An Introduction to
Mechanical Engineering

Michael Clifford (editor)
Richard Brooks
Kwing-So Choi
Donald Giddings
Alan Howe
Thomas Hyde
Arthur Jones
Edward Williams

University of Nottingham

An Introduction to Mechanical Engineering

Part 2

Michael Clifford (editor), Richard Brooks, Kwing-So Choi, Donald Giddings, Alan Howe, Thomas Hyde, Arthur Jones and Edward Williams

HODDER EDUCATION
AN HACHETTE UK COMPANY

First published in Great Britain in 2010 by
Hodder Education, An Hachette UK Company,
338 Euston Road, London NW1 3BH

Hachette UK's policy is to use papers that are natural, renewable and recyclable products and made from wood grown in sustainable forests. The logging and manufacturing processes are expected to conform to the environmental regulations of the country of origin.

British Library Cataloguing in Publication Data
A catalogue record for this book is available from the British Library

Library of Congress Cataloging-in-Publication Data
A catalog record for this book is available from the Library of Congress

ISBN: 978 0 340 93996 3

1 2 3 4 5 6 7 8 9 10

Typeset in 10.5/12pt Bembo by Tech-Set Ltd, Gateshead
Printed and bound in Italy for Hodder Education, an Hachette UK Company.

What do you think about this book? Or any other Hodder Education title? Please send your comments to educationenquiries@hodder.com

www.hoddereducation.com

Contents

Unit 4 – Electromechanical drive systems 247

Unit 5 – Feedback and control theory 317

Unit 6 – Structural vibration 376

Introduction

'I think I should understand that better', Alice said very politely, 'if I had it written down: but I can't quite follow it as you say it.'

(*Alice's Adventures in Wonderland* by Lewis Carroll)

This book builds on the experience and knowledge gained from *An Introduction to Mechanical Engineering Part 1* and is written for undergraduate engineers and those who teach them. These textbooks are not intended to be a replacement for traditional lectures, but like Alice, we see the benefit of having things written down.

In this book, we introduce material to supplement the foundational units in *Part 1* on solid mechanics, thermodynamics and fluid mechanics. In addition, the reader will encounter units on control, electromechanical drive systems and structural vibration. This material has been compiled from the authors' experience of teaching undergraduate engineers, mostly, but not exclusively, at the University of Nottingham. The knowledge contained within this textbook has been derived from lecture notes, research findings and personal experience from within the lecture theatre and tutorial sessions.

The material in this book is supported by an accompanying website: www.hodderplus.co.uk/mechanicalengineering, which includes worked solutions for exam-style questions, multiple-choice self-assessment and revision material.

We gratefully acknowledge the support, encouragement and occasional urgent but gentle prod from Stephen Halder and Gemma Parsons at Hodder Education, without whom this book would still be a figment of our collective imaginations.

Dedicated to past, present and future engineering students at the University of Nottingham.

Mike Clifford, August 2010

Acknowledgements

The authors and publishers would like to thank the following for use of copyrighted material in this volume:

Figure 1.04 reproduced with the permission of The McGraw-Hill Companies; photo of butterfly on page 10 © Imagestate Media; photo of crane on page 10 © DLILLC/Corbis; photo of dolphin on page 10 © Stockbyte/Photolibrary Group Ltd; photo of whale on page 10 © Xavier MARCHANT – Fotolia.com; Figure 1.23 © Rick Sargeant – Fotolia.com; Figure 1.24b © Vitaly Krivosheev – Fotolia.com; Figure 1.25 reproduced with permission of The McGraw-Hill Companies; Figure 1.27 D.S. Miller, 1986, *Internal Flow* Systems, 2nd edn, Cranfield: BHRA, p. 215, reproduced with permission of the author; Figure 1.28 27 D.S. Miller, 1986, *Internal Flow* Systems, 2nd edn, Cranfield: BHRA, p. 207, reproduced with permission of the author; Figure 1.29 27 D.S. Miller, 1986, *Internal Flow* Systems, 2nd edn, Cranfield: BHRA, p. 208, reproduced with permission of the author; Figure 1.31 reproduced with permission of The McGraw-Hill Companies; Figure 1.32 reproduced with permission of The McGraw-Hill Companies; Figure 1.33 reproduced with permission of The McGraw-Hill Companies; Figure 1.38a © Cambridge University Press/Courtesy of Prof. T.T. Lim; Figure 1.39 Andrew Davidhazy; Figure 1.40 ADAM HART-DAVIES/SCIENCE PHOTO LIBRARY; Figure 1.42 I.H. Abbott and A.E. von Doenhoff, 1959, *Theory of Wing Sections*, New York: Dover, pp. 462 and 463; Figure 1.45 NASA/ Sean Smith; Figure 1.48 reproduced with permission of The McGraw-Hill Companies; Figure 2.1 Drax Power Limited; Figure 2.14 reproduced by permission of the Chartered Institution of Building Services Engineers; Figure 2.22a © chukka_nc/released with a Creative Commons 2.0 licence; Figure 2.41 reprinted by permission of John Wiley & Sons Inc.; Figure 2.44 reprinted by permission of John Wiley & Sons Inc.; Figure 2.59 R.A. Bowman, A.C. Mueller and W.M. Nagle, 1930, 'Mean temperature difference in design', *Transactions of the* ASME, vol. 12, 417–422; Figure 2.62 R.A. Bowman, A.C. Mueller and W.M. Nagle, 1930, 'Mean temperature difference in design', *Transactions of the* ASME, vol. 12, 417–422; Figure 2.68 PROATES® is a registered trademark of E.ON Engineering Ltd. Reproduced with the permission of E.ON; Figure 4.27 Westend 61 GmbH/Alamy; Figure 4.67 Festo Ltd.; Figure 4.69 image reproduced with permission of Design & Draughting Solutions Ltd - www.dds-ltd.co.uk; Figure 4.70 with kind permission from Hagglunds Drives; Figure 4.71b with kind permission from Hagglunds Drives; Figure 5.1 Papplewick Pumping Station Trust; Figure 6.1 Keystone/Getty Images; Figure 6.2 © patrickw – Fotolia.com; Figure 6.15b © Ulrich Müller – Fotolia.com; Figure 6.26 IAE International Aero Engines; Figure 6.77 Siemens; Figure 6.90 Photo courtesy of Technical Manufacturing Corporation; Figure 6.101 Dave Pattison/Alamy.

Every effort has been made to trace and acknowledge the ownership of copyright. The publishers will be pleased to make suitable arrangements with any copyright holders whom it has not been possible to contact.

Unit 1

Fluid dynamics

UNIT OVERVIEW

- Introduction
- Basic concept in fluid dynamics
- Boundary layers
- Drag on immersed bodies
- Flow through pipes and ducts
- Dimensonal analysis in fluid dynamics

1.1 Introduction

Fluid dynamics is the study of the dynamics of fluid flow. Here we learn how flows behave under different external forces and conditions. In a sense this is similar to rigid body dynamics in physics, where Newton's second law is used to describe the motion of rigid bodies. Here, we must apply Newton's second law to fluid flows in a different way since fluids do not behave exactly like rigid bodies. This will be discussed in Section 1.2, where basic equations to describe fluid motion are derived and explained. Some discussions on laminar and turbulent flows are also given there, paving the way for what will follow.

The fluid that we deal with in this unit is a viscous fluid, so the velocity of fluid flow becomes zero at the solid surface. The consequence of this **no-slip condition** is that flow velocity changes from zero at the wall to the free-stream value sufficiently far away from the wall surface. This thin layer is called the **boundary layer**, an important concept in fluid dynamics, which explains how the fluid forces are generated. So, in Section 1.3, we learn the basic behaviour of boundary layers to be able to estimate the viscous drag acting on the solid surface.

The boundary layers over solid bodies behave differently depending on their shape. For example, the drag force acting on sports cars is much less than that on pickup trucks, where the boundary layer is separated from the body surface of vehicle creating a strong flow disturbance. In Section 1.4 we study the streamlining strategy to reduce the drag force of immersed bodies. We also discuss how the drag of immersed bodies is affected by the Reynolds number as well as the wall roughness.

Pipes and ducts are important engineering components used in many fluid systems. It is important, therefore, that the flow resistance can be correctly estimated for different type of ducts and pipes. In general there are two types of flow resistance. One is due to the friction drag, while the other relates to the loss of energy due to boundary layer separation. In Section 1.5, we study a method of minimizing the flow resistance similar to the streamlining strategy for immersed bodies. The discussion of pipes and ducts is extended to non-circular shapes by introducing the concept of hydraulic diameter.

The final section of this unit deals with the non-dimensional numbers of fluid dynamics. We are already familiar with the Reynolds number, but there are many other non-dimensional numbers in fluid dynamics. In Section 1.6, we learn how to identify relevant non-dimensional numbers for different types of fluid flow. We also study how these non-dimensional numbers are used to carry out model tests. Applications of the similarity principle to fluid machinery are given, emphasizing the importance of non-dimensional numbers in fluid dynamics.

1.2 Basic concept in fluid dynamics

Navier–Stokes equations

The main aim of fluid dynamics is to understand the dynamic behaviour of fluid flows. Since all fluids are continuous, we can determine the velocities and pressure of flows as a function of space and time. To achieve this, we require the governing equations to represent the fluid flows: the Navier–Stokes equations.

For simplicity, we consider only two-dimensional, **isothermal** (no thermal input or output) and **Newtonian** (the shear stress is linearly proportional to the strain rate) flows with constant density and viscosity. Therefore, u, v and p (velocities and pressure in Cartesian coordinates) are function of x, y and t. If the fluid flows are steady, they are only functions of x and y.

The Navier–Stokes equations can be derived by applying Newton's second law of motion to fluid flow. By considering a *small* control volume $dxdy$ with a unit depth, the fluid mass times acceleration is given in vector form by

$$m \cdot \vec{a} = m \cdot \frac{d\vec{V}}{dt} = \rho \cdot dxdy \cdot \frac{d\vec{V}}{dt} = \vec{F} \qquad \textbf{(1.1)}$$

The x- and y-component forces, F_x and F_y, acting on the control volume are given by

$$m \cdot a_x = \rho \cdot dxdy \cdot \frac{du}{dt} = F_x \quad m \cdot a_y = \rho \cdot dxdy \cdot \frac{dv}{dt} = F_y \qquad \textbf{(1.2, 1.3)}$$

Since, the total derivative (or material derivative) $\frac{d}{dt}$ represents

$$\frac{d}{dt} = \frac{\partial}{\partial t} + \frac{\partial x}{\partial t}\frac{\partial}{\partial x} + \frac{\partial y}{\partial t}\frac{\partial}{\partial y} = \frac{\partial}{\partial t} + u\frac{\partial}{\partial x} + v\frac{\partial}{\partial y} \qquad \textbf{(1.4)}$$

Therefore,

$$m \cdot a_x = \rho\, dxdy \cdot \frac{du}{dt} = \rho\, dxdy \cdot \left[\frac{\partial u}{\partial t} + u\frac{\partial u}{\partial x} + v\frac{\partial u}{\partial y}\right] = F_x$$

$$m \cdot a_y = \rho\, dxdy \cdot \frac{dv}{dt} = \rho\, dxdy \cdot \left[\frac{\partial v}{\partial t} + u\frac{\partial v}{\partial x} + v\frac{\partial v}{\partial y}\right] = F_y \qquad \textbf{(1.5, 1.6)}$$

Here, we shall consider only those forces acting on the surfaces of the control volume, although a body force will be introduced later. Surface forces include hydrostatic pressure p, the normal stresses τ_{xx}, τ_{yy} and the shear stresses τ_{xy}, τ_{yx}.

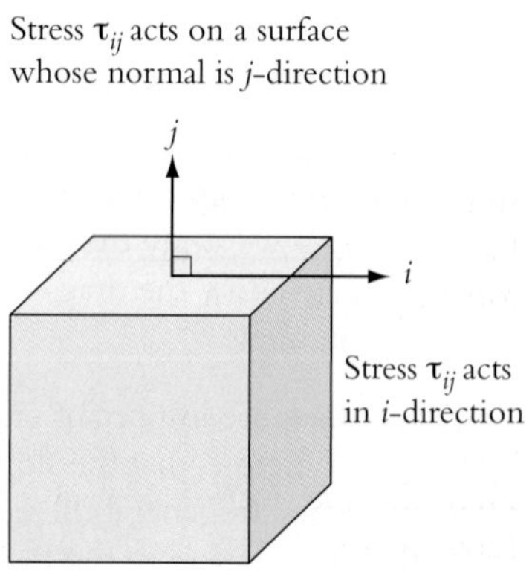
Stress τ_{ij} acts on a surface whose normal is j-direction
j
i
Stress τ_{ij} acts in i-direction

Figure 1.1 Stress tensor τ_{ij}. The first index i of the stress tensor τ_{ij} indicates the direction of the stress that is acting on the surface, whose normal direction is indicated by the second index j.

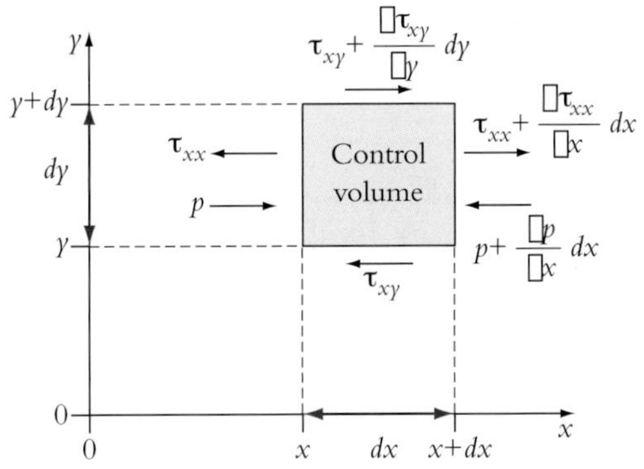

Figure 1.2 The balance of surface forces in the *x*-direction

From Figure 1.2, the total surface forces in the x-direction can be obtained as follows.

$$\left[\left(\tau_{xx} + \frac{\partial \tau_{xx}}{\partial x} dx\right) dy - \tau_{xx}\, dy\right] + \left[\left(\tau_{xy} + \frac{\partial \tau_{xy}}{\partial y} dy\right) dx - \tau_{xy}\, dx\right] - \left[\left(p + \frac{\partial p}{\partial x} dx\right) dy - p\, dy\right] \quad \textbf{(1.7)}$$

Here, the forces acting on the surface at $(x + dx)$ can be obtained from the forces on the surface at x by using Taylor's expansion.

In a similar way, the total surface force in the y-direction can be obtained from Figure 1.3 as

$$\left[\left(\tau_{yy} + \frac{\partial \tau_{yy}}{\partial y} dy\right) dx - \tau_{yy}\, dx\right] + \left[\left(\tau_{yx} + \frac{\partial \tau_{yx}}{\partial x} dx\right) dy - \tau_{yx}\, dy\right] - \left[\left(p + \frac{\partial p}{\partial y} dy\right) dx - p\, dx\right] \quad \textbf{(1.8)}$$

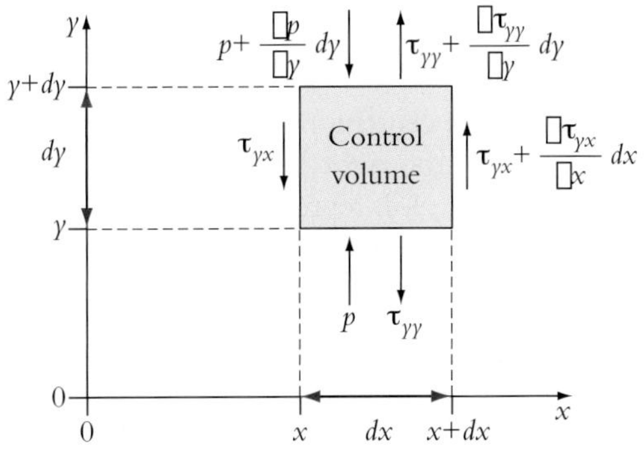

Figure 1.3 The balance of surface forces in the *y*-direction

Therefore, the *total force* acting on the surfaces of control volume can be given by

$$F_x = \left(-\frac{\partial p}{\partial x} + \frac{\partial \tau_{xx}}{\partial x} + \frac{\partial \tau_{xy}}{\partial y}\right) dx\, dy \quad F_y = \left(-\frac{\partial p}{\partial y} + \frac{\partial \tau_{yx}}{\partial x} + \frac{\partial \tau_{yy}}{\partial y}\right) dx\, dy \quad \textbf{(1.9, 1.10)}$$

So far, we have considered only the surface forces. There are flows, such as water waves around a ship, where the gravity force plays an important role. Therefore, we should also consider such a body force acting on the control volume together with the surface forces. In such a situation, the vertical force F_y in equation (1.10) must be replaced by

$$F_y = \left(-\frac{\partial p}{\partial y} + \frac{\partial \tau_{yx}}{\partial x} + \frac{\partial \tau_{yy}}{\partial y}\right) dx\, dy - \rho g\, dx\, dy \quad \textbf{(1.11)}$$

Here, the additional force due to gravity is $\rho g\, dx\, dy$, which always acts in negative y-direction (downwards).

We finally obtained the following equations of fluid motion.

$$\rho\left(\frac{\partial u}{\partial t} + u\frac{\partial u}{\partial x} + v\frac{\partial u}{\partial y}\right) = -\frac{\partial p}{\partial x} + \left(\frac{\partial \tau_{xx}}{\partial x} + \frac{\partial \tau_{xy}}{\partial y}\right)$$

$$\rho\left(\frac{\partial v}{\partial t} + u\frac{\partial v}{\partial x} + v\frac{\partial v}{\partial y}\right) = -\frac{\partial p}{\partial y} + \left(\frac{\partial \tau_{yx}}{\partial x} + \frac{\partial \tau_{yy}}{\partial y}\right) - \rho g \qquad \textbf{(1.12, 13)}$$

For the *Newtonian* flow, the stress is linearly proportional to the velocity gradient, we can write this relationship in a matrix form as follows.

$$\begin{pmatrix} \tau_{xx} & \tau_{xy} \\ \tau_{yx} & \tau_{yy} \end{pmatrix} = \begin{pmatrix} \mu\dfrac{\partial u}{\partial x} & \mu\dfrac{\partial u}{\partial y} \\ \mu\dfrac{\partial v}{\partial x} & \mu\dfrac{\partial v}{\partial y} \end{pmatrix} \qquad \textbf{(1.14)}$$

By substituting this relationship into equation (1.12) and (1.13), we have the final form of the Navier–Stokes equations after dividing all terms by the fluid density ρ.

$$\frac{\partial u}{\partial t} + u\frac{\partial u}{\partial x} + v\frac{\partial u}{\partial y} = -\frac{1}{\rho}\frac{\partial p}{\partial x} + \nu\left(\frac{\partial^2 u}{\partial x^2} + \frac{\partial^2 u}{\partial y^2}\right)$$

$$\frac{\partial v}{\partial t} + u\frac{\partial v}{\partial x} + v\frac{\partial v}{\partial y} = -\frac{1}{\rho}\frac{\partial p}{\partial y} + \nu\left(\frac{\partial^2 v}{\partial x^2} + \frac{\partial^2 v}{\partial y^2}\right) - g \qquad \textbf{(1.15, 16)}$$

where, $\nu = \mu/\rho$ is called the **kinematic viscosity** with dimension $[L^2\, T^{-1}]$.

Now, the Navier–Stokes equations consist of four parts, which are:

(Inertia force) = (Pressure force) + (Viscous force) + (Gravity force) **(1.17)**

In others words, the left-hand side of the Navier–Stokes equations represents the inertia force (i.e. mass times acceleration), while pressure force represented by the pressure gradient and viscous force led by the viscosity are on the right-hand side. The gravity force can usually be omitted from the equations unless the surface wave or the natural convection is involved.

By examining the magnitude in each term of the Navier–Stokes equations, we can see that the left-hand side of the equations of is of the order of u^2/L, while the right-hand side is of the order of $\nu u/L^2$. Here, u represents the velocity scale (either u or v), L represents the length scale (x or y) and ν is the kinematic viscosity. The ratio of these two will give a *non-dimensional* value called the **Reynolds number**:

$$Re\ \text{(Reynolds number)} = \frac{uL}{\nu} \propto \frac{\text{inertia force}}{\text{viscous force}} \qquad \textbf{(1.18)}$$

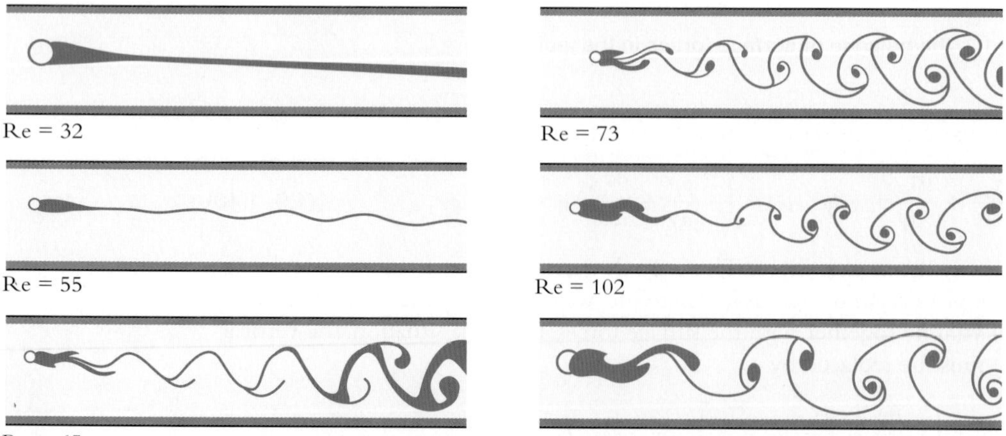

Figure 1.4 Effect of the Reynolds number on the vortex shedding from a circular cylinder (H. Schlichting, 1968, *Boundary Layer Theory*, 6th edn, New York: McGraw Hill, reproduced with permission of the McGraw-Hill Companies)

Similarly, the ratio between the magnitude of the inertia term $\left(\text{of the order of } \frac{u^2}{L}\right)$ and that of the gravity term (of the order of g) is called the **Froude number**:

$$Fr\ (\text{Froude number}) = \frac{u}{\sqrt{gL}} \propto \left(\frac{\text{inertia force}}{\text{gravity force}}\right)^{\frac{1}{2}} \tag{1.19}$$

Here, it is customary to take a square root of the ratio to define the Froude number. For example, a yacht with a length of $L = 20\,\text{m}$ travelling at $u = 10\,\text{m/s}$ would have $\text{Re} = 1.3 \times 10^8$ and $\text{Fr} = 0.7$ assuming that $\nu = 1.5 \times 10^{-6}\,\text{m}^2/\text{s}$.

Worked example

Find the acceleration of a fluid particle with the velocity field given by

$$\begin{aligned}\vec{V} &= u\vec{i} + v\vec{j} \\ &= (3t)\vec{i} + (2xy)\vec{j}\end{aligned}$$

$$\frac{Du}{Dt} = \frac{\partial u}{\partial t} + u\frac{\partial u}{\partial x} + v\frac{\partial u}{\partial y} = 3$$

$$\frac{Dv}{Dt} = \frac{\partial v}{\partial t} + u\frac{\partial v}{\partial x} + v\frac{\partial v}{\partial y} = 0 + (3t)(2y) + (2xy)(2x)$$

$$\frac{D\vec{V}}{Dt} = \frac{Du}{Dt}\vec{i} + \frac{Dv}{Dt}\vec{j} = 3\vec{i} + (6ty + 4x^2y)\vec{j}$$

Continuity equation

The **continuity equation** can be derived in a similar way as we have done to obtain the Navier–Stokes equation. Here, however, we should consider the *mass balance* in the control volume instead of the *force balance*.

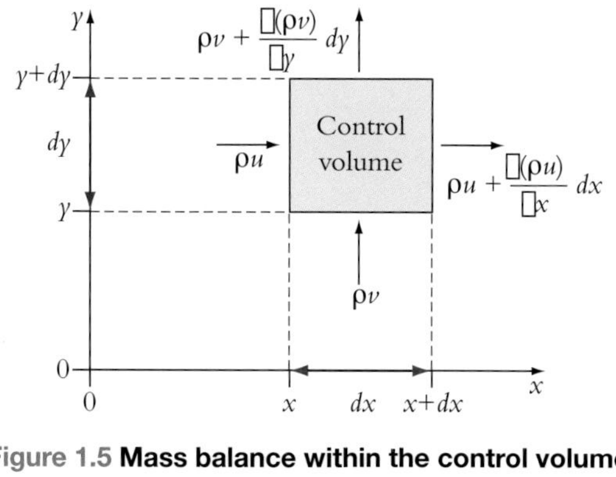

Figure 1.5 Mass balance within the control volume

Mass flux into the control volume in the x-direction is given by

$$\left[(\rho u)\cdot dy + \frac{\partial(\rho u)}{\partial x}dx\cdot dy\right] - (\rho u)\cdot dy = \frac{\partial(\rho u)}{\partial x}\cdot dx\,dy \tag{1.20}$$

while the mass flux into the control volume in the y-direction is

$$\left[(\rho v)\cdot dx + \frac{\partial(\rho v)}{\partial y}dy\cdot dx\right] - (\rho v)\cdot dx = \frac{\partial(\rho v)}{\partial y}\cdot dx\,dy \tag{1.21}$$

Here again, we have used Taylor's expansion to evaluate the mass flux out from the surface at $(x + dx)$ and $(y + dy)$ from the mass flux into the surface at x and y, respectively.

Since there must be no changes in mass within the control volume:

$$\frac{\partial(\rho u)}{\partial x} \cdot dx\,dy + \frac{\partial(\rho v)}{\partial y} \cdot dx\,dy = 0 \qquad (1.22)$$

Assuming that the fluid density ρ is constant, we have:

$$\frac{\partial u}{\partial x} + \frac{\partial v}{\partial y} = 0 \qquad (1.23)$$

This is called the continuity equation based on the conservation of mass.

Worked example

When the velocity field is given by

$$\vec{V} = (x^2 - y^2)\vec{i} + v\vec{j}$$

determine the v-velocity from the two-dimensional continuity equation.

$$u = x^2 - y^2$$

$$\therefore \quad \frac{\partial u}{\partial x} = 2x$$

Substituting into the continuity equation $\left(\frac{\partial u}{\partial x} + \frac{\partial v}{\partial y} = 0\right)$ we get

$$\frac{\partial v}{\partial y} = -2x$$

$$\therefore \quad v = -2xy + C(x)$$

Worked example

There is a two-dimensional steady flow whose velocity vector is given by

$$\vec{V} = 2xy\vec{i} - y^2\vec{j}$$

where $\vec{i}$ and $\vec{j}$ are unit vectors in the x- and y-directions, respectively. Obtain an expression for the pressure gradient of this flow if the fluid density is constant and the viscosity and gravity forces are negligible.

$$u = 2xy \qquad v = -y^2$$

$$\frac{\partial u}{\partial x} = 2y \qquad \frac{\partial u}{\partial y} = 2x$$

$$\frac{\partial v}{\partial x} = 0 \qquad \frac{\partial v}{\partial y} = -2y$$

Substituting these into the Navier–Stokes equations, we get

$$(2xy)(2y) + (-y^2)(2x) = -\frac{1}{\rho}\frac{\partial p}{\partial x}$$

$$2xy^2 = -\frac{1}{\rho}\frac{\partial p}{\partial x} \quad \therefore \quad \frac{\partial p}{\partial x} = -2\rho\, xy^2$$

$$(2xy)(0) + (-y^2)(-2y) = -\frac{1}{\rho}\frac{\partial p}{\partial y}$$

$$2y^3 = -\frac{1}{\rho}\frac{\partial p}{\partial y} \quad \therefore \quad \frac{\partial p}{\partial y} = -2\rho\, y^3$$

Laminar and turbulent flows

The flow through a pipe remains smooth and steady below the **critical Reynolds number**, given by

$$Re \equiv \frac{U \cdot d}{\nu} = 13{,}000 \qquad \textbf{(1.24)}$$

However, the flow becomes fluctuating and random when the Reynolds number exceeds this value. Figure 1.6 shows this process where the transition from laminar to turbulent flow is demonstrated with dye injected into a pipe through a needle.

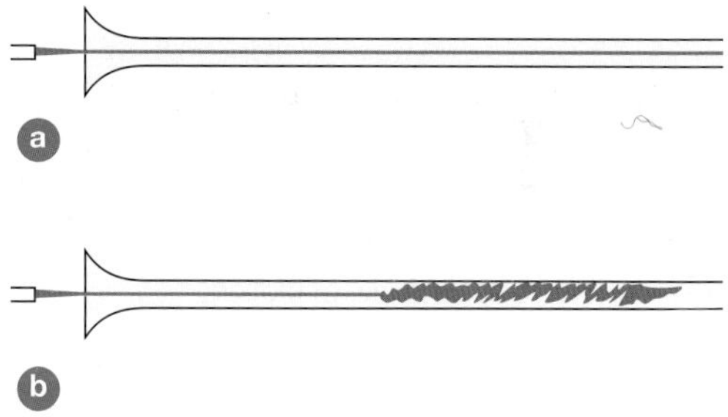

Figure 1.6 **The flow in a pipe changes from laminar flow (a) to turbulent flow (b)**

We can see the change in flow behaviour as a function of the Reynolds number. Figure 1.7(a) shows a laminar flow at subcritical Reynolds number, where the dye filament stays straight through until the end of the pipe. As the Reynolds number is increased in Figures 1.7(b) and (c) the flow develops patterns, signifying the flow transition that is taking place. Finally, turbulent flow is reached in Figure 1.7(d) where the dye patterns seem to be random and chaotic.

Although the Reynolds number is an important parameter affecting the transition process to turbulence, there are other influential factors such as the wall roughness, initial disturbance and external disturbance. For example, a sharp inlet to the pipe will disturb the flow, and therefore reduce the critical Reynolds number. Vibration of the experimental setup from the floor, as well as noise transmitted to the flow, will accelerate the process of transition.

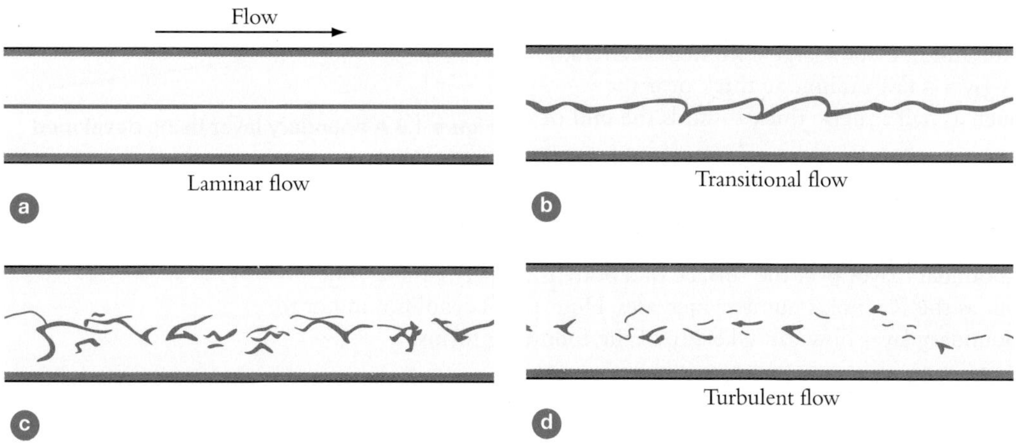

Figure 1.7 **Flow transition to turbulence in a pipe flow. Laminar flow (a), where the injected dye stays straight through the pipe; transitional flow (b), where the flow develops certain patterns; turbulent flow (c, d), where the flow becomes random and chaotic.**

There is, however, no precise definition of turbulence. Indeed, it is very difficult to determine whether a particular flow is turbulent or not. For example, random waves that can be observed over the surface of the water in a swimming pool are probably not turbulence. Therefore, we have to look at the symptoms of the flow to determine whether it is turbulent or not.

Symptoms of turbulence include:

- irregularity
- high diffusivity
- high Reynolds number
- three-dimensionality
- dissipativeness
- continuous flow.

Every turbulent flow is different, yet they have many common characteristics as listed above. We must, therefore, check whether a particular flow satisfies all of these characteristics before declaring that it is a turbulent flow.

Learning summary

By the end of this section you should have learnt:

✓ the Navier–Stokes equations are governing equations for fluid motion, which can be derived from Newton's second law of motion;
✓ the continuity equation guarantees the conservation of mass;
✓ the Reynolds number indicates a relative importance of inertial force in flow motion to viscous force;
✓ the Froude number signifies the importance of inertial force in flow motion against the gravity force;
✓ all flows become turbulent above the critical Reynolds number.

1.3 Boundary layers

The boundary layer is a thin layer created over the surface of a body immersed in a fluid (Figure 1.8), where the viscosity plays a significant role. Due to the non-slip condition of viscous flows, the velocity at a solid surface is always zero. This means that the velocity gradually increases from zero at the wall to the freestream velocity U_o at the edge of the boundary layer. This creates a thin, highly sheared region called the boundary layer, over a body surface in a moving fluid (or over a moving body in a still fluid). Over a large commercial aircraft, for example, the boundary layer a few millimetre thick near the cockpit can grow to as much as half a metre thick towards the end of the fuselage.

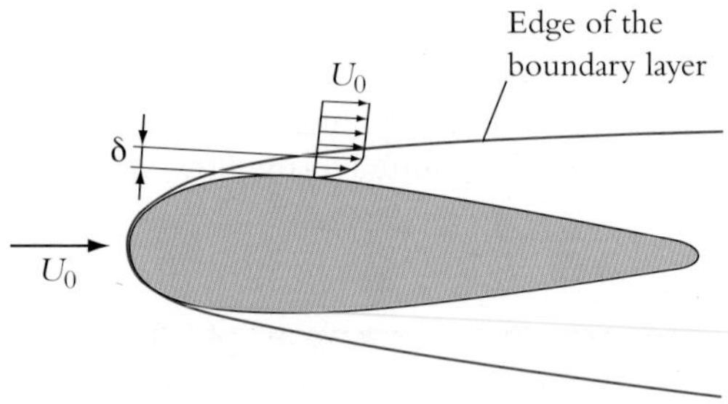

Figure 1.8 A boundary layer being developed over a wing

Reynolds number

Similar to pipe flows, the boundary layer over the surface of a body is initially laminar, but will soon become turbulent as the Reynolds number increases. Here, the Reynolds number to describe the state of the boundary layer flow can take any of the following forms.

$$Re \equiv \frac{U_o x}{\nu}, \frac{U_o \delta}{\nu}, \frac{U_o \delta^*}{\nu}, \frac{U_o \theta}{\nu} \quad (1.25)$$

While the pipe diameter is always used to define the Reynolds number of pipe flows, the length scale of boundary layer flows takes either the streamwise length x along the body surface or one of the boundary layer thicknesses such as δ, δ^* or θ. The boundary layer thickness δ is defined as the distance from the wall to the point where the boundary layer velocity reaches the freestream velocity. Discussions on other boundary layer thickness, such as δ^* or θ, will be given later in this section.

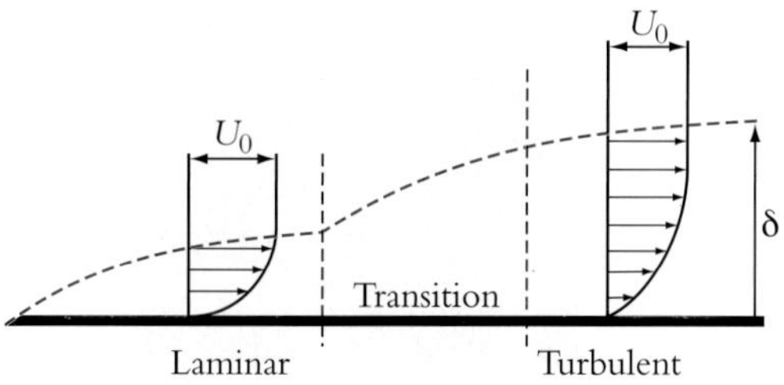

Figure 1.9 Development of the boundary layer over a flat plate parallel to the flow. Note that the boundary layer thickness grows faster when it is turbulent.

The boundary layer thickness over a flat plate at the streamwise length x from the leading edge can be obtained using the following formulae.

$$\delta = 5.0\,x\left(\frac{U_o x}{\nu}\right)^{-\frac{1}{2}} \quad \text{for laminar flow } (\mathrm{Re}_x < 3 \times 10^6) \tag{1.26}$$

$$\delta = 0.37\,x\left(\frac{U_o x}{\nu}\right)^{-\frac{1}{5}} \quad \text{for turbulent flow } (\mathrm{Re}_x > 3 \times 10^6) \tag{1.27}$$

The boundary layer growth depends on the flow condition, whether it is laminar or turbulent. The growth is much faster for turbulent boundary layers (Figure 1.9) as the diffusivity increases as a result of transition to turbulence (Section 1.1). The critical Reynolds number for the boundary layer over a flat plate is usually around 3×10^6 as indicated in equations (1.26) and (1.27). It may take quite a different value, however, depending on the initial as well as boundary conditions of the flow. For the boundary layer over a rough surface, for example, the critical Reynolds number is less than 10^6. This means that the boundary layer transition takes place much earlier over a rough surface as compared to that over the smooth surface.

Worked example

A laminar boundary layer is being developed over a flat plate with the free-stream velocity of 1.5 m/s. Assuming that the pressure gradient along the plate is zero, obtain the distance x from the leading edge where the boundary layer thickness δ becomes 10 mm. Using the critical Reynolds number R_{xc} of 10^6, determine the transition point where the boundary layer becomes turbulent. The kinematic viscosity and density of the fluid (air) are 1.5×10^{-5} m²/s and 1.2 kg/m³, respectively.

$$\delta = 5.0\,x\left(\frac{U_o x}{\nu}\right)^{-\frac{1}{2}} = 5.0\sqrt{\frac{\nu}{U_o}}\sqrt{x}$$

$$\therefore \quad x = \frac{1}{25.0}\delta^2\frac{U_o}{\nu} = \frac{1}{25.0}(0.01)^2\frac{1.5}{1.5 \times 10^{-5}} = 0.40\ [\mathrm{m}]$$

The transition point x_t can be obtained from

$$R_{xc} = \frac{x_t U_o}{\nu} = 10^6$$

$$\therefore \quad x_t = 10^6\frac{\nu}{U_o} = 10^6 \times \frac{1.5 \times 10^{-5}}{1.5} = 10\ [\mathrm{m}]$$

Reynolds number of living things

Reynolds number of living things depends on their size, flight or swim speed and the medium they live in. The following are the Reynolds numbers of some of familiar living things. Unless the body size is very small or the flight or swim speed is very low, as with butterflies, the flow around the body of living things is most likely turbulent.

Butterflies: $\mathrm{Re} \approx \dfrac{(0.3\,\mathrm{m/s}) \times (0.08\,\mathrm{m})}{(1.5 \times 10^{-5}\,\mathrm{m^2/s})} = \mathbf{1600}$

Cranes: $\mathrm{Re} \approx \dfrac{(15\,\mathrm{m/s}) \times (1.0\,\mathrm{m})}{(1.5 \times 10^{-5}\,\mathrm{m^2/s})} = \mathbf{1\,000\,000}$

Dolphins: $\mathrm{Re} \approx \dfrac{(10\,\mathrm{m/s}) \times (2\,\mathrm{m})}{(1.0 \times 10^{-6}\,\mathrm{m^2/s})} = \mathbf{20\,000\,000}$

Whales: $\mathrm{Re} \approx \dfrac{(15\,\mathrm{m/s}) \times (25\,\mathrm{m})}{(1.0 \times 10^{-6}\,\mathrm{m^2/s})} = \mathbf{375\,000\,000}$

Velocity profiles of boundary layers

The **Blasius profile** is the theoretical velocity profile of the **laminar boundary layer** over a flat plate where the static pressure does not change along the plate. Here, the Blasius profile is independent of the Reynolds number of the boundary layer, as shown in Figure 1.10. The Blasius profile can be approximated by the parabolic velocity profile given by equation (1.28).

$$\frac{u}{U_o} = \frac{y}{\delta}\left(2 - \frac{y}{\delta}\right) \qquad (1.28)$$

For the **turbulent boundary layer**, however, there is no theoretical solution to represent the velocity profile. The turbulent boundary layer profile can be approximated by the **one-seventh law** given by

$$\frac{u}{U_o} = \left(\frac{y}{\delta}\right)^{\frac{1}{7}} \qquad (1.29)$$

The atmospheric boundary layer is often simulated by the one-seventh law in a wind tunnel, where building or bridge models are tested. Indeed, this empirical law has a reasonable agreement with the actual profile of the turbulent boundary layer. However, the velocity gradient of the one-seventh law at the wall is always incorrect, since

$$\frac{d}{dy}\left(\frac{u}{U_o}\right)_{y=0} = \infty \qquad (1.30)$$

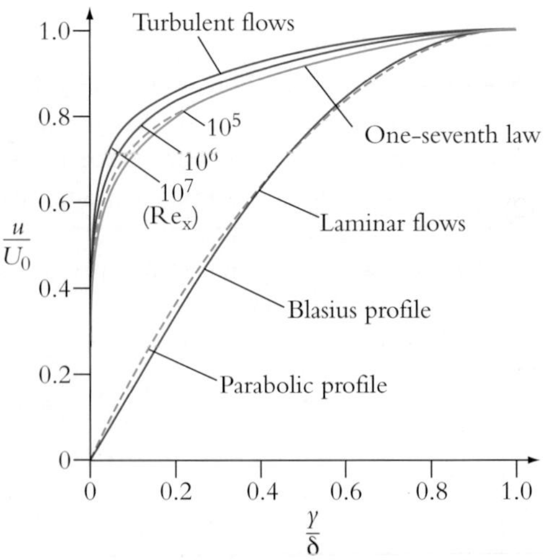

Figure 1.10 Velocity profiles of laminar and turbulent boundary layers

Therefore, it cannot be used to investigate the turbulent boundary layer close to the wall surface. Here, we should use the logarithmic velocity profile instead (Figure 1.11). With the logarithmic velocity profile (the log law, for short) a large part of the turbulent boundary layer can be represented by

$$\frac{u}{u^\star} = 5.75 \log_{10} \frac{u^\star y}{\nu} + 5.5 \qquad \textbf{(1.31)}$$

except for a very thin region near the wall (the viscous sublayer) where the velocity is given by the following linear profile.

$$\frac{u}{u^\star} = \frac{u^\star y}{\nu} \qquad \textbf{(1.32)}$$

Here, the friction velocity $u^\star$ is given by $u^\star = \sqrt{\frac{\tau_w}{\rho}}$. It should be noted that the turbulent velocity profile is dependent on the Reynolds number (Figure 1.10).

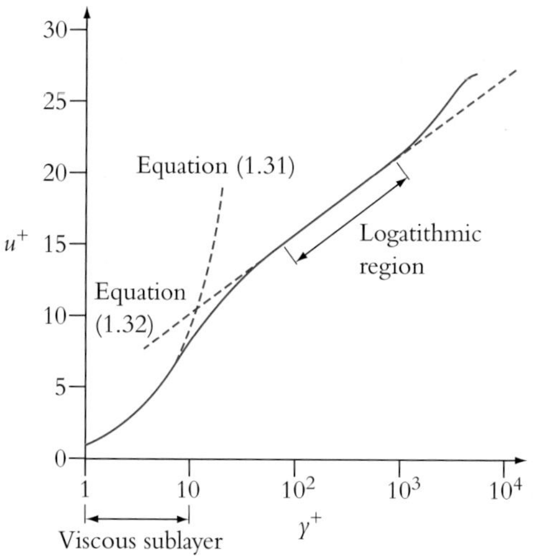

Figure 1.11 Logarithmic velocity profile of the turbulent boundary layer

Worked example

An atmospheric boundary layer ($\delta = 80$ m) with the free-stream velocity of $U_0 = 10$ m/s has the wall-shear stress of $\tau_w = 0.077$ Pa. The kinematic viscosity and density of air are $\nu = 1.5 \times 10^{-5}$ m²/s and $\rho = 1.2$ kg/m³, respectively.

(1) Estimate the thickness of the viscous sublayer.

(2) Obtain the velocity at $y = 1$ m from the ground.

(1) The thickness of the viscous sublayer is given by

$$y^+ = \frac{yu^\star}{\nu} = 10 \quad \therefore\ y = 10\frac{\nu}{u^\star}$$

$$\text{where, } u^\star = \sqrt{\frac{\tau_w}{\rho}} = \sqrt{\frac{0.077}{1.2}} = 0.25\ \text{m/s}$$

$$\therefore \qquad y = 10 \times \frac{1.5 \times 10^{-5}}{0.25} = 0.6 \times 10^{-3}\ \text{m} = 0.6\ \text{mm}$$

(2) Using the logarithmic velocity profile,

$$\frac{u}{u^\star} = 5.5 + 5.75 \log_{10}\frac{u^\star y}{\nu}$$

$$\therefore \qquad u = u^\star\left(5.5 + 5.75 \log_{10}\frac{u^\star y}{\nu}\right)$$

$$= 0.25\left(5.5 + 5.75 \log_{10}\frac{0.25 \times 1}{1.5 \times 10^{-5}}\right)$$

$$= 7.4\ \text{m/s}$$

If we use the one-seventh law instead, we have

$$\frac{u}{U_o} = \left(\frac{y}{\delta}\right)^{\frac{1}{7}} \quad \therefore\ u = U_o\left(\frac{y}{\delta}\right)^{\frac{1}{7}}$$

$$u = 10\left(\frac{1}{80}\right)^{\frac{1}{7}} = 5.3\ \text{m/s}$$

Friction velocity in turbulent boundary layers

Logarithmic velocity profile of turbulent boundary layer is given by

$$\frac{u}{u^\star} = 5.75 \log_{10} \frac{u^\star y}{\nu} + 5.5$$

Where $\frac{u}{u^\star}$ and $\frac{u^\star y}{\nu}$ are non-dimensional variables for the local velocity and the distance from wall, respectively.

$$u^+ \equiv \frac{u}{u^\star} = \left[\frac{LT^{-1}}{LT^{-1}}\right] \qquad y^+ \equiv \frac{u^\star y}{\nu} = \left[\frac{(LT^{-1})L}{L^2T^{-1}}\right]$$

Using these notations, we can write the logarithmic profiles as

$$u^+ = 5.75 \log_{10} y^+ + 5.5$$

Logarithmic velocity profile can represent a large part of the turbulent boundary layers and turbulent pipe flows except in the viscous sublayer ($y^+ < 10$), where we should use

$$\frac{u}{u^\star} = \frac{u^\star y}{\nu} \quad \text{or} \quad u^+ = y^+$$

Since $u = \frac{u^{\star 2}}{\nu} y$ in the viscous sublayer, we have

$$\frac{du}{dy} = \frac{u^{\star 2}}{\nu}$$

$$\therefore \qquad \mu \frac{du}{dy} = \mu \frac{u^{\star 2}}{\nu} = \frac{\mu u^{\star 2}}{(\mu/\rho)} = \rho u^{\star 2} = \tau_w$$

We see that the velocity gradient within the viscous sublayer is constant.

Effect of wall roughness

The wall roughness can affect the laminar boundary layer by promoting an early transition to turbulence. For the turbulent boundary layer, the wall roughness enhances the fluid mixing to increase the near-wall velocity gradient, which leads to an increase in skin-friction drag. Here, it is important to know how rough the wall should be before it starts affecting the skin-friction drag of the turbulent boundary layer.

The wall surface of the turbulent boundary layer is **hydraulically smooth** when $\varepsilon^+ \equiv \frac{\varepsilon u^\star}{\nu} < 5$.

Here, ε is the roughness height, $u^\star = \sqrt{\frac{\tau_w}{\rho}}$ is the friction velocity and ν is the kinematic viscosity of the fluid. In other words, the wall surface is smooth as far as fluid dynamics of turbulent boundary layer is concerned if the Reynolds number ε^+, based on the roughness height ε and the friction velocity $u^\star$, is less than 5. This suggests that the skin-friction drag of the turbulent boundary layer will be increased only when $\varepsilon^+ > 5$. Recalling that the thickness of viscous sublayer is given by $y^+ = 10$, we can say that the wall surface is hydraulically smooth if the roughness is completely submerged in the viscous sublayer.

If we know the roughness height ε, the **Moody chart** for pipe flows (and the analogous chart for the boundary layers) will give us the effects of roughness on the skin-friction drag in the turbulent boundary layer. Wall roughness should be considered always relative to the boundary layer thickness, so we must use the non-dimensional roughness height in studying its effect on the skin-friction drag.

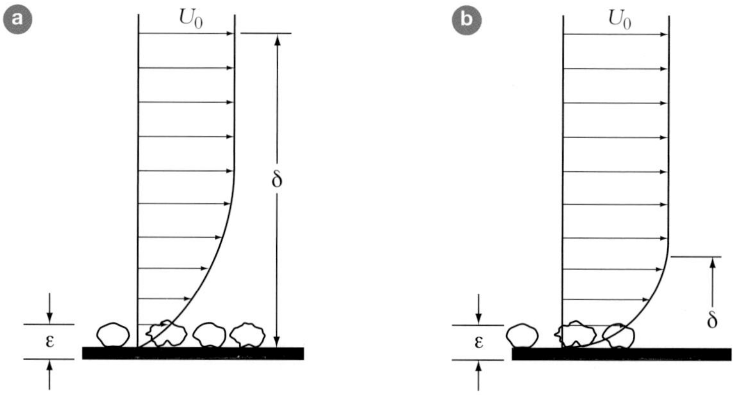

Figure 1.12 Schematic view of the boundary layer. The wall surface is considered smooth (a); it should be considered rough (b) despite the identical physical size of roughness.

Momentum integral equation

It is possible to estimate the skin-friction drag of the boundary layer using the momentum integral equation. To derive the equation, we consider a balance of forces on a control volume within the boundary layer (Figure 1.13), which is being developed over a flat plate. Here, the line 1 and line 3 indicate the entry and exit of the control volume, while the streamline just above the boundary layer (line 2) and the flat plate (line 4) indicate the upper and bottom surface, respectively.

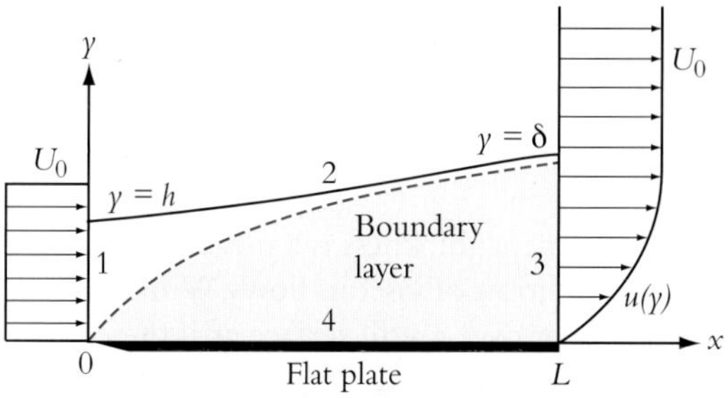

Figure 1.13 Control volume used in the derivation the momentum integral equation

Since there is no flow across the streamline or the plate wall, the mass flow rate to the control volume through line 1 must be equal to the mass flow rate out of line 3. This gives

$$\rho\, U_o h = \rho \int_0^{\delta(x)} u\, dy \qquad (1.33)$$

The only force acting on the control volume is the skin-friction drag over the flat plate since the pressure gradient over a flat plate is zero. Accordingly, the skin-friction drag must be equal to the change in the momentum flux within the control volume. Therefore, the skin-friction drag over the flat plate is given by

$$D(x) = \rho\, U_o^2 h - \rho \int_0^{\delta(x)} u^2\, dy \qquad (1.34)$$

By substituting equation (1.33) into equation (1.34), we obtain

$$D(x) = \rho \int_0^{\delta(x)} U_o u\, dy - \rho \int_0^{\delta(x)} u^2\, dy \qquad (1.35)$$

$$= \rho \int_0^{\delta(x)} u(U_o - u) dy$$

Introducing the *momentum thickness* θ, a measure of the momentum loss as a result of the boundary layer growth, which is is given by:

$$\theta = \int_0^{\delta} \frac{u}{U_o}\left(1 - \frac{u}{U_o}\right) dy \qquad (1.36)$$

Equation (1.35) can be written as

$$D(x) = \rho U_o^2 \theta \qquad (1.37)$$

Therefore, the momentum thickness θ represents the skin-friction drag D over a flat plate. A differential form of this equation can be given by

$$\tau_w(x) = \rho U_o^2 \frac{d\theta}{dx} \tag{1.38}$$

if we note that

$$D(x) = \int_0^x \tau_w(x)\,dx \tag{1.39}$$

Using the **skin-friction coefficient**, $C_f = \frac{\tau_w}{\frac{1}{2}\rho U_o^2}$, equation (1.38) can be written as

$$C_f = 2\frac{d\theta}{dx} \tag{1.40}$$

This is the **Kármán's momentum integral equation**, which is valid for both laminar and turbulent boundary layers as long as the pressure gradient is zero. The Kármán's momentum integral equation indicates that the local skin-friction coefficient is exactly twice the streamwise change of the momentum thickness.

Another important parameter in boundary-layer theory is the **displacement thickness** δ^* defined as

$$\delta^* = \int_0^\delta \left(1 - \frac{u}{U_o}\right)dy \quad \text{or} \quad U_o\delta^* = \int_0^\delta (U_o - u)\,dy \tag{1.41}$$

The displacement thickness is a measure of mass-flow deficit in the boundary layer due to the non-slip condition of viscous flows. With this concept we can treat the development of the boundary layer over a wall surface as if the on-coming flow were shifted by the amount of the displacement thickness δ^* (Figure 1.14). Therefore, we can estimate the change in flow rate though a duct, for example, without considering the change in velocity profile.

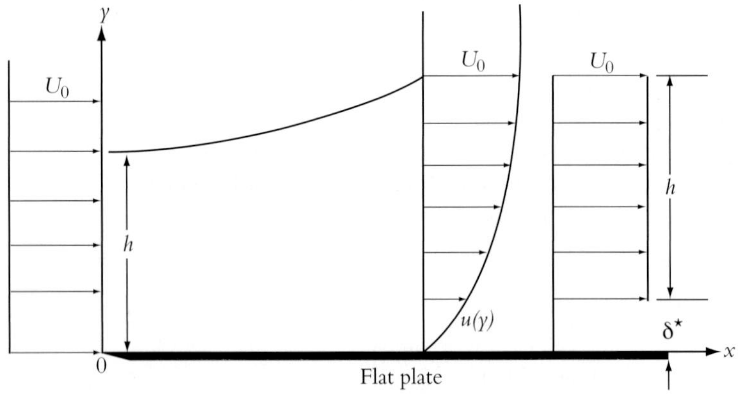

Figure 1.14 Simulated effect in the boundary layer using the displacement thickness concept

The ratio of the displacement thickness δ^* to the momentum thickness θ is called the **shape factor** H

$$H = \frac{\delta^*}{\theta} \tag{1.42}$$

This is a good indicator for the flow status, which can be used to check whether the flow is laminar or turbulent. Over a flat plate with zero pressure gradient, for example, the shape factor of the laminar boundary layer is $\boldsymbol{H \approx 2.6}$, while the shape factor of the turbulent boundary layer is $\boldsymbol{H \approx 1.4}$. For the boundary layer under transition, the shape factor takes a value between 1.4 and 2.6.

The shape factor can also be used to find out if it is close to flow separation (to be discussed next in this section). Since the velocity profile will become tall and thin as the flow separation is approached, the shape factor of the boundary layer will increase in value whether it is laminar or turbulent. Usually this increase in the H value is quite rapid, giving warning that the boundary layer flow is about to detach from the wall.

Worked example

Obtain the displacement thickness, momentum thickness and the shape factor of the boundary layer when the velocity profile is given by $\frac{u}{U_o} = \frac{y}{\delta}$ (Figure 1.15).

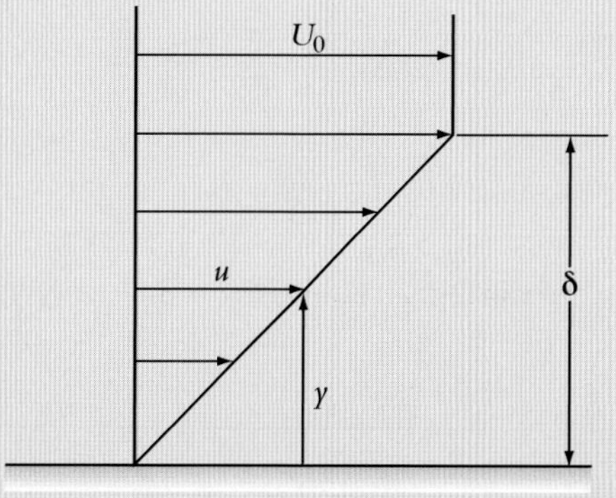

Figure 1.15

Changing the variables by setting $\eta = \frac{y}{\delta}$, we have

$$\delta^* = \int_0^\delta \left(1 - \frac{u}{U_o}\right)dy = \delta\int_0^1 \left(1 - \frac{u}{U_o}\right)d\left(\frac{y}{\delta}\right) = \delta\int_0^1 \left(1 - \frac{y}{\delta}\right)d\left(\frac{y}{\delta}\right)$$

$$= \delta\int_0^1 (1 - \eta)d\eta = \tfrac{1}{2}\delta$$

$$\theta = \int_0^\delta \frac{u}{U_o}\left(1 - \frac{u}{U_o}\right)dy = \delta\int_0^1 \frac{u}{U_o}\left(1 - \frac{u}{U_o}\right)d\left(\frac{y}{\delta}\right) = \delta\int_0^1 \left(\frac{y}{\delta}\right)\left(1 - \frac{y}{\delta}\right)d\left(\frac{y}{\delta}\right)$$

$$= \delta\int_0^1 \eta(1 - \eta)d\eta = \frac{1}{6}\delta$$

$$H = \frac{\delta^*}{\theta} = 3$$

Formulae for the boundary layer development

Laminar:

$$\left.\begin{aligned} \delta/x &= 5.0 \\ \delta^*/x &= 1.721 \\ \theta/x &= 0.664 \\ C_f &= 0.664 \\ C_D &= 1.328 \end{aligned}\right\} (U_o x/\nu)^{-1/2} \qquad (1.43)$$

Turbulent:

$$\left.\begin{aligned} \delta/x &= 0.37 \\ \delta^*/x &= 0.046 \\ \theta/x &= 0.036 \\ C_f &= 0.058 \\ C_D &= 0.0725 \end{aligned}\right\} (U_o x/\nu)^{-1/5} \qquad (1.44)$$

The drag coefficient C_D is a non-dimensional drag of a flat plate of length x as defined by $C_D = \frac{D}{\frac{1}{2}\rho U_o^2 x}$, while the skin-friction coefficient C_f is a non-dimensional, local wall-share stress at x given by $C_f = \frac{\tau_w}{\frac{1}{2}\rho U_o^2}$.

Worked example

Obtain the skin-friction drag and the corresponding drag coefficient up to the transition point of the boundary layer in the worked example on page 9.

$$D = \int_0^x \tau_w(x)dx$$

$$= \int_0^x \frac{\rho}{2} U_o^2 \cdot C_f(x)dx$$

$$= 0.664 \cdot \frac{\rho}{2} U_o^2 \int_0^x \left(\frac{U_o x}{\nu}\right)^{-\frac{1}{2}} dx$$

$$= 0.664 \times \frac{1.2}{2} \times 1.5^2 \times \left(\frac{1.5}{1.5 \times 10^{-5}}\right)^{-\frac{1}{2}} \int_0^{10} x^{-\frac{1}{2}} dx$$

$$= 0.00275 \times 2\sqrt{10} = 0.0179\,[\text{N/m}]$$

$$\therefore \quad C_D = \frac{D}{\frac{1}{2}\rho U_o^2 L} = \frac{0.0179}{\frac{1.2}{2} \times 1.5^2 \times 10} = 0.00133$$

Alternatively, we can use $C_D = 1.328Rx^{-\frac{1}{2}}$ to give $C_D = 0.00133$, where $R_x = 10^6$.

Therefore, $D = C_D \frac{1}{2}\rho U_o^2 L = 0.0180\,[\text{N/m}]$

We could also compute the skin-friction drag from Kármán's momentum integral equation,

$$\theta = 0.664x\,Rx^{-\frac{1}{2}} = 0.664 \times 10 \times (10^6)^{-\frac{1}{2}} = 0.00664\,[\text{m}]$$

Therefore,

$$D = \rho U_o^2 \theta = 1.2 \times 1.5^2 \times 0.00664 = 0.0179\,[\text{N/m}]$$

This gives the drag coefficient of $C_D = 0.00133$.

Boundary-layer equations

Assuming that the boundary-layer thickness is small as compared to the streamwise length of development, we can derive a special form of the Navier–Stokes equations, called **Prandtl's boundary-layer equations**. In order to do that, we must make a number of assumptions.

(1) The *length scale* in the vertical (normal) direction of the boundary layer is much smaller than that of the longitudinal (streamwise) scale. In other words,

$$\Delta y \ll \Delta x \quad \text{or} \quad \frac{\partial}{\partial y} \gg \frac{\partial}{\partial x}$$

(2) The *velocity scale* in the vertical direction of the boundary layer is much smaller than that in the longitudinal direction:

$$v \ll u$$

(3) The Reynolds number is very large, so that

$$\text{Re} = \frac{uL}{\nu} \sim \left(\frac{L}{\delta}\right)^2$$

Then, the Navier–Stokes equations (1.15) and (1.16) will take a very simple form

$$u\frac{\partial u}{\partial x} + v\frac{\partial u}{\partial y} = -\frac{1}{\rho}\frac{dp}{dx} + v\frac{\partial^2 u}{\partial y^2} \tag{1.45}$$

$$0 = -\frac{1}{\rho}\frac{\partial p}{\partial y} \tag{1.46}$$

Equation (1.46) suggests that $p = p(x)$, therefore the pressure is constant across the boundary layer. Since the pressure gradient of the boundary layer over a flat plate is always zero, i.e. $\partial p/\partial x = 0$, this means that the pressure is constant everywhere in the boundary layer over a flat plate.

The derivation of the boundary-layer equations can be done using the *order of magnitude* analysis. Here, we set

$x \sim L$ (x is of the same order of magnitude as the plate length L)

$y \sim \delta$ (y is of the same order of magnitude as the boundary layer thickness δ)

$p \sim \rho u^2$ (p is of the same order of magnitude as the dynamic pressure ρu^2)

$u \sim u$

$v \sim (\delta/L) \cdot u$ (using the continuity equation, we find that v is of the same order of magnitude as $(\delta/L) \cdot u$)

After replacing x, y, p, u and v with L, δ and u, we find that the first viscous term in the Navier–Stokes equations is much smaller than the second viscous term. It should also be noted that all terms in the y-equation become zero, except for the pressure gradient term.

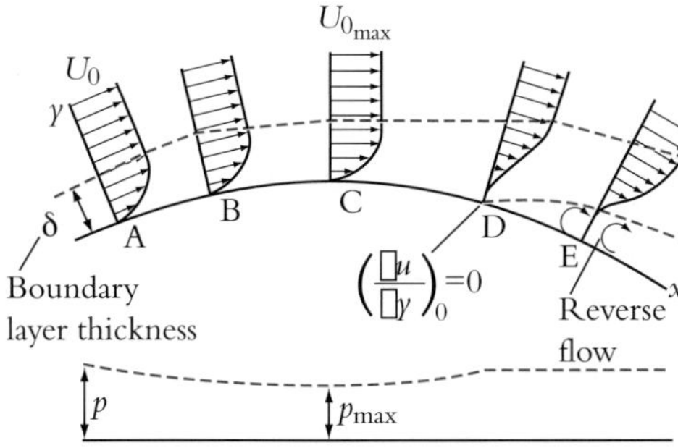

Figure 1.16 Development of the boundary layer over a convex surface, showing a retardation that leads to a flow separation with reverse flow. The flow is from left to right.

Effect of pressure gradient

So far, we have studied only the boundary layers with zero pressure gradient. However, we can extend the boundary-layer theory to cover situations with non-zero pressure gradient. Figure 1.16 shows a typical development of the boundary layer over a curved surface, where a dramatic change in the velocity profile is taking place. Where the pressure gradient is negative, or favourable, i.e. $\partial p/\partial x < 0$ (from point A to C in Figure 1.16), the lost energy of the boundary layer due to skin-friction drag can easily be replenished by the pressure force acting in the flow direction. Therefore, flow separation does not generally take place easily in such a pressure gradient condition. However, when the pressure gradient is positive, or adverse, $\partial p/\partial x > 0$ (from point C onward in Figure 1.16), the pressure force cannot easily replenish the lost momentum. This is because the pressure force acts *against* the boundary layer under an adverse pressure gradient. This leads to a separation of the boundary layer, or simply a **flow separation**, creating a region of **flow reversal** (Figure 1.16).

The flow separation point is defined as a location where the wall-shear stress becomes zero, i.e.

$$\tau_w \sim \left(\frac{\partial u}{\partial y}\right)_{y=0} = 0 \tag{1.47}$$

Therefore, the velocity gradient at $y = 0$ (at the wall) also becomes zero.

Worked example

A sports equipment company is required to design a new swimming costume for the next Olympic Games to help swimmers break the world record in the 200 m freestyle. Answer the following questions assuming that the typical swim speed is 2 m/s.

(1) Suggest how you might design a new swim cap. In doing so, you need to explain the design concept based on fluid mechanical principles.

(2) A new swimming costume must also cover the entire arms and legs. How should the design concept for these parts of the costume differ from that of a cap? Again, your answer must be based on fluid mechanical principles.

(3) What considerations should be given to the choice of fabric for the swim cap and the swimming costume? Your answer must be accompanied by clear and sound reasons.

(1) The Reynolds number of the boundary layer over a swim cap is estimated as

$$R_x = \frac{U_o x}{\nu} = \frac{2.0 \times 0.1}{1.0 \times 10^{-6}} = 2.0 \times 10^5$$

which is a subcritical Reynolds number. Therefore, the flow is laminar. In this case, there are two possible strategies in designing the swim cap.

(a) Maintain the laminar flow by making it as smooth as possible, since the skin-friction drag of laminar flows is much lower than that of turbulent flows.

(b) Promote turbulent flow by *tripping* the boundary layer, which can reduce the pressure drag of the swimmer's head by moving the separation point further downstream.

(2) To increase the thrust, the drag on arms and legs must be increased by

(a) making the surface of the swimming costume covering the arms and legs rough, thereby increasing the skin friction drag, or

(b) making the surface smooth as possible, thereby allowing the laminar flow to separate early.

(3) We must choose the fabric of the swimming costume carefully to achieve these objectives: a smooth fabric where we want laminar flow, and a rough fabric in certain parts of the costume to promote turbulence.

Learning summary

By the end of this section you should have learnt:

✓ viscous fluid does not slip at a solid wall surface. This is called the non-slip condition of flow motion;

✓ the boundary layer is a thin fluid layer near a solid wall surface, where the velocity is less than the freestream velocity;

✓ the momentum thickness signifies the loss of momentum in the boundary layer due to skin-friction drag;

✓ the displacement thickness is a measure of mass flow deficit in the boundary layer;

✓ the boundary layer equations are a simplified form of the Navier–Stokes equations;

✓ flow separation occurs over a curved surface when the static pressure increases in the flow direction.

1.4 Drag on immersed bodies

Pressure drag

While the friction drag D_{fric} results from the *viscous* action of fluids on the body surface, the pressure drag D_{pres} comes from the static pressure distribution around the body, mainly due to boundary-layer separation. The total drag acting on immersed bodies in incompressible flows, therefore, consists of the friction drag and the pressure drag. We can write

$$D_{tot}\text{ (total drag)} = D_{fric}\text{ (friction drag)} + D_{pres}\text{ (pressure drag)} \tag{1.48}$$

The relative importance of D_{pres} to D_{fric} depends on the body shape as well as the Reynolds number. When the immersed bodies are streamlined, the friction drag dominates the total drag. When the non-streamlined bodies (bluff bodies) are placed in a fluid flow, however, the total drag is dominated by the pressure drag, and the contribution of the friction drag is usually negligible.

Drags can be expressed in terms of the non-dimensional drag coefficient, which is given by

$$C_D \equiv \frac{D}{\frac{1}{2}\rho V^2 A} \tag{1.49}$$

where A (which must be specified in quoting C_D values) is either the frontal area (for thick bodies, such as motor cars) or the plan-form area (for long bodies, such as aircraft wings).

In terms of drag coefficient, equation (1.49) can be written as:

$$C_{Dtot} = C_{Dfric} + C_{Dpres} \tag{1.50}$$

Flow around a circular cylinder

For a circular cylinder with radius a and length b, the pressure drag is given by

$$D_{pres} = \int_0^{2\pi} ab(p - p_\infty)\cos\theta \, d\theta \tag{1.51}$$

whose drag coefficient is given by

$$C_{Dpres} = \int_0^{2\pi} \frac{ab(p - p_\infty)\cos\theta \, d\theta}{(2ab)\frac{1}{2}\rho V^2} = \frac{1}{2}\int_0^{2\pi} \frac{(p - p_\infty)}{\frac{1}{2}\rho V^2}\cos\theta \, d\theta = \frac{1}{2}\int_0^{2\pi} C_p \cos\theta \, d\theta \tag{1.52}$$

where the pressure coefficient C_p is given by

$$C_p = \frac{p - p_\infty}{\frac{1}{2}\rho V^2} \tag{1.53}$$

It should be noted that the frontal area of circular cylinder ($2ab$) is used to non-dimensionalize the drag to give the pressure drag coefficient. This equation suggests that C_{Dpress} can be obtained by integrating the streamwise component of C_p over the circular cylinder. The cylindrical coordinate system being used in the computation is shown in Figure 1.17, where the angle θ is measured clockwise from the frontal stagnation point.

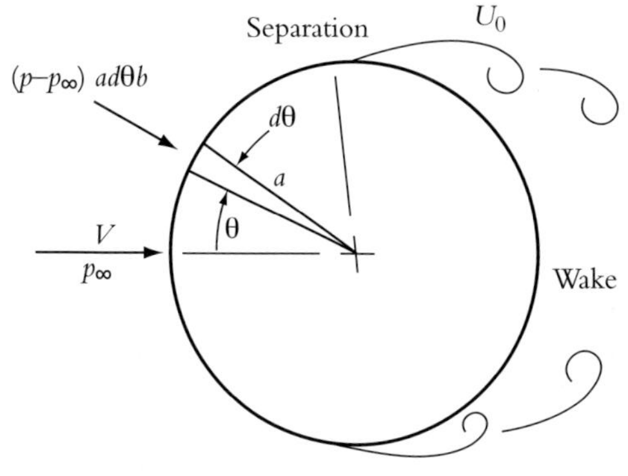

Figure 1.17 The coordinate system used for the integral of static pressure around a circular cylinder to give the pressure drag. Here, p is the static pressure over the cylinder surface and p_∞ is the freestream pressure.

Figure 1.18 compares the distribution of pressure coefficient C_p over a circular cylinder between the laminar flow and the turbulent flow. It should be noted that both C_p curves are asymmetric with respect to $\theta = 90°$, indicating that the static pressure over the front of the circular cylinder is much higher than that in the rear. The integrated pressure difference between the front and rear surfaces gives the pressure drag acting on the circular cylinder.

Figure 1.18 also shows a significant difference in the static pressure distribution between the turbulent flow and the laminar flow. While the static pressure for the laminar flow stays near the minimum value of $C_p = -1.0$ in the rear of the circular cylinder, the turbulent flow recovers to a much greater value of $C_p = -0.4$ after reaching the minimum value of $C_p = -2.1$ at around $\theta = 75°$. This reflects a small C_D value of 0.3 for the turbulent flow as compared to $C_D = 1.2$ for the laminar flow (Figure 1.19). The main reason for the smaller C_D value for turbulent flow is that the flow separation takes place much further downstream due to greater mixing capability of the turbulent flow. As a result, the wake region in the downstream of turbulent flow separation is narrower than for laminar flow (Figure 1.21).

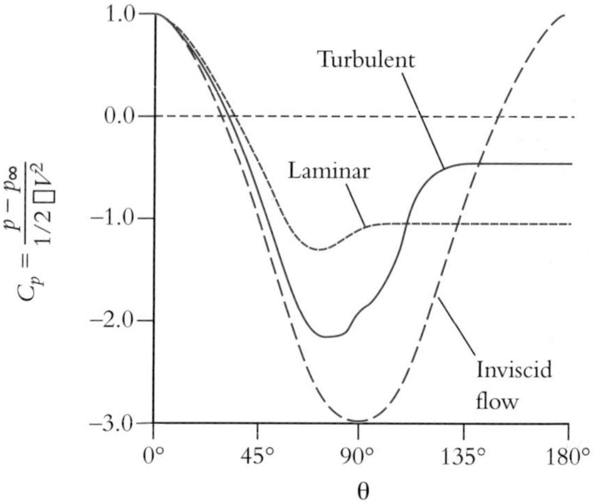

Figure 1.18 Non-dimensional pressure distribution over a circular cylinder, where the C_p curve for the laminar and turbulent flow are compared with the solution of inviscid flow.

However, inviscid theory gives a symmetric C_p curve (Figure 1.18), suggesting that the drag on a circular cylinder is zero for zero viscosity fluids. Certainly this is not a realistic assumption in calculating the drag force on immersed bodies. Indeed, inviscid theory cannot impose the non-slip condition on the wall, so there will be no boundary layer development or flow separation over the immersed bodies.

Drag of bluff bodies

As has been previously suggested, the drag coefficient $\mathbf{C_D}$ of immersed bodies is a function of the Reynolds number. The drag coefficient is gradually reduced with an increase in the Reynolds number as seen in Figure 1.19. Once the Reynolds number reaches the critical value ($Re \approx 5 \times 10^5$ for circular cylinders and spheres with smooth surface) the drag coefficient will drop suddenly. This is the transition point where the flow around the immersed bodies will become turbulent from laminar flow.

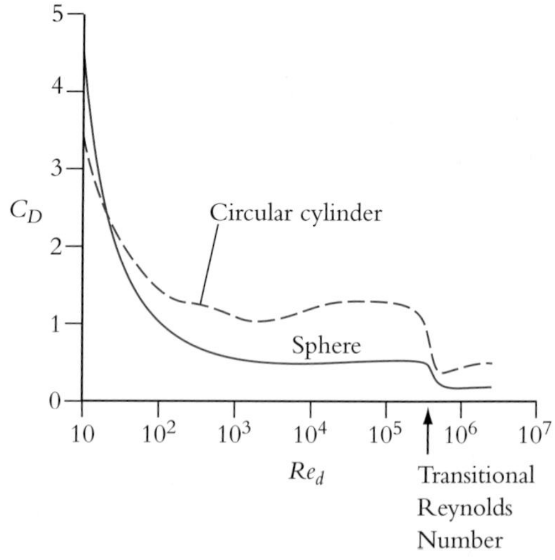

Figure 1.19 Drag coefficient of circular cylinder and sphere as a function of the Reynolds number, showing that the C_D value reduces as the transition takes place.

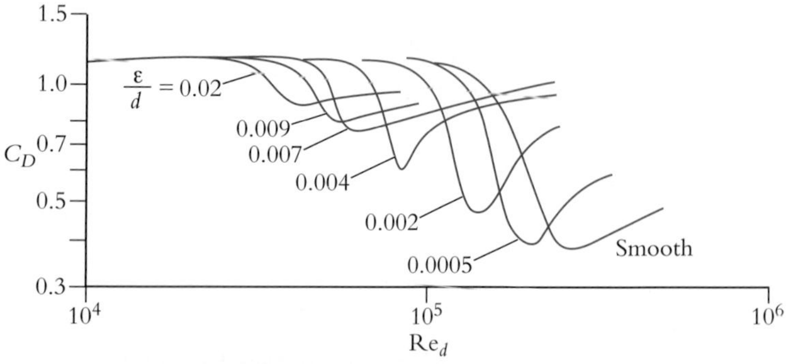

Figure 1.20 Drag coefficient of a circular cylinder, showing that the transition takes place at lower Reynolds number with an increase in the surface roughness.

If the wall surface is rough, the transition to turbulences over a circular cylinder takes place at lower Reynolds number (Figure 1.20), but the drag coefficient C_D for the turbulent flow (after transition) is greater than that with a smooth surface. Figure 1.20 also shows that the critical Reynolds number reduces with an increase in the roughness ratio ε/d. The flow over a sphere is qualitatively similar to that over a circular cylinder.

Worked example

Obtain the drag force on a baseball of 73 mm diameter at the critical Reynolds number assuming that the flow around the ball is turbulent. The density and kinematic viscosity of air are 1.2 kg/m^3 and 1.5 × 10^{-5} m^2/s, respectively.

From Table 1.2, we find that $C_D \approx 0.2$ for a sphere when the flow is turbulent, where the critical Reynolds number is $\mathrm{Re} = 3 \times 10^5$. Therefore, the drag force on the ball can be obtained by

$$D = C_D \frac{\rho}{2} U^2 A$$

where,

$$A = \frac{\pi}{4} d^2 = \frac{\pi}{4} \times (0.073)^2 = 0.0042\ [\mathrm{m}^2]$$

$$U = Re\frac{\nu}{d} = (3 \times 10^5) \times \frac{1.5 \times 10^{-5}}{0.073} = 61.6\ [\mathrm{m/s}]$$

$$\therefore \quad D = 0.2 \times \frac{1.2}{2} \times (61.6)^2 \times 0.0042 = 1.91\ [\mathrm{N}]$$

The dimples on a golf ball can reduce the pressure drag by making the ball surface rough. This reduces the transition Reynolds number by artificially forcing (tripping) the boundary layer to turbulent flow at low Reynolds numbers. As a result, the wake becomes narrower as can be seen in Figure 1.21. Although the friction drag is increased in this case, the total drag of the golf ball is reduced. This is because the golf balls are bluff (non-streamlined) bodies, where D_{pres} is much greater than D_{fric}.

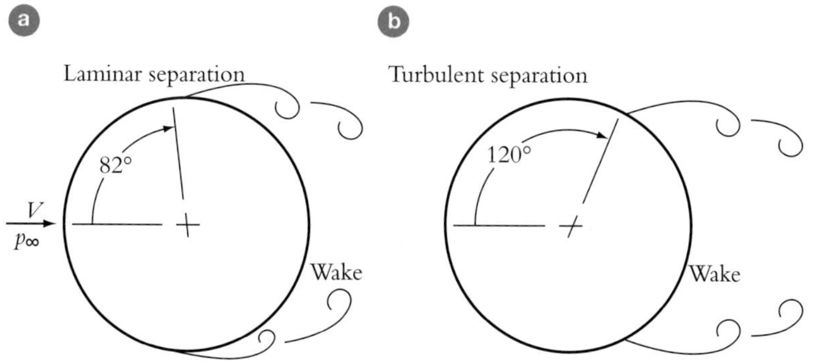

Figure 1.21 Comparison of laminar separation with turbulent separation, showing that the separation point moves further downstream when the boundary layer becomes turbulent. This reduces the pressure drag by making the wake narrower.

Tables 1.1 and 1.2 give the drag coefficient of two-dimensional and three-dimensional bodies, respectively. It should be noted here that the drag coefficient of sharp-edged bodies, such as squares and cubes, is insensitive to the Reynolds number since the flow is always separated at the sharp edges. In other words, the C_D value of sharp-edged bodies remains constant whether the flow is laminar or turbulent as long as the Reynolds number is greater than 10^4.

Shape		C_D	Shape		C_D
Plate:	⟶ │	2.0	Half-cylinder:	⟶ ◖	1.2
Square cylinder:	⟶ ☐	2.1		⟶ ◗	1.7
	⟶ ◇	1.6	Equilateral triangle:	⟶ ◁	1.6
Half tube:	⟶ (	1.2		⟶ ▷	2.0
	⟶)	2.3		Laminar	Turbulent
Elliptical cylinder:	1:1 ⟶ ◯			1.2	0.3
	2:1 ⟶			0.6	0.2
	4:1 ⟶			0.35	0.15
	8:1 ⟶			0.25	0.1

Table 1.1 Drag coefficient of two-dimensional bodies at Re $> 10^4$

Body		L/D	C_D
Cube:	⟶ ☐		1.07
	⟶ ◇		0.81
60° cone:	⟶ ◁		0.5
Disk:	⟶ │		1.17
Cup:	⟶)		1.4
	⟶ (		0.4
Parachute:	⟶ ◁		1.2
Circular cylinder:	⟶ ▭ d, L	0.5	1.15
		1	0.90
		2	0.85
		4	0.87
		8	0.99

			Laminar	Turbulent
Ellipsoid:	⟶ ⬭ d, L	0.75	0.5	0.2
		1	0.47	0.2
		2	0.27	0.13
		4	0.25	0.1
		8	0.2	0.08

Table 1.2 Drag coefficient of three-dimensional bodies at Re $> 10^4$

Worked example

The fork ball is a baseball pitch thrown like a straight ball but with little or no rotation, where the ball initially travels straight but falls sharply as it gets closer to the batter who is standing about 18 m away from the pitcher.

(1) Draw a figure showing the drag coefficient of the baseball as a function of the Reynolds number by considering the baseball as a smooth sphere.

(2) Obtain the drag force on a baseball with 73 mm diameter at the critical Reynolds number assuming that the flow around the ball is turbulent. The density and kinematic viscosity of air are 1.2 kg/m^3 and 1.5×10^{-5} m^2/s, respectively.

(3) By what percentage does the drag force change if the flow becomes laminar rather than turbulent?

(4) Explain the behaviour of the fork ball as described above using the principles of fluid mechanics.

(5) Discuss how the pitcher should adjust the delivery of the fork ball for it to remain effective if the ball surface becomes rough during a game.

(1)

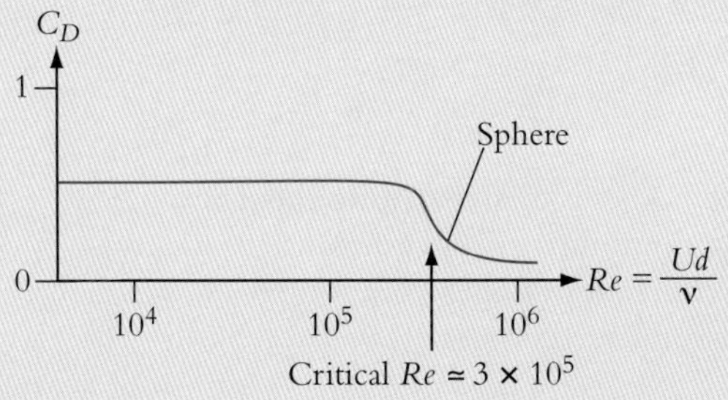

Figure 1.22

(2) $C_D \approx 0.2$ for a sphere when the flow is turbulent, where

$$C_D = \frac{D}{\frac{\rho}{2}U^2 A}, A = \frac{\pi d^2}{4} = 0.0042\,\text{m}^2 \quad \text{and} \quad U = \text{Re}\frac{\nu}{d} = 61.6\,\text{m/s}$$

Therefore,

$$D = 0.2 \times \frac{1.2}{2} \times (61.6)^2 \times 0.0042 = 1.91\,\text{N}$$

(3) When the flow around the sphere is laminar, $C_D \approx 0.5$. Therefore, the drag will be increased to 2.5 times (150% increase).

(4) The fork ball is the result of reverse transition (from turbulent to laminar rather than usual laminar to turbulent route) of flow around the ball. During this transition, the ball will experience a 150% increase in drag, resulting in a sharp drop near the batter.

(5) As shown in Figure 1.20 for a circular cylinder (similar for a sphere), the drag increase will be smaller when the surface is rough. Therefore, the amount of drop will be reduced as a result of ball roughness. The critical Reynolds number for flow transition will be lowered so that the pitcher must throw a fork ball with a lower initial speed for it to be effective.

Worked example

A man jumped from an airplane with a parachute of 7.3 m in diameter. Assuming that the total mass of the man and parachute is 80 kg, calculate the speed of descent when he reaches terminal velocity.

The drag coefficient of parachute is $C_D = 1.2$ regardless of the Reynolds number, as it is a "sharp-edged" body.

The terminal velocity will be reached when the drag of the parachute is balanced by the weight of the parachute and the man, i.e. $D = W$.

Here,

$$D = C_D \tfrac{1}{2}\rho A V^2 = 1.2 \times (1.2/2) \times (\pi/4) \times (7.3)^2\, V^2 \text{ [N]}$$

$$W = Mg = 80 \times 9.8 \text{ [N]}$$

$$\therefore \quad V^2 = \frac{80 \times 9.8}{1.2 \times (1.2/2) \times (\pi/4) \times (7.3)^2} = 26.0 \text{ [m}^2\text{/s}^2\text{]}$$

to give $V = 5.1$ [m/s]

Streamlining strategy

An important strategy in reducing pressure drag $\boldsymbol{D_{pres}}$ of immersed bodies is to **streamline** them, by shaping the bodies in such a way as to move the flow separation point further downstream. This will effectively reduce the width of wake (the area in the downstream of flow separation), leading to a reduction of the low pressure region in the rear of the immersed bodies.

Figure 1.23 shows a procedure for streamlining a rectangular cylinder that has sharp corners. The drag can be easily reduced to nearly a half by rounding the front corners of a cylinder, which reduces the drag coefficient C_D from 2.0 to 1.0. A more dramatic reduction in drag can be obtained by tapering the rear corners, resulting in a reduction of drag to nearly one seventh of its original. It is surprising to observe that a fully streamlined cylindrical body is equivalent in terms of the total drag with a circular cylinder one tenth of its width. This shows the effectiveness of streamlining strategy in reducing drag by tapering the rear of immersed bodies.

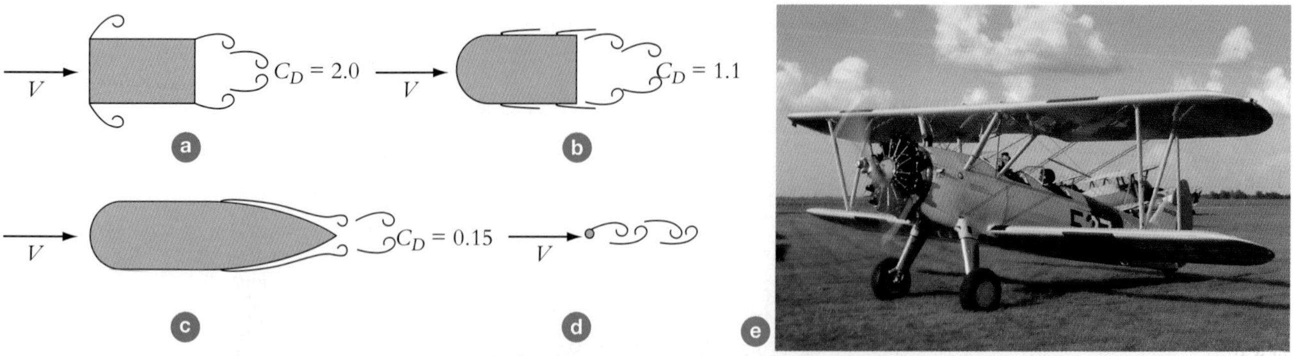

Figure 1.23 Reducing the drag coefficient through streamlining strategy. The drag coefficient of a rectangular cylinder (a) can be reduced by tapering the front (b) and rear (c) of the body. A circular cylinder, as shown in (d), has a diameter that is nearly one tenth of the width of the streamlined cylinder (c), yet both have the same drag. With its cylindrical wing struts, no wonder a biplane (e) cannot fly very fast.

The width of the wake region can be reduced if the flow separation is moved back towards the rear of the body. In practice, however, it is often difficult to do this as it may reduce the capacity (volume) of the vehicle. Instead, trucks can benefit much by attaching a deflector on top of the cab (Figure 1.24), which reduces a large separation region in front of the trailer, leading to a drag reduction of up to 20%.

Figure 1.24 A deflector can reduce the pressure drag of a truck by 20%, by steering the streamlines away from the frontal surface of the trailer. If there is no trailer attached to the cab, however, there will be a large increase in drag.

Learning summary

By the end of this section you should have learnt:

- ✓ pressure drag is a result of the boundary layer separation, where the static pressure difference is created between the front and rear of the bodies;
- ✓ drag coefficient of immersed bodies is reduced with an increase in the Reynolds number when the flow is laminar;
- ✓ drag coefficient of immersed bodies is suddenly reduced at the critical Reynolds number when the flow becomes turbulent;
- ✓ surface roughness will reduce the critical Reynolds number of immersed bodies, thereby reducing their drag at lower Reynolds number;
- ✓ streamlining is an effective strategy for reducing drag, where the immersed bodies are rounded at the front and tapered at the rear.

1.5 Flow through pipes and ducts

Friction factor

The friction factor f of pipe flows, as defined by

$$f = \frac{h_f}{\frac{L}{d}\frac{V^2}{2g}} \tag{1.54}$$

is a non-dimensional form of the frictional head loss h_f. The friction factor is a function of the Reynolds number $R_d = \frac{Vd}{\nu}$ as well as the relative surface roughness $\frac{\varepsilon}{d}$, where V is the bulk velocity, d is the pipe diameter, ν is the kinematic viscosity of fluid and ε is the typical surface roughness height.

For laminar flows ($R_d < 2 \times 10^3$) it can be shown that the friction factor is the only function of the Reynolds number, where

$$f = \frac{64}{R_d} \quad \textbf{(1.55)}$$

This is called the Darcy–Weisbach equation for laminar pipe flows. It should be noted that the surface roughness does not affect the friction factor for laminar pipe flows.

For turbulent pipe flows, Colebrook gave the following formula for f, covering a wide range of Reynolds number and surface roughness

$$\frac{1}{\sqrt{f}} = -2.0\log\left(\frac{\varepsilon/d}{3.7} + \frac{2.51}{R_d\sqrt{f}}\right) \quad \textbf{(1.56)}$$

Although accurate in presenting the friction factor for both transitional and fully turbulent pipe flows ($R_d > 4 \times 10^3$), this formula is difficult to use in practice since the friction factor is not given in a closed form. In other words, iteration is required to obtain the friction factor from this equation for a given Reynolds number and roughness ratio. It is for this reason Moody has presented a chart where the friction factor can be easily read. This is called the **Moody chart**, where the friction factor is given as a function of the Reynolds number R_d and the roughness ratio $\frac{\varepsilon}{d}$ (Figure 1.25). Following a curve of constant $\frac{\varepsilon}{d}$ value (as shown on the right-hand side of the chart) to meet a constant Reynolds number line, one can read off the friction factor on the left-hand side of the chart. Typical surface roughness ε for pipes and ducts, from iron to concrete, can be found in Table 1.3.

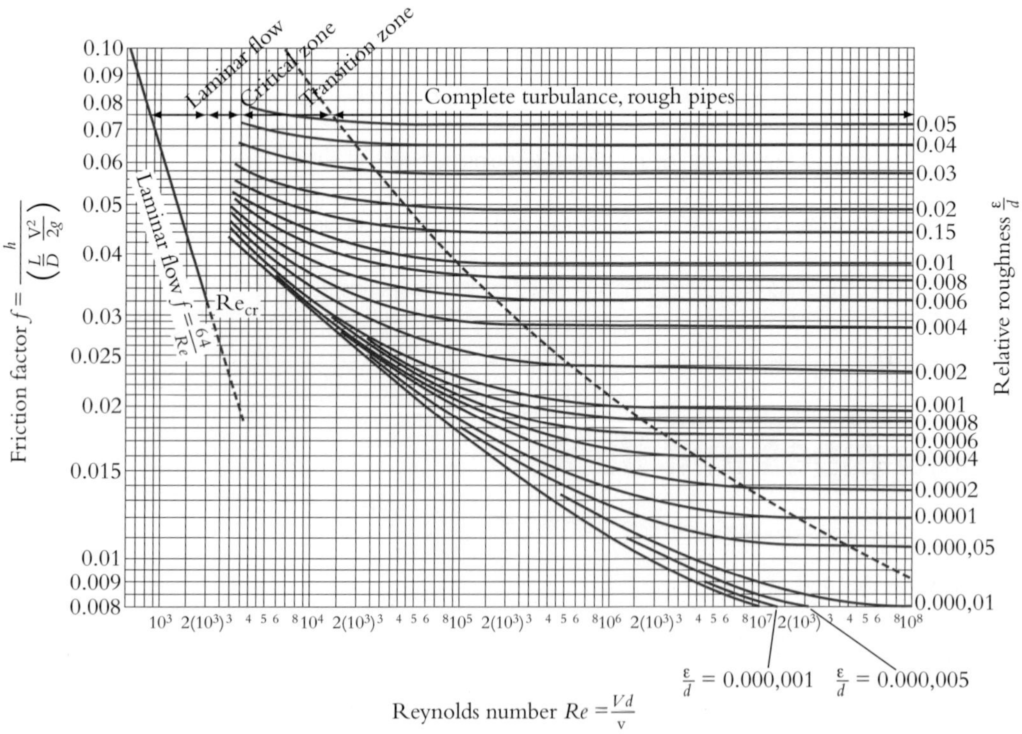

Figure 1.25 Moody chart (taken from F. M. White, 2008, *Fluid Mechanics*, New York: McGraw Hill and reproduced with permission of the McGraw-Hill Companies)

Material	Condition	ε mm	Uncertainty, %
Steel	Sheet metal, new	0.05	±60
	Stainless, new	0.002	±50
	Commercial, new	0.046	±30
Iron	Cast, new	0.26	±50
	Wrought, new	0.046	±20
	Galvanized, new	0.15	±40
Brass	Drawn, new	0.002	±50
Plastic	Drawn tubing	0.0015	±60
Glass	—	Smooth	
Concrete	Smoothed	0.04	±60
	Rough	2.0	±50
Rubber	Smoothed	0.01	±60
Wood	Stave	0.5	±40

Table 1.3 Typical surface roughness in pipe and channel flows

There are two different definitions for the friction factor. The Darcy friction factor is used throughout this textbook, which is defined by

$$f = \frac{h_f}{\frac{L}{d}\frac{V^2}{2g}} \tag{1.57}$$

Meanwhile, the Fanning friction factor is given by

$$f_F = \frac{h_f}{4\frac{L}{d}\frac{V^2}{2g}} \tag{1.58}$$

which contains a factor of 4 in its definitions. Since the friction head is given by

$$h_f = \frac{\Delta p}{\rho g}$$

where,

$$\Delta p = \frac{\pi DL}{\frac{\pi}{4}d^2} \cdot \tau_w = \frac{4L}{d} \cdot \tau_w$$

Therefore,

$$f = \frac{\frac{1}{\rho g} \cdot \frac{4L}{d} \cdot \tau_w}{\frac{L}{d}\frac{V^2}{2g}} = 4 \cdot \frac{\tau_w}{\frac{\rho}{2}V^2} = 4C_f$$

Therefore, the Darcy friction factor is related to the skin-friction coefficient through

$$f = 4C_f \tag{1.59}$$

Using this relationship, the friction velocity u^* of the pipe flow can be given in terms of friction factor f as

$$f = 8\left(\frac{u^*}{V}\right)^2 \quad \text{or} \quad u^* = \sqrt{\frac{f}{8}}\,V \tag{1.60}$$

If we use the Fanning friction factor, we have $f_F = C_f$ instead.

Minor losses

Whenever there are changes in velocity magnitude or direction in a pipe or duct system, there will be associated pressure drops, called minor losses. The minor losses are typically found at

- the entrance to the pipe or exit
- sudden expansion or contraction
- bends
- valves.

The minor losses are caused by the internal flow separation as a result of changes in the magnitude or direction of the flow through the pipes or ducts. These are similar to the pressure reductions along the immersed bodies as a result of boundary layer separation. Although they are called minor losses, the pressure drops can be a significant part of the total pressure drop when the pipe or duct has a short straight section.

The minor head loss h_m in a duct or pipe system is expressed by

$$h_m = K \cdot \frac{V^2}{2g} \qquad (1.61)$$

where, K is the minor loss coefficient or the K-factor. Therefore, the total head loss through the pipe system is given by summing the frictional head loss h_f and all the minor losses, to give

$$h_{tot} = h_f + \sum h_m \qquad (1.62)$$

The frictional head loss h_f through a pipe or duct is given from equation (1.54) as

$$h_f = f \frac{L}{d} \frac{V^2}{2g} \qquad (1.63)$$

Examples of flow causing minor losses are shown in Figure 1.26, where the flow direction changes through a circular and square bend. Both flows are separated at the bend, but the degree of flow separation and the turbulence being produced are very different. The turbulence is much greater in a flow through a square bend than through a circular bend. This explains why the minor loss is much greater for the flow around a square bend. Therefore, the use of square bends should be avoided in a pipe or duct system.

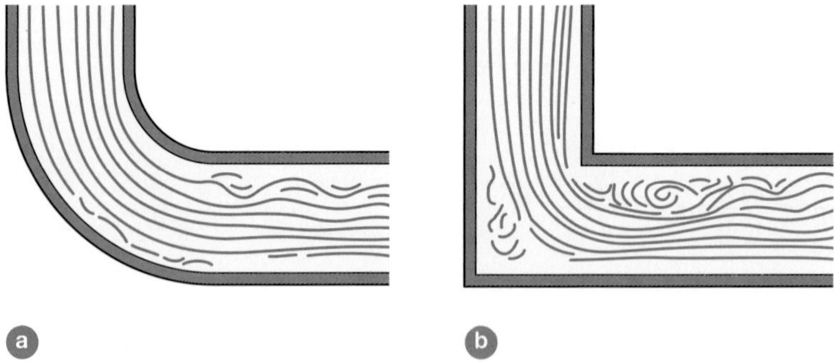

Figure 1.26 Flow through a circular (a) and square (b) bends

The minor loss coefficient for mitre bends including that of the square bend is given in Figure 1.27. There is no Reynolds number dependency on this value since the flow is always separated at the sharp corner. Figure 1.28 shows the minor loss coefficient of circular bends at $\mathrm{Re} = 10^6$ as a function of the bend-radius-to-pipe-diameter ratio $\frac{r}{d}$ and the bend angle θ_b. Here, the minor loss coefficient depends on the Reynolds number, so a correction factor given in Figure 1.29 should be applied to this value.

Other examples for large minor loss in a pipe and duct system include sudden contraction and expansion, where the flow separation takes place at the junctions (insets in Figure 1.30). The K-factors are a function of the rate of contraction or expansion.

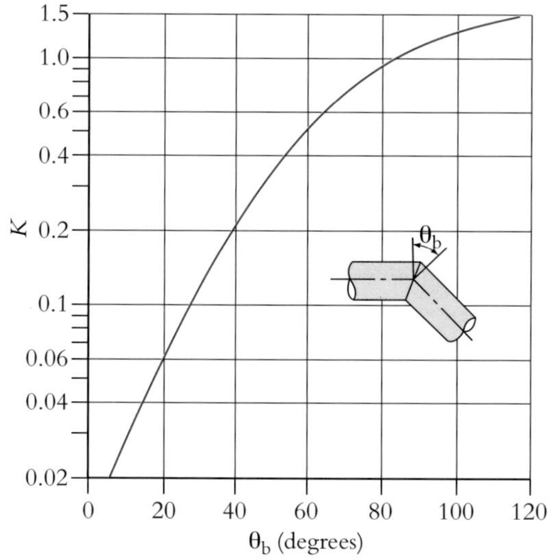

Figure 1.27 Mitre bend loss coefficient

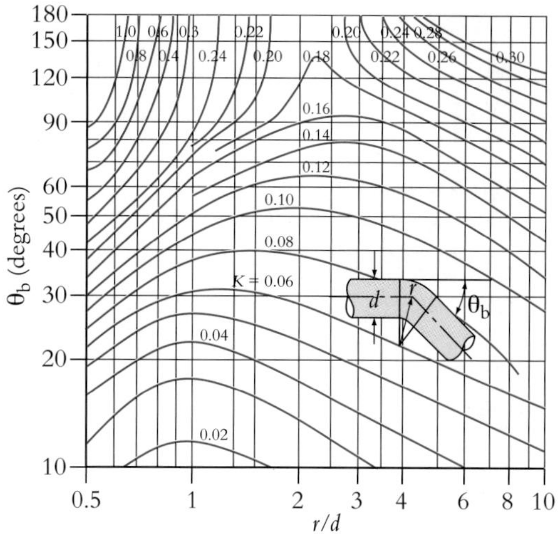

Figure 1.28 The loss coefficient of circular bends at Re = 10^6. For other Reynolds numbers this coefficient must be multiplied by the correction factor given in Figure 1.29

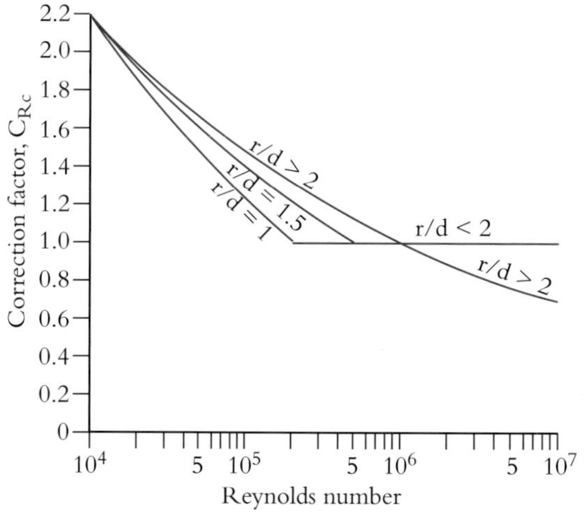

Figure 1.29 The Reynolds number correction factor for circular bend loss coefficient

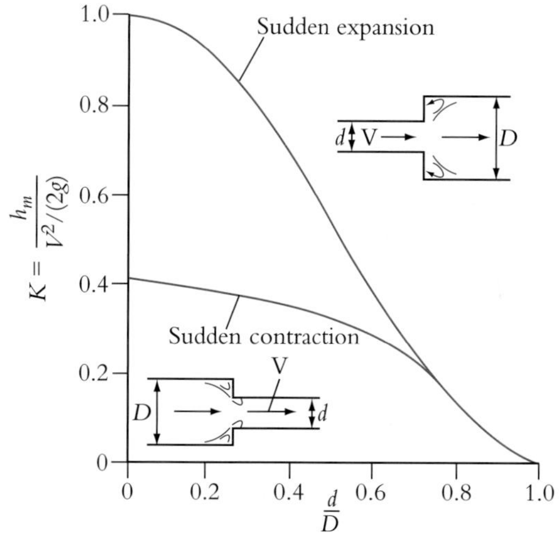

Figure 1.30 Minor losses for the flow through a sudden contraction or expansion as a function of $\frac{d}{D}$

Figure 1.31 illustrates some commercial valve geometries. Typical *K*-factor for the gate and disk valves is $K \approx 0.2$ when they are fully open, while it is $K \approx 4$ for the globe valve. Valves are the main source of minor losses in a pipe system as one can see in Figure 1.32, showing typical values of the *K*-factor when valves are partially open. It is shown that the head loss in the pipe or duct system will be increased by more than 100 times when a gate valve is closed by 75%. We must be careful, therefore, in the selection and use of the valves in a pipe and duct system.

There are further sources for minor losses in a pipe system. Figure 1.33 shows the entry losses for different entry geometry. As we would expect, the *K*-factor for the pipe entry is a function of the relative radius and length of the entry. Note that the *K*-factor is always unity for a sharp exit from the pipe.

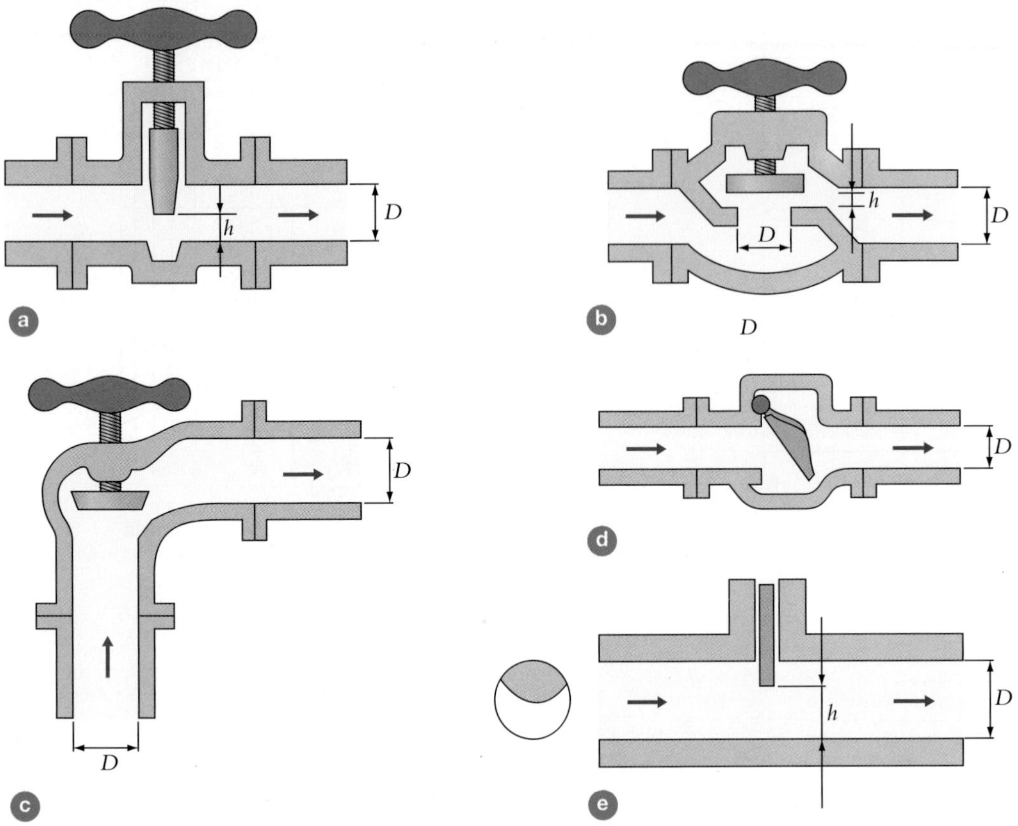

Figure 1.31 **Commercial valve geometries. (a) gate valve; (b) globe valve; (c) angle valve; (d) swing-check valve; (e) disk-type gate valve (F. M. White, 2008, *Fluid Mechanics*, New York: McGraw Hill. Reproduced with permission of The McGraw-Hill Companies)**

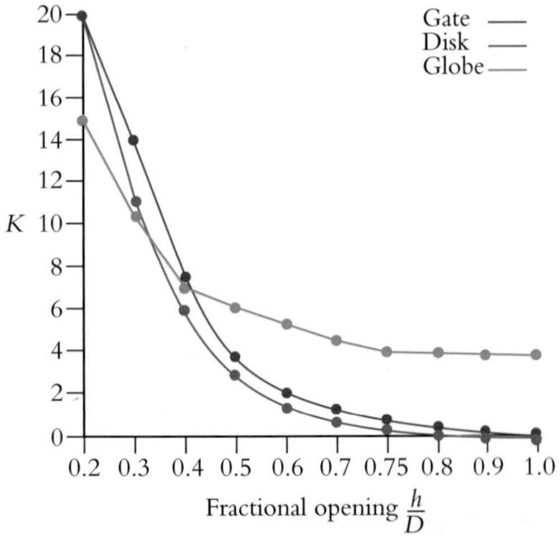

Figure 1.32 **Typical minor losses of valves when they are partially open (F. M. White, 2008, *Fluid Mechanics*, New York: McGraw Hill. Reproduced with permission of The McGraw-Hill Companies)**

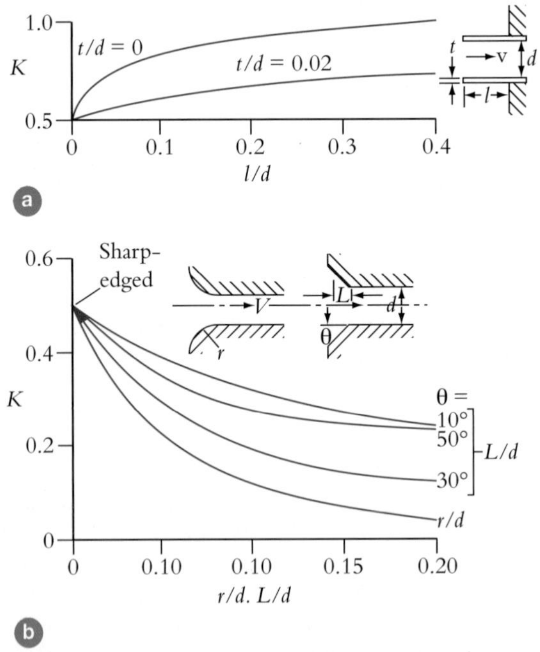

Figure 1.33 **Entry losses to re-entrant inlets (a), rounded inlets (b) and bevelled inlets (c). Note that the exit losses are K = 1 for all exit shapes. (F. M. White, 2008, *Fluid Mechanics*, New York: McGraw Hill. Reproduced with permission of The McGraw-Hill Companies)**

Hydraulic diameter

When the pipes and ducts are *not* circular, we can use the *hydraulic diameter* $\boldsymbol{D_h}$ in place for the diameter of the circular pipe to calculate pipe losses. The hydraulic diameter is defined by

$$D_h = 4 \times \frac{\text{cross-sectional areas}}{\text{wetted perimeter}} \tag{1.64}$$

With this concept, we can obtain the friction factor of non-circular pipes and ducts using the Moody chart just as we have obtained the friction factor for a circular pipe from it. Here, the Reynolds number and the relative roughness can be defined by $\frac{VD_h}{\nu}$ and $\frac{\varepsilon}{D_h}$, respectively, using the hydraulic diameter D_h instead of d, while the frictional head loss is given by

$$h_f = f\frac{L}{D_h}\frac{V^2}{2g} \tag{1.65}$$

The hydraulic diameter of a circular pipe is its physical diameter; that is $D_h = d$.

Worked example

Calculate the hydraulic diameter D_h of the following pipes and channels.

parallel plates

$$D_h = 4 \times \frac{hl}{2l}$$

$$\therefore \quad D_h = 2h$$

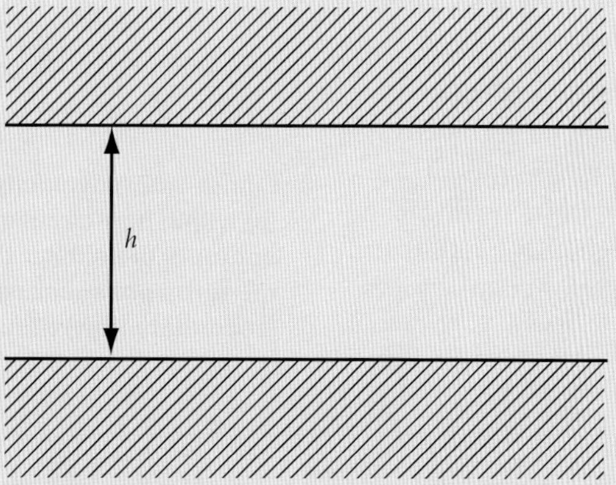

Figure 1.34

concentric cylinders

$$D_h = 4 \times \frac{\frac{\pi}{4}(D^2 - d^2)}{\pi(D + d)}$$

$$\therefore \quad D_h = D - d$$

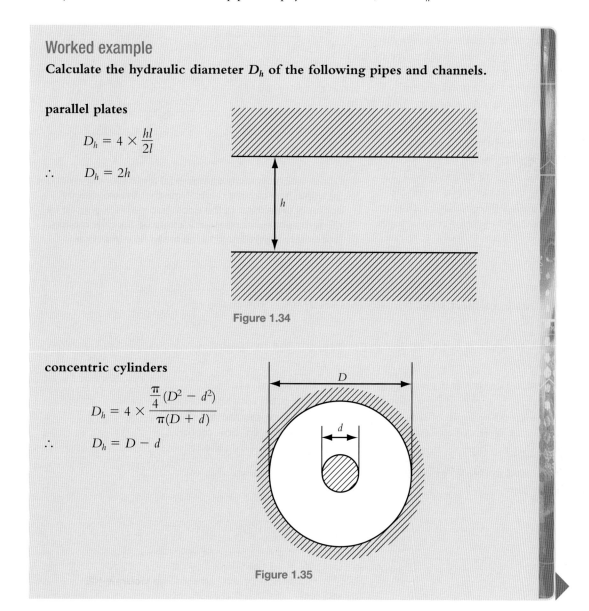

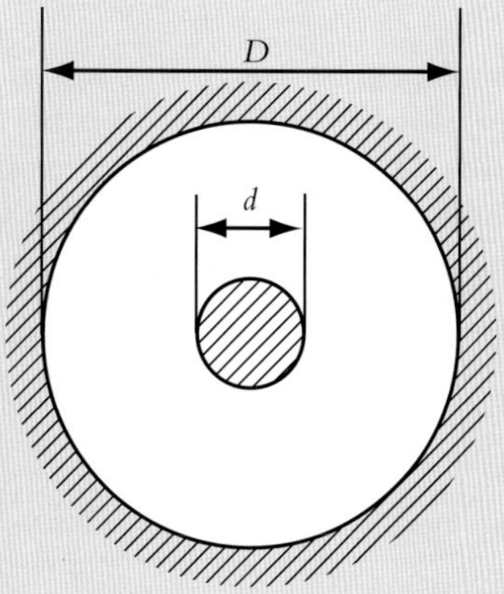

Figure 1.35

rectangle

$$D_h = 4 \times \frac{lh}{2(l + h)}$$

$$\therefore \quad D_h = \frac{2lh}{(l + h)}$$

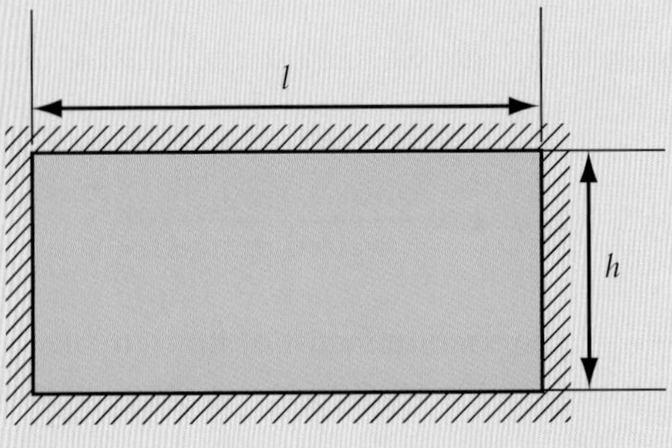

Figure 1.36

Secondary flows

Secondary flows can be observed in non-circular pipes and ducts, where the fluid near the corners has a movement across the mean flow direction. The secondary flows are driven by the *turbulent-shear stresses*, which act towards the corners of non-circular ducts. As a result, the isovelocity contours are similar in shape to the non-circular pipe or duct cross sections. This is why the hydraulic diameter concept works well for turbulent pipe or duct flows. Figure 1.37 shows the isovelocity contours (a) and secondary motions (b) across a triangular pipe section normal to the mean flow.

Secondary flows in non-circular pipes and ducts are turbulent flow phenomena, so there are no secondary motions for laminar flows. There are other types of secondary flows, which are caused by the centrifugal force acting on the flow around the circular bend. These secondary flows can be observed even for circular pipes or laminar flows.

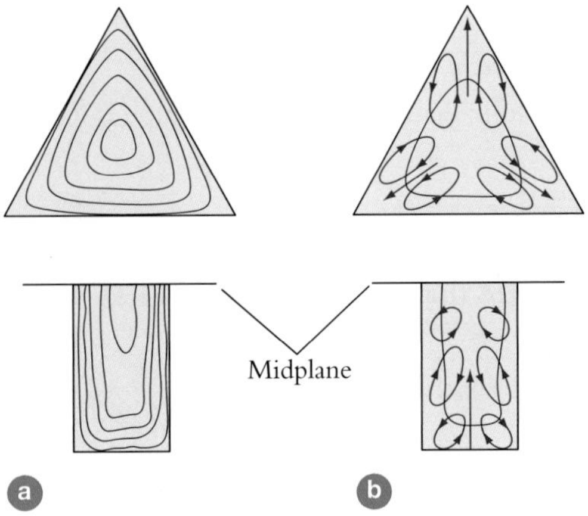

Figure 1.37 Isovelocity contours (a) and secondary motions (b) within non-circular ducts. Note that the isovelocity contours are similar to the duct shape. This is caused by the turbulent shear stresses, setting up the flow motion (secondary flow) normal to the mean flow direction.

Worked example

A section of pipe is made of sheet steel, whose cross section has the shape of an equilateral triangle.

(1) If one side of the triangle is h, obtain the expression for the hydraulic diameter of this pipe.

(2) Air of density $\rho = 1.2\,\text{kg/m}^3$ and viscosity $\mu = 1.8 \times 10^{-5}\,\text{kg/m·s}$ at atmospheric pressure is supplied by a blower through 30 m pipe with the cross section of an equilateral triangle of side 22.5 mm. Obtain an expression for the frictional head loss in terms of the volumetric flow rate and the friction factor. You can assume that the pipe section is straight and is placed horizontally.

(3) If the blower is rated at 746 W of output power, obtain the maximum flow rate through the pipe.

(1) The wetted perimeter of a triangle of side h is $3h$, and its cross-sectional area is given by

$$A_c = \frac{1}{2} \times h \times \left(\frac{h}{2} \tan 60°\right) = \frac{\sqrt{3}}{4} h^2$$

Therefore,

$$D_h = 4 \times \frac{\frac{\sqrt{3}}{4}h^2}{3h} = \frac{\sqrt{3}}{3}h^2$$

(2) Since $h = 22.5 \times 10^{-3}$ m, $A_c = 2.19 \times 10^{-4}$ m^2 and $D_h = 1.30 \times 10^{-2}$ m
The friction head loss is given by

$$h_f = f\frac{L}{D_h}\frac{V^2}{2g} = f\frac{L}{D_h}\frac{\left(\frac{Q}{Ac}\right)^2}{2g} = \frac{L}{D_h}\frac{A_c^{-2}}{2g}fQ^2$$

$$= \frac{30 \times (2.19 \times 10^{-4})^{-2}}{(1.30 \times 10^{-2}) \times 2 \times 9.8}fQ^2$$

$$= 2.45 \times 10^9 fQ^2$$

(3) The power is given by

$$P = 746 = \rho g h_f Q = 1.2 \times 9.8 \times (2.45 \times 10^9)(fQ^2)Q$$

$$fQ^3 = 2.59 \times 10^{-8} \quad \text{(i)}$$

Here, the pipe roughness ratio is given by

$$\frac{\varepsilon}{D_h} = \frac{0.05}{13.0} = 3.85 \times 10^{-3} \quad \text{(ii)}$$

The Reynolds number is

$$\mathrm{Re} = \frac{\left(\frac{Q}{A_c}\right)D_h}{\left(\frac{\mu}{\rho}\right)} = \frac{\left(\frac{1.30 \times 10^{-2}}{2.19 \times 10^{-4}}\right)}{\left(\frac{1.8 \times 10^{-5}}{1.2}\right)}Q = 4.0 \times 10^6\, Q \quad \text{(iii)}$$

The flow rate Q can be obtained by iteration through the following process.

(1) Assume an appropriate flow rate Q.
(2) Compute ε/D_h and Re using equations **(ii)** and **(iii)**, respectively.
(3) Look up f against ε/D_h and Re using the Moody chart.
(4) Check to see if f and Q satisfy equation **(i)**.
(5) If not, change the flow rate Q and repeat the above until it converges.

You should get Q = 0.0094 m^3/s.

Learning summary

By the end of this section you should have learnt:

- ✓ the friction factor of a pipe flow is a function of the Reynolds number and the surface roughness ratio, which can be obtained from the Moody chart;
- ✓ whenever there are changes in velocity magnitude or direction in a pipe or duct system, there will be associated pressure drops, called minor losses;
- ✓ the total head loss through the pipe system is obtained by adding the frictional head loss and all the minor losses;
- ✓ when the pipes and ducts are not circular, we can use the hydraulic diameter D_h to calculate the pipe losses;
- ✓ the secondary flows in non-circular pipes and ducts are driven by the turbulent-shear stresses which act towards the corners of non-circular ducts.

1.6 Dimensional analysis in fluid dynamics

Non-dimensional numbers

Non-dimensional numbers such as the Reynolds number, Re and the Froude number, Fr are important in understanding the characteristics of the flow as well as in comparing the flow behaviour with others. These non-dimensional quantities can be obtained as a ratio of two different physical quantities. For example, the Reynolds number represents the ratio of inertial force to viscous force, while the Froude number is given as a ratio of inertial force to gravity force. The definition of non-dimensional numbers appearing in many fluid flows and their physical ratios are summarized in Table 1.4.

Parameter	Definition	Ratio	Situation
Cavitation number (Euler number)	$\mathrm{Ca} = \dfrac{p - p_v}{\rho U^2}$	$\dfrac{\text{Pressure}}{\text{Inertia}}$	Cavitation
Drag coefficient	$C_D = \dfrac{D}{\frac{1}{2}\rho U^2 A}$	$\dfrac{\text{Drag force}}{\text{Dynamic force}}$	Aerodynamics, hydrodynamics
Froude number	$\mathrm{Fr} = \dfrac{U^2}{gL}$	$\dfrac{\text{Inertia}}{\text{Gravity}}$	Free-surface flow
Grashof number	$\mathrm{Gr} = \dfrac{\beta \Delta T g L^3 \rho^2}{\mu^2}$	$\dfrac{\text{Buoyancy}}{\text{Viscosity}}$	Natural convection
Lift coefficient	$C_L = \dfrac{L}{\frac{1}{2}\rho U^2 A}$	$\dfrac{\text{Lift force}}{\text{Dynamic force}}$	Aerodynamics, hydrodynamics
Mach number	$\mathrm{Ma} = \dfrac{U}{a}$	$\dfrac{\text{Flow speed}}{\text{Sound speed}}$	Compressible flow
Pressure coefficient	$C_P = \dfrac{p - p_\infty}{\frac{1}{2}\rho U^2}$	$\dfrac{\text{Static pressure}}{\text{Dynamic pressure}}$	Aerodynamics, hydrodynamics
Prandtl number	$\mathrm{Pr} = \dfrac{\mu c_p}{k}$	$\dfrac{\text{Dissipation}}{\text{Conduction}}$	Heat convection
Reynolds number	$\mathrm{Re} = \dfrac{\rho U L}{\mu}$	$\dfrac{\text{Inertia}}{\text{Viscosity}}$	Viscous flow
Roughness ratio	$\dfrac{\varepsilon}{L}$	$\dfrac{\text{Wall roughness}}{\text{Body length}}$	Turbulent, rough walls
Strouhal number	$\mathrm{St} = \dfrac{\omega L}{U}$	$\dfrac{\text{Oscillation}}{\text{Mean speed}}$	Oscillating flow
Weber number	$\mathrm{We} = \dfrac{\rho U^2 L}{\Upsilon}$	$\dfrac{\text{Inertia}}{\text{Surface tension}}$	Free-surface flow

Table 1.4 Non-dimensional numbers in fluid dynamics

The **Strouhal number**, St, is a non-dimensional frequency of flow oscillation, appearing in such a situation as Kármán's vortex street (Figure 1.38). As seen in Figure 1.38 the Strouhal number of vortex shedding from a circular cylinder is nearly constant (St $\approx$ 0.2) over a large range of the Reynolds number.

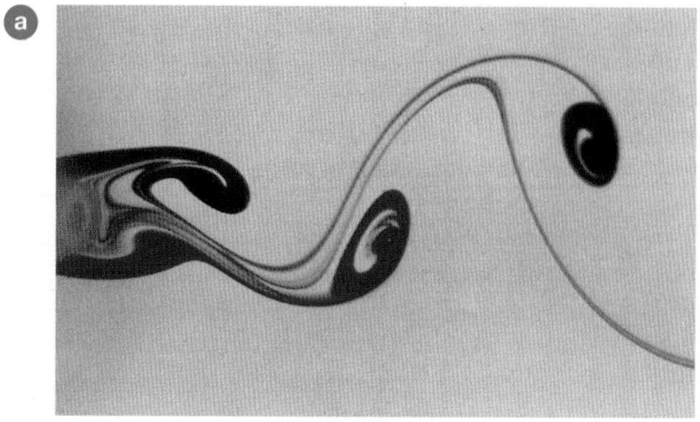

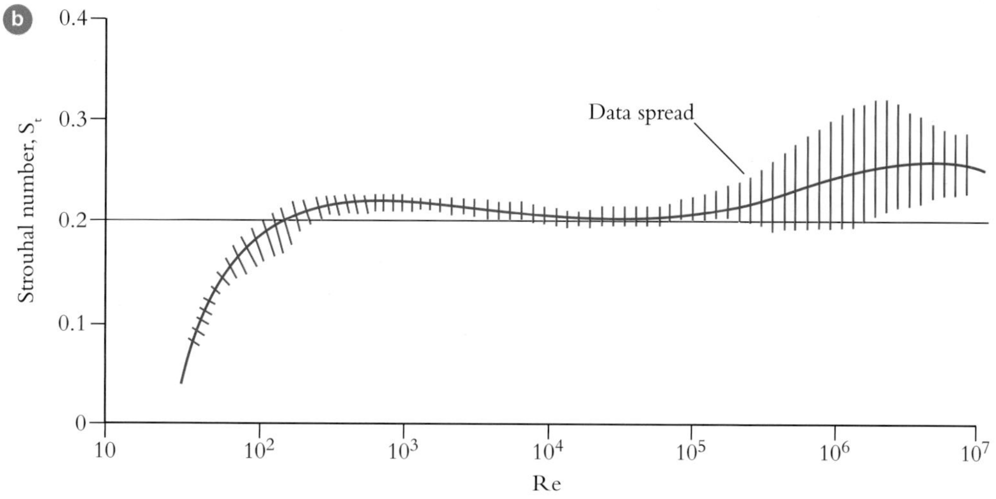

Figure 1.38 Vortex shedding from a circular cylinder. (a) Flow visualization; (b) Strouhal number (non-dimensional frequency for vortex shedding) as a function of the Reynolds number.

Worked example

A chimney of 2 m in diameter and 40 m high is subjected to 22.35 m/s storm winds, where the density and dynamic viscosity of air are 1.225 kg/m^3 and 1.78×10^{-5} Pa · s, respectively.

(1) Obtain the Reynolds number of the flow around the chimney, assuming that the velocity of the storm winds is uniform along the chimney. Is the flow laminar or turbulent at this Reynolds number?

(2) Estimate the frequency of the vortex shedding from the chimney.

(1) The Reynolds number is given by

$$\mathrm{Re} = \frac{Ud}{\mu/\rho} = \frac{22.35 \times 2}{1.78 \times 10^{-5}/1.225} = 3.08 \times 10^6$$

Therefore, the flow around the chimney is turbulent.

(2) From Figure 1.38, we get the value for the Strouhal number as

$$St = \frac{fd}{U} \approx 0.25$$

Therefore,

$$f = St\frac{U}{d} = 0.25 \times \frac{22.35}{2} = 2.8\,\mathrm{Hz}$$

Mach number, Ma, is a ratio of the flow speed to the speed of sound, indicating the compressibility effect of the fluid. Figure 1.39 presents the pictures of a flying bullet visualized by the shadowgraph technique, showing the shock waves over the bullet body at a high Mach number.

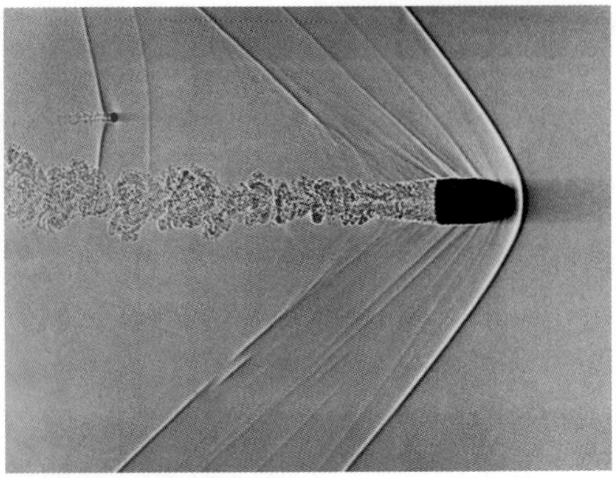

Figure 1.39 **Effect of Mach number on the shockwave around a bullet**

Weber number, We, indicates a relative magnitude of the inertial force to the surface tension. This non-dimensional number becomes important when there is significant effect of surface tension in the flow phenomena, such as the droplets and in capillary flows. Figure 1.40 shows a sequence of pictures to show the break-up of droplets, where the Weber number plays a significant role in determining its behaviour.

Figure 1.40 **The break-up of droplets is governed by the Weber number**

Grashof number, Gr, is defined as a ratio of the buoyancy force to the viscous force. The Grashof number becomes important when a strong buoyancy effect is considered in a situation such as natural convection. Figure 1.41 shows the pattern of thermal convection between two concentric cylinders visualized by cigarette smoke, where the outer cylinder is cooled by 14.5 K. This gives the Grashof number of 120 000 based on the gap distance.

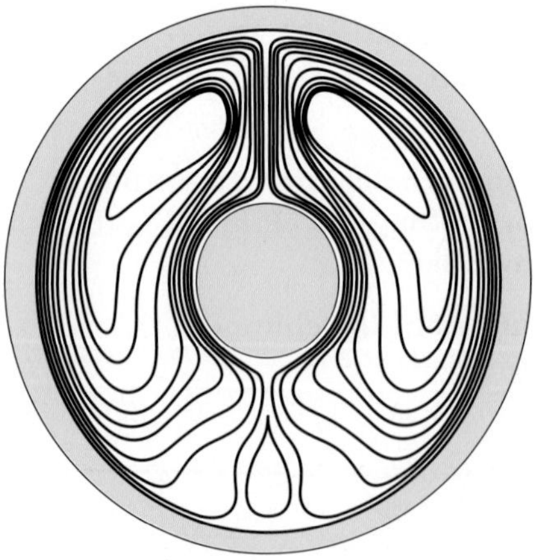

Figure 1.41 **The streamlines in convection between concentric cylinders, which is governed by the Grashof number**

Drag coefficient C_D and **lift coefficient** C_L are non-dimensional numbers used for lift-generating bodies, such as wings and propellers. They represent the drag force and the lift force, respectively, acting on the bodies, relative to the dynamic pressure of the free stream. Figure 1.42 shows the lift coefficient (a) and drag coefficient (b) of the NACA 0012 airfoil. As the angle of attack is increased beyond the stall angle (around 15°), there will be a flow separation (c) to reduce the lift coefficient dramatically with an associated increase in the drag coefficient.

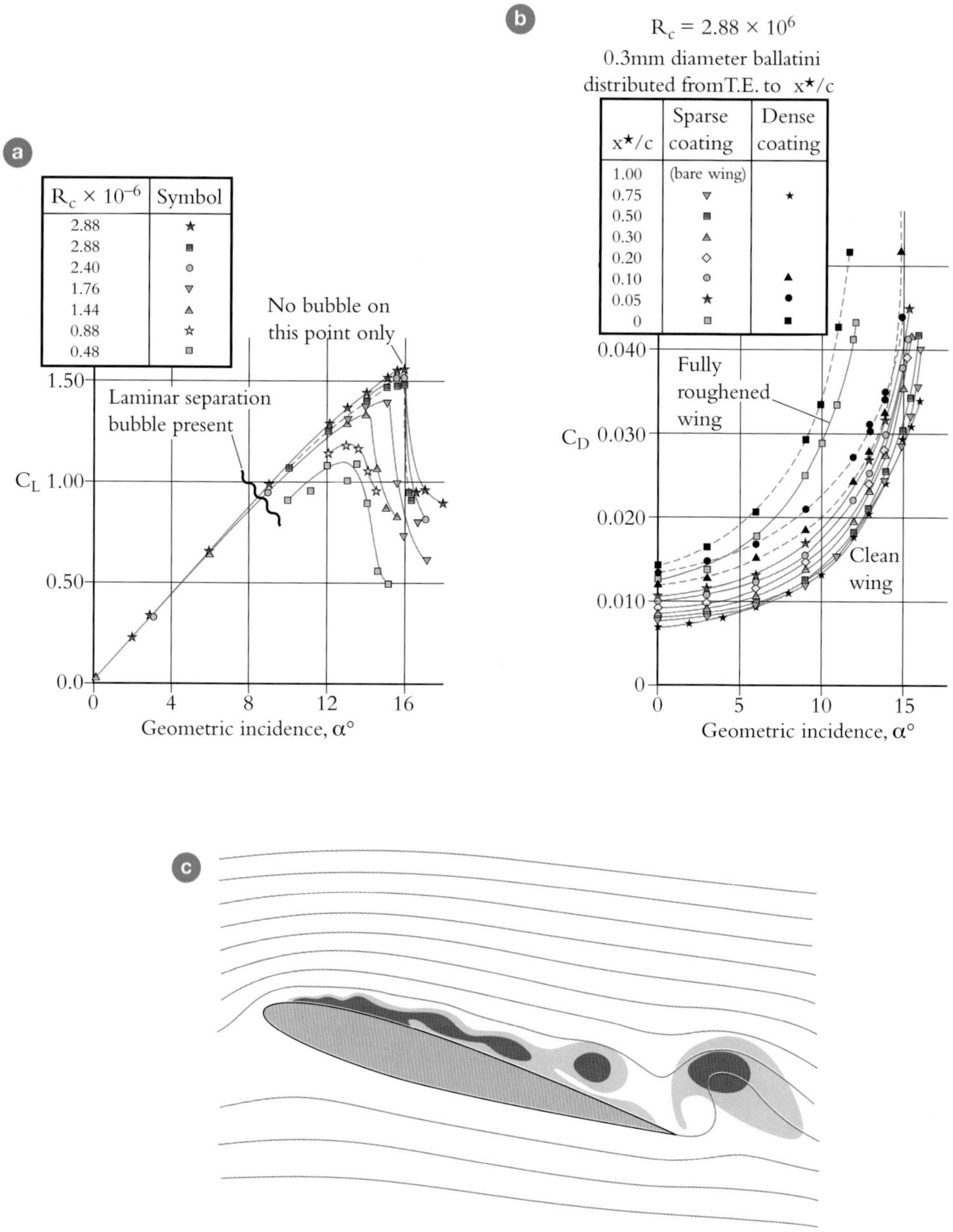

Figure 1.42 Lift (a) and drag coefficient (b) of the NACA 0012 airfoil (c)

Buckingham's theorem

The Buckingham π theorem, or the π theorem for short, is a systematic method for finding relevant non-dimensional quantities. It applies not only to the problems in fluid mechanics, but to many other areas of engineering and science. Buckingham's theorem consists of two parts. The first gives the number of non-dimensional quantities to account for a given flow problem, and the second determines each of the non-dimensional quantities involved.

Part 1

If there are n physical variables describing a physical process which contains m basic dimensions, where $n > m$. Then, this physical process can be described only with $(n - m)$ independent non-dimensional quantities:

$$\pi_1, \pi_2, \ldots, \pi_{n-m} \tag{1.66}$$

Here, the physical variables are those to indicate the characteristics of the physical quantity or status (e.g. velocity, pressure, density, etc.), while the basic dimensions are given by M (mass), L (length), T (time) or Θ (temperature).

Part 2

In order to find the non-dimensional quantities $\pi_1, \pi_2, \ldots, \pi_{n-m}$, we first pick m physical variables that do not form any of π_i themselves. Then, the product of these variables with one additional variable from $(n - m)$ remaining physical variables will give the non-dimensional quantities.

For example, if there are five physical variables v_1, v_2, v_3, v_4 and v_5 which contain three basic dimensions M, L and T, then there are two $(5 - 3 = 2)$ non-dimensional quantities to describe this physical process (Part 1). Picking v_1, v_2 and v_3 physical variables, the non-dimensional quantities π_1 and π_2 can be given by multiplying v_1, v_2 and v_3 by v_4 and v_5, respectively, in such a way that π_1 and π_2 will become non-dimensional (Part 2) by choosing the powers a, b, c, d, e, f, g and h appropriately.

$$\pi_1 = (v_1)^a(v_2)^b(v_3)^c(v_4)^d$$

$$\pi_2 = (v_1)^e(v_2)^f(v_3)^g(v_5)^h$$

It is strongly advisable to go through worked examples carefully to understand Buckingham's theorem, where a detailed procedure to find the number of non-dimensional quantities as well as to determine their non-dimensional forms are given. To start the process, we need to understand the flow problem with a view to identifying relevant physical variables. Table 1.5 lists the physical variables appearing in many flow problems together with their dimensions, which can help apply Buckingham's theorem.

Quantity	Symbol	Dimensions
Acceleration	dV/dt	LT^{-2}
Angle	θ	None
Angular velocity	ω	T^{-1}
Area	A	L^2
Density	ρ	ML^{-3}
Force	F	MLT^{-2}
Kinematic viscosity	ν	L^2T^{-1}
Length	L	L
Mass flow	m	MT^{-1}
Moment, torque	M	ML^2T^{-2}
Power	P	ML^2T^{-3}
Pressure, stress	p, σ	$ML^{-1}T^{-2}$
Speed of sound	a	LT^{-1}
Strain rate	ε	T^{-1}
Surface tension	Υ	MT^{-2}
Temperature	T	Θ
Velocity	V	LT^{-1}
Viscosity	μ	$ML^{-1}T^{-1}$
Volume	$\mathcal{V}$	L^3
Volume flow	Q	L^3T^{-1}
Work, energy	W, E	ML^2T^{-2}

Table 1.5 Dimensions of physical variables used in fluid dynamics

Worked example

Obtain the non-dimensional quantities for a pipe flow by considering the following physical variables:

$$Q\,[L^3T^{-1}]\ a\ [L]\ l\ [L]\ \Delta p\ [ML^{-1}T^{-2}]\ \mu\ [ML^{-1}T^{-1}]$$

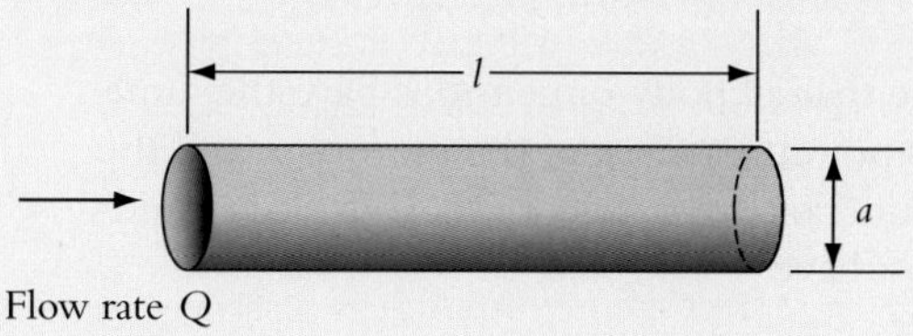

Figure 1.43

There are five ($n = 5$) physical variables Q, a, l, Δp and μ, and three ($m = 3$) basic dimensions M, L and T. Therefore, the number of non-dimensional quantities in this flow problem will be two ($n - m = 2$).

In order to find these two non-dimensional quantities π_1 and π_2, we firstly pick three ($m = 3$) physical variables Q, l and Δp. It is guaranteed that the products of these physical variables do not form any non-dimensional quantities, since only Δp contains the basic dimension M.

Therefore, we set $\pi_1 = Q^a \cdot l^b . \Delta p^c \cdot \mu^d \quad \pi_2 = Q^e \cdot l^f \cdot \Delta p^g \cdot a^h$

We now have to determine a, b, c, d, e, f, g and h to make π_1 and π_2 non-dimensional. Substituting the basic dimensions into above,

$$\pi_1 = [L^{3a} \cdot T^{-a}] \cdot [L^b] \cdot [M^c L^{-c} T^{-2c}] \cdot [M^d L^{-d} T^{-d}]$$

$$\pi_2 = [L^{3e} \cdot T^{-e}] \cdot [L^f] \cdot [M^g L^{-g} T^{-2g}] \cdot [L^h]$$

In order for these quantities to be non-dimensional, we must solve the following:

$$\text{L: } 3a + b - c - d = 0 \qquad 3e + f - g + h = 0$$

$$\text{T: } -a - 2c - d = 0 \qquad -e - 2g = 0$$

$$\text{M: } c + d = 0 \qquad g = 0$$

The solution is given by

$$g = 0, e = 0, f = -h, c = -d, a = d, b = -3d$$

Therefore, the non-dimensional quantities can be given by

$$\pi_1 = [Q \cdot l^{-3} \cdot \Delta p^{-1} \cdot \mu]^d \qquad \pi_2 = [l \cdot a^{-1}]^f$$

where d and f are arbitrary constants. We cannot determine these constants with the number of available equations. We set $d = f = 1$ for simplicity to get

$$\pi_1 = \frac{Q\mu}{\Delta p l^3} \qquad \pi_2 = \frac{1}{a}$$

In other words, π_1 is a non-dimensional flow rate and π_2 is the length to diameter ratio.

The pipe flow problem can be described by only two non-dimensional numbers, so that

$$\pi_1 = f_n(\pi_2)$$

to give

$$Q = f_n\left(\frac{l}{a}\right) \cdot \frac{l^3}{\mu} \cdot \Delta p$$

Buckingham's theorem cannot determine the functional form of $\frac{l}{a}$.

Dimensional analysis and model testing

To carry out scaled model tests, it is important that we should follow the similarity principle. This basically consists of the geometric similarity and the dynamic similarity. First it is vital that the geometric similarity is satisfied between the test model and the prototype. Only after that should we proceed with the scaled model test, ensuring that the flow around the prototype and that around a test model are dynamically similar.

To satisfy the geometric similarity, we must ensure that all body dimensions have the same linear scale in all directions. In other words, corresponding points of the prototype and the model must be related by the same linear scale ratio. For example, for a 1/10th scale wind-tunnel test of a wing (Figure 1.44) the model must have

- thickness, width and length that are 1/10th those of the prototype
- nose radius that is 1/10th of the prototype
- 1/10th of the surface roughness as compared to that of the prototype
- 1/10th the thickness of coating, if that is either painted or lacquered.

In addition, both the prototype and the model must have the same relative posture to the flow direction. This ensures that they have the same angle of attack. The dynamic similarity is guaranteed by making sure that all relevant non-dimensional numbers are the same during the scaled model tests.

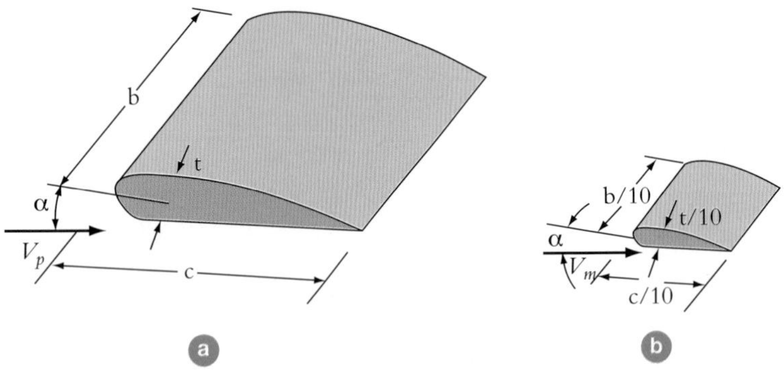

Figure 1.44 Prototype wing (a) and the 1/10th scale model (b) which are geometrically similar

When we carry out a wind tunnel test of an aircraft, for example, we need an exact scaled replica of the prototype aircraft in order to fulfil the geometric similarity. Dynamic similarity in the test will be satisfied by carrying out the model test at the same Reynolds number and Mach number as in the prototype flight. If the Mach number is small, and therefore the compressibility effect is negligible, then the only relevant non-dimensional number to consider will be the Reynolds number to satisfy the dynamic similarity.

Figure 1.45 Wind tunnel test of the X-48 blended wing body at NASA Langley Research Center

Worked example

Derive the flow speed at which a scaled model experiment should be carried out, when the only relevant non-dimensional quantity is the Reynolds number.

If the Reynolds number Re is the only non-dimensional quantity for the flow being considered, we must ensure that

$$\mathrm{Re}_p = \mathrm{Re}_m$$

Therefore,

$$\frac{V_p L_p}{\nu_p} = \frac{V_m L_m}{\nu_m}$$

So that, we have

$$\frac{V_m}{V_p} = \left(\frac{L_p}{L_m}\right)\left(\frac{\nu_m}{\nu_p}\right)$$

In order words, the velocity V_m in the model test must be determined by the scale of the model and the kinematic viscosity ratio.

For example, if the model size is a tenth of the prototype in the same working fluid ($\nu_p = \nu_m$), the test speed V_m must be 10 times faster than the prototype speed V_p in order to ensure the dynamic similarity. This is not always easy to achieve. There are special experimental facilities to attain high Reynolds numbers through a change in fluid density or viscosity.

Worked example

A test for high speed boat will be carried out in a towing tank. Here, we must consider the Froude number as the second non-dimensional quantity in addition to the Reynolds number. What are the conditions at which we should carry out the tank test to ensure the dynamics similarity?

We must have the following relationships in order to ensure the dynamic similarity:

$$\mathrm{Re}_p = \mathrm{Re}_m \quad \text{and} \quad \mathrm{Fr}_p = \mathrm{Fr}_m$$

Therefore,

$$\frac{V_p L_p}{\nu_p} = \mathrm{Re} = \frac{V_m L_m}{\nu_m} \qquad \frac{V_p}{\sqrt{g L_p}} = \mathrm{Fr} = \frac{V_m}{\sqrt{g L_m}}$$

The only way to satisfy these two conditions is

$$\frac{V_m}{V_p} = \sqrt{\frac{L_m}{L_p}} \qquad \frac{\nu_m}{\nu_p} = \frac{V_m L_m}{V_p L_p} = \left(\frac{L_m}{L_p}\right)^{\frac{3}{2}}$$

In practice, the first condition can be met relatively easily, but the second condition is difficult to realize.

Similarity rules in turbomachinery

Turbomachinery can be classified into two types – pumps and turbines. Pumps are used to carry out useful work, such as to circulate hot water in a central heating system, by adding energy to the fluid through the rotation of impellors. Turbines, on the other hand, extract energy from the moving fluid through shaft work as the blades or impellors rotate. In this section, we describe the performance of centrifugal pumps (Figure 1.46) through an application of similarity rules.

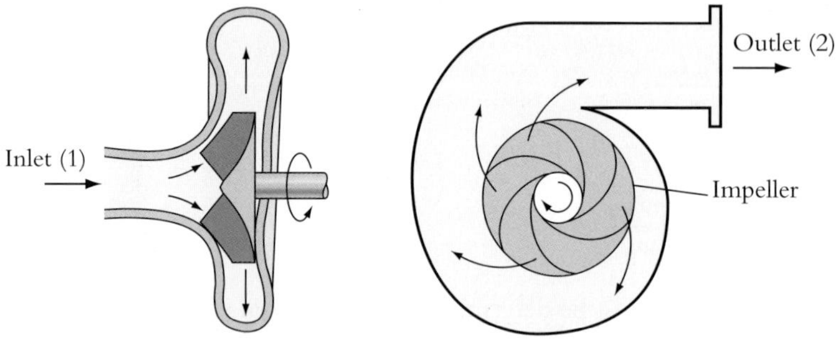

Figure 1.46 Centrifugal pump, where the flow moves in (1) and exits from (2).

If we apply the Bernoulli equation at the inlet (1) and outlet (2) of the pump, we have

$$\left(\frac{p}{\rho g}+\frac{V^2}{2g}+z\right)_1+h_s=\left(\frac{p}{\rho g}+\frac{V^2}{2g}+z\right)_2+h_f$$

Therefore,

$$\left(\frac{p}{\rho g}+\frac{V^2}{2g}+z\right)_2-\left(\frac{p}{\rho g}+\frac{V^2}{2g}+z\right)_1=h_s-h_f\equiv H \quad \textbf{(1.67)}$$

Typically inlet and outlet areas of centrifugal pumps are similar ($A_1 \approx A_2$), therefore $V_1^2 \approx V_2^2$. Also, their locations are usually very close to each other ($z_1 \approx z_2$). Therefore,

$$H \approx \frac{\Delta p}{\rho g}=\frac{p_2-p_1}{\rho g} \quad \textbf{(1.68)}$$

This is called the **net pump head**.

The power delivered to the fluid is called the **water horsepower**, given by

$$P_w=\rho g Q H \quad \textbf{(1.69)}$$

while the power required to drive the pump is given by

$$bhp=\omega T \quad \textbf{(1.70)}$$

This is called the **brake horsepower**. The efficiency of the pump is defined as a ratio of the water horsepower to the brake horsepower as

$$\eta=\frac{P_w}{bhp}=\frac{\rho g Q H}{\omega T} \quad \textbf{(1.71)}$$

Worked example

Find out non-dimensional quantities describing the performance of a centrifugal pump, where we can assume that there are four (n = 4) physical quantities given by

pump head	gH	$[L^2T^{-2}]$
flow discharge	Q	$[L^3T^{-1}]$
impeller diameter	D	$[L]$
shaft speed	n	$[T^{-1}]$

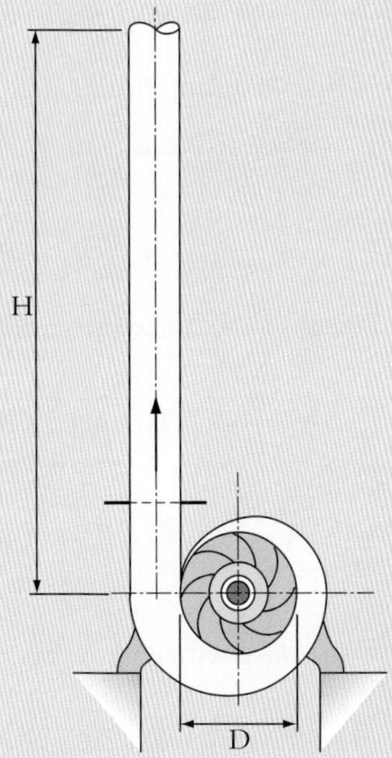

Figure 1.47

There are only two basic dimensions (m = 2) in this problem. From the π theorem, therefore, there must be only two ($n - m = 2$) non-dimensional quantities describing the pump performance.

In order to find these two quantities π_1 and π_2, we first pick two physical quantities D and n. We can easily see that these two *cannot* form a non-dimensional quantity.

Then,

$$\pi_1 = D^a \cdot n^b \cdot Q^c \qquad \pi_2 = D^d \cdot n^e \cdot (gH)^f$$

Substituting the basic dimensions into these, we have

$$\pi_1 = [L^a] \cdot [T^{-b}] \cdot [L^{3c} \cdot T^{-c}] \qquad \pi_2 = [L^d] \cdot [T^{-e}] \cdot [L^{2f} \cdot T^{-2f}]$$

After some calculation, we find that $\pi_1 = \dfrac{nD^3}{Q}$ and $\pi_2 = \dfrac{gH}{n^2D^2}$. Since the π theorem cannot give a functional form of a non-dimensional quantity, we take the liberty of making an assumption on its form. Conventionally, we use $\dfrac{Q}{nD^3}$ rather then $\dfrac{nD^3}{Q}$. Therefore, we set

$$\pi_1{}' = \frac{1}{\pi_1} = \frac{Q}{nD^3}$$

The performance of the centrifugal pump can be expressed as $\pi_2 = f_n(\pi_1{}')$:

$$\frac{gH}{n^2D^2} = f_n\left(\frac{Q}{nD^3}\right)$$

or,

$$C_H = f_n(C_Q)$$

where, C_H and C_Q are the head coefficient and the capacity coefficient, respectively.

Centrifugal pumps

In general, pump performance is dictated not only by the flow discharge, impellor diameter and shaft speed, but also by the fluid density and viscosity as well as the surface roughness of the impellor. We should, therefore, write the pump head gH as a function of all of these variables.

$$gH = f_1(Q, D, n, \rho, \mu, \varepsilon)$$

from which we can find

$$C_H = g_1\left(C_Q, R_e, \frac{\varepsilon}{D}\right)$$

using Buckingham's theorem. Likewise, the brake horse power bhp has the following functional relationship:

$$bhp = f_2\ (Q, D, n, \rho, \mu, \varepsilon)$$

from which we can find

$$C_P = g_2(C_Q, R_e, \varepsilon/D)$$

Here,

$$\text{Capacity coefficient} \ldots\ C_Q = \frac{Q}{nD^3} \quad \textbf{(1.72)}$$

$$\text{Head coefficient} \ldots\ C_H = \frac{gH}{n^2D^2} \quad \textbf{(1.73)}$$

$$\text{Power coefficient} \ldots\ C_P = \frac{bhp}{\rho n^3 D^5} \quad \textbf{(1.74)}$$

$$\text{Reynolds number} \ldots\ \text{Re} = \frac{\rho n D^2}{\mu} \quad \textbf{(1.75)}$$

$$\text{Roughness ratio} \ldots\ \frac{\varepsilon}{D} \quad \textbf{(1.76)}$$

And the efficiency of the centrifugal pump η is given by

$$\eta = \frac{\rho \cdot g \cdot H \cdot Q}{bhp} = \frac{C_H \cdot C_Q}{C_P} = g_3(C_Q)$$

If we assume that the effects of R_e and ε/D are small (i.e. the Reynolds number is sufficiently large and the internal pump surface is hydrodynamically smooth), we have

$$C_H = g_1\ (C_Q),\ C_P = g_2\ (C_Q)$$

which are exactly the same as in the previous worked example.

Unique performance curves exist for a family of pumps with similar geometries when the pump performance is plotted in terms of non-dimensional quantities such as C_Q, C_H and C_P (Figure 1.48). These save us much effort and time by using only one set of non-dimensional curves to describe the pump performance of an entire family.

For example, when the pumps from the same family are compared at the homologous points, such as at the best efficiency point (BEP), C_Q, C_H and C_P are equal. Therefore,

$$\frac{Q_2}{Q_1} = \frac{n_2}{n_1}\left(\frac{D_2}{D_1}\right)^3 \qquad \frac{H_2}{H_1} = \left(\frac{n_2}{n_1}\right)^2\left(\frac{D_2}{D_1}\right)^2$$

$$\frac{P_2}{P_1} = \frac{\rho_2}{\rho_1}\left(\frac{n_2}{n_1}\right)^3\left(\frac{D_2}{D_1}\right)^5 \qquad \eta_1 = \eta_2$$

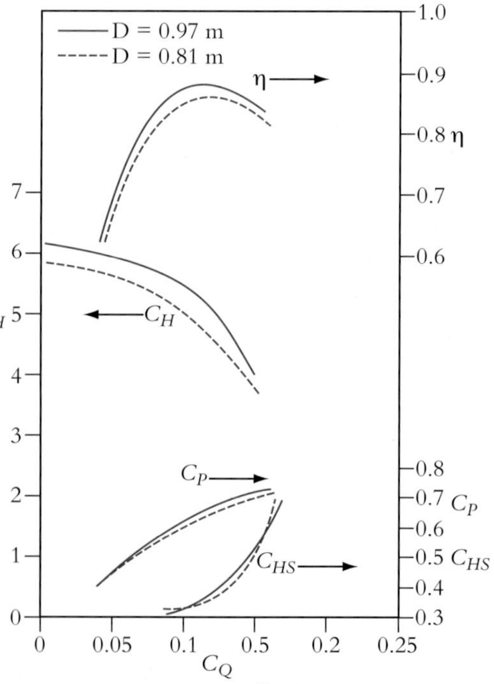

Figure 1.48 Performance of centrifugal pumps in non-dimensional quantities (F. M. White, 2008, *Fluid Mechanics*, New York: McGraw Hill. Reproduced with permission of The McGraw-Hill Companies)

Specific speed

It is useful to know the type of pumps at the early stage of design when only the required flow discharge and pump head are known together with a likely shaft speed. For this purpose, we define the non-dimensional *specific speed Ns'* as

$$Ns' = \frac{C_Q^{\star\frac{1}{2}}}{C_H^{\star\frac{3}{4}}} = \frac{n \cdot Q^{\star\frac{1}{2}}}{(gH^{\star})^{\frac{3}{4}}} \qquad (1.77)$$

where, $C_Q^{\star} = \dfrac{Q^{\star}}{n\,D^3}$ and $C_H^{\star} = \dfrac{(gH)^{\star}}{n^2\,D^2}$

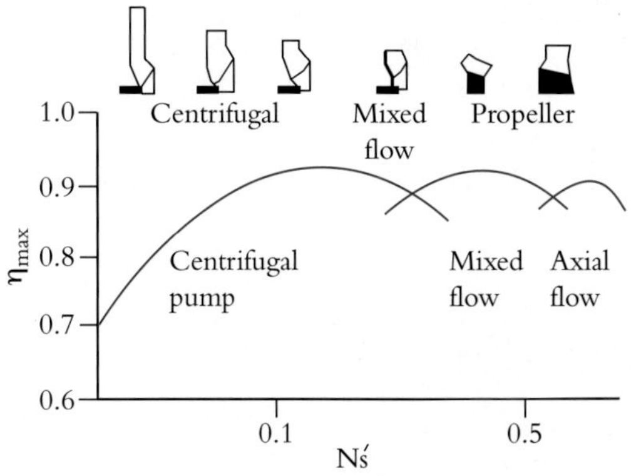

Figure 1.49 **Non-dimensional specific speed of different type of pumps**

where, (*) indicate the values at BEP (best-efficiency point). Pump designers can identify the type of required pumps by calculating the specific speed Ns' from $Q^{\star}$, $H^{\star}$ and n without requiring the size of the pump D.

As seen in Figure 1.49, centrifugal pumps with large pump head H and small flow discharge Q have relatively small specific speed $Ns' \approx 0.10$, while axial pumps (small H and large Q) tend to have larger specific speed $Ns' = 0.5 \sim 1.0$.

Learning summary

By the end of this section you should have learnt:

- ✓ non-dimensional numbers are important in understanding the characteristics of the flow as well as in comparing the type of flow with others;
- ✓ Buckingham's theorem gives not only the number of non-dimensional quantities involved, but it also determines each non-dimensional quantity;
- ✓ to carry out model tests, we need to ensure both the geometric and dynamic similarities are satisfied;
- ✓ one can identify the shape of required pumps by calculating the specific speed without knowing the size of the pump.

References

Colebrook, F., 1939, 'Turbulent flow in pipes, with particular reference to the transition between smooth and rough pipe laws', *Journal of the Institution of Civil Engineers*, vol. 11, no. 4, 133–156.

Moody, L.F., 1994, 'Friction factors for pipe flow', *Transactions of the ASME*, vol. 66, no. 8, 671–684.

Unit 2

Thermodynamics

UNIT OVERVIEW

- Introduction
- Air conditioning
- Gas mixtures
- Combustion
- Reciprocating compressors
- Heat transfer
- Heat exchangers
- Vapour power cycles
- Reciprocating internal combustion engines

2.1 Introduction

This unit continues from the foundation in thermodynamics laid out in *An Introduction to Mechanical Engineering: Part 1*. Its aim is to provide an applied emphasis to the concepts already learnt. The applications presented here concentrate mainly around power generation from oil, gas and coal by combustion to provide heat transfer to produce work via steam power. The first section reviews and refines the information that is relevant from the thermodynamics in *Part 1*, and the subsequent sections provide analysis techniques for practical engineering applications.

The material covered will address the properties of working fluids as in *Part 1*; perfect and semi-perfect gases and steam are of interest. Here **refrigerants** are introduced along with **combustion reactions** and **product gas mixtures**, with a focus on application. Our attention is on what a **working fluid** can achieve. By manipulating the working fluid, practical ends can be met:

- **air conditioning** to affect the temperature and humidity of atmospheric air;
- producing electricity by heating steam with combustion reactions, which drives generators via **steam turbines**;
- **gas and vapour compressors** increase pressure at a given flow rate of fluid;
- **combustion** gases are used directly to drive engines.

Transferring or **insulating heat** is involved in all practical thermodynamic processes, and the modes of transfer and the basic techniques for calculation of heat transfer are presented. This unit considers how classical thermodynamic machines work and how to calculate the relationships between heat and work.

The first section considers mixtures of gases, seeing how the volume and mass of each gas in a mixture relates to the others. Later sections progress on to the chemistry of combustion reactions and then to the energy release of the reactions whether using gas, liquid or solid fuels. Moisture is a life-supporting component of our environment and is contained in the atmosphere – rain, mist and humidity – and in reactants and products of combustion. Mixtures of gases with moisture have specific properties that require special treatment. The unit shows how to control atmospheric humidity.

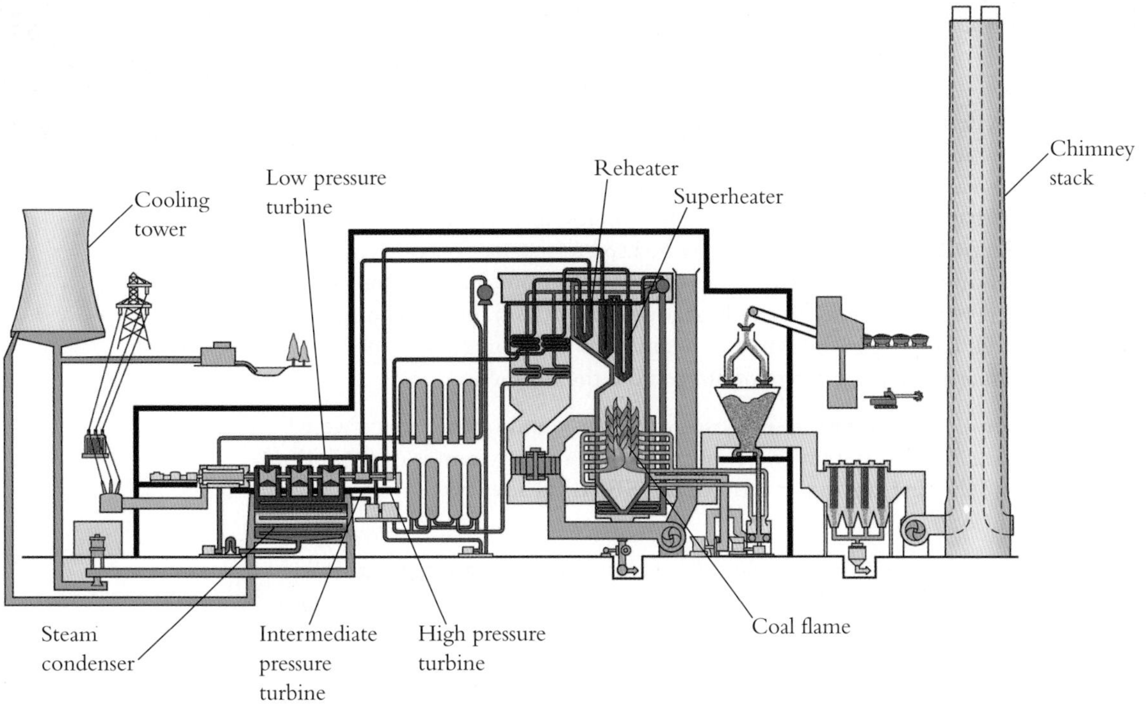

Figure 2.1 Schematic of Drax power station (Drax Power Limited)

Other characteristics of the thermodynamic machine require heat transfer, either to produce work or to achieve desired temperatures. Situations are considered in which steady heat and work transfer occur ('steady' meaning processes that do not vary over time). A particularly important thermodynamic system is the **steam power plant** as shown in the schematic in Figure 2.1 of Drax power station – the largest in the UK. This unit shows how steam can be used to produce power, and how the devices within power stations relate to each other to produce useful power output. It relies not only on the internal steam cycle, but also on the effective combustion energy release and heat transfer to the steam at the correct rate.

The law of conservation of energy states that energy can be neither created nor destroyed but only transformed from one type to another, and this is the essence of the **first law of thermodynamics** met in Part 1, and henceforward referred to as 'the first law'. Heat and work are forms of energy; they have the same units as energy. The laws of thermodynamics consider the relationship between heat and work and the effect that the transfer has on the energy of the matter that undergoes the transfer.

It is important to be able to define the **system** and the **control volume**. This means having in mind what physical part of a thermodynamic device is under consideration, in order to deal with it in isolation. A sketch of the control volume within a process is often useful. Figure 2.2 shows a crude representation of a thermodynamic system where cool stream of fluid enters a tube, is accelerated by a propeller and receives energy in the form of work, proceeds to receive energy in the form of heat from a hot source (whether combustion or some other source)

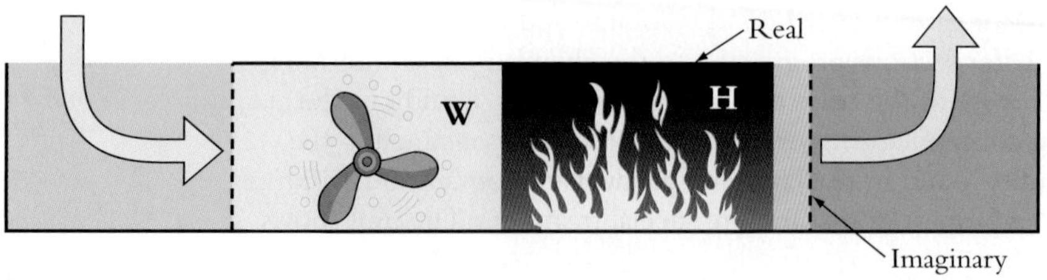

Figure 2.2 A control volume in a crude open system thermodynamic device having work and heat, real boundaries at physical walls and imaginary boundaries within the fluid, marking the extent of the fluid under consideration

before passing out of the system. A control volume within the system describes what happens in a particular part of the system where some thermodynamic action occurs and it is useful to separate out this space, breaking down the complex system into manageable parts. The **real** solid **boundaries** (perhaps moving or flexible but nonetheless impervious) are usually relatively easy to define. An **imaginary boundary** may be used to separate out internal, connected parts of the working fluid in process, by careful consideration of the separation point required to isolate the region of influence of heat, work and mass transfers. The system has to do with the fluid itself rather than the system containing it. The control volume is often specified by reference to an artefact, but it is important to note the distinction.

In *Part 1*, both closed systems – such as pistons in closed cylinders – and open systems were considered although no specific open systems were discussed. The open system is of significant importance in this unit, and the first law in the form of the **SFEE** (steady flow energy equation) dominates the analysis.

The direction of heat and work transfer is important, since maintaining an **energy audit** is vital. It is easy to lose track of an **energy budget** when the system is large and complex with many processes occurring. Even though the subsystems can be separated, e.g. considering just one cylinder in a four-cylinder engine, it is important to maintain an **energy inventory** in the same sense. It is logical to consider the transfer of heat and work to the fluid as positive, and that is the convention adopted in this unit. This is the European convention. However, for argument's sake, and for completeness, it is useful to consider the alternative convention. The American method considers useful work output from the system as positive – a convention that has particular merit since it is usually work achieved by a machine that is of most interest. Note that the terms above indicate a deliberate relationship to financial accounting – just like money, energy has to be accounted for properly.

Equilibrium is a concept that has been developed on the understanding of kinetic energy (KE) and potential energy (PE). If a system is in unstable equilibrium it is likely to lose PE to KE. In the broader thermodynamic sense, there are several forms of available PE that can be unstable. However, when and only when all are balanced, a state of equilibrium is achieved. For example, after complete combustion when no further oxygen is available and the energy state of the products is extremely adverse to a reverse reaction, or when the temperature is the same throughout, or a vapour is in contact with its liquid at constant volume and temperature. A state of equilibrium is one where measurable properties do not change with time.

In order to change from one state to another, a **process** must be followed. The idea of a sequence of processes in a **cycle** enables cyclic machines, the cycle implying a repeating sequence of events physically, but in the thermodynamics sense it has a very specific meaning. A cycle is a sequence of processes that alter the state of the working fluid and then return it to its initial state. Cyclic processes are used to produce or make use of external work. **Para-cyclic** machines are very common, in that new material of the same properties is induced from the environment at one point in the cycle and ejected to the environment at another point in the cycle, to be recycled at some stage. These are open-cycle machines such as jet engines and internal combustion engines.

Reversibility of processes is an important concept – it is important to understand how a reversible process might be possible and what the useful output from such a process is. It is only possible in very controlled conditions and with very slow processes, so it is practically unachievable, but it implies a state of ideal processing of the working fluid with least waste of effort. The reason for this will be established when we look at the **second law of thermodynamics** (henceforward referred to as 'the second law') in due course. On a typical **process diagram**, say pressure versus volume of a gas undergoing processes, we can indicate reversible and irreversible processes (Figure 2.3). This indication convention can be important and should be noted – dotted lines show irreversible processes, solid lines show reversible processes. Most lines drawn for real processes will therefore be dotted, but solid-lined reversible processes are often useful for comparison in order to calculate losses due to irreversibility.

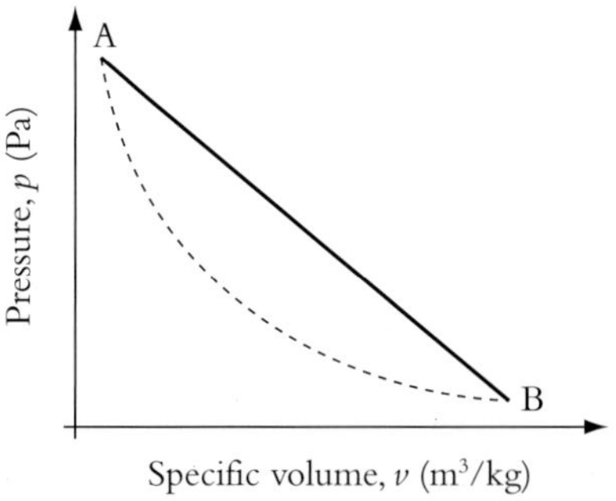

Figure 2.3 Schematic graph of pressure versus specific volume, illustrating a reversible process with a solid line and an irreversible process with a dashed line working on a gas between two states A and B

The first law of thermodynamics

The **first law** is very important – telling us about the exchange that happens between work and heat. In words:

> *When any closed system is taken through a cycle, the net work delivered to the surroundings is proportional to the net heat taken from the surroundings.*

Equation (2.1) is a statement of the first law, and it represents the conservation of energy principle:

$$Q + W = U_2 - U_1 = m(u_2 - u_1) \tag{2.1}$$

where
Q [J] is a quantity of heat transferred in the process
W [J] is the work done on the system during the process
U [J] is the internal energy of the working fluid
m [kg] is the mass of the working fluid in the system
u [J/kg] is the specific internal energy of the working fluid

States 1 and 2 are the initial and final states of the system. Internal energy is the amount of energy possessed by the molecules of matter in the system in their kinetically energized states. The formula may be expressed in differential form as:

$$dq + dw = du \tag{2.2}$$

Note that the capital letters indicate bulk quantities, but importantly, lower case indicates that only 1 kg of the working fluid is being considered. As written here it concerns the conservation of energy in a **closed system** – i.e. where there is no new mass coming in. It is valid for both reversible and irreversible processes because it is just concerned with end states and how much heat and work have been transferred to the system. This law is unaffected by what the route between the states is.

The law in words actually talks about the work delivered to the surroundings and the heat taken from them during a cycle – note that a cycle means that the state has returned to its starting point, i.e. $u_1 = u_2$. This is due to the route of the work, which can be displayed on a state diagram such as the p–v diagram, and there can be a variety of different routes. Similarly, there can be a variety of different routes that result in the same work. However, the route is usually defined by the engine that is acting on the system.

The steady flow energy equation is the first law applied to an open system, and can be expressed as:

$$\dot{Q}_{12} + \dot{W}_{12} = \dot{m}\left(h_2 - h_1 + \frac{C_2^2 - C_1^2}{2} + g(z_2 - z_1)\right) \tag{2.3}$$

where
C [m/s] is the speed of the fluid
g [m/s^2] is gravitational acceleration
z [m] is height above a fixed datum point

Here, heat, work, enthalpy, kinetic energy and potential energy all have same dimension of energy the joule, J. A superscript dot implies a rate, i.e. J/s or W and kg/s. **Enthalpy** is the flow process equivalent of internal energy. Since $h = u + pv$, it consists of the internal energy plus the '**flow work**'; the pv part. Mostly in the machines that we consider, changes in velocity (KE) and in height (PE) are negligible compared to changes in enthalpy. For example, consider that the change in KE accelerating 1 kg of water from 1 to 10 m/s is 50 J and the PE for raising the same through 10 m is approximately 100 J; the enthalpy change of water increasing in temperature from 10°C to 20°C – a relatively modest temperature change – is 42,000 J! Since KE and PE can often be neglected in reality, the SFEE can be reduced to equation (2.4) that shows a close resemblance to the first law for the closed system:

$$\dot{Q}_{12} + \dot{W}_{12} = \dot{m}(h_2 - h_1) \quad \textbf{(2.4)}$$

Properties

There are several **properties** to be aware of, and the following table shows a selection.

Extensive	**Intensive**
Volume, V [m^3]	Density, ρ [m^3/kg]
Mass, m [kg]	Pressure, p [N/m^2]
Enthalpy, H [J]	Temperature, T [K]
Entropy, S [J/K]	Specific enthalpy, h [J/kg]

In the right-hand column are **intensive** properties. In the left are **extensive** properties. Intensive properties are the same for any quantity of material being considered. Extensive properties depend on the amount of material. Intensive properties are mostly expressed per kg of material; this is a **specific** measure of an extensive property. Extensive is indicated by capital symbols, intensive by lowercase. In changing from extensive (the whole) to intensive (specific), H becomes h, S becomes s, V becomes v; T, and p are inherently intensive and a specific measure cannot be stated. h, s, v are specific enthalpy, specific entropy and specific volume.

Equation of state

The equation of state relating p, v and T is given by equation (2.5):

$$pv = RT, \quad \textbf{(2.5)}$$

or

$$pV = mRT \quad \textbf{(2.6)}$$

where R is the specific gas constant having units J/kgK. Note the alternative use of specific and total volume, v and V. These equations hold for all gases.

Specific heat capacity is defined separately for constant pressure or constant volume processes, and is directly related to the first law since specific heat capacity at constant pressure is defined as:

$$c_p = \left.\frac{\partial h}{\partial T}\right|_p$$

which leads to

$$h_2 - h_1 = c_p(T_2 - T_1) \quad \textbf{(2.7)}$$

And specific heat capacity at constant volume is defined as:

$$c_p = \left.\frac{\partial u}{\partial T}\right|_v$$

which leads to

$$u_2 - u_1 = c_v(T_2 - T_1) \quad \textbf{(2.8)}$$

Specific heat capacity is the amount of heat absorbed by 1 kg of gas when its temperature is raised by 1 K at either constant pressure or constant volume. A **semi-perfect gas** describes the behaviour of most gases except the very light ones (helium and hydrogen) – specific heat capacity varies with temperature and it is useful to take the average from tables of thermodynamic properties when calculating heat transfer to or from a quantity of gas.

For a **perfect gas**, the relationship between the constants associated with the gas is important and useful. It is important to realize that since the specific gas constant R and the ratio of specific heats, γ (gamma) are usually known, it is possible to work out the specific heat capacities from the quantities above. This is illustrated in the following derivation:

Since

$$h = u + pv, \mathrm{d}h = \mathrm{d}u + \mathrm{d}(pv)$$

so,

$$c_p \mathrm{d}T = c_v \mathrm{d}T + R\mathrm{d}T$$

$$c_p = c_v + R \tag{2.9}$$

Also

$$\gamma = \frac{c_p}{c_v}$$

Since γ and R are usually known or determinable, useful relationships with c_p and c_v can be formed, e.g.

$$\gamma\, c_v = c_v + R$$

so

$$c_v = \frac{R}{(\gamma - 1)} \tag{2.10}$$

Processes

Processes describe how a working fluid is changed from one state to another. The following list describes processes met in Part 1:

Adiabatic = no heat transfer, $pv^{\gamma} = c, q = 0$

Isothermal = constant temperature, $pv = c$

Isobaric = constant pressure, $p = c$

Isochoric = constant specific volume, $v = c$

Polytropic = unspecified process type, $pv^n = c$

Isentropic = constant entropy, $pv^{\gamma} = c, q = 0$

Isentropic implies an adiabatic process that is reversible. This is commonly achieved approximately in open system devices such as axial flow turbines and compressors.

The p–v representation can be changed to a relation between p and T or v and T by using the equation of state.

For example, in $pv^n = c$, substitute $p = \dfrac{mRT}{v}$ to give $\dfrac{mRT_1 v_1{}^n}{v_1} = \dfrac{mRT_2 v_2{}^n}{v_2}$, i.e.

$$T_1 v_1{}^{n-1} = T_2 v_2{}^{n-1}$$

Work occurs when the boundary of a system moves against a force – this can be an external atmosphere, or a mechanical device. Work can be expressed as:

$$W = -\int_1^2 p\mathrm{d}V \tag{2.11}$$

where

W [J] – work done on the working fluid

p [N/m^2 or Pa] – instantaneous pressure of the working fluid

V [m^3] – instantaneous volume of the working fluid

Equation (2.11) expresses the work done in a closed system per kg of working fluid, for a reversible process, and relies on knowledge of the pressure at every stage. A reversible process is ideal, so equation (2.11) tells us the ideal work assuming no friction, no turbulence, nothing unrecoverable. Work is negative because of the first law sign convention that we have chosen to use, such that if the gas has expanded, the energy has gone from the system in order to work against the exterior. It is useful to know how to calculate the ideal work done for a process, to compare with the actual work and hence quantify the losses of energy from the system.

A *p–v* diagram, illustrated in Figure 2.4(a), is a **state diagram**, which shows the relationship between pressure and specific volume. The limit of an individual process is bounded by constant pressure and constant volume process lines (horizontal and vertical respectively) and with curves between representing polytropic processes, including isothermal ($n = 1$) and adiabatic ($\gamma = h$) processes. n is always positive. Other process lines may be possible if the pressure–volume relationship is prescribed by an external mechanical device, such as a spring on a piston. The area under the curve represents the work done for a reversible process. In irreversible processes, work is 'lost' in overcoming friction, in the turbulent fluid and in other such resistances.

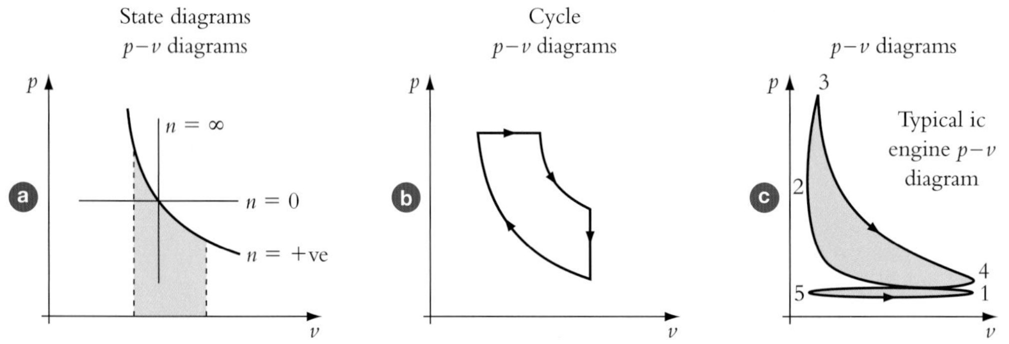

Figure 2.4 Schematic graphs of typical state diagrams using pressure and specific volume to indicate what is happening to the working fluid: (a) shows the effect of polytropic index on ideal processes from 0 (constant pressure process) to ∞ (constant volume process), (b) shows a cyclic process involving four separate processes and (c) shows a real paracyclic process, the Otto cycle for petrol engines.

For a cyclic process (Figure 2.4(b)), where the original state is returned to, the area contained in the loop of the cycle is the work output (or input) to the system. The sign depends on the directional sense of the process path. Considering a clockwise process, net work is produced by the system; positive compression of the gas is indicated in the lower curve, and heat is added to the gas in order to make it work to expand against the external system in the upper curve. The area under the upper curve is greater than the area under the lower one and net work is negative (more work is removed from the gas than is put into it, and the working fluid does work on the surrounding machinery). If the cycle were anticlockwise then the positive work (to compress the gas) is the area under the upper curve; a larger amount of work is put into the system than is taken out and the system must lose heat to do this.

Figure 2.4(c) shows a typical *p–v* diagram for a four-stroke engine – a polytropic compression from 1 to 2 as the crank drives the piston up. Combustion from 2 to 3 as the piston starts to be driven down by the expanding combustion gases. A polytropic expansion follows from 3 to 4, the pressure at 4 being still above the atmospheric in-draft pressure at 1 – the gases still have expansion to do if there were enough room to move. The combustion gases are exhausted from 4 to 5 and a new draught of fresh air is taken in as the crank drives the piston down from 5 to 1. The work done against the surroundings is the shaded area in the *p–v* graph. By controlling the form of the processes between states in this way, a significant amount of work is produced. The upper area is the useful work out; the lower slim area is the pumping work to get the exhaust out and the fresh air in and is carried by the other cylinders in the cycle and by the flywheel. Note the direction round the cycle is clockwise – net work out.

The second law of thermodynamics

It is important to consider the efficiency of processes. In the *p–v* diagram for a cycle there is no indication about the effectiveness of the relationships between heat and work. Given that work cycle, how effective was the conversion of heat to work? How much heat was required to get that quantity of work done? The second law addresses this issue; in words it can be stated as:

> *It is impossible to construct a system, which will operate in a cycle, extract heat from a reservoir, and do an equivalent amount of work on the surroundings.*

From the second law there are eight corollaries, i.e. facts that follow from it and can be proved by logical arguments as described in *Part 1*. They can be summarized as follows:

(1) Heat can't pass from cold to hot without work.

(2) Reversibility implies maximum efficiency.

(3) Temperatures of heat source and sink determine maximum efficiency.

(4) There is a universal zero value of temperature.

(5) If you exchange heat with more than two reservoirs the efficiency reduces from the maximum possible.

(6) $\int dQ/T$ for a complete cycle is zero if reversible, and less than zero if irreversible

(7) $\int dq/T$ between two states is the change in entropy, s.

(8) If $q = 0$ then s increases or remains constant.

The first corollary is part of everyday experience and is perhaps the best way for an engineer to remember the second law. The second corollary is not obvious unless the second law is considered, except that we intuitively know that friction causes energy to be lost to disorder. The third is very interesting – the extreme temperatures available to a heat engine determine the best that it can possibly perform. If it is possible to raise the upper temperature or lower the sink temperature then efficiency will increase. Number four is simple to remember, but it was a new concept since most temperature scales before the second law depended on the behaviour of a material such as water or alcohol. The fifth is reasonably obvious – increasing the number of devices that transfer heat in the system introduces more disorder. Number six is surprising, and is where we find the difference between reversible and irreversible processes. This sets the scene for the most efficient cycle possible – the Carnot cycle, introduced in a subsequent section. The seventh is the definition of entropy; this affects the system and the immediate surroundings, i.e. the level of disorder in the external reservoirs of hot and cold. The final one is a condition for a cycle – if a cycle involves no heat transfer then it is irreversible if entropy increases and reversible if entropy does not change.

Figure 5(a) shows a symbolic representation of a **heat engine**; it has some undefined machinery in the circle, which by virtue of heat exchange with the hot and cold reservoirs, is able to develop work on the surroundings. We define cycle efficiency as:

$$\eta = \frac{\text{Work done}}{\text{Heat supplied}} = \frac{W_{out}}{Q_{high}} = \frac{1 - Q_{low}}{Q_{high}} < 1$$

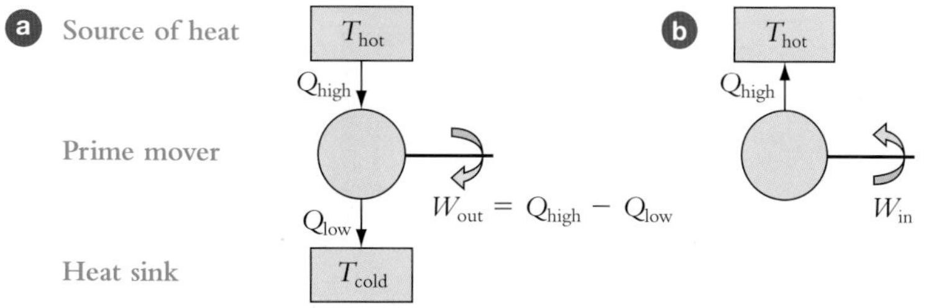

Figure 2.5 Schematic representation of heat engines; (a) shows a heat engine designed to produce work out of the working fluid, and (b) shows how the second law allows work to be entirely converted to heat.

The most important observation here is that, in order to produce work output, heat has to be lost to a cold reservoir. To produce work the engine must exchange heat with both the hot and the cold reservoirs. With only one reservoir (Figure 2.5(b)) work can be done on the system, but not produced by it.

Work and heat occur at system boundaries when the working fluid changes its state. Work implies that the boundary of the system is moved; heat implies that there is a change of temperature of the system. Heat transfer is a less useful form of energy transfer than work because all energy transfer by work can be converted to energy transferred as heat without having to lose any energy in the exchange, but not all heat transfer can be converted to energy transferred by work; in using heat transfer to produce work, some heat supplied must be lost, or transferred to a low temperature, and hence low energy content reservoir as part of the process.

Real processes

Figure 2.6(a) shows various flow devices that are used in thermodynamic machines to relate the working fluid to the outside world. A **compressor** does work by sucking gas axially into a series of disks with aerofoil blades on the periphery. As the gas is compressed it passes through ever narrower passages with smaller blades, until it exits at high pressure. These are commonly used in the inlet of a turbofan jet engine. The **turbine** is a compressor in reverse – it takes a hot, highly compressed fluid and expands it through ever-widening passages to get work out while reducing the pressure and the energy of the working fluid; a turbine is found at the rear of a jet engine. A common symbol for a compressor or turbine is the quadrilateral shape at top right of Figure 2.6(a). **Pumps** are used to deliver a volume of pressurized liquid, in which there is relatively little energy increase in the liquid because liquids are virtually incompressible. The **throttle** is a constriction on the flow, the most common example being a tap on a kitchen sink. A throttle involves a pressure drop, but is usually at a sufficiently low flow rate so that the kinetic energy change is insignificant. In this case, the change of enthalpy is negligible – there is an exchange between u and pv internally.

Placing these flow processes in a cycle, as in Figure 2.6(b), we see a schematic of a realistic heat engine, the **vapour power cycle**. The hot source of heat and the cold heat sink are external to the system – they transfer heat across pipes that contain the working fluid – a liquid being pushed through the system by a pump is energized by the heating process and converted to a vapour, producing work through a turbine. The fluid is then condensed back to a liquid in the cold sink before repeating the cycle. The elements of this machine are a simplified description of what occurs in a power-generation station.

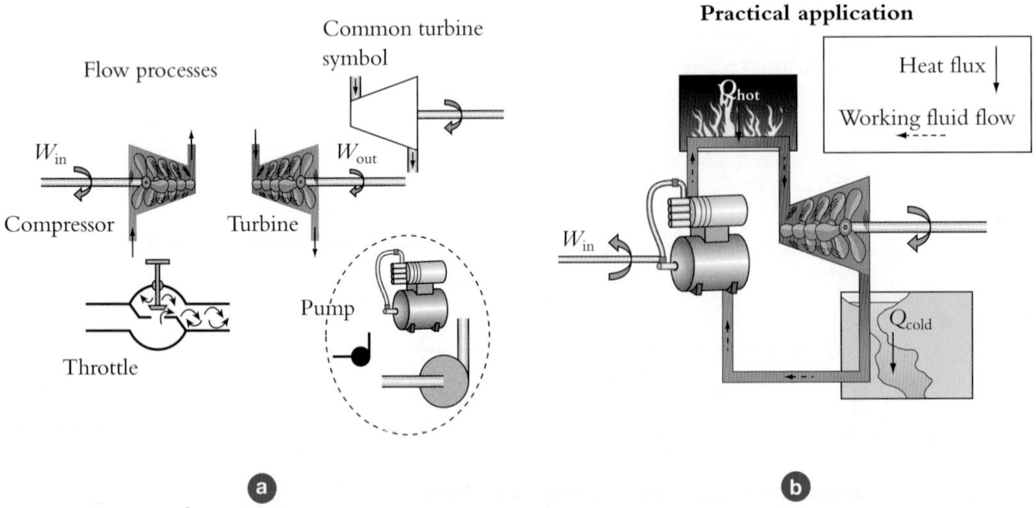

Figure 2.6 (a) Flow processes that alter the state of a working fluid in an open process; (b) a practical implementation of the heat engine.

Having a basic heat engine to analyse, we now consider its efficiency. The **Carnot efficiency** is a very important measure. Referring to the heat engine of Figure 2.5, the cycle efficiency is the ratio of the work obtained to the heat required. It can be seen from the heat engine schematic that work out is the difference between the heat in and heat out. This is determined by the temperature of each of the reservoirs:

$$\eta_{\text{Carnot}} = \frac{\text{Work done}}{\text{Heat supplied}} = \frac{W_{\text{out}}}{Q_{\text{high}}} = \frac{1 - Q_{\text{low}}}{Q_{\text{high}}} = \frac{1 - T_{\text{low}}}{T_{\text{high}}} < 1$$

Q [W] – rate of heat supplied to a reservoir of heat energy
T [K] – temperature of a reservoir of heat energy

This shows that the higher the temperature difference, the better the efficiency. The most efficient machine will exchange heat with a sink at absolute zero. For all real cycles, efficiency is lower than the Carnot Efficiency because a small amount of work is required to drive the process (the pump in the vapour power cycle) and because heat is added at more than one temperature.

Entropy is a measure of disorder; the state of agitation of a system; the energy allocated to maintaining a state of disorder over the ordered state. This *property* is defined from the second law:

$$dq_{\text{reversible}} = Tds$$

where
q [J/kg] – specific heat transfer
T [K] – instantaneous working fluid temperature
s [J/kgK] – instantaneous specific entropy of the working fluid

this is the heat transfer equivalent of $dw_{\text{reversible}} = -pdv$. It is the minimum heat transfer required to get from one state to another. This applies for any reversible process undergone by a closed system.

Isentropic efficiency is used to characterize the performance of many open-system machines. Since open-system machines (for example, axial flow turbines and compressors) operate under near adiabatic conditions, and a reversible adiabatic process is the most efficient, it is useful to compare the actual machine work with the ideal machine work. The **isentropic work** output of a turbine is greater than the energy transferred out of the fluid in practice, and the isentropic work input to a compressor is less than the theoretical input power required to drive the compressor, i.e. the useful increase in energy to the fluid is less than the isentropic amount. The work is measured in terms of change of working fluid enthalpy, and can be determined from the steady flow energy equation.

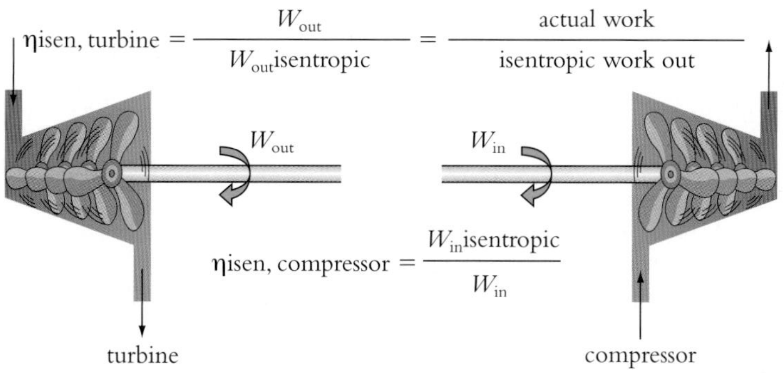

Figure 2.7 Isentropic efficiency is defined as the ratio of ideal work to actual work in an axial flow compressor or turbine.

The state diagram in Figure 2.8 describes the properties of steam. It is useful to notice the characteristics of the chart. The standard notation is to use f to indicate saturated liquid and g for saturated gas. If the temperature is 278 K (5°C) and the pressure is 1.013 25 bar (i.e. the standard atmospheric pressure at sea level) then the vapour is a sub-cooled liquid. If we

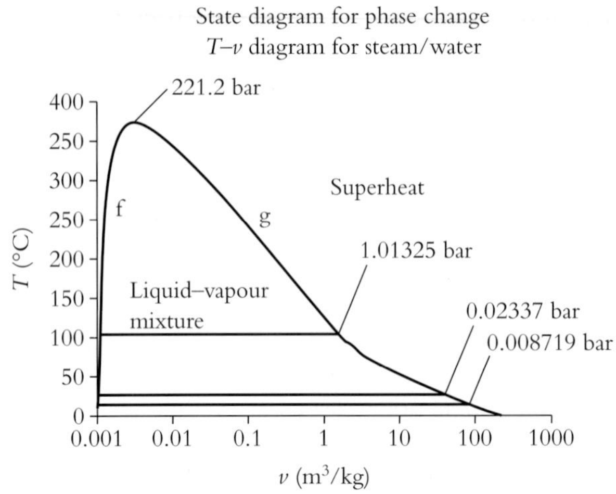

Figure 2.8 State diagram for water in terms of pressure and specific volume. The area under the curve is the region where vapour and liquid exist together, the left-hand side labelled 'f' indicates the line separating saturated liquid from the mixture region, and the right-hand side labelled 'g' indicates saturated vapour, separating superheated vapour from the mixture region. Note that the constant pressure lines, which are horizontal in the mixture region, follow approximately the 'f' line and rise sharply in the superheated steam region.

reduce the pressure slowly to 0.008 719 bar, boiling will occur. This happens on solid surfaces within the fluid, and bubbles rise and expand within the fluid. 0.008 719 bar is p_S, or saturation pressure, for 5°C; similarly 0.023 37 bar is p_S for 20°C. Other points to note from this chart are that the saturated liquid region is very narrow, i.e. liquid volume doesn't change a great deal with temperature, certainly up to 100°C at atmospheric pressure. The volume difference between f and g grows as the temperature falls. Phase change diagrams, regardless of the properties compared, all have this approximate form, with division into the three regions. The shape of the diagrams is always a dome containing the vapour–liquid mixture with the f–g dividing line. The position of the dome and its width vary with properties of the particular fluid.

Just as the area on the *p–v* graph represents the work done for a reversible process, Figure 2.9 shows the *T–s* diagram, in which the area represents the heat transfer for a reversible process. The heat transfer for the reversible case is useful because it is the limiting heat transfer, to which actual heat transfer can be compared. Although it has no meaning for irreversible processes, it shows the ideal for a particular type of process. It is particularly useful for liquid–gas materials such as water and refrigerants.

The *T–s* diagram for steam in Figure 2.9 has constant pressure lines drawn on it. The pressure lines are horizontal through the central vapour–gas mixture region under the dome, and then follow the contour of the saturated liquid side of the dome (the left-hand side from the critical point at the apex). Entropy for water is assigned a value of zero at zero Celsius for convenience. By the third law of thermodynamics the absolute value of entropy of any substance is zero at zero K. Despite this, use of an alternative reference state is helpful. This diagram is particularly useful for the forthcoming vapour power cycle section of this unit and for refrigeration. This is very useful regarding the Carnot efficiency, against which the efficiency of all plant is compared. Turbines are considered as constant entropy devices.

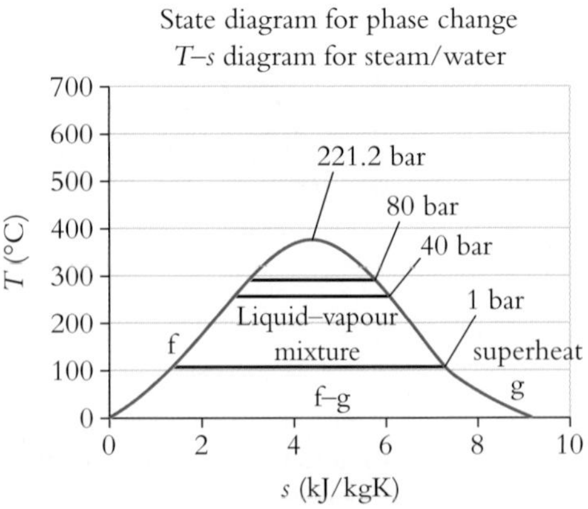

Figure 2.9 State diagram for water in terms of temperature and specific entropy. The characteristics of this chart are similar to the p–v chart, except that the dome is more spread out on the horizontal axis.

Learning summary

By the end of this section you will have learnt:

- ✓ there are principal descriptive terms used in the previous volume that are now to be used with an emphasis placed on how they find application in thermodynamic processes. In particular, the ideas of cycles, processes and irreversibility are useful for considering practical thermodynamic systems;
- ✓ the first law of thermodynamics is directly represented by the steady flow energy equation in most engineering processes because the fluids move through the system. The effects of potential and kinetic energy terms are considered to be negligible in many practical thermodynamic processes. The equation then reduces to a form that considers the interchange between the enthalpy of the fluid and the energy transferred to it in the forms of heat and work. *This is the most important and straightforward relationship to remember for applied thermodynamics – it is the cornerstone of applied thermodynamic machines*;
- ✓ the properties of thermodynamic working fluids are important in determining thermodynamic state, and the equation of state for perfect and semiperfect gases is used in most calculations involving machines where a gas is the working fluid;
- ✓ thermodynamic processes change the state of a working fluid, and cyclic processes and heat and work transfers in a series of processes are commonly used to produce heat and work;
- ✓ the work done on the fluid in a closed process is $W = -\int_1^2 p\,dV$ for a *reversible* process;
- ✓ state diagrams and the representation of processes on them indicate what is going on in heat engines;
- ✓ the second law of thermodynamics describes *how* heat and work can and cannot be converted in a thermodynamic machine;
- ✓ the eight corollaries of the second law are principles of real thermodynamic processes and demonstrate pragmatic observations relating to the second law;
- ✓ cycle efficiency describes how well a machine takes ingoing energy and converts it to useful output;
- ✓ the practical components of real thermodynamic machines are turbines for work out, pumps and compressors for work in, and throttles for reducing pressure;
- ✓ the Carnot efficiency is the measure of the maximum achievable performance of a heat engine;
- ✓ isentropic efficiency is a comparison of the real process power output to the ideal, reversible process power output.

2.2 Air conditioning

Air conditioning is mainly concerned with controlling temperature and humidity. It is important for controlling the comfort of living organisms and for keeping complex electronic circuitry within suitable climatic conditions.

Comfort air conditioning for people:

- people have a limited comfort zone due to the requirement to maintain a steady core body temperature of 37°C;
- people produce heat and release moisture into the atmosphere;
- people generate heat at ~80W resting, 120W office work, and up to 400W during physical work;
- people produce sweat at varying rates, and 100% humid air during respiration.

Control conditioning for computers:

- computers have a limited 'comfort' zone requirement for steady core temperature and dry conditions, which are cooler and dryer than for people.

A so-called 'swamp cooler' is a good example of air conditioning – blowing air through a wet cloth in hot dry climates causes the air to emerge on the other side sensibly colder – typically in British indoor conditions, this will cause a 3 or 4°C drop in the air temperature. This is akin to the common experience of climbing out of a swimming pool and feeling cold in air that is actually warmer than the water. The temperature difference is due to evaporation – otherwise known as evaporative cooling. This difference increases with increasing temperature and decreasing **humidity** – a term of primary importance in air conditioning.

Figure 2.10 shows an experimental apparatus for air conditioning tests. A duct of approximately 2 m length has an in-draught fan, a 5 kW steam boiler for moisture addition to the air stream, a 2 kW air heater, a cooler unit (refrigeration circuit) capable of approximately 3 kW of cooling, another heater of 1 kW and an orifice plate to measure the air flow rate (section 3.3 of *Part 1*). All of these components affect the condition of the air.

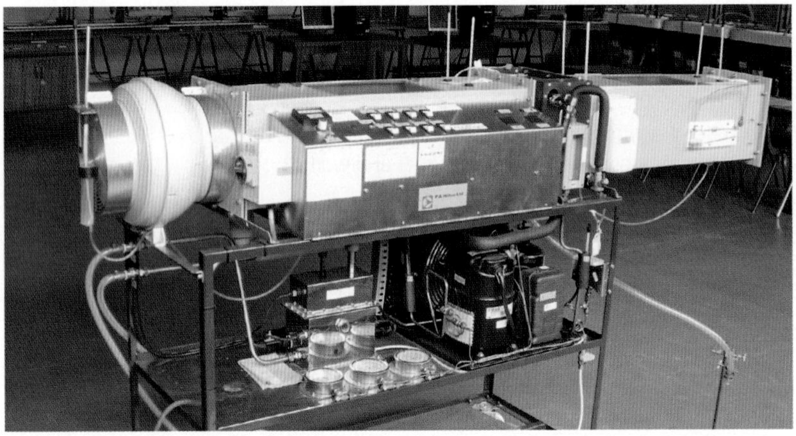

Figure 2.10 A basic air conditioning unit – fan and refrigerator circuit

When it is very dilute, water vapour is considered to be a perfect gas. Figure 2.11 shows the principle of mixed gases in which gas molecules of different gases have different colours to illustrate that they are intimately mixed and are undergoing molecular motion and collisions with a level of kinetic energy (dependent on temperature) that gives internal energy to the system and results in an overall pressure. These gases can be imagined separately filling the entire space. They would then give a pressure in the space on their own. The **law of partial pressures** observes that when the gases are mixed, their individual, separate pressures add up to make the total pressure of the combined gas mixture. The pressure ratio is directly proportional to the ratio of the number of **moles** as described in equation (2.12).

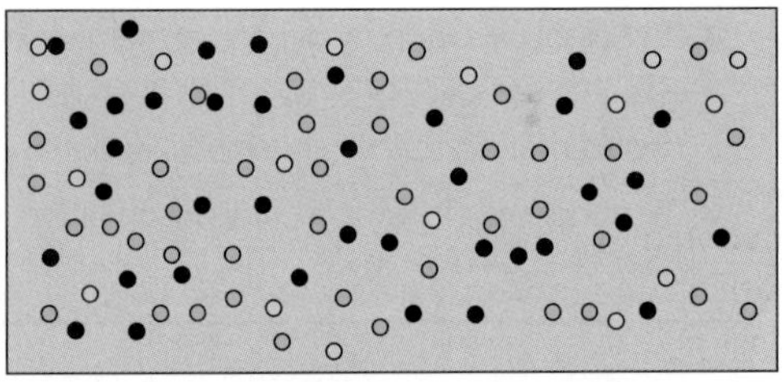

Figure 2.11 Perfect gases have partial pressures – the pressure of that amount of gas if it alone occupied the entire volume. Here molecules of three gases are inter-mixed.

Gibbs–Dalton law (law of partial pressures)

$$p = \sum p_i \qquad (2.12)$$

where
p [Pa] mixture pressure
p_i [Pa] – the partial pressure due to component i

We will see later that the partial pressure is related to the number of moles or mols (n) of gas in the mixture:

$$\frac{p_i}{p} = \frac{n_i}{n} \qquad (2.13)$$

where
n [mols] – number of moles of a species (subscript i) or of the mixture (no subscript).

Note that the number of moles indicates the quantity of a substance. Every molecular substance has a molecular mass; that is the mass of a specific number of molecules of the substance. The specific number of molecules in a mole is the **Avogadro number – 6.23 × 10^{23}** molecules per mole. Since molecules generally have differing sizes, they have different molecular masses. The molecular mass is defined in the units g/mol or kg/kmol.

Air is classified either with or without water vapour as **atmospheric air** or **dry air** respectively. The properties of dry air are available in tables, and are just that – air without any moisture whatsoever. Dry air is a mixture of gases, which can be regarded as existing in a universally constant ratio across the globe at sea level, mainly nitrogen and oxygen. The partial pressures are determined by the number of moles of each gas present in a given volume, which is in the same ratio as the volumetric ratio. Water vapour can be supported in the air by virtue of its partial pressure – or its saturation pressure at the temperature of the atmospheric air.

Worked example

Atmospheric air consists of dry air and water vapour.

Dry air, composed of N_2 and O_2, by mass the proportions are approximately 76.7% nitrogen and 23.3% oxygen, and by volume, 79% nitrogen and 21% oxygen.

What are the partial pressures, assuming atmospheric pressure is one standard atmosphere, 1.01325 bar?

nitrogen – 0.79 × 1.01325 = 0.80047 bar

oxygen – 0.21 × 1.01325 = 0.21278 bar

The practice of boiling water is familiar, and it occurs when the **saturation pressure** of a fluid is equal to the atmospheric pressure. Water boils at 100°C (the boiling point) at a pressure of 1 atmosphere (1.01325 bar). This is the **saturation temperature**, T_s, and the corresponding saturation pressure, usually denoted p_s. The chart of saturation pressure vs. saturation temperature in Figure 2.12 shows that p_s drops sharply when the temperature falls below 100°C.

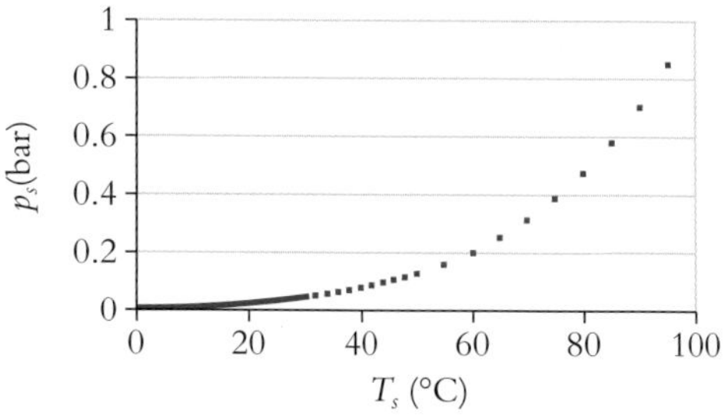

Figure 2.12 Water vapour and its saturation temperature and pressure

Saturation pressure shows the pressure that water vapour exerts on a free water surface because of the energetic molecules it consists of. When the saturation pressure is equal to the local atmospheric pressure, boiling occurs; it can occur anywhere on a solid surface submerged in the liquid, the surface providing nucleation sites for vapour bubble growth. Initially, small bubbles grow as they rise through the liquid before bursting through the free surface.

Apart from boiling, liquid can evaporate from a free surface and condense at a free surface. In equilibrium, evaporation and condensation occur at equal rates. The partial pressure of the vapour in contact with its liquid will be the saturation pressure corresponding to the temperature of the atmosphere. When water vapour is not in contact with a free liquid water surface, the partial pressure will be less than or equal to the saturation pressure at that temperature. The level of **humidity** will be determined by the saturation pressure of the liquid at the prevailing temperature.

There are two measures of humidity, which are useful in different ways. **Absolute humidity**, ω, describes the mass of water vapour contained by a given mass of air.

$$\omega = \frac{m_s\,(\text{vapour})}{m_a\,(\text{air})} = \frac{m_s/V}{m_a/V} = \frac{v_a}{v_s}$$

Note here that $m_s/V = 1/v_s$ and $m_a/V = 1/v_a$, and hence the relationship in terms of specific volume also.

The following derivation shows how this formula can be converted from mass fractions to partial pressures:

$$p_s V = m_s R_s T \Rightarrow \frac{m_s}{V} = \frac{p_s}{R_s T}$$

and

$$p_a V = m_a R_a T \Rightarrow \frac{m_a}{V} = \frac{p_a}{R_a T}$$

Therefore,

$$\omega = \frac{R_a p_s}{R_s p_a}$$

Since R_a is 287 J/kgK and R_s is 461 J/kgK and $P_{\text{atmos}} = p_a + p_s$, this equation becomes:

$$\omega = \frac{287 p_s}{461 (p_{\text{atmos}} - p_s)} = \frac{0.622 p_s}{(p_{\text{atmos}}\ 2\ p_s)}$$

This relationship is very useful in the analysis of air conditioning systems, as will become clear in the subsequent parts of this section.

The second measure is **relative humidity,** φ, which gives a measure of how far from saturation the atmospheric air is, and is defined by equation (2.14):

$$\phi = \frac{p_s}{p_g} \qquad \textbf{(2.14)}$$

Note carefully the redefinition of p_s for this equation only – here p_s is partial pressure of water vapour actually in the air, not the saturation pressure, and p_g is the partial pressure of vapour if the mixture is saturated at the temperature, T, of the mixture, i.e. p_g is the saturation pressure. Note that this use of p_s is only to be used for air conditioning and not to be confused with the saturation pressure also denoted p_s in the tables. Here, and only for air conditioning, p_g means saturation pressure. This is the standard notation for air-conditioning engineers, which is unfortunately at odds with the standard notation used elsewhere in thermodynamics.

A typical range of ω is from zero up to about 30 g H_2O per kg of dry air at normal earth sea level conditions and φ varies between 0 and 100%.

If the temperature of the air falls until the saturation point or 100% relative humidity occurs, the air is at the **dew point** temperature; this is the temperature at which air becomes saturated when cooled at constant pressure (since ω = constant, P_s is constant during cooling). This can be verified by reference to Figure 2.8 or Figure 2.9.

Examples

(1) A sample of atmospheric air contains 12 g of water in 1.2 kg of dry air, what is the absolute humidity?

In this case there is 12 g per 1.2 kg, which is a specific or absolute humidity, ω, 0.01. This is typical of the standard atmospheric condition.

(2) What is the specific volumetric ratio of moisture to air?

The inverse of ω is the volumetric ratio – i.e. partial volume ratio. The specific volume ratio of air to vapour follows, i.e. 0.01 m³/kg of air per m³/kg of vapour. This shows the significant difference in specific volume of the water vapour and dry air.

(3) If dry air specific volume is 1 m³/kg, and partial pressure of water vapour is 0.016 bar, what is the specific volume ratio and, hence, what is the specific humidity?

Air specific volume, v is 1 m³/kg and specific volume of vapour is (from Figure 2.8 or, more accurately, from thermodynamic property tables) 83 m³/kg, specific volume ratio is $\frac{1}{83} = 0.012$, which is the specific humidity.

(4) Given a specific humidity of 0.02, and atmospheric pressure of 1.01325 bar, what is the partial pressure of water vapour?

$\omega = 0.02$, and $p_{atmos} = 1.01325$ bar, $p_s = \dfrac{\omega . p_{atmos}}{(0.622 + \omega)}$; therefore, $p_s = 0.03156$ bar

(5) The humidity is 50% in atmospheric air at 20°C; what is the partial pressure of the water vapour in the air?

At 20°C, p_g is 0.02337 bar. Given the relative humidity is 50%, the vapour pressure, p_s must be $0.02337 \times 0.5 = 0.0117$ bar.

(6) If the temperature of the atmospheric air is 35°C and the relative humidity is 50%, what is the dew point?

Find p_s from the relative humidity, φ, formula with p_g at 35°C. Then identify the temperature that corresponds to this, that is the dew point. $p_g = 0.056$ bar (from thermodynamic property tables). $p_s = 0.5 \times 0.056 = 0.028$ bar. From the tables the temperature that corresponds to this as its p_g is 23°C. This means that when the temperature falls to 23°C, moisture will condense out of the air, either as a cloud if the bulk air temperature decreases, or more usually on sky-facing surfaces which cool down due to losing heat to the night sky by thermal radiation.

Hygrometry or psychrometry

The analysis of air condition is called **hygrometry** and the two measures just introduced provide the basis for the work of air-conditioning engineers. In order to control air condition, we must be able to measure it easily. The basic tools of the trade are **wet and dry bulb thermometers** as shown in Figure 2.13 The wet bulb differs from the dry one only in that it has a muslin sock around the bulb that is soaked in water from a reservoir. If the local atmosphere has humidity less than 100% then water will evaporate from the sock and cause the temperature of the bulb to drop, owing to the latent heat of evaporation, and therefore the thermometer will register a lower temperature. This temperature enables us to calculate the relative humidity.

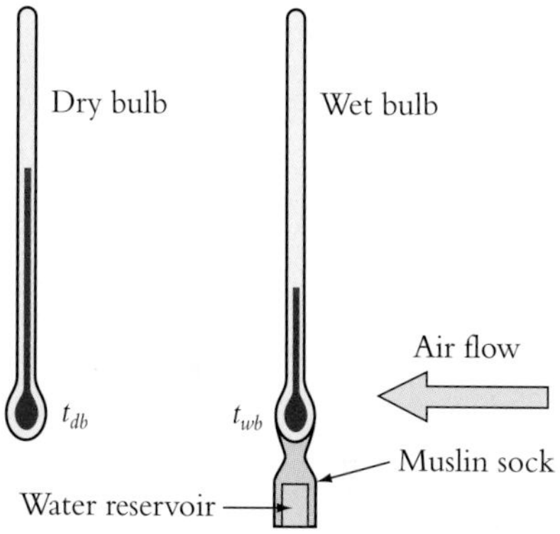

Figure 2.13 Wet and dry bulb thermometers

A spreadsheet can be calculated based on this theory to determine the relative and specific humidity from the wet and dry bulb temperatures. The chart in Figure 2.14 is called a **psychrometric chart** and it shows the condition of the air for a given pair of wet and dry bulb temperatures. Data represented includes relative and specific humidity, enthalpy of the mixture and specific volume of the mixture.

Examples of reading the psychrometric chart

(1) What is the relative humidity if T_{DB} is 30°C and T_{WB} is 20°C?

The dry bulb temperature is read from the bottom axis, and constant temperature lines rise vertically. The wet bulb temperature is read from the 100% relative humidity curve on the left-hand side. Constant wet bulb temperature lines are inclined to the horizontal. The intersection of the two lines defines the air condition, and the relative humidity is indicated on the curves at 10% intervals. The approximate value for this particular condition is 40%.

(2) What is the specific humidity if it is 100% humid at 30°C? What does this tell you about the relative masses of air and water at 30°C?

Taking the position on the 100% relative humidity curve, and marking the point when the dry bulb temperature is 30°C defines the air condition. Now read horizontally across to the vertical axis on the right-hand side to show that the absolute humidity is 0.025.

(3) What is the enthalpy for a 50% relative humidity at 20°C dry bulb?

The intersection of the 50% relative humidity curve with the vertical 20°C dry bulb temperature defines the air condition. The enthalpy is found by placing a rule on the air condition point, and rotating about that point until the enthalpy on the upper enthalpy scale and the enthalpy on the lower scale are the same. In this case it is 39 kJ/kg.

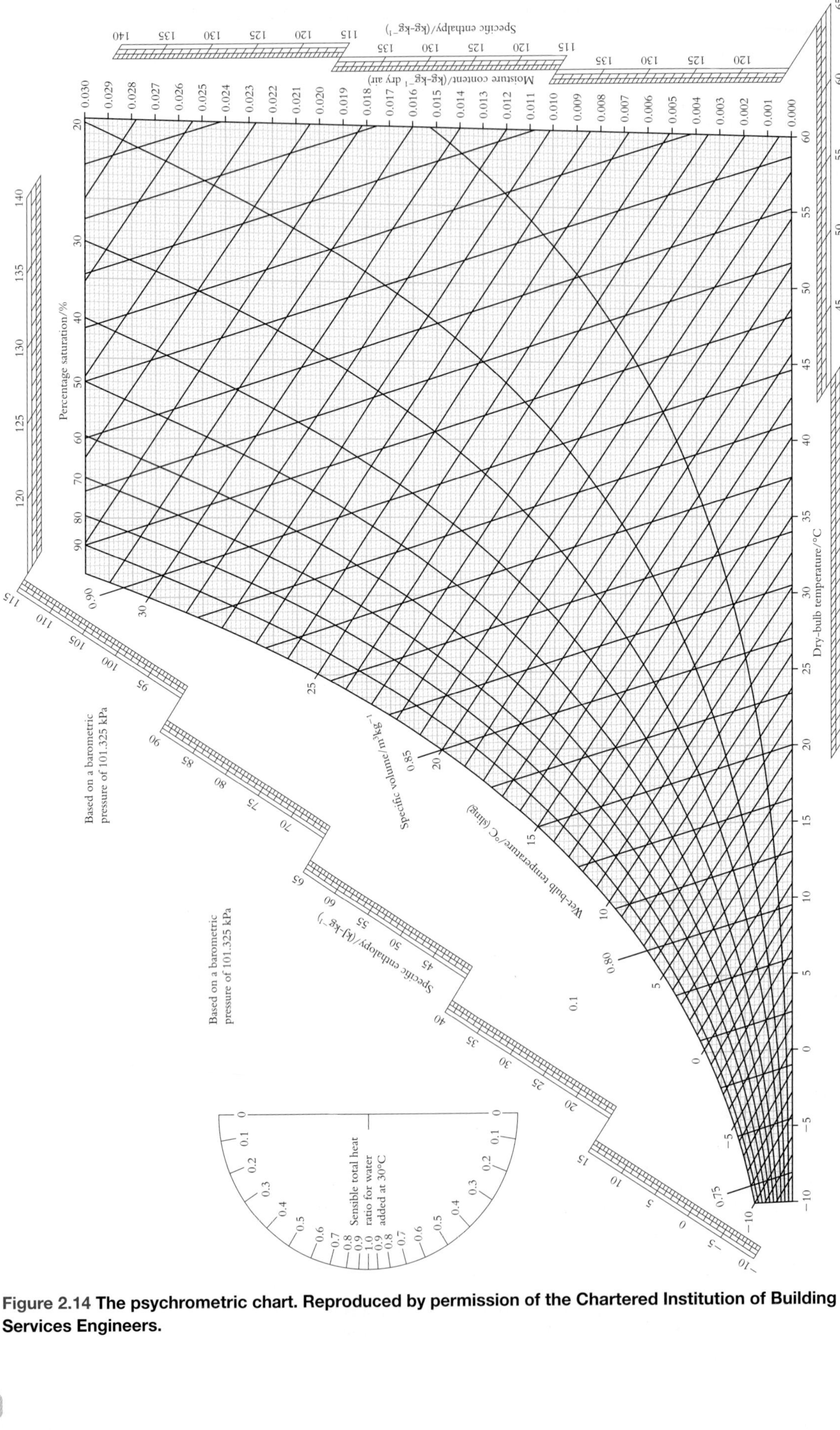

Figure 2.14 The psychrometric chart. Reproduced by permission of the Chartered Institution of Building Services Engineers.

Figure 2.15 shows the basic principles of operation of an air-conditioning unit. Air to be conditioned is drawn in by a fan, cooled down either to simply reduce the temperature or, more commonly, to also decrease humidity by reaching dew point and causing condensation. Air is heated up after the cooler when condensation has occurred in order to make the temperature comfortable before it is issued to the room. The power of the heating and cooling is determined from the desired changes in enthalpy of the dry air/water vapour mixture between each of the sections. This must be equal to the energy required to be drawn from the external source. The steady flow energy equation is used, applying conservation of mass and energy. With reference to Figure 2.15:

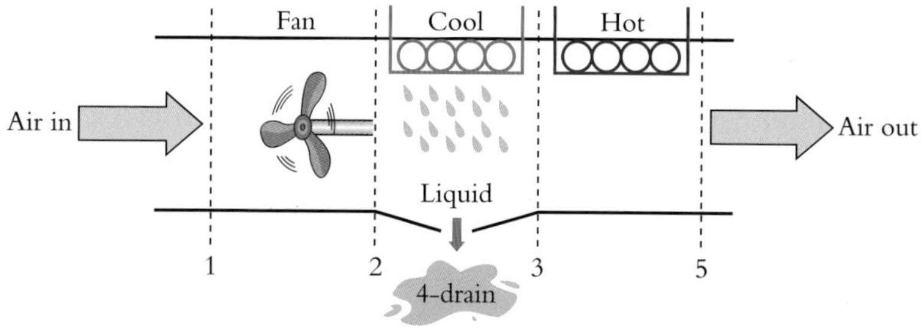

Figure 2.15 Enthalpy exchanges in the cooling section

Mass conservation:

$$\dot{m}_{s2} = \dot{m}_{s3} + \dot{m}_{w4}$$

$$\dot{m}_{a1} = \dot{m}_{a2} = \dot{m}_{a3} = \dot{m}_{a4}$$

At stage 3, there is the same mass of dry air, but some moisture has fallen out of the cooling unit (refrigerator coils) to the drain. So we have the same mass of dry air with different enthalpy (calculable by dry air temperature and specific heat capacity at constant pressure, c_p), and we have a (smaller) amount of water vapour with the air. We also have an amount of condensate – the condensed water vapour on the cooling coils.

Then $m_a h_{a1}$ is ingoing air enthalpy, m_{s2} is from specific humidity and h_{s2} from enthalpy of water vapour at t_2; m_{w4} is the condensate at the temperature of the cooling coils with h_{w4} from tables as h_{fw4}.

Worked example

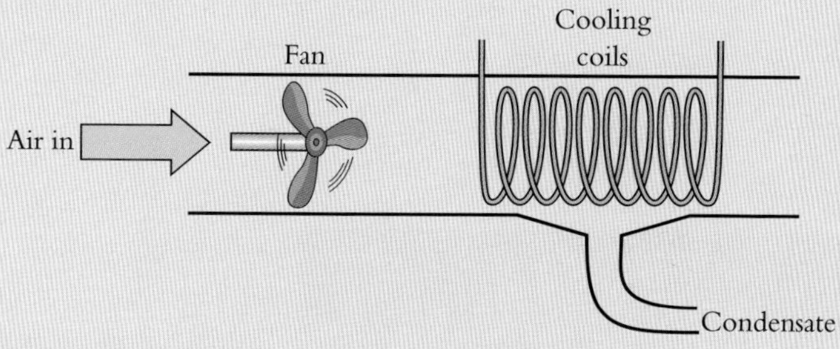

Figure 2.16

An air-conditioning unit has a fan that delivers a mass of atmospheric air. The mass flow rate of the dry air in this is m_a, which is 0.5 kg/s. The specific humidity is 0.01 kg/kg air, and it can be concluded that the mass flow rate of the water vapour, m_{s2}, is 0.005 kg/s. The ingoing air temperature is 30°C.

The unit causes a drop in specific humidity to 0.006 kg/kg. What is the rate of heat taken out of the air in the cooler section of the air conditioning unit, $\dot{q}$? What is the pressure of the dry air, p_a, and of the water vapour, p_s, at exit?

(1) Calculate p_{s2} and p_{s3} using specific humidity–vapour pressure formula.

(2) Calculate the mass flow rate of dry air and of water vapour at entry to, and at the exit from, the cooler.

(3) What is the dew point temperature corresponding to $p_g = p_{s3}$?

(4) Find c_{pair} from the tables at the appropriate mean temperature.

(5) Find, from the tables, the specific enthalpy of liquid water at the dew point temperature, and of the vapour at points 2 and 3.

(6) Use the SFEE across the cooler to find $\dot{Q}$ the rate of heat delivery to the air stream.

(1) using $\omega = \dfrac{0.622p_s}{p - p_s} = \dfrac{0.622p_s}{1.01325 - p_s} 0.01 \Rightarrow p_{s2} \Rightarrow 0.01 \times 1.01325 - 0.01p_{s2} = 0.622p_{s2}$

$\Rightarrow 0.01013 = 0.632p_{s2} \Rightarrow p_{s2} = 0.016\,\text{bar}$

p_{s3} must be 100% ϕ, $\quad \therefore\ p_g = p_{SAT} = p_{s3}$

$$\omega = \frac{0.622p_{s3}}{p - p_{s3}} \Rightarrow 0.006 = \frac{0.622p_{s3}}{1.01325 - p_{s3}} \Rightarrow 0.00608 - 0.006p_{s3} = 0.622p_{s3}$$

$\Rightarrow 0.00608 = 0.628p_{S3} \Rightarrow p_{S3} = 0.0097\,\text{bar}$

(2) Atmospheric air mass flow rate at entry is given as 0.5 kg/s. We know

$$\omega \Rightarrow \frac{\dot{m}_W}{\dot{m}_{DRYAIR}} = 0.01 \text{ and } \dot{m}_{DRYAIR} + \dot{m}_W = 0.5\,\text{kg/s} \ \therefore\ \dot{m}_W = 0.005\,\text{kg/s}$$

and

$$\dot{m}_{DRYAIR} = 0.495\,\text{kg/s}.$$

At exit, we must have $\dot{m}_{DRYAIR} = 0.495$ kg/s still – it can only go out one way. We know that $\omega = 0.006$, and therefore we say

$$\omega = \frac{\dot{m}_W}{0.495} = 0.006 \Rightarrow \dot{m}_W = 0.003\,\text{kg/s}$$

(3) at p_{s3}, $t_{SAT} \sim 6.5°\text{C}$ from tables of saturated steam and water.

(4) The mean temperature of 6.5°C and 30°C is 18.25°C or 291 K. Using the data for dry air at low pressure,

T [K]	c_p [KJ/kgK]	c_v [KJ/kgK]	γ [–]	$\mu \times 10^{-5}$ [kg/m.s.]	$K \times 10^{-5}$ [kW/m.K.]	Pr [–]
175	1.0023	0.7152	1.401	1.182	1.593	0.74
200	1.0025	0.7154	1.401	1.329	1.809	0.73
225	1.0027	0.7156	1.401	1.467	2.020	0.72
250	1.0031	0.7160	1.401	1.599	2.227	0.72
275	1.0038	0.7167	1.401	1.725	2.428	0.71
300	1.0049	0.7178	1.400	1.846	2.624	0.70
325	1.0063	0.7192	1.400	1.962	2.816	0.70

Table 2.1 Dry air at low pressure (Rogers and Mayhew, 1995)

$c_{p,air})_{291K}$ could be interpolated for accuracy, but inspection shows it is 1.0045 kJ/kgK.

(5) The saturated water and steam table is required again.

T [°C]	p_s [bar]	v_g [m³/kg]	h_f [kJ/kg]	h_{fg} [kJ/kg]	h_g [kJ/kg]
0.01	0.006112	206.1	0	2500.8	2500.8
1	0.006566	192.6	4.2	2498.3	2502.5
2	0.007054	179.9	8.4	2495.9	2504.3
3	0.007575	168.2	12.6	2493.6	2506.2
4	0.008129	157.3	16.8	2491.3	2508.1
5	0.008719	147.1	21.0	2488.9	2509.9
6	0.009346	137.8	25.2	2486.6	2511.8
7	0.01001	129.1	29.4	2484.3	2513.7
8	0.01072	121.0	33.6	2481.9	2515.5
9	0.01147	113.4	37.8	2479.6	2517.4

Table 2.2 Saturated water and steam (Rogers and Mayhew, 1995)

From this it can be seen that h_f of liquid water at 6.5°C is about 27 kJ/kg, h_g at 6.5°C is 2512 kJ/kg and at 30°C, h_g is 2555 kJ/kg.

29	0.04004	34.77	121.5	2432.4	2553.9
30	0.04242	32.93	125.7	2430.0	2555.7
32	0.04754	29.57	134.0	2425.3	2559.3

Table 2.3

(6) Now use the SFEE, which is the statement of the first law (conservation of energy) for moving fluids, and ignore the terms from kinetic and potential energy as being negligible, $Q + W = \Delta H$. In the air-conditioning unit between 2 and 3, there is no fan or other working device, so $W = 0$. We need the change in enthalpy of the fluids in section 2 to 3 in order to work out the heat transfer to cause it.

Referring to the schematic, the enthalpy flows coming in at 2 and out at 3 and through the condensate collection chute, are all indicated in terms of mass flow rate and specific enthalpy. The flow of enthalpy can then be compared in the SFEE:

$$\dot{Q} = \dot{m}_{DRYAIR} h_{DRYAIR,2} - \dot{m}_{DRYAIR} h_{DRYAIR,3} + \dot{m}_{W,2} h_{g,2} - \dot{m}_{W,3} h_{g,3} - \dot{m}_{COND} h_{f,COND}$$

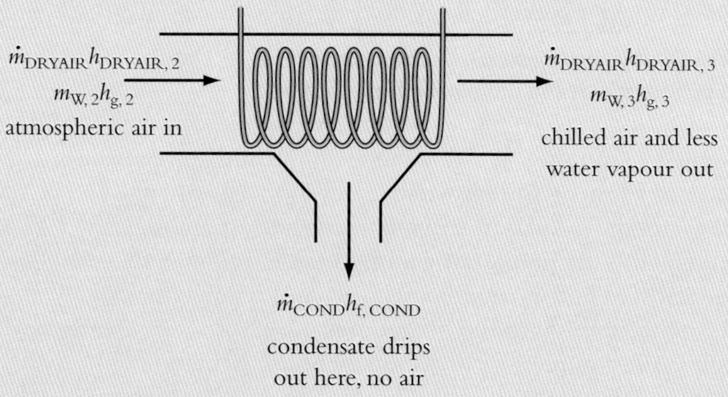

Figure 2.17

We can use $\Delta H = mCp\Delta T$ for the enthalpy change of dry air, which is

$$0.495 \times 1.004 \times (30 - 6.5) = 11.68\,\text{kW}$$

just to cool the dry air. The change in enthalpy of the water vapour carried in the air is mainly due to the loss of mass as vapour that condenses. Therefore, with the data for h_g and mass flow rate of vapour going in at 2 and out at 3, we have

$$0.006 \times 2555 - 0.003 \times 2485 = 5.32\,\text{kW}$$

The condensate is assumed to leave at 6.5°C, and it has a mass rate of 0.003 kg/s, therefore enthalpy flow rate is

$$0.003 \times 27 = 1.27\,\text{kW}$$

Putting all this in the SFEE, we have

$$\dot{Q} = 11.68 + 5.32 - 1.27 = 15.73\,\text{kW}$$

Refrigeration and heat pumps

In considering a practical air-conditioning plant, the cooling duty must be supplied by a refrigeration unit, and the application of refrigeration and heat pumps is therefore considered here. Figure 2.18 shows a heat pump in which work is converted into heat to produce a flow of heat from cold to hot. It is a heat engine operating in reverse. The process is indicated by arrows, essentially this is equivalent to running a reverse of a vapour power cycle with the reverse a **condensing hot reservoir**, an **evaporating cold reservoir** – a compressor and a throttle.

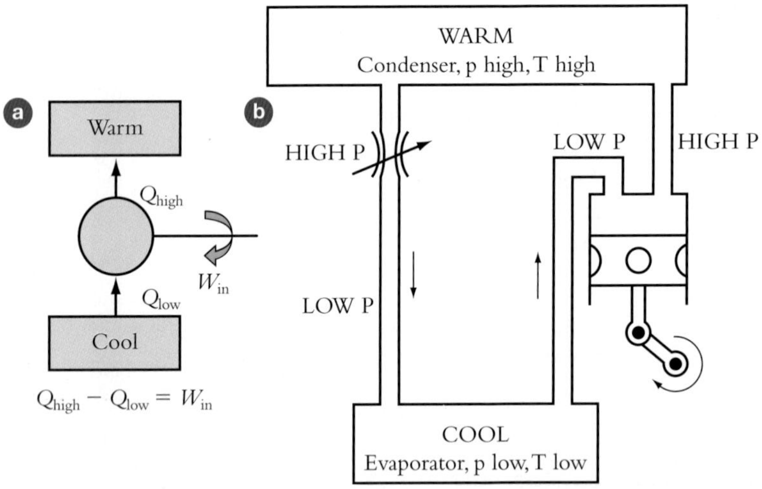

Figure 2.18 Refrigeration or heat pump circuit: (a) schematic of the reversed heat engine, (b) schematic of the physical circuit required

Superheated **refrigerant** vapour is pumped round the circuit by the reciprocating compressor. The compressor maintains the evaporator at a low pressure by drawing evaporated vapour out of it; since the vapour pressure is low and the saturation temperature is correspondingly low – lower than the surrounding cold region – the liquid refrigerant in the evaporator tubes evaporates as heat is absorbed from the relatively hot surroundings at the cool temperature. The compressor drives the evaporated vapour through to the higher pressure condenser tubes at the top. The high pressure means high saturation pressure and therefore high saturation temperature, and the vapour is forced to condense since the saturation temperature is higher than the hot reservoir temperature. This is a refrigerator too, if the cool side is the useful output. Note the symbol for the throttle on the left, which reduces pressure at constant enthalpy.

The pressure, p, vs specific enthalpy, h, properties chart is the easiest chart to deal with for refrigeration cycles since it yields the enthalpy for a process directly. The enthalpy is the energy content of the working fluid, and hence energy balances can very easily be performed with operations occurring at two pressure levels. The ***p–h* diagram** for the commonly used refrigerant R134a is displayed in simplified form in Figure 2.19 Note the examples of constant temperature lines drawn on in blue (10°C) and pink (40°C). On the superheated side of the chart (the right-hand side of the domed line), the temperature lines descend in a curve; on the subcooled liquid side the lines rise vertically because the liquid is incompressible over the range of pressures represented. A typical heat pump process is plotted in Figure 2.19, showing a cycle that runs in the anti-clockwise direction. We know from the earlier description of cycles that this direction means that net work is being put into the cycle.

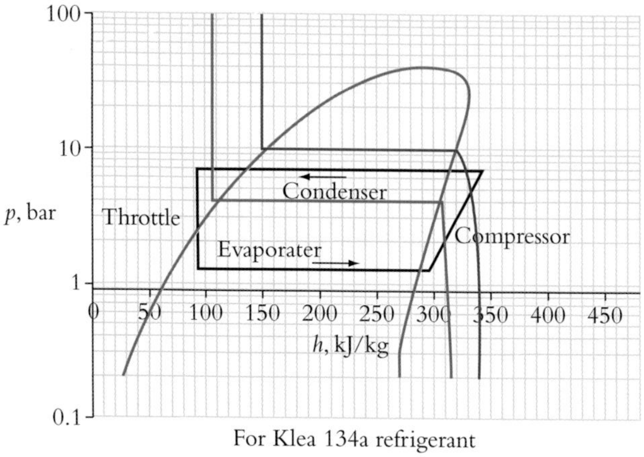

Figure 2.19 *p–h* diagram of a typical modern refrigerant

The chart is plotted as follows:

- horizontal high pressure and low pressure lines across;
- find the temperature of the evaporator exit at the pressure in the evaporator, on the superheated side of the chart;
- find the temperature of the compressor outlet, which is in the superheated region, at the high pressure of the condenser;
- draw a line showing the process of the compressor between evaporator exit and compressor exit;
- find the temperature of the outlet from the condenser, which is in the subcooled liquid region;
- a vertical line through the saturated liquid point corresponding to the condenser exit temperature defines the process in the throttle, assuming that it is a perfectly constant enthalpy device;
- the change in enthalpy across the three main components gives the heat/work transferred in that component.

The reason for the throttle having no change in enthalpy may not be immediately obvious, but in practice the reason is basically as follows: we know that the throttle is short and should be well insulated, and so $\dot{Q}$ is nearly zero. There is obviously no work in the throttle – no paddles or pistons to move the fluid. Use the SFEE $\dot{Q} + \dot{W} = \dot{m}\Delta h$, where $\dot{Q}$ and $\dot{W}$ are rates of heating and power input and $\dot{m}$ is the mass flow rate. It is easy to see that if $\dot{Q}$ and $\dot{W}$ are zero, then the enthalpy entering is the same as that leaving.

The heat pump circuit relies on vapour–liquid behaviour. Vapour enters the condenser tube (Figure 2.20(a)) at a temperature above T_{high} and exchanges heat, Q out of the condenser to cool the vapour. The vapour condenses when it reaches T_{sat} until all is condensed. T_{sat} needs to be a few degrees above T_{hot} for it to work properly. The condensed and slightly subcooled liquid passes through the throttle (Figure 2.20(b)), in which the SFEE causes no change in enthalpy as noted above. This reduces the pressure of the fluid and therefore the T_{sat} reduces to a point just below the

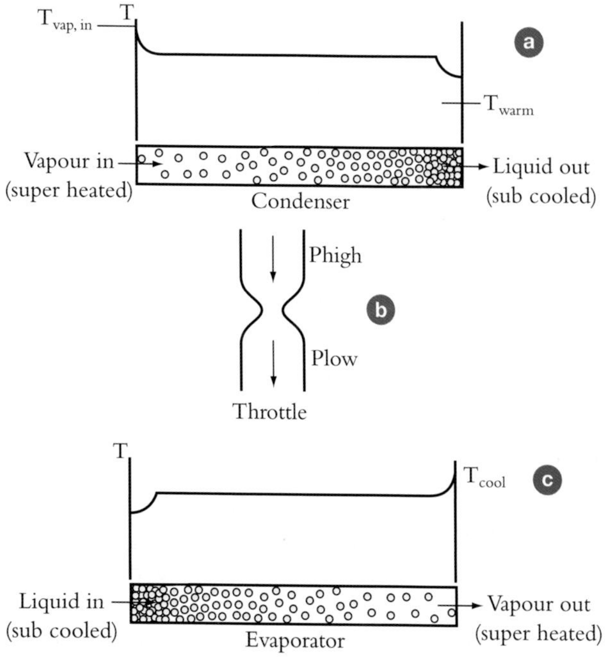

Figure 2.20 How the refrigeration circuit works

temperature of the cool reservoir (which contains the evaporator). As it enters the evaporator, liquid heats up to T_{sat} and starts to boil (Figure 2.20(c)). By the time it reaches the end of the evaporator, all is evaporated and the vapour gains a little superheat.

The heat pump is used not only for refrigeration: the hot reservoir can be the useful heat output of the unit. This is the case in **ground source heat pumps** used for domestic heating. This is an effective way of providing heating because the coefficient of performance ensures that a significantly higher amount of heat is transferred than the energy supplied to the compressor.

Coefficient of performance

The heat pump effectiveness is measured in terms of how much heat is transferred either out of the cold reservoir or into the hot reservoir, depending on whether it is being used as a refrigeration unit or a heat pump. In either case, the cost of pumping the heat is the power into the compressor. A measure of how well the heat pump works is therefore the ratio of the heat transferred to the work input, and these are given by the enthalpy changes in the evaporator and condenser for the heat transfers and by the enthalpy change in the compressor. From Figure 2.20, the change in enthalpy is given on the horizontal axis in kJ/kg of refrigerant. The difference is obviously far greater in the evaporator and condenser than in the compressor. For the cycle shown these enthalpy changes are approximately 200 kJ/kg, 250 kJ/kg and 50 kJ/kg respectively. The ratio is therefore for refrigeration $\frac{200}{50} = 4$ and for the condenser $\frac{250}{50} = 5$. These are far greater than unity and therefore cannot be stated as an efficiency. This ratio is known as the **coefficient of performance** and is the standard way of quantifying the effectiveness of heat pumps.

Learning summary

By the end of this section you will have learnt:

- ✓ atmospheric air is a mixture of dry air and water vapour;
- ✓ air condition defines the temperature and humidity of atmospheric air;
- ✓ the law of partial pressures and partial internal energy determines what proportion of a gas pressure or internal energy is due to each individual gas component;
- ✓ there are two measures of humidity: specific (or absolute) humidity – the mass of water vapour per unit mass of dry air; and relative humidity – the ratio of the partial pressure of water vapour in the air, p_s, to the maximum partial pressure of the water vapour at that temperature, p_g, which is the water saturation pressure;
- ✓ gases may be quantified in moles, which is a specific number of molecules (the Avogadro number, 6.023×10^{26} per kmol);
- ✓ dew point is the temperature at which atmospheric air becomes 100% saturated with water vapour (i.e. $p_s = p_g$), such that if the temperature is further reduced, water condenses out of the atmosphere;
- ✓ the psychrometric chart is the key tool of air condition monitoring together with the wet and dry bulb thermometers, and the terms psychrometry and hygrometry are introduced to describe the study of atmospheric air;
- ✓ the principles of operation of an air-conditioning unit require enthalpy balances to determine the heat power required by heating and cooling in the unit to produce a particular condition;
- ✓ the heat pump is the generic refrigeration unit that is required to produce cooling in the air conditioning unit, and that can also be used to provide heating energy from a cold source;
- ✓ the p–h diagram is used to plot heat pump processes, since it quickly yields enthalpy changes between points, which represent the heat and work transfers required to produce the changes of state;
- ✓ the coefficient of performance is introduced as the measure of heat pump effectiveness.

2.3 Gas mixtures

There are two ways of analysing gas mixtures – by mass fractions and by volume fractions. The law of partial pressures and internal energy leads to analysis by mass – **gravimetric analysis**. Using it enables calculation of properties of a mixture, i.e. h, c_v, c_p, R, M, s from the individual properties combined according to their mass fractions.

Gravimetric analysis

The proportions by mass are useful for determining the fluid properties of the mixture. The Gibbs–Dalton law of partial pressures presented in Section 2.2 gives the method for doing this as follows:

Deductions from the Gibbs–Dalton law of partial pressures:

Enthalpy

$$H = U + pV$$

$$H = \Sigma U_i + \Sigma p_i V$$

$$H = \Sigma H_i \tag{2.15}$$

Here, the subscript i indicates a particular species of gas in a mixture of gases, and it is comprised as shown here by use of the Gibbs-Dalton Law, of a contribution from each of the constituent gas enthalpies.

Specific heat at constant volume, c_v

From the summation (conservation) of internal energy

$$m(u_2 - u_1) = \Sigma m_i (u_{2i} - u_{1i})$$

$$mc_v (T_2 - T_1) = \Sigma m_i c_{vi} (T_2 - T_1)$$

$$c_v = \Sigma \frac{m_i}{m} c_{vi} \tag{2.16}$$

Specific heat at constant pressure, c_p

From the summation (conservation) of enthalpy

$$m(h_2 - h_1) = \Sigma m_i (h_{2i} - h_{1i})$$

$$mc_p(T_2 - T_1) = \Sigma m_i c_{pi} (T_2 - T_1)$$

$$c_p = \Sigma \frac{m_i}{m} c_{pi} \tag{2.17}$$

Gas constant R for the mixture

From Joule's law

$$R = c_p - c_v$$

$$R = \Sigma \frac{m_i c_{pi}}{m} - \Sigma \frac{m_i c_{vi}}{m}$$

$$R = \Sigma \frac{m_i}{m} (c_{pi} - c_{vi})$$

$$R = \Sigma \frac{m_i}{m} R_i \tag{2.18}$$

Apparent molar mass $\widetilde{m}$

$$R = \Sigma \frac{m_i R_i}{m}$$

$$R = \frac{\widetilde{R}}{\widetilde{m}}$$

$$\frac{\widetilde{R}}{\widetilde{m}} = \Sigma \frac{m_i}{m} \frac{\widetilde{R}}{\widetilde{m}_i}$$

$$\frac{1}{\widetilde{m}} = \Sigma \frac{m_i}{m} \frac{1}{\widetilde{m}_i} \quad (2.19)$$

Entropy

This can be calculated by treating the mixture as a perfect gas.
From the first law, $dq = -dw + du$
Substituting from the second law and reversible work in a closed system:

$$Tds = pdv + du$$

$$ds = \frac{p}{T}dv + \frac{du}{T}$$

Now, $du = c_v dt$ and $p/T = R/v$

$$ds = R\frac{dv}{v} + c_v \frac{dt}{T} \quad (2.20)$$

Hence:

$$s_2 - s_1 = R\ln\left(\frac{v_2}{v_1}\right) + c_v \ln\left(\frac{T_2}{T_1}\right) \quad (2.21)$$

This can be converted to alternative forms, i.e. p–T and p–v forms, by using the equation of state.

Since $pv = mRT$, to get the p–T form: $v_2/v_1 = (p_1/p_2)(T_2/T_1)$

Substitute into the previous entropy formula to give:

$$s_2 - s_1 = R\ln\left(\frac{p_1}{p_2}\right) + R\ln\left(\frac{T_2}{T_1}\right) + c_v \ln\left(\frac{T_2}{T_1}\right)$$

$$s_2 - s_1 = R\ln\left(\frac{p_1}{p_2}\right) + (R + c_v)\ln\left(\frac{T_2}{T_1}\right)$$

$$s_2 - s_1 = R\ln\left(\frac{p_1}{p_2}\right) + c_p\ln\left(\frac{T_2}{T_1}\right) \quad (2.22)$$

Therefore for mixtures

$$m(s_2 - s_1) = \Sigma m_i(s_{i,2} - s_{i,1})$$

$$= \Sigma m_i\left[c_{pi}\ln\left(\frac{T_2}{T_1}\right) - R\ln\left(\frac{p_{i,2}}{p_{i,1}}\right)\right]$$

$$m(s_2 - s_1) = \Sigma m_i(s_{i,2})_{V,T} - \Sigma m_i(s_{i,1})_{V,T} \quad (2.23)$$

Note: Even for a perfect gas mixture, the entropy depends on state rather than temperature alone (see h and u, which are just temperature dependent for a perfect gas).

Worked example

Calculate c_p for the following fuel and air mixture at 300 K: methane 0.1 kg, ethane 0.25 kg, nitrogen 0.75 kg and oxygen 0.23 kg

Using the tables of thermodynamic properties, the c_p for each of the gases can be read at 300 K as methane, 2.226 kJ/kgK, ethane 1.766 kJ/kgK, nitrogen 1.040 kJ/kgK and oxygen 0.918 kJ/kgK. Using the formula from above the c_p of the mixture is

$$\frac{(0.1 \times 2.226 + 0.25 \times 1.766 + 0.75 \times 1.040 + 0.23 \times 0.918)}{1.33} = 1.245\ \text{kJ/kgK}$$

Volumetric and molar analysis

Volumetric analysis is directly related to **molar analysis**. So if we have the proportions of volumes of the gases reacting, we know the proportions of moles reacting. The analysis examined so far is gravimetric, that is the analysis of gas mixtures by the separation of the constituents and their estimation by mass. However, in some cases the composition based on volumes (or volumetric composition) of gases in the mixture is known.

Example: The volumetric composition of air is approximately 79% nitrogen, N_2, 21% oxygen, O_2, by volume. That is in 1 m^3 of air, the partial volumes of nitrogen, N_2 and oxygen, O_2 are 0.79 m^3 and 0.21 m^3.

Law of partial volumes: Amagat's law of partial volumes

The volume of a mixture of gases is equal to the sum of the volumes of the individual constituents when each exists alone at the pressure and temperature of the mixture

$$V = \sum_i V_i$$

V = mixture volume at p, T.

V_i = partial volume of gas component i at p, T.

$V_{i)p,T}$ refers to the volume occupied by gas 'i' at the mixture pressure, p, and mixture temperature, T.

It is important to understand the subscripts that identify the various components.

Formula for the relation between partial pressure and partial volume

Partial pressures and partial volumes are alternative ways of describing the mixture. Either a component occupies p_i at V or V_i at P but not V_i at p_i. Using the gas law, considering a perfect gas:

By partial pressure, $p_i V = m_i R_i T$

By partial volume, $p V_i = m_i R_i T$

Divide to give: $\dfrac{V_i}{V} = \dfrac{p_i}{p}$ **(2.24)**

Equation (2.24) shows that proportions by volume are equal to proportions by pressure. So if a gas occupies 10% of volume of a mixture, its partial pressure will be 10% of the mixture pressure.

Molar analysis

By definition, this is the same as the volumetric analysis but using mole fractions. The mole was presented as a measure of substances in the previous section on air conditioning.

Definitions:

(1) 1 mol of gas has a mass in kg equal to its molecular weight.

(2) 1 mol of any perfect gas occupies *the same volume* V at a given p and T as 1 mol of any other perfect gas (e.g. the standard atmosphere condition, $V = 22.4\,\text{m}^3$ at $T = 0°\text{C}$ and $p = 1$ atmosphere).

(3) 1 mol of a substance contains an Avogadro number of molecules, $N_A = 6.025 \times 10^{23}\,\text{mol}^{-1}$.

(4) Usually it is more convenient to quantify in kmol.

Part (2) above tells us that the volume occupied by the same number of molecules of any gas is the same.

The atomic masses of all elements can be found in the periodic table of the chemical elements. To get the molecular mass of a molecule made up of more than one atom, add the atomic masses of the atoms. For example, carbon dioxide is CO_2, so for every single molecule of this there is one atom of carbon, atomic mass 12 kg/kmol, and two atoms of oxygen (indicated by subscript 2) of 16 kg/kmol each, so the total molecular mass of CO_2 is 12 + 16 + 16 = 44 kg/kmol.

Difference between mass and volume fractions

Consider 1 kmol of hydrogen and 2 kmol of oxygen mixing without reacting in Figure 2.21, pictured by volume, at constant pressure and temperature. It is known that at the same p and T, 1 kmol of any gas occupies the same volume – regardless of mass – as 1 kmol of any other gas. Since the molecular mass of hydrogen is so much less than oxygen, this shows us the effect of molecular weight, that vastly different volume and mass fractions of gases can occur in a mixture.

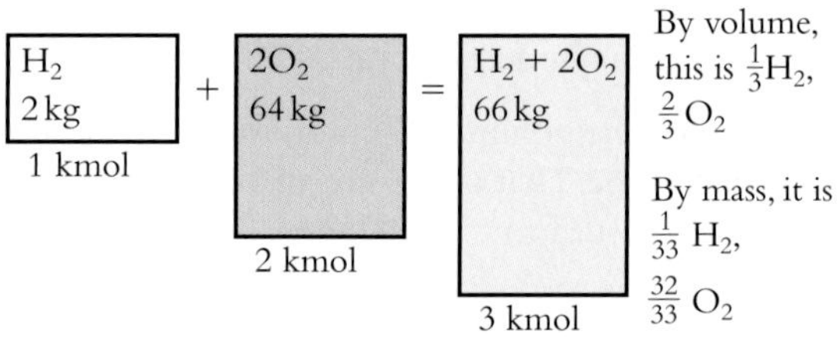

Figure 2.21 Proportions by mass compared to volume

The equation of state for perfect gas is:

$$p_i V \text{ (or } p V_i) = mRT \qquad \textbf{(2.25)}$$

To calculate by number of mole instead of by mass:

$n = \dfrac{m}{M}$ where n is the number of kmol, m is the mass of the sample [kg] and M is molecular mass [kg/kmol]. Knowing that the specific gas constant, R is the molar gas constant $\tilde{R}$, 8,314 [J/kmolK], divided by the molecular mass of the gas:

$$mR = \frac{\tilde{R}}{M} = n\tilde{R}$$

Substituting in equation (2.25) gives:

$$p_i V \text{ (or } p V_i) = n_i \tilde{R} T \qquad \textbf{(2.26)}$$

Molar analysis can be used to determine partial pressure and partial volumes and vice versa.

Formula for the relation between partial pressure and partial volume and molar fraction.

For the mixture as a whole, considered as a perfect gas, we have:

$$pV = n\tilde{R}T$$

(Note the molar form of the gas law). Using equation (2.26) gives:

$$\frac{p_i V}{pV} = \frac{ni\tilde{R}T}{n\tilde{R}\text{T}}$$

$$\therefore \quad \frac{p_i}{p} = \frac{n_i}{n} \qquad \textbf{(2.27)}$$

Combining with equation (2.24) leads to:

$$\frac{p_i}{p} = \frac{V_i}{V} = \frac{n_i}{n} \tag{2.28}$$

in which

$$\Sigma p_i = p, \Sigma V_i = V \text{ and } \Sigma n_i = n$$

Equation (2.28) shows us the important fact that volume fraction is equal to molar fraction. Consider a reaction $2H_2 + O_2 \rightarrow 2H_2O$. The molar fraction of H_2 in the reactants is 2/3. This is also its volume fraction. Usually we have a volumetric analysis and a reaction to consider in combustion-related calculations.

Formula for apparent molar mass of mixture, M

This is calculated from

$$\widetilde{m} = \frac{m}{n}$$

and therefore

$$m = \sum_i n_i \widetilde{m}_i$$

by conservation of mass.

Therefore the molar mass of the mixture is

$$\widetilde{m} = \sum_i \frac{n_i \widetilde{m}_i}{n} \tag{2.29}$$

Compare this with the relation derived by partial pressure analysis:

$$\frac{1}{\widetilde{m}} = \sum_i \frac{m_i}{m} \frac{1}{\widetilde{m}_i} \tag{2.30}$$

Apparent gas constant of mixture, R

Since the specific gas constant is defined as: $R = \dfrac{\widetilde{R}}{\widetilde{m}}$

Therefore,

$$\frac{\widetilde{R}}{R} = \sum \frac{n_i}{n} \frac{\widetilde{R}}{R_i}$$

and

$$\frac{1}{R} = \sum \frac{n_i}{n} \frac{\widetilde{R}}{R_i} \tag{2.31}$$

Conversions between mass and volume fractions

To convert from gravimetric to volumetric proportions, consider volumetric analysis:

of constituent, i:

$$pV_i = \frac{m_i \widetilde{R}}{\widetilde{m}_i} T \qquad \text{(equation A)} \tag{2.32}$$

of mixture:

$$pV = \frac{mR}{\widetilde{m}} T \qquad \text{(equation B)} \tag{2.33}$$

dividing equation (2.32) by (2.33) gives $\dfrac{V_i}{V} = \left(\dfrac{m_i}{m}\right)\left(\dfrac{\widetilde{m}}{\widetilde{m}_i}\right)$ or $\dfrac{m_i}{m} = \left(\dfrac{V_i}{V}\right)\left(\dfrac{\widetilde{m}_i}{\widetilde{m}}\right)$ (2.34)

Note that $pV_i = n_i \widetilde{R} T$ and $n_i = \dfrac{m_i}{\widetilde{m}}$.

The conversion can be done in tabular form.

Example of conversion from volume fraction to mass fraction

Typically, a volumetric analysis is stated as, for example, 10% methane, 10% ethane (C_2H_6), 65% nitrogen and 15% oxygen, i.e. $\frac{V_i}{V} = \frac{n_i}{n} = 0.1, 0.1, 0.65$ and 0.15 respectively for each gas.

Component	$V_i/V = n_i/n$	$\tilde{m}_i$	$(n_i/n)\,\tilde{m}_i$	$\frac{m_i}{m} = \frac{n_i}{n}\tilde{m}_i \cdot \frac{1}{\tilde{m}}$
A methane	0.1	16	1.6	0.058
B ethane	0.1	30	3.0	0.109
C nitrogen	0.65	28	18.2	0.659
D oxygen	0.15	32	4.8	0.174

$$\tilde{m} = \sum \frac{n_i \tilde{m}_i}{n} = 27.6\,\text{kg/kmol}$$

Alternatively for mass to volume fraction:

Component	m_i/m	$\tilde{m}_i$	$(m_i/m)(1/\tilde{m}_i)$	$\frac{n_i}{n} = \frac{m_i}{m}\frac{1}{\tilde{m}_i}.\tilde{m}$
A carbon	0.09	12	0.0075	0.192
B O_2	0.21	32	0.0066	0.169
C N_2	0.70	28	0.0250	0.639

$$\frac{1}{\tilde{m}} = \sum \frac{m_i}{m}\frac{1}{\tilde{m}_i} = 0.0391\,\text{kmol/kg}$$

For 0.1 kg of carbon combusting in 1 kg of air. Mass ratio of air is $O_2 = 0.233$, $N_2 = 0.767$.

Total mass is 1.1 kg. This is useful because $C + O_2 \rightarrow CO_2$, which we shall consider in the section on combustion.

Work and heat transferred

The equations given in this volume and in the first volume for the work done and heat transferred during various processes undergone by a single, perfect gas can be used for a mixture of perfect gases. That is, the appropriate values of specific heats and gas constants are calculated from the values applicable to individual constituents. Therefore knowing c_v, c_p, R and γ for the mixture, all other equations can be used.

Learning summary

By the end of this section you will have learnt:

✓ gas mixtures have several gases intimately mixed, which can be quantified in proportion by mass or by volume;

✓ the Gibbs–Dalton law of partial pressures leads to proportion analysis by mass, or gravimetric analysis;

✓ Amagat's law of partial volumes leads to analysis by volume proportions;

✓ combining the two with the gas law leads to a useful relationship between partial volumes, molar proportions and partial pressures of the gases: $\frac{p_i}{p} = \frac{V_i}{V} = \frac{n_i}{n}$ which can be used to make conversions between gravimetric and volumetric analysis.

2.4 Combustion

Fuels produce energy by reacting with **oxidizers**. The oxidizer is commonly the oxygen in atmospheric air but may be oxygen embedded in a compound that has high oxygen content, e.g. nitrates with NO_3 (i.e. one nitrogen atom per molecule with three oxygen atoms) as seen in fireworks in Figure 2.22(a), which contain gunpowder: potassium nitrate (75% mass), charcoal (15% mass) and sulphur (10% mass). The fuel may be a solid (e.g. coal), a liquid (e.g. petrol, which may be considered as octane, C_8H_{18}, or a gas (e.g. natural gas, which may be considered as methane, CH_4, as shown burning in Figure 2.22(b). Since fuels mainly comprise hydrogen and carbon, they are called **hydrocarbons**. If sufficient starting energy is provided to initiate a reaction, then the oxygen will react with the hydrocarbon fuel in combustion. This section outlines how to calculate the amount of air required to burn a given amount of fuel, and the composition of the products of combustion formed.

Solid and liquid fuels are analysed by mass (gravimetric analysis). Gaseous fuels are analysed by volume (volumetric analysis). In this section, all fuels are assumed to be burnt in air, and it is assumed that air contains 23.3% O_2, 76.7% N_2 by mass or 21 % O_2 , 79% N_2 by volume as described in the introductory part of Section 2.2 on air conditioning. The ratio of nitrogen to oxygen by volume is $\frac{79}{21} = 3.76$. The ratio by mass is $\frac{76.7}{23.3} = 3.29$. These analyses correspond to a mean molar mass $\tilde{m}$ of 28.85 kg/kmol and a corresponding mean specific gas constant R of 0.287 kJ/kgK. If the small percentage of CO_2 and Ar are included, then $\tilde{m} = 28.96$ and $R = 0.287$ kJ/kg K. This assumes approximate air data from tables (Rogers and Mayhew, 1995).

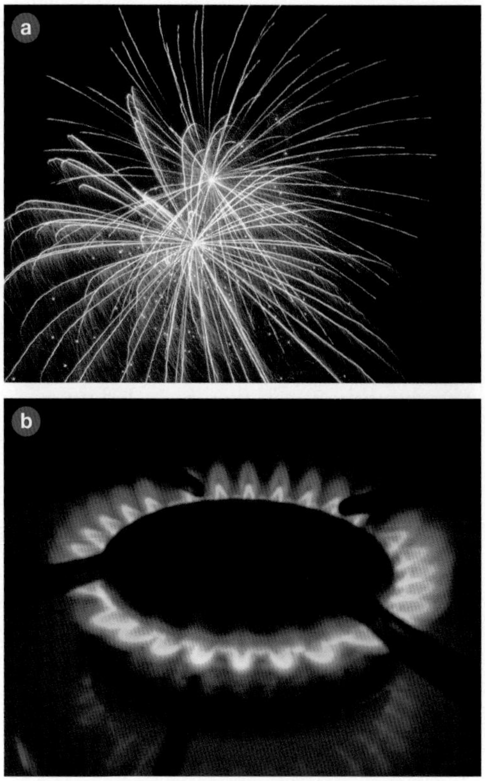

Figure 2.22 Combustion characteristics

Stoichiometric combustion

Molar reaction equations are used for the analysis of a combustion process, which begins by the formulation of the chemical equation that shows how the atoms of the reactants are combined to form the products.

How to balance equations:

e.g. butane $C_4H_{10} + xO_2 \rightarrow aCO_2 + bH_2O$

The unknown quantities are labelled by the unknown number of moles for each, and the combustion of only 1 mole (if working in grams) or 1 kmol (if working in kg) of the fuel is considered. We know that since matter cannot be created or destroyed, the atoms on the reactant side must be present on the products side and so a count of atoms reveals that:

C: $4 = a$

H: $10 = 2b; b = 5$

O: $2x = 2a + b; x = 13$

Therefore:

$C_4H_{10} + 13O_2 \rightarrow 4CO_2 + 5H_2O$

By the law of conservation of mass, the number of atoms of each element is the same at the beginning as at the end of any chemical reaction:

the number of atoms of reactants = the number of atoms of products

The reaction equations are molar, i.e. in terms of molecules reacting, and therefore we can make use of molar masses to find the masses of each involved in the reaction, and molar quantities are equivalent to volume for gases as described in section 2.3 in the subsection on molar analysis. Equations conserve:

(1) number of atoms of each element from reactants to products of combustion

(2) mass of reactants = mass of products.

Usually the number of moles (alternatively called mols) of reactants will not equal the number of mols of products.

In order to convert to mass fractions from mole fractions, use $n = \frac{m}{\tilde{m}} \Rightarrow m = n\tilde{m}$ as shown in the example below in which:
n = number of kmol in sample
m = mass of sample
$\tilde{m}$ = molecular mass of the sample

Example: forming water.

$$\mathbf{H_2 + \tfrac{1}{2}\,O_2 \rightarrow H_2O}$$

$$\mathbf{1\,kmol + \tfrac{1}{2}\,kmol \rightarrow 1\,kmol}$$

$$1\,\text{kmol} \times 2\,\text{kg/kmol} + \tfrac{1}{2}\,\text{kmol} \times 32\,\text{kg/kmol} \rightarrow 1\,\text{kmol} \times 18\,\text{kg/kmol}$$

$$2\,\text{kg} + 16\,\text{kg} \rightarrow 18\,\text{kg}$$

The proportions can then be scaled up or down depending on mass of fuel used. Similarly, if we have the masses, we can convert to molar equations.

Note: Incidentally, at this point it is worth noting that the naming of hydrocarbon fuels gives away their chemical formula. The prefixes, meth-, eth-, prop-, but-, pent-, hex-, hept-, oct- etc. denote the number of carbon atoms in the molecules increasing from one to eight in this list. The suffixes -ane and -ene denote the bonding of the carbon atoms to each other within the molecules. A carbon atom can be considered as having four links with which to connect to other atoms. Therefore methane has one carbon atom with its four links connected one each to a hydrogen atom, hence CH_4. Ethane has two carbon atoms with one link connected between them and then each one having three links connected to a hydrogen, hence C_2H_6. Ethene (Figure 2.23(a)) has two carbon atoms with two links connected between them – a stronger bond – and each with only two links each to hydrogen atoms, hence the C_2H_4. molecule. The schematic representation of butane (Figure 2.23(b)) gives the idea of the single carbon links and the single hydrogen links.

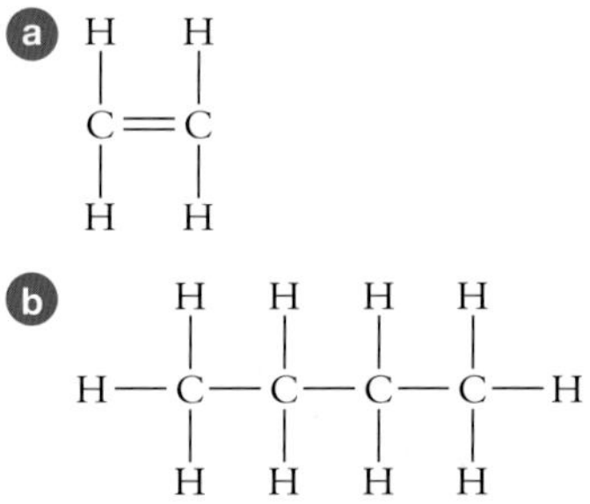

Figure 2.23 Schematic diagram of the structure of (a) ethene and (b) butane

Stoichiometric combustion describes a combustion reaction in which all of the fuel and oxidizer is used to make the products, so there is no leftover fuel or oxidizer.

A combustion process can be:

Complete: where sufficient O_2 is available to convert all carbon and all hydrogen in the hydrocarbon fuel to CO_2 and H_2O. There may be excess oxygen in the products.

Incomplete: where not enough O_2 is available, and other products such as CO appear. Oxygen prefers to react with hydrogen, so all hydrogen is consumed before carbon starts to be converted to CO or CO_2.

A stoichiometric reaction is one where all the oxygen is used up and all the fuel is burnt to the **ultimate** products (CO_2, H_2O), which are the most stable combination of those atoms at atmospheric conditions. Stoichiometric combustion describes the most efficient use of oxygen and fuel, and it is useful to use the ratio of oxygen to fuel in the stoichiometric case in order to compare actual oxygen–fuel ratios. This can be done by volume or by mass. For example, if ethane is burned in oxygen, the equation is described by:

$$C_2H_6 + 3\tfrac{1}{2}\,O_2 \rightarrow 2\,CO_2 + H_2O$$

The stoichiometric oxygen–fuel ratio of C_2H_6 (by volume) is oxygen moles ÷ fuel moles $= \frac{3.5}{1} = 3.5$ – i.e. by number of moles. This can be done because of $\frac{V_i}{V} = \frac{n_i}{n}$, i.e. molar proportions represent volume proportions.

Air–fuel ratio

Most combustion applications use the oxygen in atmospheric air and therefore draw in significant amounts of nitrogen to the reaction. Nitrogen is not reactive at atmospheric conditions, but can react at the high temperatures typically found in flames. Although there can be reaction of nitrogen it does not happen in significant quantity compared to the more aggressive carbon and hydrogen reactions, and can be neglected with only a very small effect on the volume and mass fractions of the other reactions. It is a significant issue for air pollution since parts per million are sufficient to cause acid rain.

The stoichiometric air–fuel ratio (**AFR**) by volume is found from the stoichiometric oxygen–fuel ratio equation, but includes the nitrogen in the air as an inert gas. This is important not for the reaction, but for the temperature of the flame – the more gas there is involved in the reaction, which has to be heated up from atmospheric conditions to the flame temperature, the lower the flame temperature.

Nitrogen is brought in with oxygen at a fixed ratio of:

$3.76 \times (O_2 \text{ volume})$ – from approximate air data.

Total air volume includes the O_2, i.e. air volume is $(1 + 3.76) \times (O_2 \text{ volume})$.

The stoichiometric air–fuel ratio (AFR) in the case of the ethane combustion before is:

$$\begin{aligned} AFR_{\text{volume}} &= (O_2/\text{fuel}) \text{ vol ratio} \times 3.76 + 1 \\ &= 3.5 \times 4.76 = 16.7 \text{ (by volume)} \end{aligned}$$

This can be converted to stoichiometric AFR by mass.

Since $n = \dfrac{m}{\tilde{m}}$

$$\begin{aligned} AFR_{\text{mass}} &= \frac{m_{\text{air}}}{m_{\text{fuel}}} = AFR_{\text{volume}} \times \frac{\tilde{m}_{\text{air}}}{\tilde{m}_{C_2H_6}} \\ &= 16.7 \times \frac{29}{30} = 16.14 \end{aligned}$$

In practice, the complete reaction equation for the air and ethane stoichiometric case will include N_2 terms where N_2 is treated as inert as in the following reaction equation:

$$C_2H_6 + 3\tfrac{1}{2}\,O_2 + 3\tfrac{1}{2} \times 3.76N_2 \rightarrow 2CO_2 + 3H_2O + 3\tfrac{1}{2} \times 3.76N_2$$

Non-stoichiometric combustion

This can either be **fuel rich** (insufficient O_2) or **air rich** (excess O_2). In these cases, combustion products can contain incomplete products such as CO and free oxygen as well as CO_2, H_2O, N_2 etc.

In the case of excess fuel the incomplete combustion of the fuel leads to CO_2 and CO products (this is in the case of hydrocarbon fuels). The hydrogen is always the first to burn off because it is a much more aggressive reaction than the carbon reactions, so it seizes any oxygen present before the carbon can get to it. In real processes, there may be both the oxygen (O_2) and CO in the products, because of incomplete mixing of fuel and air.

In the case of excess O_2, the products will include free oxygen, but the fuel is burnt to the ultimate products. The % excess air supplied is a useful indicator for mixture quality. This is defined as:

$$\% \text{ excess air} = \frac{\text{air supplied} - \text{min air for stoichiometric combustion}}{\text{min air for stoichiometric combustion}} \times 100$$

This is the same by volume and by mass because:

$$\text{by volume: } \frac{\dfrac{V_a}{V_f} - \dfrac{V_{a,\text{stoich}}}{V_f}}{\dfrac{V_{a,\text{stoich}}}{V_f}} \qquad \text{But } \frac{V_a}{V_f} \times \frac{\tilde{m}_a}{\tilde{m}_f} = \frac{m_a}{m_f} \text{ because } \frac{V_a}{V_f} = \frac{n_a}{n_f}$$

$$\therefore \text{ Multiply throughout by } \frac{\dfrac{\tilde{m}_a}{\tilde{m}_f}\ \dfrac{V_a}{V_f} \times \dfrac{\tilde{m}_a}{\tilde{m}_f} - \dfrac{\tilde{m}_a}{\tilde{m}_f} \times \dfrac{V_{a,\text{stoich}}}{V_f}}{\dfrac{V_{a,\text{stoich}}}{V_f} \times \dfrac{\tilde{m}_a}{\tilde{m}_f}} = \frac{\dfrac{m_a}{m_f} - \dfrac{m_{a,\text{stoich}}}{m_f}}{\dfrac{m_{a,\text{stoich}}}{m_f}} \qquad \textbf{(2.35)}$$

Wet and dry products

Since the products of hydrocarbon combustion contain H_2O, i.e. steam, which condenses when the product temperature is reduced to the atmospheric condition, it is useful to consider the products as wet when the water is contained as a vapour and dry when the vapour is considered to have condensed out of the products. This gives rise to the terms **wet products** and **dry products** of combustion. It also affects the energy value of the fuel, such that less energy is used when the products escape to the atmosphere containing the water as vapour. Notably, modern domestic *condensing* boilers make use of the energy in the water vapour by condensing it in the exhaust stream and capturing some of its energy. A gas analyser can be used to measure the component gases of a combustion exhaust stream in order to test how complete the combustion is, and it informs about the state of mixture of the incoming gases.

Worked example

Determine the stoichiometric AFR:

(1) by mass (2) by volume

for propane, C_3H_8. Calculate the volumetric composition by wet and dry analysis of the products. If the fuel is burnt with 30% excess air, what are the volumetric and gravimetric composition of wet products?

Aim:
To realize stoichiometric air–fuel ratio by mass and by volume.
To realize the importance of wet and dry analysis of product gases.

Stoichiometric air to fuel ratio (AFR):

(1) by mass

Reaction equation of the fuel:

$$C_3H_8 + 5O_2 \rightarrow 3CO_2 + 4H_2O$$
$$44\,\text{kg} + 5(32\,\text{kg}) \rightarrow 3(44\,\text{kg}) + 4(18\,\text{kg})$$

The oxygen–fuel ratio by mass is therefore $5 \times \frac{32}{44} = 3.636$. The air–fuel ratio can be worked out using the approximate ratio of air to oxygen in atmospheric air, which is 1:0.233 by mass:

$$\frac{\text{air[kg]}}{\text{fuel[kg]}} = \frac{\text{oxygen[kg]}}{\text{fuel[kg]}} \times \frac{1}{0.233} = 3.636 \times 4.29$$

The AFR by mass is therefore 15.6.

(2) by volume

The oxygen–fuel ratio by volume is 5:1 from the reaction equation above. To find the AFR by volume, use the approximate ratio of air to oxygen in atmospheric air, which is 1:0.21 by volume:

$$\frac{\text{air[kmol]}}{\text{fuel[kmol]}} = \frac{\text{oxygen[kmol]}}{\text{fuel[kmol]}} \times \frac{1}{0.21} = 5 \times 4.76$$

The AFR by volume is therefore 23.8.

The complete representative equation with all mixture components of the air–fuel stream includes nitrogen in the approximate ratio for atmospheric air. Including this in the reaction equation gives:

$$C_3H_8 + 5\left(O_2 + \frac{0.79}{0.21}N_2\right) \rightarrow 3CO_2 + 4H_2O + 5 \times \frac{0.79}{0.21}N_2$$

In order to get a volumetric analysis of wet and dry products, use a table for compactness. Wet includes the water in the reaction; dry does not include the water.

	Wet		Dry	
Component	V_i	$\frac{V_i}{V}$	V_i	$\frac{V_i}{V}$
CO_2	3	0.116	3	0.138
H_2O	4	0.155	0	0
N_2	18.81	0.729	18.81	0.862
Σ	25.81	1	21.81	1

When there is 30% excess air in the reacting stream of air and fuel, there will be 30% more O_2 and N_2 in the representative equation:

$$C_3H_8 + 5 \times 1.3(O_2 + {}^{0.79}/_{0.21}N_2) \rightarrow 3CO_2 + 4H_2O + 0.30_2 + 5 \times 1.3 \times {}^{0.79}/_{0.21}N_2$$

The list of products can be made for gravimetric and volumetric analysis:

Component	V_i	$\frac{V_i}{v}$	$\tilde{m}_i$	$m_i = \tilde{m}_i \frac{V_i}{V}$	$\frac{m_i}{m} = \frac{\tilde{m}_i}{\tilde{m}} \frac{V_i}{V}$
	[kmol]	[–]	[kg/kmol]	[kg/kmol]	
CO_2	3	0.091	44	4	0.141
H_2O	4	0.122	18	2.187	0.077
O_2	0.3	0.046	32	1.456	0.051
N_2	24.45	0.742	28	20.78	0.731
Σ	32.95	1		$\Sigma = \tilde{m} = 28.4$	1

Note that the column for the individual mixture species components has units kg/kmol and not just kg. This happens because, when multiplying by the volume fraction, we are considering per unit volume, and that is equivalent to per kmol of the total product gases. The sum of this column strictly speaking makes up the molecular mass of the product gases as an incidental by-product because we use the kmol as a convenient total sample size of the product gases. The final column calculating the ratio of each gas to the total mass of product gases uses the molecular mass of the previous column.

Worked example

Benzene C_6H_6 is burned in air and dry products of combustion contain 2% CO by volume and no free oxygen. Determine the stoichiometric air–fuel ratio by mass and the actual air ratio by mass.

Aim:

To calculate excess air supplied.

To show how to use the air–fuel ratio to indicate substoichiometric combustion.

The benzene molecule is complex, in the form of a ring with a mixture of double and single links between the carbon, as shown.

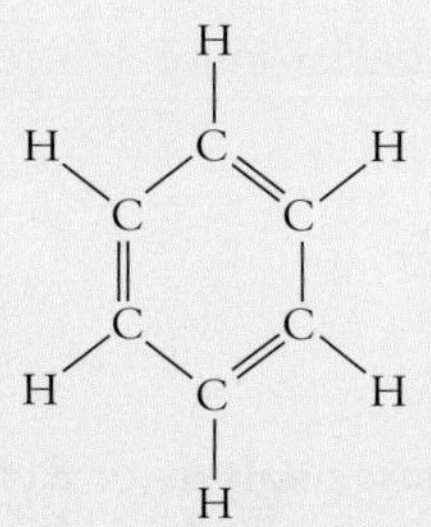

Figure 2.24 Schematic diagram of the structure of benzene

For the stoichiometric combustion case, the reaction of the fuel with oxygen can be written:

$$C_3H_6 + 7\tfrac{1}{2}O_2 \rightarrow 6CO_2 + 3H_2O$$

$$78\,\text{kg} + 7\tfrac{1}{2}(32\,\text{kg}) \rightarrow 6(44\,\text{kg}) + 3(18\,\text{kg})$$

The oxygen–fuel ratio by mass is therefore $7\frac{1}{2} \times 32{:}78 = 3.077$.

The AFR by mass can be calculated using the approximate ratio of air to oxygen in atmospheric air by mass, which is 1:0.233 = 4.29:

$$\frac{\text{air[kg]}}{\text{fuel[kg]}} = \frac{\text{oxygen[kg]}}{\text{fuel[kg]}} \times \frac{\text{air[kg]}}{\text{oxygen[kg]}} = 3.077 \times 4.29$$

The AFR by mass is therefore 13.2.

For the actual combustion, insufficient oxygen is supplied to burn the fuel completely. We assume that the order of reaction of the atoms in the fuel is hydrogen first completely, since it is the fastest and by far the most aggressive. Subsequent to this, the carbon burns, firstly all to carbon monoxide and then some CO reacts to form CO_2 to consume the remaining oxygen. The lack of sufficient air therefore leads to some CO remaining in the exhaust gas stream.

We assume in the representative reaction that there is a proportion of the stoichiometric O_2 required, and write:

$$C_6H_6 + y(O_2 + {}^{0.79}/_{0.21}N_2) \rightarrow aCO_2 + bCO + cH_2O + {}^{0.79}/_{0.21}N_2$$

The first step is to conduct an atomic balance in terms of y, such that when the proportion y is known, the equation balances identical quantities of elements on each side.

Carbon: $6 = a + b$
Hydrogen: $6 = 2c$, i.e. $c = 3$.
Oxygen: $2y = 2a + b + c$ and substituting c, and

$$a = 6 - b,\ 2y = 12 - 2b + b + 3 \text{ or } b = 15 - 2y.$$

The second piece of information is the proportion of CO in the dry products by volume, which is 2%. We therefore create the proportion from the above equation and equate to 0.02:

$$\frac{b}{a + b + 3.76y} = 0.02$$

And substituting $b = 2y - 15$ and $a + b = 6$:

$$\frac{15 - 2y}{6 + 3.76y} = 0.02$$

Therefore, $y = (15 - 0.12)/(2 + 0.0752) = 7.16$.

The AFR by mass for the actual combustion process is:

$$\frac{\text{air[kg]}}{\text{fuel[kg]}} = \frac{7.16 \times (32\,\text{kg/kmol} + 3.76 \times 28\,\text{kg/kmol})}{78\,\text{kg}} = \frac{983}{78} = 12.6$$

Worked example

A particular coal has a mass analysis of 81% C, 5% H, 5% O, 9% ash. The volumetric analysis of the dry products is:

10% CO_2, 1% CO, 8% O_2, 81% N_2

Determine the minimum air required for complete combustion of 1 kg of coal and the percentage of excess air required to produce the product gas proportions.

Aims:

To realize the ash content effect on the air–fuel ratio.
To determine the stoichiometric reaction.
To realize the effect of oxygen in coal on the stoichiometric reaction.
To calculate excess air supplied.

In this case it is useful to determine the required oxygen by working out for each component how much oxygen is required to burn it individually. So instead of a representative reaction equation, it is more obvious to create a table:

Constituent	Reaction	Mass per kg of coal	Mass of O_2 per kg	Oxygen required
carbon	$C + O_2 \rightarrow CO_2$	0.81	32/12 = 2.67	2.16
hydrogen	$H_2 + \frac{1}{2}O_2 \rightarrow H_2O$	0.05	16/2 = 8	0.4
oxygen		0.05	–1	–0.05
ash		0.09	0	0

Adding the oxygen required in the last column, per kg of coal there is 2.51 kg of O_2 required. For complete combustion, the air required for 1 kg of coal is this mass multiplied by the air:oxygen ratio by mass in the atmosphere, which is 1:0.233 = 4.29.

Therefore complete combustion requires $4.29 \times 2.51 = 10.77$ kg.

In order to find the actual air supplied, we have to take account of the information presented by the constituents of the product gas stream. This contains nitrogen that can only have come from the air supplied. The carbon came in a fixed proportion of the coal, and if we can work out the nitrogen to carbon ratio, we can infer the air:carbon ratio, and hence the air:coal ratio. The other useful information is that the ratio of products is given by volume, and we therefore know the molar proportions, which then give not only the mass proportions of each species molecule, but also the mass proportions of the species atoms.

For 1 kmol of product gases, there is 0.81 kmol of N_2, with mass $0.81 \times 28 = 22.68$ kg. This is carried in with a proportion of air by mass of 1:0.767 or $\frac{22.68}{0.767} = 29.57$ kg.

There is 0.1 kmol of CO_2, containing 0.1 kmol of carbon, with mass $0.1 \times 12 = 1.2$ kg, and there is 0.01 kmol of CO, containing 0.01 kmol of carbon, with mass $0.01 \times 12 = 0.12$ kg. The total amount of carbon in 1 kmol of product gas is therefore 1.32 kg, which is carried in the proportion 1:0.81 of coal to carbon. Therefore the mass of coal to carry 1.32 kg of carbon is $\frac{1.32}{0.81} = 1.63$ kg.

The air to coal ratio is therefore 29.57:1.63 = 18.14 by mass.

Therefore percentage excess air is:

$$\text{excess air \%} = \frac{18.14 - 10.77}{10.77}$$

or 68.4%.

How solids burn

Coal, plastics and wood all burn in an oxygen environment, and the pattern of combustion is very familiar if not formally recognized. The combustion has characteristic stages as follows: initial heating with evaporation of any moisture content up to the temperature at which polymers melt and begin to evaporate. This stage is called devolatilization involving the evaporation of **volatiles** – low boiling point hydrocarbons in the fuel. The volatile molecules burn in the surrounding gas giving the familiar flame of airborne combustion. There then follows smouldering combustion of the **char** content that is left behind, until only the ash is left. Char is carbon, which is solid, does not evaporate and relies on diffusion of oxygen coming in through the pores to burn. As long as the diffusion rate is sufficiently fast to produce a reasonable amount of combustion, whilst insufficient cooling occurs, then the char remains at high temperature sufficient to initiate combustion at new sites penetrating deeper into the body. This type of solid-bound combustion has a characteristic glow dependent on the temperature of the char. Figure 2.25 shows the early stages of combustion of a piece of rubber, showing a sooty flame as volatile molecules are released and burnt, and some are cooled faster than they are burnt, forming carbon-heavy particles or soot.

Figure 2.25 Initial stage of combustion of a (a) ignition; (b) 30s; (c) 60s; (d) 120s piece of car tyre rubber in a 900°C furnace

Combustion energy

This section describes how to calculate the energy released by combustion, and the temperature of the products. The temperature affects the mechanical strengths of materials of the combustion vessel as well as thermal efficiency. Combustion, just like other processes considered before, can be a closed system process as in an internal combustion engine, or an open process as in a jet engine or boiler furnace. As before, it is the first law that we are interested in. The two forms are the direct first law and the derived SFEE for each of the processes. Combustion releases chemical energy, which is converted to internal energy in the closed system and enthalpy in the open system.

Internal energy (ΔU_0) of combustion

The closed system is the easiest to consider to start with, that is for non-flow processes. To use the first law of thermodynamics, $Q + W = \Delta U$, we need to know, when energy is released from combustion, how much is converted to heat and work, and how much is taken to increase the internal energy of the working fluid. ΔU_0 is the **internal energy of combustion** at reference conditions of temperature T_0 (usually 25°C) and of pressure $p_\ominus$ (usually 1 bar). This is determined from the first law applied to a constant volume, closed system process, as shown in the following.

For an analysis of closed system combustion, refer to schematic diagram of the process in Figure 2.26.

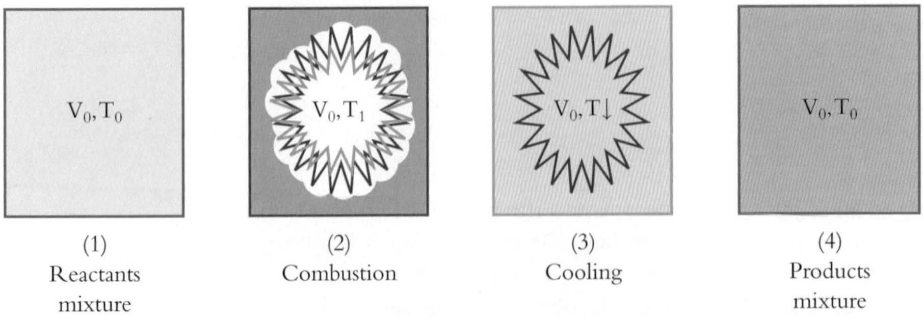

Figure 2.26 Schematic diagram of combustion in a closed vessel

(1) **Reactants** are brought together; at V_0, T_0, the fuel and air mixture is ignited and burnt.

(2) **Combustion** then takes place; this releases chemical energy which is transformed into 'sensible' internal energy, that is kinetic energy of the molecules, which dictates the temperature of the mixture. Some changes to properties will also occur and the chemical species will change. Still at V_0, but temperature will have increased.

(3) and (4) **Products** are cooled back to T_0. Notice that bringing the products back to temperature T_0 is a standardizing procedure – it allows us to consider, in a standard way, the quantity of heat available from a combustion reaction.

Apply the first law to the processes (a) to (c):

$$Q = -W + U_C - U_A \quad \textbf{(2.36)}$$

But there is no work because it is a constant volume process.

$$Q = U_{p0} - U_{r0} \quad \textbf{(2.37)}$$

where subscript p indicates product species mixture and r indicates reactant species mixture.

$$Q = U_{p0} - U_{r0} = \Delta U_0 \quad \textbf{(2.38)}$$

Since Q is negative, because heat is transferred from the system to surroundings, then $U_{r0} > U_{p0}$ and the internal energy of combustion is negative. Internal energy of combustion tells us how much heat had to be removed to convert the reactants to the products and bring the products back to the standard state temperature. So Q in equation (2.38) is the heat released by the reaction.

Enthalpy (ΔH_0) of combustion

For steady flow process combustion, the situation is similar except the steady flow energy equation $Q + W = \Delta H$ (neglecting KE and PE) is used. In the same way as for the closed system, in which a standard state internal energy of combustion is used, in the open system, ΔH_0, the **enthalpy of combustion** at reference temperature T_0 (25°C) and pressure p_0 (1 bar) is used. Note the subscript zero, means at standard condition, i.e. at 25°C.

Analysis of open system combustion

ΔH_0 can be determined from the SFEE for a flow process with no work, and with heat transfer to restore the temperature of the products to initial temperature of the reactants, T_0. Consider the situation shown in Figure 2.27, in which the combustion takes place from the standard state condition at 1 (i.e. temperature is T_0), releasing energy. The combustion energy released in the flame is removed through the walls of the system by heat transfer in order to reduce the hot combustion product gases to the initial temperature of the reactants. Writing the steady flow energy equation including kinetic energy for now:

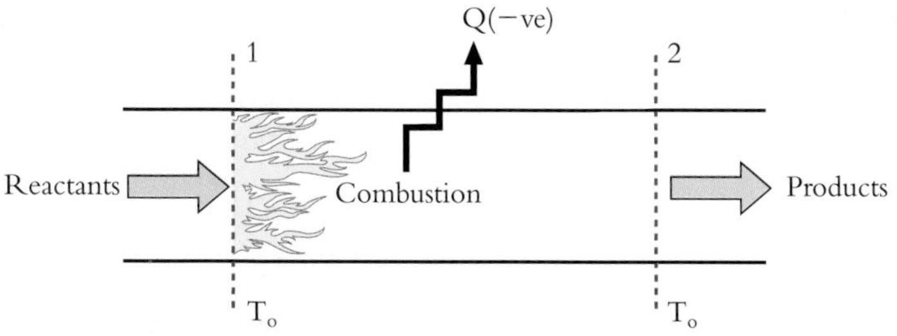

Figure 2.27 Open system combustion model. Note that the beginning and end conditions are forced to be at the standard state.

$$H_1 + m\frac{C_1^2}{2} + Q(-ve) = H_2 + m\frac{C_2^2}{2}$$

$$Q = (H_2 - H_1) + m\left(\frac{C_2^2}{2} - \frac{C_1^2}{2}\right) \quad \textbf{(2.39)}$$

The kinetic energy terms are often negligibly small compared to the enthalpy and equation (2.39) simplifies to

$$Q = H_{P_o} - H_{R_o} = \Delta H_o \quad \textbf{(2.40)}$$

For a combustion reaction, Q is negative, i.e. the system produces heat that is extracted from the working fluid via the containing system walls, in order to bring the products back to the standard condition temperature. Since the flow is in a duct of some kind, it is often the case that pressure does not change significantly across the flame.

Converting between ΔH_0 and ΔU_0 from $H = U + pV$

Values of enthalpy of reaction are stated in data tables, but internal energy of reaction is not necessarily stated. If we want to know the internal energy of reaction for a closed process, then we must be able to work it out. It is therefore useful to be able to convert from one to the other, and this is based on the formula for enthalpy as follows:

$$\Delta H_0 = H_{p0} - H_{R0} = (\Delta U_{p0} + \Delta p_{p0} V_{p0}) - (\Delta U_{R0} + p_{R0} V_{R0})$$

$$\Delta H_0 = (\Delta U_{p0} - U_{R0}) + (\Delta p_{p0} V_{p0} - p_{R0} V_{R0})$$

$$\Delta H_0 = \Delta U_0 + (\Delta p_{p0} V_{p0} - p_{R0} V_{R0}) \quad \textbf{(2.41)}$$

– For solid and liquid components, PV is negligible
– For perfect gas components, $pV = nRT$, therefore

where $$\Delta H_0 = \Delta U_0 + (\Delta n_p - n_R) R_{T0} \quad \textbf{(2.42)}$$

n_p = kmols of products per kmol of fuel burnt.
n_R = kmols of reactants per kmol of fuel burnt.
ΔH_0 and ΔU_0 are in kJ/kmol of fuel burnt.

Usually energy contributions are considered in terms of mass, and so conversion from kJ/kmol to kJ/kg must be done before use in the steady flow energy equation. Enthalpy of reaction is stated for common reactions in the data tables. There is a very easy way to work them out from **enthalpy of formation** for the molecules involved in the reaction which are given in the tables in order to determine reaction enthalpy for a wider range of reactions. Formation enthalpies are stated in kJ/kmol for molecules, and represent the energy required to form the molecule from the constituent atoms. Atoms have formation enthalpies dependent on their structure, but are often zero. In particular, all diatomic molecules (e.g. N_2, O_2) have formation enthalpies of zero. In order to calculate the energy released by a chemical reaction, write the reaction down, write the formation enthalpy of each molecule below the reaction equation, multiplying by the number of moles involved in the reaction. The difference between the sum of reactant formation enthalpies and the sum of the product formation enthalpies is the reaction energy. For example considering the previous example of propane:

$$C_3H_8 + 5O_2 \rightarrow 3CO_2 + 4H_2O$$

$$44\,\text{kg} + 5(32\,\text{kg}) \rightarrow 3(44\,\text{kg}) + (18\,\text{kg})$$

$$-118\,910 + 5 \times 0 \rightarrow 3 \times -393\,520 + 4 \times -241\,830$$

The enthalpy on the reactant side is –118 910 kJ and the enthalpy of the products is –2 147 880 kJ. Therefore for the combustion of 1 kmol of propane, the reaction energy released is

$$-2\,147\,880 - (-118{,}910) = -2\,028\,970\,\text{kJ}$$

Where formation enthalpies are taken from data tables in kJ/kmol.

This is a lot of energy, but it is important to remember that it comes from 1 kmol of propane, which weighs 3×12 kg + 8×1 kg = 44 kg, which is a substantial amount of fuel.

Calorific values

Δh_0 and Δu_0 are usually given at 25°C, per kg or kmol ($\tilde{h}$) of burnt fuel. In the tables, Δh_0 is given and stated as $\Delta \tilde{h}^{\theta}$ kJ/kmol at $T = 25°C$. The phase of the reactants and products, which could be vapour or liquid (e.g. H_2O) should be specified. Alternatively, calorific values are quoted instead of Δh_0 and Δu_0. These are defined according to British Standard.

Gross (or higher) CV at constant volume ($Q_{gr,v}$), $\Delta U_{25°C}$, with H_2O in products as liquid.

Net (or lower) CV at constant volume ($Q_{net,v}$), $\Delta U_{25°C}$, with H_2O in products as vapour.

Gross (or higher) CV at constant pressure ($Q_{gr,p}$) $\Delta H_{25°C}$, with H_2O in products as liquid.

Net (or lower) CV at constant pressure ($Q_{net,p}$) $\Delta H_{25°C}$, with H_2O in products as vapour.

By convention, calorific values are stated as positive values. However, we know that according to the European first law convention, positive energy transfer is into the system. Since a combustion reaction produces heat out of the system, combustion energy is indicated as negative for use in the first law equations. The difference between gross and net is due to the state of the water in the product gases. Look at tables of formation enthalpies (Rogers and Mayhew, 1995): H_2O (liquid) is −285 820 kJ/kg and H_2O (vapour) is −241 830 kJ/kg. The difference is the enthalpy difference between liquid and vapour water at 25°C. This is the contribution of the H_2O content to the overall product enthalpy. Hence the higher calorific value (i.e the more negative the heat removed to restore to the standard temperature) is with the water as liquid because it reclaims the energy of condensation, which in turn must be removed from the system by negative heat transfer.

Gas mixture conditions when inlet and outlet are not at standard conditions

Consider the closed system combustion, similar to the situation in Figure 2.26, but this time with beginning and end conditions that are different from the standard state. If initial and final states are not at the reference state temperature T_0, then artificial processes are considered in order to adjust conditions such that combustion takes place at the reference temperature T_0. The reactants and products are assumed to be perfect gases (plus liquids and solids) so that h and u depend on T and *only* T. With these assumptions, we can use $\Delta u = c_v \Delta T$ and $\Delta h = c_p \Delta T$ for gas mixtures heating and cooling, or more accurately the absolute values of h and u directly from tables, to calculate the difference in enthalpy and internal energy between reference and actual states. Values of the molar enthalpy or energy content (note that now it is just 'enthalpy' not 'formation enthalpy' or 'combustion enthalpy') and the molar internal energy are given for common gases in the tables over a large range of temperatures, so they can be read from there. The specific values, in order to consider fractions by mass rather than volume, are calculated by dividing by the molecular mass, e.g. $\tilde{h}$ [kJ/kmol]/$\tilde{m}$ [kg/kmol] $= h$ [kJ/kg].

Worked example

What is the specific enthalpy change for carbon dioxide between 25°C and 327°C?

For CO_2 molar enthalpy at 327°C or 600K is 12 916 kJ/kmol. At 25°C or 298.15K it is zero.

So $\tilde{h}_{600K} - \tilde{h}_{298.15K} = 12\,916$ kJ/kmol, and since $\tilde{m}_{CO2}$ is 44 kg/kmol, specific enthalpy change is $h_{600K} - h_{298K} = 293$ kJ/kg.

The alternative method of computation is to use specific heat capacity. $c_p = 0.846$ kJ/kgK at 300K and $c_p = 1.075$ kJ/kgK at 600K.

$h_{600K-298K} = mc_p \Delta T = 1 \times ((0.846 + 1.075)/2) \times 300 = 288.15$ kJ/kg.

The first method is more accurate since it makes no assumptions. The second method assumes that the change of c_p with temperature is linear – it is not, quite.

Closed system combustion

Having determined that internal energy and enthalpy changes of gases between different temperatures can be determined, we can now return to the question at hand, which is to determine the enthalpy difference (i.e. the internal energy difference for the closed system) from the standard state for the beginning and end states of combustion. Figure 2.28 shows a schematic of combustion in a closed cylinder with a moveable piston for extracting work, similar to an internal combustion engine. In the analysis of the combustion, we know what the energy release is at standard conditions, so our analysis breaks the process from state 1 to state 2 into a number of artificial steps:

(1) start with the gas at the initial conditions v_1, T_1;

(2) make the gas go from the initial conditions v_1, T_1 to the standard conditions v_0, T_0 and note the heat and work to do this;

(3) as in the analysis for Figure 2.24, combust;

(4) as in analysis for Figure 2.24, cool back to T_0 noting the energy release to do so;

(5) make v_0 T_0, standard conditions go to final conditions v_2, T_2 noting the heat and work to do this.

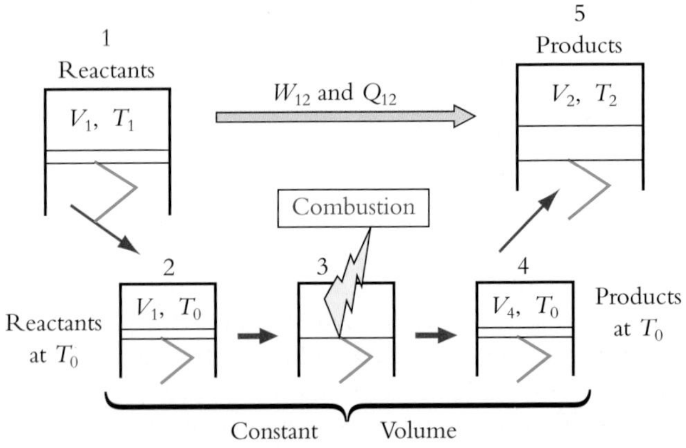

Figure 2.28 Closed system combustion case with conditions at beginning and end that are different from the standard state. A cylinder is depicted similar to an internal combustion engine, and three intermediate artificial conditions are inserted to aid in the analysis of the process.

Note that there are only two real states: at the beginning conditions 1, and at the end conditions 2. Three intermediate artificial conditions are inserted between the beginning and end states to simplify the analysis because we know what happens with energy release at standard conditions. This assumes that we know the start and end conditions and wish to know the energy released.

From applying the first law to the states (1) to (2)

$$Q_{12} + W_{12} = U_2 - U_1 \text{ but}$$

$$U_2 - U_1 = U_{p2} - U_{R1}$$

$$U_2 - U_1 = (U_{p2} - U_{p0}) - (U_{R1} - U_{R0}) + (U_{p0} - U_{R0})$$

$$U_2 - U_1 = \sum_{\text{PRODUCTS}} m_i c_{vi}(T_2 - T_0) - \sum_{\text{REACTANTS}} m_i c_{vi}(T_1 - T_0) + \Delta U_0 \qquad (2.43)$$

Where U, with capital, indicates extensive internal energy for the whole quantity of gases considered.

Or with tables for values of internal energy:

$$U_2 - U_1 = \sum_{\text{PRODUCTS}} m_i(u_{i2} - u_{i0}) - \sum_{\text{REACTANTS}} m_i(u_{i1} - u_{i0}) + m_{\text{fuel}}\Delta u_0 \tag{2.44}$$

Note that internal energy does not depend on pressure for a perfect gas. Given the conditions that occur between states 1 and 2, heat and work exchanges with the surroundings can then be calculated. For example, if we know that the system is perfectly insulated, then the energy change must be the work done by the surroundings on the working fluid; alternatively, if there is no machinery to extract work then the energy must be exchanged as heat with the surroundings.

Open system combustion

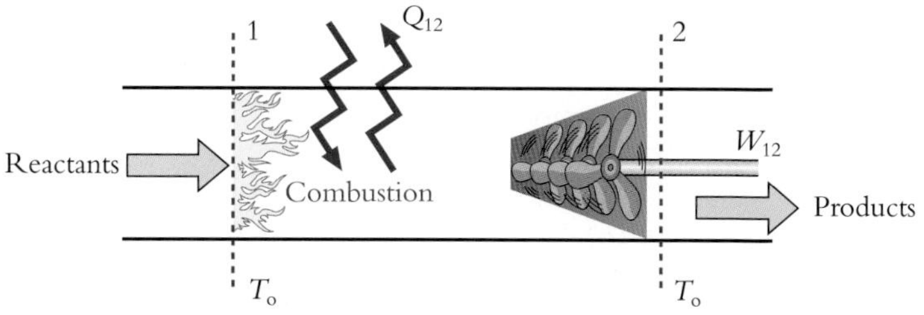

Figure 2.29 Open system combustion schematic in which the beginning and end states are at the standard condition

The open system can be handled in the same way as the closed system: apply the same idea as for the closed system of bringing reactants to the standard state T_0, and the products returned from T_0 to actual state 2.

Apply the steady flow energy equation, noting the superscript dots to indicate flow rates:

$$H_1 + \dot{m}\frac{C_1^2}{2} + \dot{Q}_{12} = -\dot{W}_{12} + H_2 + \dot{m}\frac{C_2^2}{2}$$

$$\dot{Q}_{12} + \dot{W}_{12} = (H_2 - H_1) + \dot{m}\frac{C_2^2}{2} - \frac{C_1^2}{2}$$

The kinetic energy is negligible compared to the enthalpy terms:

$$\dot{Q}_{12} + \dot{W}_{12} = H_2 - H_1$$

Note that condition 2 is for the products and condition 1 is for the reactants:

$$\dot{Q}_{12} + \dot{W}_{12} = H_{P2} - H_{R1}$$

Now, similar to the treatment of the closed system case, reduce the product condition to standard state, and the same for the reactants:

$$\dot{Q}_{12} + \dot{W}_{12} = (H_{P2} - H_{P0}) - (H_{R1} - H_{R0}) + (H_{P0} - H_{R0})$$

Note now that the last enthalpy difference term is the combustion enthalpy at standard conditions, and using the expression for enthalpy from specific heat capacity at constant pressure:

$$\dot{Q}_{12} + \dot{W}_{12} = \Sigma m_i c_{pi}(T_2 - T_1) - \Sigma m_i c_{pi}(T_1 - T_0) + \Delta H_0$$

Alternatively, use the enthalpy of the gases from tables for a more accurate solution, which gives:

$$\dot{Q}_{12} + \dot{W}_{12} = \Sigma m_i(h_{i2} - h_{i0}) - \Sigma m_i(h_{i1} - h_{i0}) + \Delta H_0 \tag{2.45}$$

Worked example

A mixture of CO and air is 10% fuel rich by volume and at 8.28 bar and 555K. The mixture burns adiabatically and at constant volume. Calculate the products temperature neglecting dissociation.

The representative stoichiometric reaction equation gives the ideal reaction, which shows the required oxygen and nitrogen to thoroughly combust all the fuel:

$$CO + \tfrac{1}{2}O_2 + \tfrac{1}{2} \times 3.76N_2 \rightarrow CO_2 + \tfrac{1}{2} \times 3.76N_2$$

We know that the reaction is not stoichiometric, because we are told that the mixture is 10% rich – i.e. there is 10% more CO than can be burnt with the stoichiometric air in the above equation. Therefore, the true representative reaction equation will have 1.1CO, thus:

$$1.1CO + \tfrac{1}{2}O_2 + \tfrac{1}{2} \times 3.76N_2 \rightarrow CO_2 + \tfrac{1}{2} \times 3.76N_2 + 0.1CO$$

Alternatively, divide by 1.1 to get the reaction in terms of combustion of 1 kmol of fuel:

$$CO + 0.91(\tfrac{1}{2}O_2 + \tfrac{1}{2} \times 3.76N_2) \rightarrow 0.9CO_2 + 1.71N_2 + 0.0CO$$

In order to determine the final temperature, we need to analyse the closed system with the method described above:

$$U_2 - U_1 = \sum_{\text{PRODUCTS}} m_i cv_i(T_2 - T_1) - \sum_{\text{REACTANTS}} m_i cv_i(T_2 - T_1) + \Delta U_0$$

The internal energy of combustion can be obtained from the tabulated values of enthalpy of combustion and the equation to convert between the two:

$$\Delta H_0 = \Delta U_0 + \Delta\,(n_p - n_R)\,RT_0$$

$\Delta\tilde{h}_0$ for CO is –283 000 kJ/kmol for the reaction $CO + \tfrac{1}{2}O_2 \rightarrow CO_2$.

$$\Delta\tilde{u}_0 = \Delta\tilde{h}_0 - (n_p - n_R)\tilde{R}T$$

$$\Delta\tilde{u}_0 = -283\,000 - (1 - 1.5)8.314 \times 298$$

Therefore the internal energy of combustion per kmol of CO is –281 761 kJ/kmol.

To put this in specific terms, divide by the molecular mass of CO, 28 kg/kmol, to give –10 063 kJ/kg.

The first law can now be applied, for adiabatic ($Q = 0$) and constant volume ($W = 0$) to give:

$$U_2 - U_1 = 0$$

Where U_2 is the total internal energy of the products at the end of the process and U_1 is the total internal energy of the reactants before combustion. Knowing the temperature of the reactants, we can work out the internal energy of the reactants before combustion. We don't know the final temperature of the products – this is what is asked for. Using equation (2.44):

$$U_2 - U_1 = \sum_{\text{PRODUCTS}} m_i c_{vi}(T_2 - T_0) - \sum_{\text{REACTANTS}} m_i c_{vi}(T_1 - T_0) + \Delta U_0$$

And reading values of molar internal energy from the tables (Rogers and Mayhew, 1995), to complete the following table of reactants:

Species	n_i kmol	$\tilde{u}_{i,555K}$ kJ/kmol	$n_i\tilde{u}_{i,555K}$ kJ/kmol
CO	1	2984	2984
O_2	0.455	3233	1471
N_2	1.71	2944	5034
Σ			9489

where the internal energy at 555K is found from the tables by interpolation. Values are stated at 400K and 600K for a number of gaseous species. The interpolation is done thus:

$$\tilde{u}_{i,555K} = \frac{555 - 400}{600 - 400}(\tilde{u}_{i,600K} - \tilde{u}_{i,400K}) + \tilde{u}_{i,400K}$$

The reader can easily confirm that the values in the table are correct.

There is no direct way to obtain the product temperature since it is determined by balancing the equation. The method used is to guess a temperature and see how far out of balance the equation is. In this case, guess 3000K.

Making the same table as above for the products

Species	n_i kmol	$\tilde{u}_{i,3000K}$ kJ/kmol	$n_i\tilde{u}_{i,3000K}$ kJ/kmol
CO_2	0.91	127 920	116 407
CO	0.09	68 598	6174
N_2	1.71	67 795	115 929
Σ			238 510

Inserting the values in the equation:

$$U_2 - U_1 = 238\,510 - 9489 + (-281\,761) = -52\,740 \text{ kJ}$$

This is negative, and shows that the products were not hot enough. Therefore try the next temperature up in the table, i.e. 3200K:

Species	n_i kmol	$\tilde{u}_{i,3200K}$ kJ/kmol	$n_i\tilde{u}_{i,3200K}$ kJ/kmol
CO_2	0.91	138 720	126 235
CO	0.09	74 391	6695
N_2	1.71	73 555	125 779
Σ			258 709

And the equation now is

$$U_2 - U_1 = 258\,709 - 9489 + (-281\,761) = -32\,540 \text{ kJ}$$

This is still too cool so the process is repeated until the answer is acceptably close to zero.

Learning summary

By the end of this section you will have learnt:

- ✓ combustion involves mixing a fuel (hydrocarbon) and an oxidizer, in order to produce heat whilst converting the chemicals involved in the reaction into reaction products;
- ✓ in this unit, the hydrocarbon fuels considered only have hydrogen and carbon composition, and produce the ultimate product gases carbon dioxide and water vapour;
- ✓ molecular reaction equations can be balanced by counting the atoms of all molecules on the reacting side of the equation, and matching the number of atoms of all molecules on the product side of the equation;

- ✓ a stoichiometric combustion reaction is ideal: every carbon atom combines with exactly one diatomic oxygen molecule to produce a carbon dioxide molecule, and every hydrogen atom combines with one other hydrogen atom and one oxygen atom to produce water vapour;
- ✓ when there is either too much or too little oxygen, incomplete products of reaction appear including carbon monoxide and oxygen. The reaction is then non-stoichiometric;
- ✓ most practical combustion installations use atmospheric air, which carries oxygen as the primary oxidizer. The oxygen in the air brings with it a fixed proportion of nitrogen. The combined requirement for oxygen with its associated nitrogen gives an air requirement that leads to the air-to-fuel ratio by mass or by volume;
- ✓ for non-stoichiometric combustion, the excess air ratio is defined as the ratio of excess air supplied to the air required for stoichiometric reaction;
- ✓ the product gases may be considered by proportion with or without moisture, since the moisture will condense out of the gases when they cool down. The alternatives are known as wet and dry products of combustion;
- ✓ each reaction has an enthalpy of reaction at standard conditions, which is determined by the difference between the sum of formation enthalpies of the reactants and products;
- ✓ standard conditions are 1 atmosphere pressure and 25°C;
- ✓ the enthalpy of reaction is referred to as the calorific value of the fuel, and may be considered as gross, in which case the water is considered as condensed out of the gases, or as net, in which case the water is considered as vapour in the gases;
- ✓ enthalpy of reaction is directly used in flow process combustion but must be converted to internal energy of reaction for closed process combustion;
- ✓ the final temperature of a combustion process can be determined by an enthalpy balance (or internal energy balance for closed processes).

2.5 Reciprocating compressors

Reciprocating piston machines are typically needed to provide a supply of compressed air, but are also used for other positive displacement requirements such as refrigerator pumps and vacuum pumps. They can be described as **single-acting** or **double-acting** (compression on one or both sides of piston), **single-stage** or **multistage** (number of stages of compression before air is delivered). They are **positive displacement** pumps and usually have **self-acting valves** (cylinder inlet and outlet valves needed to prevent backflow, which are opened and closed by pressure differentials across the valves). A typical example can be seen in Figure 2.30. The range of sizes and pressures is from the order of millimetres of water gauge pressures (e.g. 5 mm water gauge is $p = \rho gh = 1000\ [\text{kg/m}^3] \times 9.81\ [\text{m/s}^2] \times 0.005\ [\text{m}] = 49.05\ \text{Pa}$), as might be used for medical purposes, to several meters of water gauge pressure (e.g. 20 m water gauge is $p = \rho gh = 1000 \times 9.81 \times 20 = 196\,200\ \text{Pa}$ or 1.9 bar gauge or approximately 2.9 bar absolute) and up to hundreds of bar pressure, as might be used for compressing carbon dioxide to be buried in disused oil and gas wells for carbon dioxide sequestration.

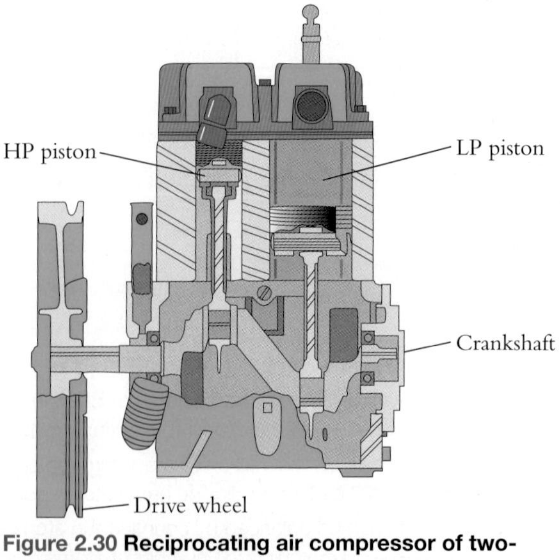

Figure 2.30 Reciprocating air compressor of two-stage type, with a big lower pressure piston, and a smaller, higher pressure piston

There is some basic terminology that must be remembered for a proper discussion of reciprocating compressors to describe their operation fully. These descriptions are related to Figure 2.31:

stroke – distance from bottom dead centre (**BDC**) to top dead centre (**TDC**)

clearance volume – approximately 10 per cent of swept volume allows the valves to be clear of the piston

swept volume – volume swept through from BDC to TDC

machine cycle – the gas goes through a machine cycle rather than a full thermodynamic cycle because it is drawn in and exhausted part way through the cycle

automatic valves – spring return valves

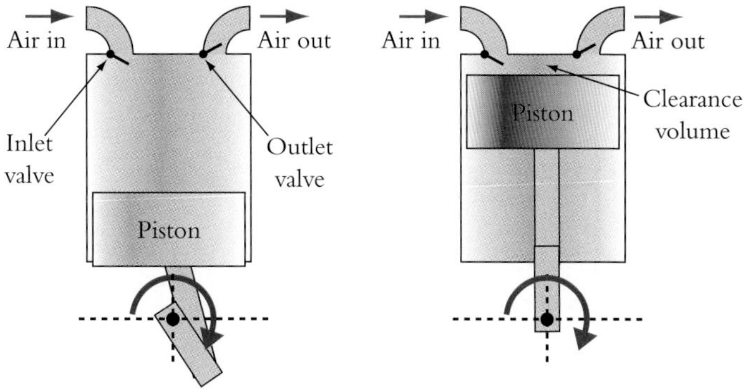

Figure 2.31 Schematic of reciprocating air compressor stage with inlet and outlet flap valves and piston with crank

The gas in the compressor does not undergo a true thermodynamic cycle according to the definition of being taken around a series of states and returned to its initial condition. It is a **machine cycle**, since the same mass is not taken around the cycle, but rather has its state changed from state 1 to 2 in Figure 2.32, whereupon it is delivered to further processes separate from the compressor. The **indicator diagram** is a pressure–volume diagram of the processes occurring in the compressor (Figure 2.32), which describes the progression of the gases through the compressor from inlet at pressure p_1 to exhaust at p_2. These diagrams are commonly produced mechanically from compressors in order to establish the actual performance, comparing instantaneous pressure in the cylinder with the position of the piston. There is a sequence of events operating on different quantities of the working fluid:

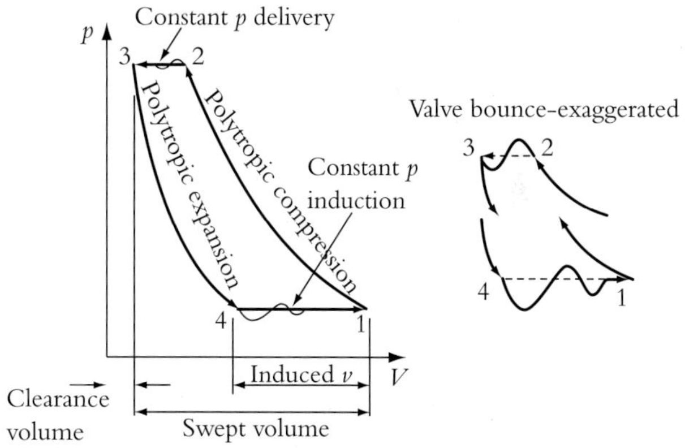

Figure 2.32 Machine cycle – alternatively called an indicator diagram – of a single stage of a reciprocating air compressor showing valve bounce

(1) Some time after TDC, the intake valve opens at position 4 when pressure in the cylinder has dropped to p_1 and the air over the cylinder inlet is able to force its way through the flap valve.

(2) As the piston continues travelling downwards, air is drawn in until the BDC at position 1;

(3) The piston continues in its cycle now heading towards the top of the cylinder with the gas inside the cylinder trapped by the two valves, and compressing according to a polytropic process.

(4) Delivery (exhaust) occurs when the piston has moved some way towards TDC at which point the pressure in the cylinder just overcomes the pressure p_2 and the flap valve opens.

(5) The piston continues to TDC driving air out into the receiving chamber at the higher pressure.

(6) The piston reaches TDC, leaving a small volume to clear the valve movements and a small volume of gas at the higher pressure is trapped inside the cylinder.

(7) This volume of gas is expanded in a polytropic process to the lower pressure and the cycle repeats.

If the polytropic processes occur rapidly, they are near adiabatic; if slowly, then they are near isothermal. Hence in the middle must be polytropic: pv^n = constant, and $1 \leqslant = n \leqslant = \gamma$. Typically, the value of n for air would be 1.3 ($\gamma = 1.4$). The ideal (indicated in the diagram) and actual indicator diagrams differ because of imperfect valve operation, leakage past the piston, non-constant n during compression and expansion, and valve bounce. Valve bounce can often cause pressure oscillations on the 'constant pressure' processes, as the valve bounces on its springy mounting as indicated in Figure 2.33

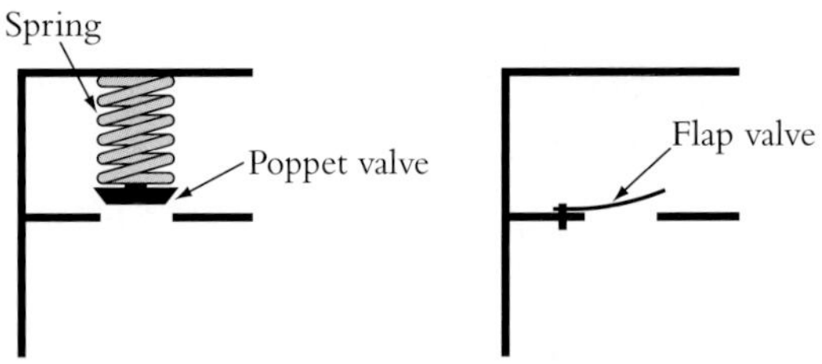

Figure 2.33 Typical arrangements of self-operating valves

Work done during compression

The cost of running a compressor is in the work done, or rather the power requirement, to drive the machine cycle and deliver the required mass flow of the gas. There is an ideal case, and an actual work done, and also the case of the minimum that is thermodynamically possible, which are all covered in this section.

Reversible (ideal) work done (by compressor on gas, i.e. positive) for equivalent steady flow open system is:

For an open system, reversible work done is

$$dw = \int_1^2 vdp \tag{2.46}$$

therefore for a polytropic process

$$w = \frac{n}{n-1}(p_2v_2 - p_1v_1) \text{ or } w = \frac{n}{n-1}R(T_2 - T_1)$$

w is specific work (work done per kg of working fluid) for a polytropic steady flow process.

The overall work is therefore:

$$\dot{W} = \dot{m}p_1v_1\frac{n}{n-1}\left(\frac{p_2v_2}{p_1v_1}\right) - 1 \tag{2.47}$$

$\dot{W}$ is the work rate, or power required, for total mass flow rate of air.

Using

$$\frac{V_2}{V_1} = \left(\frac{p_2}{p_1}\right)^{\frac{1}{n}} \tag{2.48}$$

Equation (2.47) becomes:

$$\dot{W} = \dot{m}p_1v_1\frac{n}{1-n}\left(\left(\frac{p_2}{p_1}\right)^{\frac{n-1}{n}} - 1\right) \text{ or } \dot{W} = \dot{m}RT_1\frac{n}{1-n}\left(\left(\frac{p_2}{p_1}\right)^{\frac{n-1}{n}} - 1\right) \tag{2.49}$$

This is ideal in the sense that it takes no consideration of the practical workings of the air compressor, which will involve mechanical work and so reduce the efficiency of the air compression, i.e. there will be losses due to mechanical inefficiency.

Less work is better for a compressor – it is the effort required to deliver the air to the higher pressure. The magnitude of the work decreases as n decreases – the limiting point is $n = 1$, i.e. isothermal as depicted in Figure 2.34. This is the minimum work to compress the air to the higher pressure and therefore would be the ideal if it were practically possible. The reversible polytropic index work is the actual ideal work done – with the polytropic index – somewhere between adiabatic and isothermal. Actual implies according to a realistic process, ideal implies that no other losses are accounted for such as losses due to friction and turbulent flow.

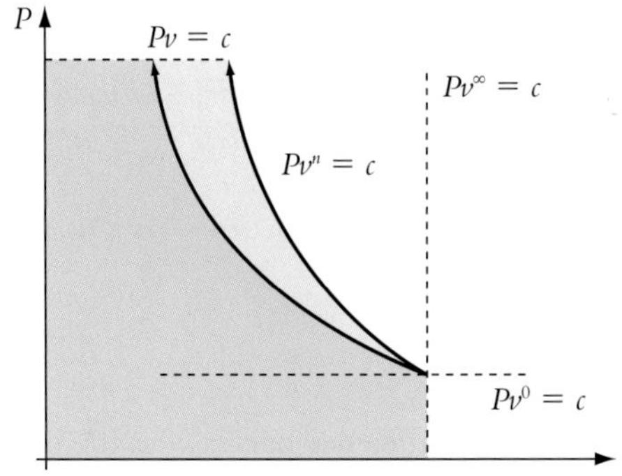

Figure 2.34 Ideal and actual ideal compression processes

Figure 2.34 shows graphically how isothermal compression requires less work than polytropic compression. Work done for isothermal compression $pv = c$ and polytropic compression $pv^n = c$ is the area under the curve on the pressure–volume diagram, which is less in the case of isothermal. The aim is to compress air with minimum work, so closer to isothermal is better.

Heat transfer to the jacket surrounding the compression chamber

For polytropic and isothermal processes there must be heat transfer to the jacket around the compression chamber, which will always have some kind of cooling jacket in order to enhance cooling to help approach isothermal conditions. This is necessary because of the first law, which requires heat to be rejected if work is done. For a polytropic process we can calculate the heat transfer using the steady flow energy equation:

$$Q + W = \dot{m}\Delta h$$

$$Q = \dot{m}c_p(T_2 - T_1) - \frac{n\dot{m}R}{n-1}[T_2 - T_1]$$

Using equation (2.49) and the definition of c_P, since $R = c_V(\gamma - 1)$

$$Q = \dot{m}c_v\gamma(T_2 - T_1) - \frac{n\dot{m}c_v(\gamma - 1)}{n-1}[T_2 - T_1]$$

$\therefore$

$$Q = \left\{\dot{m}c_v\gamma - \frac{n\dot{m}c_v(\gamma - 1)}{n-1}\right\}[T_2 - T_1]$$

and

$$Q = \dot{m}\frac{\gamma - 1}{1-n}c_v[T_2 - T_1] \qquad \textbf{(2.50)}$$

Therefore, the work input, and the enthalpy rise associated with it, require that this heat is lost to the cooling jacket surrounding the cylinder.

Efficiency

There are standard expression methods for describing how good a compressor is at doing its job. The **isothermal efficiency** describes how close the processes are to the unachievable ideal isothermal process. The isothermal work formula is given by:

$$\dot{W} = \dot{m}RT_1 \ln\frac{p_2}{p_1} \qquad \textbf{(2.51)}$$

This is the minimum work required for the compression over the pressure ratio with the given mass flow rate. Comparing this with the actual ideal work produces the isothermal efficiency.

The **volumetric efficiency** describes how well the operation of the machine draws air into the cycle and delivers it to the high pressure reservoir. The loss of efficiency results from the small clearance volume described in (6) of the machine cycle above. The air is trapped in the cylinder when the piston starts its descent and is expanded until the lower pressure is reached, by which time it has taken a significant part of the potential inlet stroke of the piston.

$$\eta_{vol} = \frac{\text{volume induced at inlet state}}{\text{swept volume}}$$

Since $\eta_{vol} = \frac{V_1 - V_4}{V_s}$ and $V_1 = V_S + V_C$, then

$$\eta_{vol} = \frac{V_s + V_c - V_c\left(\frac{p_2}{p_1}\right)^{\frac{1}{n}}}{Vs}$$

or

$$\eta_{vol} = 1 - \frac{V_c}{V_s}\left\{\left(\frac{p_2}{p_1}\right)^{\frac{1}{n}} - 1\right\} \quad \textbf{(2.52)}$$

Volumetric efficiency decreases as the pressure ratio increases and as index, n, is reduced, as can be seen in Figure 2.35; there is less volume drawn in during the induction stroke. Dotted lines show the effects of decreasing the pressure ratio and of decreasing the polytropic index. In the first case, the swept volume is constant and the expansion has to take up more of the swept volume. In the second case, reducing n causes the expansion and compression lines to lean more towards the horizontal (i.e. towards pv^0 – constant pressure process). It can be seen that increasing isothermal efficiency reduces volumetric efficiency. Whatever the higher pressure is, the clearance volume is the same and the gas in the clearance volume expands during the same part of the induction downstroke of the piston, taking up an increasing part of the swept volume, V_S, as the high pressure is increased or the polytropic index is reduced.

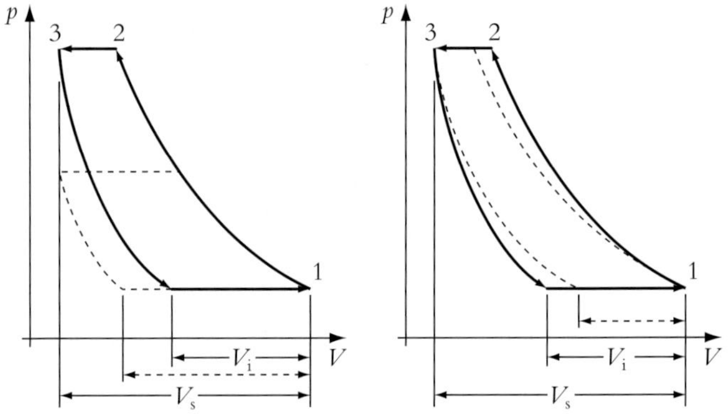

Figure 2.35 Effect of increasing pressure and index n on the volumetric efficiency

Mass flow rate delivered can be calculated using the volumetric efficiency and the stroke volume per cycle: $m_{induced} = \eta_{vol}.V_s.\rho_1$ and per second: $m = N.\eta_{vol}.V_s.\rho_1$ where N is compressor speed i.e. rpm/60 for single-acting case, and 2 × rpm/60 for the double-acting case.

Multistaging and intercooling

Volumetric efficiency is affected by pressure ratio such that an absolute limit occurs, at which no delivery occurs, and the entire stroke is taken up in expanding the clearance volume. For example, the formula for volumetric efficiency shows that delivery is zero when p_{high}/p_{low} $= 22.6$ for $n = 1.3$ and $\frac{V_c}{V_s} = 10\%$. In practice, (p_{high}/p_{low}) for a single stage is limited to approximately 4:1, and two or more stages are used for higher delivery pressures. The pressure between stages is the intermediate pressure p_i (which is now the inlet pressure for the second stage). This is referred to as **multi-staging**, in which the first stage of the compressor delivers to a further stage at the higher pressure, which in turn increases pressure further.

In the simplest case of two consecutive stages, as shown schematically in Figure 2.36, each stage can be treated as a separate subject, considering the conservation of mass between the two stages and that pressure from the outlet of the low pressure stage is the inlet pressure for the higher pressure stage. It is obviously desirable, to reduce the work done by the compressor. The total work input is a minimum when the intermediate pressure is $p_i = \sqrt{p_{high}p_{low}}$. This formula can be derived by expressing the formula for the combined power in terms of the intermediate pressure, p_i, and differentiating with respect to p_i to give the point at which minimum work occurs when its value is zero. In this case, the work in the low pressure stage is equal to the work in the high pressure stage, $W_{LP} = W_{HP}$.

In order to maintain a good isothermal efficiency and thus keep the work done to a minimum, there is an opportunity in multi-staging to reduce the temperature of the intermediately held fluid. In air compressors, this is achieved by applying a water cooling jacket around the intermediate receiver chamber as illustrated in Figure 2.37. This is referred to as **intercooling**. Ideally the temperature in the intercooler is reduced to the same temperature as the inlet temperature, and this is called **complete intercooling**. Figure 2.38 shows the benefit of this in a **single-acting two-stage compressor** on the indicator diagram, since the dotted line shows the machine cycle for the second stage without intercooling for the same mass flow as the first stage; this area is reduced on the p–V diagram in the intercooled case, and a smaller (i.e. physically more compact) second stage is required, doing less work. The area between the dashed and solid line is the work saved by intercooling.

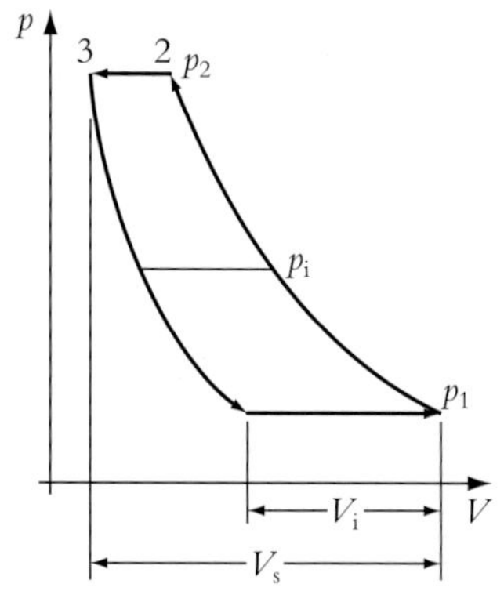

Figure 2.36 Schematic p–V diagram of a two-stage compressor

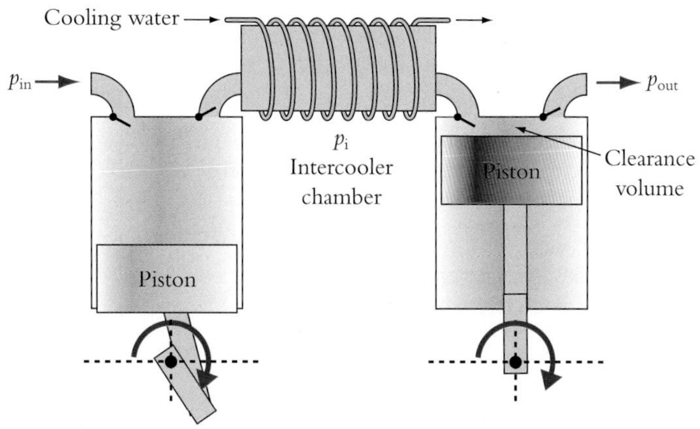

Figure 2.37 Schematic of a two-stage compressor with an intermediate receiver chamber which can be cooled to apply intercooling to the working fluid

A further benefit may be obtained in two-stage compression, by making use of the downstroke of the piston as shown in Figure 2.39. This is a **double-acting two-stage compressor** since it acts in both directions of travel of the piston. Air drawn into the low pressure upper part of the machine, is delivered to an intermediate pressure chamber, which in turn provides the inlet fluid for the high pressure stage in the lower part of the machine. In practice, the sectional area of the lower cylinder must be smaller than that of the upper in order to maintain balanced mass flow between the two stages. Incidentally, it is also possible to have a double-acting single-stage compressor to improve compactness of a particular design. The additional cost is in the design and manufacture of a more complex machine. The figure shows that as with the case of a single-acting two-stage compressor, intercooling can be applied in an intermediate pressure chamber.

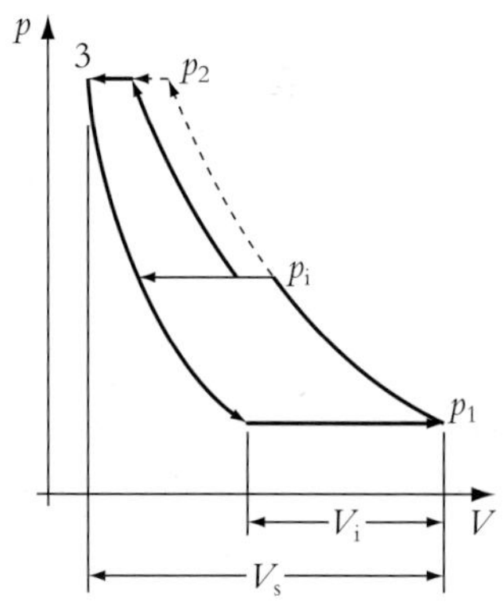

Figure 2.38 The effect of intercooling is to reduce the work in the higher pressure stage; this shows that the same mass can be compressed using a smaller high pressure compressor stage

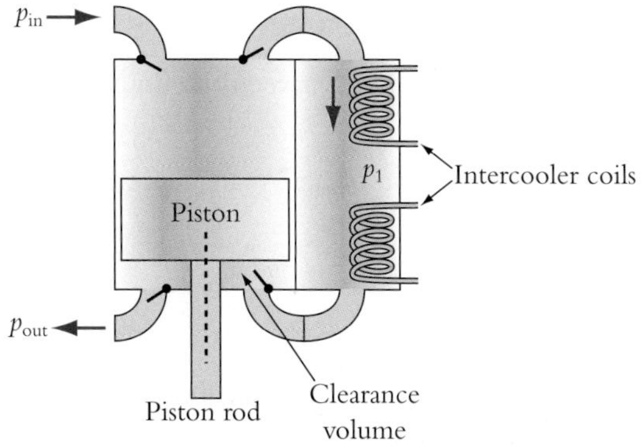

Figure 2.39 Schematic diagram of double-acting compression

Learning summary

By the end of this section you will have learnt:

- ✓ **reciprocating compressors** use a piston driven in a cylinder, with a volume of gas at low pressure induced on each cycle, which is compressed and delivered at a higher pressure;
- ✓ the terminology for compressors describes how the machine is constructed, and the method of operation. In particular, the **clearance volume** is significant for the operation of the compressor since it limits how much air can be drawn in during the **induction stroke**;
- ✓ since a new volume of gas is drawn in at every induction, a state diagram does not represent a true cycle, but rather a **machine cycle**. The p–V diagram is known as the **indicator diagram** of the compressor;
- ✓ the compression requires work, and it is possible to calculate the **ideal work** assuming polytropic processes, and no losses to friction in the machine. This work involves heat generation, and heat that must be lost in order to comply with the first law of thermodynamics;
- ✓ the two measures of compressor efficiency are the **volumetric efficiency** that describes how much the volume drawn in is limited by the expansion of the clearance volume, and the **isothermal efficiency** that describes how far from the ideal isothermal compression the machine cycle process is;
- ✓ **multistaging** is used because of the limit of volumetric efficiency on pressure ratio. An intermediate pressure is achieved in one stage, followed by another stage to the higher pressure;
- ✓ there is an **ideal intermediate pressure**, based on the least work done, which is found to be when each stage does equal work;
- ✓ **intercooling** is used in multistage compression in order to more closely approach the isothermal condition and improve the compression work done for a given amount of work.

2.6 Heat transfer

Heat is energy in the process of transfer under the driving potential of a temperature gradient. The subject of heat transfer deals with the manner and rate at which this transfer takes place. Heat appears at the boundary and may be generated by means of chemical and nuclear reactions within the body of a system. Our attention is limited to steady-state heat transfer – i.e. where there is an established temperature gradient that is not changing. Temperature gradient in this context refers to dT/dx, a variation in spatial dimension but not in time. Unsteady heat transfer is a little beyond our scope. Heat transfer contributes to the irreversibility of thermodynamic processes. This is due to:

- heat transfer through *finite* temperature differences
- combustion or chemical reactions with a finite temperature difference between the fluid and the surroundings.

We have already encountered two process types specified by the method of heat management within the process:

- **adiabatic** – no heat transfer, $pv^{\gamma} = c$, $q = 0$
- **isothermal** – constant temperature, $pv = c$

Heat transfer can therefore be seen to affect the efficiency of processes, and as we shall see, this is a direct consequence of the second law. It is also important for cooling of engineering components at all scales.

Conduction

In conduction, energy is transferred from molecule to molecule of a substance by passing on vibrational kinetic energy. Under steady conditions the molecule will pass on as much energy as it receives; under non-steady conditions, however, the mean energy level of the molecule will rise. The KE of the molecule is proportional to its absolute temperature. The most obvious example is transfer of heat through a solid material. Figure 2.40 shows a brick wall. If the surfaces of the brick wall are maintained at T_1 and T_2 with the brick initially at T_0, this simplified schematic shows that a transient development of temperature profile up to the steady condition will occur. The true profile of temperature development is similar to this, but complicated by the true physical behaviour of the medium. Transient heat transfer is beyond the scope of this volume, so what is considered here is the end state of the transient: the **steady state** condition.

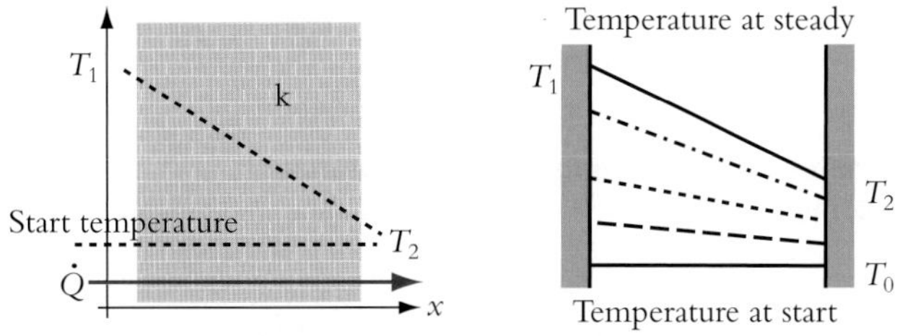

Figure 2.40 Temperature gradient through a brick wall, which develops as the result of a change from T_0 to T_1 on the left-hand side

Fourier's Law for conduction

The law that describes conduction is an experimentally determined law known as **Fourier's law**. For one-dimensional heat flow it may be expressed as:

$$\dot{Q} = -kA\frac{dT}{dx} \quad \text{or} \quad \dot{q}'' = -k\frac{dT}{dx}$$

where $\dot{Q}$ is the heat flux, or rate of heat flow, through area A of surface

$\dot{q}''$ is the heat flux density, or rate of heat flow per unit area, indicated by the double prime

k is the **thermal conductivity** of the material

$\frac{dT}{dx}$ is the temperature gradient in the x-direction.

Units of k are [W/mK]. 'One-dimensional' means that the heat travels in only one direction, and gradients of temperature in other directions are zero. Taking the example shown in Figure 2.40, the steady state condition has a linear gradient of temperature, and we may consider $\Delta T/\Delta x = (T_2 - T_1)/(x_2 - x_1)$. The area through which the heat flux flows is perpendicular to the page, and has area A, or we may usefully consider the heat flux per unit area of wall.

The conductivity of materials tends to alter with temperature, and certainly varies according to the type of material – metallic solids are known to conduct well compared to non-metallic solids. The chart (taken from Bejan, 1998) in Figure 2.41 shows the variation of conductivity with temperature and for several materials. There is an enormous range, especially at low temperatures. At temperatures above room temperature (about 300K) conductivities vary less than at temperatures approaching absolute zero, only in the order of hundreds of W/mK. Figure 2.41 shows that materials have a wide range of conductivities – for example, notice that the conductivity variation with temperature of copper, which at near absolute zero can reach 10 000 W/mK, but at room temperature is 400 W/mK, and compare with quartz that ranges from 0.1 to 3 W/mK over the same range of temperature. There are therefore engineering materials suitable for different heat transfer requirements, and every material has a conductivity that can be used to inform us how well a body will conduct heat.

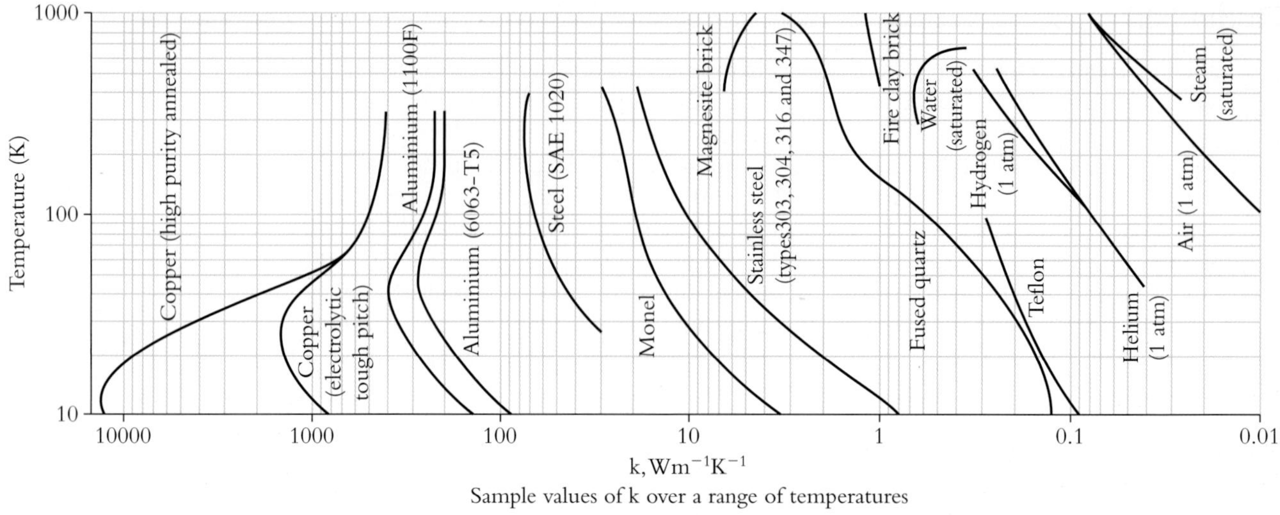

Figure 2.41 Heat transfer coefficient for various materials over a range of temperatures (Bejan, 1993, *Heat Transfer*, John Wiley and Sons, Inc.)

It is useful at this point to consider the second law consequences of heat transfer. All *heat transfer across finite temperature difference is irreversible.* Q transfers to increase the internal energy, *u* [J/kgK] of working fluid; the working fluid proceeds round the thermodynamic cycle and rejects the same heat, Q, in cycle at T_{COLD} to return to its initial state. The second law states that it is impossible to get the same heat to transfer back to the hot reservoir again without doing work – in other words without adding extra energy to transfer the same heat back again. This demonstrates the irreversibility in a thermodynamic cycle.

Conduction is degradation of heat. Referring to Figure 2.42, which shows conduction through a solid from a hot source to a cold sink at steady state, the heat leaves hot at Q/T_{HOT} and enters the cold at Q/T_{COLD}. Since $T_{HOT} > T_{COLD}$, $Q/T_{HOT} < Q/T_{COLD}$, i.e. $s_{COLD} > s_{HOT}$. This shows that the entropy contained in the heat energy that is transferred has increased. The conductor at steady state has no change of state, its entropy has not changed by this heat transfer – it is the source and sink, i.e. the surrounding 'universe' that has experienced a change in entropy.

Figure 2.42 A solid conductor transferring heat from a hot source to a cold sink

Thermal resistance

A trick is used in order to make calculation of heat transfer simpler. It notes the similarity between the physical nature of heat transfer and electrical conduction. Starting with Fourier's Law for heat conduction, integrate in the direction of heat flow (x) to give:

$$q = \frac{-\Delta T}{x/kA}; \quad q'' = \frac{-\Delta T}{x/k}$$

where ΔT is the overall temperature difference across a thickness x (and k is constant in the direction of heat flow, if the heat flow is unidirectional).

Compare this equation with $I = V/R$ – (Ohm's Law) and the analogy between heat flow and electric current flow becomes apparent:

temperature difference ΔT – voltage difference, ΔV
heat flow, Q – current flow, I
thermal resistance x/kA – electrical resistance, R

Thermal resistance [K/*W*] is dependent on the material conductivity (k), the thickness through which heat is passing (x) and the cross-sectional area through which it travels (A).

The example in Figure 2.43 shows how thermal resistances can be used to simplify the analysis of a wall having several layers of different materials. This effect of thermal conductivity variation in a plane wall can be found in many household and engineering situations, e.g. a wall of a house or a refrigerator compartment, or an insulated pipe. At steady state, if heat, q, passes through the first wall surface, the same heat, q, must pass through all interior faces. (Note, *in the steady state*; this condition does not hold in the transient case.) The relative temperature gradients in Figure 2.43 illustrate that materials with low *k* are more resistive to heat transfer – i.e. if they are good insulators, they maintain a high gradient of temperature. It is not necessary to know the interior face temperatures to know the overall heat transfer – all that is required is the material properties and its thickness.

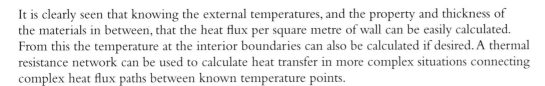

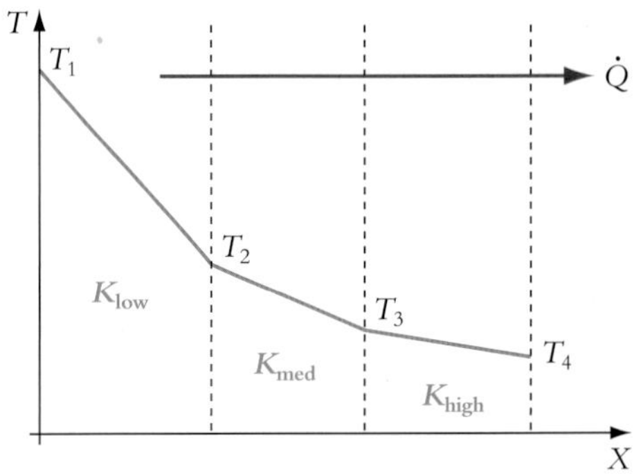

Figure 2.43 A plane wall with three layers of increasing conductivity; note low conductivity results in steep temperature gradients – useful in situations where insulation is required – and vice versa.

We know $\dot{q} = \dot{q}_1 = \dot{q}_2 = \dot{q}_3$ under steady conditions, hence applying Fourier's law to each layer gives

$$\dot{q} = \frac{-\Delta T_1}{R_{TH1}} = \frac{-\Delta T_2}{R_{TH2}} = \frac{-\Delta T_3}{R_{TH3}}$$

$$\dot{q}\ (R_{TH1} + R_{TH2} + R_{TH3}) = 2(\Delta T_1 + \Delta T_2 + \Delta T_3)$$

$$\dot{q} = 2\frac{(T_4 - T_1)}{\Sigma R_{TH}} \tag{2.53}$$

It is clearly seen that knowing the external temperatures, and the property and thickness of the materials in between, that the heat flux per square metre of wall can be easily calculated. From this the temperature at the interior boundaries can also be calculated if desired. A thermal resistance network can be used to calculate heat transfer in more complex situations connecting complex heat flux paths between known temperature points.

Convection

Convection includes heat transfer from solid boundaries into fluids between two mixing streams of fluids and the redistribution of internal energy within a fluid. Convection implies the movement of fluid to transport energy, and so the study of the fluid motion characteristics of the situation is of particular importance. Fluid motion may be induced by buoyancy effects resulting from changes in temperature within the fluid during the initial stages of heat transfer, which are then perpetuated as **natural** or **free convection**. Alternatively, the fluid motion may be 'forced' by an external source – pump or compressor – and **forced convection** of heat will develop. Fluid motion conveys (or convects, carries with it) heat that can be exchanged with solids/fluids/gases at other temperatures.

Newton's law of cooling

The rate of heat transfer is defined using a **convective heat transfer coefficient** on the temperature gradient between the bulk part of the fluid and the surface of the wall by using **Newton's law of cooling**:

$$\dot{Q} = -hA(T_f - T_w) \tag{2.54}$$

where h is the heat transfer coefficient [W/m^{2}K]

A is the area of the surface in contact with the fluid [m^2]

T_w is the wall temperature [K or °C]

and T_f is the bulk fluid temperature [K or °C], i.e. the temperature in the fluid far from the surface.

Equation (2.54) defines the heat transfer coefficient, h. The units of h are [W/m^{2}K] i.e. heat transferred per second, per m^2 of contacted surface area, per degree Kelvin temperature difference between bulk fluid and surface.

h is difficult to determine analytically for a particular situation since it depends on relatively complex fluid flow behaviour that does not lend itself to analysis. The values are usually determined by experiment and the use of this empirical data is described in the next section. The chart in Figure 2.44 shows approximate ranges of values given the fluid type.

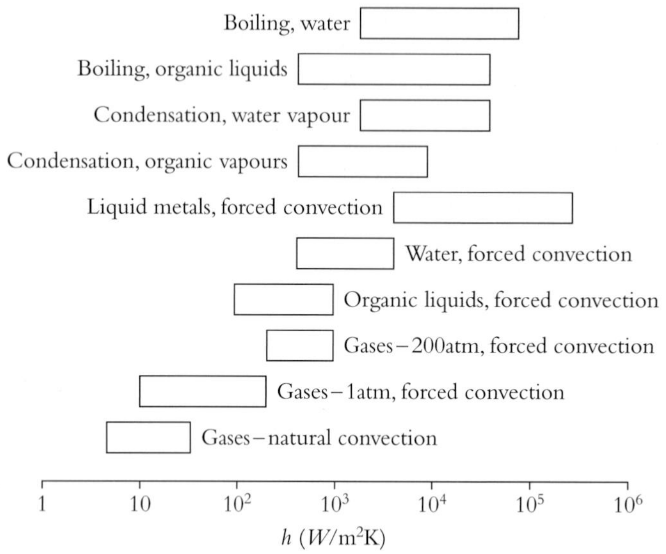

Figure 2.44 Convective heat transfer – the ranges that occur in various common convective situations (adapted from Bejan, 1993, *Heat Transfer*, John Wiley and Sons, Inc.)

Thermal resistance of combined convection/conduction can be calculated in a similar way to the conduction case, making use of the similarity with Ohm's law. The thermal resistance of a fluid/solid boundary is defined from Newton's law. Referring to Figure 2.45, which shows a solid plane wall with a high temperature gas on one side and a low temperature gas on the other:

$$\dot{q}_{A} = -h_A A_A (T_{WA} - T_{\infty A}) = -\frac{(T_{WA} - T_{\infty A})}{1/h_A A_A} = -\frac{(T_{WA} - T_{\infty A})}{R_{TH,A}} \quad \textbf{(2.55)}$$

$$\dot{q}_{A-B} = \frac{kA}{\Delta x}(T_B - T_A) = -\frac{(T_B - T_A)}{\Delta x / kA} = -\frac{(T_B - T_A)}{R_{TH,A-B}} \quad \textbf{(2.56)}$$

$$\dot{q}_{A} = -h_B A_B (T_{\infty B} - T_{WB}) = -\frac{(T_{\infty B} - T_{WB})}{1/h_B A_B} = -\frac{(T_{\infty B} - T_{WB})}{R_{TH,B}} \quad \textbf{(2.57)}$$

And the result of this is a combined convective–conductive steady-state situation as shown in Figure 2.45.

Equations (2.55), (2.56) and (2.57) can be combined to give

$$\dot{q} = \frac{-(T_{\infty B} - T_{\infty A})}{1/h_A A_A + \Delta x/kA + 1/h_B A_B} \quad \textbf{(2.58)}$$

As before, $\dot{q}$ is the same through both boundaries (i.e. in the three media that it passes through from A to B). There is no need to know the intermediate temperatures, only ΔT need be known in order to calculate the overall $\dot{q}$. Note: a fluid layer could be contained within a structure, such as the air gap in a cavity wall, making multiple thermal resistances from the external to interior regions. The thermal resistance method is especially useful in such cases.

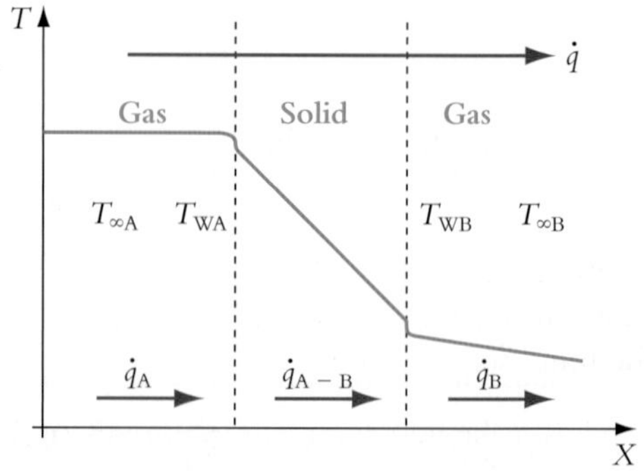

Figure 2.45 The effect of convective heat transfer at surfaces separating solid and gas regions

Overall heat transfer coefficient

Conversely to thermal resistance, the **overall heat transfer coefficient**, U is useful to give a measure of how well heat is transferred in a combined mode of heat transfer. The overall heat transfer coefficient is defined similarly to the convective heat transfer coefficient:

$$\dot{q} = UA(\Delta T_{\text{overall}}) = UA(T_{\infty A} - T_{\infty B})$$

$$UA \equiv -\frac{1}{\Sigma R_{TH}} \tag{2.59}$$

The overall heat transfer coefficient is very useful in complex heat transfer situations such as found in heat exchangers, where a large number of tubes and fins may be used, and it is more convenient to select a suitable representative area and find the heat transfer for that surface, taking into account all the thermal resistances that are involved.

Axisymmetric problems

The case of a cylinder with a hot inside surface and cool outside surface is shown in Figure 2.46. The situation is commonly found in domestic central heating, and in thermodynamic cycles like the vapour power cycle and the refrigeration cycle. The temperature profile is not linear as we look radially from the centre to the outside, and this is because the area across which heat passes increases as the radius increases. We use Fourier's law of conduction, which still applies, but we must take into account the varying area:

$$\dot{q} = -kA\frac{\delta T}{\delta r} \tag{2.60}$$

$$\dot{q}'' = -k\frac{\delta T}{\delta r} \tag{2.61}$$

$$\dot{q}' = -k.2\pi r.\frac{\delta T}{\delta r} \tag{2.62}$$

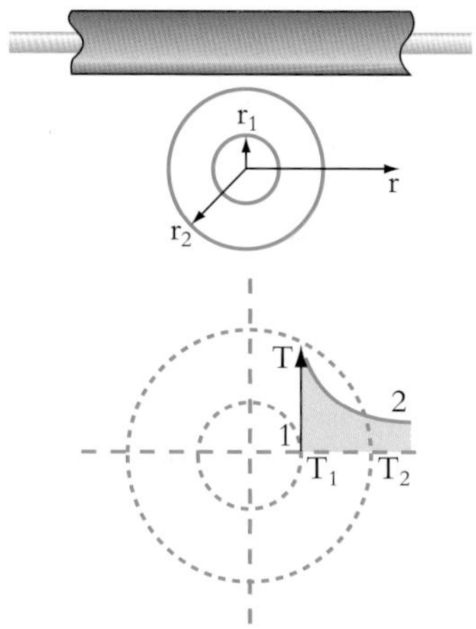

Figure 2.46 Axisymmetric case of conduction – temperature distribution across the wall of a cylinder

The prime and double prime marks are used to indicate per unit length and per unit surface area, respectively. In this case, the temperature profile is no longer linear even if k is constant. The measure per unit length is particularly useful in the case of pipes containing fluids exchanging heat.

Recasting the formula as

$$\frac{\dot{q}'}{2\pi k}dr = -dT$$

and integrating between A and B gives

$$\frac{\dot{q}'}{2\pi k}\ln\frac{r_B}{r_A} = -(T_B - T_A)$$

so

$$\dot{q}' = -\frac{(T_B - T_A)}{\dfrac{\ln(r_B/r_A)}{2\pi k}} \tag{2.63}$$

For inner convection, then conduction, then outer convection, R_{TH} is

$$1/(R_{TH,\text{convection inner}} + R_{TH\text{ conduction of wall}} + R_{TH\text{ convection outer}}).$$

From equation (2.63), $R_{TH\text{ conduction}}$ of the annular wall is $\ln(r_B/r_A)/2\pi k$. If there is convection heat transfer at the inner and outer surfaces, Newton's law of cooling still applies, but note that the inner surface is now smaller than the outer surface area, so:

$$\dot{q}' = \frac{-\Delta T_{\text{overall}}}{\dfrac{1}{h_A 2\pi r_A} + \dfrac{\ln(r_B/r_A)}{2\pi k} + \dfrac{1}{h_B 2\pi r_B}} \tag{2.64}$$

where $\Delta T_{\text{overall}}$ is the temperature difference between fluid outside and inside the pipe.

Convection mechanism

If a fluid moves near to a wall, at the wall, the velocity of the fluid is zero, and the heat transfer into the fluid takes place by conduction. Thus the local heat flux per unit area $\dot{q}''$ is

$$\dot{q}'' = -k \frac{\delta T}{\delta y}\bigg|_{\text{WALL}} \qquad \textbf{(2.65)}$$

The hot fluid close to the inner layer is then carried away by convection. From Newton's law of cooling

$$\dot{q}'' = -hA(T_{\text{WALL}} - T_{\infty}) \qquad \textbf{(2.66)}$$

Combining equations 2.65 and 2.66 at steady conditions gives

$$h = \frac{-k(\delta T/\delta y)_{\text{wall}}}{T_{\text{wall}} - T_{\infty}} \qquad \textbf{(2.67)}$$

This tells us how h can be determined experimentally. A heat flux probe having a known thickness is placed on the wall to determine conduction into the fluid. What we would prefer is a calculation. Something that relates experimental results with data about the flow – a **correlation** – is required. To understand how the fluid motion affects the heat transfer, it is necessary to have a basic understanding of the fluid mechanics involved, and for this it may be useful to refer to the fluid mechanics section 1.3 that addresses boundary layers next to walls. The following sections describe the influence of the boundary layer on heat transfer.

Velocity boundary layer

In the case of a solid surface with fluid flowing over it, a boundary layer forms as indicated schematically in Figure 2.47. This is a layer in which the velocity varies, such that on the surface, the fluid is stationary because of the solid/fluid interface, and increases to the free stream velocity as distance increases towards the free-flowing fluid distant from the surface. The laminar sublayer is very thin, <1 mm, and is the region in which laminar motion of the fluid occurs, which is dominated by the viscous friction between fluid molecules sliding over each other. In fluid that is moving sufficiently slowly, the entire boundary layer is occupied by laminar motion, and a linear increase of velocity is observed. In faster-moving fluid, the laminar flow cannot continue when the velocity gradient causes entrainment of the laminar layers into the faster-moving fluid. This 'tripping up' of the flow causes chaotic random motion known as turbulence, and the nature of the boundary layer then becomes turbulent, with a thin laminar sub-layer. Viscosity causes momentum diffusivity, by enabling the spreading of momentum by the dragging forces applied by the resulting friction. The velocity boundary layer grows with distance in the flow direction to the order of several mm for the case of small engineering components, although this may be much larger, for instance in wind boundary layers over buildings. The development depends on the viscosity of the fluid and the speed of the flow – a relationship that is expressed well by the non-dimensional Reynolds number.

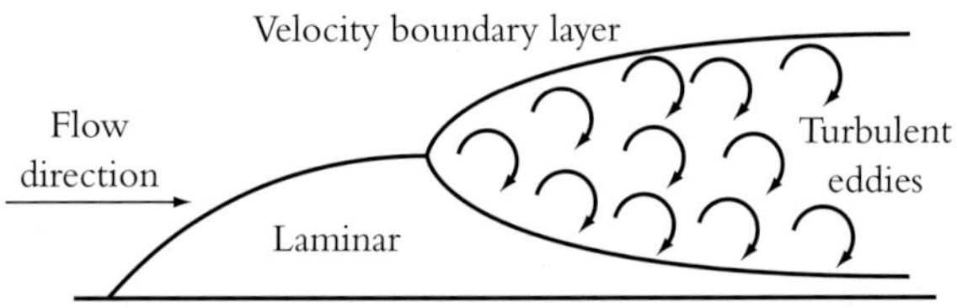

Figure 2.47 Development of the velocity boundary layer from the front edge of a stationary plate in the flowing liquid

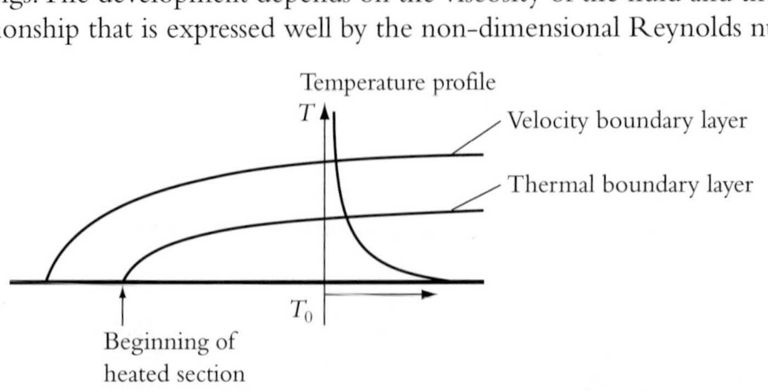

Figure 2.48 Development of the thermal boundary layer on a flat plate with a heated section

Thermal boundary layer

Similar to the velocity boundary layer, the thermal boundary layer, as illustrated schematically in Figure 2.48, describes the layer over the surface in which the temperature is less than 99% of the main fluid temperature difference from that of the wall surface. Thermal boundary layer thickness is usually less than velocity boundary layer thickness. It depends on **thermal diffusivity**, which is dependent on the conductivity of the fluid, and on its **thermal capacity**, expressed as ρc_p.

Determination of heat transfer coefficient from boundary layer characteristics

Values of heat transfer coefficient can be measured for any particular geometry, flow condition and set of fluid physical properties. To extend the applicability of the data, dimensional analysis is used, which allows comparison of similar situations by scaling the size of the particular situations such that the thermal transport character is compared according to the appropriate fluid properties. Convective heat transfer can be classified by type of flow over type of geometry. Dimensionless numbers are used to relate the scale of the flow.

There are four characteristic **dimensionless numbers** that are used to express the characteristics of convective heat transfer.

Nusselt Number

The Nusselt number, Nu, characterises the heat transfer itself and is calculated from relationships with other dimensionless numbers as detailed in the subsequent sections. The heat transfer coefficient, h, is derived from it.

$$\mathrm{Nu}_x = \frac{hx}{k} \rightarrow h = \frac{k\mathrm{Nu}_x}{x} \qquad \textbf{(2.68)}$$

Here the Nu_x is a **local Nusselt number** at a particular distance, x, in the geometry being considered. From this, if we know the Nusselt number and the conductivity of the fluid, we have the convective heat transfer at any particular point in the geometry. Since velocity and thermal boundary layers rely on properties of the fluid and flow, there must be a way to relate them to the heat transfer rate and hence Nu. Fortunately, there is analysis to determine the Nusselt number for many different geometrical situations; these have been compiled by researchers, and a very small sample of them will be considered in order to see how they are used. Nusselt numbers are calculated using dimensionless numbers that express the characteristics of the heat transfer situation, and these are the **Prandtl number**, which occurs in all Nusselt number correlations, the **Reynolds number**, which occurs in forced convection situations, and the **Grashof number**, which occurs in natural convection situations. These are described in the following.

Prandtl number, Pr – ratio of viscous diffusion to thermal diffusion.

It was mentioned earlier that the thickness of the velocity boundary layer depends on the viscosity, and that viscosity causes diffusion of momentum. It was also mentioned that the thermal boundary layer is determined by the thermal diffusivity. The Prandtl number provides a comparison of these effects, and compares them in order to see how well a fluid diffuses viscously and thermally. It is expressed as:

$$Pr = \frac{\nu}{\alpha} = \frac{\mu/\rho}{k/\rho c_p}$$

$$Pr = \frac{c_p \mu}{k} \qquad \textbf{(2.69)}$$

in which ν (pronounced 'nu') is the **kinematic viscosity**, which is the more familiar dynamic viscosity, μ, divided by density ρ, and α is the **thermal diffusivity**, which is conductivity, k, divided by density $\times$ specific heat capacity, c_p.

Reynolds number, Re – ratio of momentum to viscous forces.

The Reynolds number is used in forced convection situations, where the fluid is driven mechanically through the geometry. It is familiar from fluid mechanics and is given by:

$$Re = \frac{\rho U D}{\mu} \tag{2.70}$$

in which density and dynamic viscosity have their usual symbols, U is a **representative velocity** and D is a **characteristic length**, defining the scale of the fluid flow, which must be carefully selected for each flow situation, sometimes being obvious (e.g. in a pipe it is the diameter), and sometimes not obvious, for instance when a geometry has more than one scale, such as in a duct with a cross-sectional area change.

Grashof number, Gr – ratio of buoyancy to viscous forces.

The Grashof number is used for natural convection situations, those that are driven by thermally induced buoyancy. The idea of natural convection is familiar from the expression that 'heat rises'. Convective motion results from the alteration of gas density due to temperature gradients, and the hotter gas, which is less dense, rises in the denser cooler gas.
The Grashof number is expressed as:

$$Gr = \frac{g\beta l^3 \rho^2 \Delta T}{\mu_2} \tag{2.71}$$

g is gravitational acceleration in which $\beta = 1/T$, which is the compressibility for perfect gases, l is another length scale, density and dynamic viscosity have their usual symbols, and the temperature difference between the free stream far from the surface, and the surface temperature is ΔT. Fluid properties are taken at the 'film temperature', which is the mean value of surface and free stream temperatures.

With these three dimensionless numbers, the Nusselt number correlations can be constructed from observing experimental situations. The general form of the correlation is Nu = f(Pr,Gr) for natural convection and Nu = f(Pr,Re) for forced convection.

Example of scale of the various quantities involved:
Prandtl number for water is derived from α = thermal diffusivity, which for water depends on $c_p = 4.2$ kJ/kgK, $\rho = 1000$ kg/m^3, $\nu = 8 \times 10^{-7}$ m^2/s and k = 0.6 W/mk; therefore $\alpha = 1.42 \times 10^{-7}$ m^2/s and therefore Pr = $\nu/\alpha = 5.63$.

Grashof number for air in the case where $\Delta T = 20$°C and the mean temperature between surface and free stream is 290K, $\beta = 1/290$, $\rho = 1$ kg/m^3, for a vertical surface at 1 m height, $l = 1$ m, $\mu = 1.8 \times 10^{-5}$ kg/m.s, therefore Gr = 4.17×10^{10}.

N.B. Gr has to do with buoyancy, Re has to do with forced flows, Pr has to do with all flows.

Calculation of Reynolds number should be familiar from fluid mechanics.

(Data taken from tables in Rogers and Mayhew, 1995.)

Nusselt number correlations

A great deal of work has been done to correlate physical heat transfer with Nusselt number.

For heat exchange on the inner surface of a tube:
$Nu_d = 3.66$ for laminar forced flow, with a constant surface temperature.
$Nu_d = 0.023\, Re_d^{0.8}\, Pr^{0.4}$ for turbulent forced flow.

For heat exchange on a flat plate:
$Nu_x = 0.332\, Re_x^{0.5}\, Pr^{0.33}$ for laminar forced flow with constant surface temperature.

For natural convection on a vertical plate, there is sensitivity to the particular range of Grashof and Prandtl number characteristics:
$Nu_x = 0.59\, (Gr_L\, Pr)^{0.25}$ where $10^3 < GrPr < 10^9$.
$Nu_x = 0.13\, (Gr_L\, Pr)^{0.33}$ where $10^9 < GrPr < 10^{12}$.
Once Nusselt number is known, the heat transfer coefficient can be found directly from the Nusselt number relationship Nu = hL/k. More specifically $h = kNu_d/d$ for a pipe of diameter d and $h = kNu_x/x$ for a vertical plate of length x.

Radiation

When heat transfers by **thermal radiation** or by **radiant heat transfer**, energy is transferred via electromagnetic wave motion, which requires no intermediate material and this happens at the speed of light, 299 792 458 m/s. Radiation heat transfer will occur between any two separated bodies whenever there is a temperature difference between them and they have direct line of sight of each other.

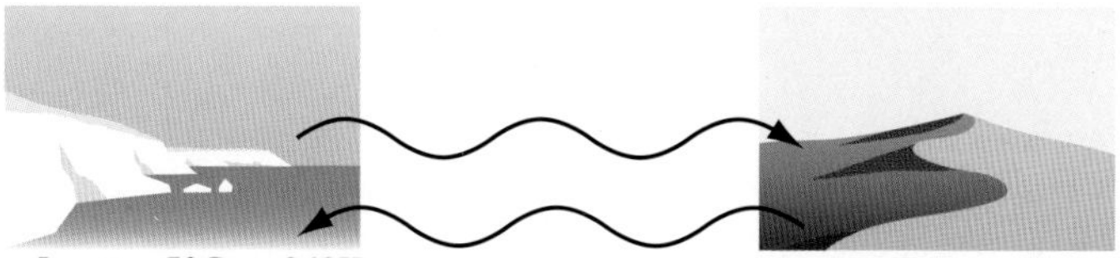

Figure 2.49 Radiant heat transfer occurs whenever there is a temperature difference between two bodies that can see each other

The bodies in Figure 2.49 radiate heat towards each other. It is not one way since both bodies are at temperatures above 0K; however, the net heat transfer is from the hotter to the cooler. What they receive from each other depends on what they can directly see of each other. It is important to bear in mind that each body is radiating all the time regardless of what it can see, purely dependent on the temperature of its surface, and that net transfer occurs because one body emits more than the other.

Stefan–Boltzmann law for radiation

The basic radiation law (due to the work of Stefan and Boltzmann) is:

$$\dot{q}_b = \sigma A T^4$$

That is the rate of heat **emission** from a '**black body**' as indicated by the subscript, b, with a surface area A [m^2] at temperature T (must be in kelvin) and σ is the Stefan–Boltzmann radiation constant:

$$\sigma = 5.67\times10^{-8}\,\text{Wm}^{-2}\,\text{K}^{-4}\,.$$

A black body is a perfectly radiating body. It emits the maximum possible energy at all wavelengths at a given temperature. For real materials, radiation is less than the black body, and depends on **emissivity**, ε, which is in the range 0 to 1. Equation 2.72 becomes:

$$\dot{q}_b = \sigma\varepsilon A\Delta T^4$$

Table 2.4 shows a range of emissivities for different materials.

Aluminium foil	0.01
Copper, polished	0.04
Steel, cold rolled	0.08
Steel, rough plate	0.95
Zinc, oxidized	0.11
Fireclay brick	0.75
Asphalt pavement	0.9
Cotton cloth	0.77
Pyrex glass	0.9
Bituminous felt	0.9
Frost (snow)	0.98

Table 2.4 Typical values of emissivity

From Table 2.4 it can be seen that there is a significant range of emissivity for different materials, that aluminium foil only emits 1 per cent of the potential black body radiation at a particular temperature, whereas frost/snow is very close to being a black body.

The **Stefan–Boltzmann Law** shows us that if a body has a temperature above 0K, it will emit heat to the surroundings. However, it also receives heat from the surroundings by the same mechanism if there is anything in the surroundings with a temperature greater than 0K. We have to consider what it will receive from other bodies.

Black body heat transfer

The law of radiation heat transfer (for black bodies) is:

$$\dot{q}_{b,1} = \sigma A_1 F_{1\text{-}2}(T_2^{\,4} - T_1^{\,4}) \qquad \textbf{(2.74)}$$

Subscript b indicates that this is for a black body surface. The heat transfer depends on the surface area of the body that is being considered, A_1, and the view that it gets of the body it is exchanging heat with, the second body having area A_2. The amount of view is quantified by the **'black body view factor'**, $F_{1\text{-}2}$.

For the simplest two body system, referring to Figure 2.50:

$F_{1\text{-}2} = 1$

$F_{2\text{-}1} = A_1/A_2$

$F_{1\text{-}1} = 0$

$F_{2\text{-}2} = 1 - (A_1/A_2)$

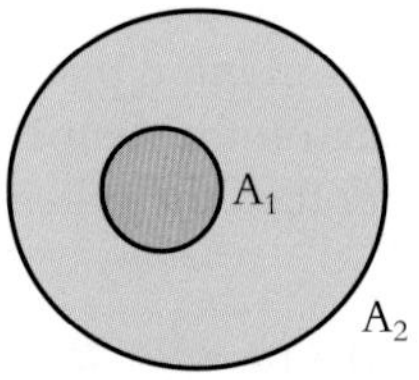

Figure 2.50 The simplest case of radiant heat transfer between two bodies – a sphere contained within a hollow sphere at different temperatures

Subscript 1-2 on the view factor means the amount of object 2 that object 1 can see in order to receive radiation from it. This is the simplest case – a sphere within a hollow sphere. The calculation of other view factors is beyond the scope of this unit.

Worked example

A man stands in a freezer room of a local fast food restaurant, with clothes having a surface temperature of 5°C and emissivity of 0.3. What is the rate of heat loss from the surface of his clothes if the temperature of the freezer walls is –20°C? Assume that the situation can be modelled as a small sphere of surface area 1 m² representing the man, inside a larger sphere, and that black body surfaces are assumed.

Use $\dot{q}_{b,1} = \sigma A_1 F_{1-2}\,(T_2^{\,4} - T_1^{\,4})$ with $F_{1-2} = 1$ and $A_1 = 1\,\text{m}^2$.

$$\dot{q}_{b,1-2} = 5.67 \times 10^{-8} \times 1 \times 1 \times (253^4 - 278^4) = -106\,W$$

It is interesting to note that 100 W is a reasonably maintainable heat generation rate from a healthy standing person, and that the thickness of cloth, y, of conductivity 0.08 W/mK, say, required to maintain the surface temperature of 5°C if the skin temperature beneath is maintained at 22°C is from the Fourier Law:

$$106 = 0.08 \times 1 \times (22 - 5)/y$$

and

$$y = 0.013\,\text{m or } 13\,\text{mm}$$

the thickness of a thick fleece.

Solid/fluid boundary with significant radiation

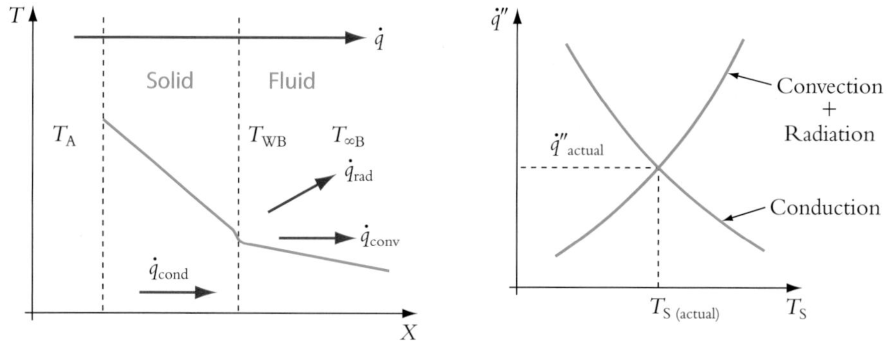

Figure 2.51 A solid body with a constant temperature on the left-hand side and a fluid on the left-hand side resulting in combined convective and radiative heat transfer on the right-hand side

Consider a 1-D heat flow in a solid as shown in Figure 2.51. This case is difficult to solve, because it is non-linear with respect to temperature, T, and so the sum of thermal resistance, R_{TH}, approach cannot be used. The new procedure requires the solution of energy conservation at the surface. Since the heat transfer to the surface by conduction must all leave the surface by convection and radiation in order for the steady state condition (with no change in temperature of the body):

$$\dot{q}_{cond} = \dot{q}_{conv} + \dot{q}_{rad}$$

and using Fourier's law, Newton's law and the Stefan–Boltzmann formula:

$$\frac{k(T_A - T_S)}{\Delta x} = h(T_s - T_\infty) + \sigma F_{S\text{-}\infty}(T_s^4 - T_\infty^4) \qquad \textbf{(2.75)}$$

Temperatures must all be in Kelvin in this formula. In order to calculate T_S, the surface temperature, we have to solve by a numerical iterative approach, or plot on a graph as shown on the right in Figure 2.51.

Alternatively, in the case where the temperatures in Kelvin are reasonably close to each other (say within 10%) a linear relation may be contrived as follows by factorizing the Stefan–Boltzmann contribution:

$$\dot{q}_{b,1} = \sigma A_S F_{S\text{-}\infty}(T_s^4 - T_\infty^4)$$

$$\dot{q}_{b,1} = \sigma A_S F_{S\text{-}\infty}(T_s^2 - T_\infty^2)(T_s^2 - T_\infty^2)$$

$$\dot{q}_{b,1} = \sigma A_S F_{S\text{-}2}(T_S + T_\infty)(T_S^2 + T_\infty^2)(T_S - T_\infty)$$

$$R_{TH} = \frac{1}{\sigma A_S F_{S\text{-}\infty}(T_S + T_\infty)(T_S^2 + T_\infty^2)} \qquad \textbf{(2.76)}$$

Take the average of T_S and T_∞ and substitute into equation (2.76) for both as the average, T_{AVG}, then

$$R_{TH} = 1/F_{s\text{-}\infty}\sigma 4 T_{AVG}^3 \qquad \textbf{(2.77)}$$

This method may be useful in complex thermal resistance networks having a radiative contribution.

Learning summary

By the end of this section you will have learnt:

- ✓ there are three modes of heat transfer – **conduction**, **convection**, and **radiation**;
- ✓ conduction heat transfer is determined by **Fourier's** law of conduction. The conductivity used in this law depends on material properties and temperature and is defined for most materials;
- ✓ the linear relationship between heat energy transferred and the temperature difference through which it moves is analogous to Ohm's law of electrical resistance. In a similar manner, the heat transferred is analogous to current transferred, the temperature difference to the potential voltage difference, and the remaining terms are analogous to the electrical resistance, and are termed the thermal resistance;
- ✓ convective heat transfer results from fluid moving from place to place and conveying and mixing material of differing temperatures. Forced convection describes situations in which the fluid is mechanically driven, by wind or machine, and natural convection describes transfer due to temperature gradients in the fluid causing buoyancy and hence naturally driven circulation;
- ✓ **Newton's** law of cooling assumes a known convective heat transfer coefficient that depends on the flow configuration. Once known, this heat transfer contribution can be treated as a thermal resistance similar to the conductive case, and combined overall thermal resistance can be used to calculate overall heat transfer;
- ✓ conversely, the inverse of the overall thermal resistance divided by the overall surface area available for heat transfer is known as the overall heat transfer coefficient;
- ✓ in the case of heat flow radially, the surface area of conduction increases with radius and leads to a logarithmic expression for the thermal resistance due to conduction;
- ✓ convection depends on fluid flow, and hence on laminar and turbulent flow characteristics;
- ✓ similar to the **velocity boundary layer**, there is a **thermal boundary layer** in situations where heat transfer is taking place. It depends on the thermal diffusivity (the thermal equivalent of viscosity), and on thermal capacity, $m.c_p$;
- ✓ the **Nusselt** number is a dimensionless parameter that represents convective heat transfer coefficient and, if known, provides the convective heat transfer coefficient directly. Correlations of experimental data with Nusselt number are made with other dimensionless numbers: the **Prandtl** number is used for all Nusselt number correlations and is the ratio of kinematic viscosity, ν, to **thermal diffusivity**, α; **Reynolds** number is used for Nusselt number correlations in forced convection situations; **Grashof** number is a dimensionless ratio that represents buoyant and viscous forces, and it is used for Nusselt number correlations in natural convection situations;
- ✓ Nusselt number correlations are available in heat transfer texts from experimental evidence;
- ✓ radiant heat transfer is by electromagnetic radiation due to the release of energy from molecules excited by heat energy. It is determined according to the **Stefan–Boltzmann** Law of radiant heat release. For radiative heat transfer, two bodies at different temperatures having direct sight of each other release and receive heat dependent on their temperatures;
- ✓ the **emissivity** of a surface limits the amount of radiant energy released and modifies the Stefan–Boltzmann Law for surfaces that are not thermally black. A truly thermally black surface is not necessarily coloured black, but has the property that it releases all its radiant energy due to its temperature;
- ✓ it is possible to calculate **combined mode** heat transfer with conduction, convection and radiation, but in cases where radiant heat transfer is present, the calculation is not directly solvable due to the fourth power of T in the Stefan–Boltzmann Law.

2.7 Heat exchangers

Heat exchangers allow exchange of heat between two fluid streams without mixing them – perhaps the most familiar is the car radiator, which cools engine cooling water with a forced draught of cool air. There are two main types of heat exchanger – the **recuperator** and the **regenerator** as shown schematically in Figure 2.52

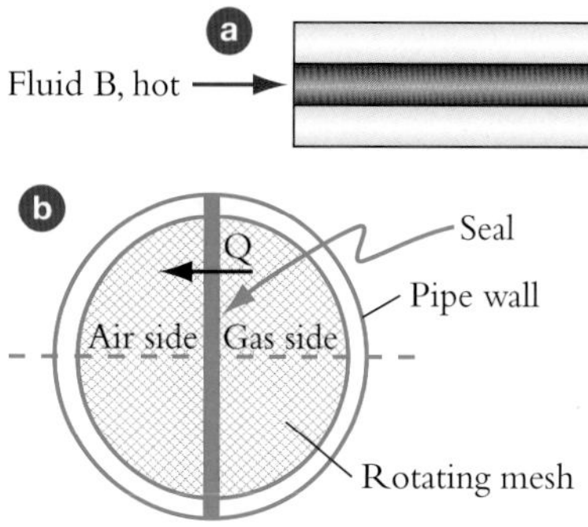

Figure 2.52 Schematic diagram of (a) a recuperator; and (b) a regenerator

The **recuperator** (Figure 2.52(a)) is the most familiar and common type of heat exchanger; it is used in condensers, evaporators and boilers, and for cooling liquids and gases. A hot stream is contained in tubes that pass through a vessel containing the cooler fluid, or vice versa in terms of hot and cold. The fluids do not make physical contact with each other. They exchange heat through the walls of the heat exchanger across a temperature gradient that varies as progress is made through the recuperator.

The **regenerator** (Figure 2.52(b)) is commonly used for preheating atmospheric air for combustion in the furnace of a power station using the exhaust gases at approximately 200°C as they travel to the chimney stack. There is a dividing wall centrally down the tube in which the regenerator is mounted, with the hot gases on one side of the wall and the cold gases on the other. The rotating mesh, or frame of a material with a significant heat capacity, has elements that absorb heat on the hot side, and release heat to the cold side, alternately taking heat from one side to the other as the mesh rotates. In this way, in the power station, the exhaust gases leave at a lower temperature, and the combustion air has been warmed, which means that more heat has been used to generate power. Any heat loss in a power station is a corresponding loss in power generation, so any recovery of potentially lost heat is useful.

This section demonstrates how to calculate the heat exchange in recuperators and from this point they will be referred to simply as **heat exchangers**. The simplest type of heat exchanger is the **shell-and-tube** configuration, which implies an inner tube containing one of the two fluids, contained within an outer shell (which is often just a bigger tube) as shown in Figure 2.52(b). These are the simplest type to analyse, because we can estimate reasonably well the temperature profile, from our knowledge of heat transfer in convective–conductive modes from the previous section, and hence the local and overall heat transfer coefficient. Figure 2.53 shows a simplified geometry of the shell-and-tube; there are obviously two directions that each fluid can pass in, and this gives rise to a very important distinguishing characteristic – whether the flow is **parallel flow**, in which the fluids travel in the same direction along the geometry, or **counter flow**, in which the fluids travel in opposite directions. From the figure, it can be seen that the corresponding temperature profiles are significantly affected by the flow directions.

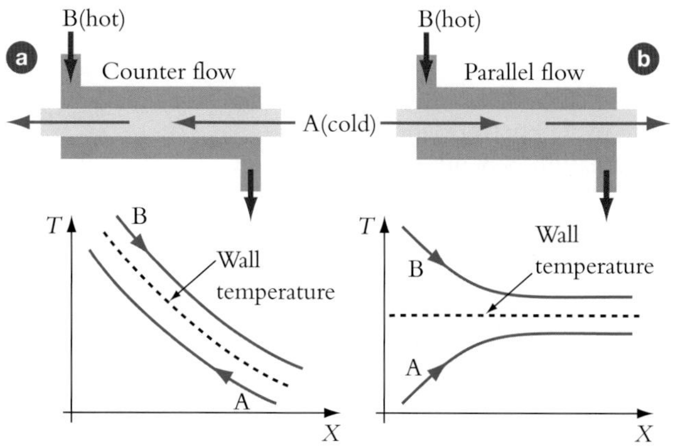

Figure 2.53 Schematic of the geometry and temperature profile in a shell-and-tube heat exchanger with counter (a) and parallel (b) flow

Energy balance

In Figure 2.54, a single-pass shell-and-tube heat exchanger is seen, which is the simplest possible configuration: an inner fluid is contained within the tube and an outer fluid is contained within the shell. By the principle of conservation of energy, heat transfer from the hot fluid = heat transfer to cold fluid:

$$\dot{Q} = \{\dot{m}c_p(T_{\text{in}} - T_{\text{out}})\}_{\text{hot}} = \{\dot{m}c_p(T_{\text{out}} - T_{\text{in}})\}_{\text{cold}} \quad \textbf{(2.78)}$$

where $\dot{m}c_p$ is the **thermal capacity rate** of the fluid, that is, its capability of storing heat per degree of temperature rise per second.

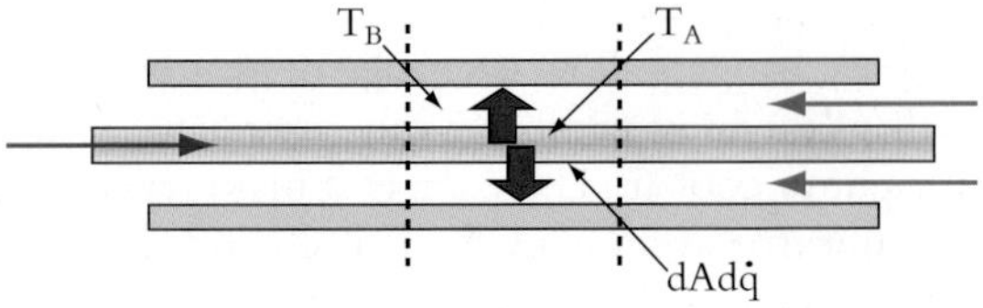

Figure 2.54 An elemental length of a shell-and-tube heat exchanger for the consideration of the heat transfer in it

Heat transfer calculation

With the developed formulae from heat transfer theory we can, in the steady state, set up an Ohm's law analogous formula for the heat transfer in terms of the overall temperature difference between the hot and cold fluids at a particular position on the heat exchanger, and the thermal resistance at that point, which is dependent on the convective heat transfer from the internal fluid to the wall of the tube, the conduction of heat from the inside of the wall to the outside of the wall, and the convection from the outside of the wall to the external fluid. Hence,

$$\dot{q} = \frac{-\Delta T_{\text{overall}}}{\dfrac{1}{h_A A_A} + \dfrac{\ln r_o/r_i}{2\pi k L} + \dfrac{1}{h_B A_B}} \quad \textbf{(2.79)}$$

Subscript A represents the inner (tube bound) fluid and subscript B represents the outer (shell bound) fluid. Since the wall of the tube is usually thin compared to the pipe diameter, the pipe inner surface area is very nearly the same as the pipe outer surface area, so we can say that $A_A = A_B = A$, and the thermal resistance can be represented instead as an overall heat transfer coefficient for that area:

$$\dot{q} = UA(\Delta T_{\text{overall}}) = UA(T_{\infty A} - T_{\infty B}) \quad \textbf{(2.80)}$$

in which

$$UA = \frac{1}{\Sigma R_{TH}}$$

and where the ∞ symbol indicates bulk temperature of the inner or outer fluid, i.e. the temperature of each fluid far enough from the surfaces to be independent of temperature gradients. Considering the element shown in Figure 2.54, there is the contribution of heat transfer for the elemental length of the heat exchanger:

$$d\dot{q} = -dA.U(T_B - T_A)$$

And the total heat transfer is

$$\dot{q} = \Sigma d\dot{q} \quad \textbf{(2.81)}$$

summing over all elements on length. This energy balance gives the overall performance of the heat exchanger, the transfer of heat between the two fluids. The calculation depends on the performance of the heat exchanger, which is due to the heat transfer. Theory tells us how to calculate heat transfer across the surfaces, but the situation is complicated by the stream-wise gradients of temperature.

Note: Even if U is constant, $(T_B - T_A)$ varies along the heat exchanger. This must be taken into account. There are two approaches available for analysis of heat exchangers – **logarithmic mean temperature difference (LMTD)** and **effectiveness-number of transfer units** (ε-NTU) – which will be elaborated on in the following sections.

LMTD method for sizing heat exchangers

In a parallel- and counter-flow single-pass tubular heat exchanger, as shown in Figure 2.54, the temperature difference between the two fluid streams varies as the flow proceeds from inlet to outlet. The LMTD (log mean temperature difference) analysis gives the equivalent constant temperature difference for calculation of heat transfer, such that it is the same as the actual heat transfer with the temperature varying with progression through the heat exchanger. LMTD finds a representative temperature dependent on the overall temperature differences at inlets and outlets, which requires the heat exchange between the two separated fluids. It is then used to find the area of a heat exchanger that is required to achieve the desired heat transfer. Alternatively, if a heat flux is required and temperatures are not known, these can be found by iteration.

Referring to Figure 2.55, the LMTD formula is expressed in terms of the temperature difference between the fluid in the tube and in the shell at position a and that at position b:

$$\text{LMTD} = \Delta T_m = \frac{\Delta Ta - \Delta Tb}{\ln(\Delta Ta/\Delta Tb)} \quad \textbf{(2.82)}$$

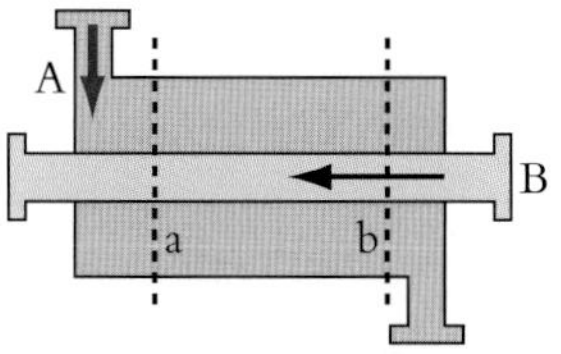

Figure 2.55 Single-pass shell and tube heat exchanger

where 'a' refers to inlet and 'b' to outlet conditions or to any two far apart points along the heat exchange surface:

$$\Delta T_a = (T_B - T_A)_a \quad \textbf{(2.83)}$$

$$\Delta T_b = (T_B - T_A)_b \quad \textbf{(2.84)}$$

These are the overall temperature differences between the two streams. Equations (2.83) and (2.84) apply to, and are the same for, parallel and counter-flow arrangements. The LMTD formula is derived from consideration of elemental heat exchange and can be easily verified.

For **sizing heat exchangers** (i.e. working out required size for given flow rate and temperature change), the LMTD approach is used to give the area required from:

$$A = \frac{\dot{Q}}{U\Delta T_m}$$

Rating of heat exchangers (i.e. working out the temperature change in a given heat exchanger) can be achieved by iteration using the LMTD approach when there is a known overall heat transfer coefficient, U, which must be determined by calculation or from tabulated data.

Worked example

Use of LMTD to rate and size a heat exchanger.

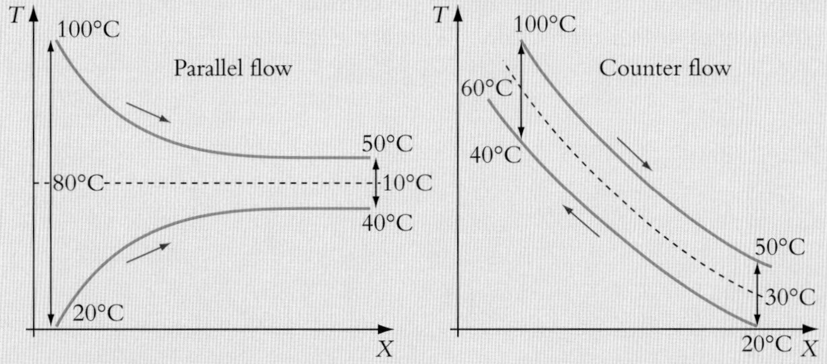

Figure 2.56 Calculation of LMTD

Calculate the LMTD for parallel- and counter-flow cases that have the same inlet and outlet stream temperatures. The cold enters at 20°C, and exits at 40°C.

Referring to Figure 2.56, the temperature profile is distinctly different in the counter-flow and parallel-flow operation. In the parallel case,

$$\text{LMTD} = \frac{(80 - 10)}{[\ln(80/10)]}$$

$$\text{LMTD} = 33.7°\text{C}$$

and in the counter flow case,

$$\text{LMTD} = \frac{(60 - 30)}{[\ln(60/30)]}$$

$$\text{LMTD} = 43.3°\text{C}$$

Note: The parallel arrangement has the lower LMTD – therefore needs a larger surface area for the same heat transfer, i.e. a larger heat exchanger. This shows the advantage of the counter-flow arrangement.

Calculation of heat transfer rate

$$Q = U\,.\,A\,.\,\text{LMTD}$$

The overall heat transfer coefficient (HTC) is determined from

$$UA = \frac{1}{\dfrac{1}{2\pi r_o h_o L} + \dfrac{\ln\dfrac{r_o}{r_i}}{2\pi k L} + \dfrac{1}{2\pi r_i h_i L}}$$

The 'i' and 'o' refer to the inside and outside surfaces of the tube. If the heat transfer coefficient variations along the tube are large, then an average U value is used.

This is usually unnecessary. To a first approximation, evaluate U at

$$T = \frac{(T_a + T_b)}{2}$$

The variation is because the heat transfer coefficient may vary with fluid temperatures since the fluid properties may vary with temperature. In a shell-and-tube heat exchanger, with the tube carrying hot fluid and the shell cold fluid, the conduction term can usually be neglected if the thickness of the wall is small and the conductivity high.

The LMTD analysis requires the outlet temperatures of the fluid streams to be known. This may necessitate an iterative solution involving guessed outlet values. This is the main disadvantage of the approach.

Procedure:

(1) guess outlet temperatures;

(2) calculate LMTD;

(3) calculate heat transfer rate from step 2;

(4) calculate outlet temperature for step 1;

(5) repeat until converged.

Comparison of the advantages of parallel and counter flow

The counter-flow type is the most compact for a given heat transfer rate due to the temperature profile.

The parallel-flow type gives a lower T_{max} for the tube wall temperature.

In more complex designs of similar type, the 'tube' flow passes through the heat exchanger more than once, and the 'shell' side flow can be directed across the tube to increase turbulent heat transfer. Generally, heat exchangers of this type are called **shell-and-tube heat exchangers**.

Figure 2.57 shows a typical shell-and-tube heat exchanger. As shown in the schematic, the tube is the section between the two end caps, and the end caps themselves contain fluid that passes through the tubes. There are two passes of the tube fluid through the shell fluid in this case, and the central mark on the opened end of the cap shows how the two halves of the cap are separated to allow for this.

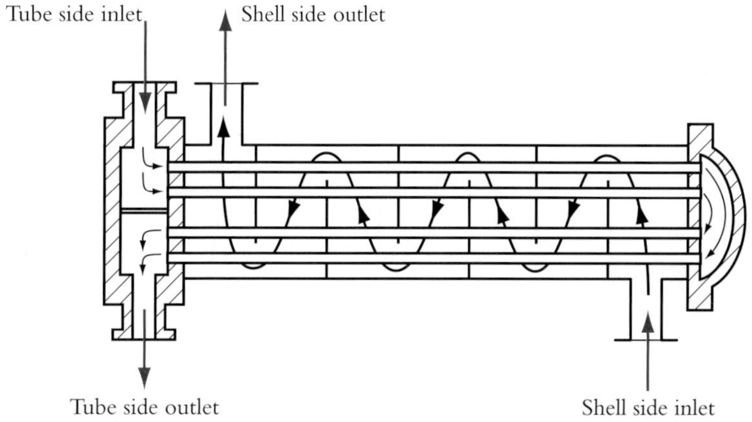

Figure 2.57 **A typical shell-and-tube heat exchanger used in the hot water system at the University of Nottingham**

Cross-flow and multipass

The stream temperature differences are more complicated in cross-flow or multipass configurations. The effects of cross-flow and multipass are accounted for by introducing a **correction factor** F such that:

$$Q = UAF\Delta T_m \quad \textbf{(2.85)}$$

The factor F is derived from calculations similar to the derivation of LMTD or read from graphs. Figure 2.58 shows a matrix heat exchanger with cross-flowing fluids; the resulting temperature profiles are complex. The LMTD method relies on prior experimental data.

A simpler and more direct approach for finding the temperatures is the ε-NTU method.

Correction factor – LMTD

Kays and London (1984) catalogues the many experiments about heat exchangers conducted by the US Navy in the 1950s, forming the basis for modern heat exchanger design and theory. They present methods for calculating the behaviour of many different heat exchanger types and configurations. The methods described in this section come from that text, where you can read the full description.

The correction factor allows us to use the LMTD type of calculation for more complex geometries. We must introduce two new variables:

$$R = \frac{T_1 - T_2}{t_2 - t_1} \text{ and } P = \frac{t_2 - t_1}{T_1 - T_2}$$

T_1 and T_2 are the temperatures of one stream and t_1 and t_2 are the other stream.

Using this method from Kays and London, we can read the correction factor from charts in the book, knowing P on the horizontal axis, a selection of curves for various values of R, and corresponding correction factor, F from the vertical axis. Three cases are illustrated in Figure 2.59 from Bowman, Mueller and Nagle (1940) in Bejan (1993).

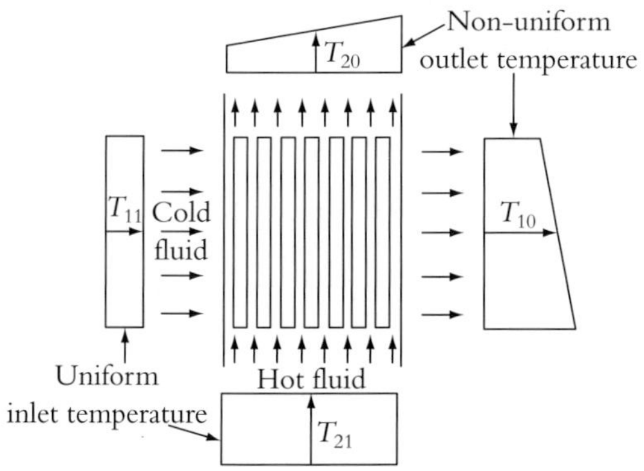

Figure 2.58 **Cross-flow heat exchanger and the resulting temperature profiles at outlet of the two streams**

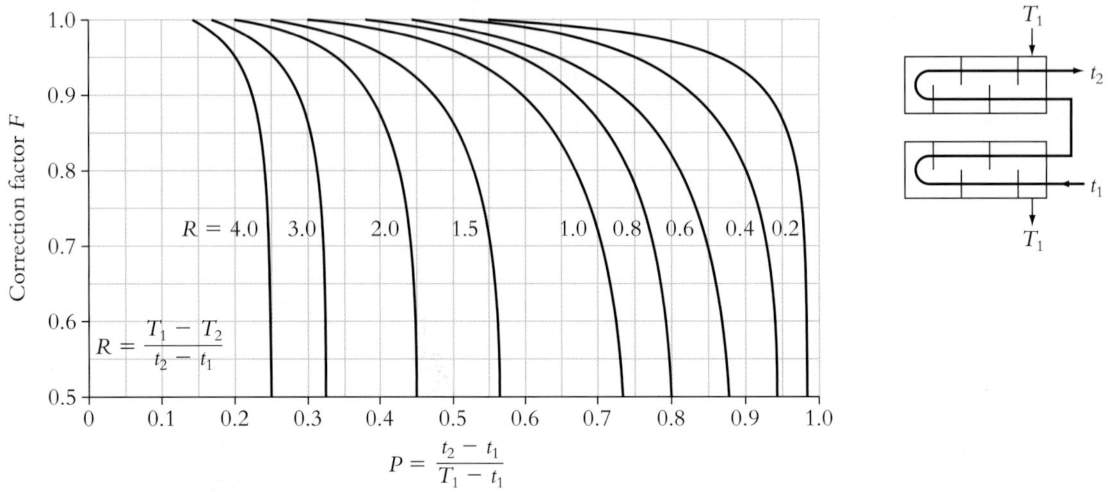

This is for two shell passes, four tube passes and multiples of them

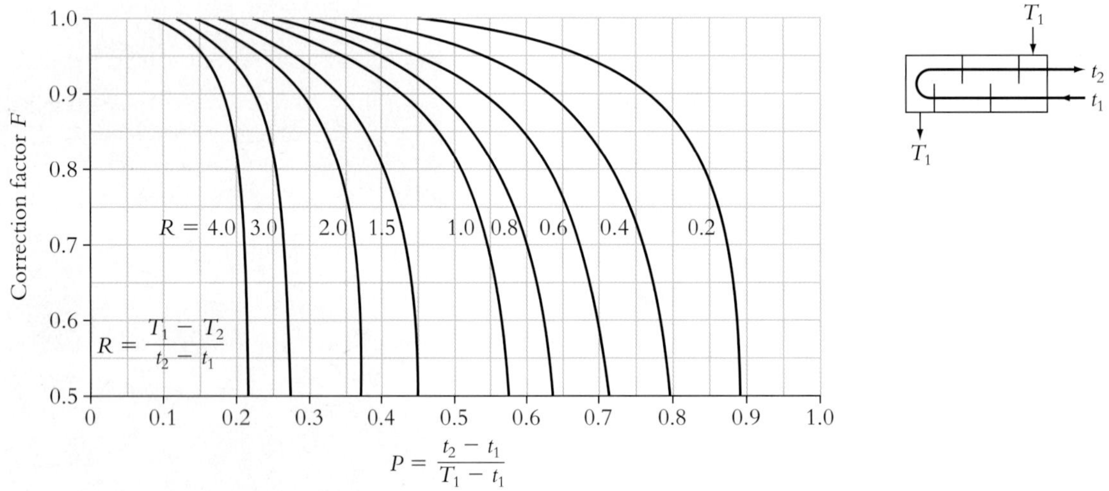

One shell pass, two tube passes and multiples of them

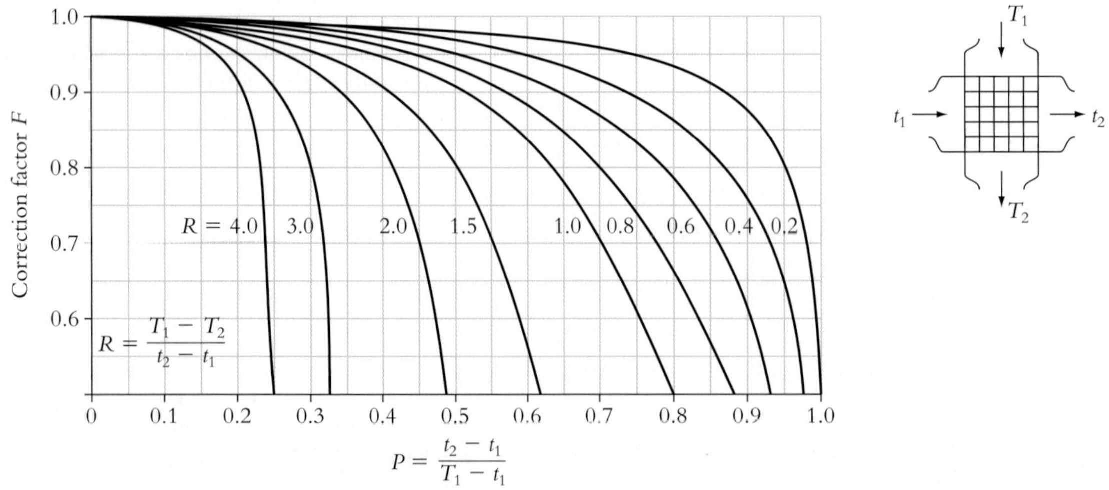

Figure 2.59 Cross-flow, both streams unmixed (i.e. the fluid passing in each direction is contained in a matrix of passages) (R.A. Bowman, A.C. Mueller and W.M. Nagle, 1930, 'Mean temperature difference in design', *Transactions of the ASME*, vol. 12, 417–422)

If factor P tends to zero, the stream having temperatures t_1 and t_2 has a change of phase, i.e. $t_1 \rightarrow t_2$.

If R tends to zero, the stream having temperatures T_1 and T_2 has a change of phase.

If a stream has a phase change, the capacity rate is effectively infinite, because the fluid will absorb heat without changing temperature. These are the two limits on the charts.

Figure 2.60 shows what is meant by the terms used with cross-flow heat exchanger: the corrugations keep fluid moving as an **unmixed stream** as it passes through the channel, the open passage has no corrugations and fluid is free to move across the full face of the channel and is called a **mixed stream**.

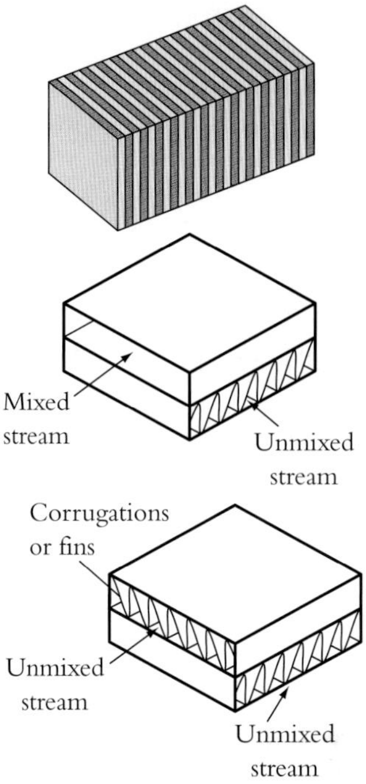

Figure 2.60 Examples of what is meant by cross-flow heat exchangers (Bejan, 1993)

Care using LMTD

It is important to understand the **capacity rate** of the fluids in the heat exchanger, i.e.

$$C_{cold} = \dot{m}_{cold}c_{p,cold} \text{ and } C_{hot} = \dot{m}_{hot}c_{p,hot}$$

These determine the magnitude of temperature changes for a certain heat transfer. It is important to always use the formula for LMTD:

$$\dot{q} = UA\,\Delta T_m \quad (2.86)$$

together with the heat capacity formulae:

$$\dot{q} = \{\dot{m}c_p(T_{in} - T_{out})\}_{hot} = \{\dot{m}c_p(T_{out} - T_{in})\}_{cold}$$

Using equation (2.86) on its own suggests that the heat transfer does not depend on the capacity rate of the fluids. The limit of the heat exchanger capacity is when condensation of the hot fluid or evaporation of the cold fluid occur across the heat exchanger, as in a boiler for example.

Effectiveness ε-NTU method of rating heat exchangers

This is an alternative method to LMTD for working out the temperature change in a heat exchanger, which examines the thermal conductance, UA, as well as the capacity rates, C_{hot} and C_{cold}; this method introduces two dimensionless groups, the **Number of heat Transfer Units**:

$$\text{NTU} = \frac{UA}{(mc_p)_{min}} \quad (2.87)$$

where $(mCp)_{min}$ is the smaller capacity rate and the effectiveness, ε:

$$\varepsilon = \frac{\text{actual heat transfer rate}}{\text{maximum heat transfer rate}} = \frac{\dot{q}}{\dot{q}_{max}}$$

in which

$$\dot{q}_{max} = C_{min}\Delta T_{max} \quad (2.88)$$

What is the maximum possible heat transfer rate?

This can best be considered in graphical terms as it asks the question: if the UA could be increased to whatever we wanted, what is the maximum achievable heat exchange between the two fluids?

As the size of the exchanger increases, the fluid with the smaller capacity rate, which has the steeper gradient (in this case illustrated in Figure 2.61; this is C_{cold}), exchanges heat with the hot fluid until the limit when the size of the heat exchanger is such that it leaves at the temperature of the hot fluid. This is the limit of the heat transfer. The minimum fluid may be either the hot or the cold fluid. In this case

$$q_{max} = C_{cold}(T_{hot,in} - T_{cold,in})$$

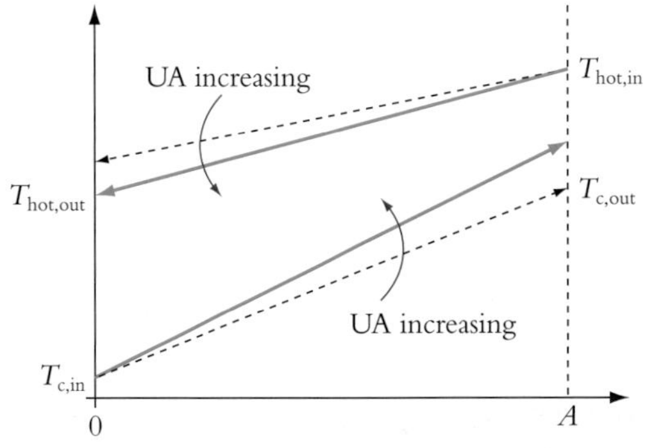

Figure 2.61 Demonstrating what the maximum possible heat transfer between two fluids in a heat exchanger is

Calculations

These usually entail evaluating effectiveness, ε from knowledge of the capacity rates, C, and NTU values, and then evaluating q_{actual}. No iteration is required. Expressions for NTU in terms of ε can be derived by analysis of the rearranged LMTD heat transfer equation:

$$\ln(\Delta T_a/\Delta T_b) = \frac{1}{q}(\Delta T_a - \Delta T_b)UA \tag{2.89}$$

in terms of effectiveness and NTU, by manipulation of equations 2.87–2.89. For example, a parallel flow gives the formula

$$NTU_{\text{parallel flow}} = -\frac{\ln[1 - \varepsilon(1 + C_{min}/C_{max})]}{1 + C_{min}/C_{max}} \tag{2.90}$$

which can be manipulated to give effectiveness:

$$\varepsilon = \frac{1 - \exp[-NTU(1 + C_{min}/C_{max})]}{1 + C_{min}/C_{max}} \tag{2.91}$$

A formula is required for each configuration of heat exchanger. Fortunately, rather than using the formulae, data are presented in charts that can be read. If $C_{min}/C_{max} = 1$, the capacity rates are balanced – a balanced heat exchanger. If $C_{min}/C_{max} = 0$, one stream, C_{max}, has phase change at nearly constant pressure, and effectively has infinite capacity rate.

Worked example

A heat exchanger has an overall heat transfer coefficient of 120 W/m2K. Superheated steam enters the tubes at 250 °C and exits the exchanger at 400 °C. Hot gases on the shell side enter at 900 °C and exchange heat with the steam. The capacity rate of steam is 200 000 W/K and that of the gases is 300 000 W/K. What is the effectiveness? Using the formula for NTU for a counter-current shell and tube, determine the required surface area of the heat exchanger. What is the temperature of the gases leaving the exchanger?

$$NTU = \frac{\ln\left(\dfrac{1 - \varepsilon\, C_{min}/C_{max}}{1 - \varepsilon}\right)}{1 - C_{min}/C_{max}}$$

The actual heat transfer is $q = C_{steam} \times (400 - 250) = 30$ MW. $q_{min} = C_{min}(900 - 250) = 130$ MW. The effectiveness, $\varepsilon = q/q_{min} = 30/130 = 0.23$. We know that $NTU = UA/C_{min} = 120 \times A/200\,000$. From the formula for a single counter-flow tube and shell exchanger:

$$NTU = \frac{\ln\left(\dfrac{1 - 0.23 \times 20\,000/30\,000}{1 - 0.23}\right)}{1 - 20\,000/30\,000}$$

Therefore, $A = 0.285 \times 200\,000/120 = 475\ m^2$. The gases exit temperature is determined by: $q = C_{gases}(900 - T_{exit})$, $30 \times 106 = 300\,000 \times (900 - T_{exit})$, and $T_{exit} = 800$ °C.

Knowing

$$UA/C_{min} = NTU$$

the chart gives effectiveness, which then yields q_{actual} from knowing q_{max}. Inspection of the chart in Figure 2.62(a) shows that if the area, C_{min}, and overall heat transfer coefficient are known, then the effectiveness of the heat exchanger is given from the experimental data represented by the chart. This is for two shell passes, four tube passes and multiples of them.

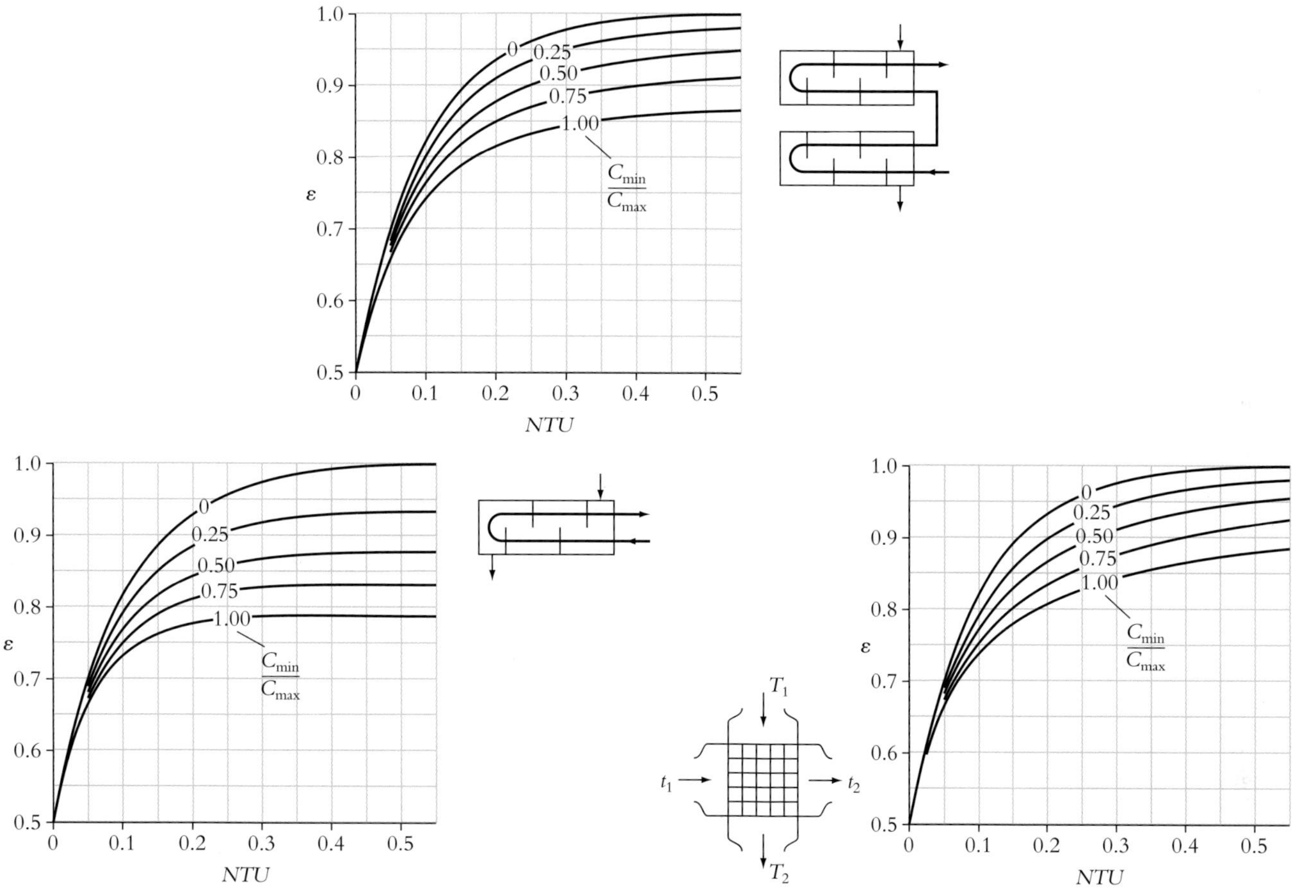

Figure 2.62 Effectiveness-NTU charts for the same geometries as the LMTD illustrated in Figure 2.59 (R.A. Bowman, A.C. Mueller and W.M. Nagle, 1930, 'Mean temperature difference in design', *Transactions of the ASME*, vol. 12, 417–422)

Calculation of the pressure drop

Calculation of the pressure drop depends on detailed knowledge of the path that the fluid follows in flowing through the heat exchanger. Expansions, entries, and bends all need to be included.

There is a pump factor that needs to be taken into account. For a liquid it is:

$$\dot{W} = \frac{1}{\eta_p}\frac{\dot{m}}{\rho}(p_{in} - p_{out}) \qquad \textbf{(2.92)}$$

in which η_p is the isentropic efficiency of the pump.

For a gas it is

$$\dot{W} = \frac{1}{\eta_p} mc_p T_{in}\left[\left(\frac{p_{out}}{p_{in}}\right)^{R/cp} - 1\right] \qquad \textbf{(2.93)}$$

Intensification of heat transfer is usually accompanied by an increase in pressure drop. The Δp across the heat exchanger can be an issue with complex flow paths. Work must be supplied to deliver fluid through this pressure gradient by a pump.

Fouling factors

Over time, fluids tend to leave residual traces on the surfaces that they flow over. The thin layer of deposits from a particular fluid flow in the heat exchanger has a conductive resistance on the shell (s) and tube (t) sides. This contributes to overall U_s (heat transfer coefficient).

$$\dot{q} = \frac{\Delta T}{\frac{1}{h_s A_s} + \frac{1}{h_t A_t}} \text{ and } \dot{q} = U_s A_s LMTD \tag{2.94}$$

to give

$$U_s A_s = \frac{1}{\frac{1}{h_s A_s} + \frac{1}{h_t A_t}}$$

Over time, the boundary surfaces corrode and acquire a scale coat, while the fluids gather impurities, so fouling factors (r) are used for both shell-and-tube, giving:

$$\frac{1}{U_s} = \left(\frac{1}{h_s} + \frac{1}{h_t}\frac{A_s}{A_t}\right) + \left(r_s + r_t\frac{A_s}{A_t}\right) \tag{2.95}$$

Fouling factor is additive to ideal heat transfer coefficients, and the effect of the fouling depends on fluids involved.

Typical fouling factors are illustrated in Table 2.5 showing the effect of various fluids.

Temperature of heating medium water temperature	Up to 115°C 50°C or less		115–205°C Above 50°C	
Water velocity	1 m/s and less	Over 1 m/s	1 m/s and less	Over 1 m/s
Water types				
Distilled water	0.0001	0.0001	0.0001	0.0001
Sea water	0.0001	0.0001	0.0002	0.0002
Brackish water	0.0004	0.0002	0.0005	0.0004
City or well water	0.0002	0.0002	0.0004	0.0004
River water, average	0.0005	0.0004	0.0007	0.0005
Hard water	0.0005	0.0005	0.0009	0.0009
Treated boiler feed water	0.0002	0.0001	0.0002	0.0002
Liquids				
Liquid gasoline, oil and liquefied petroleum gases				0.0002–0.0004
Vegetable oils				0.0005
Caustic solutions				0.0004
Refrigerants, ammonia				0.0002
Methanol, ethanol and ethylene glycol solutions				0.0004
Gases				
Natural gas				0.0002–0.0004
Acid gas				0.0004–0.0005
Solvent vapours				0.0002
Steam (non-oil bearing)				0.0001
Steam (oil bearing)				0.0003–0.0004
Compressed air				0.0002
Ammonia				0.0002

Table 2.5 Representative fouling factors [m^2K/W] related to various fluid flows. From Chenoweth (1990) in Bejan (1993).

Learning summary

By the end of this section you will have learnt:

- ✓ heat exchangers are classified firstly as **regenerators** and **recuperators**. The recuperator, which is the type involving tubes carrying one fluid surrounded by another fluid at a different temperature, is the concern of this unit;
- ✓ recuperators are defined secondly by direction of flow as counter-flow or parallel-flow of the two fluids involved;
- ✓ thermal **capacity rate** is the specific heat capacity of a fluid multiplied by the amount of mass of the fluid considered, which is the mass flow rate of the fluid. It determines the temperature rise for a given heat input;
- ✓ heat transfer theory shows how an elemental approach could be taken to the analysis of a particular heat exchanger, but there are two methods used with overall heat transfer coefficients, which allow for simplified analysis in general cases;
- ✓ since the temperature varies throughout a heat exchanger, the **logarithmic mean temperature difference** (LMTD) is used with a known overall heat transfer coefficient to work out either the heat transferred for a given sized heat exchanger or the size of a heat exchanger given the heat transfer required. It can be shown that the LMTD is the correct mean to calculate the overall heat transfer;
- ✓ the collection of experimental results by Kays and London (1984) provides a method for modifying the LMTD for more complex heat exchangers than the simple shell-and-tube configuration;
- ✓ it is important to remember that the capacity rate of each fluid determines its temperature and heat transfer;
- ✓ the **effectiveness (ε-NTU) method** relates the known data of overall heat transfer coefficient and minimum of the two capacity rates in the NTU and relies on experimental data from heat transfer texts to relate this to the ratio of actual heat transfer to maximum possible heat transfer between the hottest and coldest temperatures available. Again this relies on the catalogue of data by Kays and London;
- ✓ the cost of heat exchangers is the pressure drop for a given size. This must be considered in a practical heat exchanger and is determined from fluid mechanics calculations;
- ✓ **fouling** often occurs in practical installations and the alteration to heat transfer coefficient must be considered;

2.8 Vapour power cycles

The idea of a vapour power cycle is to use a condensable vapour in a heat engine to create power from a source of heat. Referring to Figure 2.63 we can see that it is similar to a heat pump running in reverse:

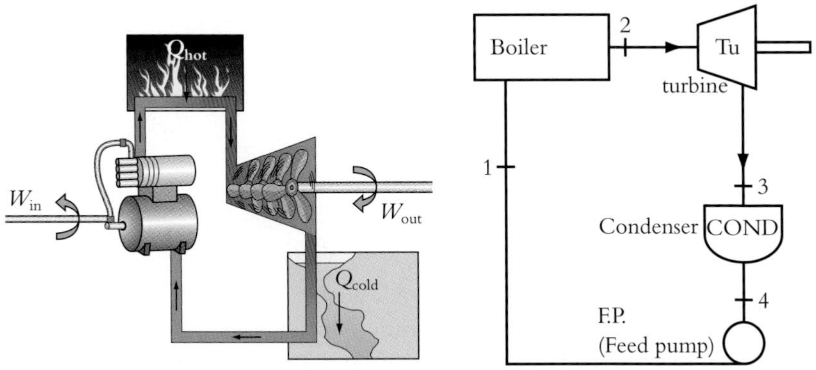

Figure 2.63 Schematic diagram of a simplified vapour power cycle

- **4–1** work is put in by a pump (at the position of the throttle in the heat pump) to drive the liquid-fluid into a high-pressure part of the circuit (rather than expanding into the low-pressure part of the circuit);
- **1–2** significant heat is added in the high-pressure side to convert the liquid into a vapour;
- **2–3** the vapour passes through a turbine (at the position of the compressor in the heat pump), driving the blades round and producing work out of the turbine shaft, and in the process losing pressure back to the low-pressure side of the circuit;
- **3–4** before running around the circuit again, the vapour-fluid is cooled in a heat exchanger to convert it fully back into a liquid.

The process is used for power production and requires that

(1) the working fluid is a condensable vapour;

(2) the power cycle consists of a series of steady flow processes.

For the remainder of this section steam will be considered as the condensable vapour. The analysis is simplified by the assumption that kinetic and potential energy changes are negligible compared with the change in enthalpy. The steady flow energy equation for change of state between points is therefore

$$Q + W = H_2 - H_1$$

Since the purpose of the cycle is to produce usable work, it is useful to see where the work is coming from and where the expended energy used to create the work is being applied. The useful measures are **thermal efficiency**, **work ratio**, and **specific steam consumption**.

Thermal efficiency

Closed cycle plants used for power generation have a thermal efficiency of between about 35% and 50%, where:

$$\eta_{th} = \frac{\text{net work output}}{\text{external heat supplied}} = \frac{\dot{w}_{out} - \dot{w}_{in}}{\dot{q}_{hot}} = \frac{(h_2 - h_3) - (h_1 - h_4)}{h_2 - h_1}$$

The net work output is measured in MW. It is useful to quantify the total energy output as electrical energy from the generator set that is mounted at the end of the turbine shaft, and this is often denoted as MW(e) to signify electrical power output. In practice it is very close to the work output from the shaft of the turbine since the efficiency of electrical generators is very high (usually over 99%).

Work ratio (r_w)

The efficiency of the cycle is not useful on its own – we want to know how much we can get out of the process. Work occurs at two points in the cycle – driving the liquid-fluid up to the high pressure, which is an energy cost in the system, and releasing energy from the high pressure vapour-fluid on its passage through the turbine to deliver work to the electrical generator set. The ratio of these is expressed as the work ratio:

$$r_w = \frac{\text{net work output}}{\text{gross work output}} = \frac{(h_2 - h_3) - (h_1 - h_4)}{h_2 - h_3} \qquad \textbf{(2.96)}$$

Equation 2.96 describes how much work we get out for work we put in – there is no consideration of heat here. High r_w values are desirable, which indicates low sensitivity of efficiency to irreversibilities. Modern steam plants have $r_w \approx 0.98$, while for gas turbines cycles $r_w \approx 0.45$, the major difference being due to high compression work involved with gases.

Specific steam consumption (SSC)

The efficiency and work ratio do no inform us of the amount of steam delivered to the turbine, which indicates the size of the plant. We use the specific steam consumption (SSC):

$$SSC = \frac{\dot{m}_{steam}}{\dot{W}_{net}} \quad \textbf{(2.97)}$$

which is usually expressed in kilograms of steam per hour per kilowatt of power produced (kg/hr/kW = kg/kWhr). A low SSC is good and implies the need for a smaller plant (that is lower capital cost).

Cycles

The simplified cycle shown in Figure 2.63 can be modified to improve performance on all three metrics by application of aspects of the second law of thermodynamics. The following applications are used routinely to improve power station performance and are regular parlance, so much so that the names of the inventors who first considered the implications of the process changes are synonymous with the processes. The technology applied in power stations increases the work produced for the size of plant and the amount of work that can be produced from the heat available. The theory explains the Carnot cycle, the Rankine cycle, superheating, reheat and regeneration.

Carnot cycle – the most efficient thermodynamic cycle

The Carnot cycle comprises four reversible processes: two isothermal and two adiabatic as shown in Figure 2.64, and these are:

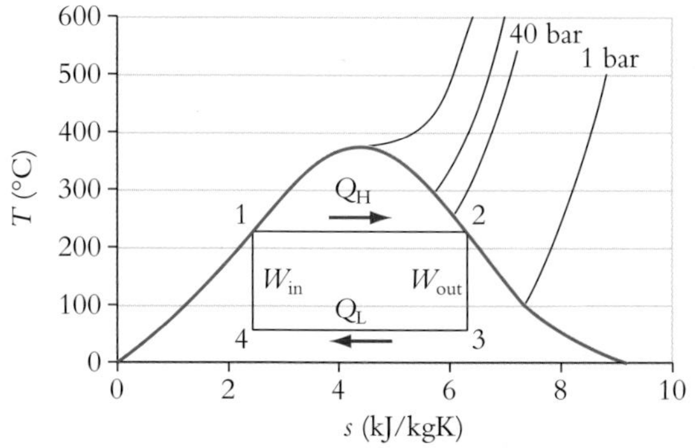

Figure 2.64 Representation of the Carnot cycle on the temperature–entropy diagram for steam

1–2 isothermal expansion by adding heat, Q_H. In this part of the cycle, heat is added from the surroundings to convert the water to low quality steam (i.e. high liquid droplet content or cloudy steam) and then to high quality steam (low liquid content), and eventually to saturated steam (i.e. no water content, or dry steam) by evaporating the liquid and thus maintaining a constant temperature in the steam. The heat input is denoted Q_H, since the heat is added in the high pressure, or (relatively) hot side of the circuit

2–3 adiabatic expansion by passing the steam through a turbine, producing work. This assumes that the turbine is perfect, i.e. it has no mechanical-friction or fluid-friction losses. The steam temperature drops from high on the high pressure side, T_H to low on the low pressure side, T_L

3–4 isothermal compression by removing heat, Q_L. Heat is removed from the system at this point in the cycle to the surroundings, and the temperature is constant as steam gradually condenses as it travels, forming small droplets, which may condense on the pipe walls of the system. Note that a high water by mass content in the steam is not necessarily a high water by volume content since the density is high (about 1000 kg/m^3, compared to steam at about 0.5 kg/m^3)

4–1 adiabatic compression by passing the low quality steam through a pump, until the point at which the steam-water mixture in the system becomes saturated water with no steam content. Temperature increases from T_L to T_H.

The constancy of the temperature as heat is added in the high-pressure side and heat is removed in the low-pressure side is, thermodynamically speaking, a good thing because we know that the Carnot efficiency improves with increasing the difference between the hottest and coldest reservoirs (surroundings), and we know that if heat is exchanged in a cycle with more than one reservoir at different temperatures, then the efficiency falls from the maximum available from the highest temperature and lowest temperature reservoirs to somewhere in between. The only way to add heat at constant temperature is with phase change and it is achieved in steam by working in the mixture region.

The usefulness of the *T-s* diagram is apparent from this process as shown in Figure 2.64 in which it can be seen that the constant entropy processes are vertical lines, and variation from the ideal will be seen if the lines incline to the vertical.

The Carnot cycle is the basic cycle to aspire to, however, since it is an impracticable cycle, and although the thermal efficiency is highest for a given high temperature, higher temperatures are achievable by working outside the mixture region. The Carnot cycle is impractical for the following reasons: the turbine will physically struggle to expand the vapour in the mixture region since at high speeds the droplets of steam that form as the quality of the steam falls will impact on the blades and cause erosion, which is very costly to repair; it is difficult to ensure that condensation reaches the exact point required for the saturated liquid at the point entering the compressor on the left side of the process, and the compressor work is quite high since it is not pumping a very dense liquid but a significantly less dense, and therefore higher specific volume, fluid, reducing efficiency, which decreases the work ratio. As a point of interest, if the cycle is reversed it becomes the Carnot refrigeration cycle.

Worked example

Find the SSC, work ratio and thermal efficiency of the Carnot cycle operating between 100 bar and 0.3 bar.

For SSC, we require the work done to compress the fluid, and the net work done. For the Carnot cycle, we need the enthalpy change between 4 and 1 on the diagram in Figure 2.64, as follows:

$$h_{f,100\text{bar}} = 1{,}408\ \text{kJ/kg}$$

$$s_{f,100\text{bar}} = 3.36\ \text{kJ/kgK}$$

Dryness fraction at 4 is from entropy $h_4 = h_1$:

$$s_{4f}(1 - x) + s_{4g}x = s_1$$

$$0.944(1 - x) + 7.767x = 3.36$$

$$x = 0.35$$

Therefore enthalpy at 4 is:

$$h_4 = 0.35h_{g0.3bar} + 0.65h_{f0.3bar}$$

$$h_4 = 0.35 \times 2{,}625 + 0.65 \times 289 = 1{,}106\ \text{kJ/kg}.$$

Therefore

$$W_{4\text{–}1} = 1408 - 1106 = 302\ \text{kJ/kg}.$$

The work out of the turbine between 2 and 3 is:

$$\Delta h = h_{g100bar} - h_3$$

h_3 is found by first getting the dryness fraction as above for constant entropy, which is 0.685, and then using that to get the mixture enthalpy, which is 1,889 kJ/kg. Therefore

$$\Delta h = 2725 - 1889 = 836\ \text{kJ/kg}.$$

SSC for Carnot is

$$\frac{1 \times 3600}{(836 - 302)} = 6.742\ \text{kg/kWh}.$$

r_W for Carnot is

$$\frac{(836 - 302)}{836} = 64\%.$$

For thermal efficiency, the external heat added is required. For the Carnot cycle, this is $h_{fg,100\text{bar}} = 1317$ kJ/kg. Therefore $\eta_{TH,Carnot} = \dfrac{(836 - 302)}{1317} = 40.5\%$.

(Using data from tables in Rogers and Mayhew, 1995.)

Rankine cycle – reducing compression work

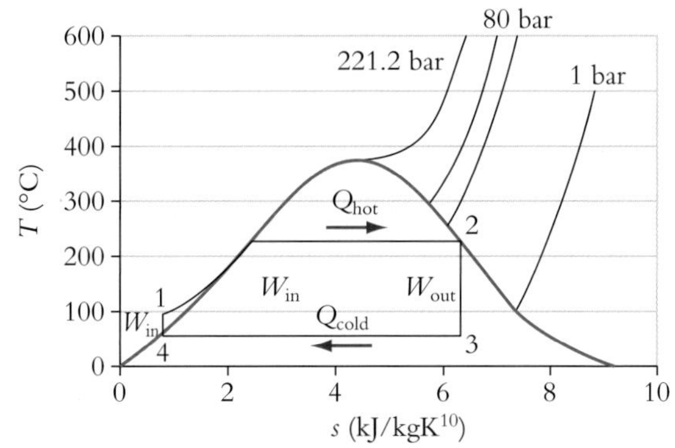

Figure 2.65 The Rankine cycle showing the small pump work in the liquid phase, which results in reduced loss

The Rankine cycle, shown on the *T-s* diagram in Figure 2.65, extends the condensation to the saturation line (process 3–4) before using a pump instead of a compressor, to increase pressure of liquid water (process 4–1) from the condenser pressure (typically a fraction of atmospheric pressure) to the boiler pressure (typically in the order of 100 bar). The mixture at outlet from the condenser is saturated liquid, so the feed pump work is greatly reduced since compressing liquid requires less work than for a liquid and vapour mixture. The work ratio is increased, because 1 and 4 are virtually coincident and, as a direct result of this, W_{in} is very small. For a flow process, the reversible process work is:

$$W = \int_4^1 v.dp$$

and since v [m^3/kg] is very nearly constant and not a function of pressure, the work is $v[p_1 - p_4]$. With v being the inverse of density, and knowing that water density is approximately 1,000 kg/m^3, an estimate of the work in changing the pressure from 0.3 bar to 100 bar is

$$0.001 \times [100 - 0.3] \times 10^5 = 9970\,\text{J/kg}$$

or 10 kJ/kg. In the Rankine cycle, thermal efficiency is reduced, because not all heat is added at the higher temperature, rather a significant amount is added at a temperature lower than the saturation temperature, but this is unimportant with regard to the increase in work output. This can be illustrated by using typical enthalpy values. SSC is also lower (i.e. improved) due to the increased work out per kg of steam during the cycle. These effects are seen in the following example.

Worked example

Compare the SSC, work ratio and thermal efficiency of the Rankine and Carnot cycles operating between 100 bar and 0.3 bar.

In the case of the Rankine cycle, the compression work is 10 kJ/kg of water. Using some of the results from the turbine work in the Carnot calculation above:

SSC for Rankine is

$$1 \times 3600/(836 - 10) = 4.891\,\text{kg/kWh}$$

Carnot was 6.742 kg/kWh.

r_W for Rankine is

$$\frac{(836 - 10)}{836} = 98.8\%$$

For Carnot it is 64%.

For thermal efficiency, the external heat added is required. For the Rankine cycle it is

$$h_{g100\text{bar}} - (h_{f,0.3bar} + 10)\ \text{kJ/kg} = 2725 - (289 + 10)$$
$$= 2426\ \text{kJ/kg}.$$

Therefore

$$\eta_{\text{TH},Rankine} = \frac{(836 - 302)}{2426} = 22\%.$$

The comparison shows that Rankine is a significant improvement over Carnot for SSC and r_W, but that the thermal efficiency is significantly affected.

Rankine cycle with superheat

To improve the cycle efficiency, the saturated steam leaving the boiler is **superheated**. Heat is applied at nearly constant pressure between 2 and 3 by passing the steam through pipes that pass through the furnace flue gases at the top of the furnace.

It is useful to note at this point where a practical implementation is considered, that the boiling of water to saturated steam occurs in tubes built into the walls of the furnace that completely fill the wall surface area, receiving mostly radiant heat transfer from the hot flame in the vicinity of the burners, before rising to a steam drum, where any liquid is passed back to the furnace wall pipes and the saturated steam proceeds into the superheating tubes. This raises the steam state into the superheated region.

Figure 2.66(a) gives some indication of the layout in the furnace. The boiler is actually in the walls – a large number of vertical pipes line the entire wall surface, and in these pipes the water is heated from state 1 to state 2 in Figure 2.65. The superheater circuit consists of yet more pipes suspended above the flame of the furnace in the middle of the hot rising gases. The gases having a higher temperature than the steam in the pipes means that there is a temperature gradient between the hot gases and the steam in the pipes of a few hundred degrees, and significant heat transfer occurs mainly by forced convective heat transfer.

Figure 2.66(b) shows the process diagram on the *T-s* chart. The use of modern materials allows T_3 values up to 650°C. The Carnot cycle is limited to $T_{max} \leqslant 374.15$°C (the critical temperature for H_2O), and superheating surpasses this temperature considerably. Some of the heat is supplied at $T > T_{sat}$, raising the average temperature above that of the purely Rankine cycle, and thus increasing the thermal efficiency. The other advantage of superheating is that the dryness fraction at exit from the turbine (4) is higher than in the Rankine cycle and consequently the turbine suffers less erosion by droplet impact.

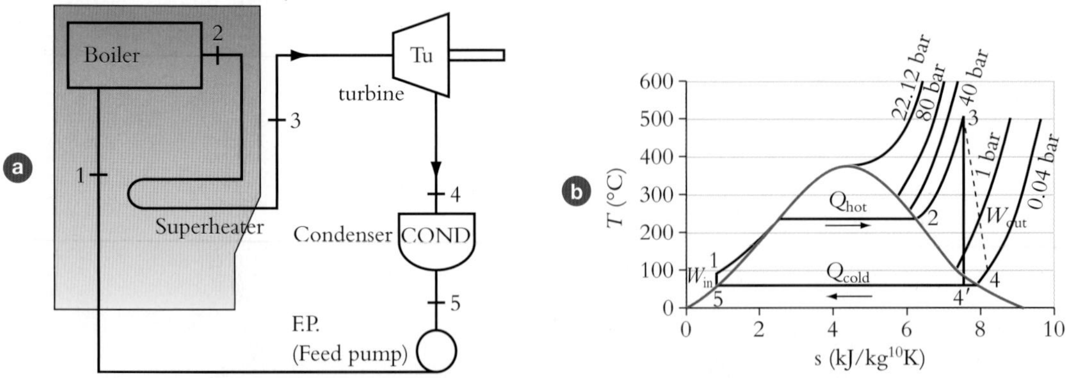

Figure 2.66 Schematic of the Rankine cycle with a superheater circuit

The real process in the turbine becomes important at this point as indicated in Figure 2.66, and the isentropic (constant entropy, reversible adiabatic) process 3–4 suffers an increase in entropy due to the second law and arrives at the same pressure line as 4, but with increased entropy and hence dryness fraction. The dotted line shows this irreversible process.

Worked example

Calculate r_W, η_{TH} and SSC for a Rankine-superheat cycle working between 100 bar and 0.3 bar, with 139°C of superheat.

The calculation of the liquid work is the same as in the previous example. At 100 bar $T_{SAT} = 311°C$, therefore superheated temperature is 450°C. The isentropic work out of the turbine requires enthalpy in:

$$h_{100\text{bar},450°\text{C}} = 3241\ \text{kJ/kg}$$

and the enthalpy at exhaust.

Calculate the exhaust enthalpy using the mixture fraction at constant entropy. Entropy at 100 bar and 450°C is 6.419 kJ/kgK. Using h_f and h_g at 0.3 bar as 289 and 2625 kJ/kg respectively, results in dryness fraction:

$$6.149 = 0.944(1 - x) + 7.767x$$

$$x = 0.762$$

and hence exhaust enthalpy:

$$h_3 = 2625x + 289(1 - x)$$

$$h_3 = 2069\ \text{kJ/kg}$$

hence work out is 3,241 – 2,069 = 1172 kJ/kg.

$$r_W = \frac{(1172 - 10)}{1172} = 99.1\%$$

External heat supplied is

$$h_{100\text{bar},450°\text{C}} - (289 + 10) = 3241 - 299 = 2942\ \text{kJ/kg}$$

$$\eta_{TH} = \frac{(1172 - 10)}{2942} = 39.5\%.$$

$$\text{SSC} = \frac{3600}{(1172 - 10)} = 3.098\ \text{kg/kWh}.$$

Use of reheaters

A 'reheater' supplies heat at a point part way through the turbine expansion (between turbines). The reheater is a tube bank that passes steam back through the furnace flue gases in the same way as superheaters, as shown in Figure 2.67. This avoids turbine expansion into low-dryness fraction conditions, which cause high turbine erosion by droplet impact.

The specific steam consumption (SSC) is reduced, hence a smaller power station is required. Figure 2.67(b) shows the effect of the reheat, which is similar to the superheater, but on a lower pressure line.

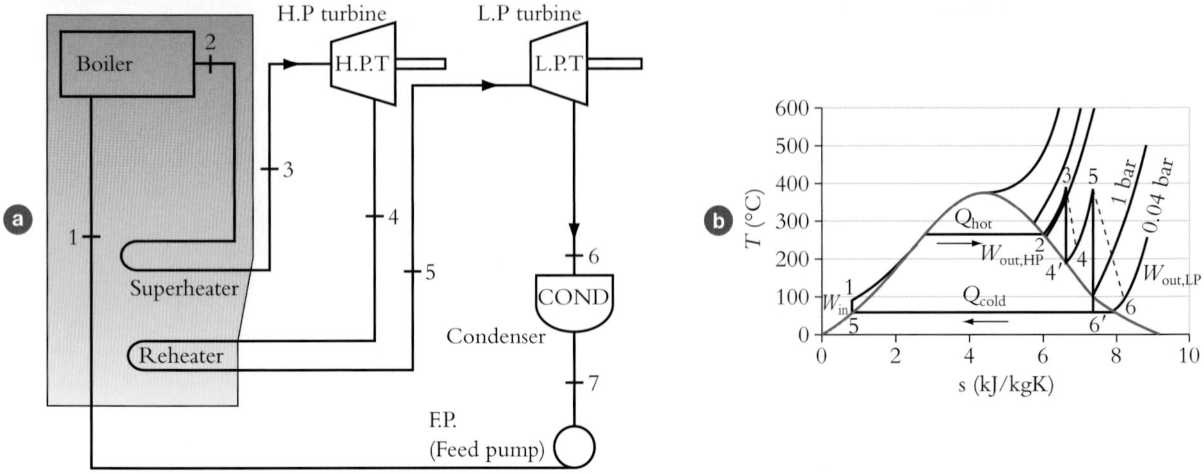

Figure 2.67 Schematic diagram of the process with a reheater circuit

The Mollier chart or enthalpy–entropy diagram for steam

This chart, a section of which is shown in Figure 2.68, is used to find the state of steam at entry to and exit from turbines. It is useful because the ideal process is isentropic – constant entropy – which means there is a vertical line from one pressure and temperature state to a lower, known pressure. When the **Mollier diagram** is used with **isentropic efficiency**, the true exit state from the turbine can be easily determined by calculating the actual exit enthalpy and hence temperature. The change in specific enthalpy that is revealed by this is used to find the actual work produced by the turbines in the steam cycle. The chart is a quick and reasonably accurate method for finding the change in properties, avoiding the need for interpolating values in the tables of steam data, which is the alternative method.

Use of feed heaters – regenerative cycle

Feed heaters are used to reduce the external heat supplied (i.e. the heat drawn into the circuit external to the steam pipes, that is heat taken out of the furnace) at temperatures below T_{max}. This is known as **regenerative heating**. They therefore increase the thermal efficiency. The reduction is achieved by bleeding off steam part way through the turbine expansion and using it to heat the feed water flow to the boiler.

There are costs associated with this mode of operation since turbine work is reduced and specific steam consumption is increased. The plant is more complex (more expensive).

There is an advantage in regenerative heating in that the LP turbine doesn't have to be so big because it handles less steam.

Two types of feed heaters are used:

- **open** in which bleed steam and feed water are mixed irreversibly and adiabatically in a constant pressure process;
- **closed** in which a recuperative heat exchanger is used with no mixing of the separate streams.

In both cases, mass flow conservation and the steady flow energy equation (SFEE) are applicable, that is, inlet and outlet conditions are related by mass flow continuity and SFEE.

(Note: recuperative is analogous to a recuperative heat exchanger – where the fluids exchange heat through a wall with no mixing.)

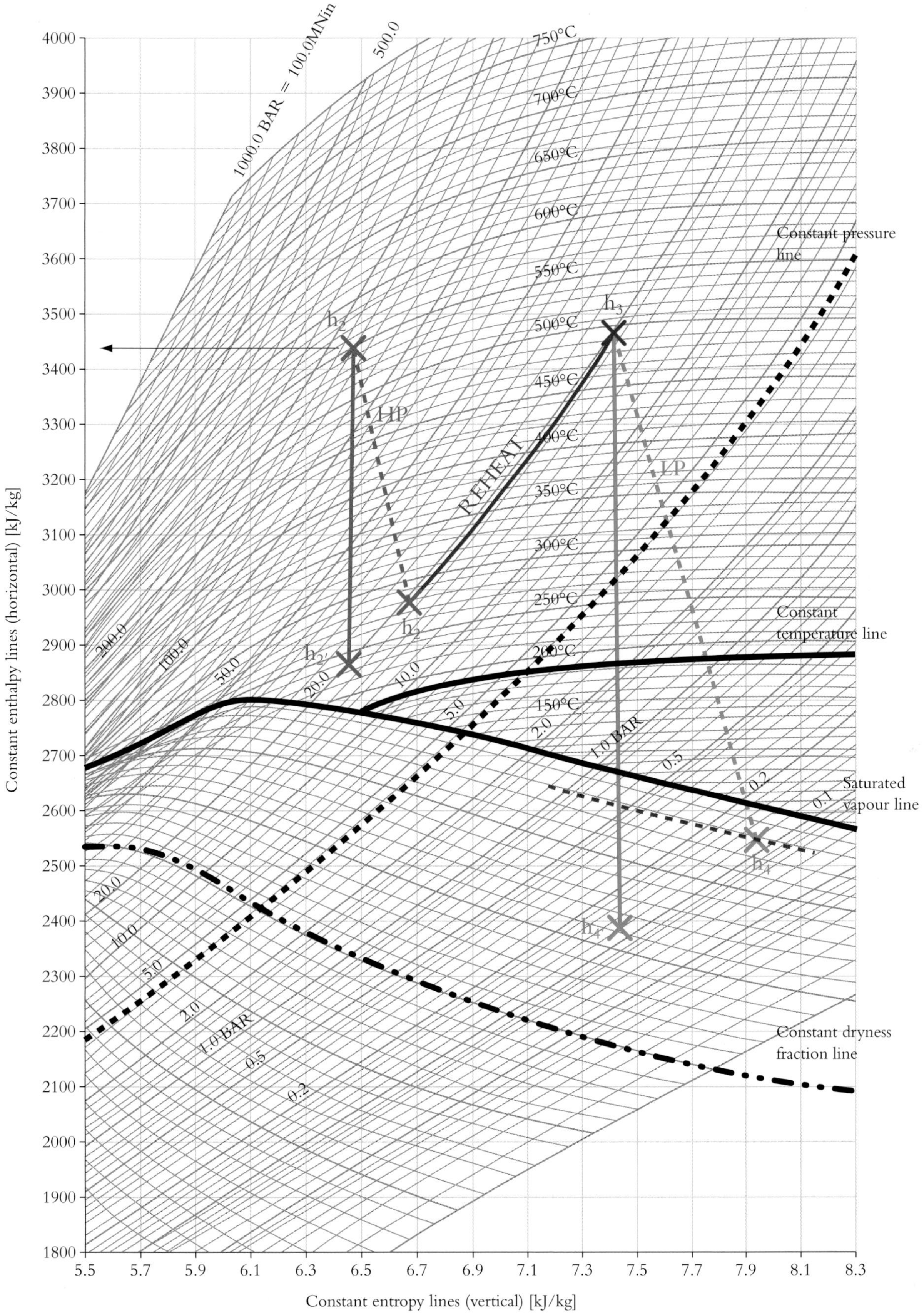

Figure 2.68 Section of the Mollier enthalpy vs. entropy diagram for steam process analysis for thermal energy systems. (PROATES® is a registered trademark of E.ON Engineering Ltd. Reproduced with the permission of E.ON)

Open feed heater

The mass flow of steam is split after the HP part of the turbine so that only a proportion of it goes on to the low pressure (LP) turbine.

The energy equation in the feed heater, assuming a perfectly insulated (i.e. adiabatic) chamber and referring to the state points indicated in Figure 2.69, becomes

$$y\dot{m}h_4 + (1 - y)\dot{m}h_7 = \dot{m}h_8$$

in which y is the diverted proportion of the total steam flow rate, $\dot{m}$, and h is specific enthalpy at the indicated locations 4, 7 and 8 in the circuit. States 4 and 7 mix irreversibly, but at constant pressure, to produce a fluid at state 8, which is usually in the saturated liquid state. The liquid reaching the boiler in the saturated state then has the requirement of receiving sufficient heat to change phase rather than having to heat up to saturation first. Since the pressure is below the boiler pressure, an extra feed pump is needed after the open feed heater.

$$p_4 = p_7 = p_8$$

$$\dot{m}_4 + \dot{m}_7 = \dot{m}_8$$

The principle of energy conservation gives

$$(\dot{m}h)_4 + (\dot{m}h)_7 = (\dot{m}h)_8$$

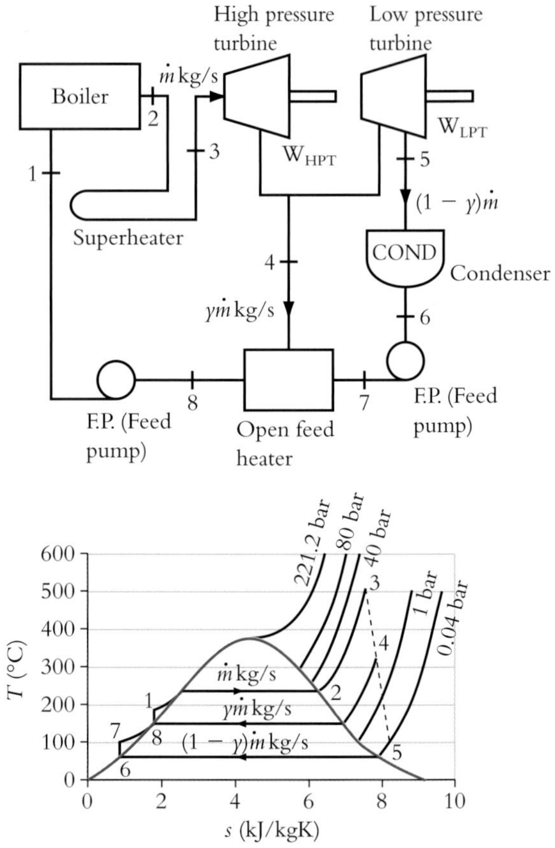

Figure 2.69 A steam circuit with open feed heater

The assumption of saturated liquid at the point before the pump is usually a reasonably good one to calculate the required bleed flow, $\dot{m}_4$. It can be shown that, for optimum efficiency, the pressure for bleeding is p_{sat} for the temperature halfway between the average boiler and condenser temperatures, but the proof is beyond the scope of this section. With the assumption of reaching saturation temperature, the process diagram appears as shown schematically in Figure 2.69. Note that this is not a true process diagram because the mass carried round each part is not the same. The feed pump between 6 and 7 increases pressure almost at constant entropy, before the liquid reaches the feed heater chamber, where it mixes with the wet steam, which has been bled out of the high pressure turbine, and the combined water leaves at saturated condition before being pumped up to the boiler steam pressure at 1. When calculating the work out it is important to remember that the low pressure turbine does not have the full flow rate of steam from the boiler but a quantity less by the bled steam.

Closed feed heater.

In this case, as indicated in Figure 2.70, some of the expanding steam is bled from the HP turbine and passed through the feed heater, which raises the feed water temperature partially towards the saturation temperature at p_1. The bled steam at 4 is cooled by heat exchange to state 8 and then throttled down to the condenser pressure. During the throttling process the expansion may lead to condensed water flashing off as steam at the lower pressure, for which a flash chamber is provided. The feed water temperature is raised from T_7 to T_1 during the process of internal heat transfer. Note that 1 is still liquid (subcooled liquid). State 8 is condensed liquid. The throttling process can form some steam, which can lead to erosion.

Flash chambers are used to avert erosion by slowing down the large volume of steam. There is no mixing between streams.

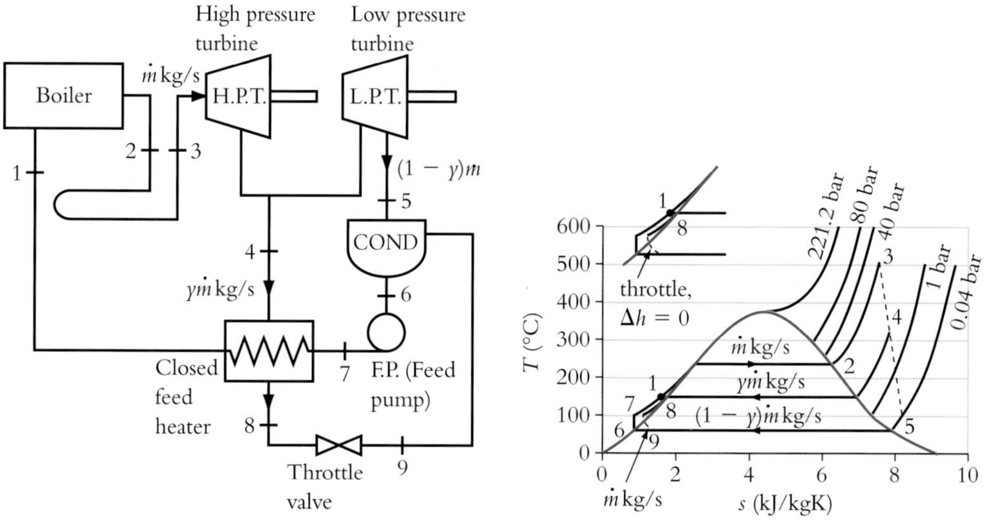

Figure 2.70 Schematic of regenerative feed water heating using a closed feed heater

The energy equation gives (assuming adiabatic conditions):

$$(\dot{m}h)_4 - (\dot{m}h)_8 = (\dot{m}h)_1 - (\dot{m}h)_7$$

Since enthalpy is constant from 8 to 9:

$$(\dot{m}h)_8 = (\dot{m}h)_9$$

Note: as before, this is not a true process diagram because the mass carried round each part is *not* the same. The inset picture in Figure 2.70 shows the detail at point 8 – the bleed steam condenses as it heats up the feed water from state 7 to state 1 before it enters the boiler. The spent bleed steam then goes through a throttle to the pressure of the condenser and joins the feed water. The throttle process is indicated by a dashed line since it is an irreversible process.

Cycles with power and process steam/heat as useful outputs – combined heat and power

On chemical process sites, there is often the requirement for steam and heat to be supplied in addition to there being a need for electrical power. Steam plant can be designed to meet all these demands. Steam is required at a much lower temperature than power steam. It would be wasteful to have a heating-only plant.

Efficiency is redefined as, for example:

$$\eta = \frac{\text{net work} + \text{process heat}}{\text{heat supplied}}$$

Efficiencies are higher than for conventional plant (70% typical) since the heat used for the process can be included in the benefit of the system.

It is convenient to use plant for power and then extract heat from the condenser. The condenser is maintained at a relatively high pressure in order that $T_{SAT} = T_{REQUIRED}$. The turbine exhausting to the relatively high pressure condenser is called a 'back pressure turbine'. This is practical when heat and power are fairly steady and well matched. When higher power is required, a 'pass-out turbine' can be used. This bleeds steam at an intermediate pressure to a heat exchanger before being fed back to the boiler. The amount of bled steam can be controlled.

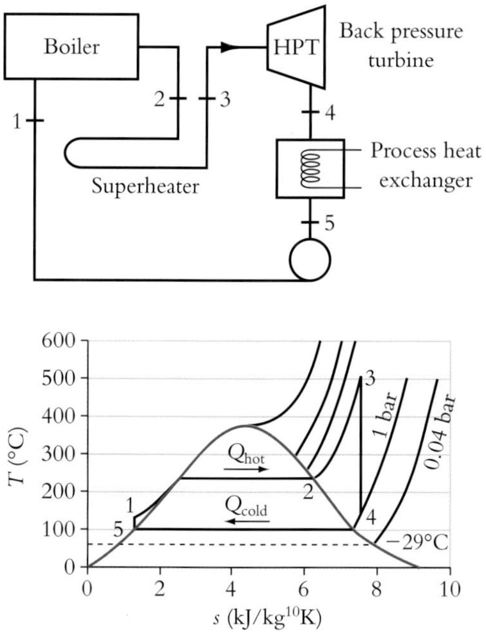

Figure 2.71 Back pressure turbine

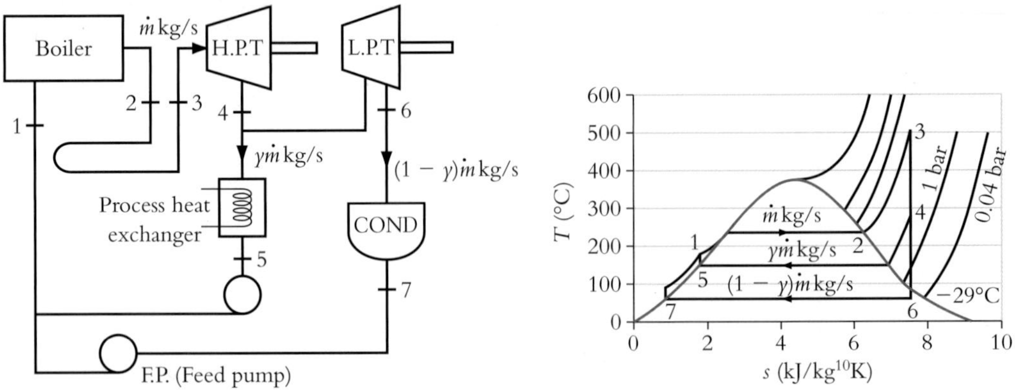

Figure 2.72 Pass-out turbine

Learning summary

By the end of this section you will have learnt:

- ✓ a vapour power cycle operates on a condensable fluid in a closed cycle, receiving heat from a hot reservoir, usually a boiler, to evaporate the fluid before expanding through a turbine to a lower pressure in order to produce work; it is then condensed to remove heat at a low pressure to a cold reservoir. A pump then increases the pressure to the high pressure side of the circuit prior to the heating part of the cycle;
- ✓ enthalpy changes between state points on the cycle are used to calculate heat and work from the first law in the form of the steady flow energy equation;
- ✓ since vapour power cycles are widely used for power production, there are measures of the effectiveness of employment of the energy input to obtained output power. **Thermal efficiency** is the ratio of work output to externally supplied heat, i.e. the heat supplied from the hot reservoir. **Work ratio** is the ratio of net work (work output less work in) to work output, and gives an indication of the loss to compression. In the case of steam vapour power cycles, the **specific steam consumption** is the mass flow rate of steam per unit of power output, measured in kg per kW-hour;
- ✓ the basic cycle has been improved by altering it to improve the effectiveness. The basic cycle is the **Carnot cycle**, which uses evaporation between saturated liquid and saturated vapour and which has expansion and compression within the vapour-liquid mixture region. This is the most thermally efficient cycle given any maximum and minimum temperature available, and is based on the idea of the **Carnot efficiency**. The **Rankine cycle** is an improvement to the Carnot cycle, by simplifying the condensation part of the cycle such that all fluid is in the saturated liquid state, and the pumping work is thus significantly reduced. The average upper temperature can be increased by employing a **superheat** part of the circuit after initial boiling, thus increasing the overall thermal efficiency; this also improves the turbine operation by reducing expansion into the mixture region. **Reheat** in the circuit is included after the initial turbine expansion, in order to avoid entering the mixture region significantly and is used in coal-fired power stations;
- ✓ the **Mollier chart** or h–s diagram is a convenient representation of the processes in the vapour power cycle since it directly yields enthalpy changes, which are the heat and work exchanges with the working fluid. The process in the turbine is close to isentropic, and it is a simple exercise to compare the isentropic performance with the real performance by use of the **isentropic efficiency**;
- ✓ **feed heating** is often employed, using some of the heat in the steam to heat the water being fed from the condenser to the pump. This improves the thermal efficiency by reducing the heat required from the boiler;
- ✓ feed heaters can either mix the steam with the feed water, in an open feed heater or pass the fluids through a recuperator, in which the fluids are separate but exchange heat, in a closed feed heater;
- ✓ **combined heat and power** is used where a quantity of heat is required for chemical processing in a factory, or where domestic heating is required for long periods of time. The thermal efficiency is artificially improved by including some of the heat lost as useful output since it is put to good use. In this case, steam is taken at higher temperature out of the final turbine, at a point where there is still useful heat in the steam.

2.9 Reciprocating internal combustion engines

Internal combustion engines were developed in the 19th century, producing petroleum- and diesel-fuelled engines that would subsequently find their most common and well-known application in the automotive industry, for power units for motor vehicles, as well as for small-scale power generation. In this section, an overview of reciprocating engines is given.

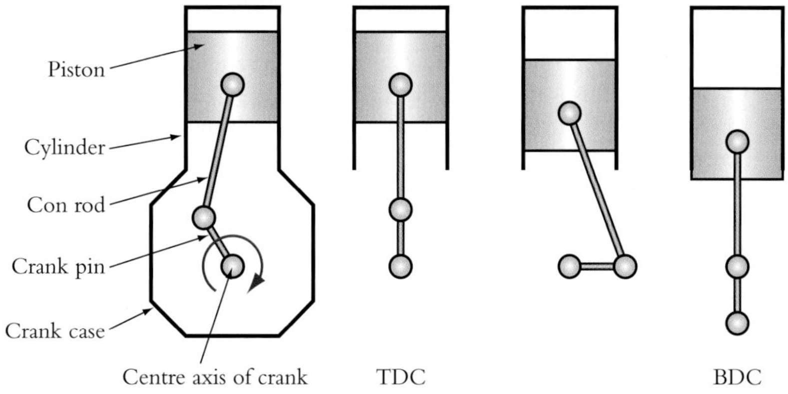

Figure 2.73 Schematic of a reciprocating engine

The reciprocating engine mechanism, shown schematically in Figure 2.73, consists of a piston that moves in a cylinder and forms a movable gas-tight plug, a connecting-rod and a crankshaft, similar to the construction of the reciprocating compressor. If the engine has more than one cylinder then the cylinders and pistons are identical, and all the connecting rods are fastened to a common crankshaft. The angular positions of the crank pins are such that the cylinders contribute their power strokes in a selected and regular sequence. By means of this arrangement the reciprocating motion of the piston is converted to a rotary motion at the crankshaft. There are many types and arrangements of engines and their classification defines them as described in the following.

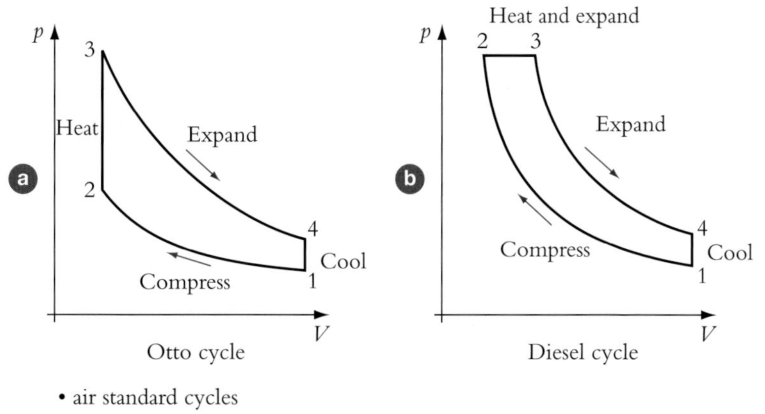

- air standard cycles
- compress and expand isentropically

Figure 2.74 Air standard cycles representing the Otto and Diesel cycles

Otto and Diesel cycles

Petrol and **combustible gases** have high volatility and are appropriate for **spark ignition** (SI) engines and a thermodynamic cycle based on the **Otto cycle**. Petrol is a complex mixture of distilled organic oil, the closest chemical to which is **octane**, or $\mathbf{C_8H_{18}}$, which leads to the definition of the **octane number**, defining the quality of the fuel. The combustion is started by spark ignition and allows for controlling timing of fuel detonation. The **air standard** Otto cycle is shown in Figure 2.74(a). An air standard cycle considers the processes in an ideal state, in which the heat is supplied and removed externally (rather than generated within the cylinder as it is in practice) and a single working fluid in a closed cycle is considered. The cycle consists of **isentropic compression** (1–2) assuming no heat transfer through the cylinder walls and a rapid process; **constant volume heating** (2–3); the piston moves down and **isentropic expansion** occurs (3–4); **heat removed** (4–1), and the cycle repeats. The power out of the cycle is the area in the cycle loop.

Diesel and **fuel oils** have relatively low volatility and are appropriate for **compression ignition** (CI) and a thermodynamic cycle based on the **Diesel cycle**. Since the fuel is a complex mixture of distilled organic oil, the chemical composition is not precisely definable, but a close approximation is **cetane**, or $\mathbf{C_{16}H_{34}}$, and this is the ideal diesel fuel leading to the definition of **cetane number** defining the quality of the fuel. The combustion is started by compression ignition, which works by increasing the pressure to such a high level that the temperature is significantly raised, sufficient to commence a rapid deflagration or combustion, as compared to the explosive reaction occurring with more volatile fuels. The **air standard Diesel cycle** is illustrated in Figure 2.74(b). It consists of **isentropic compression** (1–2); **constant pressure heat addition** (2–3); **isentropic expansion** (3–4) and **constant volume heat removal** (4–1).

The practical implementation of these cycles are the spark ignition engine, in which the air and fuel are mixed before compression and the compression ignition engine, in which the air only is compressed, and the fuel is injected into the compressed air that is then at a sufficiently high temperature to initiate spontaneous combustion.

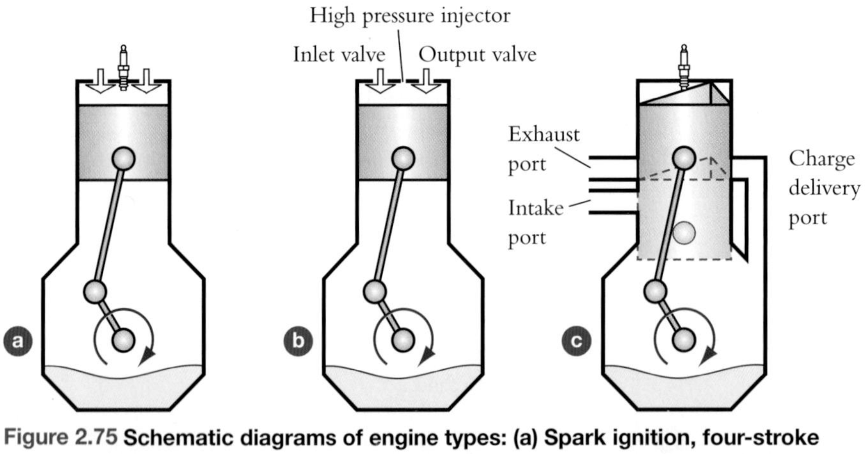

Figure 2.75 Schematic diagrams of engine types: (a) Spark ignition, four-stroke engine, (b) compression ignition, four-stroke engine, (c) spark ignition, two-stroke engine with BDC position indicated to show clearance of charge delivery and exhaust ports

Cycle processes arrangement

Figure 2.75 shows the practical differences between three engine types, which between them represent the common engine cycles used. The stroke of the piston is the distance it moves from the position most extreme from the crankshaft to that nearest it. This takes place over half a revolution of the crankshaft. In petrol engine practice, the extreme positions of the piston are referred to as top dead centre (TDC), and bottom dead centre (BDC). In oil-engine practice they are referred to as outer dead centre (ODC) and inner dead centre (IDC) respectively. An engine that requires four strokes of the piston (that is two revolutions of the crankshaft) to complete its cycle is called a **four-stroke** cycle engine. An engine that requires only two strokes of the piston (that is one crankshaft revolution) is called a **two-stroke** cycle engine. In all reciprocating internal combustion (IC) engines the gases are induced into, and exhausted from, the cylinder through ports, the opening and closing of which are related to the piston position. In a two-stroke engine (Figure 2.75(c)) the ports can be opened or closed by the piston itself, but in the four-stroke engine a separate shaft, called the **camshaft**, is required; this is driven from the crankshaft through a two-to-one speed reduction. The two-stroke engine has a piston head that is shaped to ensure that the charge delivery is directed upward into the cylinder away from the exhaust port to prevent a short circuit to the exhaust. The four-stroke petrol engine requires inlet and outlet valves. The diesel engine requires a high pressure injection port, and outlet valves. There may be a number of valves and ports in each cylinder to improve fuel distribution and combustion.

Control of power output

Spark ignition engines are **quantity governed**. Air and fuel are mixed outside the cylinder, and the quantity of mixture induced is controlled by the throttle plate position. The throttle plate controls the air induced by restricting the inlet pipe, which may deliver to a distribution manifold or directly to each individual cylinder. Diesel engine are **quality governed**. They have minimum restriction to air flow into the engine (no throttle plate). Fuel is injected directly into the cylinder, under pressure, towards the end of the compression stroke. The amount of fuel injected controls the power output.

Combustion initiation

In the case of the **spark ignition engine** the air-to-fuel mixture is **ignited by spark**, typically at an angle of 10–40° before top dead centre (TDC) during the compression stroke. The **compression ratio** is approximately 8:1, such that at end of compression the gases in the piston are at eight times atmospheric pressure. The mixture ratio must be near stoichiometric (~14.5:1 by mass) for ignition by spark to occur. Current engines run at stoichiometric under most conditions, for efficiency of the emissions control system. The air–fuel ratio affects the power and economy – slightly richer causes increased power, slightly leaner causes improved fuel economy. A **throttle** or restriction on the air flow adjusts the air–fuel ratio.

In the case of the **compression ignition engine** fuel injection takes place near the end of the compression stroke. Combustion begins, as the fuel mixes with the air in the cylinder, by self-ignition when compression heating raises the mixture temperature to 800–900K. The compression ratio in a CI engine is greater than 12:1 in order to achieve the required temperature. These engines normally run at air–fuel ratios (AFR) in the range 20:1 to 25:1. They can run out to approximately 40:1 AFR. The air flow is not restricted or otherwise controlled upstream, the mixture being control by injection of the fuel at an adjustable rate.

Spark ignition engine

Referring to Figure 2.76, the characteristics of a real SI engine cycle can be identified as follows:

1–2 Induction stroke. Air/fuel mixture is drawn into the cylinder through the intake valve. The cylinder pressure is less than atmospheric due to losses across the valve, the intake manifold and the throttle plate.

2–3 Compression stroke. The mixture is compressed, and ignited by spark before TDC. Combustion proceeds at near constant volume at approximately 50° of crank rotation. Peak cylinder pressure occurs just after TDC, and is typically in the order of 25 bar for the **wide open throttle** condition (WOT).

3–4 Power stroke. The hot and high pressure combustion products do work on the piston, giving useful work output.

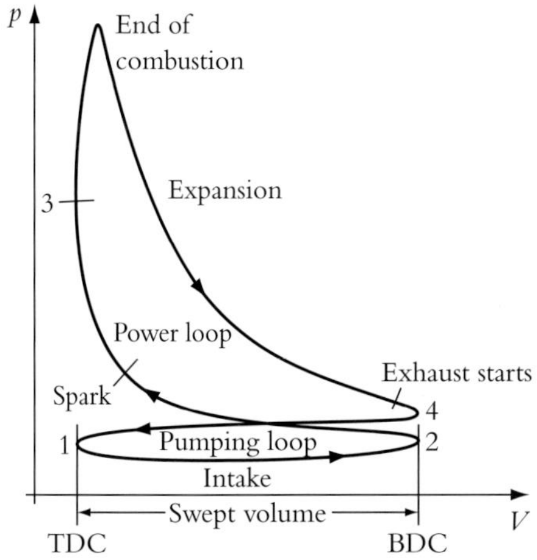

Figure 2.76 **Spark ignition cycle on the *p-V* diagram**

4–1 Exhaust stroke. The energy-depleted combustion products are expelled through the exhaust port into the exhaust manifold. Cylinder pressure is slightly above atmospheric, due to the restrictions on the gas flow of the valves and pipes downstream from the engine. Notice that energy is lost when the exhaust starts as a free expansion from a pressure significantly higher than atmospheric.

Overall, the cycle is distinctly different from the ideal Otto cycle, mainly because of the combustion process used to produce the heat internal to the closed volume, and the need to replenish the volume gases. The process is also distinctly different from the standard air Otto cycle.

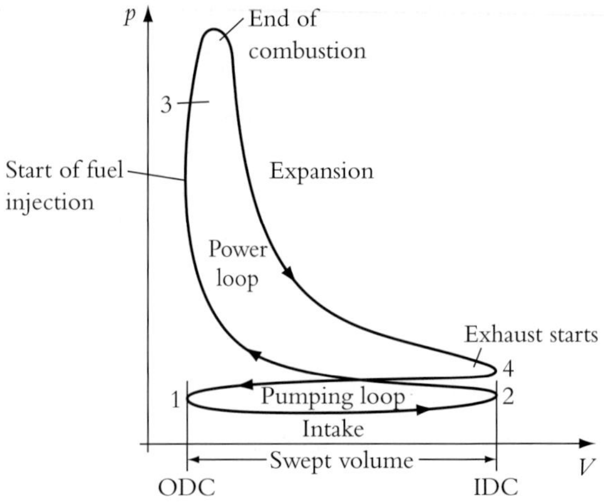

Figure 2.77 Compression ignition cycle on the *p-V* diagram

Diesel engine

1–2 Induction stroke. Air is induced into the cylinder – no fuel is present in the air at this stage.

2–3 Compression. The air only is compressed and fuel injection begins just before ODC and continues for about 20–30° of crank rotation. The injection is done at very high pressure in order to form a fine and quick atomized spray. The rail pressure has increased to the point where most efficient diesel engines now have up to 2000 bar. The combustion duration is longer than for an SI engine because it is nearer to constant pressure than constant volume. The rate of burn is controlled by the rate at which fuel evaporates and mixes with air in the cylinder. Peak cylinder pressure is in the order of 40 bar at full load.

3–4 Power stroke. The combusting mixture expands against the load applied to the engine.

4–1 Exhaust stroke. A stroke to remove the spent gases. Again notice the immediate loss of pressure at exhaust.

Notice that the cycle is again distinctly different from the ideal Diesel cycle due to the real combustion processes used to provide the heat and due to the induction and exhaust strokes.

Performance assessment

For spark ignition engines (and similarly for diesels), the main concerns are fuel economy at idle and part load, wide open throttle power output, and emission levels. Also important, but less easy to quantify, are the characteristics of the engine response to changes in demand (throttle opening or closing).

(i) **Indicated power** or **ip**: Rate of work done on the piston by the gas as determined from the *p-V* diagram. The top loop of the diagram is the power loop (positive work done by gases); the lower loop is the pumping loop (negative work done by gases). The indicated power is given by:

$$\text{ip} = (\text{imep}) \times LAN\ [\text{W}]$$

where L = stroke, A = piston area, N = cycles/sec, that is, (engine speed/2 for four-stroke engine), imep is the indicated mean effective pressure. From the diagram:

$$\text{imep} = \frac{\text{power loop area} + \text{pumping loop area}}{\text{diagram base length equivalent to swept volume}}$$

(imep values for spark ignition engines are ~5 bar at high load conditions).

The above is sometimes referred to as the net imep.

The gross imep is given by neglecting the pumping loop area.

(ii) **Brake power** or **bp**: Measured power output, Tω, where T = brake torque, ω = engine speed (rad/s)

(iii) **Mechanical efficiency**: $\eta_m = \dfrac{\text{Brake Power}}{\text{Indicated Power}}$ This is typically ~90% at full load. Losses are due to friction in the valve drive train, piston friction, bearings etc.

(iv) **Brake specific fuel consumption**: bsfc. $= \dfrac{\dot{m}_f}{\text{brake power}}$, where $\dot{m}_f$ is the fuel mass flow rate.

This is a measure of engine efficiency. It is more commonly used than **brake thermal efficiency**, $\eta_m = \dfrac{\text{Brake Power}}{\text{Indicated Power}}$

where Q_{net} is the lower calorific value of the fuel. (η_b is around 35% for modern engines).

$$Q_{net} = \Delta U_o$$

(v) **Volumetric efficiency**: $\eta_v = \dfrac{\text{Volume induced/cycle}}{\text{Swept Volume}}$ where the volume induced is referred to the conditions at inlet to the intake manifold. (WOT (wide open throttle) η_v is typically >80%).

Compression ratio and uncontrolled combustion

Thermal efficiency increases with increasing compression ratio (r):

$$r = \frac{V_{max}}{V_{min}} = \frac{V_{BDC}}{V_{TDC}} = \frac{V_{clearance} + V_{displaced}}{V_{clearance}}$$

This reflects the increase in the upper temperature of the cycle, which increases thermal efficiency (recall that $\eta_{Carnot} = 1 - \frac{T_{MIN}}{T_{MAX}}$). In addition, there is less residual gas left in the cylinder at the end of the exhaust stroke to dilute the fresh change for the next cycle. Spark ignition engines have *r* values in the range 9–10. These are unlikely to increase dramatically for the following reasons:

(1) **Production tolerances**. High *r* values mean larger variations from build-to-build for given tolerances on dimensions.

(2) **Spontaneous ignition**. If the *r* value is increased progressively, temperatures during compression will increase, and eventually the charge will start to burn spontaneously before spark ignition.

(3) **Knock**. This is a problem even on current engine designs, and is more likely to occur when *r* is increased. This is an uncontrolled combustion phenomenon. Combustion starts normally by spark ignition but unburnt gas ahead of the flame front is compressed. If the unburnt gas temperature is raised sufficiently it will self-ignite as in (2), producing a characteristic knocking sound.

In some circumstances, the flame front is accelerated and propagates across the combustion chamber with a shock wave. This detonation wave is reflected back and forth giving rise to high frequency noise. Knock phenomena can cause serious overheating and loss of efficiency.

(4) **Pre-ignition.** One effect of knock is to produce local hot spots on the combustion chamber wall that can initiate combustion before the spark on following cycles. Similarly, deposits on spark plugs and walls can also produce hot spots.

In CI engines, r is higher, but similar issues occur, due to the nature of the combustion. There is initial delay as fuel and air mix, followed by initial spread of the flame, and finally by combustion as the fuel is fully injected. **Diesel knock** may occur if the initial stages occur too rapidly.

Spark timing

In SI engines, maximum work output is achieved when combustion occurs at TDC, at constant volume. In practice, this is not achieved. The best spark timing gives peak pressures, which occur at approximately 15° after TDC. To achieve this, spark ignition occurs before TDC.

At a given nominal load, speed and AFR (air to fuel ratio), the optimum timing is normally taken as the 'minimum advance for best torque' (MBT timing). Usually this is determined experimentally, and timing requirements mapped out as a calibration for engines of a given design. Flame speeds are approximately proportional to engine speed, so although engine speed varies by an order of magnitude between idle and maximum engine speed, the duration of combustion requires a similar crank angle interval. Even so, timing is advanced from about 10° before TDC at idle and 30° before TDC at high speed, full load.

In addition, timing is adjusted to compensate for changes in load. Timing is advanced under light load conditions by up to about 20°. This compensates for charge dilution by residual combustion products, and low cylinder pressures and temperatures, which are not conducive to good combustion.

Modern electronic ignition systems perform adjustments electronically, but still use engine speed and manifold vacuum (or air flow rate) to identify timing requirements.

Fuelling systems

On spark ignition engines, fuel injection systems have largely replaced carburettors, although the basic function is the same. The carburettor meters fuel into the throat of a venturi, such that the AFR is approximately constant for all running conditions. The air mass flow rate through the venturi depresses the throat gauge pressure, and fuel flow has the same functional dependence on this pressure. This is because, as the constriction occurs, the air flow speeds up and the static pressure drops correspondingly, as seen in Part 1 in Section 3.3 on venturis. Mixture strength is increased at idle (to compensate for poor combustion conditions due to residual dilution in particular) and at wide open throttle (to give maximum power output, which is then limited by the availability of air).

Current fuel injection systems are multipoint (most expensive) or single-point (still more expensive than a carburettor). In multipoint systems, fuel is injected into each tract of the intake manifold near to the intake valve. One fuel injector per cylinder is required. Single-point system inject fuel into the air flow before it is split into separate routes to individual cylinders.

Multipoint systems (either electronic, EFI, or mechanical, MFI) are most common. Fuel is injected in discrete shots lasting approximately 10 msec. Fuel supply pressure (3 bar) is much lower than diesel fuel injection systems, and the injection timing is not critical. For diesel engines, fuel injection is inherently necessary. Each cylinder has an injector that fires fuel into the combustion chamber directly (direct injection diesels) or into a prechamber connected to the combustion chamber (indirect injection diesels). The injection system is sequential (each cylinder receives the fuel charge at the same point in its cycle). Because the injection is into the cylinder, late in the compression stroke, the fuel supply pressure is much higher (200 bar and now much higher pressures are available) than for spark ignition engine systems. The injection spray characteristics (droplet size, distribution, direction etc.) are also more critical since this directly influences the way in which combustion progresses.

Supercharging

The loss in the exhaust due to the sudden expansion when exhaust commences, can be used to benefit the charging part of the cycle. This is done generally on diesel engines to very useful effect in improving engine efficiency. The exhaust gases are passed over a turbine, which is directly linked to a compressor that delivers the air to the inlet manifold. The compressor turns at speeds in the order of tens of thousands of rpm This is called a **turbocharger**, and the basic idea is indicated on the p-V diagram in Figure 2.78. It improves the volumetric efficiency since the air in to the cylinder is at a higher pressure.

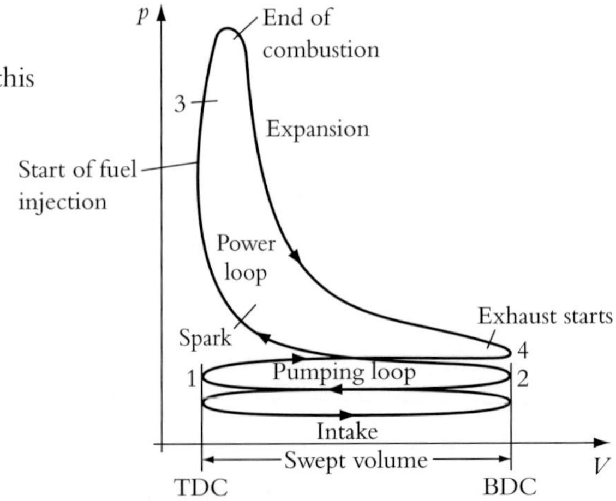

Figure 2.78 Principle of the use of a turbocharger to improve volumetric efficiency

Learning summary

By the end of this section you will have learnt:

- ✓ internal combustion engines are either operated on the Otto cycle, in which case they work with petrol and are called spark ignition engines, or on the Diesel cycle, in which case they work with diesel fuel and are called compression ignition engines;
- ✓ diesel fuel may be mainly considered as cetane, $C_{16}H_{34}$. Petrol fuel may be mainly considered to be octane, C_8H_{10};
- ✓ both cycles are based on the ideal standard air cycle versions of the cycle, which assume external heat supply and ideally enacted processes. In reality the cycles vary from this, mainly due to the nature of heat addition by internal combustion;
- ✓ SI engines have air–fuel ratios of approximately 14.5:1 (near stoichiometric) and a compression ratio of approximately 8. They are quantity governed, in that the rate of air flow is adjusted to alter power and economy;
- ✓ CI engines have air–fuel ratios between 20:1 and 25:1 and compression ratio in excess of 12. They are quality governed, in that the air flow is unrestricted, and the injection of fuel controls the power and economy;
- ✓ there are several measures of engine performance: indicated power is the power directly from the p–V diagram; brake power is the measured power at the engine shaft; mechanical efficiency is the ratio of brake power to indicated power; brake specific fuel consumption, which is the fuel required for a specific brake power; volumetric efficiency, which is volume induced / swept volume, similar to that for air compressors;
- ✓ the compression ratio affects the combustion performance and can lead to poor combustion and hence poor mechanical behaviour if handled badly. In particular, it can lead to a noisy combustion behaviour known as knock or diesel knock;
- ✓ spark timing is important for SI engines in order to provide the maximum pressure at the right moment in the cycle. The spark occurs a few degrees in advance of top dead centre position in order to create maximum pressure at top dead centre;
- ✓ fuelling systems for SI engines rely on injection of a premixed air and fuel mixture. Traditionally, carburretors were used, but recently these have been superceded by fuel injection. For CI engines, injection is necessary for the cycle to work, and the improvements have come from injection strategies due to electronic control and higher pressures available for fuel injection;
- ✓ the exhaust gases are ejected significantly above atmospheric pressure. This expansion can be used in a turbocharger to drive a turbine, which in turn drives a compressor at a very high rotational speed. The compressor delivers air to the cylinder significantly above atmospheric pressure, thus increasing the volumetric efficiency.

References

Bejan, A., 1993, *Heat Transfer*, Canada: John Wiley and Sons, Inc.

Bowman, R.A., Mueller, A.C. and Nagle, W.M., 1930, *Mean temperature difference in design*, Trans. ASME, vol. 12, 417–422.

Chenoweth, J.M., 1990, 'Final report of HTRI/TEMA joint committee review: the fouling section of the TEMA Standards'. *Heat Transfer Engineering*, vol. 11, no. 1, 73–107.

Kays, W.M. and London, A.L., 1984, *Compact Heat Exchangers*, New York: McGraw-Hill.

Unit 3

Solid mechanics

UNIT OVERVIEW

- Introduction
- Combined loading
- Yield criteria
- Deflection of beams
- Elastic–plastic deformations
- Elastic instability
- Shear stresses in beams
- Thick cylinders
- Asymmetrical bending
- Strain energy
- Fatigue
- Fracture mechanics
- Thermal stresses

3.1 Introduction

An Introduction to Mechanical Engineering: Part 1 covered the basic principles of solid mechanics (Unit 1) including basic mechanics and design analysis, stress and strain and the analysis of simple engineering loading situations such as uniaxial loading, beam bending, multi-axial stresses and torsion. This basic introduction assumed that all materials behaved in a linear elastic way and focused on establishing fundamental principles and simplified methods of analysis to solve for stresses and strains within components and structures.

In this unit, the application of solid mechanics to engineering problems is taken to a greater depth. Further elastic analysis is considered including such aspects as combined loading applied to components; further bending analyses, such as shear stresses in beams; the calculation of bending deflections; and the bending of beams with asymmetric sections. Powerful methods of analysing elastic deformations in more complex shaped structures using the concept of stored strain energy is also included. The analysis of more detailed stress distributions, in thick cylinders for instance, is also explained. The purpose of these further elastic analyses is to determine stresses, strains and deformations under more complex loading and where geometry varies in a more complex way.

Stresses and strains are calculated for a reason: to design components fit for purpose. In this sense, it is important that we consider how components might fail under loading. The unit therefore also includes several sections on failure, including yield criteria to assess when failure might occur, analysis beyond the elastic region, where elastic–plastic behaviour becomes important and the analysis of particular types of failure other than yielding, including brittle fracture and fatigue (under cyclic loading). We learn how to analyse components under such conditions and how to design against such failures. Finally, failure of components is not necessarily restricted to excessive mechanical stresses but may be caused by other factors such as temperature or excessive deformations. The unit therefore also covers the analysis of thermal stresses and situations where elastic instability – buckling – might occur.

All of the above topics are covered in depth by firstly introducing the basic concepts and theory, secondly developing standard methods for analysis and finally illustrating these methods with many worked examples. At the end of the unit students will have learnt a wealth of easily usable analytical methods for solving many practical engineering problems.

3.2 Combined loading

Many engineering problems can be analysed as simple loading situations e.g. uniaxial loading, beam bending, torsion etc. However, it is also very common in the real world for engineering components and structures to be subjected to several loads simultaneously. This is a combined loading situation and can be analysed by superposing the effects of the individual loads.

Mohr's circle – recap

Mohr's circle for plane stress was introduced in *Introduction to Mechanical Engineering: Part 1* as a useful graphical technique for analysing plane stresses acting on an element in a material or structure. For combined loading situations, it is common to reduce the problem to such a plane stress problem and analyse using Mohr's circle. The analysis will give the principal stresses, the maximum shear stresses and the angles of the principal planes for the element. Figure 3.1 shows a shaft subjected to combined loading of a torque, T, and a compressive axial load, P. Let us assume that such loading gives rise to an axial stress of −12 MPa (i.e. compressive) and a shear stress of −6 MPa (i.e. causes the element to rotate clockwise) acting on a surface element as shown in the figure. The Mohr's circle analysis is then as follows:

The known stresses on the element are:

$$\sigma_x = -12\,\text{MPa}$$
$$\sigma_y = 0\,\text{MPa}$$
$$\tau_{xy} = -6\,\text{MPa}$$
$$\tau_{yx} = 6\,\text{MPa}$$

Figure 3.2 shows Mohr's circle for this stress system. To draw the circle, firstly draw the point B which represents stresses on the x-plane (coordinates: −12, −6). Next draw point E which represents stresses on the y-plane (coordinates: 0, +6). Join the two points with the line BE that intersects the x-axis at the centre of the circle, C. The circle can now be drawn and the following quantities measured:

$$\sigma_1 = 2.5\,\text{MPa}$$
$$\sigma_2 = -14.5\,\text{MPa}$$
$$\tau_{max} = 8.5\,\text{MPa}$$
$$2\theta = 45°$$

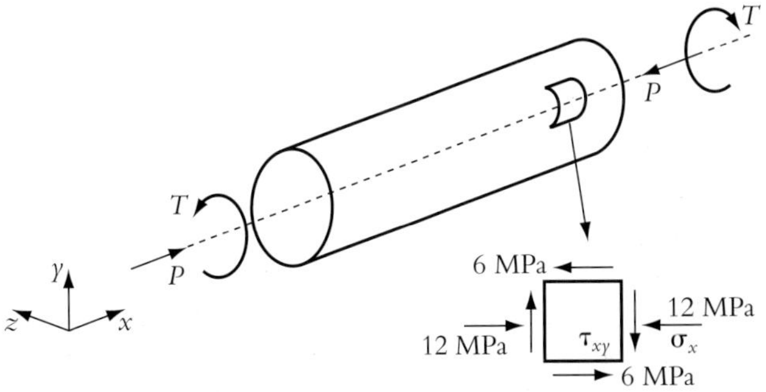

Figure 3.1 Shaft subjected to combined torque and compressive axial loading

On the element, the angle of the principal plane (P1) from the y-plane is $\theta = 22.5°$ anticlockwise as shown in Figure 3.2.

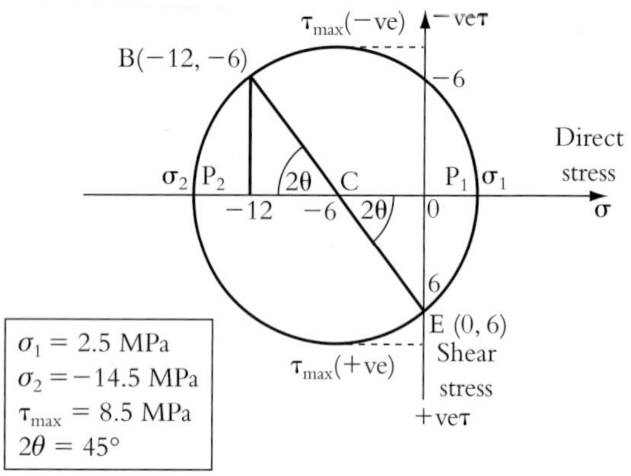

Figure 3.2 Mohr's circle for combined torsion and compression

Alternatively, the important parameters in the circle can be calculated analytically as follows:

The centre of the circle is given by $C = (\sigma_x + \sigma_y)/2 = -6$

The radius of the circle is given by $R = \sqrt{\left(\frac{\sigma_x - \sigma_y}{2}\right)^2 + \tau_{xy}^2} = 8.5$

The principal stresses are:

$$\sigma_1 = C + R = 2.5\,\text{MPa}$$
$$\sigma_2 = C - R = -14.5\,\text{MPa}$$
$$\tau_{max} = R = 8.5\,\text{MPa}$$

The angle of the principal planes: $\tan 2\theta = \frac{\tau_{xy}}{\left(\frac{\sigma_x - \sigma_y}{2}\right)} = 1$

$$2\theta = 45°$$
$$\theta = 22.5°$$

If the analytical approach is taken (which does give the more accurate results), then it is always advisable to sketch Mohr's circle in order to gain a clear understanding of the orientation of the principal planes and the maximum shear planes with respect to the *x*- or *y*-planes.

Superposition of combined loads

The **principal of superposition** states that:

$$\begin{bmatrix}\text{The total effect of } \textit{combined} \\ \text{loads applied to a body}\end{bmatrix} = \sum \begin{bmatrix}\text{The effects of the individual} \\ \text{loads applied } \textit{separately}\end{bmatrix}$$

Thus, when a body or structure is subjected to a combination of different types of loading simultaneously, we can consider separately the effects of each load on the local stresses that act on an element. Stresses on the element can then be summed to determine the effect of the combined loading. A number of combined loading examples can be used to illustrate:

Combined bending and axial loading

Figure 3.3 shows a beam carrying a uniformly distributed load (UDL) along its span, while simultaneously being subjected to an axial compressive force, F. The figure shows how the effect of the combined loading on the stress distribution through the thickness of the beam at the centre of its span is determined. The effect of the UDL and the axial force are obtained separately and then summed to give the combined stress distribution in the beam. The symmetrical bending stress distribution about the neutral axis is essentially skewed to more compression by the effect of the axial stress.

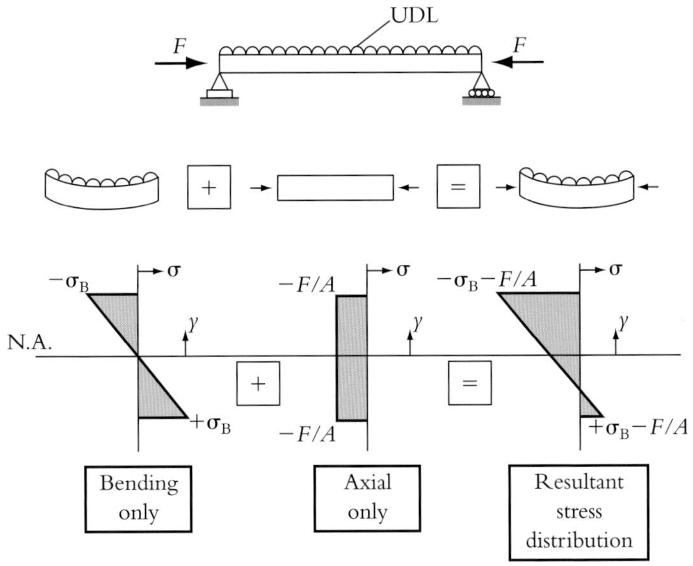

Figure 3.3 **Combined bending and axial loading**

Combined bending and torsion

Figure 3.4 shows a similar beam to the above example, except now the beam carries a torque instead of the axial load. This loading situation is typical of a shaft with self-weight (UDL) transmitting a torque. In this case, the beam cross section can be assumed to be a solid circle with diameter d. The stresses at the centre of the span, at the bottom surface of the beam, are given by the usual bending and torsion equations as follows.

Arising from the UDL, bending stress (σ_B):

$$\sigma_B = \frac{My}{I}$$

where $$y = d/2$$

Arising from the torque, torsional shear stress (τ)

$$\tau = \frac{Tr}{J}$$

where $$r = d/2$$

These two stresses can be superposed and illustrated as acting on an element at the surface of the beam, as shown in Figure 3.4.

Mohr's circle can now be used for this element to obtain the principal stresses and maximum shear stress at this position.

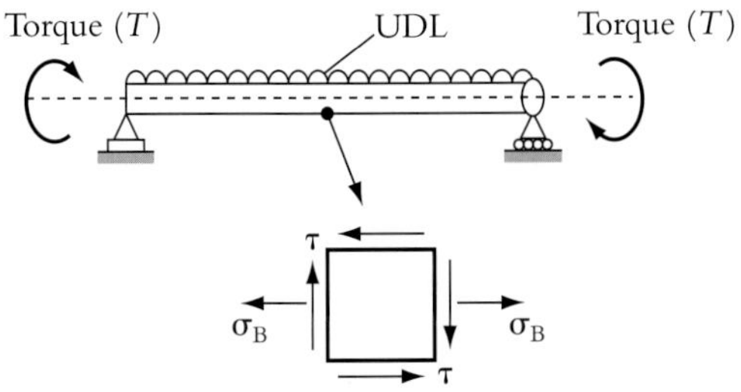

Figure 3.4 **Combined bending and torsion**

Combined pressure, axial and torsional loading

A combination of three loads can be illustrated by considering a thin-walled cylinder, as shown in Figure 3.5, subjected to an internal pressure, P, an axial tensile force, F, and a torque, T. Figure 3.5 shows the stresses arising from each load separately acting on a surface element in the plane of the cylinder wall. Superposition of these three stresses are also shown on the element. Mohr's circle can again be used to obtain the principal stresses and maximum shear stress for the element.

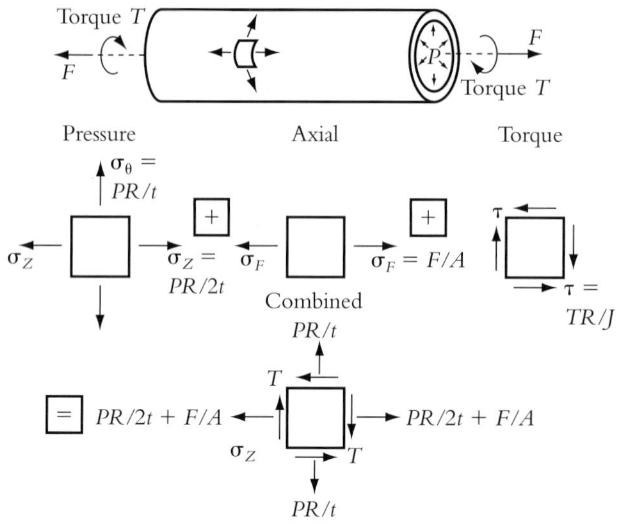

Figure 3.5 Combined internal pressure, axial tensile force and torque

Methodology for combined loading

The methodology for analysing components or structures under combined loading can now be summarized:

(i) Identify a 2D element at the location of interest in the component.

(ii) Determine the stresses acting on the element arising from each individual load.

(iii) Superpose the stresses from each individual load to obtain the combined stresses on the element.

(iv) Use Mohr's circle to determine the principal stresses and the maximum shear stress on the element.

Worked example

Combined bending and torsion – offset loading on a cantilever

Figure 3.6 shows a solid circular cross section cantilever beam, of length *L* and diameter *d*, fixed at one end. Attached at the free end of the beam is a crank arm which allows a vertical load, *P*, to be applied at an offset distance, a, from the axis of the cantilever.

Determine the maximum shear stress on the upper surface at the fixed support of the cantilever beam (position A).

The following load and dimensions apply:

$P = 1\,\text{kN}$
$L = 200\,\text{mm}$
$a = 120\,\text{mm}$
$d = 30\,\text{mm}$

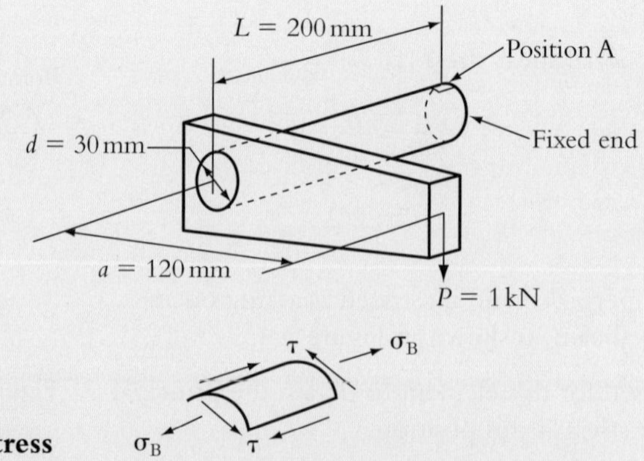

Figure 3.6 Combined bending and torque acting on a cantilever beam

We consider the stresses acting on a small surface element at position A. The load gives rise to a bending moment and torsional moment at the cross section at position A as follows:

Bending moment $M = P \cdot L$

Torsional moment $T = P \cdot a$

These moments give rise to separate bending and shear stresses acting on the element at position A which can be superposed to give the total effect of the combined loading as shown in Figure 3.6. The stresses are:

$$\text{Bending stress:} \qquad \sigma_B = \frac{My}{I} = \frac{PL\frac{d}{2}}{\frac{\pi d^4}{64}} = \frac{32PL}{\pi d^3} = 75.45\,\text{MPa}$$

$$\text{Torsional shear stress} \qquad \tau = \frac{Tr}{J} = \frac{Pa\frac{d}{2}}{\frac{\pi d^4}{32}} = \frac{16Pa}{\pi d^3} = 22.64\,\text{MPa}$$

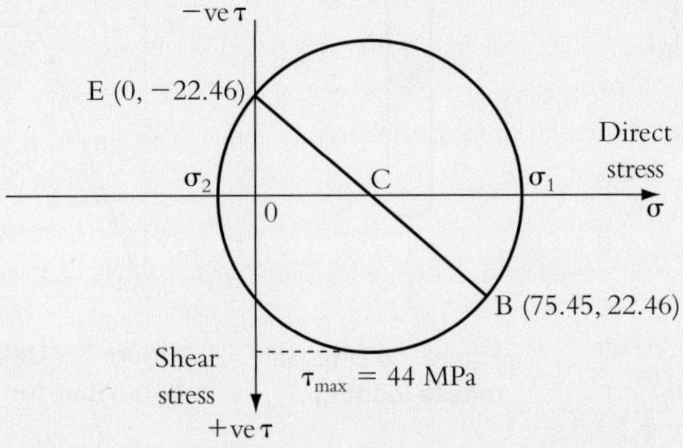

Figure 3.7 Mohr's circle for combined bending and torsion

Mohr's circle can now be drawn for the element to determine the maximum shear stress as shown in Figure 3.7 The coordinates of point B on the circle (σ_B,τ) correspond to the stresses on the element in the longitudinal direction, i.e. along the cantilever. Point E corresponds to the stresses ($0, -\tau$) in the transverse direction to this. The line joining these two points defines the diameter of the circle and, where it crosses the σ-axis, the centre, C. The circle can now be drawn and its radius measured to give the maximum shear stress as follows:

$\tau_{max} = \text{radius} = 44\,\text{MPa}$

Alternatively, by calculation:

Given the element stresses

$$\sigma_x = 75.45\,\text{MPa}$$
$$\sigma_y = 0\,\text{MPa}$$
$$\tau_{xy} = 22.64\,\text{MPa}$$

$$\text{The maximum shear stress, } \tau_{max} = \text{radius} = \sqrt{\left(\frac{\sigma_x - \sigma_y}{2}\right)^2 + \tau_{xy}{}^2}$$

$$= \sqrt{\left(\frac{75.45}{2}\right)^2 + 22.64^2}$$

$$= 44\,\text{MPa as before}$$

Learning summary

By the end of this section you should have learnt:

- ✓ the basic use of Mohr's circle for analysing the general state of plane stress;
- ✓ how the effect of combined loads on a component can be analysed by considering each load as initially having an independent effect;
- ✓ how to use the principle of superposition to determine the combined effect of these loads.

3.3 Yield criteria

Elastic–plastic deformations

Uniaxial tensile tests (Figure 3.8) are used to obtain some important material properties. Typical 'stress–strain curves' are shown in Figure 3.9.

A linear elastic range is exhibited by a number of materials. In this region the strain is proportional to the stress (Hooke's Law) and on removal of the stress, the strain returns to zero. The constant of proportionality is the Young's Modulus, E, for the material. If the stress exceeds a certain value, i.e. the yield stress, σ_y, the strain does not return to zero on removal of the load. Table 3.1 gives typical values of E and σ_y for some common engineering materials.

A tensile specimen, with a circular cross-section, made from mild steel, which is a ductile material, breaks with a 'cup and cone' mode of failure, see Figure 3.10, the cone angle is ~45°.

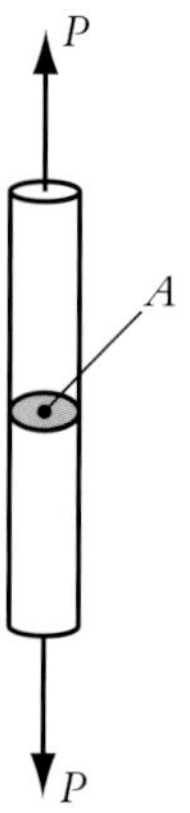

Figure 3.8 Uniaxial tensile loading

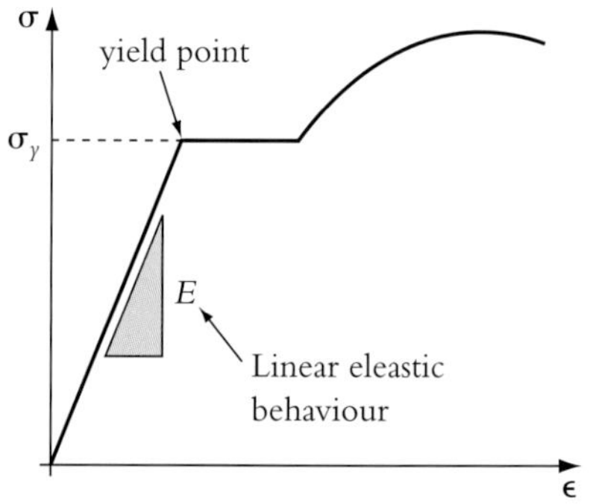

Figure 3.9 Uniaxial stress–strain behaviour for mild steel

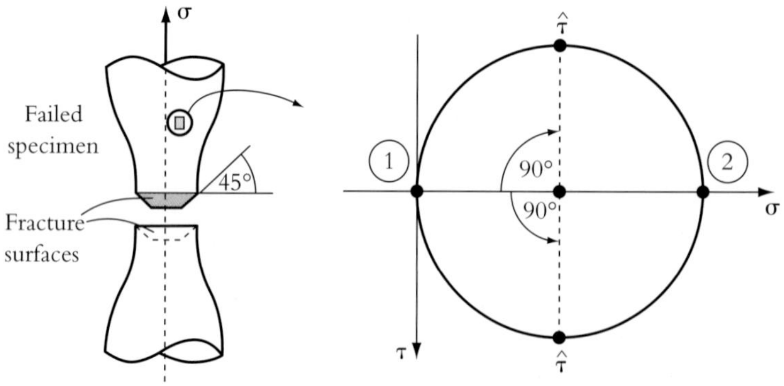

Figure 3.10 Cup and cone failure for mild steel and Mohr's circle for uniaxial tensile loading

Material	E(GPa)	σ_y(MPa)
mild steel	200	350
copper	120	310
brass	100	390
aluminum alloys	70	140–500

Table 3.1 Typical values of E and σ_y for common engineering materials at 20°C

A Mohr's circle for the loading condition, see Figure 3.10, indicates that the 45° plane is that on which the maximum shear stress, $\hat{\tau}$, occurs.

If a circular cross section mild steel bar is subjected to torsional loading (Figure 3.11), the torque angle response is as shown in Figure 3.12. For $T \leqslant T_y$, $T \propto \theta$, and the torque–angle behaviour is reversed on removal of the torque. If the torque is continuously increased until failure occurs, the fracture plane is transverse to the axis of the specimen (Figure 3.13). A Mohr's circle for the torsion loading case, Figure 3.13, indicates that failure occurs on the maximum shear stress plane.

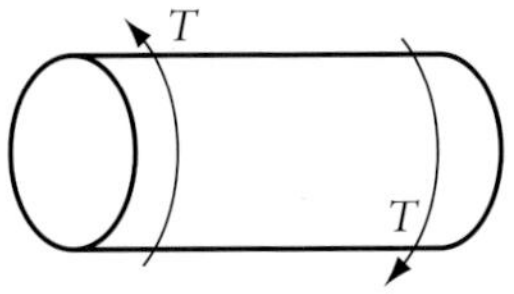

Figure 3.11 Pure torsional loading for a circular cross section bar

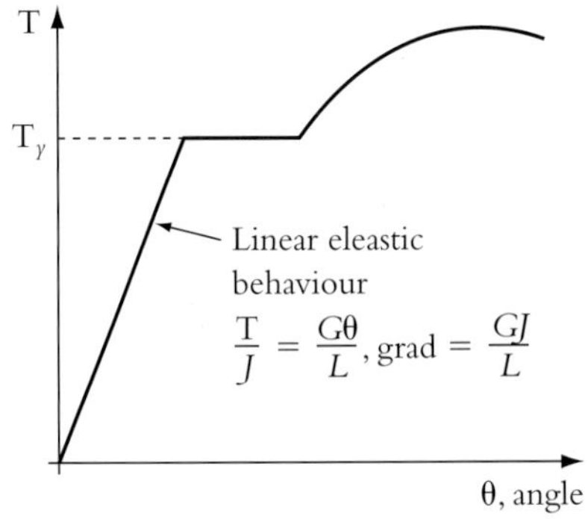

Figure 3.12 Torque–angle behaviour for mild steel

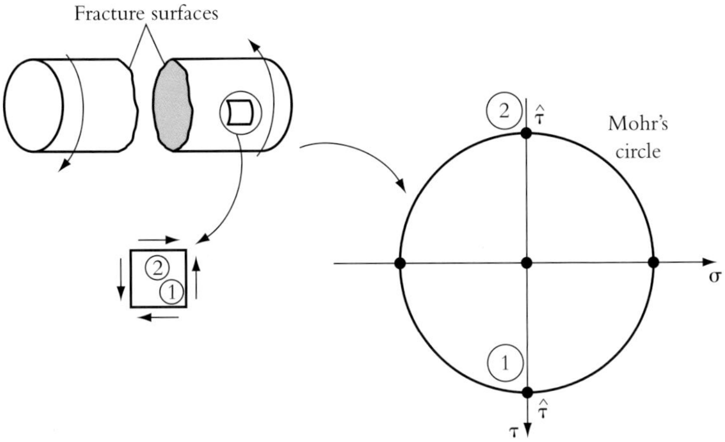

Figure 3.13 Transverse failure mode exhibited by mild steel and pure torsional loading and corresponding Mohr's circle

Therefore, the results of both the tension and the torsion tests indicate that failure occurs on the planes that contain the maximum shear stresses. This behaviour is similar to that of many 'ductile' materials.

Similar torsion tests on 'brittle' materials (e.g. cast iron) give different behaviour. The torque-angle response is as shown in Figure 3.14 and the fracture occurs on a 45° helix, as shown in Figure 3.15.

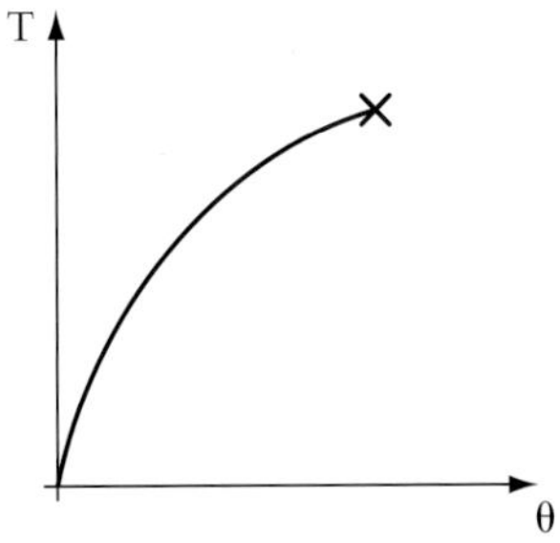

Figure 3.14 Torque–angle behaviour for cast iron

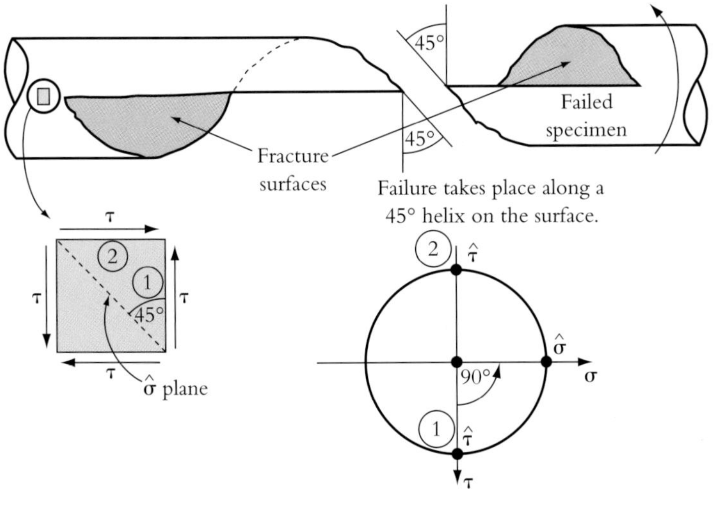

Figure 3.15 The 45° helix failure mode exhibited by mild steel under pure torsional loading and the corresponding Mohr's circle

A Mohr's circle for this loading condition shows that the point on the Mohr's circle associated with the maximum principal stress is 90° (ccw) from point 1. Therefore, this represents a plane at 45° from plane 1 on the element. This maximum principal stress, $\hat{\sigma}$, plane corresponds to the 45° helix angle of the fracture on the surface. This indicates that failure in this material has occurred on a plane on which the maximum principal stress exists.

The tests and stress states used to come to the above conclusions for 'ductile' and 'brittle' failures are very simple and it would therefore be unwise to base failure criteria on this evidence alone.

Yielding of ductile materials

The topic of yield criteria is limited to the prediction of the initiation of yielding in ductile materials. A yield criterion closely related to the maximum shear stress (Tresca) criterion is the maximum shear strain energy (von Mises) criterion; these two criteria generally provide a good indication of yield and are widely used in elastic–plastic analysis.

Maximum shear stress (Tresca) yield criterion

The Tresca yield criterion states that the material will yield when the maximum shear stress in the material exceeds a limiting value.

If σ_1, σ_2 and σ_3 are the three principal stresses ($\sigma_1 > \sigma_2 > \sigma_3$) then, Figure 3.16 shows that

$$\tau_{max} = \frac{\sigma_1 - \sigma_3}{2}$$

The limiting value can be related to the uniaxial yield stress, σ_y, obtained from a uniaxial tensile test. In this case,

$$\sigma_1 = \sigma_y \text{ and } \sigma_2 = \sigma_3 = 0$$

$$\tau_{max} = \frac{\sigma_y - 0}{2} = \frac{\sigma_y}{2}$$

The τ_{max} criterion therefore states that the material will yield if:

$$\sigma_1 - \sigma_3 \geqslant \sigma_y \quad (\sigma_1 > \sigma_2 > \sigma_3)$$

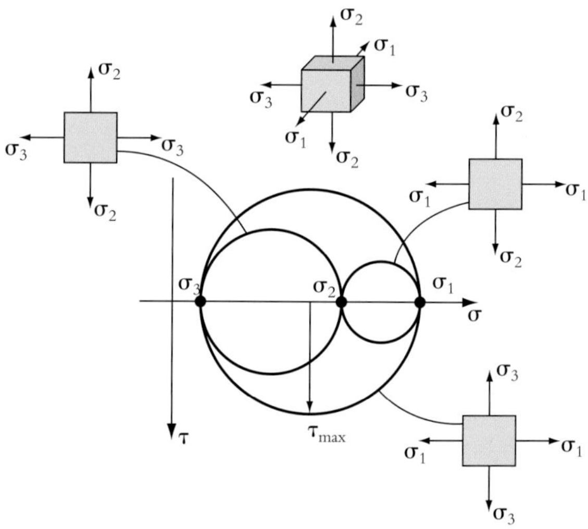

Figure 3.16 Mohr's circle representation of a three dimensional stress state

Maximum shear strain energy (von Mises) yield criterion

The von Mises yield criterion states that the material will yield when the maximum shear strain energy (per unit volume) exceeds a limiting value. If σ_1, σ_2 and σ_3 are the three principal stresses ($\sigma_1 > \sigma_2 > \sigma_3$) then:

$$\frac{\text{shear strain energy}}{\text{unit volume}} = \frac{1}{12G}\{(\sigma_1 - \sigma_2)^2 + (\sigma_2 - \sigma_3)^2 + (\sigma_3 - \sigma_1)^2\}$$

Again, the limiting value can be related to the uniaxial yield stress, σ_y, obtained from a uniaxial tensile test. Thus, at yield $\sigma_1 = \sigma_y$ and $\sigma_2 = \sigma_3 = 0$:

$$\frac{\text{shear strain energy}}{\text{unit volume}} = \frac{1}{12G}\{2\sigma_y^2\}$$

The von Mises yield criterion can thus be expressed as follows:

$$\frac{1}{12G}\{(\sigma_1 - \sigma_2)^2 + (\sigma_2 - \sigma_3)^2 + (\sigma_3 - \sigma_1)^2\} \geqslant \frac{1}{12G}\{2\sigma_y^2\}$$

which can be reduced to the following, more common expression for the onset of yield, according to the von Mises yield criterion:

$$(\sigma_1 - \sigma_2)^2 + (\sigma_2 - \sigma_3)^2 + (\sigma_3 - \sigma_1)^2 \geqslant 2\sigma_y^2$$

Two-dimensional stress systems (i.e. $\sigma_3 = 0$)

For plotting purposes here, σ_1 and σ_2 can take on any values, i.e. σ_1 is not necessarily always greater than σ_2. The yield boundaries on the σ_1–σ_2 plane are shown in Figure 3.17.

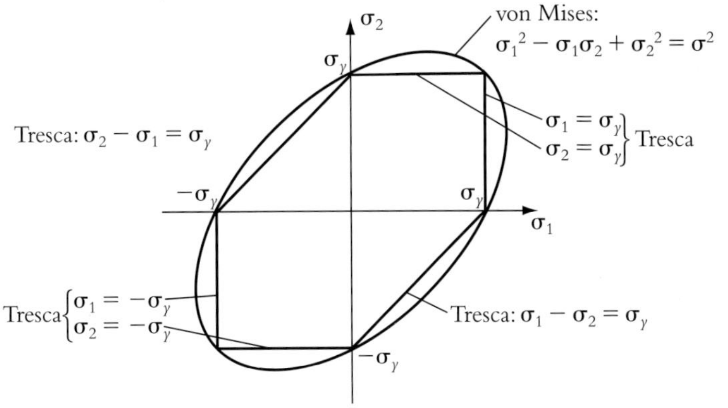

Figure 3.17 The yield boundaries, for a two-dimensional stress state, on the $\sigma_1 - \sigma_2$ plane ($\sigma_3 = 0$).

In general, the von Mises yield criterion is easier to handle analytically because it is continuous. This is particularly important for the calculation of incremental plastic strains, since the plastic strains are related to the normals to the yield surface, and at the corners of the Tresca yield locus there is ambiguity about the directions of the normals, whereas there is no such ambiguity about the von Mises yield locus.

Three-dimensional stress systems

The Tresca yield criterion

$$\sigma_1 - \sigma_3 = \sigma_y \quad (\sigma_1 > \sigma_2 > \sigma_3)$$

and the von Mises yield criterion

$$(\sigma_1 - \sigma_2)^2 + (\sigma_2 + \sigma_3)^2 + (\sigma_3 - \sigma_1)^2 = 2\sigma_y^2$$

are not altered if a constant stress component, say σ, is added to each stress component, i.e.

$$(\sigma_1 + \sigma) - (\sigma_3 - \sigma) = \sigma_1 - \sigma_3 = \sigma_y$$

and

$$((\sigma_1 + \sigma) - (\sigma_2 + \sigma))^2 + ((\sigma_2 + \sigma) - (\sigma_3 + \sigma))^2 + ((\sigma_3 + \sigma) - (\sigma_1 + \sigma))^2$$
$$= (\sigma_1 - \sigma_2)^2 + (\sigma_2 - \sigma_3)^2 + (\sigma_3 - \sigma_1)^2 = 2\sigma_y^2$$

This implies that the addition of a 'hydrostatic stress state', i.e. $\sigma_1 = \sigma_2 = \sigma_3 = \sigma$ does not change the shapes of the yield surfaces shown in the section on two-dimensional stress systems.

The mean principal stress $\sigma_h = \frac{1}{3}(\sigma_1 + \sigma_2 + \sigma_3)$, which is known as the **hydrostatic stress** for a given stress state $(\sigma_1, \sigma_2, \sigma_3)$, is the stress that causes volume change. Now, the independence of the yield criteria with respect to hydrostatic stress means that the three-dimensional yield criteria are prismatic surfaces with the axes of the prisms in each case being the line $\sigma_1 = \sigma_2 = \sigma_3$. This is called the **hydrostatic line** in 3D stress space (Haigh–Westergaard stress space) and it has direction cosines $\left(\frac{1}{\sqrt{3}}, \frac{1}{\sqrt{3}}, \frac{1}{\sqrt{3}}\right)$. The yield boundaries can thus move any distance in the direction $\sigma_1 = \sigma_2 = \sigma_3$. The yield surfaces for both the von Mises and the Tresca yield criteria therefore have a constant oblique section and hence a constant perpendicular cross section, whose true shape can be seen in the view along the line $\sigma_1 = \sigma_2 = \sigma_3$. Any arbitrary stress 'vector' $(\sigma_1, \sigma_2, \sigma_3)$, e.g. $\overrightarrow{OB}$ and $\overrightarrow{OD}$, (Figure 3.18), in the stress space can be decomposed into two components, one parallel to the hydrostatic line, e.g. $\overrightarrow{OA}$ and $\overrightarrow{OC}$, and one perpendicular to the hydrostatic line, e.g. $\overrightarrow{AB}$ and $\overrightarrow{CD}$. The oblique planes that are perpendicular to the hydrostatic line are called **deviatoric planes** and are given by equations of the form $\sigma_1 + \sigma_2 + \sigma_3 =$ constant, each representing a different level of hydrostatic stress. The deviatoric plane with $\sigma_1 + \sigma_2 + \sigma_3 = 0$ is known as the π-plane. It can be shown that the component of $(\sigma_1, \sigma_2, \sigma_3)$ parallel to the hydrostatic line is $(\sigma_h, \sigma_h, \sigma_h)$, e.g. $\overrightarrow{OA}$ and $\overrightarrow{OC}$, while the component parallel to the deviatoric planes is $(\sigma_1 - \sigma_h, \sigma_2 - \sigma_h, \sigma_3 - \sigma_h)$, $\overrightarrow{AB}$ and $\overrightarrow{CD}$. Only the latter component of stress is important in determining yield according to the von Mises and Tresca criteria.

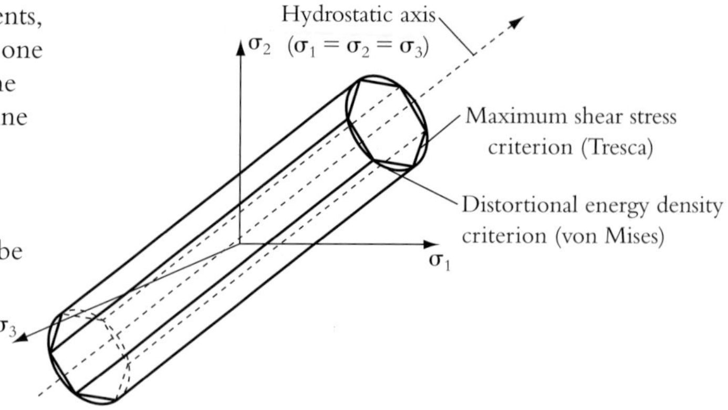

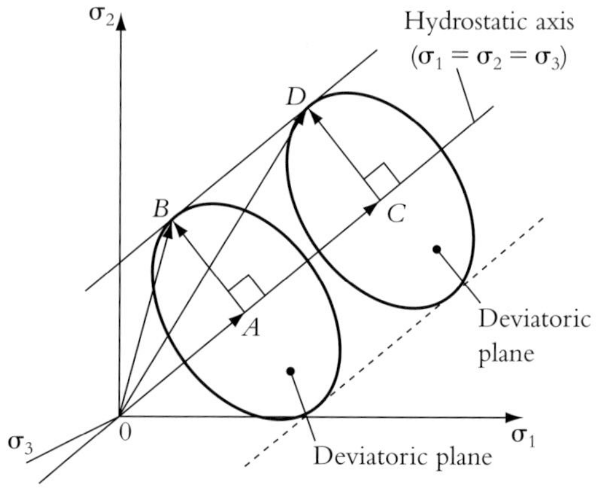

Figure 3.18 (a) Representation of yield surfaces for a three-dimensional stress state and (b) the decomposition of the stress into hydrostatic and deviatoric components

The view along the $\sigma_1 = \sigma_2 = \sigma_3$ line of the von Mises and Tresca yield criteria is an isometric view that shows the three axes included at 120° intervals. This is sometimes called a view on the π-plane, on which the Tresca yield surface is a hexagon and the von Mises yield surface is a circle. Therefore, large principal stresses do not necessarily result in yield; it is the stress differences and the route to the final stress state that govern whether yielding will occur.

Figure 3.19 can be used, instead of the equations, to decide whether a certain stress state will be safe. Simply plot on the diagram each of the three principal stresses parallel to each of the three axes and see whether the final point lies inside the appropriate yield surface.

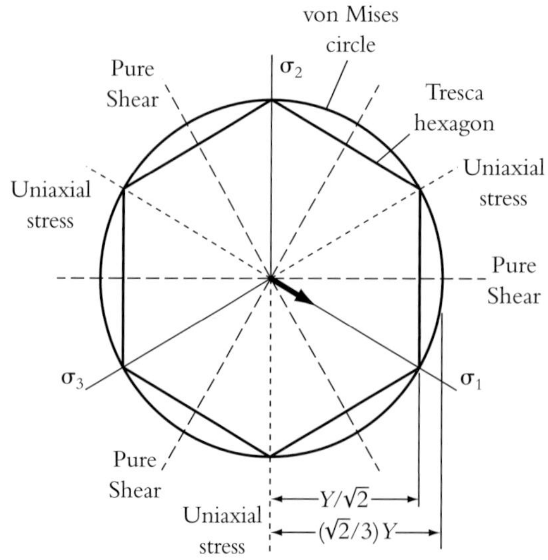

Figure 3.19 View of the yield surfaces on the π-plane

Note:

(i) The yield condition can be examined by either using the appropriate equation, i.e.

$$\text{Tresca: } \sigma_1 - \sigma_3 = \sigma_y \quad (\sigma_1 > \sigma_2 > \sigma_3)$$

$$\text{von Mises: } (\sigma_1 - \sigma_2)^2 + (\sigma_2 - \sigma_3)^2 + (\sigma_3 - \sigma_1)^2 = 2\sigma_y{}^2$$

or by plotting principal stresses on the π-plane.

(ii) All three principal stresses are important. At free surfaces the normal stress is usually zero, but it may be important, particularly if the other two principal stresses are of the same sign.

Learning summary

By the end of this section you should have learnt:

- ✓ the difference between ductile and brittle failure, illustrated by the behaviour of bars subjected to uniaxial tension and torsion;
- ✓ the meaning of yield stress and proof stress, in uniaxial tension, for a material;
- ✓ the Tresca (maximum shear stress) yield criterion and the 2D and 3D diagrammatic representations of it;
- ✓ the von Mises (maximum shear strain energy) yield criterion and the 2D and 3D diagrammatic representations of it.

3.4 Deflection of beams

In this section we address the important topic of beam deflections. The design of engineering structures and components is very often dictated by the strength of the materials used and consequently the stresses within the structure, but often the limiting factor is the allowable deflection. This is particularly important for engineering artefacts made from materials of lower stiffness, such as aluminium, plastics or composites, but may also be critical for high stiffness steel structures comprising slender flexible members. It is therefore important as part of the design process to be able to calculate maximum deflections in a structure in addition to the position at which they occur.

This section focuses on deflections in beam structures. Following the derivation of the fundamental deflection equation for a beam, a flexible procedure is introduced, called Macaulay's method, which allows deflections to be calculated at any position along a beam span. In particular, the method allows us to deal with different types of loading such as point loads, uniformly distributed loads (UDLs) and point moments, including discontinuities in these loads. Although not the only method for calculating deflections, as we will see later in the strain energy section, it is a particularly powerful and flexible method.

Equation of the elastic line

The section of span of a beam, shown in Figure 3.20, is under pure bending, i.e. there is a constant bending moment along this section and no shear force. Under pure bending conditions, the neutral surface of the beam is a circular arc with radius of curvature, R, as shown. The transverse deflection of the neutral surface is given by the coordinate y of any position along this surface.

(N.B. Do not confuse this 'y' definition of deflection with the 'y' denoting distance from the neutral axis/surface in the beam bending equation – as detailed in *An Introduction to Mechanical Engineering: Part 1*.

The line denoting the neutral surface in Figure 3.20 is known as the **elastic line** or the **deflection curve** of the beam.

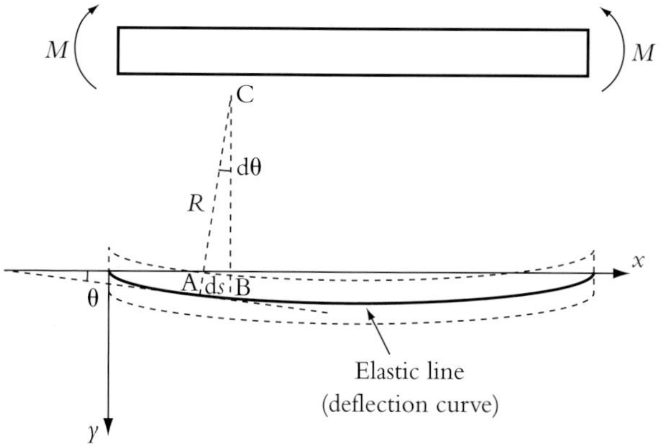

Figure 3.20 The elastic line (i.e. deflection curve) of a beam under pure bending

Referring to Figure 3.20, the angle $d\theta$ defines an element AB, ds in length, of the elastic line, and, because $d\theta$ is small,

$$ds = R\,d\theta$$

in other words, the **curvature**,

$$\frac{1}{R} = \frac{d\theta}{ds} \tag{3.1}$$

For a beam bent in the elastic range, the elastic line is a very flat curve i.e. deflections are small. In this case,

$$ds \approx dx$$

where dx is the increment of x between A and B, and equation (3.1) becomes

$$\frac{1}{R} = -\frac{d\theta}{dx} \tag{3.2}$$

Note the introduction of the **minus (−)** sign because θ decreases as x increases for positive bending, i.e. sagging. This maintains a positive radius of curvature for positive bending.

Also, θ is the slope of the elastic line at position A,

$$\theta = \frac{dy}{dx} \tag{3.3}$$

Combining equations 3.2 and 3.3 gives,

$$\frac{1}{R} = -\frac{d^2y}{dx^2} \tag{3.4}$$

And finally, combining equation (3.4) with the part of the beam bending equation relating radius of curvature to the bending moment (M), Young's modulus (E) and second moment of area (I), i.e.,

$$\frac{M}{I} = \frac{E}{R} \text{ or } \frac{1}{R} = \frac{M}{EI} \tag{3.5}$$

gives,

$$-EI\frac{d^2y}{dx^2} = M \tag{3.6}$$

Equation (3.6) is the **differential equation of the elastic line**, relating the deflection, y, to the applied bending moment, M, the Young's modulus, E, and second moment of area. I. The product of E and I, i.e. EI, is termed the **flexural rigidity** of the beam.

Successive integration of equation (3.6) will yield the slope, dy/dx, and the deflection, y, at all points along the beam.

Equation (3.6) has been derived for the case of pure bending, i.e. constant bending moment along the section, and does not take into account deflections due to shear. For long slender beams, shear deflections can be neglected.

Equation (3.6) may also be integrated, and solutions found, if the bending moment, M, is a continuous function of x. This condition applies only to beams under specific simple loading conditions. A complication arises where **discontinuities** in M exist, such as where there are point loads and point moments or where there is an abrupt change in distributed loading.

Various methods have been developed to solve such problems with discontinuities. In the next section we will introduce and develop the procedure called **Macaulay's method**, a versatile solution procedure that can handle most discontinuities that we are likely to come across.

Macaulay's method (also termed the method of singularity functions)

Named after the mathematician W.H. Macaulay, this method uses a mathematical technique to deal with discontinuous loading. The moment expression $M(x)$, i.e. M as a function of x, is replaced with the step function $M\langle x - a\rangle$, in which a defines the points where discontinuities arise. The procedure will now be developed from first principles.

Figure 3.21 shows a simply supported beam carrying three point loads W_A, W_B and W_C, along its span. Each of the three loads gives rise to a discontinuity in the bending moment. Knowing the applied loads, the end reactions R_1 and R_2 can be calculated from equilibrium conditions for the whole beam. Now, considering each part of the span between the loads separately:

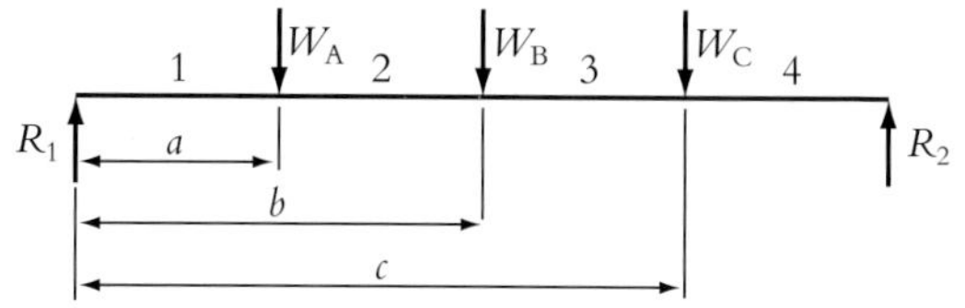

Figure 3.21 Simply supported beam carrying point loads

Span 1

Figure 3.22(a) shows the free-body diagram of the left-hand end of the beam, cut at a section in span 1. The unknown bending moment, M, and shear force, S, at this section are shown in the diagram. Considering the bending moment only and taking moments about the cut section,

$$M - R_1 x = 0$$

and combining with the equation of the elastic line,

$$EI\frac{d^2y}{dx^2} = -M = -R_1 x \qquad (3.7)$$

This equation applies to span 1 only. Now consider span 2.

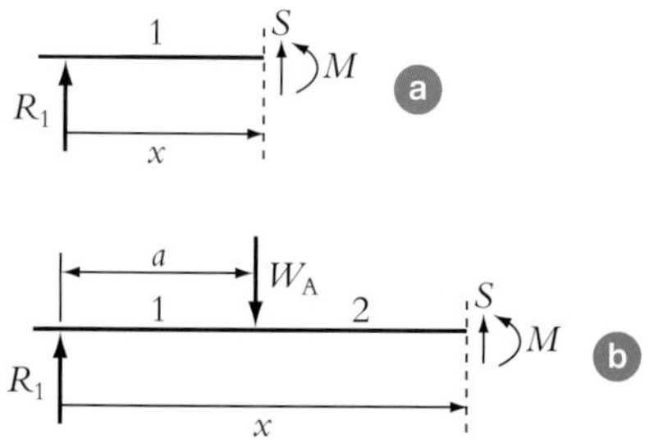

Figure 3.22 Free-body diagrams of beam cut at different sections

Span 2

Figure 3.22(b) shows the free-body diagram of the left-hand end of the beam, now cut at a section in span 2. As before, the unknown bending moment, M, and shear force, S, at this section are shown in the diagram. Again, considering the bending moment only and taking moments about the cut section,

$$M + W_A(x - a) - R_1 x = 0$$

And, again, combining with the equation of the elastic line,

$$EI\frac{d^2y}{dx^2} = -M = -R_1 x + W_A(x - a) \qquad (3.8)$$

This equation now applies to span 2 only.

Similar equations can be derived for spans 3 and 4 as follows.

Span 3

$$EI\frac{d^2y}{dx^2} = -M = -R_1 x + W_A(x - a) + W_B(x - b) \qquad (3.9)$$

Span 4

$$EI\frac{d^2y}{dx^2} = -M = -R_1 x + W_A(x - a) + W_B(x - b) + W_C(x - c) \qquad (3.10)$$

It is interesting to note that the forms of equations (3.7), (3.8), (3.9) and (3.10) are very similar, in that extra terms are added to take account of the point loads as we move towards the right-hand end of the beam. In fact, due to this similarity, equation (3.10) can be used to apply to all spans by rewriting it in a slightly different form:

$$EI\frac{d^2y}{dx^2} = -M = -R_1x + W_A < x - a > + W_B < x - b > + W_C < x - c > \qquad \textbf{(3.11)}$$

Note the change of bracket shape in equation (3.11). The $< >$ brackets are termed **Macaulay brackets**, and equation (3.11) is now applicable to any point in the beam if we adopt what is called Macaulay's convention.

Macaulay's convention: 'Whenever a bracketed term becomes negative it must be ignored.'

Adopting this convention, the general expression for M in equation (3.11) can be integrated twice to give the slope, dy/dx, and the deflection, y, at any point along the beam. During the integration, boundary conditions for the beam are used to determine the integration constants. It should also be noted that the bracketed terms *must not* be multiplied out in any general expression for y, dy/dx or d^2y/dx^2, otherwise the Macaulay convention will be lost.

Worked example

Macaulay's method example – point load on a beam

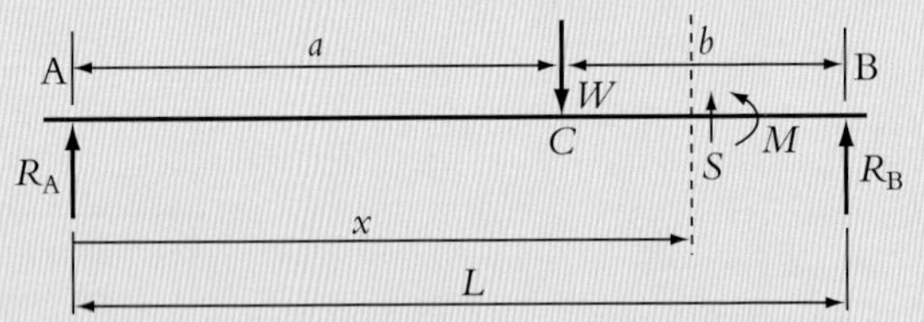

Figure 3.23 Simply supported beam carrying a single point load

Figure 3.23 shows a simply supported beam carrying a point load, distance a from the left-hand end and distance b from the right-hand end.

Derive an equation for the deflection of this beam at *any* point along its length.

Firstly, determine the reaction force, R_A, by taking moments for the whole beam about B,

$$R_AL - Wb = 0 \qquad \therefore \quad R_A = \frac{Wb}{L}$$

Now, cutting the beam at a section in the extreme right-hand part of the span, distance x from the left-hand end, the left-hand part of the beam can be considered as a free body with an unknown bending moment, M, and shear force, S, on the cut section. Taking moments about this cut section for the free body, we have,

$$M + W < x - a > - R_Ax = 0 \qquad \textbf{(3.12)}$$

Here we have used Macaulay brackets to indicate that the expression applies anywhere along the beam on condition that the Macaulay convention is adhered to. The second term in equation 3.12, $W< x - a >$, is specific to a point load and is termed a 'singularity function'. Different forms of singularity function appear for different types of loading as we will see overleaf.

From (3.12) and the equation of the elastic line (i.e. (3.6)),

$$EI\frac{d^2y}{dx^2} = -M = -R_A x + W<x - a> = -\frac{Wb}{L}x + W<x - a> \quad (3.13)$$

Integrating:

$$EI\frac{dy}{dx} = -\frac{Wb}{2L}x^2 + W\frac{<x - a>^2}{2} + A \quad (3.14)$$

where A is an integration constant.

Integrating again:

$$EIy = -\frac{Wb}{6L}x^3 + W\frac{<x - a>^3}{6} + Ax + B \quad (3.15)$$

where B is also an integration constant.

We use the boundary conditions to solve for the two integration constants, A and B:

$y = 0$ at $x = 0$ $\therefore$ from (3.15)

$$0 = 0 + 0 + 0 + B$$

$$\therefore \qquad B = 0$$

(Note that the second term after the equal sign is zero because the quantity in the Macaulay bracket, $<x - a>$, is negative at $x = 0$ – 'The Macaulay Convention'.)

$y = 0$ at $x = L$ $\therefore$ from (3.15)

$$0 = \frac{-WbL^2}{6} + \frac{W(L-a)^3}{6} + AL$$

therefore:

$$A = \frac{-WbL^2}{6} + \frac{Wb^3}{6L} = \frac{Wb}{6}\left(L - \frac{b^2}{L}\right)$$

Substituting for A and B in (3.15), a general expression for y can be obtained as follows,

$$y = \frac{1}{EI}\left[-\frac{Wbx^3}{6L} + \frac{W<x - a>^3}{6} + \frac{Wbx}{6}\left(L - \frac{b^2}{L}\right)\right] \quad (3.16)$$

N.B. The Macaulay bracketed term in equation (3.16) is ignored when $x < a$.

The slope, dy/dx can be obtained from (3.14) or by differentiating (3.16).

A *special case* for this problem occurs when the load is applied at the centre of the beam span. The deflection at point C, the load point, is then given by putting $b = a$ and $x = a$ in (3.16), giving,

$$y = \frac{WL^3}{48EI}$$

which is the well-known result for the central deflection of a simply supported beam with the load applied at centre span.

Other loading conditions

Uniformly distributed load

Consider a uniformly distributed load (UDL), q N/m, acting over part of a beam's span as shown in Figure 3.24. The UDL runs from a position C, distance a from the left-hand end of the beam all the way to the right-hand end of the beam. A discontinuity occurs where the UDL commences at position C.

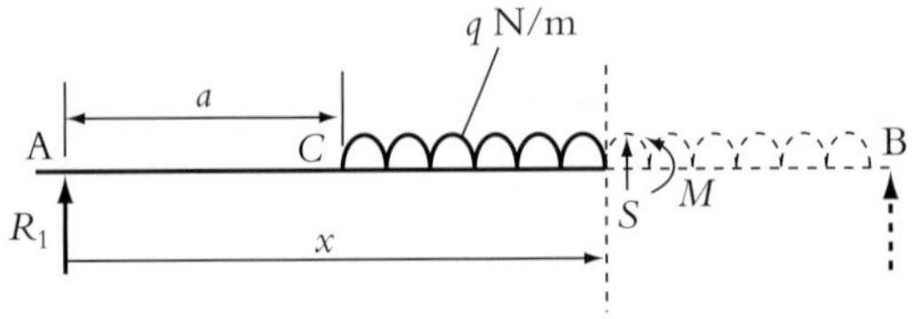

Figure 3.24 Simply supported beam carrying a uniformly distributed load (UDL)

Cutting a section to the right of the discontinuity and considering the part of the beam to the left of this section as a free body, moments can be taken about the section, giving the following equilibrium equation,

$$M + \frac{q<x-a>^2}{2} - R_1x = 0 \qquad \textbf{(3.17)}$$

Combining equation (3.17) with the equation of the elastic line,

$$EI\frac{d^2y}{dx^2} = -M = -R_1x + \frac{q<x-a>^2}{2}$$

As for a point load, this equation can be integrated twice to give the deflection at any position along the beam. Integration constants are determined from the boundary conditions as before. The procedure is identical to the point load case except that the singularity function, $\frac{q<x-a>^2}{2}$, is different. The Macaulay convention applies equally in this case, with the Macaulay brackets in the singularity function indicating whether the term is included or not depending on the sign of the bracketed term. It is excluded if $<x - a>$ is negative, i.e. if $x < a$.

How do we deal with a UDL which only acts over part of the beam span as shown in Figure 3.25(a)?

In this case, there are two discontinuities at distances a and b from the left-hand end of the beam. The UDL can be extended to the right-hand end of the beam and an additional negative counterbalance UDL superimposed over the newly extended part, also shown in Figure 3.25(b). This gives a statically equivalent system to the partially extended UDL.

Cutting a section in the newly extended part and looking at equilibrium of moments on the left-hand part of the beam we have,

$$M + \frac{q<x-a>^2}{2} - \frac{q<x-b>^2}{2} - R_1x = 0$$

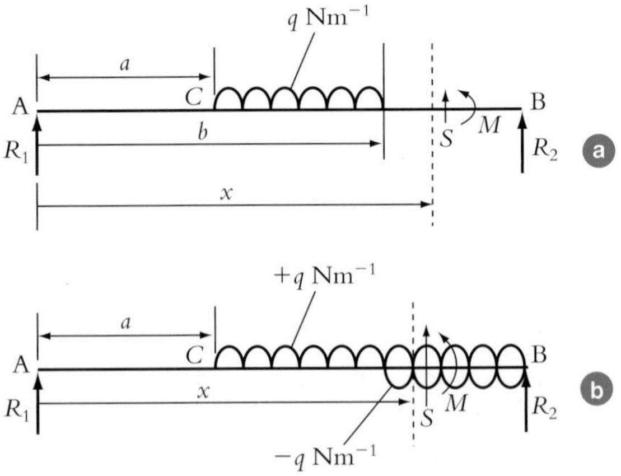

Figure 3.25 Dealing with a UDL acting over part of the beam's span – 'counterbalancing' the load

which, combining with the equation of the elastic line gives,

$$EI\frac{d^2y}{dx^2} = -M = -R_1x + \frac{q<x-a>^2}{2} - \frac{q<x-b>^2}{2}$$

As before, this equation can be integrated twice to give the deflection at any position along the beam. There are now two singularity functions corresponding to the discontinuities in UDL at positions a and b. The negative sign for the second function arises from the negative counterbalance UDL.

Point bending moment

Figure 3.26 shows both a point load, W_A, acting at position a, and a point bending moment, M_B, acting at position b on a simply supported beam. This problem has a combination of two loads (point load and point moment) acting on the beam and shows the versatility of Macaulay's method.

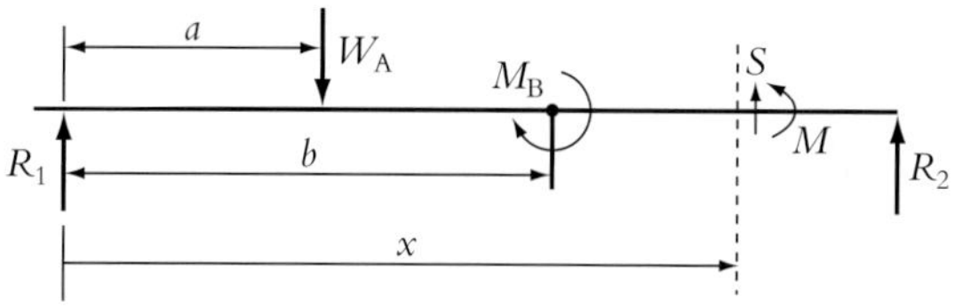

Figure 3.26 Simply supported beam carrying a point bending moment and a point load

Cutting a section to the right of the point bending moment discontinuity and considering the part of the beam to the left of this section as a free body, moments can be taken about the section, giving the following equilibrium equation,

$$M - R_1x + W_A<x - a> - M_B<x - b>^0 = 0 \qquad \textbf{(3.18)}$$

Note the form of the discontinuity function for the point bending moment, i.e. $M_B<x - b>^0$. The zero exponent is a special form that allows subsequent integration to give higher level exponents when deriving the expressions for slope and deflection as shown below. Combining equation (3.18) with the equation of the elastic line,

$$EI\frac{d^2y}{dx^2} = -M = -R_1x + W_A<x - a> - M_B<x - b>^0$$

Integrating gives,

$$EI\frac{dy}{dx} = -\frac{R_1x^2}{2} + W_A\frac{<x - a>^2}{2} - M_B<x - b> + A$$

The point bending moment singularity function has integrated to have an exponent of order 1 and A is an integration constant.

Further integration gives,

$$EIy = -\frac{R_1x^3}{6} + W_A\frac{<x - a>^3}{6} - M_B\frac{<x - b>^2}{2} + Ax + B \qquad \textbf{(3.19)}$$

The final form of the applied bending moment singularity has an exponent 2. There are two integration constants, A and B, which can be solved for by using the boundary conditions. As before, equation (3.19) can be used to determine the deflection at any position along the beam on condition that Macaulay's convention is used when evaluating the bracketed terms.

Summary of singularity functions

We have seen that Macaulay's method can be used to find deflections at any position along a beam where point loads, UDLs and point moments produce discontinuities in the bending moment distribution. The method can also be used where there is a combination of these loads acting on a beam.

When developing the bending moment equation for the beam with load discontinuities, the following singularity functions are used for each different type of load:

Load	Singularity function
Point load (W)	$W<x - a>$
UDL (q) with single discontinuity	$\frac{q<x - a>^2}{2}$
UDL (q) with double discontinuity	$\frac{q<x - a>^2}{2} - \frac{q<x - b>^2}{2}$
Point moment (M_B)	$M_B<x - b>^0$

It is interesting to note that in these singularity functions the exponent is 1 for a point load, 2 for a UDL and 0 for a point moment.

Summary of Macaulay's method for beam deflections

(1) Take the origin at the left-hand end of the beam.

(2) When necessary, extend and counterbalance uniformly distributed loading to the right-hand end of the beam

(3) Obtain an expression for M in the extreme right-hand part of the beam span, i.e. beyond the final discontinuity.

(4) Use Macaulay brackets in relevant singularity functions for discontinuities.

(5) Use the equation of the elastic line, i.e. $EI\,d^2y/dx^2 = -M$.

(6) Integrate to give dy/dx and y. Do not multiply out bracketed terms during these integrations.

(7) Use boundary conditions to determine integration constants.

(8) Evaluate slope (dy/dx) and deflection (y) at required positions. Bracketed terms are ignored when the value within brackets is negative.

Worked example

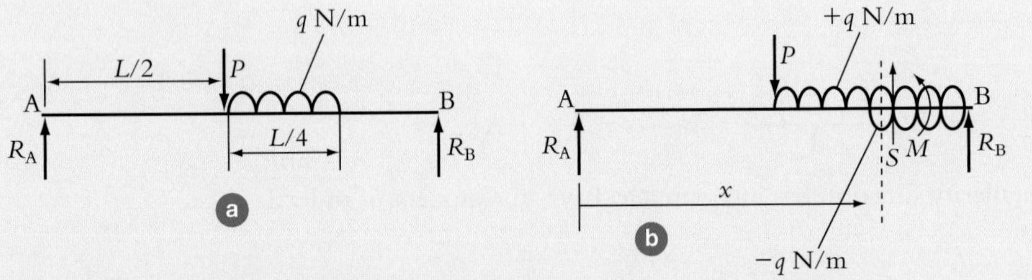

Figure 3.27 **Point load and UDL discontinuities acting on the same beam**

Figure 3.27(a) shows a simply supported beam carrying a point load, $P = 2\,\text{kN}$, at centre span and a uniformly distributed load, $q = 4\,\text{kN/m}$ over part of the right-hand half of the span. The length of the beam is 2 m and the flexural rigidity of its cross section is $EI = 10^5\,\text{Nm}^2$.

Use Macaulay's method to determine the deflection of the beam at centre span below the point load.

Reaction forces

Considering equilibrium of the full beam,

Vertical forces:

$$R_A + R_B = P + \frac{qL}{4} \tag{3.20}$$

Moments about A

$$\frac{PL}{2} + \frac{qL}{4}\frac{5L}{8} - R_B L = 0$$

$$R_B = \frac{P}{2} + \frac{5qL}{32} = 2250\,\text{N}$$

and from (3.19)

$$R_A = \frac{P}{2} + \frac{3qL}{32} = 1750\,\text{N}$$

Moment expression and integration

The UDL is extended to the right-hand end of the beam and a counterbalancing negative UDL is added, as shown in Figure 3.27(b). Taking moments about a section in the extreme right-hand end of the beam beyond the last discontinuity, we have,

$$M = R_A x - P<x - \frac{L}{2}> - \frac{q<x - \frac{L}{2}>^2}{2} + \frac{q<x - \frac{3L}{4}>^2}{2}$$

and combining with the equation of the elastic line,

$$EI\frac{d^2y}{dx^2} = -M = -R_A x + P<x - \frac{L}{2}> + \frac{q<x - \frac{L}{2}>^2}{2} - \frac{q<x - \frac{3L}{4}>^2}{2}$$

Integrating twice gives,

$$EI\frac{dy}{dx} = -\frac{R_A x^2}{2} + \frac{P<x - \frac{L}{2}>^2}{2} + \frac{q<x - \frac{L}{2}>^3}{6} - \frac{q<x - \frac{3L}{4}>^3}{6} + A$$

$$EIy = -\frac{R_A x^3}{6} + \frac{P<x - \frac{L}{2}>^3}{6} + \frac{q<x - \frac{L}{2}>^4}{24} - \frac{q<x - \frac{3L}{4}>^4}{24} + Ax + B \qquad \textbf{(3.21)}$$

where A and B are integration constants.

Boundary conditions

Setting the boundary conditions in equation (3.21) we have,

When $x = 0, y = 0 \quad \therefore\; B = 0$

N.B. Negative Macaulay bracketed terms are ignored, i.e. set to zero

When $x = L, y = 0$

$$\therefore \qquad 0 = \frac{R_A L^3}{6} + \frac{PL^3}{48} + \frac{qL^4}{16.24} + \frac{qL^4}{256.24} + AL$$

Substituting values for R_A, P, q and L, and solving for A gives,

$$A = 922\,\text{Nm}^2$$

Deflection at centre span

Equation (3.21) can now be used to evaluate the deflection at centre span. As $x = L/2$, the three Macaulay bracketed terms are zero (using Macaulay convention). Therefore,

$$EIy = -\frac{R_A x^3}{6} + Ax = -\frac{1750.1^3}{6} + 922.1 = 630$$

$$\therefore \qquad y = \frac{630}{10^5} = 6.3\,\text{mm}$$

Statically indeterminate beams

Macaulay's method can also be used to solve for deflections and slopes of a *statically indeterminate* beam, i.e. a beam where the reaction forces and moments cannot be determined by the equations of statics alone. An example is a clamped–clamped beam subjected to a point load, as shown in Figure 3.28(a).

Drawing the free-body diagram for this beam, Figure 3.28(b), the end reactions comprise moments M_A and M_B, which restrain rotation, and the vertical reaction forces R_A and R_B. There are therefore four unknowns that cannot be solved for by equilibrium conditions alone. They have to satisfy both equilibrium (two conditions) and the boundary conditions, i.e. slope and deflection at both ends (four conditions). Equilibrium is incorporated in the Macaulay bending moment equation, and, as we have seen, boundary conditions are also an integral part of the method.

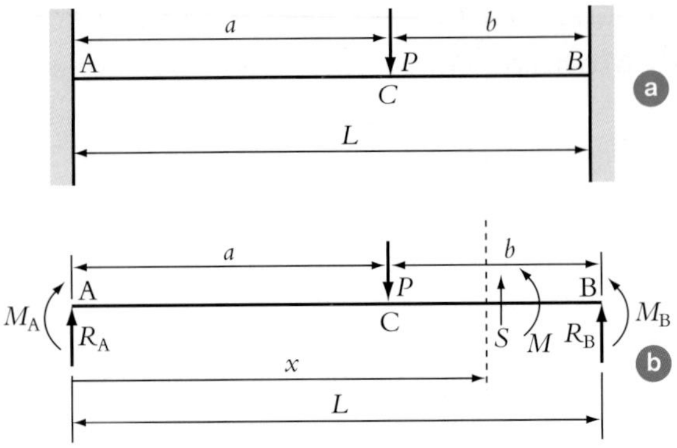

Figure 3.28 **Statically indeterminate beam: clamped–clamped beam subjected to a point load**

Taking a section, X–X, at distance x from the left-hand support, beyond the load discontinuity, and considering the left-hand part of the beam as a free body, the equation for the unknown moment, M, on the section is,

$$M = -EI\frac{d^2y}{dx^2} = M_A + R_A x - P<x - a>$$

Integrating this expression twice gives,

$$-EI\frac{dy}{dx} = M_A x + \frac{R_A x^2}{2} - \frac{P<x - a>^2}{2} + A$$

and

$$-EIy = \frac{M_A x^2}{2} + \frac{R_A x^3}{6} - \frac{P<x - a>^3}{6} + Ax + B \qquad (3.22)$$

This expression for the deflection, y, includes four unknowns, namely M_A and R_A, and the integration constants A and B. We can use the boundary conditions to solve for these unknowns.

Boundary conditions

When $x = 0, y = 0$

$$\therefore \qquad B = 0$$

When $x = 0, dy/dx = 0$

$$\therefore \qquad A = 0$$

N.B. Negative Macaulay bracketed terms are ignored, i.e. set to zero

When $x = L, y = 0$

$$\therefore \qquad 0 = \frac{M_A L^2}{2} + \frac{R_A L^3}{6} - \frac{P(L - a)^3}{6} \qquad (3.23)$$

When $x = L, dy/dx = 0$

$$\therefore \qquad 0 = M_A L + \frac{R_A L^2}{2} - \frac{P(L - a)^2}{2} \qquad (3.24)$$

N.B. Macaulay bracketed terms are converted back to normal brackets at this stage.

Equations (3.23) and (3.24) can be solved simultaneously to give R_A and M_A as follows,

$$R_A = \frac{P(L - a)^2(L + 2a)}{L^3} \qquad (3.25)$$

$$M_A = -\frac{Pa(L - a)^2}{L^2} \qquad (3.26)$$

Substituting for R_A and M_A from (3.25) and (3.26) into (3.22) gives the required general expression for deflection, y, at any position along the beam,

$$y = \frac{1}{EI}\left[\frac{Pa(L-a)^2x^2}{2L^2} - \frac{P(L-a)^2(L+2a)x^3}{6L^3} + \frac{P<x-a>^3}{6}\right]$$

Special case – centrally loaded beam

For a beam loaded at centre span, i.e. $a = L/2$, Equations (3.25) and (3.26) give,

$$R_A = \frac{P}{2}$$

$$M_A = -\frac{PL}{8}$$

And the maximum deflection at the centre of the span, under the load, is,

$$y_{max} = -\frac{1}{EI}\left[\frac{M_A L^2}{8} + \frac{R_A L^3}{48} - \frac{P(0)^3}{6}\right]$$

$$= \frac{PL^3}{192EI}$$

which is the well-known expression given in many textbooks.

Comparing this result with a simply supported beam, where the central deflection is given by $y_{max} = PL^3/48EI$, it can be seen that clamping the ends of the beam results in a deflection which is 25% of the deflection of a simply supported beam.

Learning summary

By the end of this section you should have learnt:

- ✓ to derive the differential equation of the elastic line (i.e. deflection curve) of a beam;
- ✓ to solve this equation by successive integration to yield the slope, dy/dx, and the deflection, y, of a beam at any position along its span;
- ✓ to use Macaulay's method, also called the method of singularities, to solve for beam deflections where there are discontinuities in the bending moment distribution arising from discontinuous loading;
- ✓ to use different singularity functions in the bending moment expression for different loading conditions including point loads, uniformly distributed loads and point bending moments;
- ✓ to use Macaulay's method for statically indeterminate beam problems.

3.5 Elastic–plastic deformations

Elastic–plastic material behaviour models

Elastic–perfectly plastic (EPP)

In this case there is assumed to be no hardening, i.e. the yield stress, σ_y, is assumed to remain constant at $\pm\sigma_y$, regardless of any previous plastic deformation, see Figures 3.29 and 3.30. Therefore, the yield surface doesn't change in either shape or position in the principal stress-space, see Figure 3.31.

This is a good model for mild steel with moderate plasticity, but is also used very generally for materials without well-defined yield, i.e. $\sigma_y = \sigma_{0.2\%}$, for moderate plasticity.

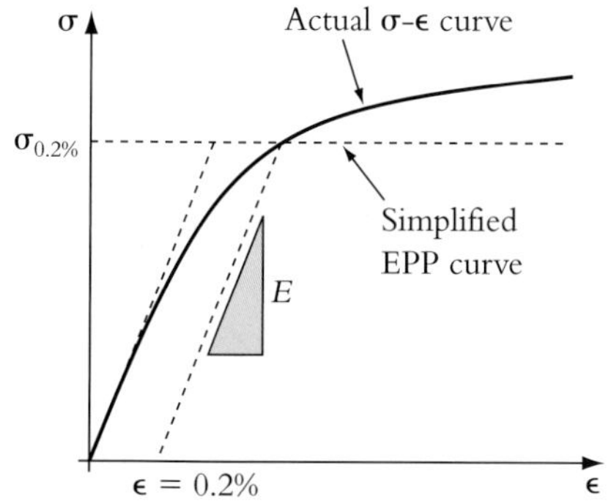

Figure 3.29 EPP stress–strain curve

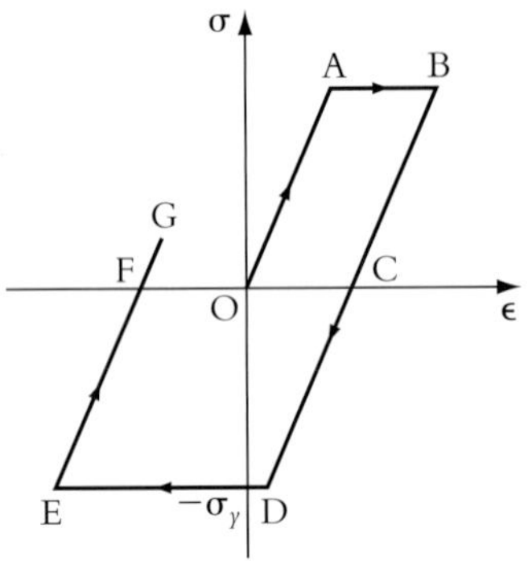

Figure 3.30 **Uniaxial cyclic σ–ε behaviour for an EPP material model**

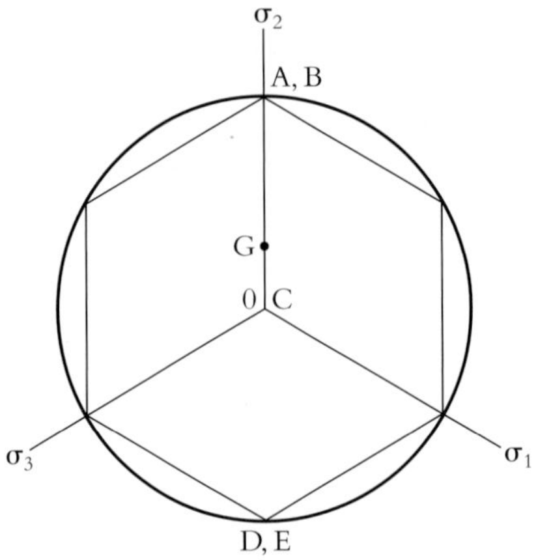

Figure 3.31 **Representation of uniaxial stress–strain behaviour (Figure 3.30) in principal stress space**

Isotropic hardening

For an isotropically hardening material, any plastic deformation causes an increase in the yield stress, and the yield surface continuously expands isotropically (i.e. by the same amount in all directions) with loading, reverse loading and reloading involving plastic deformation, as indicated in Figures 3.32 and 3.33. Thus, cyclic tension–compression loading, i.e. varying between equal and opposite fixed values of applied stress, which initially gives yield (in tension) will quickly 'shake down' to elastic behaviour, since the subsequent compressive stress will not be large enough to cause yield in compression.

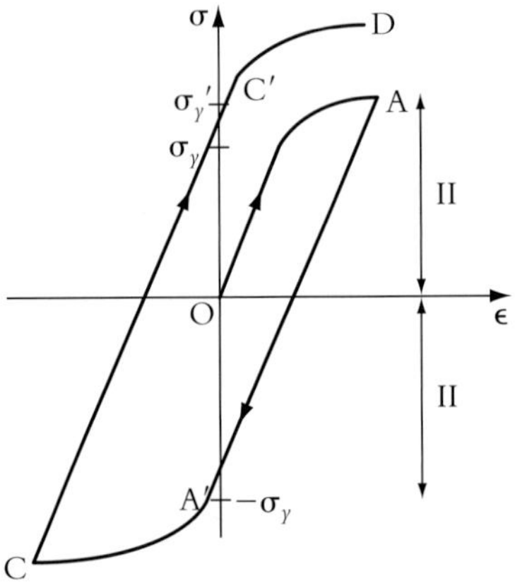

Figure 3.32 **Uniaxial cyclic σ–ε behaviour for an isotropic material model**

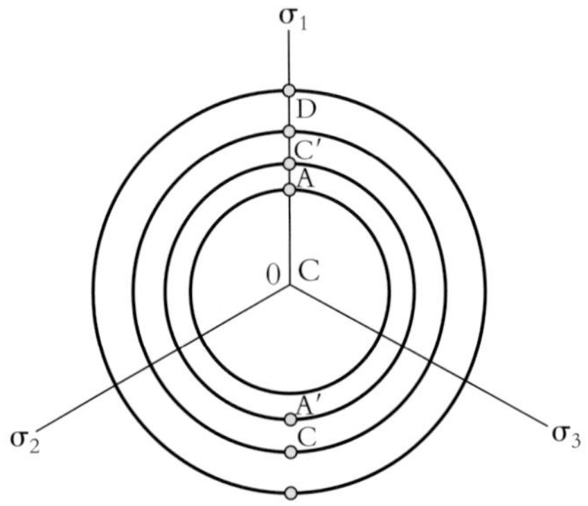

Figure 3.33 **Representation of isotropic stress–strain behaviour in principal stress space**

Kinematic hardening (assuming linear plasticity for simplicity)

For a kinematically hardening material, the yield stress in compression following yield in tension is reduced due to the tensile yield, as indicated in Figures 3.34 and 3.35.

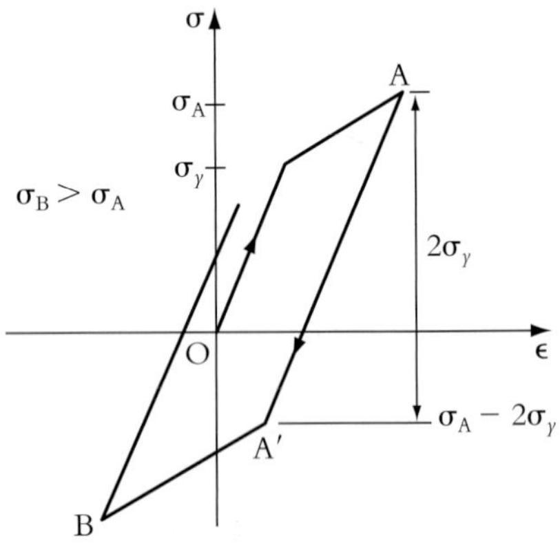

Figure 3.34 **Uniaxial cyclic σ–ε behaviour for a kinematically hardening material**

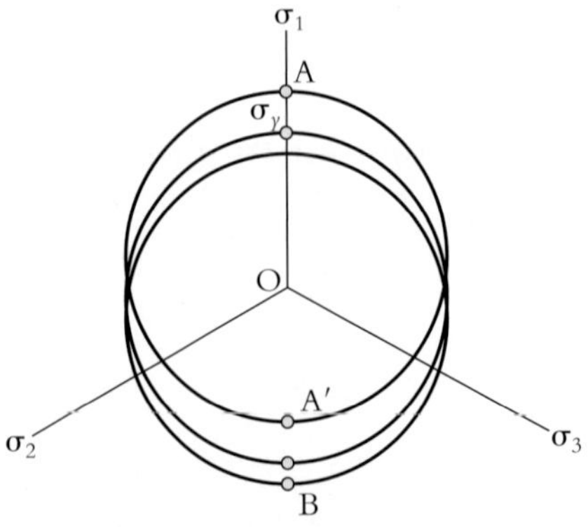

Figure 3.35 **Representation of kinematic stress-strain behaviour in principal stress space**

Kinematic hardening assumes that the yield range remains constant at a value of $2\sigma_y$, so that the subsequent yield in compression will be the highest tensile stress with plasticity, $-2\sigma_y$, as shown in Figure 3.34. This can be represented in principal stress space as a translation of the yield surface without changing shape or size, as shown in Figure 3.35. In this case, cycling between tensile and compressive stresses of equal and opposite values, which initially gives yield (in tension) will also give yield due to the subsequent compressive stress since the compressive yield stress will have been reduced. Thus, the kinematic hardening model predicts that constant alternating plastic strains will occur after the first loading cycle.

In the following analyses related to the elastic–plastic deformation of components (e.g. beams in bending, and torsion of shafts) only EPP material behaviour models will be considered.

Bending of a beam with rectangular cross section, assuming elastic–perfectly plastic material behaviour model

The following examples will illustrate the basic approach.

Worked example

A rectangular section beam (100 mm wide × 200 mm deep) is made from an elastic–perfectly plastic material with $E = 200\,\text{GN/m}^2$ and $\sigma_y = 250\,\text{N/mm}^2$. Calculate the radius of curvature and the bending stress distribution when a pure moment of 190 kNm is applied *and* after the moment is removed.

Check whether or not yield has occurred. The applied moment at first yield, M_y, is that which causes the maximum elastic bending stress to become equal to the yield stress. Since the maximum elastic stress occurs at the top and bottom (extreme) fibres of the cross section, first yield occurs when $\sigma = \pm\sigma_y$ at the positions $y = \pm 100$ mm. Thus, M_y is given by

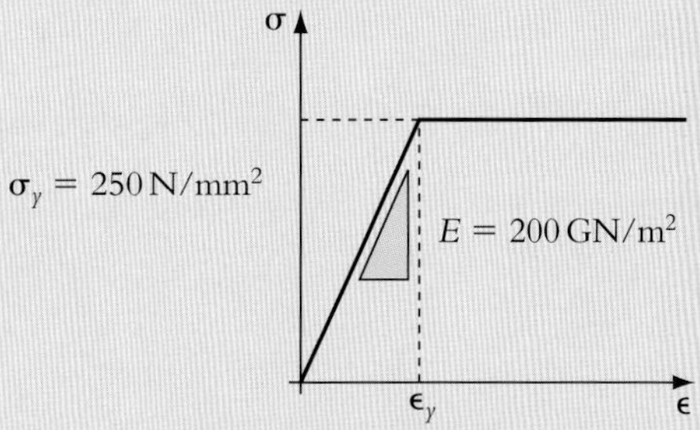

Figure 3.36 The EPP material behaviour model

$$M_y = \frac{\sigma_y I}{y} = \frac{250\,\text{N/mm}^2 \times \left(\dfrac{100 \times 200^3}{12}\right)\text{mm}^4}{100\,\text{mm}}$$

i.e. $M_y = 166.7 \times 10^6\,\text{Nmm} = 166.7\,\text{kNm}$

The applied moment (190 kNm) exceeds M_y and therefore yielding will occur.

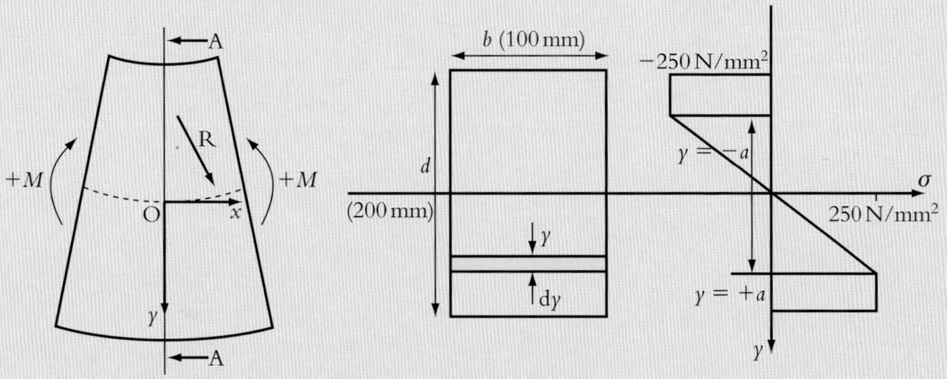

Figure 3.37 A section of the beam in bending, the beam cross section and the variation of stress through the beam cross section

Assume that yielding occurs at $y \geqslant a$ and $y \leqslant -a$.

Variation of stress with y:

$$\text{For } -a < y < a,\ \sigma = +250\frac{y}{a}\,\text{N/mm}^2$$

$$\text{For } a < y < \frac{d}{2},\ \sigma = +250\,\text{N/mm}^2$$

and

$$\text{for } -\frac{d}{2} < y < -a,\ \sigma = 250\,\text{N/mm}^2$$

Moment equilibrium:

$$\therefore \qquad M = \int_A y(\sigma dA) = \int_{y=-d/2}^{y=+d/2} \sigma y(bdy)$$

$$\text{i.e.} \qquad M = 2\int_{y=0}^{y=d/2} \sigma ybdy$$

$$\therefore \qquad M = 2\left\{\int_{y=0}^{y=+a}\left(-250\frac{y}{a}\right)ybdy + \int_{y=+a}^{y=+d/2} 250ybdy\right\}$$

$$= 2 \times 250b\left\{-\left[\frac{y^3}{3a}\right]_0^{+a} + \left[\frac{y^2}{2}\right]_{+a}^{+d/2}\right\}$$

$$= 500b\left\{\frac{a^2}{3} + \frac{d^2}{8} - \frac{a^2}{2}\right\}$$

$$\therefore \qquad M = 250b\left\{\frac{d^2}{4} - \frac{a^2}{3}\right\}$$

$$190 \times 10^6 = 250 \times 100\left(10^4 - \frac{a^2}{3}\right)$$

which gives

$$a = 84.85\,\text{mm}$$

Compatibility requirement:

$$\varepsilon = \frac{y}{R}$$

At $y = +84.85$ mm, $\sigma = 250\,\text{N/mm}^2$ and since this point is within the elastic range, use of the linear elastic stress–strain equation gives:

$$\varepsilon = \frac{\sigma}{E} = \frac{250}{200 \times 10^3} = \frac{84.85}{R} = \left(\frac{y}{R}\right)$$

$$R = 67.9 \times 10^3\,\text{mm} = 67.9\,\text{m}$$

On unloading, assume that the stress change which occurs is elastic, but check the resulting solution to establish whether or not reverse yielding has occurred in order to validate this assumption.

Assuming elastic unloading, the maximum stress change will occur at $y = +\frac{d}{2}$ and is given by

$$\Delta\sigma^{el}_{max} = -\frac{My}{I} = -\frac{-190 \times 10^6\,\text{Nmm} \times (+100\,\text{mm})}{\left(\frac{100 \times 200^3}{12}\right)\text{mm}^4}$$

$$= -285\,\text{N/mm}^2$$

$$\therefore \qquad \Delta\sigma^{el}_{max} = -285\,\text{N/mm}^2 \text{ at } y = +\frac{d}{2}$$

with a corresponding stress change of $+285\ \text{N/mm}^2$ at $y = -\frac{d}{2}$

By adding the stresses on the cross section when loaded to the stress changes that occur on unloading, the residual stresses can be obtained, as shown in Figure 3.38.

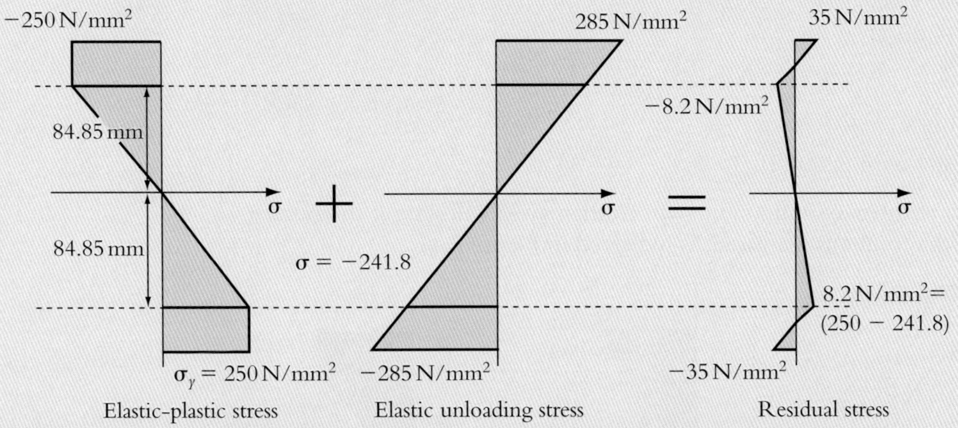

Figure 3.38 The stress distribution on loading, the stress change on unloading and the residual stress distribution

It is clear that the residual stress is well below the yield stress of $\pm 250\,\text{N/mm}^2$, so reverse yielding does not occur.

There will also be a residual curvature corresponding to the residual stresses. At $y =$ 84.85 mm, there has been no plastic deformation. Therefore, the residual strain is given by

$$\varepsilon\ (\text{residual}) = \frac{\sigma}{E} = \frac{8.2}{200 \times 10^3} = \frac{84.85}{R} = \left(\frac{y}{R}\right)$$

$$\therefore \qquad R = 2.070 \times 10^6\,\text{mm}$$

i.e. the residual radius of curvature is 2070 m.

On releasing the moment, the radius of curvature changes from 67.9 m to 2070 m. This change of curvature is called 'spring back' and is particularly important when bending bars of metal to specified radii of curvature.

Torsion of circular shafts, assuming an elastic–perfectly plastic material behaviour model

Figure 3.39 shows a circular shaft subjected to pure torque, T. This results in a twist, θ, and the rotation of an initially axial line by an angle γ. This causes the distortion of an element of material on the surface of the shaft, as indicated in Figure 3.39.

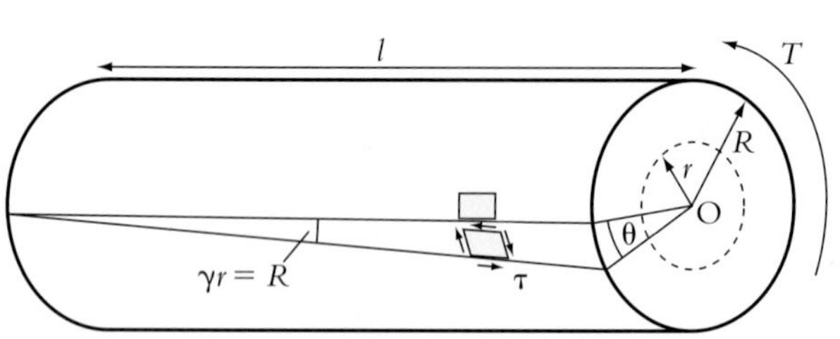

Figure 3.39 **A circular shaft subjected to a pure torque, *T***

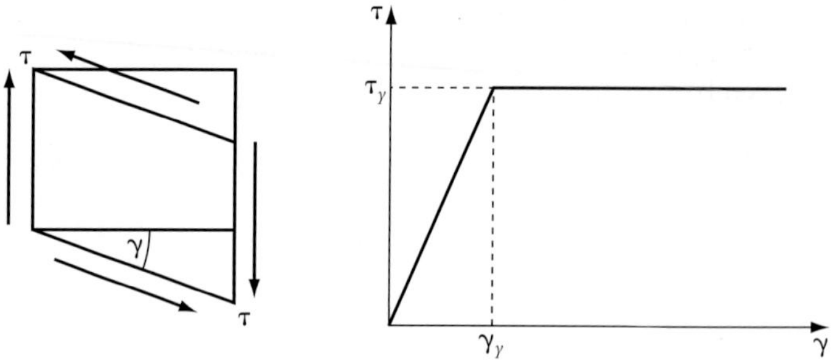

Figure 3.40 **Distortion of the initially rectangular element and τ–γ material behaviour**

Compatibility requirement:
Assuming plane sections remain plane and radii remain straight, then at $r = R$

$$l \times (\gamma_{r=R}) = R\theta \Rightarrow \gamma_{r=R} = \frac{R\theta}{l}$$

and hence at a general radial position, r,

$$\gamma = \frac{r\theta}{l} \qquad \textbf{(3.27)}$$

i.e. the shear strain varies linearly with radius, r.

Since the highest strains occur at the maximum radius $r = R$, then yield will start at the outer surface and spread inwards as the torque is increased, as indicated in Figure 3.41.

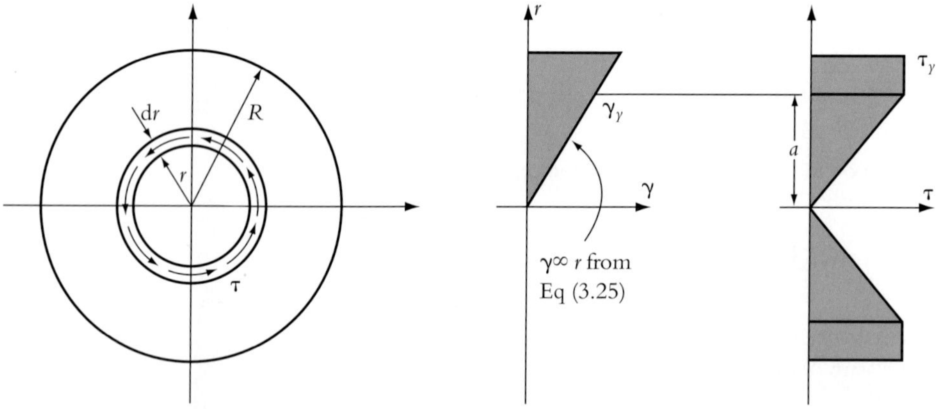

Figure 3.41 **Shear stress on the shaft cross section, variation of shear strain with radius and variation of shear stress with radius**

Equilibrium (moments about the axis):

$$T = \int_{r=0}^{r=R} r \times \tau \times 2\pi r dr \qquad \textbf{(3.28)}$$

Torque arm (r); stress × area ($\tau \times 2\pi r dr$) = force

i.e.

$$T = 2\pi \int_{r=0}^{r=R} \tau r^2 dr$$

As with the case of beam bending, residual stresses can be obtained by determining (elastically) the change of stress that occurs when the load is removed. Remember that it is necessary to check that the residual stresses are not great enough to cause reverse yield. For *elastic unloading*, the following elastic torsion formulae can be used:

$$\frac{T}{J} = \frac{\tau}{r} = \frac{G\theta}{l}$$

The relationship between τ–γ and the more commonly available uniaxial $\sigma - \varepsilon$ data depends on the yield criterion. The Mohr's circle for this case, Figure 3.42, indicates that $\sigma_1 = -\sigma_3 = \tau$ and $\sigma_2 = 0$. Hence, for a material obeying the von Mises criterion.

$$\sigma_{vm} = \frac{1}{\sqrt{2}}\sqrt{(\sigma_1 - \sigma_2)^2 + (\sigma_2 - \sigma_3)^2 + (\sigma_3 - \sigma_1)^2}$$
$$= \sigma_y \text{ at yield}$$
$$= \frac{1}{\sqrt{2}}\sqrt{(2\tau_y)^2 + (\tau_y)^2 + (\tau_y)^2} = \sqrt{3}\tau_y = \sigma_y \text{ at yield}$$
$$\Rightarrow \tau_y = \frac{\sigma_y}{\sqrt{3}} = 0.577\sigma_y$$

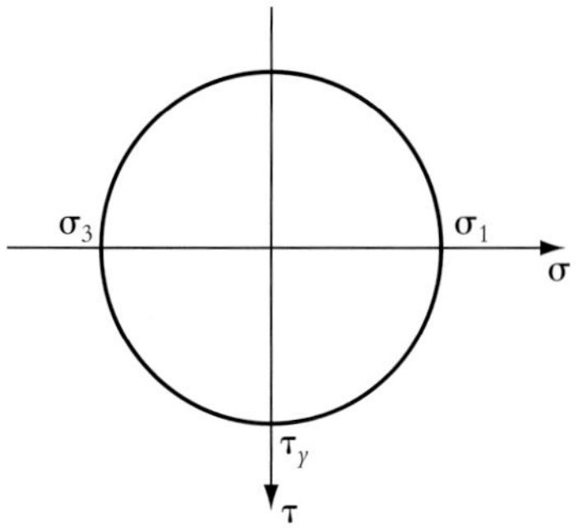

Figure 3.42 **Mohr's circle**

And for a material obeying the Tresca yield criterion

$$\sigma_y = \sigma_1 - \sigma_3 = \tau_y - (-\tau_y) = 2\tau_y = \sigma_{vm}$$

$\therefore$
$$\tau_y = \frac{\sigma_y}{2} = 0.5\sigma_y$$

Worked example

A solid circular shaft of diameter 50 mm and length 1000 mm is subjected to a torque, *T*. The shaft is made from an elastic–perfectly plastic material with $\tau_y = 100\,\text{N/mm}^2$, $G = 70\,\text{GN/m}^2$. Determine the magnitude of the torque required to cause yielding to occur at a radius of 15 mm (and greater) and the angle of twist.

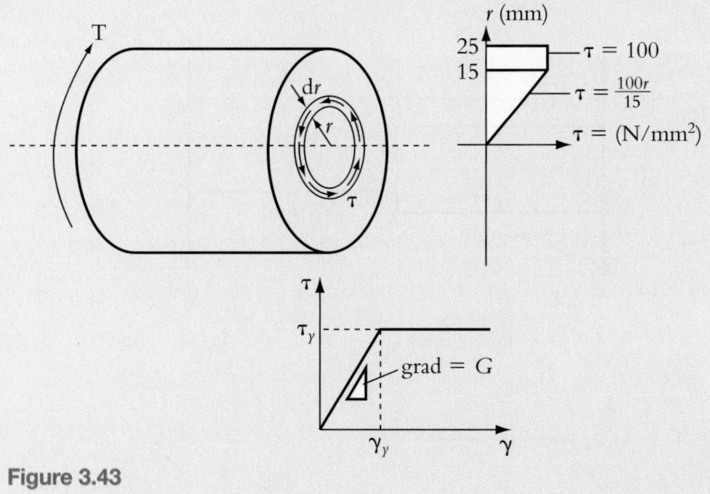

Figure 3.43

Equilibrium (moments about the axis)

$$T = \int_{r=0}^{r=25} r \times \tau \times 2\pi r \mathrm{d}r$$

Torque arm × force   stress × area

$$T = \int_0^{15} 2\pi\frac{100}{15}r^3\mathrm{d}r + \int_{15}^{25} 200\pi r^2\mathrm{d}r$$
$$= 200\pi\left\{\left[\frac{r^4}{15 \times 4}\right]_0^{15} + \left[\frac{r^3}{3}\right]_{15}^{25}\right\}$$
$$= 200\pi\left[\frac{15^4}{15 \times 4} + \frac{25^3}{3} - \frac{15^3}{3}\right]$$

$\therefore$
$$T = 3.096 \times 10^6\,\text{Nmm} = 3.096\,\text{kNm}$$

Relationship between τ_y and γ_y: at the outermost elastic point, $\tau = \tau_y$ and $\gamma = \gamma_y$ and the elastic relation $G = \dfrac{\tau_y}{\gamma_y}$ is applicable. It should be noted that outside this outermost elastic point,

i.e. $r = 15$ mm, the strain, which will be larger, will be elastic–plastic and consequently will not be governed by $\tau = G\gamma$, which is the elastic relation. Nonetheless, at $r = 15$ mm, we have

$$\gamma_y = \frac{\tau_y}{G} = \frac{100}{70\,000} = 1.4286 \times 10^{-3}\,\text{rad}$$

by invoking the compatibility requirement,

$$r_y\theta = \gamma_y l$$

and hence

$$\theta = \frac{\gamma_y l}{r_y}$$

i.e.

$$\theta = \frac{1.4286 \times 10^{-3} \times 1000\,\text{mm}}{15\,\text{mm}} \times \left(\frac{360}{2\pi}\frac{\text{deg}}{\text{rad}}\right)$$

$\therefore$

$$\theta = 5.456°$$

Worked example

A stepped circular shaft is built in at both ends; the two shaft segments are the same length, *l*. If the stepped shaft is made from an elastic–perfectly plastic material with $\tau_y = 100\text{N/mm}^2$, find the magnitude of the torque applied at the step to just cause yield in the smaller shaft.

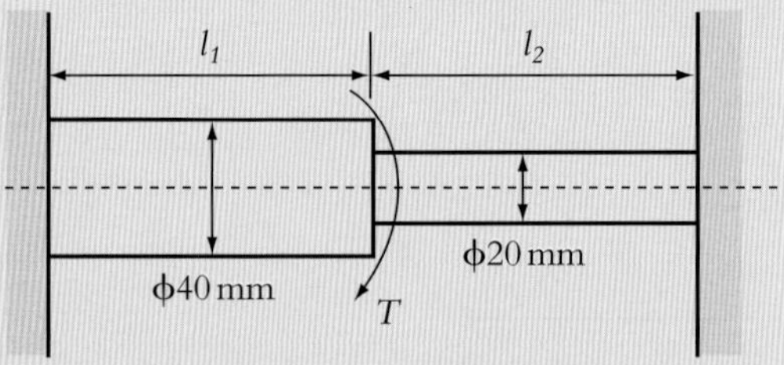

Figure 3.44

Stress–strain behaviour

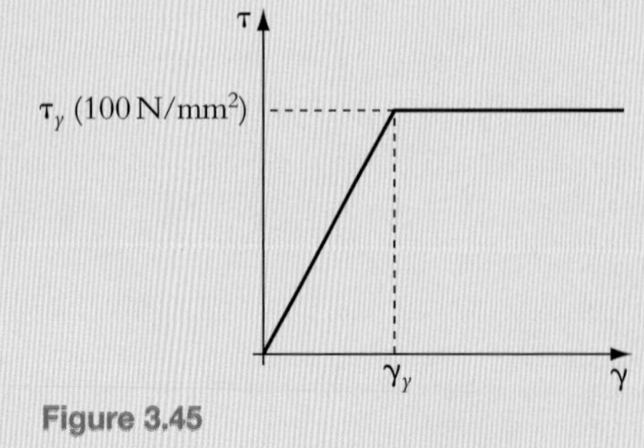

Figure 3.45

Equilibrium

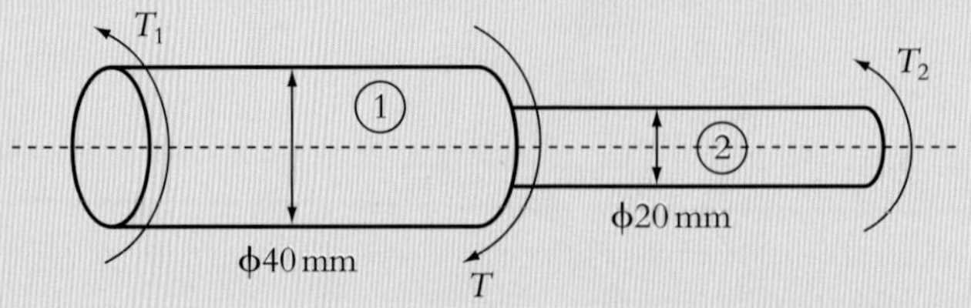

Figure 3.46

$$T = T_1 + T_2$$

The small shaft just reaches yield, so the linear elastic torsion equation can be used for that shaft, i.e.

$$\frac{T}{J} = \frac{\tau}{r} = \frac{G\theta}{l}$$

$$\Rightarrow \qquad \frac{T_2}{\frac{\pi \times 20^4}{32}\,\text{mm}^4} = \frac{100\,\text{N/mm}^2}{10\,\text{mm}}$$

$$\Rightarrow \qquad T_2 = 157.1 \times 10^3\,\text{Nmm} = 157.1\,\text{Nm}$$

For the large shaft, θ and l are the same as the small shaft, and since $\gamma = \frac{r\theta}{l}$ (elastic *or* plastic), then

$$\gamma_1|_{r=10\,\text{mm}} = \gamma_2|_{r=10\,\text{mm}}$$

Therefore, shaft 1 must also be elastic for $r < 10$ mm and will be plastic for $r > 10$ mm.

Thus, for the large shaft,

for $\qquad r < 10\,\text{mm}, \tau = \tau_y \frac{r}{10} = 10r$

for $\qquad r > 10\,\text{mm}, \tau = \tau_y = 100\,\text{N/mm}^2$

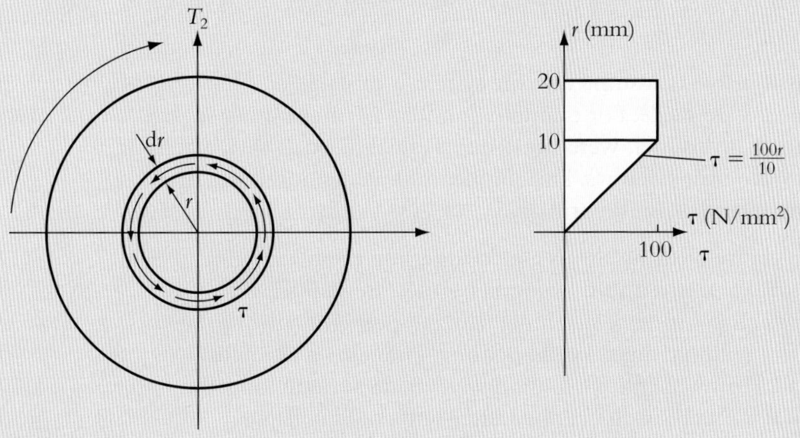

Figure 3.47

Equilibrium

$$T_1 = \int_{r=0}^{r=10} r \times \underbrace{10r}_{\text{stress}} \times \underbrace{2\pi r dr}_{\text{area}} + \int_{r=10}^{r=20} r \times \underbrace{100}_{\text{stress}} \times \underbrace{2\pi r dr}_{\text{area}}$$

Torque arm × force $\qquad$ Torque arm × force

$$T_1 = 20\pi\int_0^{10} r^3 dr + 200\pi\int_{10}^{20} r^2 dr$$

$$= 20\pi\left[\frac{r^4}{4}\right]_0^{10} + 200\pi\left[\frac{r^3}{3}\right]_{10}^{20}$$

$$= 20\pi \times \frac{10^4}{4} + 200\pi\left[\frac{20^3}{3} - \frac{10^3}{3}\right]$$

i.e. $\qquad T_1 = 1623 \times 10^3\,\text{Nmm} = 1623\,\text{Nm}$

i.e. $\qquad \text{total torque} = T = T_1 + T_2$

$$= (1623 + 157.1)\,\text{Nm}$$

$\therefore \qquad T = 1780\,\text{Nm}$

Learning summary

By the end of this section you should have learnt:

✓ the shapes of uniaxial stress–strain curves and the elastic–perfectly plastic approximation to uniaxial stress–strain curves;

✓ the kinematic and isotropic material behaviour models used to represent cyclic loading behaviour;

✓ the elastic–plastic bending of beams and the need to use equilibrium, compatability and σ–ε behaviour to solve these types of problems;

✓ the elastic–plastic torsion of shafts and the need to use equilibrium, compatability and τ–γ behaviour to solve these types of problems;

✓ to determine residual deformations and residual stresses.

3.6 Elastic instability

Introduction

For many structural problems, it is reasonable to assume that the system is in stable equilibrium. However, not all structural arrangements are stable. For example, consider a 1 metre-long stick with the cross-sectional area of a pencil. If this stick were stood on its end, the axial stress would be small, but the stick could easily topple over sideways. This simple example demonstrates that in some configurations, stability considerations can be primary.

This section is concerned with the stability of struts. Struts are compression members with cross-sectional dimensions that are small compared to the length, i.e. they are slender. If a circular rod of, say, 5 mm diameter, which has its ends machined flat and perpendicular to the axis, were made 10 mm long to act as a column, there would not be a problem of instability and it could carry a considerable load. However, if the same rod were made a metre long, the rod would become laterally unstable at a much smaller applied force and could collapse.

Buckling also occurs in many other situations with compressive forces. Examples include thin sheets, which have no problem carrying tensile loads, narrow beams unbraced laterally, and vacuum tanks, as well as submarine hulls. Thin-walled tubes can wrinkle like paper when subjected to torque.

Buckling phenomenon

Consider the response of a marble when subjected to disturbances from an initial equilibrium position on different types of surfaces, as shown in Figure 3.48. If the surface is concave, the marble will return to its original equilibrium position and the marble is said to be in a stable equilibrium position. If the surface is flat the marble will move to another equilibrium position and the marble is said to be in a neutral equilibrium position. Finally, if the surface is convex, the marble will roll off uncontrollably in an unstable fashion and the marble is said to be in an unstable equilibrium position.

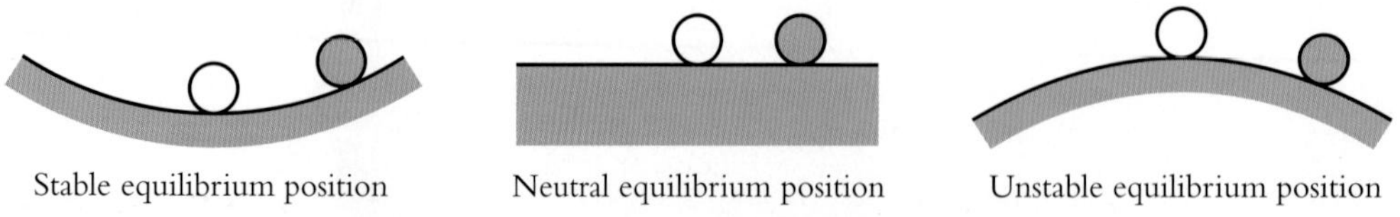

Figure 3.48 Equilibrium states for a marble on various surfaces

This analogy is useful for understanding the energy approach to buckling problems. Every deformed structure has a potential energy associated with it, which depends on the strain energy stored in the structure and the work done by the external loads. A concave potential energy function at equilibrium gives a stable equilibrium while a convex potential energy function gives unstable equilibrium.

Alternatively, buckling problems may be treated as bifurcation problems. Referring to Figure 3.49(a), it is clear that the tensile force will tend to restore the bar to equilibrium if there is a slight displacement to the right. However, the same bar under the action of a compressive force, Figure 3.49(b), will continue to fall when subjected to a slight displacement. This illustrates unstable equilibrium.

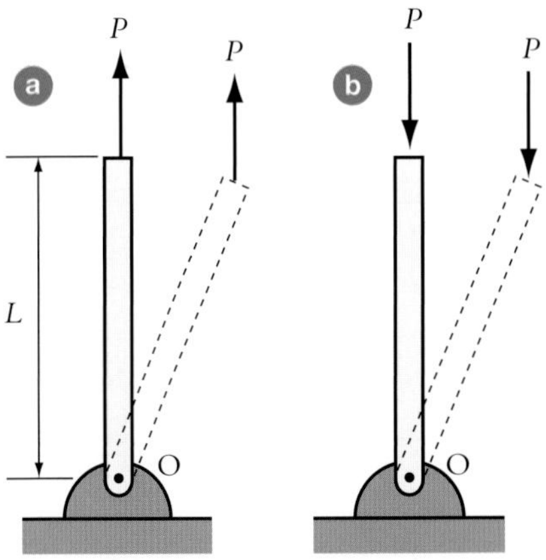

Figure 3.49 Examples of stable and unstable equilibrium

Figures 3.50(a) and (b) illustrate a slightly more complicated example of the same phenomenon. The vertical bar is supported horizontally by two springs of stiffness, k. If the bar of length L, is displaced a small amount, x, horizontally, there is a displacing moment of Px about O and a restoring moment $2kxL$. Hence we get $Px < 2kxL$ for stable equilibrium and $Px > 2kxL$ for unstable equilibrium.

The critical condition occurs when

$$Px = 2kxL$$

or

$$P_c = 2kL$$

where P_c is termed the critical load between stable and unstable equilibrium.

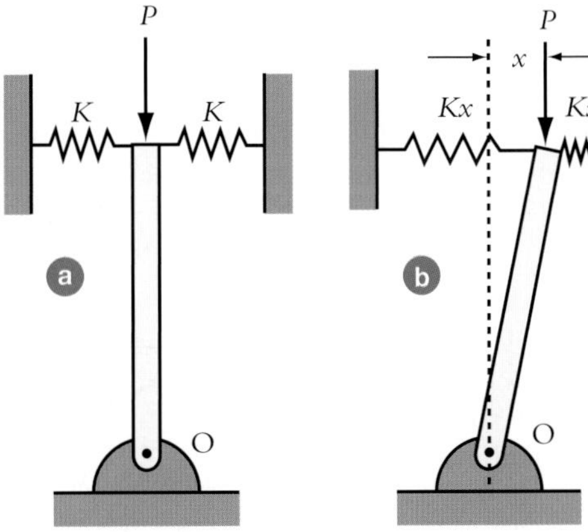

Figure 3.50 Axially loaded rigid bar with transverse springs

Figure 3.51 shows a rigid bar subjected to a compressive axial load with a torsional spring at its base; a free-body diagram of the problem is also shown. Taking moments about the point O, gives.

$$PL \sin \theta = K_\theta \theta$$

$$\frac{PL}{K_\theta} = \frac{\theta}{\sin \theta}$$

Figure 3.52 shows a graph of $\frac{PL}{K_\theta}$ versus θ. There is a stable region for low loads and an unstable region for high loads. Below point A, the bar will return to its equilibrium position if rotated slightly to either the right or the left. Once the load

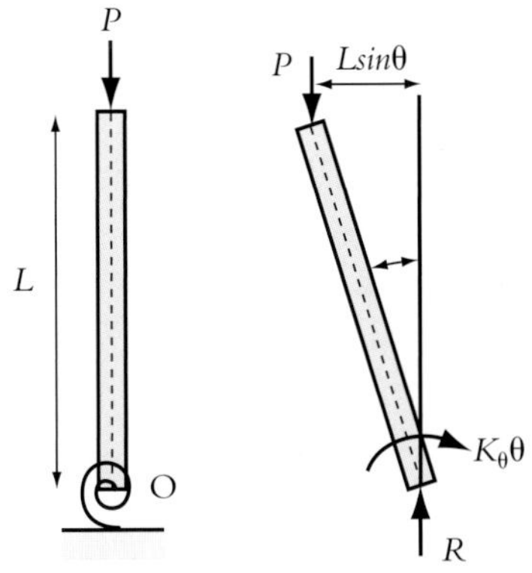

Figure 3.51 Rigid bar supported by a torsional spring

exceeds the value at point A, then any disturbance will cause the bar to rotate along either the right branch or the left branch of the bifurcation curve. Point A is called the bifurcation point, at which there are three possible solutions. The associated load at point A is called the critical (buckling) load. $\theta = 0°$ is the trivial solution. Only non-trivial solutions are generally of interest, i.e. for $\theta \neq 0°$.

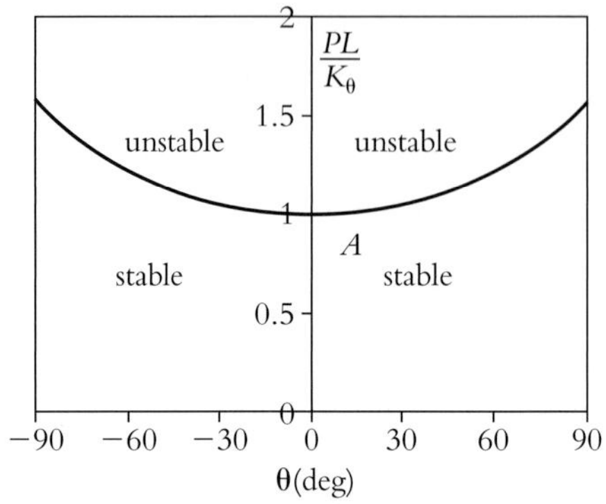

Figure 3.52 Variation of PL/K_σ with θ, indicating stable and unstable regions

Ideal struts

Ideal struts are assumed to be initially perfectly straight and of uniform section, and subjected to purely axial loading. Expressions will be developed relating the critical buckling load to the applied load, the material properties and the member dimensions, for different support conditions of the struts. At a critical load, members that are circular or tubular in cross section could buckle sideways in any direction. Often, compression members do not have equal flexural rigidity, EI, in all directions and there will be one axis about which the flexural rigidity is a minimum, depending on the dimensions. The member will therefore buckle about this axis and the I-value (second moment of area) referred to here is assumed to be the minimum value, based on the nominal dimensions of the member.

Case 1: Hinged–hinged

Consider an initially straight strut with its ends free to rotate around frictionless pins, as shown in Figure 3.53, which will be referred to as a hinged–hinged case. The dashed line represents the initially straight strut. The strut is now considered to be perturbed, from its initially straight position, as shown in Figure 3.53. This perturbation is equivalent to the movement of the marble in Figure 3.48.

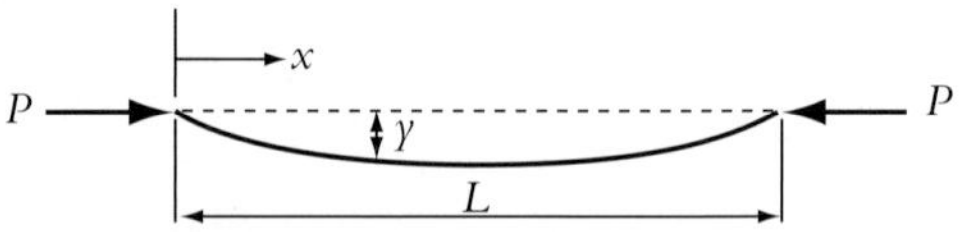

Figure 3.53 A hinged–hinged strut

The bending moment, M, depends on the deflection, y, and is hence a function of position, x.

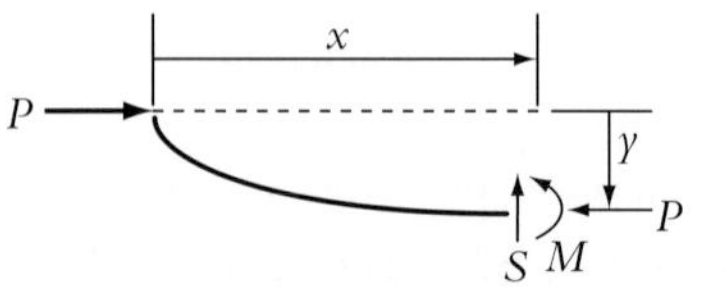

Figure 3.54 Free-body diagram for the left-hand portion of the strut

The deflection, y, is related to the moment M, by the relationship

$$EI\frac{d^2y}{dx^2} = -M$$

and

$$M = Py$$

Therefore,

$$\frac{d^2y}{dx^2} + \frac{P}{EI}y = 0$$

or

$$\frac{d^2y}{dx^2} + \alpha^2 y = 0$$

where

$$\alpha^2 = \frac{P}{EI}$$

Hence,

$$y = A \sin \alpha x + B \cos \alpha x$$

In order to determine A and B, we need two boundary conditions, i.e. at $x = 0$, $y = 0$ and at $x = l$, $y = 0$.

Therefore, B $= 0$ and A $\sin(\alpha l) = 0$

The condition $A = 0$ results in a trivial solution, i.e. $y = 0$, which is the case for an undeflected strut. Hence, the non-trivial solution is $\sin(\alpha l) = 0$, which gives $\alpha l = n\pi$, where $n = 0, 1, 2, \ldots$

$$\therefore \quad \alpha^2 l^2 = n^2\pi^2$$

$$\frac{P}{EI} l^2 = n^2\pi^2$$

or

$$P = \frac{n^2\pi^2 EI}{l^2}$$

$n = 0$ gives another trivial solution, i.e. $P = 0$,
$n = 1$ gives

$$P_c = \frac{\pi^2 EI}{l^2}$$

This is called the **Euler buckling** (or **crippling**) **load**, P_c; it is the lowest load at which buckling can occur (Euler solved this problem in 1757).
For $n = 2$ we have

$$P = \frac{4\pi^2 EI}{l^2}$$

and this corresponds to a different deflected (buckling) shape of the strut.

For $n = 1$, $y = y_{max}$ at $x = \frac{l}{2}$ and therefore $A = y_{max}$ and the deflected shape of the strut is given by the following expression:

$$y = y_{max} \sin(\alpha x) = y_{max} \sin\left(\frac{n\pi x}{l}\right)$$

The magnitude of y_{max} cannot be determined from the boundary conditions and it can become arbitrarily large, leading to elastic instability of the structure. The first three buckling mode shapes are shown in Figure 3.55. If buckling mode I is prevented from occurring by installing a restraint (support), then the column would buckle at the next highest mode at a critical load value that is higher than for the lower mode. The inflexion points, I, for each deflection curve have zero deflection. Recalling that the curvature $\frac{d^2y}{dx^2}$ at an inflexion point is zero indicates that the internal moment at these points is zero. If roller supports are put at any other point than point I, the boundary value problem must be solved for new eigenvalues (buckling loads) and eigenvectors (mode shapes).

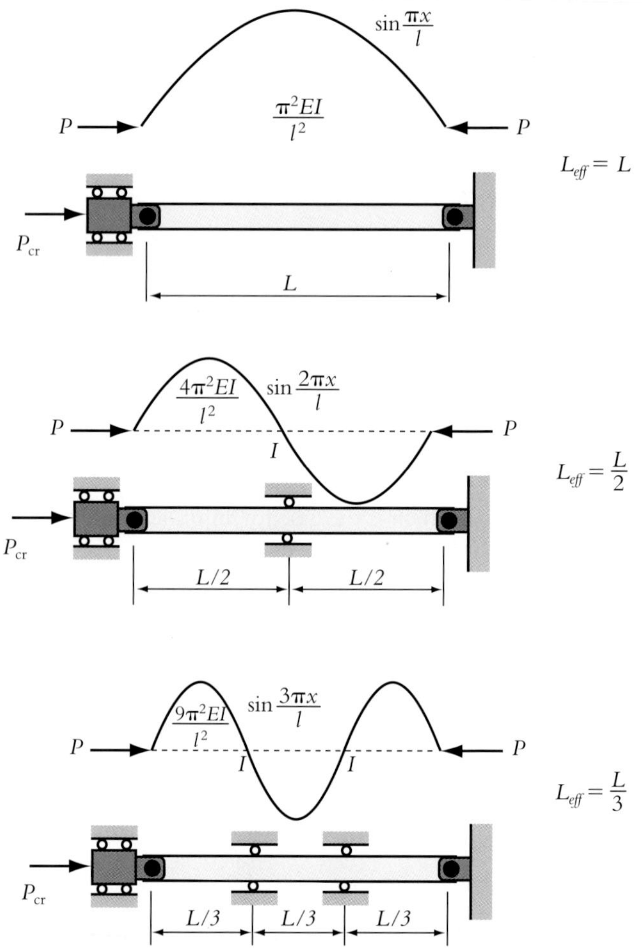

Figure 3.55 Buckling mode shapes for a hinged–hinged strut with $n = 1$, $n = 2$ and $n = 3$

Case 2: Free–fixed

Figure 3.56 shows the deflected shape and free-body diagram for a fixed-free strut.

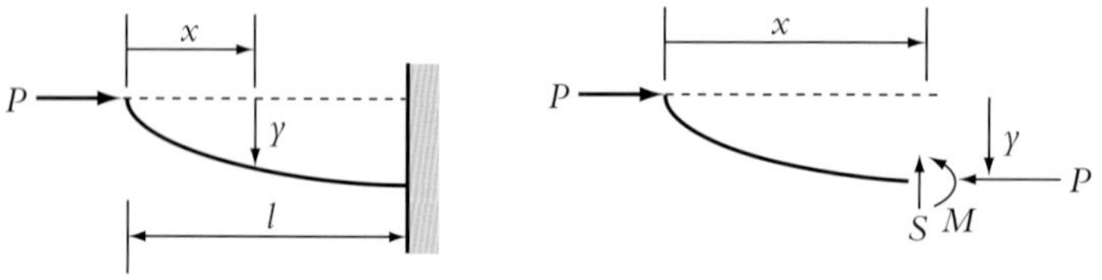

Figure 3.56 A fixed–free strut

$$EI\frac{d^2y}{dx^2} = -M$$

Where $$M = Py$$

$\therefore$ $$EI\frac{d^2y}{dx^2} + Py = 0$$

or $$\frac{d^2y}{dx^2} + \alpha^2 y = 0$$

where $$\alpha^2 = \frac{P}{EI}$$

The solution to this differential equation is:

$$y = A \sin \alpha x + B \cos \alpha x$$

At $x = 0$, $y = 0$, and therefore $B = 0$.

At $x = l$, $\frac{dy}{dx} = 0$, and therefore $A\alpha \cos(\alpha l) = 0$

So far, the mathematical solution is practically identical to that of a free–free strut. However, the boundary conditions are different, i.e. in this case, $A = 0$ or $\alpha = 0$ or $\cos(\alpha l) = 0$, leading to trivial solutions (as before). The non-trivial solution results from taking $\cos(\alpha l) = 0$, which implies that $\alpha l = \frac{n\pi}{2}$

i.e. $$\frac{P}{EI} l^2 = \frac{n^2\pi^2}{4} \text{ where } n = 1, 3, \ldots$$

The smallest, non-trivial, value of P occurs with $n = 1$, i.e.

$$P_c = \frac{\pi^2 EI}{4l^2}$$

By comparison with the hinged–hinged case, it can be seen that the solution is the same except that 'l' is replaced by '$2l$', i.e. the fixed–free case can be treated as the hinged–hinged case for a strut with an equivalent length of $2l$.

Case 3: Fixed–fixed

The solution procedure is the same as that for cases 1 and 2, leading to:

$$P_c = \frac{4\pi^2 EI}{l^2}$$

The fixed–fixed case shows a significant increase in the buckling capacity relative to the hinged–hinged case.

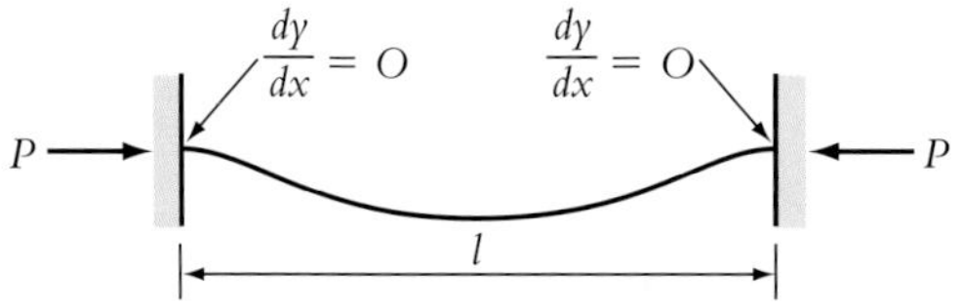

Figure 3.57 A fixed–fixed strut

Case 4: Fixed–hinged

$$P_c = \frac{2.045\pi^2 EI}{l^2} \left(\approx \frac{2\pi^2 EI}{l^2}\right)$$

This case differs from the previous examples in that a transverse force, R, is necessary to create this mode of deformation.

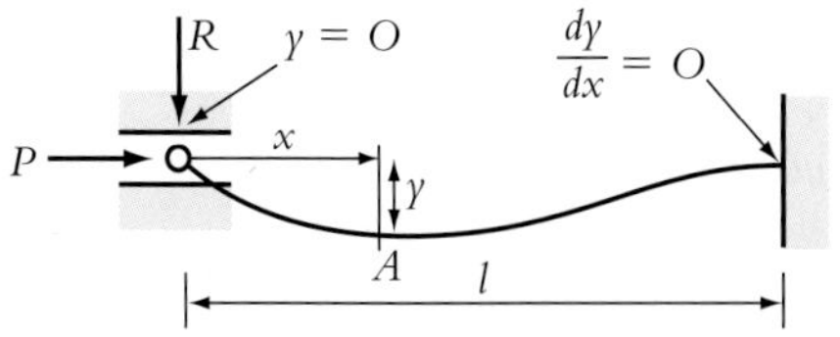

Figure 3.58 A fixed-hinged strut

$$EI\frac{d^2y}{dx^2} = -M$$

but $$M + Rx = Py$$

$$\frac{d^2y}{dx^2} + \frac{P}{EI}y = \frac{R}{EI}x$$

i.e. $$\frac{d^2y}{dx^2} + \alpha^2 y = \frac{Rx}{EI}$$

where $$\alpha^2 = \frac{P}{EI}$$

The solution to this type of differential equation consists of two parts, a homogenous solution and a particular integral. The homogeneous solution is given by

$$y = A \sin \alpha x + B \cos \alpha x$$

The particular integral for such second order differential equations is generally obtained by taking

$$P \cdot I = y = C \cdot f(x)$$

where

$$\frac{d^2y}{dx^2} + \alpha^2 y = \text{const} \cdot f(x)$$

And substituting for y in the differential equation gives the solution for C. In this particular case, $f(x) = x$, so that

$$\alpha^2 C \cdot f(x) = \frac{R}{EI} f(x)$$

$$C = \frac{R}{\alpha^2 EI} = \frac{R}{P}$$

$$\text{P.I} = y = \frac{R}{P} \cdot f(x) = \frac{R}{P} x$$

Hence, the complete solution is:

$$y = \underbrace{A \sin \alpha x + B \cos \alpha x}_{\text{Homogeneous solution}} + \underbrace{\frac{R}{P} x}_{\text{Particular integral}}$$

There are three unknowns this time, A, B and R. Therefore, we need three boundary conditions, i.e. at $x = 0$, $y = 0$ and so $B = 0$,

at $\quad x = l, y = 0, \text{and so, } 0 = A \sin \alpha l + \frac{R}{P} l$

Therefore

$$A = -\frac{Rl}{P \sin \alpha l}$$

Also, at $x = 1$,

$$\frac{dy}{dx} = 0,$$

$$-\frac{Rl\alpha}{P \sin \alpha l} \cos \alpha l + \frac{R}{P} = 0$$

Therefore

$$\tan \alpha l = \alpha l$$

The smallest non-trivial root to this equation is

$$\alpha l = 4.493 (\approx 1.43\pi)$$

$\therefore$

$$\alpha^2 l^2 = \frac{P}{EI} l^2 = 1.43^2 \pi^2$$

i.e.

$$P_c = \frac{2.045\pi^2 EI}{l^2}$$

Example problem

A steel framework with four vertical legs of solid rectangular cross section (20 mm × 30 mm) each of length 400 mm is to be used to support a platform carrying equipment. Determine the maximum weight that the framework can support using the following assumptions:

(a) hinged–hinged connections at both ends

(b) fixed–fixed connections at both ends

(c) hinged–fixed connections.

Assume $E = 200$ GPa and a yield stress of 500 MPa for the steel.

Case 5: Hinged–hinged with initial curvature

Figure 3.59 shows a hinged–hinged strut with initial curvature, y_0, and the deflected shape, y.

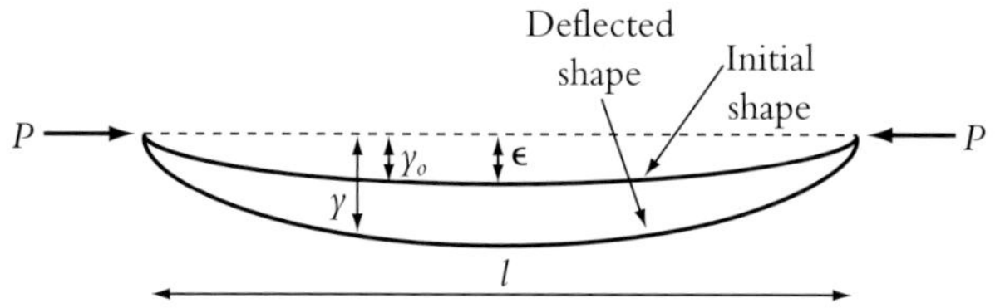

Figure 3.59 A hinged–hinged strut with initial curvature

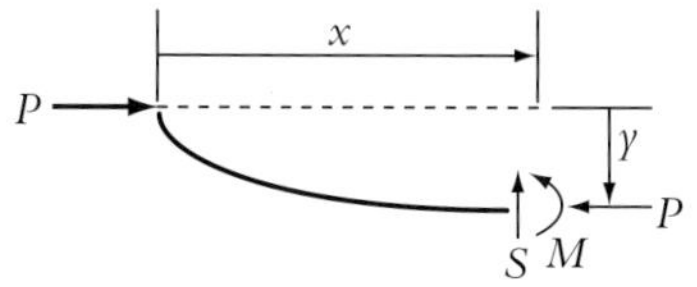

Figure 3.60 Free-body diagram for a hinged–hinged strut with initial curvature

Suppose that $y_0 = \varepsilon \sin\left(\frac{\pi x}{l}\right)$ describes the initial shape.

From the free-body diagram (Figure 3.60),

$$EI\frac{d^2(y - y_0)}{dx^2} = -Py$$

$$\therefore \quad \frac{d^2y}{dx^2} + \alpha^2 y = -\frac{\pi^2}{l^2}\varepsilon \sin\left(\frac{\pi x}{1}\right)$$

Where

$$\alpha^2 = \frac{P}{EI}$$

In this case there is a particular integral, as well as a homogeneous function.

Assume the particular integral is obtained by taking $y = C\sin\left(\frac{\pi x}{l}\right)$, where C is an unknown constant, and

substitute this into the differential equation:

$$-C\frac{\pi^2}{l^2}\sin\left(\frac{\pi x}{l}\right) + \alpha^2 C\sin\left(\frac{\pi x}{l}\right) = -\frac{\pi^2}{l^2}\varepsilon\sin\left(\frac{\pi x}{l}\right)$$

$$\Rightarrow \quad C = \frac{-\frac{\pi^2}{l^2}\varepsilon}{\left(\alpha^2 - \frac{\pi^2}{l^2}\right)}$$

Thus the overall solution, i.e. the homogeneous function and the particular integral, is

$$y = A\sin(\alpha x) + B\cos(\alpha x) - \frac{\frac{\pi^2}{l^2}\varepsilon\sin\left(\frac{\pi x}{l}\right)}{\left(-\frac{\pi^2}{l^2} + \alpha^2\right)}$$

The A and B values are obtained by substituting the boundary conditions

At $x = 0$, $y = 0$

$\therefore \quad B = 0$

At $x = l$, $y = 0$

$\therefore \quad 0 = A \sin(\alpha l)$, i.e. $A = 0$

$$\therefore \quad y = \frac{\frac{\pi^2}{l^2}\varepsilon\sin\frac{\pi x}{l}}{-\frac{\pi^2}{l^2} + \alpha^2}$$

When $\frac{\pi^2}{l^2} = \alpha^2$, $y = \infty$ (except at $x = 0$ or l)

i.e.

$$\frac{\pi^2}{l^2} = \frac{P}{EI}$$

$$\therefore \quad P = \frac{\pi^2 EI}{l^2}$$

i.e. the initially curved strut buckles when

$$P(=P_c) = \frac{\pi^2 EI}{l^2}$$

and
$$y = \frac{\varepsilon \sin\left(\frac{\pi x}{l}\right)}{\left(\frac{l^2 P}{\pi^2 EI} - 1\right)} = \frac{\varepsilon \sin\left(\frac{\pi x}{l}\right)}{\left(\frac{P}{P_c} - 1\right)}$$

Therefore, the assumed initial shape has little effect on the solution, i.e.

$$P_c = \frac{\pi^2 EI}{l^2}$$

and the deflected shape, y, would be practically the same.

Eccentrically loaded struts

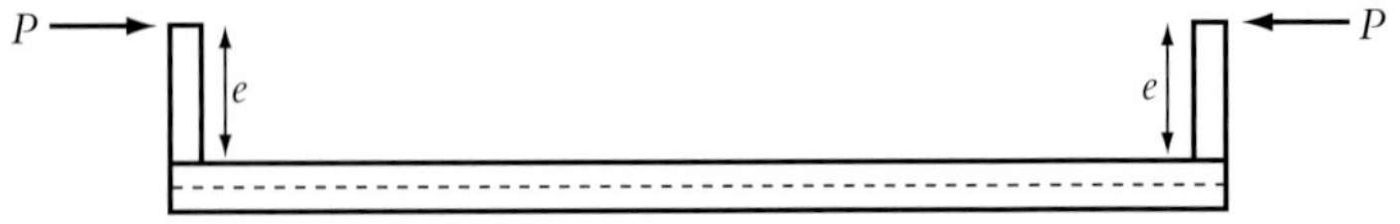

Figure 3.61 **An eccentrically loaded strut**

Consider a long, slender member subjected to a slightly eccentric compressive load as shown in Figure 3.61; the axial load is P and the eccentricity of the load is e. This form of loading will cause the member to curve so that at a distance x from the left-hand end, the eccentricity of the load becomes $(e + y)$. The deflection, y, is related to the moment, M, by the relationship

$$EI\frac{d^2y}{dx^2} = -M$$

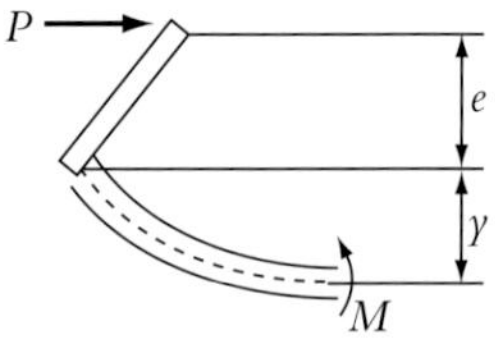

Figure 3.62 **Free-body diagram for an eccentrically loaded strut**

and, x from the left-hand-end,

$\therefore$
$$M = P(e + y)$$

$$EI\frac{d^2y}{dx^2} = -P(e + y)$$

or

$$\frac{d^2y}{dx^2} + \alpha^2 y = -\alpha^2 e$$

where

$$\alpha^2 = \frac{P}{EI}$$

The solution to this differential equation consists of two parts, a homogeneous solution and a particular integral. The homogeneous solution is given by

$$y = A\sin\alpha x + B\cos\alpha x$$

In general the particular integral is obtained by taking

$$P\cdot I = y = C\cdot f(x)$$

where

$$\frac{d^2y}{dx^2} + \alpha^2 y = \text{const.} f(x)$$

and then substituting into the differential equation for y and solving for C. In this case, the particular integral is taken to be C, since $f(x)$ is 1 and hence $C = -e$. Thus, the total solution to the differential equation is:

$$y = A\sin\alpha x + B\cos\alpha x - e$$

In order to determine A and B, we use two boundary conditions.

At $x = 0$, $y = 0$ and at $x = \frac{l}{2}$, $\frac{dy}{dx} = 0$

Thus, $0 = B - e$, i.e. $B = e$, and $0 = A\alpha\cos\left(\frac{\alpha l}{2}\right) - B\alpha\sin\frac{\alpha l}{2}$

i.e.
$$A = e\tan\left(\frac{\alpha l}{2}\right)$$

So,
$$y = e\tan\left(\frac{\alpha l}{2}\right)\sin(\alpha x) + e\cos(\alpha x) - e$$

$\therefore$
$$y = e\left(\tan\left(\frac{\alpha l}{2}\right)\sin(\alpha x) + \cos(\alpha x) - 1\right)$$

The maximum deflection, $\hat{y}$, will occur at $x = \frac{l}{2}$, so that,

$$\hat{y} = e\left(\tan\left(\frac{\alpha l}{2}\right)\sin\left(\frac{\alpha l}{2}\right) + \cos\left(\frac{\alpha l}{2}\right) - 1\right)$$

or
$$\hat{y} = e\left(\frac{\sin\left(\frac{\alpha l}{2}\right)\sin\left(\frac{\alpha l}{2}\right)}{\cos\left(\frac{\alpha l}{2}\right)} + \frac{\cos^2\left(\frac{\alpha l}{2}\right)}{\cos\left(\frac{\alpha l}{2}\right)} - 1\right)$$

$$\hat{y} = e\left(\frac{\sin^2\left(\frac{\alpha l}{2}\right) + \cos^2\left(\frac{\alpha l}{2}\right)}{\cos\left(\frac{\alpha l}{2}\right)} - 1\right)$$

$$\hat{y} = e\left(\sec\left(\frac{\alpha l}{2}\right) - 1\right)$$

Therefore, when $\frac{\alpha l}{2} = \frac{\pi}{2}$, then $\hat{y} = \infty$

However, $\alpha^2 = \frac{P}{EI}$, so that when $\frac{P}{EI} = \frac{\pi^2}{l^2}$, $\hat{y} = \infty$

The value of P which causes $\hat{y} = \infty$ is the buckling load (or the **Euler crippling load**), P_c, so $\hat{y} = \infty$, when

$$P = P_c = \frac{\pi^2 EI}{l^2}$$

It is important to note that P_c is independent of e, so that the buckling load is the same no matter how large or small the eccentricity, so even very small eccentricities will give the same buckling load.

Figure 3.63 shows the relationship between transverse deflection $\hat{y}$ and the applied load ratio $\frac{P}{P_c}$ for a range of eccentricity values, e.

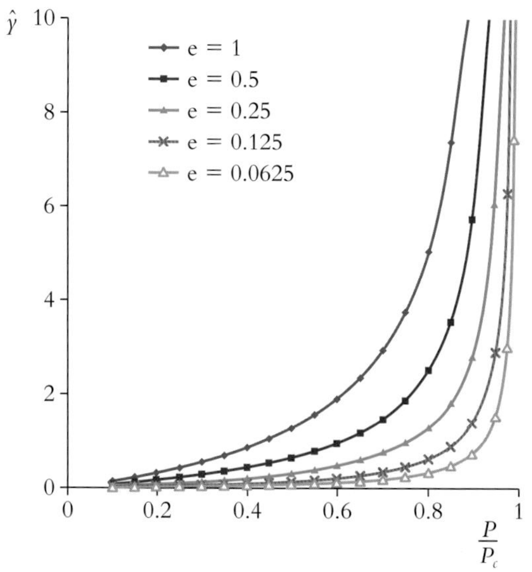

Figure 3.63 Variation of $\hat{y}$ with P/P_c for a range of eccentricity values

General formula: $P_c = \dfrac{\pi^2 EI}{L_{eff}^2}$

Description	Schematic	Critical buckling load, P_c	Effective length, L_{eff}
Free–fixed	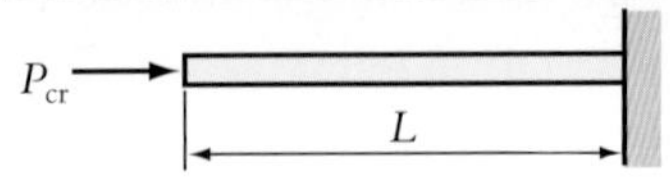	$P_c = \dfrac{\pi^2 EI}{4l^2}$	$2l$
Hinged–hinged	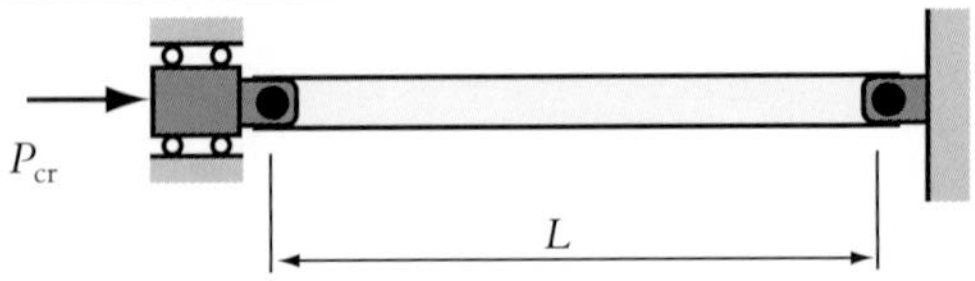	$P_c = \dfrac{\pi^2 EI}{l^2}$	l
Hinged–hinged, initially curved		$P_c = \dfrac{\pi^2 EI}{l^2}$	l
Fixed–hinged	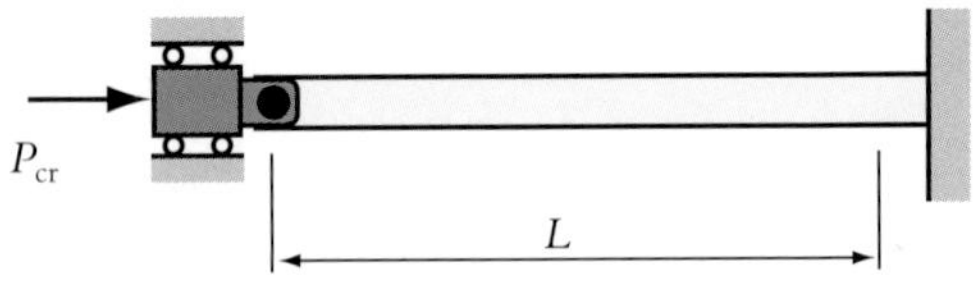	$P_c = \dfrac{2.045\pi^2 EI}{l^2}$	$0.7l$
Fixed–fixed	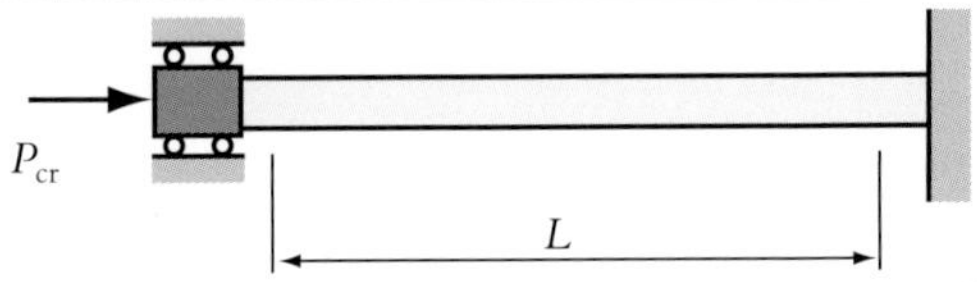	$P_c = \dfrac{4\pi^2 EI}{l^2}$	$\dfrac{l}{2}$

Table 3.2 Summary of Euler buckling loads of struts

The effective length, L_{eff}, is a measure of how much longer (and thus more unstable) a given strut configuration appears to be in terms of critical buckling load, relative to the hinged–hinged case. Thus, the fixed–fixed case, for example, has a shorter effective length because it is more stable and thus appears to be shorter with respect to buckling.

Beams with both axial and transverse loads

A beam which is acted on by a compressive axial force in addition to transversely applied loads is referred to as a beam–column. In this section, the buckling load for such a member, under the action of a point load at the mid span, is derived.

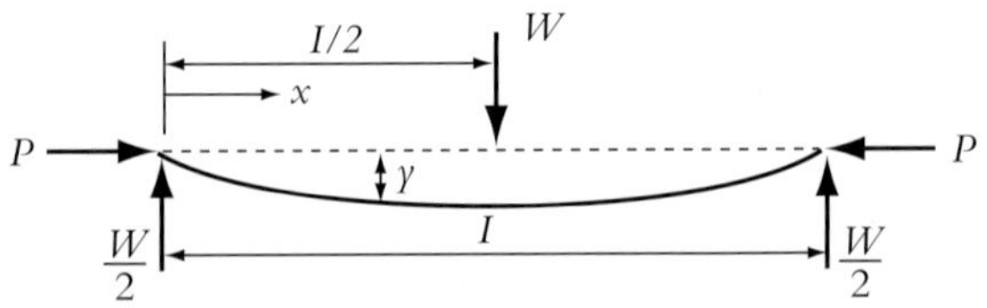

Figure 3.64 A beam–column with a central transverse load

For $0 \leqslant x \leqslant \dfrac{l}{2}$, $M = Py + \dfrac{W}{2}x$ and $M = -EI\dfrac{d^2y}{dx^2}$

$$\therefore \quad \frac{d^2y}{dx^2} + \alpha^2 y = -\frac{W}{EI}\frac{x}{2} \quad \text{where} \quad \alpha^2 = \frac{P}{EI}$$

The solution to this differential equation, including using both the homogeneous function and the particular integral, is

$$y = A\sin(\alpha x) + B\cos(\alpha x) - \frac{Wx}{2P}$$

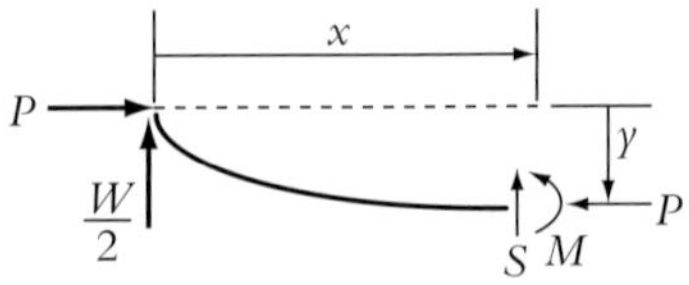

Figure 3.65 A free-body diagram for the left-hand section of the beam–column

The constants A and B are obtained by using the boundary condition:

at $x = 0, y = 0$

$\therefore \quad B = 0$ and at $x = \dfrac{l}{2}, \dfrac{dy}{dx} = 0$

$$\therefore \quad 0 = A\alpha\cos\left(\frac{\alpha l}{2}\right) - \frac{W}{2P}$$

giving

$$A = \frac{W}{2P}\frac{1}{\alpha\cos\left(\frac{\alpha l}{2}\right)}$$

Therefore, the solution is

$$y = \frac{W}{2P}\left(\frac{\sin(\alpha x)}{\alpha\cos\left(\frac{\alpha l}{2}\right)} - x\right)$$

The maximum deflection, y_{max}, occurs at $x = \frac{l}{2}$, i.e.

$$y_{max} = \frac{W}{2P}\left(\frac{\sin\left(\frac{\alpha l}{2}\right)}{\alpha\cos\left(\frac{\alpha l}{2}\right)} - \frac{l}{2}\right)$$

i.e.

$$y_{max} = \frac{W}{2P\alpha}\left(\tan\left(\frac{\alpha l}{2}\right) - \frac{\alpha l}{2}\right)$$

The maximum absolute bending moment occurs at the mid-span:

$$M_{max} = \left|-\frac{Wl}{4} - Py_{max}\right| = \frac{W}{2\alpha}\tan\frac{\alpha l}{2}$$

These expressions for y, y_{max} and M_{max} become infinite when $\frac{\alpha l}{2}$ is a multiple of $\frac{\pi}{2}$, i.e. when

$$\frac{\alpha l}{2} = \sqrt{\frac{P}{EI}}\frac{l}{2} = \frac{n\pi}{2}$$

where n is an integer.

When $n = 1$,

$$P_c = \frac{\pi^2 EI}{l^2}$$

i.e. the critical buckling load is unaffected by the presence of the transverse load, W.

The expression for y_{max} can be rewritten as

$$y_{max} = \frac{W}{2EI\alpha^3}\left(\tan\left(\frac{\alpha l}{2}\right) - \frac{\alpha l}{2}\right)$$

$$y_{max} = \left[\frac{Wl^3}{48EI}\right]\left\{\frac{3\left(\tan\left(\frac{\alpha l}{2}\right) - \frac{\alpha l}{2}\right)}{\left(\frac{\alpha l}{2}\right)^3}\right\}$$

Since $\frac{Wl^3}{48EI}$ would be the deflection without P, it is clear that the presence of the compressive force has magnified the deflection by a factor of $3\left(\tan\left(\frac{\alpha l}{2}\right) - \left(\frac{\alpha l}{2}\right)\right)/\left(\frac{\alpha l}{2}\right)^3$ which is approximately equal to 1.33 for $\frac{\alpha l}{2} = \frac{\pi}{4}$, for example. By contrast, a tensile axial load would have the effect of reducing the transverse deflections.

It also follows that the bending moments in slender members can be substantially increased by the presence of compressive axial forces.

Some important notes

(1) In contrast to the classical cases considered here, actual compression members are seldom truly pinned or completely fixed against rotation at the ends. Because of this uncertainty regarding the fixity of the ends, struts or columns are often assumed to be pin-ended. This procedure is conservative.

(2) The above equations are not applicable in the inelastic range, i.e. for $\sigma > \sigma_y$, and must be modified.

(3) The critical load formulae for struts or columns are remarkable in that they do not contain any strength property of the material and yet they determine the load-carrying capacity of the member. The only material property required is the elastic modulus, E, which is a measure of the stiffness of the strut.

Compressive loading of rods

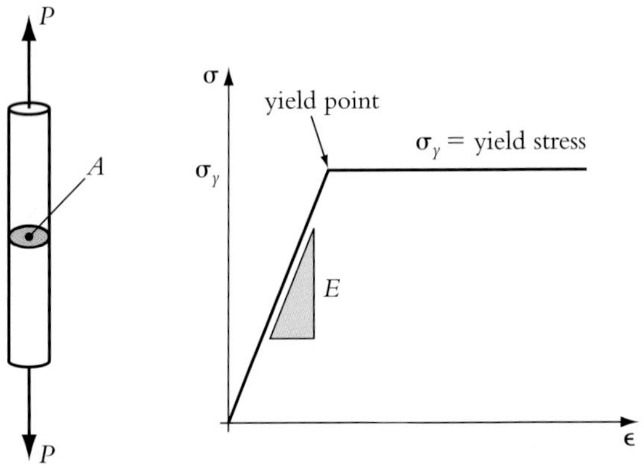

Figure 3.66 Tensile test specimen and elastic perfectly plastic stress–strain behaviour

If we assume that the rod loading is perfectly axial, and the material can be represented by an elastic–perfectly plastic stress–strain curve, then the plastic collapse failure would occur in compression if $\sigma\left(=\frac{-P}{A}\right)$ reaches $-\sigma_y$ before the buckling load is reached. Now

$$P_c = \frac{\pi^2 EI}{l^2}$$

and defining the second moment of area, I, as

$$I = Ak^2$$

where k is the radius of gyration, gives

$$P_c = \frac{\pi^2 EAk^2}{l^2}$$

and

$$\sigma = \frac{P_c}{A} = \frac{\pi^2 Ek^2}{l^2} = \frac{\pi^2 E}{(l/k)^2}$$

and l/k is the slenderness ratio.

Therefore, buckling will occur if $\sigma = \dfrac{\pi^2 E}{(l/k)^2}$,

whereas plastic collapse will occur if $\sigma > \sigma_y$.

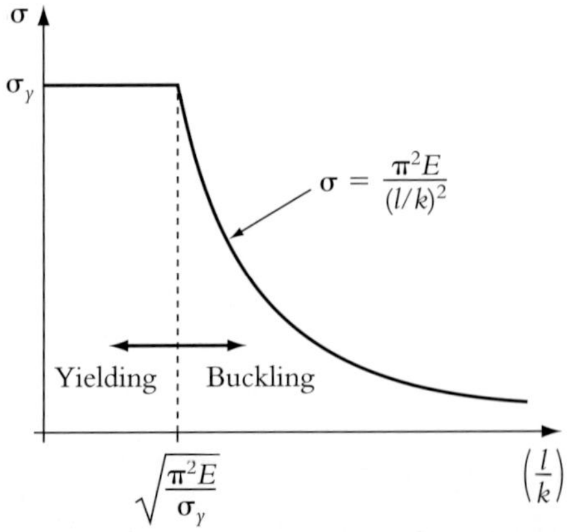

Figure 3.67 Plot of σ versus l/k indicating the buckling and plastic collapse regions

Assuming that the load is to be applied eccentrically as shown in Figure 3.68, this can be represented diagrammatically as shown in Figure 3.69.

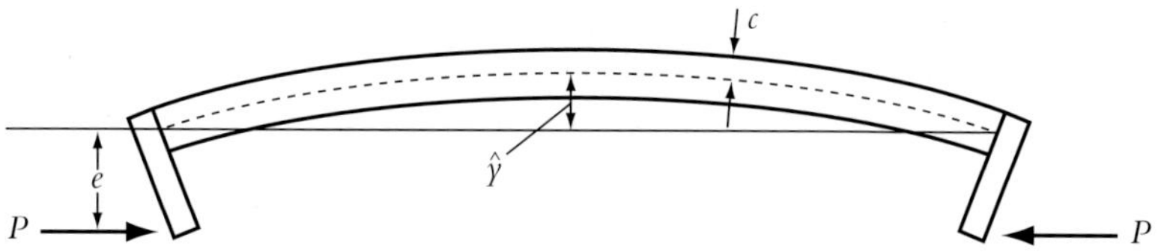

Figure 3.68 Eccentrically loaded bar

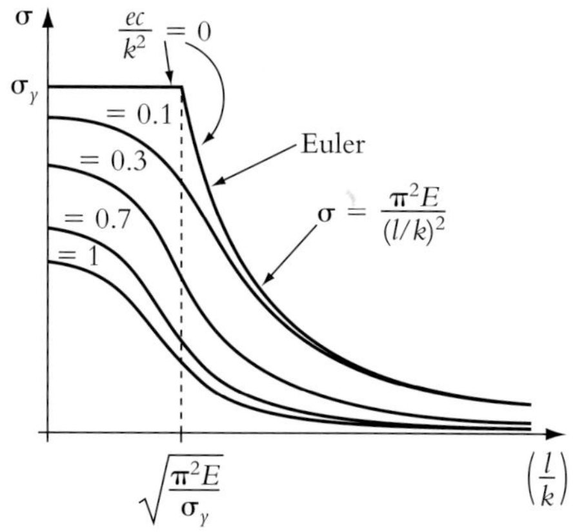

Figure 3.69 Effect of ec/k^2 on the buckling characteristics of an eccentrically loaded strut

We find that:

$$\hat{\sigma}_c = -\left(\frac{P}{A} + \frac{P(e + \hat{y})c}{Ak^2}\right)$$

i.e.

$$\hat{\sigma}_c = -\frac{P}{A}\left(1 + \frac{(e + \hat{y})c}{k^2}\right)$$

Recall $\hat{y} = e\left(\sec\left(\frac{\alpha l}{2}\right) - 1\right)$ for an eccentrically loaded strut

$\therefore$

$$\hat{\sigma}_c = -\frac{P}{A}\left(1 + \frac{ec}{k^2}\left(1 + \sec\left(\frac{\alpha l}{2}\right) - 1\right)\right)$$

i.e.

$$\hat{\sigma}_c = -\frac{P}{A}\left(1 + \frac{ec}{k^2}\sec\left(\frac{\alpha l}{2}\right)\right)$$

giving

$$-\frac{P}{A} = \frac{\hat{\sigma}_c}{1 + \frac{ec}{k^2}\sec\left(\frac{\alpha l}{2}\right)}$$

It should be noted that the latter equation is a non-linear equation in P since α also contains P. The plotted results were obtained by substituting in the material yield stress for $\hat{\sigma}_c$ and then solving for an allowable load P using known values of eccentricity e, thus:

$$\sigma_y = -\frac{P}{A}\left(1 + \frac{ec}{k^2}\sec\left(\sqrt{\frac{P}{EI}}\frac{l}{2}\right)\right)$$

It should also be noted that the buckling load for an eccentrically loaded strut is the same as that for a hinged–hinged ideal strut. Therefore, the plotted results show that yield failure will always occur first except for small eccentricities at high slenderness ratios where the yield curve approaches the Euler buckling curve.

Worked example

A hoist is constructed using two wooden bars (E = 12.4 GPa) as shown in Figure 3.70 and Figure 3.71. The allowable normal stress is 13.8 MPa. Determine the maximum permissible weight W that can be lifted using the hoist for the two cases: (a) L = 1.2 m; (b) L = 1.5 m.

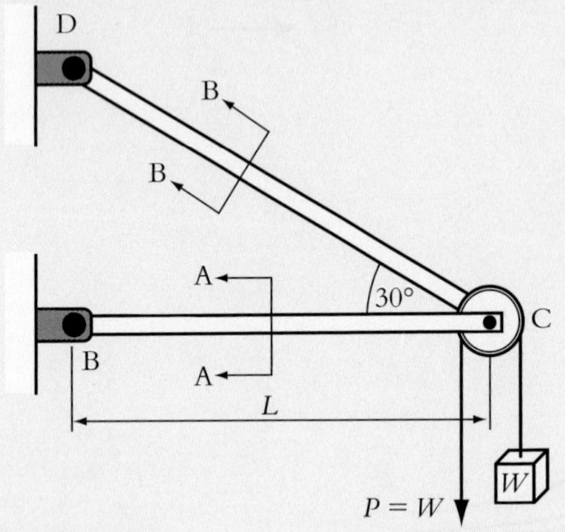

Figure 3.70

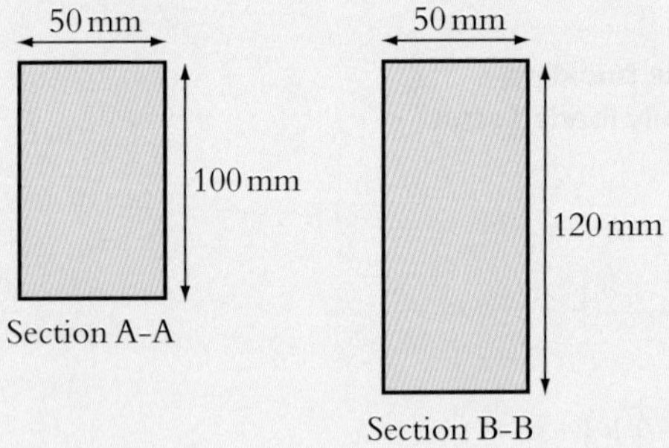

Figure 3.71

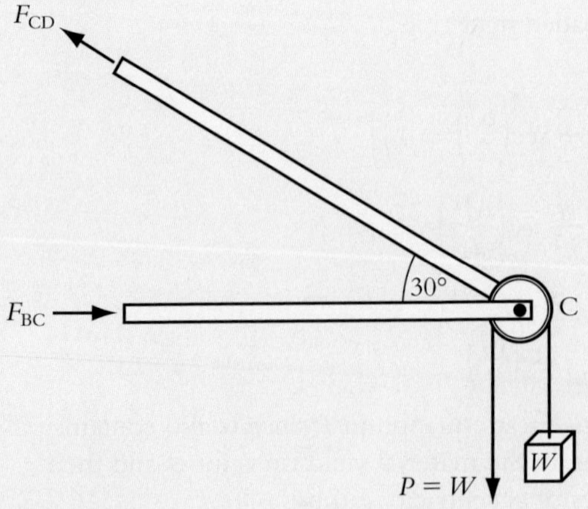

Figure 3.72

By inspection, we see that member BC will be in compression. Thus, to determine the maximum value of W, we need to consider buckling failure of member BC and strength failure of both members due to the axial stress exceeding the given allowable stress. The internal forces in members BC and CD can be found in terms of W. The axial stresses in the members are then compared with the given allowable values to determine one set of limits on W. To determine the critical buckling load, the smaller second moment of area about the perpendicular axes through the section centroid should be used and the upper limit on W to prevent buckling failure can be found. The maximum value of W that satisfies the strength and buckling criteria can now be determined.

The free-body diagram for the pulley is shown in Figure 3.72, with F_{BC} drawn compressive and F_{CD} as tensile. The internal axial forces are found from

$$\sum F_y = 0 \quad \Rightarrow \quad F_{CD} \sin 30 = 2W \quad \Rightarrow \quad F_{CD} = 4W$$

$$\sum F_x = 0 \quad \Rightarrow \quad F_{BC} = F_{CD} \cos 30 \quad \Rightarrow \quad F_{BC} = 3.464W$$

The cross-sectional areas for the two members are $A_{BC} = 5000\,\text{mm}^2$ and $A_{CD} = 6000\,\text{mm}^2$. Thus the axial stresses in terms of W can be found and these must be less than 13.8 MPa, giving two limits on W:

$$\sigma_{CD} = \frac{F_{CD}}{A_{CD}} = \frac{4W}{6000} \leqslant 13.8\,\text{MPa} \quad \text{or} \quad W \leqslant 20.7\,\text{kN}$$

$$\sigma_{BC} = \frac{F_{BC}}{A_{BC}} = \frac{3.464W}{5000} \leqslant 13.8\,\text{MPa} \quad \text{or} \quad W \leqslant 17.9\,\text{kN}$$

For cross section A-A, we note that

$$I_{min} = \frac{1}{12} \times 100 \times 50^3 = 1041\,666.7\,\text{mm}^4$$

Thus, for length (a),

$$P_{crit} = \frac{\pi^2 EI}{L^2} = \frac{\pi^2 \times 12.4 \times 10^3 \times 1041\,667}{1200^2} = 56658.8\,\text{N}$$

Thus, a second limit on F_{BC} is $F_{BC} < 88\,529$ N, thus $3.464W < 88\,529$ N or $W < 25.6$ kN. Since this is greater than the value that causes yielding, the yielding value will occur first, so that the maximum permissible value of W for $L = 1.2$ m is 17.9 kN.

Then, for length (b),

$$P_{crit} = \frac{\pi^2 EI}{L^2} = \frac{\pi^2 \times 12.4 \times 10^3 \times 1041\,667}{1500^2} = 88\,529.4\,\text{N}$$

Thus, a second limit on F_{BC} is $F_{BC} < 56\,658.8$ N, thus $3.464W < 56\,658.8$ N or $W < 16.36$ kN. Since this is less than the value that causes yielding, the buckling value becomes limiting, so that the maximum permissible value of W for $L = 1.5$ m is 16.4 kN.

Note: In case (a) the design was governed by material strength whereas in case (b) buckling governed the design. If there were several bars of different lengths and cross sectional dimensions, it would save significant time to calculate the slenderness ratio that separates long columns from short columns, i.e. the buckling failure regime from the yielding failure regime. Using $\sigma_{cr} = 13.8$ MPa in $\sigma = \dfrac{\pi^2 E}{(L/k)^2}$ gives $L/k = 94.2$ as the threshold ratio separating

long columns from short columns. For case (a) the slenderness ratio is 83.1, so that it is classified as a short column and material strength governs W_{max}. For case (b) the slenderness ratio is 103.9 so that buckling governs W_{max}.

Learning summary

By the end of this section you should have learnt:

✓ Macaulay's method for determining beam deflection in situations with axial loading;

✓ the meanings of and the differences between stable, unstable and neutral equilibria;

✓ how to determine the buckling loads for ideal struts;

✓ the effects of eccentric loading, initial curvature and transverse loading on the buckling loads;

✓ how to include the interaction of yield behaviour with buckling and how to represent this interaction graphically.

3.7 Shear stresses in beams

Whereas bending stresses in beams arising from transverse loading are important, **transverse** (i.e. through-thickness) **shear stresses** due to these same loads also exist. For long slender beams, the shear stresses can generally be neglected, and it is only necessary to do a bending calculation for the beam. However, as the beam-span-to-depth ratio reduces, i.e. if the beam is shorter and thicker, shear stresses become more important and should be calculated in any design evaluation. This can be particularly important for laminated beams, e.g. plywood or composite beams, where the transverse shear can cause failure between individual layers (plies) making up the beam.

In this section we will derive a general formula for calculating the shear stress distribution through the thickness of a beam. We will then introduce the concept of **shear centre** which is the point through which the resultant of the shear stresses always act. The shear centre becomes important for beam sections that have low torsional rigidity, i.e. can twist easily, such as thin-walled sections. For such beams, if the resultant of the applied transverse loads does not act through the shear centre, it can cause twisting of the beam, i.e. there is a bending–twisting interaction in the system. The designer should avoid this situation if possible or, at least, evaluate the degree of twisting that might take place.

Transverse shear stress distribution

The through-thickness shear force in a beam is the integral of the shear stresses over the cross-section. In this section we will determine an expression for the shear stress distribution (transverse i.e. through-thickness) at a section as a function of the shear force at that position. Consider an element of beam length, δx, as shown in Figure 3.73. The bending moment at x, section AC, is M and at $x + \delta x$, section BD, is $M + \delta M$. The direct bending stresses on AC are,

$$\sigma_{AC} = \frac{My}{I}$$

where y = distance from the neutral surface
I = second moment of area of the section
and on BD, the bending stresses are,

$$\sigma_{BD} = \frac{(M + \delta M)y}{I}$$

Thus, when the bending moment varies along the length of the beam on an element such as ABEF, also shown in Figure 3.73, there is a net axial force due to change in the bending stresses. The force on the face FB is the integral of the bending stresses over the area FB,

$$F_{FB} = \int_A \frac{(M + \delta M)}{I} y\,dA$$

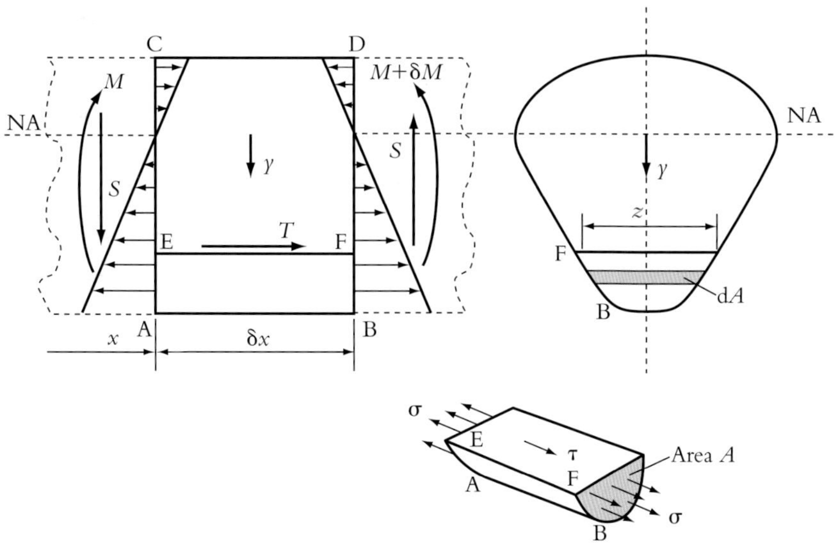

Figure 3.73 **Element of beam length subjected to a varying bending moment**

Similarly the force on the face EA is

$$F_{EA} = \int_A \frac{M}{I} y\mathrm{d}A$$

The net force to the right acting on element ABEF is the difference between these:

$$\text{Net force (bending) } F_{EA} = \int_A \frac{\delta M}{I} y\mathrm{d}A \qquad \textbf{(3.29)}$$

In order to maintain the equilibrium of ABEF, shear stresses must act on the plane EF, of average value τ, as shown in Figure 3.74. These shear stresses are complementary to the *transverse* shear stresses. For positive transverse shear stresses, as shown, the complementary shear stresses act in the positive x-direction. The net force to the right due to these complementary shear stresses is

$$\text{Net force(shear)} = \tau \cdot z \cdot \delta x$$

where z is the width of the section at that depth.

Now, the equilibrium of ABEF requires the net force due to bending to balance the net force due to the complementary shear. Thus,

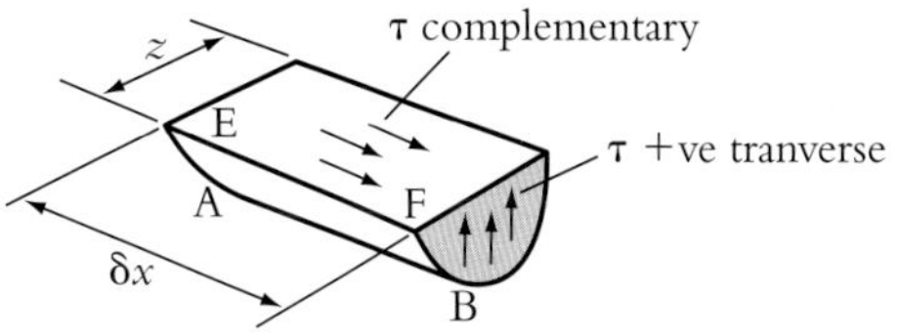

Figure 3.74 **Equilibrium of stresses on ABEF**

$$\tau z \delta x + \int_A \frac{\delta M}{I} y\mathrm{d}A = 0$$

$$\therefore \qquad \tau = -\frac{1}{Iz}\frac{\delta M}{\delta x}\int_A y\mathrm{d}A$$

But, in the limit,

$$\frac{\lim}{\delta x \to 0}\frac{\delta M}{\delta x} = \frac{\mathrm{d}M}{\mathrm{d}x} = -S$$

where S = shear force at the section

$$\therefore \qquad \tau = \frac{S}{Iz}\int_A y\mathrm{d}A \qquad \textbf{(3.30)}$$

This is the general expression for transverse shear stress at any position y through the thickness. The integral can also be written in discrete form as follows,

$$\tau = \frac{S}{Iz} A\bar{y} \tag{3.31}$$

where A is the area of the part of the cross-section outside the position, y, at which τ is determined, and $\bar{y}$ is the distance of the centroid of this area from the neutral axis, as shown in Figure 3.75.

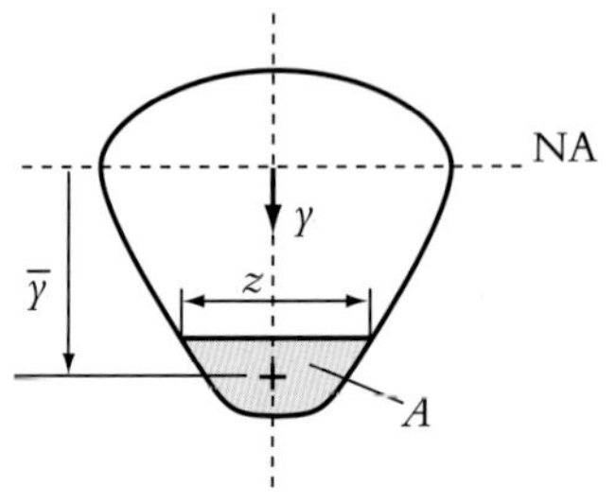

Figure 3.75 Definition of shear stress parameters in a cross-section

Determination of shear stress distribution for different cross-sectional shapes

Rectangular section

Referring to Figure 3.76 and using the discrete form for shear stress distribution, i.e. equation (3.31) above, we have,

$$A = \left(\frac{d}{2} - y\right)b \quad \text{and} \quad \bar{y} = \left(\frac{d}{2} + y\right)\frac{1}{2}$$

$$\tau = \frac{S}{(bd^3/12) \cdot b}\left(\frac{d}{2} - y\right)b\left(\frac{d}{2} + y\right)\frac{1}{2}$$

$$\therefore \qquad \tau = \frac{6S}{bd^3}\left[\left(\frac{d}{2}\right)^2 - y^2\right] \tag{3.32}$$

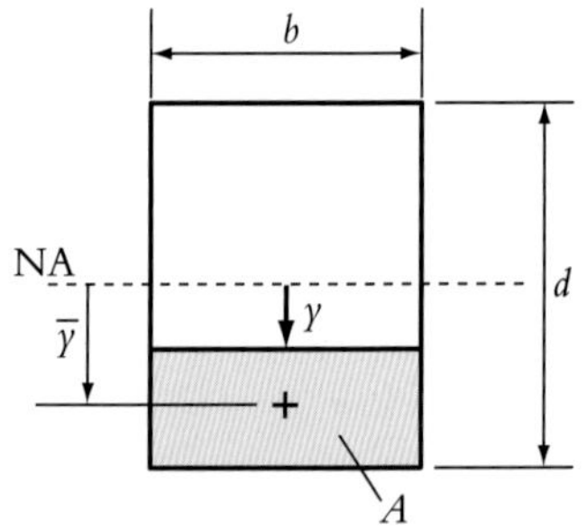

Figure 3.76 Rectangular cross section

Note the parabolic distribution of shear stress (i.e. τ varies with y^2), illustrated in Figure 3.77. Also, at the top and bottom of the section, where $y = \pm d/2$, equation (3.32) gives $\tau = 0$. As expected, there is no complementary shear stress on the top and bottom free surfaces, therefore the transverse shear stress is also zero at these positions.

At the neutral axis, i.e. where $y = 0$, equation (3.32) gives,

$$\tau = \frac{6S}{bd^3}\frac{d^2}{4} = 1.5\frac{S}{bd}$$

This is the position of maximum shear stress whose magnitude is 1.5× the average shear stress S/bd.

Note also that, in this analysis, the shear stress does not vary across the width of the section.

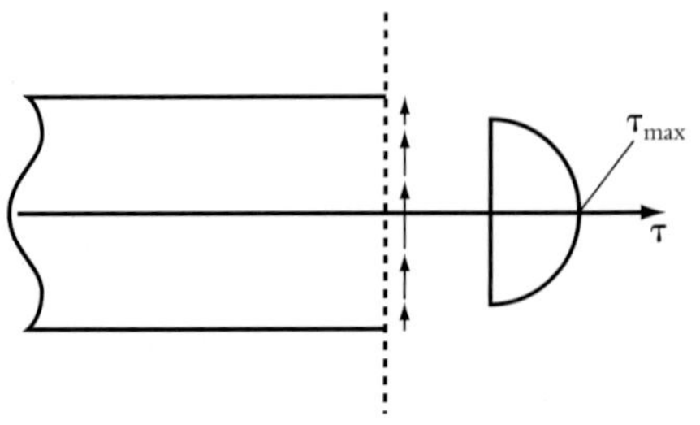

Figure 3.77 Shear stress distribution in a rectangular section

Circular section

Figure 3.78 shows a solid circular cross-section of a beam. To calculate the transverse shear stress distribution in this section, we use the integral form of the shear equation, i.e. equation (3.30).

However, because of the circular shape, it is convenient to change the variables y and z in this equation to polar variables, R and θ. Referring to Figure 3.78 we have

$$y_1 = R\sin\theta_1$$
$$dy_1 = R\cos\theta_1 d\theta_1$$
$$z_1 = 2R\cos\theta_1$$
$$z = 2R\cos\theta$$

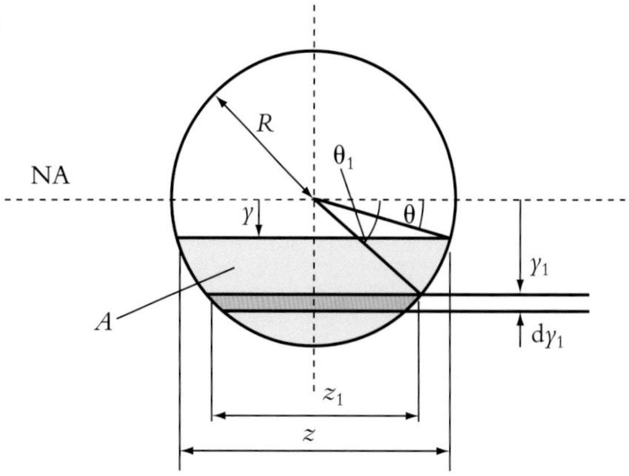

Figure 3.78 **Solid circular cross-section**

and the second moment of area, $I = \pi D^4/64 = \pi R^4/4$

The shear equation now becomes,

$$\tau = \frac{S}{Iz}\int y dA = \frac{S \cdot 4}{\pi R^4 \cdot 2R\cos\theta}\int_y^R y_1 z_1 dy_1$$

$$= \frac{S \cdot 4}{\pi R^4 \cdot 2R\cos\theta}\int_\theta^{\pi/2} R\sin\theta_1 \cdot 2R\cos\theta_1 \cdot R\cos\theta_1 d\theta_1$$

$$= \frac{4SR^3}{\pi R^5\cos\theta}\int_\theta^{\pi/2}\cos^2\theta_1 \cdot \sin\theta_1 d\theta_1$$

$$= \frac{4S}{\pi R^2\cos\theta}\left[\frac{-\cos^3\theta_1}{3}\right]_\theta^{\pi/2}$$

$$\therefore \qquad \tau = \frac{4S}{3\pi R^2}\cos^2\theta$$

But

$$\cos^2\theta = 1 - \sin^2\theta = 1 - \left(\frac{y}{R}\right)^2$$

$$\therefore \qquad \tau = \frac{4S}{3\pi R^2}\left[1 - \left(\frac{y}{R}\right)^2\right] \qquad \textbf{(3.33)}$$

Again, a parabolic distribution and the maximum value of τ, at the neutral axis, when $y = 0$ is

$$\therefore \qquad \tau_{max}(at\ y = 0) = \frac{4S}{3\pi R^2} = \frac{4}{3}\tau_{average}$$

In this case, τ must vary across the width of the section. As can be seen in Figure 3.79, at the free surface the shear stress must be zero. Therefore, the complementary shear on the cross section, *normal to the boundary*, is also zero. Thus, shear must be tangential to the boundary as drawn.

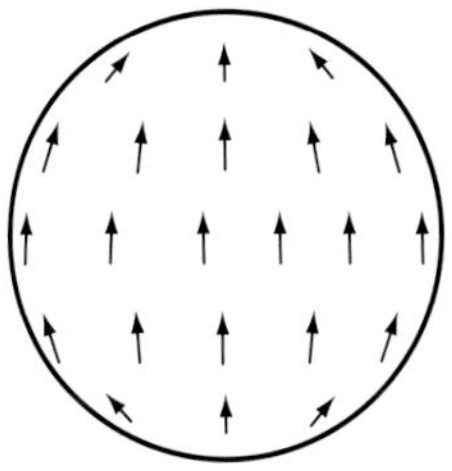

Figure 3.79 **Shear stress distribution in a solid circular section**

I-section

To determine the transverse shear stress distribution in an I-section, we need to consider the web and flange areas separately.

Transverse shear in the web

Figure 3.80(a) shows an I-section and the position y where we wish to determine the shear stress. Using the discrete form of the shear stress equation we have

$$\tau = \frac{S}{Iz}A\bar{y} = \frac{S}{Iz}[A_1\bar{y}_1 + A_2\bar{y}_2]$$

$$= \frac{S}{Ib}\left[\left(\frac{d}{2} - y\right)\cdot b \cdot \frac{1}{2}\cdot\left(\frac{d}{2} + y\right) + B\left(\frac{D}{2} - \frac{d}{2}\right)\cdot\frac{1}{2}\cdot\left(\frac{D}{2} + \frac{d}{2}\right)\right]$$

$$= \frac{S}{Ib}\left[\frac{b}{2}\cdot\left(\frac{d^2}{4} - y^2\right) + \frac{B}{2}\cdot\left(\frac{D^2}{4} - \frac{d^2}{4}\right)\right]$$

and

$$I = \frac{BD^3}{12} - \frac{(B - b)d^3}{12}$$

The maximum τ at $y = 0$:

$$\tau_{max} = \frac{S}{Ib}\left[\frac{BD^2}{8} - \frac{(B - b)d^2}{8}\right]$$

At the bottom and top of the web, where $y = \pm d/2$:

$$\tau = \frac{S}{Ib} \cdot \frac{B}{8} \cdot (D^2 - d^2) \qquad (3.34)$$

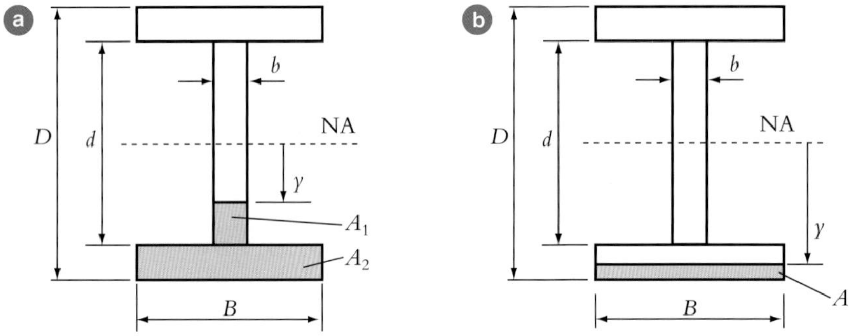

Figure 3.80 I-section

Transverse shear in the flange

Figure 3.80(b) shows the position y where we wish to determine the shear stress in the flange. Again, using the discrete form of the shear stress equation we have

$$\tau = \frac{S}{Iz} A\bar{y}$$

$$= \frac{S}{Ib}\left[B \cdot \left(\frac{D}{2} - y\right) \cdot \frac{1}{2} \cdot \left(\frac{D}{2} + y\right)\right]$$

$$= \frac{S}{2I}\left(\frac{D^2}{4} - y^2\right)$$

At $y = D/2$, $\tau = 0$, as expected, i.e. zero shear complementary to the free surface.

At $y = d/2$

$$\tau = \frac{S}{8I}(D^2 - d^2)$$

Comparing this expression with equation (3.34), there is a step change in τ from the web to the flange of magnitude B/b, i.e. the ratio of the flange width to the web width.

Figure 3.81 shows the transverse shear stress distribution down the centre line of the section and illustrates the step change discussed above. The shear in the flanges is small compared to the web and the shear stress in the web is approximately uniform with vertical position. Because of the small shear in the flanges, the average shear stress in the web is $\approx S/bd$, i.e. the shear force divided by the area of the web.

The above distribution only applies down the centre line of the section. The shear stresses in the flanges are small and non-uniform across the width. This must be the case as they must be zero at the top and bottom surfaces (i.e. free surfaces) of the flanges. There are, however, more significant shear stresses in the flanges, which act parallel to the flanges, i.e. horizontally. These can be determined by a similar analysis.

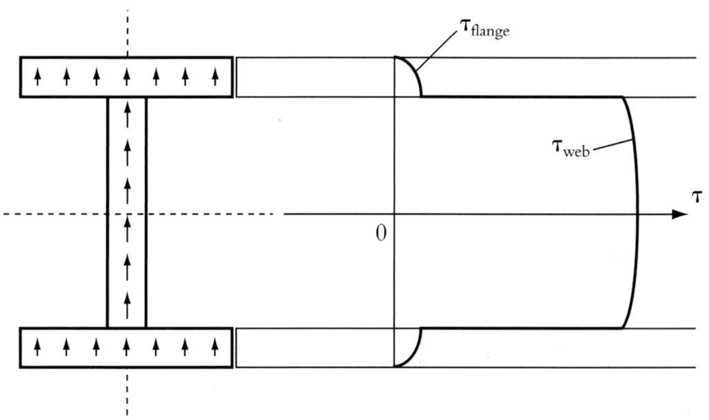

Figure 3.81 **Transverse shear stress distribution in an I-section**

Horizontal shear in the flange

Figure 3.82 shows a small element, δx, of I-beam over which the bending moment changes from M to $M + \delta M$. To determine the hidden horizontal shear stress, τ, at distance a from the edge of the flange, the equilibrium of an element of the flange is considered. Equilibrium of stresses acting on the element gives

$$\int_A \frac{(M + \delta M)}{I} y\mathrm{d}A - \int_A \frac{M}{I} y\mathrm{d}A + \tau z\mathrm{d}x = 0$$

$$\therefore \qquad \tau = -\frac{1}{Iz}\frac{\delta M}{\delta x}\int_A y\mathrm{d}A$$

In the limit as $\delta x \rightarrow 0, \dfrac{\delta M}{\delta x} = \dfrac{\mathrm{d}M}{\mathrm{d}x} = -S$

$$\therefore \qquad \tau = -\frac{S}{Iz}\int_A y\mathrm{d}A = \frac{S}{Iz}A\bar{y}$$

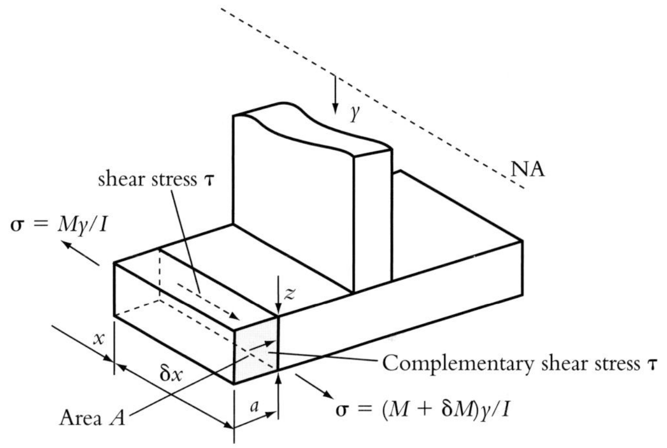

Figure 3.82 **Determining horizontal shear in the flange of an I-section**

This is the same shear equation as before except that the interpretation of the quantities A, $\bar{y}$ and z is different as shown in Figure 3.82. At a distance a from the edge of the flange, the horizontal shear stress is given by,

$$\therefore \qquad \tau = \frac{S}{Iz}\cdot(az)\cdot\frac{1}{2}\left(\frac{D}{2} + \frac{d}{2}\right)$$

$$= \frac{Sa}{4I}(D + d)$$

τ therefore varies linearly with a from zero at the flange edge to a maximum value at the flange centre ($a = B/2$),

$$\tau_{max} = \frac{SB}{8I}(D + d)$$

τ is also parallel to the flange, i.e. horizontal.

We can now draw the dominant shear stresses in both the flanges and the web. Figure 3.83 shows the distribution of these horizontal and vertical (transverse) shear stresses. The critical stress position is likely to be at the join of the web and flange where both the shear and bending stresses are high.

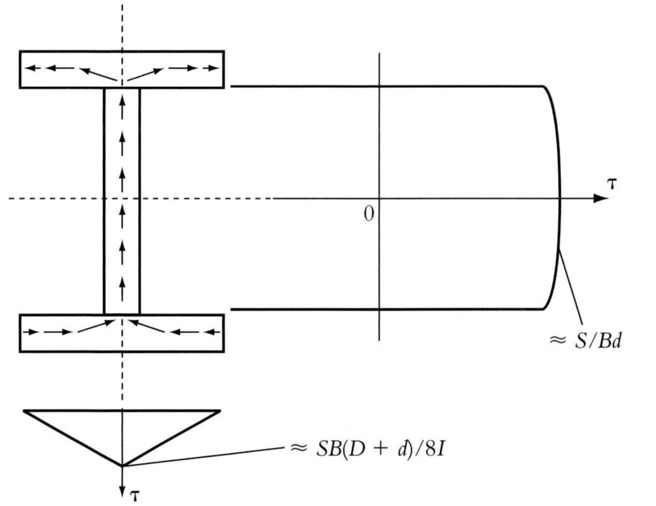

Figure 3.83 **Shear stress distribution in the flange (horizontal) and the web (vertical-transverse) of an I-section**

Shear centre

Consider the shear stress distribution in a symmetric, thin-walled channel section bending in the plane of the web as shown in Figure 3.84.

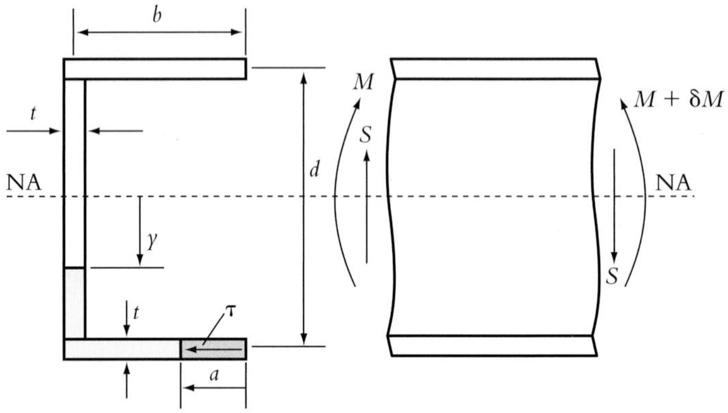

Figure 3.84 Determining shear stress distribution in a thin-walled asymmetric channel section

For the flange at distance *a* from the edge, the horizontal shear stresses are,

$$\tau = \frac{S}{Iz}A\bar{y} = \frac{S}{It}(at)\cdot\left(\frac{d}{2}\right) = \frac{S\cdot d\cdot a}{2I}$$

analysed as above for the flange in an I-section.

For the web at distance *y* from the neutral axis, the transverse shear stresses are

$$\tau = \frac{S}{Iz}A\bar{y} = \frac{S}{It}\left[bt\frac{d}{2} + \left(\frac{d}{2} - y\right)t\left(\frac{d}{2} + y\right)\frac{1}{2}\right]$$

$$= \frac{S}{2I}\left(bd + \left(\frac{d}{2}\right)^2 - y^2\right)$$

We can now draw the shear stress distribution in both the flanges and the web, as shown in Figure 3.85(a). For this shear stress distribution, note that the shear stress in the upper flange is in the opposite sense to that in the lower flange, i.e. there is no horizontal resultant. Also, as there are no shear stresses on the free surfaces, the shear stresses act along the walls, i.e. horizontal in the flanges and vertical in the web.

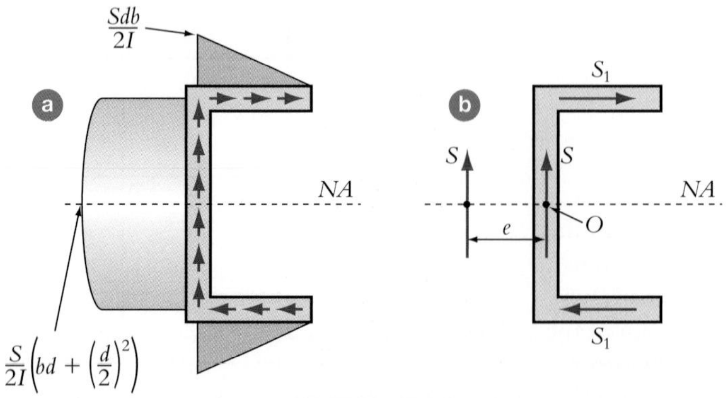

Figure 3.85 Shear stress distribution in the flanges (horizontal) and the web (vertical-transverse) of an asymmetric channel section

We can now look at the resultant forces arising from this shear stress distribution as shown in Figure 3.85(b).

The total shear force in the lower flange, S_1, is the integral of the shear stresses in this flange:

$$S_1 = \int_0^b \tau t da = \int_0^b \frac{Sda}{2I} t da$$

$$= \frac{Sdtb^2}{4I}$$

An equal and opposite shear force acts in the upper flange.

The shear force in the web is approximately S, i.e. the total vertical shear load assuming thin flanges carry negligible vertical shear load.

The resultant of all the shear stresses must be the vertical shear force S, and its line of action is distance e outside the web. Now taking moments about O in the web,

$$S \cdot e = 2S_1 \frac{d}{2}$$

$$\therefore \quad e = \frac{S_1 d}{S} = \frac{d^2 tb^2}{4I}$$

It can be shown that the resultant of the shear stresses for a section, for bending in any plane, always act through one point, the **shear centre**. The shear centre always lies on an axis of symmetry. For sections with two axes of symmetry, the shear centre is at the centroid.

If the applied vertical loads do not act in the plane of the resultant of the shear stresses, i.e. through the shear centre, then there is a torsional load on the section as shown in Figure 3.86. For arbitrary solid sections, the location of the shear centre is a complicated problem. However, it is not usually important to determine the shear centre for solid sections because such sections usually have a considerable torsional rigidity and twist very little due to bending loads. However, for thin-walled open sections, which have low torsional rigidity, the position of the shear centre may be very important.

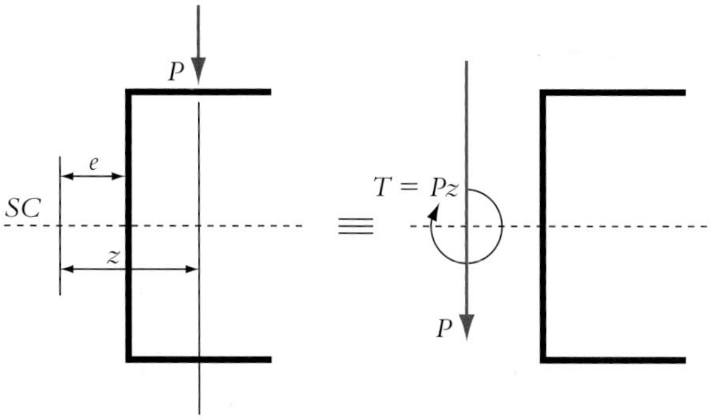

Figure 3.86 Location of the shear centre in an asymmetric channel section

Worked example

Shear stresses in a beam

The section shown in Figure 3.87 is subjected to a vertical shear force, $S = 50$ kN, acting down the vertical centre line, i.e. the y-axis. The second moment of area of the section, about the x-axis, which passes through the centroid of area, G, is $I_{xx} = 2.31 \times 10^6$ mm^4. G is positioned 14 mm below the flange.

(a) Determine the magnitude of the transverse (i.e. vertical) shear stress at positions A, B, G and C on the vertical centre line.

(b) Sketch the variation of the transverse shear stress down the vertical centre line.

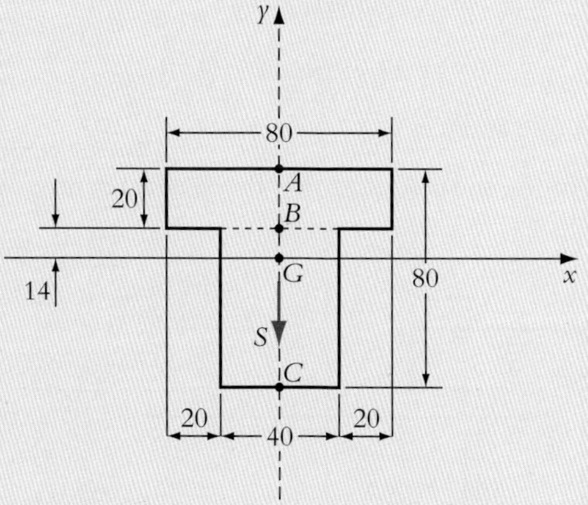

Figure 3.87 Worked example: a T-section

In the top flange

Consider a position in the top flange, vertical distance y from the centroid, G, as shown in Figure 3.88(a). Using the discrete form of the shear formula,

$$\tau = \frac{S}{Iz}A\bar{y} = \frac{S}{2.31x10^6 \cdot 80} \cdot (80 \cdot (34 - y)) \cdot \frac{(34 + y)}{2}$$

$$= \frac{S}{4.62x10^6}(1156 - y^2)$$

$$= 0.0108(1156 - y^2)$$

At position A, $y = 34$ $\quad \therefore \quad \tau = 0$

At position B, $y = 14$ $\quad \therefore \quad \tau = 10.4\,\text{N/mm}^2$ (MPa)

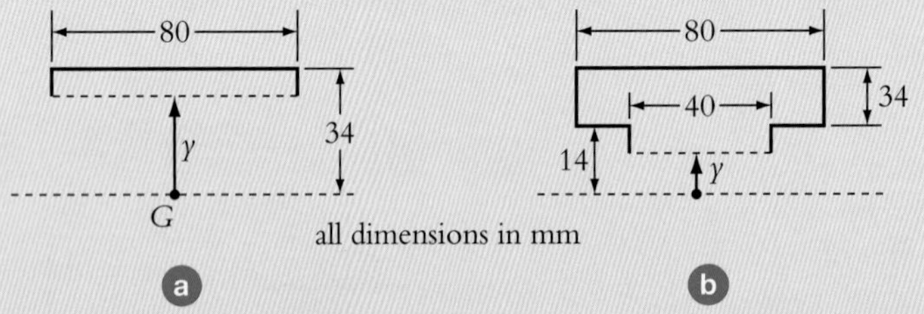

Figure 3.88 **Determining the shear stress distribution in a T-section**

In the lower section

Consider a position in the lower section, again vertical distance y from the centroid, G, as shown in Figure 3.88(b).

At position B, there is a step change in the shear stress given by the ratio of the section widths at this point:

$$\tau = \frac{10.4 \cdot 80}{40} = 20.8\,\text{N/mm}^2\ (\text{MPa})$$

At position G, i.e. the neutral axis, we can use the discrete formula for the shear stress. In this case, to simplify calculation, the relevant area can be regarded as the area below the neutral axis:

$$\tau = \frac{S}{Iz}A\bar{y} = \frac{S}{2.31x10^6 \cdot 40} \cdot (46 \cdot 40) \cdot 23$$

$$= 22.91\,\text{N/mm}^2\ (\text{MPa})$$

At position C, $\tau = 0$ i.e. a free surface.

A sketch of the variation of the shear stress down the vertical centre line is now given in Figure 3.89.

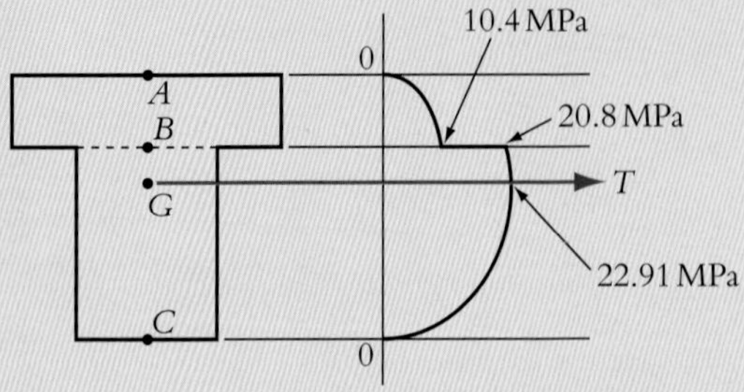

Figure 3.89 **Transverse (vertical) shear stress distribution in a T-section**

Shear centre

For the thin-walled semicircular cross-section shown in Figure 3.90, determine the position of the shear centre (assume bending about the axis of symmetry X-X)

Shear stress distribution

To solve this problem it is necessary to change from a rectangular coordinate system (x-y) to a polar coordinate system (r-θ). Referring to Figure 3.90 and using the integral form of the shear stress formula, we obtain a general expression for the shear stress distribution parallel to the wall of the section. Thus,

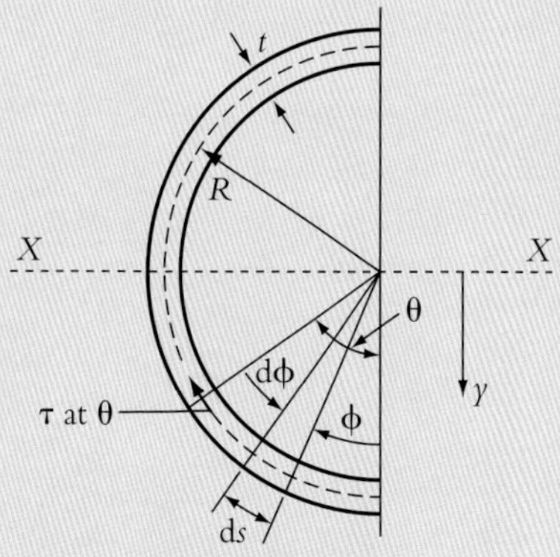

Figure 3.90 A thin-walled semicircular cross-section

$$\tau = \frac{S}{Iz}\int_A y\mathrm{d}A$$

with $y = R\cos\phi$

$$\mathrm{d}A = \mathrm{d}s \cdot t = Rt\mathrm{d}\phi$$

$$z = t$$

giving,

$$\tau = \frac{S}{Iz}\int_0^\theta R\cos\phi \cdot Rt\mathrm{d}\phi$$

$$= \frac{SR^2}{I}\int_0^\theta \cos\phi\,\mathrm{d}\phi$$

$$\therefore \quad \tau = \frac{SR^2\sin\theta}{I} \qquad (3.35)$$

Now,

$$I = \int_A y^2\mathrm{d}A = \int_0^\pi (R\cos\theta)^2 \cdot R\mathrm{d}\theta \cdot t$$

$$= \int_0^\pi R^3 t\cos^2\theta\,\mathrm{d}\theta$$

$$= \int_0^\pi \frac{R^3 t}{2}(1 + \cos 2\theta)\mathrm{d}\theta$$

$$= \frac{R^3 t}{2}\left[\theta + \frac{\sin 2\theta}{2}\right]_0^\pi$$

$$\therefore \quad I = \frac{\pi R^3 t}{2} \qquad (3.36)$$

From (3.35) and (3.36),

$$\tau = \frac{SR^2\sin\theta \cdot 2}{\pi R^3 t}$$

$$\therefore \quad \tau = \frac{2S\sin\theta}{\pi Rt}$$

The twisting moment (torque) associated with the above shear stress distribution for the whole cross-section is found by taking moments about O:

$$\text{Torque} = \int_0^{\pi} \tau \cdot (R d\theta) t \cdot R = \frac{R^2 t \cdot 2S}{\pi R t} \int_0^{\pi} \sin\theta \, d\theta$$

$$= \frac{2SR}{\pi} [-\cos\theta]_0^{\pi} = \frac{4SR}{\pi}$$

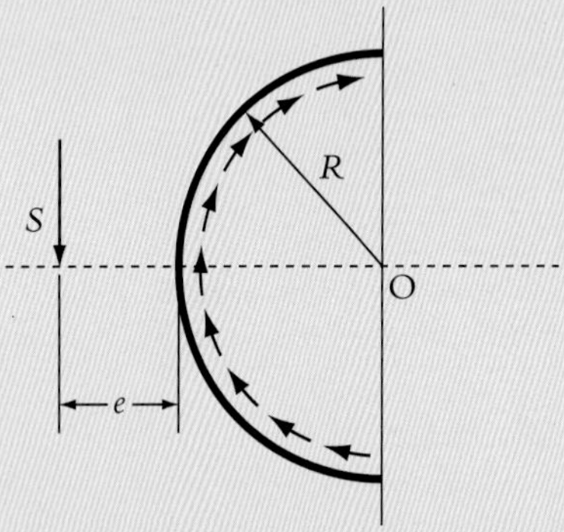

Figure 3.91 Location of the shear centre for a thin-walled semicircular section.

To counteract this twisting moment, as shown in Figure 3.91, the shear force, S, must be applied at the *shear centre*, a distance e, given by,

$$S(e + R) = \frac{4SR}{\pi}$$

$$\therefore \quad e = \frac{4R}{\pi} - R = 0.273R$$

Learning summary

By the end of this section you should have learnt:

- ✓ in addition to longitudinal bending stresses, beams also carry transverse shear stresses arising from the vertical shear loads acting within the beam;
- ✓ how to derive a general formula, in both integral and discrete form, for evaluating the distribution of shear stresses through a cross section;
- ✓ how to determine the distribution of the shear stresses through the thickness in a rectangular, circular and I-section beam;
- ✓ in an I-section, in addition to transverse vertical shear stresses in the flange and the web, more dominant horizontal shear stresses also occur in the flange;
- ✓ the resultant of the shear stresses for a section always act through one point, the shear centre;
- ✓ how to calculate the position of the shear centre;
- ✓ if the applied loads do not act through the shear centre, then there is a resultant torsional load which can result in twisting of the section if the torsional rigidity is low, e.g. in thin-walled sections.

3.8 Thick cylinders

Introduction

Thick cylinders differ from thin cylinders in that the variation of stress through the wall thickness is significant in thick cylinders when subjected to internal and/or external pressure, but for thin cylinders, the variation of stress is negligible. Figure 3.92 includes drawings of thick cylinders with closed ends and with pistons. For closed-ended, internally pressured cylinders the axial force on the inside of the end closures produces a distribution of axial stress in the cylinder while for cylinders with pistons the resultant axial force in the cylinder, and hence also the axial stress, is zero.

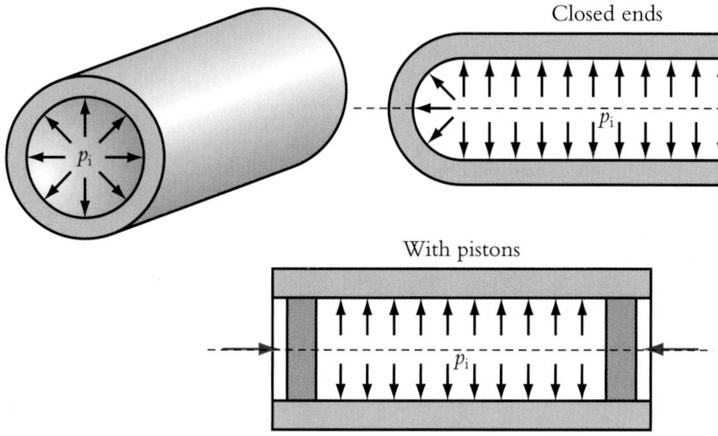

Figure 3.92 Thick cylinders subjected to internal pressure

Analysis of thin cylinders

For an internally pressurized thin cylinder situation, it is reasonable to assume that the variations of the stresses through the wall thickness are negligible. This results in the problem being *statically determinate*, i.e. expressions for the stresses can be obtained by consideration of equilibrium alone, as described below.

$$2\sigma_\theta tl = pdl \qquad p\frac{\pi d^2}{4} = \sigma_a \pi dt$$

$$\therefore \quad \sigma_\theta = \frac{pd}{2t} = \frac{pR}{t} \qquad \therefore \quad \sigma_a = \frac{pd}{4t} = \frac{pR}{2t}$$

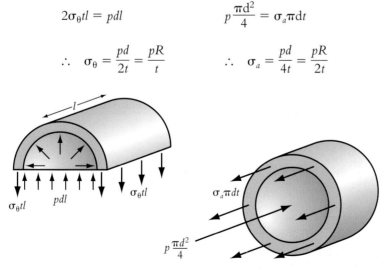

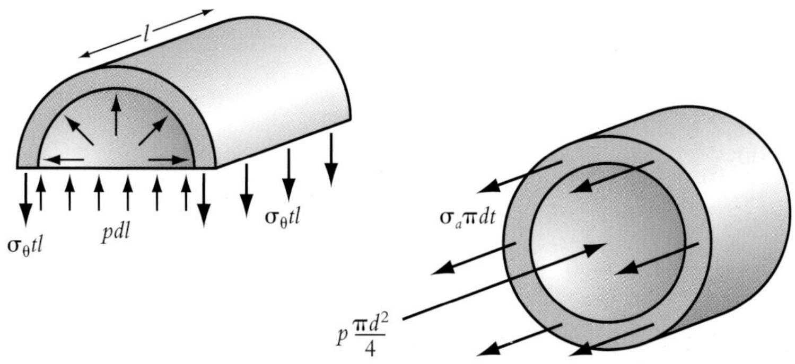

Figure 3.93 Free-body diagrams used to obtain hoop and axial stresses in thin cylinders

Analysis of thick cylinders

Thick cylinder problems are *statically indeterminate*. Therefore, in order to obtain a solution it is necessary to consider equilibrium, compatibility and the material behaviour (stress–strain relationship).

Assumptions:

- plane transverse sections remain plane (this is true remote from the ends);
- deformations are small;
- the material is linear elastic, homogenous and isotropic;

Equilibrium

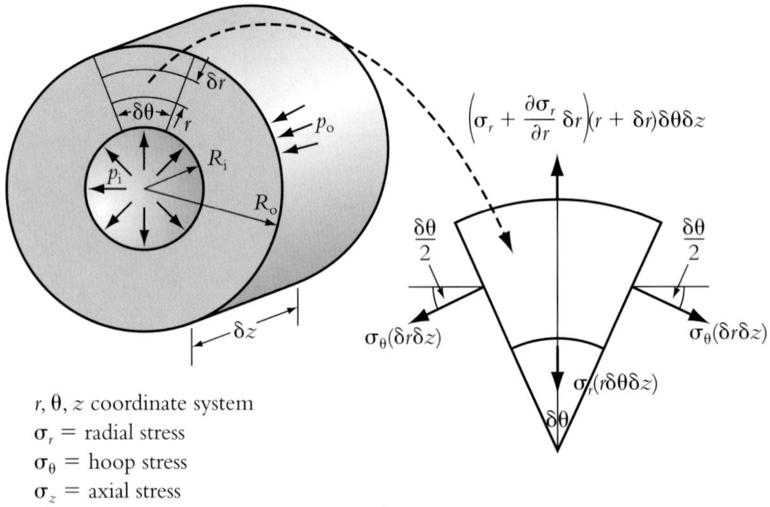

Figure 3.94 A thick cylinder and a free-body diagram of an element of material within the cylinder

$$\left(\sigma_r + \frac{d\sigma_r}{dr}\delta r\right)(r + \delta r)\delta\theta\delta z = \sigma_r(r\delta\theta\delta z) + 2\sigma_\theta(\delta r\delta z)\sin\left(\frac{\delta\theta}{2}\right)$$

For small δθ

$\therefore$

$$\sin\left(\frac{\delta\theta}{2}\right) \approx \left(\frac{\delta\theta}{2}\right)$$

Therefore:

$$\sigma_r(r + \delta r)\delta\theta + \frac{d\sigma_r}{dr}\delta r(r + \delta r)\delta\theta = \sigma_r r\delta\theta + \sigma_\theta\delta r\delta\theta$$

$$r\sigma_r + \sigma_r\delta r + \frac{r\,d\sigma_r}{d_r}\delta r + \frac{d\sigma_r}{dr}\delta r^2 = \sigma_r r + \sigma_\theta\delta r$$

As

$$\delta r \to 0, \frac{d\sigma_r}{dr}dr^2 \to 0$$

Therefore:

$$\sigma_\theta - \sigma_r = r\frac{d\sigma}{dr} \qquad (3.37)$$

Compatibility

$$\varepsilon = \frac{\text{extension}}{\text{original length}}$$

$$\text{Hoop strain, } \varepsilon_\theta = \frac{(r+u)\delta\theta - r\delta\theta}{r\delta\theta} = \frac{u}{r} \quad (3.38)$$

$$\text{Radial strain, } \varepsilon_r = \frac{\left(u + \frac{\mathrm{d}u}{\mathrm{d}r}\delta r\right) - u}{\delta r} = \frac{\mathrm{d}u}{\mathrm{d}r} \quad (3.39)$$

$$\text{Axial strain, } \varepsilon_z = \text{constant} \quad (3.40)$$

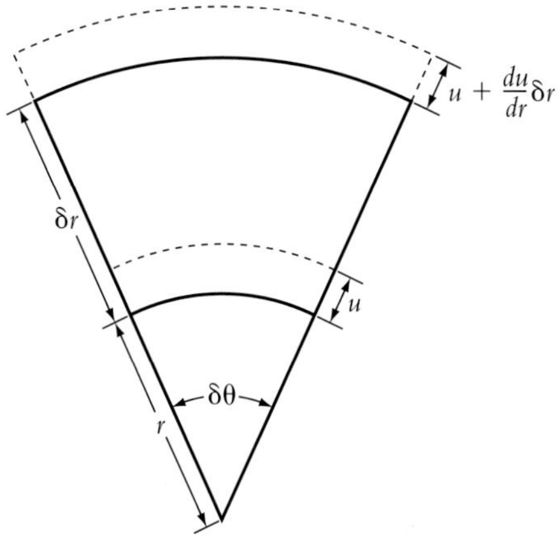

Figure 3.95 Initial and deformed shape of an element of material

Material behaviour (stress–strain relationships)

Generalized Hooke's law (linear elastic and isotropic)

$$\varepsilon_\theta = \frac{1}{E}(\sigma_\theta - \nu(\sigma_r + \sigma_z)) \quad (3.41)$$

$$\varepsilon_r = \frac{1}{E}(\sigma_r - \nu(\sigma_\theta + \sigma_z)) \quad (3.42)$$

$$\varepsilon_z = \frac{1}{E}(\sigma_z - \nu(\sigma_r + \sigma_\theta)) \quad (3.43)$$

Equations (3.37) to (3.43) have seven unknowns, i.e. u, σ_θ, σ_r, σ_z, ε_θ, ε_r and ε_z which are all functions of r, p_o, p_i, R_o, R_i, ν and E.

Substituting $u = r\varepsilon_\theta$ from equation (3.38) into equation (3.39) gives $\varepsilon_r = \frac{\mathrm{d}}{\mathrm{d}r}(r\varepsilon_\theta)$

i.e.
$$\varepsilon_r = \varepsilon_\theta + r\frac{\mathrm{d}\varepsilon_\theta}{\mathrm{d}r} \quad (3.44)$$

Using equation (3.41) and equation (3.42) in equation (3.43) gives

$$\frac{1}{E}(\sigma_r - \nu(\sigma_\theta + \sigma_z)) = \frac{1}{E}(\sigma_\theta - \nu(\sigma_r + \sigma_z)) + \frac{r}{E}\left(\frac{\mathrm{d}\sigma_\theta}{\mathrm{d}r} - \nu\frac{\mathrm{d}\sigma_r}{\mathrm{d}r} - \nu\frac{\mathrm{d}\sigma_z}{\mathrm{d}r}\right) \quad (3.45)$$

Using equation (3.40), $\frac{\mathrm{d}\varepsilon_z}{\mathrm{d}r} = 0$, then equation (3.43) gives:

$$0 = \frac{\mathrm{d}\sigma_z}{\mathrm{d}r} - \nu\frac{\mathrm{d}\sigma_r}{\mathrm{d}r} - \nu\frac{\mathrm{d}\sigma_\theta}{\mathrm{d}r}$$

$$\therefore \quad \frac{\mathrm{d}\sigma_z}{\mathrm{d}r} = \nu\frac{\mathrm{d}\sigma_r}{\mathrm{d}r} + \nu\frac{\mathrm{d}\sigma_\theta}{\mathrm{d}r} \quad (3.46)$$

Substituting (3.46) in (3.45)

$$(1 + \nu)\sigma_r = (1 + \nu)\sigma_\theta + r\frac{d\sigma_\theta}{dr} - r\nu\frac{d\sigma_r}{dr} - r\nu^2\frac{d\sigma_r}{dr} - r\nu^2\frac{d\sigma_\theta}{dr}$$

i.e. $$(1 + \nu)\sigma_r = (1 + \nu)\sigma_\theta + r(1 - \nu^2)\frac{d\sigma_\theta}{dr} - r\nu(1 + \nu)\frac{d\sigma_r}{dr}$$

∴ $$\sigma_\theta - \sigma_r = r\nu\frac{d\sigma r}{dr} - r(1 - \nu)\frac{d\sigma_\theta}{dr} \quad (3.47)$$

Substituting into equation (3.37) from equation (3.47) gives

$$r\nu\frac{d\sigma_r}{dr} - r(1 - \nu)\frac{d\sigma_\theta}{dr} = r\frac{d\sigma_r}{dr}$$

i.e. $$r(1 - \nu)\left[\frac{d\sigma_r}{dr} + \frac{d\sigma_\theta}{dr}\right] = 0$$

∴ $$\frac{d}{dr}(\sigma_r + \sigma_\theta) = 0$$

i.e. $$\sigma_r + \sigma_\theta = \text{constant} = 2A\text{, say} \quad (3.48)$$

$$\sigma_\theta - \sigma_r = r\frac{d\sigma_r}{dr}$$

i.e. $$2\sigma_r = 2A - r\frac{d\sigma_r}{dr}$$

i.e. $$r\frac{d\sigma_r}{dr} + 2\sigma_r = 2A$$

∴ $$\frac{1}{r}\frac{d}{dr}(r^2\sigma_r) = 2A$$

Hence:

$$r^2\sigma_r = \frac{2Ar^2}{2} - B$$

i.e. $$\sigma_r = A - \frac{B}{r^2}$$

and using equation (3.48) leads to

$$\sigma_\theta = A + \frac{B}{r^2}$$

Note that, since ε_z = const and $\sigma_r + \sigma_\theta$ = const, then equation (3.43) shows that σ_z = const, i.e. it is independent of r. The value of σ_z can therefore be obtained from a consideration of axial equilibrium.

$$\sigma_r = A - \frac{B}{r^2}$$

and

$$\sigma_\theta = A + \frac{B}{r^2}$$

The constants, A and B, are the so-called Lame's constants, which are the constants of integration, can be obtained from the boundary conditions, i.e.

at $$r = R_i, \sigma_r = -p_i$$

at $$r = R_o, \sigma_r = -p_o$$

∴ $$-p_i = A - \frac{B}{R_i^2}$$

and $$-p_o = A - \frac{B}{R_o^2}$$

Hence, A and B can be determined

For closed-ended cylinders:

$$\pi(R_o^2 - R_i^2)\sigma_z + \pi R_o^2 p_o = \pi R_i^2 p_i$$

i.e.

$$\sigma_z = \frac{R_i^2 p_i - R_o^2 p_o}{(R_o^2 - R_i^2)}$$

For a solid cylinder, i.e. $R_i = 0$

$$\sigma_{r(r = Ri)} = A - \frac{B}{0^2} = \infty$$

unless $B = 0$.

Therefore, B must be zero, since the stresses cannot be infinite, and so, for a solid cylinder, the radial and hoop stresses are equal to each other and they are constant,

i.e.

$$\sigma_r = \sigma_\theta = A$$

Also, since a solid cylinder can only have external pressure, the constant A must equal the external pressure.

Displacements are most conveniently obtained by using equations (3.41) and (3.43) together with equations (3.38) and (3.40), i.e.

$$\varepsilon_\theta = \frac{u}{r} = \frac{1}{E}(\sigma_\theta - \nu(\sigma_r + \sigma_z))$$

$$\varepsilon_z = \frac{\Delta l}{l} = \frac{1}{E}(\sigma_z - \nu(\sigma_r + \sigma_\theta)) = \text{constant}$$

where l is the cylinder length, Δl is the increase in cylinder length and u is the radial displacement at radius r.

Analysis of rotating discs

Rotating components such as flywheels and turbine discs can be regarded as **thick cylinders with body forces**, as well as possible pressure loads, and represent an extension of the thick cylinder theory discussed in the previous section.

Equilibrium

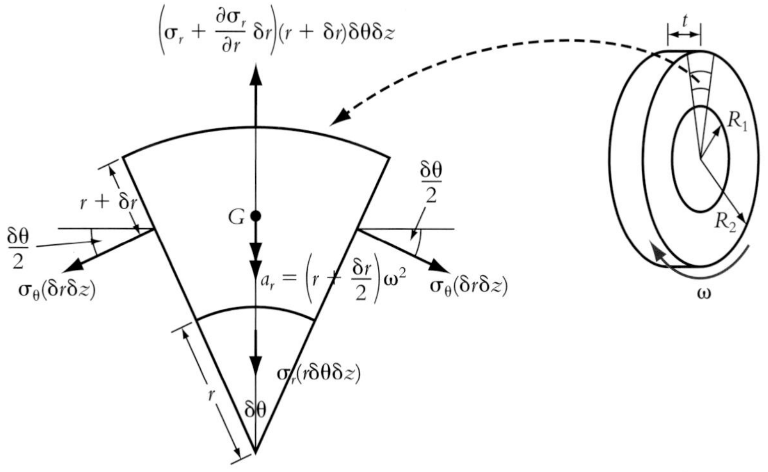

Figure 3.96 Free-body diagram of an element of material within a disc

At a radius r, $v = \omega r$ and $a_r = \frac{v^2}{r} = \omega^2 r$, directed towards the centre of rotation.

$$\sigma_r r\delta\theta\delta z + 2\sigma_\theta \delta r\delta z\left(\frac{\delta\theta}{2}\right) - \left(\sigma_r + \frac{d\sigma_r}{dr}\delta r\right)(r + \delta r)\delta\theta\delta z = \left[\rho\left(r + \frac{\delta r}{2}\right)\delta\theta\delta z\delta r\right]\left(r + \frac{\delta r}{2}\right)\omega^2$$

$$\sigma_r r + \sigma_\theta \delta r - r\sigma_r - \sigma_r \delta r - r\frac{d\sigma_r}{dr}\delta r - \frac{d\sigma_r}{dr}(\delta r)^2 = \rho\left(r + \frac{\delta r}{2}\right)^2 \delta r\omega^2$$

$$\therefore \quad \sigma_\theta - \sigma_r - r\frac{d\sigma_r}{dr} - \frac{d\sigma_r}{dr}\delta r = \rho r^2\omega^2 + \rho\left(\frac{\delta r}{2}\right)^2\omega^2 + \rho r\delta r\omega^2$$

Neglecting small terms, i.e. those containing δr and $(\delta r)^2$

$$\sigma_\theta - \sigma_r = r\frac{d\sigma_r}{dr} + \rho r^2\omega^2 \tag{3.49}$$

also

$$\sigma_z = 0$$

Compatibility and stress–strain relationships (linear elastic)

$$\varepsilon_\theta = \frac{u}{r} = \frac{1}{E}(\sigma_\theta - \nu\sigma_r) \tag{3.50}$$

$$\varepsilon_r = \frac{du}{dr} = \frac{1}{E}(\sigma_r - \nu\sigma_\theta) \tag{3.51}$$

Substituting for u from equation (3.50) into to equation (3.51) gives:

$$\frac{d}{dr}\left(\frac{r}{E}(\sigma_\theta - \nu\sigma_r)\right) = \frac{1}{E}(\sigma_r - \nu\sigma_\theta)$$

i.e.,

$$\sigma_\theta - \nu\sigma_r + r\left(\frac{d\sigma_\theta}{dr} - \nu\frac{d\sigma_r}{dr}\right) = \sigma_r - \nu\sigma_\theta$$

$$\therefore \quad (\sigma_\theta - \sigma_r)(1 + \nu) + r\left(\frac{d\sigma_\theta}{dr} - \nu\frac{d\sigma_r}{dr}\right) = 0 \tag{3.52}$$

Substituting for $\sigma_\theta - \sigma_r$ from equation (3.49) into equation (3.52) gives:

$$\left(r\frac{d\sigma_r}{dr} + \rho r^2\omega^2\right)(1 + \nu) + r\frac{d\sigma_\theta}{dr} - r\nu\frac{d\sigma_r}{dr} = 0$$

$$\therefore \quad r\frac{d\sigma_r}{dr} + r\nu\frac{d\sigma_r}{dr} + (1 + \nu)\rho r^2\omega^2 + r\frac{d\sigma_\theta}{dr} - r\nu\frac{d\sigma_r}{dr} = 0$$

i.e.,

$$\frac{d}{dr}(\sigma_\theta + \sigma_r) = -(1 + \nu)\rho\omega^2 r$$

$$\therefore \quad \sigma_\theta + \sigma_r = -(1 + \nu)\rho\omega^2\frac{r^2}{2} + 2A \tag{3.53}$$

Subtracting equation (3.49) from equation (3.53) gives:

$$2\sigma_r + r\frac{d\sigma_r}{dr} = -(1 + \nu)\rho\omega^2\frac{r^2}{2} - \rho\omega^2 r^2 + 2A$$

$$\frac{1}{r}\frac{d}{dr}(r^2\sigma_r) = \frac{\rho\omega^2 r^2(3 + \nu)}{2} + 2A$$

$$r^2\sigma_r = -\rho\omega^2(3 + \nu)\frac{r^4}{8} + Ar^2 - B$$

where B is a constant of integration. Therefore,

$$\sigma_r = A - \frac{B}{r^2} - \frac{\rho\omega^2(3 + \nu)}{8}r^2$$

and from equation (3.53):

$$\sigma_\theta = A + \frac{B}{r^2} + \frac{\rho\omega^2(3+\nu)}{8}r^2 - \frac{\rho\omega^2(1+\nu)}{2}r^2$$

$$\sigma_\theta = A + \frac{B}{r^2} - \frac{\rho\omega^2(1+3\nu)}{8}r^2$$

Worked example

Thick cylinder with pistons

A cylinder with 50 mm bore and 100 mm OD is subjected to an internal pressure of 400 bar. The end loads are supported by pistons which seal without restraint. Determine the distributions of stress across the cylinder wall.

$$p = 400 \text{ bar} = 400 \times 100\,\text{kPa}$$
$$= 40 \times 1000\,\text{kPa}$$
$$= 40\ \text{N/mm}^2$$

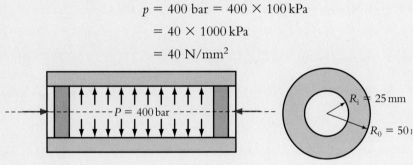

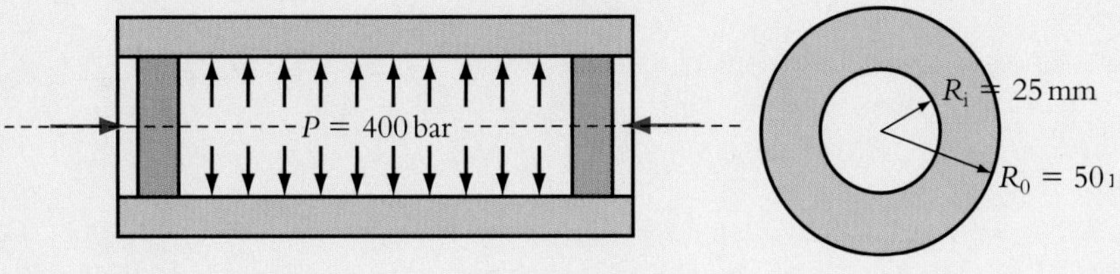

Figure 3.97

Since there is no axial load on the cylinder, then $\sigma_z = 0$.
For a thick cylinder,

$$\sigma_r = A - \frac{B}{r^2}$$

and

$$\sigma_\theta = A + \frac{B}{r^2}$$

At $r = 25$ mm, $\sigma_r = -40\ \text{N/mm}^2$

$$\therefore \qquad -40 = A - \frac{B}{625} \tag{3.54}$$

At $r = 50$ mm, $\sigma_r = 0$

$$0 = A - \frac{B}{2500} \tag{3.55}$$

Eliminating A from equations (3.54) and (3.55) gives:

$$40 = B\left(\frac{1}{625} - \frac{1}{2500}\right)$$
$$= B\left(\frac{4-1}{2500}\right)$$

$$\therefore \qquad B = \frac{40 \times 2500}{3}$$

Substituting for B into equation (3.55) gives:

$$0 = A - 40 \times \frac{2500}{3 \times 2500}$$

$$\therefore \qquad A = \frac{40}{3}$$

Hence,

$$\sigma_\theta = \frac{40}{3} + \frac{40 \times 2500}{3r^2} = \frac{40}{3}\left(1 + \frac{2500}{r^2}\right)$$

and

$$\sigma_r = \frac{40}{3} - \frac{40 \times 2500}{3r^2} = \frac{40}{3}\left(1 - \frac{2500}{r^2}\right)$$

At $r = 25$ mm,

$$\sigma_\theta = \frac{40}{3} \times 5\,\text{N/mm}^2 = 66.7\,\text{N/mm}^2$$

and

$$\sigma_r = \frac{40}{3} \times (-3)\,\text{N/mm}^2 = -40\,\text{N/mm}^2$$

At $r = 50$ mm,

$$\sigma_\theta = \frac{40}{3} \times 2\,\text{N/mm}^2 = 26.7\,\text{N/mm}^2$$

and

$$\sigma_r = 0$$

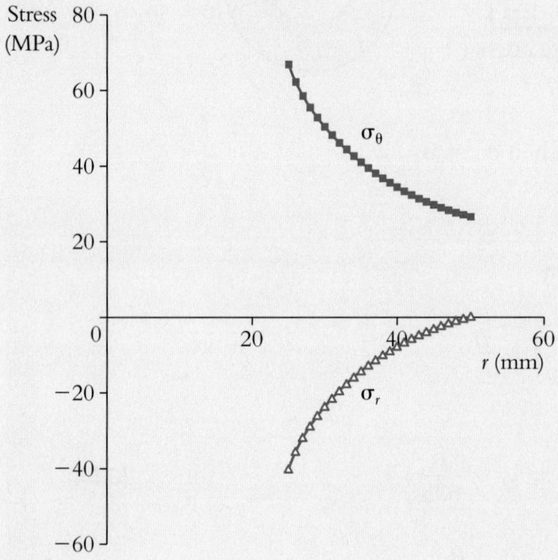

Figure 3.98

Worked example

Shrink/interference fit

A pair of mild steel cylinders (E = 200 GPa) of equal length have the following dimensions:

(1) 40 mm bore and 80.06 mm outside diameter
(2) 80 mm bore and 120 mm outside diameter

i.e. there is a diametral interference of 0.06 mm. The larger cylinder is heated, placed around, and allowed to shrink onto, the smaller cylinder. Calculate the stresses after assembly.

Conditions

- After assembly, the radial interference pressure, p, will be the same on both cylinders, i.e. cylinder 1 will have an external pressure, p, and cylinder 2 will have an internal pressure, p, as indicated in Figure 3.99.
- The decrease in the outside radius of cylinder 1, i_1, plus the increase in the inside radius of cylinder 2, i_2, will be equal to the radial interference, i.e. $i = i_1 + i_2$.
- Axial stresses are assumed to be zero (or negligible).

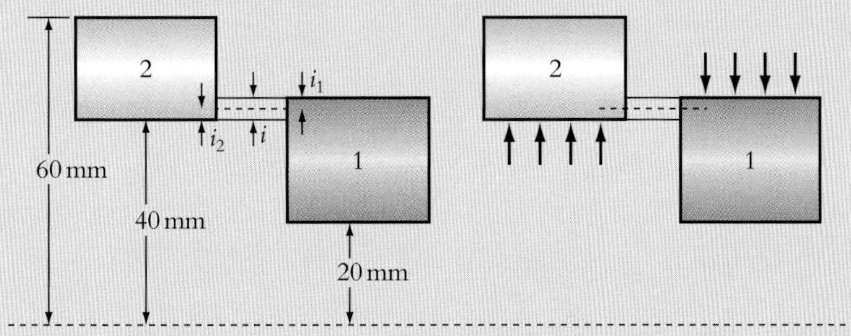

Figure 3.99

For cylinder 1:

$$\sigma_r = A_1 - \frac{B_1}{r^2}$$

and

$$\sigma_\theta = A_1 + \frac{B_1}{r^2}$$

at $r = 20\,\text{mm}$, $\sigma_r = 0$,

$$\therefore \quad B_1 = 400A_1$$

at $r = 40\,\text{mm}$ (no significant difference with 40.03 mm), $\sigma_r = -p$

$$\therefore \quad -p = A_1 - \frac{20^2}{40^2}A_1 = A_1 - \frac{400}{1600}A_1$$

$$\therefore \quad A_1 = -\frac{4}{3}p$$

and

$$B_1 = -\frac{1600}{3}p$$

Thus

$$(\sigma_r)_1 = -\frac{4p}{3}\left(1 - \frac{400}{r^2}\right)$$

and

$$(\sigma_\theta)_1 = -\frac{4p}{3}\left(1 + \frac{400}{r^2}\right)$$

$$\varepsilon_\theta = \frac{u}{r} = \frac{1}{E}(\sigma_\theta - \nu(\sigma_r + \sigma_z)) = \frac{1}{E}(\sigma_\theta - \nu\sigma_r)$$

At the outside of cylinder 1, $r = 40\,\text{mm}$,

$$\frac{-i_1}{40} = \frac{1}{200\,000}(\sigma_\theta - \nu\sigma_r)$$

i.e. $$\frac{-i_1}{40} = \frac{1}{200\,000}\left(-\frac{4p}{3}\right)\left(1 + \frac{400}{1600} - \nu\left(1 - \frac{400}{1600}\right)\right)$$

$$i_1 = \frac{8p}{30\,000}\left(\frac{5}{4} - \frac{3\nu}{4}\right)$$

∴ $$i_1 = \frac{2p}{30\,000}(5 - 3\nu)$$

For cylinder 2

$$\sigma_r = A_2 - \frac{B_2}{r^2}$$

and

$$\sigma_\theta = A_2 + \frac{B_2}{r^2}$$

At $r = 60\,\text{mm}$, $\sigma_r = 0$

∴ $$B_2 = 3600\,A_2$$

At $r = 40\,\text{mm}$, $\sigma_r = -p$

∴ $$-p = A_2 - \frac{60^2}{40^2}A_2 = A_2 - \frac{3600}{1600}A_2$$

i.e. $$A_2 = \frac{4}{5}p$$

and

$$B_2 = 3600 \times \frac{4}{5}p$$

Thus,

$$(\sigma_r)_2 = \frac{4p}{5}\left(1 - \frac{3600}{r^2}\right)$$

and

$$(\sigma_\theta)_2 = \frac{4p}{5}\left(1 + \frac{3600}{r^2}\right)$$

$$\varepsilon_\theta = \frac{u}{r} = \frac{1}{E}(\sigma_\theta - \nu(\sigma_r + \sigma_z)) = \frac{1}{E}(\sigma_\theta - \nu\sigma_r)$$

At the inside of cylinder 2, $r = 40\,\text{mm}$,

$$\frac{+i_2}{40} = \frac{1}{200\,000}\left(\frac{4p}{5}\right)\left(1 + \frac{3600}{1600} - \nu\left(1 - \frac{3600}{1600}\right)\right)$$

i.e. $$i_2 = \frac{8p}{50\,000}\left(\frac{13}{4} + \frac{5\nu}{4}\right)$$

∴ $$i_2 = \frac{2p}{50\,000}(13 + 5\nu)$$

But $i_1 + i_2 = i = 0.03\,\text{mm}$

∴ $$\frac{2p}{30\,000}(5 - 3\nu) + \frac{2p}{50\,000}(13 + 5\nu) = 0.03$$

$$\frac{10p}{30\,000} - \frac{2\nu p}{10\,000} + \frac{26p}{50\,000} + \frac{2\nu p}{10\,000} = 0.03$$

$$\frac{50p + 78p}{150\,000} = 0.03$$

i.e. $$p = \frac{4500}{128}\,\text{N/mm}^2 = 35.2\,\text{N/mm}^2$$

For cylinder 1,

$$(\sigma_r)_1 = -46.9\left(1 - \frac{400}{r^2}\right)$$

$$(\sigma_\theta)_1 = -46.9\left(1 + \frac{400}{r^2}\right)$$

and for cylinder 2,

$$(\sigma_r)_2 = 28.2\left(1 - \frac{3600}{r^2}\right)$$

$$(\sigma_\theta)_2 = 28.2\left(1 + \frac{3600}{r^2}\right)$$

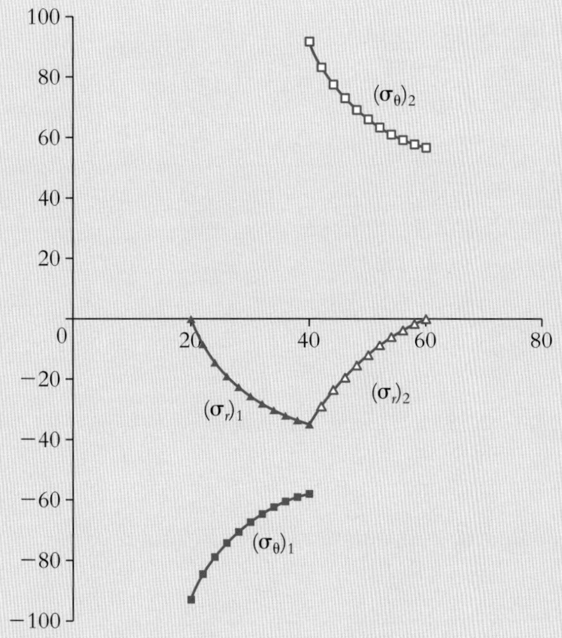

Figure 3.100

Worked example

A turbine rotor disc with an angular velocity of 4000 rpm has an external diameter of 1.2 m and has a 0.1 m diameter hole bored along its axis. Determine the stress distributions in the disc.

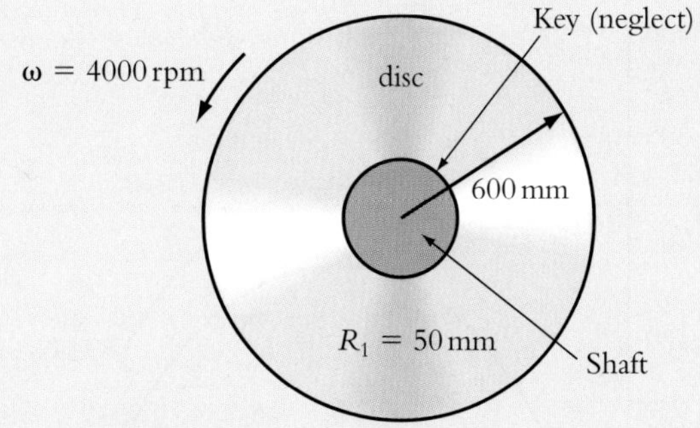

Figure 3.101

Take $\rho = 7850\,\text{kg/m}^3$ and $\nu = 0.3$.

$$\sigma_r = A - \frac{B}{r^2} - \frac{\rho w^2(3+\nu)r^2}{8}$$

$$\sigma_\theta = A + \frac{B}{r^2} - \frac{\rho w^2(1+3\nu)r^2}{8}$$

at $r = 50\,\text{mm}$, $\sigma_r = 0$

$$\therefore \quad O = A - \frac{B}{50^2} - \frac{7850 \times 10^{-9}}{8} \times \left(4000 \times \frac{2\pi}{60}\right)^2 (3 + 0.3) \times 50^2 \times 10^{-3}\,\text{N/mm}^2$$

i.e.

$$O = A - \frac{B}{2500} - 1.4204 \qquad \textbf{(3.56)}$$

at $r = 600\,\text{mm}$, $\sigma_r = 0$

$$\therefore \; O = A - \frac{B}{600^2} - \frac{7850 \times 10^{-9}}{8} \times \left(4000 \times \frac{2\pi}{60}\right)^2 \times (3 + 03) \times 600^2 \times 10^{-3}\,\text{N/mm}^2 \qquad \textbf{(3.57)}$$

Subtracting equation (3.56) from equation (3.57)

$$\frac{B}{2500} - \frac{B}{600^2} = 204.54 - 1.42$$

$\therefore$

$$B = 511\,350.0$$

and

$$A = 205.95$$

Hence,

$$\sigma_r = 205.95 - \frac{511\,350}{r^2} - 0.000\,568r^2$$

and

$$\sigma_\theta = 205.95 + \frac{511\,350}{r^2} - 0.000\,327r^2$$

at

$$r = 50\,\text{mm}, \sigma_r = 0 \text{ and } \sigma_\theta = 409.7\,\text{N/mm}^2$$

at

$$r = 600\,\text{mm}, \sigma_r = 0 \text{ and } \sigma_\theta = 89.6\,\text{N/mm}^2$$

The maximum σ_r value could be obtained by

$$\frac{d\sigma_r}{dr} = 0$$

to find the r-value and then substituting this value of r into expression for σ_r.

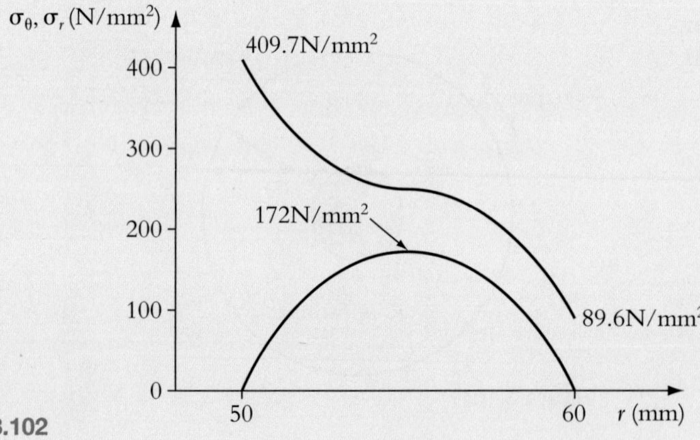

Figure 3.102

Learning summary

By the end of this section you should have learnt:

- ✓ the essential differences between the stress analysis of thin and thick cylinders, leading to an understanding of statically determinate and statically indeterminate situations;
- ✓ how to derive the equilibrium equations for an element of material in a solid body (e.g. a thick cylinder);
- ✓ the derivation of Lame's equations;
- ✓ to determine stresses caused by shrink-fitting one cylinder onto another;
- ✓ to include 'inertia' effects into the thick cylinder equations in order to calculate the stresses in a rotating disc.

3.9 Asymmetrical bending

The beam bending equation, $\frac{M}{I} = \frac{\sigma}{y} = \frac{E}{R}$, has been derived and is generally used to determine stresses in a beam with a *symmetrical* cross section. The symmetry is usually about an axis perpendicular to the neutral axis of the section. For a section where this symmetry does not apply, i.e. asymmetric sections, a complication arises, making bending analysis more difficult. In these cases, applying a bending moment about the neutral axis will, in general, result not only in bending about that axis but also in simultaneous bending about the perpendicular axis; there is an interaction effect. To analyse such sections we introduce a new geometric quantity called the **product moment of area** and this leads to the concept of **principal second moments of area** and **principal axes** for the section. These are axes for which the product moment of area is zero and the above interaction effect during bending does not occur. Thus, it is convenient to analyse the bending of asymmetric sections about these axes. In this section, we will look at the theory behind this effect and develop a general procedure for dealing with asymmetrical bending situations.

Second moments of area of a complex shaped cross section

Second moments of area about parallel axes

Consider an arbitrary shaped cross section, as shown in Figure 3.103. The centroid of area, G, is at the origin, O, of the O-x-y axes set. A parallel axes set, O′-x'-y', also exists, distance a and b from the O-x-y, as shown in the figure. The centroid of area, G, is positioned at coordinates $(x', y') = (a, b)$ in this parallel axes set.

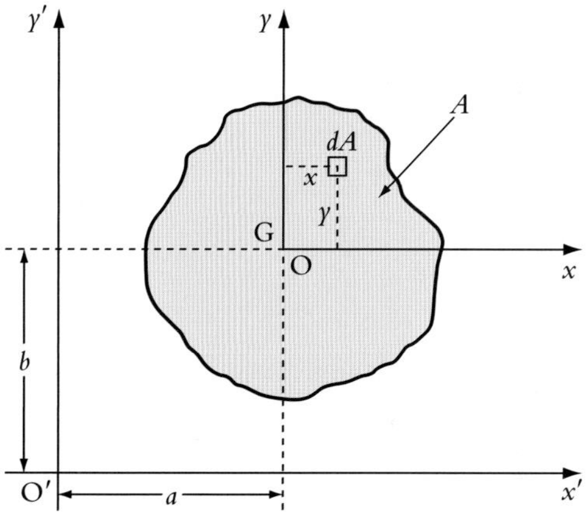

Figure 3.103 **Parallel axis theorem**

We know that the second moments of area, I_x and I_y, of the section with respect to the x- and y-axes are given by,

$$I_x = \int_A y^2 \mathrm{d}A$$

and

$$I_y = \int_A x^2 \mathrm{d}A$$

i.e. the product of an element of area, dA, and its distance squared from the particular axis (x or y), integrated over the full cross-sectional area, A.

The **parallel axis theorem** allows the calculation of the second moments of area, $I_{x'}$ and $I_{y'}$, with respect to the x'- and y'-axes as follows,

$$I_{x'} = I_x + Ab^2 \quad \textbf{(3.58)}$$

and

$$I_{y'} = I_y + Aa^2 \quad \textbf{(3.59)}$$

I_x and I_y are the second moments of area about a set of axes through the centroid and are always the minimum second moments. $I_{x'}$ and $I_{y'}$ will always be greater because the second terms in equations (3.58) and (3.59) are always positive as the distances between the axes, a and b, are squared.

Product moment of area

We now introduce a new quantity, the **product moment of area, I_{xy}**, which is defined as,

$$I_{xy} = \int_A xy \mathrm{d}A$$

I_{xy} is the summation of the elements of area times the product of their coordinates. We can now develop the parallel axis theorem for the product moments of area as follows,

$$I_{x'y'} = \int_A x'y' \mathrm{d}A = \int_A (x + a)(x + b)\mathrm{d}A$$

$$= \int_A xy\mathrm{d}A + a\int_A y\mathrm{d}A + b\int_A x\mathrm{d}A + ab\int_A \mathrm{d}A$$

but $\int_A y\mathrm{d}A$ and $\int_A x\mathrm{d}A$ are both zero because the origin of axes Oxy is at the centroid of area, G.

Thus,

$$I_{x'y'} = I_{xy} + abA \qquad \textbf{(3.60)}$$

This is the **product parallel axis theorem**. Again, I_{xy} is the product moment of area about a set of axes through the centroid. In this case, $I_{x'y'}$, can be either positive or negative, depending on the signs of a and b.

Symmetric sections

Figure 3.104 shows a section where one axis (the y-axis in this case) is an axis of symmetry. The sum of the contributions to the product moment of area from elements of area, dA, on opposite sides of the axis of symmetry will cancel out because of the change of sign of the x-coordinate. Thus, in general, if a section has an axis of symmetry, then I_{xy} is zero.

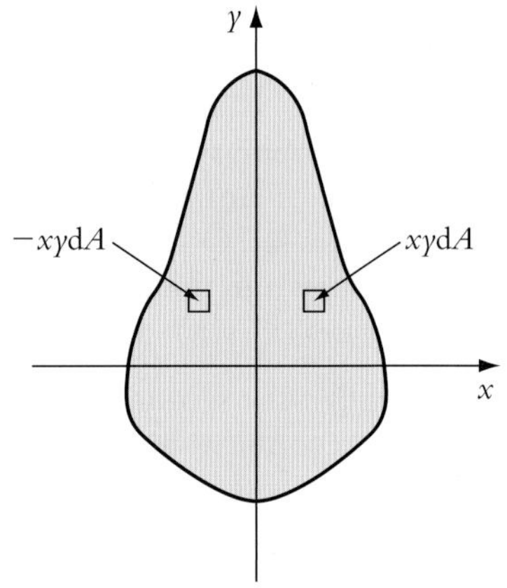

Figure 3.104 Symmetric section

Rotation of axes

Referring to Figure 3.105, the question arising is that, given I_x, I_y and I_{xy} for a particular section, what are the second moments of area I_u, I_v and I_{uv} about a set of axes, Ouv, through the same origin but rotated through an angle, θ.

The coordinate transformation equations for rotation through an angle, θ, are

$$u = x\cos\theta + y\sin\theta$$

$$v = -x\sin\theta + y\cos\theta$$

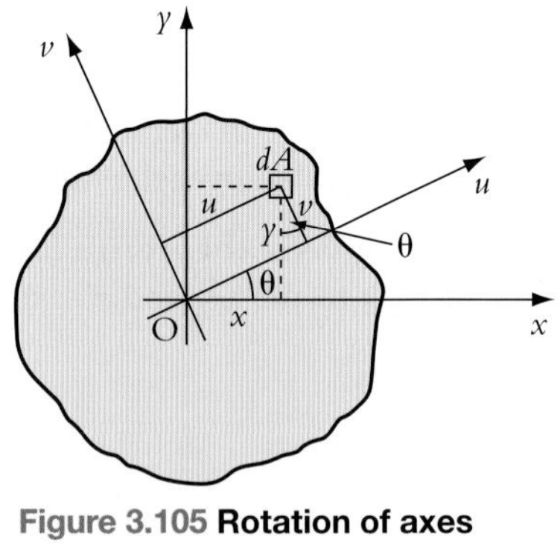

Figure 3.105 Rotation of axes

using these equations, we can now derive the second moment of area about the u-axis, I_u, as follows,

$$I_u = \int_A v^2\mathrm{d}A = \int_A (x^2\sin^2\theta + y^2\cos^2\theta - 2xy\sin\theta\cos\theta)\mathrm{d}A$$

$$= I_y\sin^2\theta + I_x\cos^2\theta - I_{xy}\sin 2\theta$$

but

$$\cos^2\theta = \tfrac{1}{2}(1 + \cos 2\theta)$$

and

$$\sin^2\theta = \tfrac{1}{2}(1 - \cos 2\theta)$$

$$\therefore \quad I_u = \tfrac{1}{2}(I_x + I_y) + \tfrac{1}{2}(I_x - I_y)\cos 2\theta - I_{xy}\sin 2\theta \qquad \textbf{(3.61)}$$

Similarly, we can show that

$$I_v = \tfrac{1}{2}(I_x + I_y) - \tfrac{1}{2}(I_x - I_y)\cos 2\theta + I_{xy}\sin 2\theta \qquad \textbf{(3.62)}$$

Equations (3.61) and (3.62) differ only by the signs of the last two terms, so adding them together gives

$$I_u + I_v = I_x + I_y = \int_A (x^2 + y^2)\mathrm{d}A = \int_A r^2\,\mathrm{d}A = \mathrm{J}$$

This is a statement of the **perpendicular axis theorem**, i.e. the sum of the second moments of area about two perpendicular axes is equal to the second moment of area about the third perpendicular axis. In this case, the latter second moment of area is the polar second moment of area, J.

Now considering the product moment of area, I_{uv}, we have,

$$I_{uv} = \int_A uv\mathrm{d}A = \int_A (x\cos\theta + y\cos\theta)(-x\sin\theta + y\cos\theta)\mathrm{d}A$$

$$= \int_A [-x^2\cos\theta\sin\theta + y^2\sin\theta\cos\theta + xy(\cos^2\theta - \sin^2\theta)]\mathrm{d}A$$

$$\therefore \quad I_{uv} = \tfrac{1}{2}(I_x + I_y)\sin 2\theta + I_{xy}\cos 2\theta \qquad \textbf{(3.63)}$$

Equations (3.61), (3.62) and (3.63) enable us to calculate I_u, I_v and I_{uv}, knowing I_x, I_y and I_{xy} and the angle of rotation, θ.

Principal second moments of area

Now, looking more closely at equation (3.63). If θ changes by 90°, 2θ changes by 180° and I_{uv} changes its sign but not its magnitude. Thus, since it is a continuous function of θ, I_{uv} must have zero value.

The axes which give rise to a zero product moment are called the **principal axes**.

For simplicity, let us assume that x and y are the principal axes. Then,

$$I_x = I_p,\ I_y = I_q \text{ and } I_{xy} = I_{pq} = 0$$

where I_p and I_q are called the **principal second moments of area**

Substituting I_p and I_q into equations (3.61), (3.62) and (3.63) we have,

$$I_u = \tfrac{1}{2}(I_p + I_q) + \tfrac{1}{2}(I_p - I_q)\cos 2\theta$$

$$I_v = \tfrac{1}{2}(I_p + I_q) - \tfrac{1}{2}(I_p - I_q)\cos 2\theta$$

$$I_{uv} = \tfrac{1}{2}(I_p - I_q)\sin 2\theta$$

These equations can be represented by a Mohr's circle as shown in Figure 3.106. Second moments are plotted on the x-axis and the product moments are plotted on the y-axis.

Note The y-axis for the circle is positive upwards, unlike Mohr's circle for stress, which is positive downwards for shear.

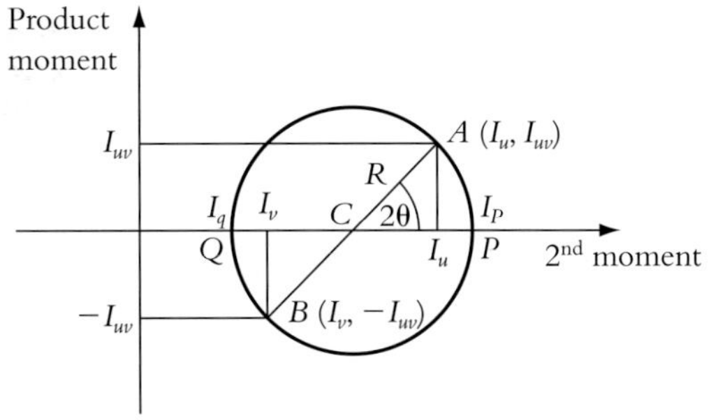

Figure 3.106 Mohr's circle

Point A on the circle has coordinates that correspond to the second moment and product moment, i.e. (I_u, I_{uv}), relevant to the u-axis. Point B on the circle has coordinates which correspond to the second moment and product moment, i.e. $(I_v, I_{vu} = -I_{uv})$, relevant to the v-axis. These two points enable the circle to be drawn.

The centre of the circle, C, and radius, R, are given by,

$$\text{centre } C = \frac{I_u + I_v}{2} \tag{3.64}$$

$$\text{radius, } R = \sqrt{\left(\frac{I_u - I_v}{2}\right)^2 + I_{uv}^2} \tag{3.65}$$

The points P and Q on the circle correspond to the principal planes for which the product moment of areas are zero and the second moments are the principal second moments of area, I_p and I_q. Their magnitudes are given by

$$I_p = \text{centre} + \text{radius}$$

and

$$I_q = \text{centre} - \text{radius}$$

where the centre and radius are given by equations (3.64) and (3.65).

Thus knowing I_u, I_v and I_{uv}, the principal second moments of area, I_p and I_q, can be determined.

The angle of the principal axes with respect to the u-v-axes is the angle θ, where 2θ is shown in Figure 3.106 and is given by,

$$\sin 2\theta = \frac{I_{uv}}{R}$$

or alternatively

$$\tan 2\theta = \frac{I_{uv}}{\left(\frac{I_u - I_v}{2}\right)}$$

Key points about the Mohr's circle for second moments of area:

(1) The positive upward direction for the product moment ensures that rotation in the Mohr's circle has the same sense as the rotation of the axes in space.

(2) I_p and I_q are both positive.

(3) If $I_p = I_q$, all product moments are zero and all axes in all directions are principal axes, e.g. this is the case for a circular section.

(4) The sign of the product moment is important. $I_{uv} = \int_A uv\mathrm{d}A$ is associated with the u-axis and can be positive or negative. The product moment associated with the v-axis is $I_{vu} = -I_{uv}$.

Summary of procedure for calculating the principal second moments of area and the directions of the principal axes:

(1) Divide the cross section into subsections for which their centroid of areas and second moments of area about their own axes can be determined.

(2) Choose a convenient set of orthogonal axes with its origin at the centroid of the full cross section.

(3) Use the parallel axis theorem to determine the second moment of area for the full cross section.

(4) Use a Mohr's circle construction to determine the principal second moments of area and the directions of the principal axes.

Worked example

Principal second moments of area
Figure 3.107 shows an asymmetric angle cross-section. Determine:

(a) the principal second moments of area
(b) the directions of the principal axes.

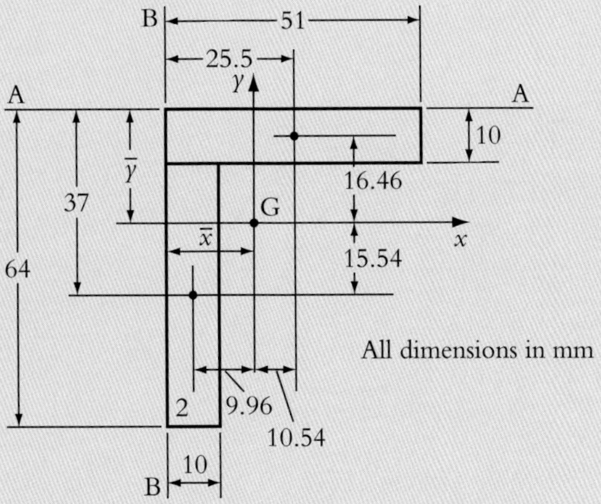

Figure 3.107 **Second moments of area of an angle section**

The section is divided into two rectangular subsections 1 and 2.

Position of the centroid:

$$\text{total area} = 51 \times 10 + 54 \times 10 = 1050\,\text{mm}^2$$

Taking moments of the areas about the datum AA,

$$1050 \times \bar{y} = (51 \times 10) \times 5 + (54 \times 10) \times 37$$

$$\therefore \quad \bar{y} = 21.46\,\text{mm}$$

Taking moments of the areas about the datum BB,

$$1050 \times \bar{x} = (51 \times 10) \times 25.5 + (54 \times 10) \times 5$$

$$\therefore \quad \bar{x} = 14.96\,\text{mm}$$

Second moments of area about a convenient set of axes:
The x and y axes are drawn as a convenient set of axes through the centroid. Using the parallel axis theorem,

$$I_x = \left(\frac{51 \times 10^3}{12} + 51 \times 10 \times 16.46^2\right) + \left(\frac{10 \times 54^3}{12} + 10 \times 54 \times 15.54^2\right)$$

$$= 404\,051\,\text{mm}^4$$

$$I_y = \left(\frac{10 \times 51^3}{12} + 10 \times 51 \times 10.54^2\right) + \left(\frac{54 \times 10^3}{12} + 54 \times 10 \times 9.96^2\right)$$

$$= 225\,268\,\text{mm}^4$$

And the product parallel axis theorem,

$$I_{xy} = (0 + 51 \times 10 \times 10.54 \times 16.46) + (0 + 54 \times 10 \times (-9.96) \times (-15.54))$$

$$= 172\,059\,\text{mm}^4$$

Note that, in the product moment of area calculation above, the product moment of each subsection about its own axis is zero due to the symmetry of each subsection. It is also important that the correct sign for the coordinates of each subsection centroid with respect to the full cross-sectional centroid are taken. Thus, for subsection 1, the coordinates are both positive (10.54 and 16.46), while for subsection 2, they are both negative (−9.96 and −15.54).

Mohr's circle:

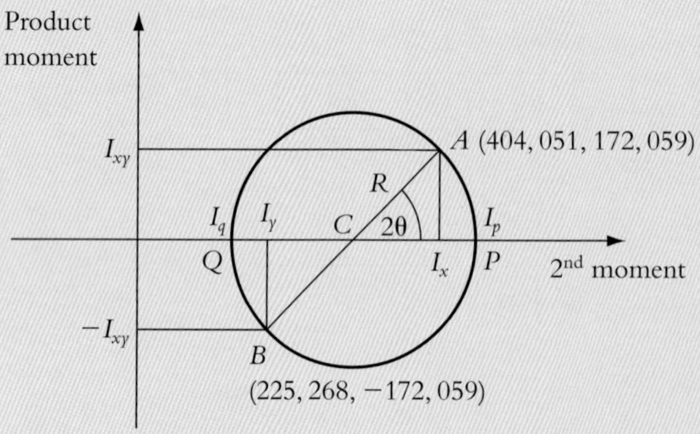

Figure 3.108 Mohr's circle

A Mohr's circle can now be drawn to represent the axes about which I_x, I_y and I_{xy} act, as shown in Figure 3.108. The centre and radius are calculated as follows:

$$\text{centre } C = \frac{I_x + I_y}{2} = 314\,659\ \text{mm}^4$$

$$\text{radius } R = \sqrt{\left(\frac{I_x - I_y}{2}\right)^2 + I_{xy}^{\ 2}} = 193\,895\ \text{mm}^4$$

Principal second moments of area:

The principal second moments of area can now be determined from the circle as follows:

$$I_p = C + R = 508\,554\ \text{mm}^4$$

$$I_q = C - R = 120\,764\ \text{mm}^4$$

and the angle, θ, of the principal axes with respect to the x-axis is given by

$$\sin 2\theta = \frac{I_{xy}}{R} = \frac{172\,059}{193\,895} = 0.887$$

$$\therefore \qquad \theta = 31.27°$$

From the Mohr's circle it can be seen that the principal axis 1, i.e. the p-axis is 31.27° clockwise from the x-axis. The principal axes can now be drawn on a sketch of the element as shown in Figure 3.109.

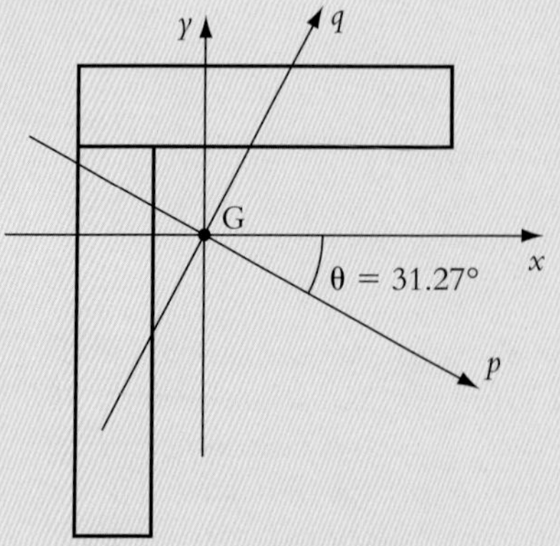

Figure 3.109 Angle of principal axes

Bending of beams with asymmetric sections

Figure 3.110 shows an arbitrary cross section of a beam subjected to a bending moment, M, acting at an angle θ to the x-axis. The origin of the x-y-axes coincides with the centroid of the section. The bending moment has two components, M_x and M_y, as shown, acting about the x- and y-axes respectively.

Note The bending moment and its components are drawn in vector form with a double arrowhead. The right-hand screw rule defines the sense of the bending moment component as shown in the figure inset.

Assume that bending takes place about the x-axis, i.e. O-x is the neutral axis. Then, the bending stress, σ, is proportional to the distance, y, from the neutral axis, or alternatively

$$\sigma = c \cdot y$$

where c is an arbitrary constant.

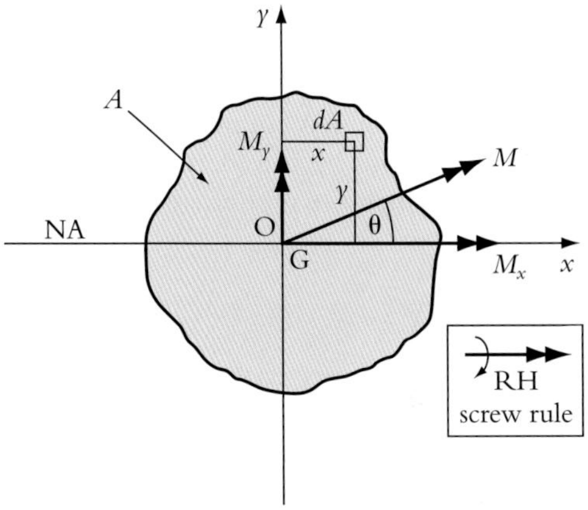

Figure 3.110 Bending about an arbitrary cross-section

The resultant moment about the x-axis is given by the sum of moments of the forces acting on each elemental area in the cross section. In the limit, this sum can be written as an integral:

$$M_x = \int_A \sigma \cdot y \mathrm{d}A$$

$$= \int_A c \cdot y \cdot y \mathrm{d}A$$

$$= c \cdot I_x$$

where I_x = second moment of area about the x-axis.

But

$$c = \sigma / y$$

$$\therefore \qquad \sigma = \frac{M_x}{I_x} y$$

which is the beam bending equation as expected.

However, there is also a resultant moment about the y-axis:

$$M_y = -\int_A \sigma \cdot x \mathrm{d}A$$

$$= -\int_A c \cdot y \cdot x \mathrm{d}A$$

$$= -c \cdot I_{xy}$$

where $I_{xy} = \int xy\mathrm{d}A$ is the product moment of area

Note The negative sign arises because a positive stress results in a negative moment about the y-axis.

Thus, in general, a moment has to be applied about the y-axis as well as the x-axis to produce bending about the x-axis only. A positive moment is required about the y-axis to counterbalance the negative moment set up by the stresses arising from M_x. This is not the case if I_{xy} is zero, i.e. for sections that are symmetric about the y-axis.

To ensure bending about the x-axis only, a resultant moment

$$M = \sqrt{M_x^2 + M_y^2}$$

must be applied at an angle, θ, given by

$$\theta = \tan^{-1}\left(\frac{M_y}{M_x}\right) = \tan^{-1}\left(\frac{-I_{xy}}{I_x}\right)$$

The resultant moment is only applied about the x-axis when $I_{xy} = 0$.

Figure 3.111 illustrates the effect for a z-section. If a bending moment is applied about the x-axis only, then the stresses in the flanges will create a resulting moment about the y-axis. Consequently, bending will take place about both the x- and y-axes. This is a consequence of I_{xy} not being zero for this asymmetric section.

To avoid this moment coupling effect, it is usually convenient to solve bending problems by considering bending about the principal axes of a section for which the product moment of area is zero.

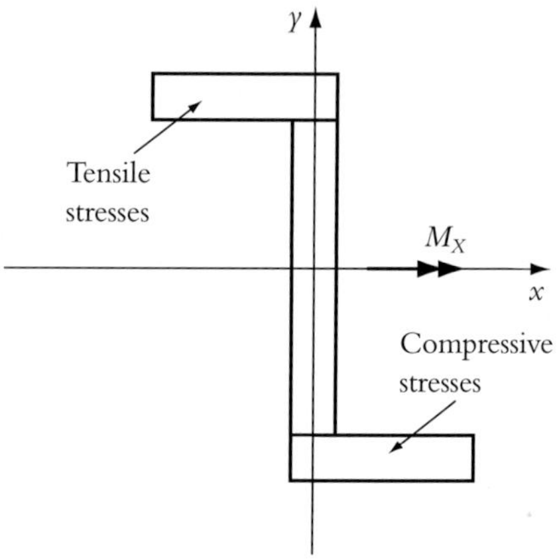

Figure 3.111 *z*-section

Solving asymmetrical bending problems

Consider the arbitrary asymmetric section shown in Figure 3.112(a). O is the centroid and O-p and O-q are the principal axes of the section. The principal axes are inclined at an angle θ to the x-y-axes. Components of an applied moment M, i.e. M_x and M_y, act about the O-x and O-y axes respectively. Firstly, M_x and M_y are resolved onto the principal directions, as illustrated in Figure 3.112(b), giving,

$$M_p = M_x \cos\theta + M_y \sin\theta$$

and

$$M_q = -M_x \sin\theta + M_y \cos\theta$$

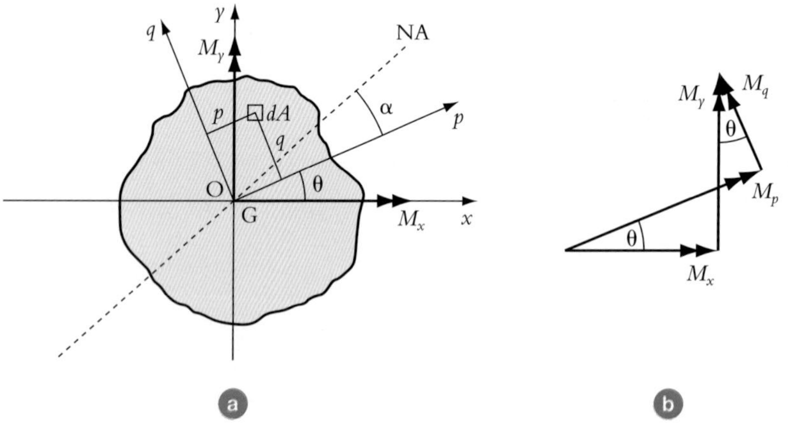

Figure 3.112 Asymmetrical section

We can now calculate the total bending stress at any position, (p,q), which arises from these two bending moment components and is given by

$$\sigma_b = \frac{M_p \cdot q}{I_p} - \frac{M_q \cdot p}{I_q} \qquad (3.66)$$

Note When p and q are both positive, i.e. in the first quadrant of the p-q axes set, a positive, M_p gives rise to a positive bending stress while a positive M_q gives rise to a negative bending stress.

The maximum stress in the section will occur at the extreme distance from the neutral axis. We therefore need to determine the position/orientation of the neutral axis which can be found by setting the bending stress, i.e. equation (3.66), to zero. Thus, the neutral axis occurs where,

$$\sigma_b = \frac{M_p \cdot q}{I_p} - \frac{M_q \cdot p}{I_q} = 0$$

$$\therefore \quad \frac{M_p \cdot q}{I_p} = \frac{M_q \cdot p}{I_q}$$

$$\therefore \quad \frac{q}{p} = \frac{M_q}{M_p} \cdot \frac{I_p}{I_q}$$

This value for q/p defines the angle, α, of the neutral axis, with respect to the p-axis, shown in Figure 3.112(a), as follows,

$$\alpha = \tan^{-1}\left(\frac{q}{p}\right) = \tan^{-1}\left(\frac{M_q \cdot I_p}{M_p \cdot I_q}\right) \tag{3.67}$$

Equation (3.66) can therefore be used to determine the magnitude of the stress at any position (p,q), and equation (3.67) can be used to determine the orientation of the neutral axis and hence the position of the maximum stress, which is at the extreme distance from the neutral axis.

Summary of the procedure for solving asymmetrical bending problems:

(1) Determine the principal axes of the section, p and q, about which $I_{xy} = 0$.

(2) Consider bending about the principal axes, i.e. resolve bending moments onto these axes.

(3) Knowing M_p, M_q, I_p and I_q, determine the general expression for the bending stress at position (p, q) as follows,

$$\sigma_b = \frac{M_p \cdot q}{I_p} - \frac{M_q \cdot p}{I_q}$$

(4) Determine the angle of the neutral axis with respect to the p-axis:

$$\alpha = \tan^{-1}\left(\frac{q}{p}\right) = \tan^{-1}\left(\frac{M_q \cdot I_p}{M_p \cdot I_q}\right)$$

(5) Evaluate the bending stress at any position in the section including the extreme positions from the neutral axis which give the maximum bending stresses.

Worked example

Asymmetrical bending

The angle section, shown in Figure 3.113, with principal axes and principal second moments of area indicated, is subjected to a bending moment of 300 Nm about the x-axis.

Determine:

(a) the position/orientation of the neutral axis

(b) the bending stresses at positions A, B and C.

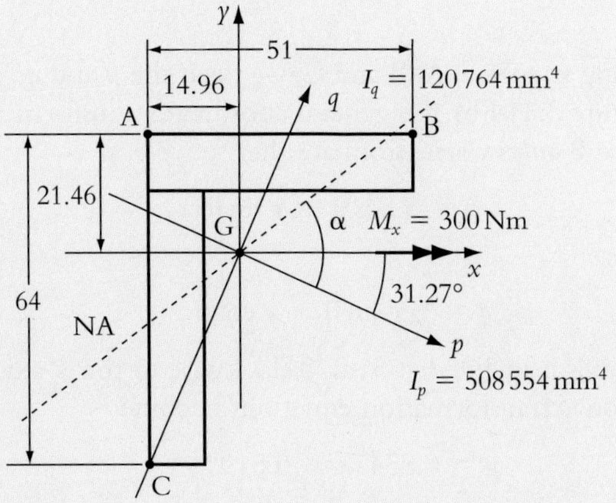

Figure 3.113 Bending about an asymmetrical angle section

Resolving the applied moment:

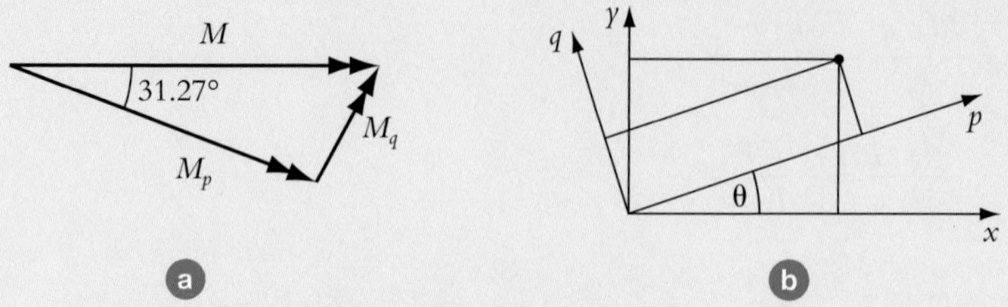

Figure 3.114 Transformation onto principal axes

Referring to Figure 3.114(a), the components of the applied bending in the p- and q-directions are

$$M_p = M\cos(31.27) = 0.855M$$

$$M_q = M\sin(31.27) = 0.519M$$

Bending stress equation:
The general expression for bending stress at position (p,q) is,

$$\sigma_b = \frac{M_p \cdot q}{I_p} - \frac{M_q \cdot p}{I_q}$$

$$= \frac{300.10^3\ 0.855q}{508\,554} - \frac{300.10^3\ 0.519p}{120\,764}$$

$$= 0.5042q - 1.2894p$$

Note that for p and q in mm, this expression gives bending stress in N/mm^2, i.e. MPa.

Orientation of the neutral axis, α, with respect to the p-axis:

$$\alpha = \tan^{-1}\left(\frac{q}{p}\right) = \tan^{-1}\left(\frac{M_q \cdot I_p}{M_p \cdot I_q}\right)$$

$$= \tan^{-1}\left(\frac{1.2894}{0.5042}\right) = \tan^{-1}(2.558)$$

$$= 68.64°$$

Orientation of the neutral axis with respect to the x-axis $= 68.64 - 31.27 = 37.37°$. This orientation is illustrated in Figure 3.113.

Bending stresses:
To determine the bending stresses at A, B and C, we need the p and q coordinates of these points. Referring to Figure 3.114(b), the general coordinate transformation equations for a set of axes, p-q, inclined θ *anticlockwise* from another set, x-y, are,

$$p = x\cos\theta + y\sin\theta$$

and

$$q = -x\sin\theta + y\cos\theta$$

For this problem, the p-axis is inclined at 31.27° clockwise to the x-axis. Thus, $\theta = -31.27°$ and the above transformation equations become

$$p = 0.8547x - 0.5191y$$

and

$$q = 0.5191x + 0.8547y$$

We can now draw a table for calculating the coordinates of A, B and C as follows

Position	x	y	p	q
A	−14.96	21.46	−23.92	10.88
B	36.04	21.46	19.66	37.05
C	−14.96	−42.54	9.3	−44.12

and the stresses follow from the general equation $\sigma_b = 0.5042q - 1.2894p$, as follows:

at A

$$\sigma_A = 36.33\,\text{MPa}$$

at B

$$\sigma_B = -6.67\,\text{MPa}$$

at C

$$\sigma_C = -34.24\,\text{MPa}$$

Learning summary

By the end of this section you should have learnt:

- ✓ an asymmetric cross section, in addition to its second moments of area about the x- and y-axes, I_x and I_y, possesses a geometric quantity called the product moment of area, I_{xy}, with respect to these axes;
- ✓ how to calculate the second moments of area and the product moment of area about a convenient set of axes;
- ✓ an asymmetric section will have a set of axes at some orientation for which the product moment of area is zero and that these axes are called the principal axes;
- ✓ the second moments of area about the principal axes are called the principal second moments of area;
- ✓ how to determine the second moments of area and the product moment of area about any oriented set of axes, including the principal axes, using a Mohr's circle construction;
- ✓ it is convenient to analyse the bending of a beam with an asymmetric section by resolving bending moments onto the principal axes of the section;
- ✓ how to follow a basic procedure for analysing the bending of a beam with an asymmetric cross section.

3.10 Strain energy

We saw in Section 3.4 how deflections of a beam can be determined by solving the differential equation of the elastic line, using Macaulay's method. This method is not appropriate for more complex shaped structures or bodies or where combined loading is applied. In this section we introduce the concept of **strain energy** which will enable us to calculate deflections of complex structures. When an elastic body is subject to loading, strain energy is stored within the body. An Italian railway engineer, called Carlo Alberto Castigliano, derived a theorem and procedure for using the strain energy to determine deflections in structures or bodies. Castigliano's theorem is a powerful and flexible method for solving deflection problems.

Strain energy expressions

The strain energy in an elastic body is equal to the work done on the body by the applied loads. Thus, when an elastic body is subjected to a single applied load P, causing it to displace by a distance u at the position of load application, as illustrated by Figure 3.115, the strain energy (denoted by the symbol capital U) is given by,

$$\text{strain energy, } U = \text{work done} = \int_0^u P\,\mathrm{d}u$$

For a linear elastic body,

$$U = \tfrac{1}{2} P \cdot u \qquad \textbf{(3.68)}$$

where P = final load
and u = final displacement

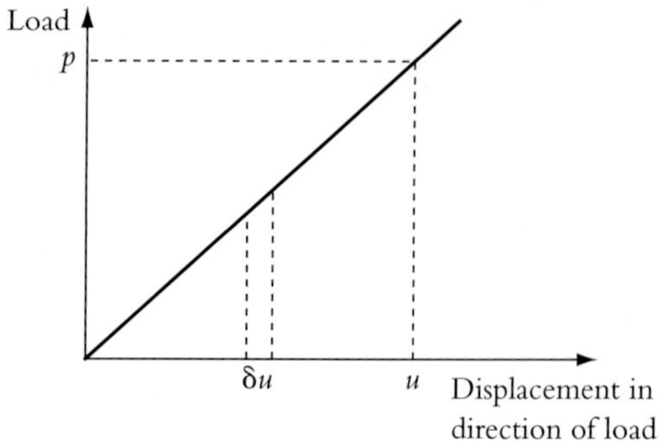

Figure 3.115 Strain energy given by area under load–displacement curve

Similar expressions can be derived for moments, torques and couples. Thus, when a moment is applied to an elastic body, causing it to rotate through an angle θ, the strain energy is given by,

$$U = \tfrac{1}{2} M \cdot \theta \qquad \textbf{(3.69)}$$

where M = final moment
and θ = final angle of rotation

We will now use equations (3.68) and (3.69) to derive expressions for the strain energy in a bar under tension, torsion and bending.

Tension in a bar

Figure 3.116 shows an element of the length of a bar, length δs, cross-sectional area, A, which is extended δu due to a tensile load P.

The strain energy for the element, δU, is given by,

$$\delta U = \text{work done} = \tfrac{1}{2} P \delta u \qquad \textbf{(3.70)}$$

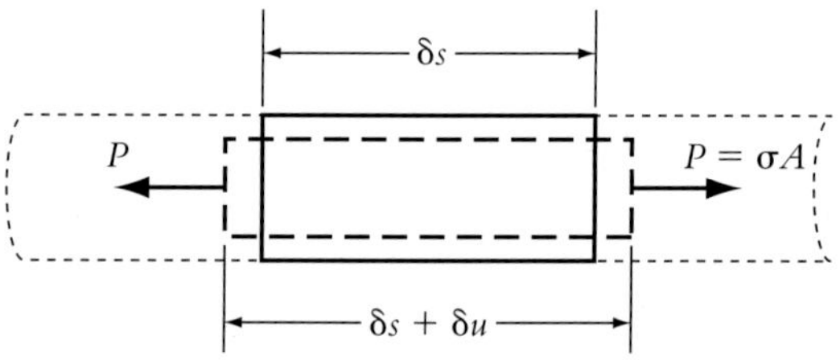

Figure 3.116 Strain energy in a bar under tension

Note: There are transverse strains/displacements due to Poisson's effect but no transverse stresses/loads. Thus, there is no work done in the transverse direction.

Now,

$$\text{strain } \varepsilon = \text{extension/original length} = \frac{\delta u}{\delta s}$$

Also, by Hooke's law

$$\varepsilon = \frac{\sigma}{E} = \frac{P}{EA}$$

$$\therefore \quad \frac{\delta u}{\delta s} = \frac{P}{EA}$$

$$\delta u = \frac{P}{EA}\delta s \qquad \textbf{(3.71)}$$

Substituting (3.71) into (3.70) gives

$$\delta U = \frac{P^2}{2AE}\delta s \qquad \textbf{(3.72)}$$

This is the expression for strain energy in an element of bar. Thus, for a bar of length L, with possibly varying A and E, integrating (3.72) gives

$$U = \int_0^L \frac{P^2}{2AE}\mathrm{d}s \qquad \textbf{(3.73)}$$

which is the general expression for strain energy in a bar under tension. This expression can also be used for a bar under compression.

For constant A and E,

$$U = \frac{P^2 L}{2AE} \tag{3.74}$$

So, knowing the load, geometry and material stiffness, we can use expression (3.72) or (3.73) to determine the strain energy in the bar.

Strain energy per unit volume

It is interesting to note that, from (3.68), the strain energy per unit volume is given by.

$$\frac{\delta U}{\delta V} = \frac{1}{2}\frac{P\delta u}{A\delta s} = \frac{1}{2}\sigma \cdot \varepsilon$$

i.e. $\frac{1}{2} \times$ stress $\times$ strain (for uniaxial stress)

Torsion in a shaft

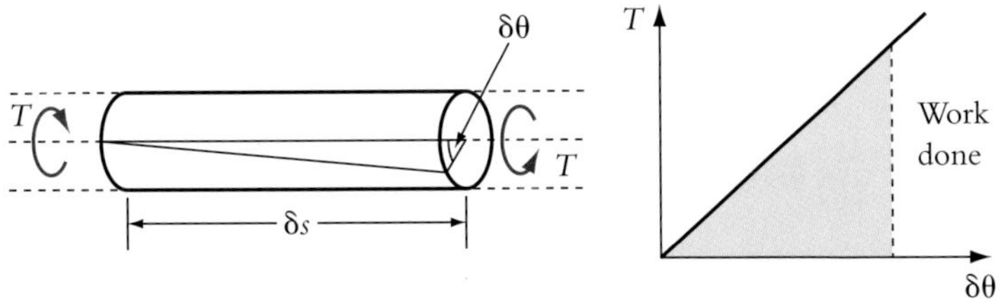

Figure 3.117 Strain energy in a shaft under torsion

Figure 3.117 shows an element of the length of a shaft, length δs, that is twisted through angle $\delta\theta$ due to a torque T.

The strain energy for the element, δU, is given by

$$\delta U = \text{work done} = \tfrac{1}{2} T\delta\theta \tag{3.75}$$

Now for a circular shaft of polar second moment of area, J, and shear modulus, G, the torsion equation, gives

$$\delta\theta = \frac{T}{GJ}\delta s \tag{3.76}$$

Substituting (3.76) into (3.75) gives,

$$\delta U = \frac{T^2}{2GJ}\delta s \tag{3.77}$$

This is the expression for strain energy in an element of the shaft. Thus, for a shaft of length L, integrating (3.77) gives

$$U = \int_0^L \frac{T^2}{2GJ}\,\mathrm{d}s \tag{3.78}$$

which is the general expression for strain energy in a shaft under torsion, and you can see the similarity in form with equation (3.74) for a bar under tension.

Bending of a bar

Figure 3.118 shows an element of a straight beam, length δs, which bends to curvature R, due to an applied bending moment M. The angle subtended by the element of beam is $\delta\phi$, also equal to the change in slope of the beam over δs.

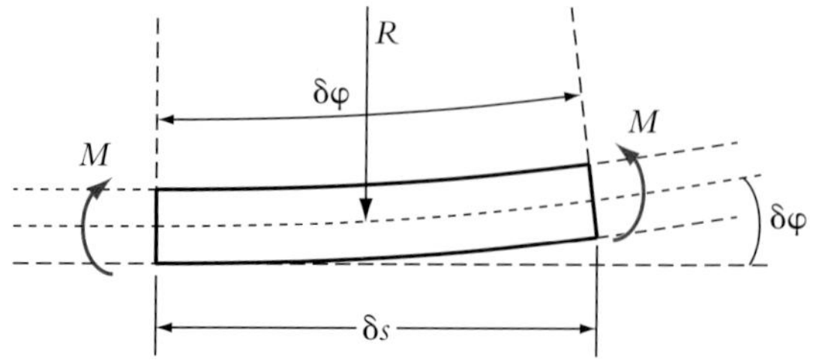

Figure 3.118 Strain energy in a beam under bending

The strain energy for the element, δU, is given by,

$$\delta U = \text{work done} = \tfrac{1}{2} M \delta\phi \qquad (3.79)$$

Now, $R \cdot \delta\phi = \delta s$ and $\dfrac{M}{I} = \dfrac{E}{R}$, the beam bending equation.

$$\therefore \qquad \delta\phi = \frac{\delta s}{R} = \frac{M}{EI}\delta s \qquad (3.80)$$

Substituting (3.80) into (3.79) gives,

$$\delta U = \frac{M^2}{2EI}\delta s \qquad (3.81)$$

Thus, for a beam of length L, integrating (3.81) gives,

$$U = \int_0^L \frac{M^2}{2EI} \mathrm{d}s \qquad (3.82)$$

i.e. the general expression for strain energy in a beam under bending, and again you can see the similarity in form to equations (3.73) and (3.78).

The three expressions for calculating strain energy can now be summarized as follows,

$$U_{\text{tension/compression}} = \int_0^L \frac{P^2}{2AE}\mathrm{d}s \quad U_{\text{torsion}} = \int_0^L \frac{T^2}{2GJ}\mathrm{d}s \quad U_{\text{bending}} = \int_0^L \frac{M^2}{2EI}\mathrm{d}s \qquad (3.83)$$

In practical engineering structures, where members are relatively long and slender, strain energy due to tension and compression can usually be neglected. Bending is usually dominant. Also, strain energy due to shear deflections can exist but, again, it can normally be neglected.

Castigliano's theorem

Consider an *elastic* body loaded by forces, P_i, as shown in Figure 3.119(a). The corresponding displacements of the load points in the direction of the loads are u_i. For the general *non-linear elastic case*, the load–displacement curve for the ith load is shown in Figure 3.119(b).

If all forces are applied simultaneously and proportionately, the total stored *strain energy*, U, in the body, is equal to the work done by the forces as follows,

$$U = \sum_i U_i = \sum_i \int P_i \mathrm{d}u_i$$

where U_i is the area under the load–displacement curve for one loading.

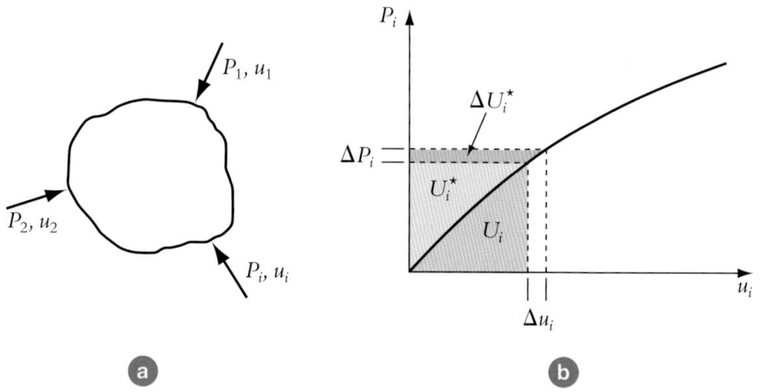

Figure 3.119 (a) Elastic body subject to several loads and (b) load–displacement curve for the *i*th load acting on the body; U_i is the strain energy and $U_i^\star$ is the complementary energy

We can also define the *complementary energy*, $U^\star$, as follows,

$$U^\star = \sum_i U_i^\star = \sum_i \int u_i \mathrm{d}P_i$$

where $U_i^\star$ is the area above the load–displacement curve for one loading.

Now, consider a small increment of the load P_i, ΔP_i, while the other loads remain constant. This causes an increment of the complementary strain energy, shown in Figure 3.119(b) and given by

$$\Delta U_i^\star = \int u_i \mathrm{d}P_i \approx u_i \Delta P_i$$

For the other loads, $P_j, j \neq i$, P_j is constant and

$$\Delta U_j^\star = \int u_j \mathrm{d}P_j = 0$$

The total change in complementary strain energy is therefore,

$$U^\star = \sum_i \Delta U_i^\star = \Delta U_i^\star = u_i \Delta P_i$$

In the limit,

$$\Delta P_i \Rightarrow 0$$

and

$$u_i = \frac{\lim}{\Delta P_i = 0} \frac{\Delta U^\star}{\Delta P_i} = \frac{\partial U^\star}{\partial P_i} \tag{3.84}$$

Note that this is a partial derivative because all other loads except P_i were kept constant.

Equation (3.84) is the general form of **Castigliano's theorem**, which, as stated in the introduction, is named after the Italian railway engineer, Carlo Alberto Castigliano (1847–1884), who developed the method. His theorem states that the deflection, u_i, at a given load point i, may be obtained by differentiating the complementary strain energy, $U^\star$, with respect to the load, P_i, acting at the point.

Note that u_i is the deflection at the point of application of the load, P_i, in the direction of P_i, and P_i is independent of other loads.

Equation (3.84) applies to any non-linear elastic body. For linear elastic bodies, strain energy is equal to the complementary strain energy as shown in Figure 3.120. Thus, for a linear elastic body,

$$U = U^\star$$

and

$$u_i = \frac{\partial U}{\partial P_i} \quad (3.85)$$

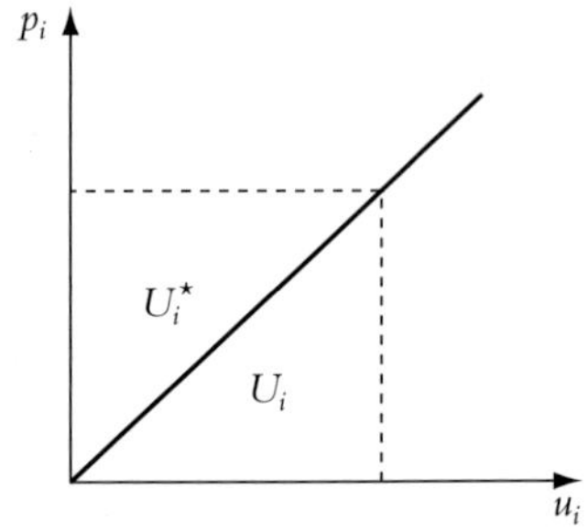

Figure 3.120 Load–displacement line for a linear elastic body

Equation (3.85) is the well-known form of Castigliano's theorem and states that **the partial derivative of the strain energy, *U*, of a linear elastic system with respect to any independent force is equal to the displacement of the structure at the point of application of the force in the direction of the force.**

Castigliano's theorem may be extended to include rotations due to moments and torques:

$$\theta_i = \frac{\partial U}{\partial M_i} \quad (3.86)$$

where θ_i is the rotation at the point of application of the moment M_i about the axis of M_i.

The above forms of Castigliano's theorem (equations (3.85) and (3.86)) can be used to determine deflections/displacements and rotations at a point in a body or structure where a load or moment is applied. The procedure is as follows:

(1) Obtain expressions for the total strain energy within a structure or body in terms of the applied loads or moments using equations (3.83).

(2) Differentiate the strain energy expressions with respect to the applied loads or moments, as in equations (3.85) and (3.86), to determine expressions for deflections or rotations respectively.

(3) Evaluate numerically the deflections and/or rotations.

Notes on further use of Castigliano's theorem

(1) To determine the deflection or rotation at a point where a load or moment is *not* applied, include a dummy load or moment at the point. Obtain and differentiate a strain energy expression incorporating the dummy load/moment, and set the latter to zero when numerically evaluating the deflections/rotations. Thus, to determine the horizontal tip deflection for the structure shown in Figure 3.121(a), with an applied vertical load, *P*, it is necessary to add a dummy load *F* at the tip. Then,

$$u_F = \left(\frac{\partial U}{\partial F}\right)_{(F=0)}$$

(2) If the displacement or rotation at a point is restrained, e.g. at a support as shown in Figure 3.121(b), and the reactions are R_o and M_o, then,

$$\frac{\partial U}{\partial R_o} = 0 \quad \text{and} \quad \frac{\partial U}{\partial M_o} = 0$$

These conditions can be used to determine the reactions R_o and M_o.

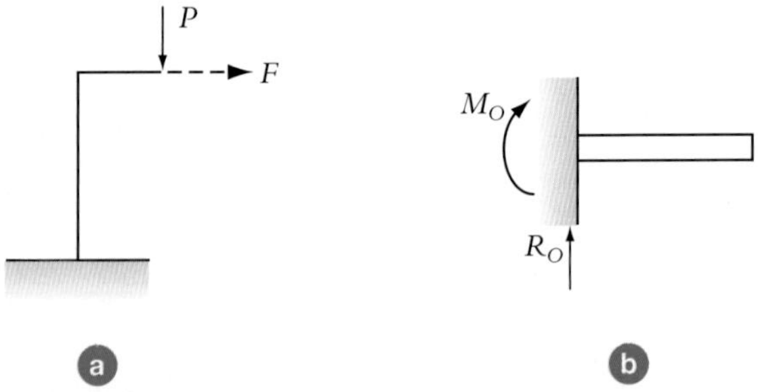

Figure 3.121 Castigliano's theorem: (a) dummy load, *F*, added to determine the horizontal deflection and (b) use of the zero displacement condition to solve for unknown reaction loads and moments

Worked examples

Beam deflection

Figure 3.122 shows a simply supported beam, ABC, with a point load at the centre of its span.

Use strain energy to derive an expression for the central deflection of the beam (assuming bending strain energy only, no shear).

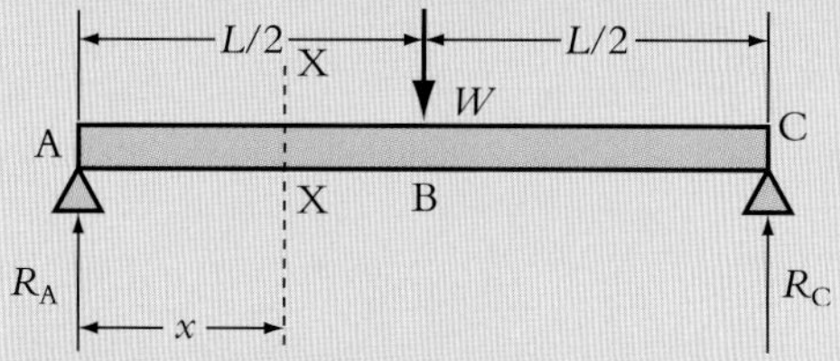

Figure 3.122 Simply supported beam with point load at centre span

Reaction forces:
From symmetry,

$$R_A = R_C = W/2$$

Bending moment distribution:
Taking a section, X-X, in the part of the span, AB, and considering the left-hand part of the beam as a free body. Moments about X-X give,

$$M - \frac{Wx}{2} = 0$$

$\therefore$

$$M = \frac{Wx}{2}$$

Strain energy:
The strain energy stored in AB is given by,

$$U_{AB} = \int \frac{M^2}{2EI} ds = \int_0^{L/2} \frac{W^2x^2}{4.2EI}$$

$$= \frac{W^2}{8EI}\left[\frac{x^3}{3}\right]_0^{L/2}$$

$$= \frac{W^2L^3}{192EI}$$

From symmetry,

$$U_{BC} = U_{AB}$$

The total strain energy

$$U_{tot} = 2U_{AB} = \frac{W^2L^3}{96EI}$$

Deflection:
The central deflection, i.e. deflection at the load, is therefore given by,

$$y_B = \frac{\partial U}{\partial W} = \frac{WL^3}{48EI}$$

Dummy load

A curved beam, whose geometric form is a quadrant in the vertical plane, as shown in Figure 3.123(a), has cross-sectional dimensions small compared with the quadrant radius.

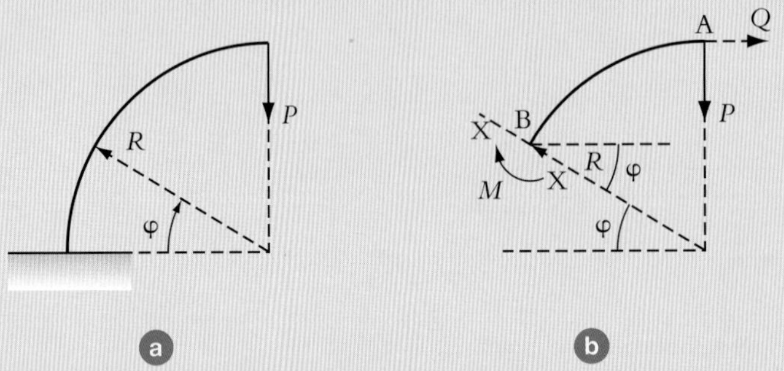

Figure 3.123 **Curved cantilever beam with point load applied at the free end**

Use strain energy to derive an expression for the horizontal and vertical deflections of its tip due to a vertical load P.

Because of the slender nature of the curved beam, direct stress and transverse shear strain energies may be neglected compared to bending strain energy.

To determine the horizontal deflection, a horizontal dummy load, Q, is added at the tip of the beam as shown in Figure 3.123(b).

Bending moment expression:

As also shown in Figure 3.123(b), to determine the bending moment distribution, a section X-X is taken at position B, angle φ from the horizontal, and equilibrium of a section AB of the beam is considered. Taking moments about X-X,

$$M + PR\cos\phi + QR(1 - \sin\phi) = 0$$

$$\therefore \quad M = -PR\cos\phi - QR(1 - \sin\phi) \tag{3.87}$$

Note that the bending moment is a function of both the applied load, P, and the dummy load, Q. Also, the sign of the bending moment is unimportant as it is squared in the strain energy integral (see below).

Strain energy:

The strain energy expression is now

$$U = \int \frac{M^2}{2EI} ds = \int_0^{\pi/2} \frac{[-PR\cos\phi - QR(1 - \sin\phi)]^2}{2} EI\, Rd\phi \tag{3.88}$$

Note that $ds = Rd\varphi$ and that the limits of integration are 0 to $\pi/2$.

Vertical displacement:

The vertical displacement is now given by differentiating equation (3.88) with respect to the applied load P, as follows,

$$u_v = \frac{\partial U}{\partial P} = \frac{1}{EI}\int_0^{\pi/2} [-PR\cos\phi - QR(1 - \sin\phi)] \cdot (-R\cos\phi) \cdot Rd\phi$$

To simplify matters, the dummy load, Q, can be set to zero before the integration is performed. We then have,

$$u_v = \frac{PR^3}{EI}\int_0^{\pi/2} \cos^2\phi \, d\phi$$

$$= \frac{PR^3}{EI}\int_0^{\pi/2} \frac{1}{2}(1 + \cos 2\phi)d\phi$$

$$= \frac{PR^3}{2EI}\left[\phi + \frac{1}{2}\sin 2\phi\right]_0^{\pi/2}$$

$\therefore$

$$u_v = \frac{\pi PR^3}{4EI}$$

Horizontal displacement:
The horizontal displacement is now given by differentiating equation (3.88) with respect to the dummy load, Q, as follows,

$$u_h = \frac{\partial U}{\partial Q} = \frac{1}{2EI}\int_0^{\pi/2} 2[-PR\cos\phi - QR(1 - \sin\phi)]\cdot[-R(1 - \sin\phi)]\cdot R\,d\phi$$

Again, setting the dummy load, Q, equal to zero before integrating we have,

$$u_h = \frac{PR^3}{EI}\int_0^{\pi/2} \cos\phi(1 - \sin\phi)]d\phi$$

$$= \frac{PR^3}{EI}\int_0^{\pi/2} \cos\phi - \frac{1}{2}\sin 2\phi \, d\phi$$

$$= \frac{PR^3}{EI}\left[\sin\phi + \frac{1}{4}\cos 2\phi\right]_0^{\pi/2}$$

$\therefore$

$$u_h = \frac{PR^3}{2EI}$$

Note that the vertical deflection is $\pi/2 \times$ horizontal deflection.

Combined strain energy

A circular cross section, offset cantilever, ABC, as shown in Figure 3.124(a) is loaded at its free end by a load, *P*.

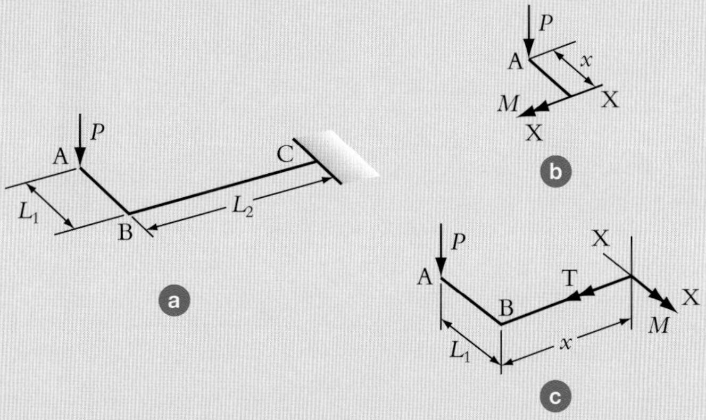

Figure 3.124 **Offset cantilever beam with a point load applied at the free end**

Neglecting strain energy due to bending shear, derive an expression for the vertical deflection at the load point.

In this example, section AB of the cantilever stores bending strain energy only, while section BC stores both bending and torsional strain energy.

Strain energy in AB:
Referring to Figure 3.124(b), which shows a section, X-X, taken distance x from A, the bending moment is obtained by considering equilibrium:

$$M + Px = 0 \qquad \therefore \quad M = -Px$$

The strain energy, U_{AB}, is given by,

$$U_{AB} = \int \frac{M^2}{2EI} ds$$

$$= \int_0^{L_1} \frac{P^2 x^2}{2EI} dx$$

$$= \frac{P^2 L_1^3}{6EI}$$

(note the limits 0 to L_1 for the length)

Strain energy in BC:
Referring to Figure 3.124(c), which shows a section, X-X, in BC, taken distance x from B, the bending moment and torque are obtained by considering equilibrium:

$$M + Px = 0 \qquad \therefore \quad M = -Px$$

$$T + PL_1 = 0 \qquad \therefore \quad T = -PL_1$$

The strain energy, U_{BC}, is now given by

$$U_{BC} = \int \frac{M^2}{2EI} ds + \int \frac{T^2}{2GJ} ds$$

$$= \int_0^{L_2} \frac{P^2 x^2}{2EI} dx + \int_0^{L_2} \frac{P^2 L_1^2}{2GJ} dx$$

$$= \frac{P^2 L_2^3}{6EI} + \frac{P^2 L_1^2 L_2}{2GJ}$$

So, the total strain energy in the cantilever is

$$U = U_{AB} + U_{BC} = P^2 \left[\frac{L_1^3 + L_2^3}{6EI} + \frac{L_1^2 L_2}{2GJ} \right] \qquad \textbf{(3.89)}$$

Deflection of the tip:
The deflection of the tip is now given by differentiating equation (3.89) with respect to the applied load, P, as follows,

$$u_v = \frac{\partial U}{\partial P} = 2P \left[\frac{L_1^3 + L_2^3}{6EI} + \frac{L_1^2 L_2}{2GJ} \right]$$

Learning summary

By the end of this section you should have learnt:

- ✓ the basic concept of strain energy stored in an elastic body under loading;
- ✓ to calculate strain energy in a body/structure arising from various types of loading, including tension/compression, bending and torsion;
- ✓ Castigliano's theorem for linear elastic bodies, which enables the deflection or rotation of a body at a point to be calculated from strain energy expressions.

3.11 Fatigue

Introduction

Fatigue failure of components and structures results from cyclic or repeated loading and from the associated cyclic stresses and strains, as opposed to failure due to monotonic or static stresses or strains, such as buckling or plastic collapse due to excessive plastic deformation. The topic of fatigue is extremely important in mechanical engineering, since machines have moving parts, which in turn give rise to stresses and strains that may vary with time, typically in a repetitive fashion. For example, the axle of a car which will transmit a time-varying torque, that changes from zero to some finite value when the car is put into gear and driven (and back to zero again when the car is taken out of gear).

An important design consideration, with respect to fatigue, is the fact that fatigue failure can occur at stresses that are well below the ultimate tensile strength of the material and often below the yield strength (or 0.2% proof stress).

Basic phenomena

The failure mechanism for an initially uncracked component with a smooth (polished) surface can be split into three parts, namely crack initiation, crack propagation and final fracture, as follows:

Stage I crack initiation: The microstructural phenomenon that causes the initiation of a fatigue crack is the development of persistent slip bands at the surfaces of the specimen. These persistent slip bands are the result of dislocations moving along crystallographic planes, leading to both slip band intrusions and extrusions on the surface. These act as excellent stress concentrations and can thus lead to crack initiation. Crystallographic slip is primarily controlled by shear stresses rather than normal stresses so that fatigue cracks initially tend to grow in a plane of maximum shear stress. This stage leads to short cracks, usually only of the order of a few grains.

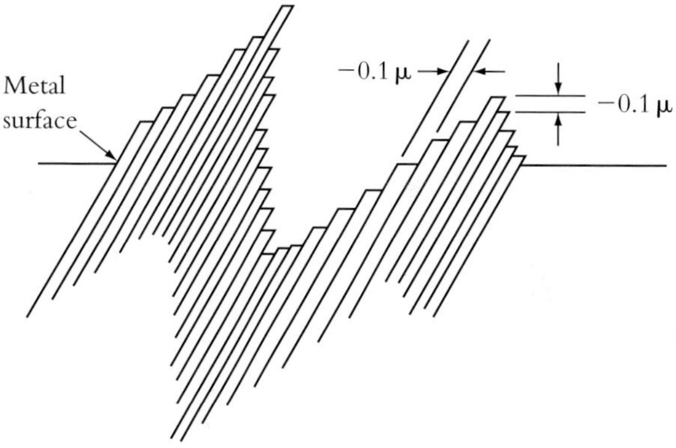

Figure 3.125 Persistent slip bands in ductile metals subjected to cyclic stress

Stage II crack propagation: As cycling continues, the fatigue cracks tend to coalesce and grow along planes of maximum tensile stress range.

Final fracture: This occurs when the crack reaches a critical length and it results in either ductile tearing (plastic collapse) at one extreme, or cleavage (brittle fracture) at the other.

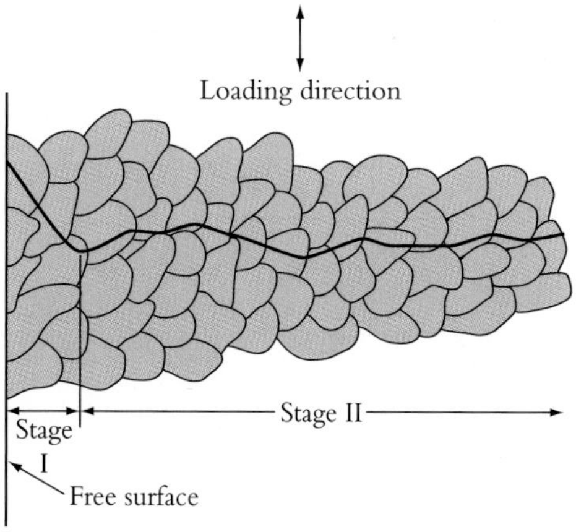

Figure 3.126 Schematic of stages I and II transcrystalline microscopic fatigue crack growth

Fatigue life analysis

In order to allow for fatigue in the analysis and design of components, a number of different approaches are adopted, two of which are described here. The more traditional approach is what is now referred to as the **total life approach**, based on laboratory tests, which are carried out under either stress- or strain-controlled loading conditions on idealized specimens. These tests furnish the number of loading cycles to the initiation of a 'measurable' crack as a function of applied stress or strain parameters. The 'measurability' is dictated by the resolution accuracy of the crack detection method employed. A typical 'measurable' crack is about 0.75 mm to 1 mm. The challenge of fatigue design is then to relate these test results to actual component lives under real loading conditions.

The second approach is known as the **damage tolerant approach**. This approach is based on the inclusion of fatigue as a crack growth process, taking account of the fact that all components have inherent flaws or cracks. The development of fracture mechanics techniques to predict crack growth has facilitated this approach as a competing technique to the total life approach. Both of the approaches have advantages and disadvantages; the former has more appeal to design engineers while the latter is more often used by material scientists and researchers. Nonetheless, even in routine design, the damage tolerant approach is gaining popularity.

Total life approach

The total life approach is based on the results of stress- and strain-controlled cyclic testing of laboratory test specimens of material, in order to obtain the numbers of cycles to failure as a function of the applied alternating stress, for example. Figure 3.127 shows a rotating bending test machine set-up. This is a constant load amplitude machine since the load doesn't change even with crack growth. The specimens usually have finely polished surfaces to minimize surface roughness effects, which would particularly affect stage I growth. In this approach, no distinction is made between crack initiation and propagation. Stress concentration effects can be studied by machining in grooves, notches or holes.

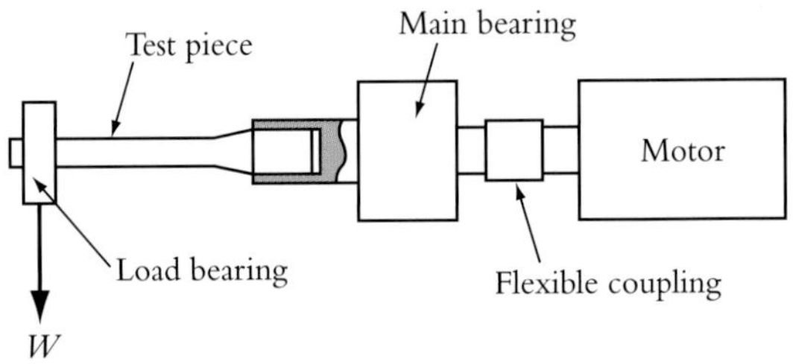

Figure 3.127 Rotating bending moment test apparatus for fully reversed fatigue loading

Traditionally, most fatigue testing was based on fully reversed (i.e. zero mean stress, $S_m = 0$), stress-controlled conditions, and the fatigue design data was presented in the form of *S–N* curves, which are either semilog or log–log plots of alternating stress, S_a, against the measured number of cycles to failure, *N*, where failure is defined as fracture. Some of the important stress parameters for cyclic loading are shown in Figure 3.128.

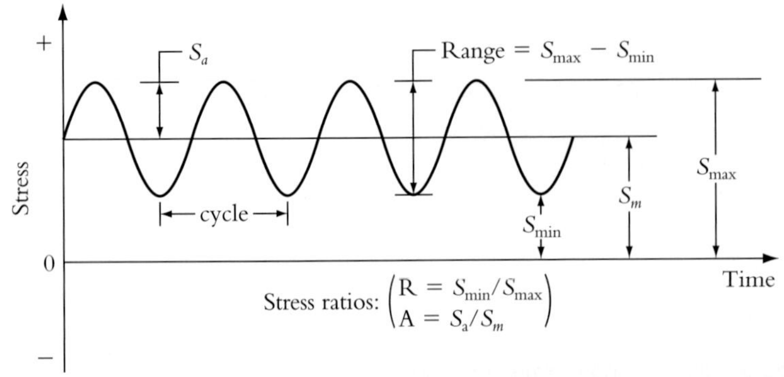

Figure 3.128 Notation used to describe constant load fatigue test cycles

Figure 3.129 contains schematic representations of two typical *S–N* curves obtained under load- or stress-controlled tests on smooth specimens. Figure 3.129(a) shows a continuously sloping curve while Figure 3.129(b) shows a discontinuity or 'knee' in the curve. A knee is only found in a few materials (notably low-strength steels) between 10^6 and 10^7 cycles under non-corrosive conditions. The curves are normally drawn through the median life value (of the scatter in *N*) and thus represent 50 per cent expected failure. The **fatigue life**, *N*, is the number of cycles of stress or strain range of a specified character that a given specimen sustains before failure of a specified nature occurs. **Fatigue strength** is a hypothetical value of stress range at failure for exactly *N* cycles as obtained from an *S–N* curve. The **fatigue limit** (sometimes called the **endurance limit**) is the limiting value of the median fatigue strength as *N* becomes very large, e.g. $>10^8$ cycles.

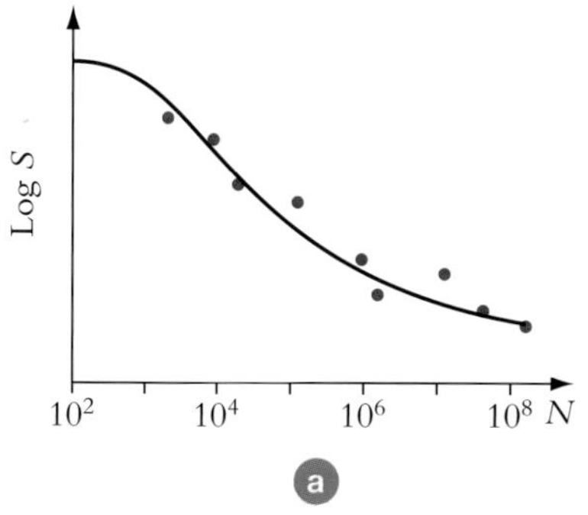

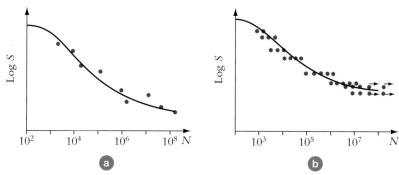

Figure 3.129 Typical S–N diagrams

Effect of mean stress

The alternating stress, S_a, and the mean stress, S_m, are defined in Figure 3.128. Early investigators of fatigue assumed that only the alternating stress affected the fatigue life of a cyclically loaded component. However, it has since been established that the mean stress has a significant effect on fatigue behaviour, as shown in Figure 3.130. It can be seen that tensile mean stresses are detrimental while compressive mean stresses are beneficial.

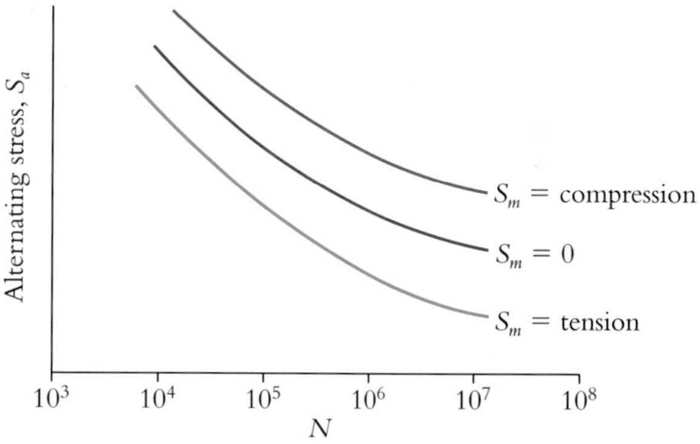

Figure 3.130 The effect of mean stress on fatigue life

The effect of mean stress is commonly represented as a plot of S_a versus S_m for a given fatigue life. Attempts have been made to develop this relationship into general relations. Three of these common relations between allowable alternating stress for a given life as a function of mean stress are shown in Figure 3.131. The modified Goodman line assumes a linear relationship between the allowable S_a and the corresponding mean stress S_m, where the slope and intercepts are defined by the fatigue life, S_e, and the material ultimate tensile strength (UTS), S_u, respectively. The Gerber parabola employs the same end-points but, in this case, the relation is defined by a parabola. Finally, the Soderberg line again assumes a linear relation, but this time the mean stress axis end-point is taken as the yield stress, S_y. The modified Goodman line, for example, can be extended into the compressive mean stress region to give increasing allowable alternating stress with increasing compressive mean stress, but this is normally taken to be horizontal for design purposes and for conservatism.

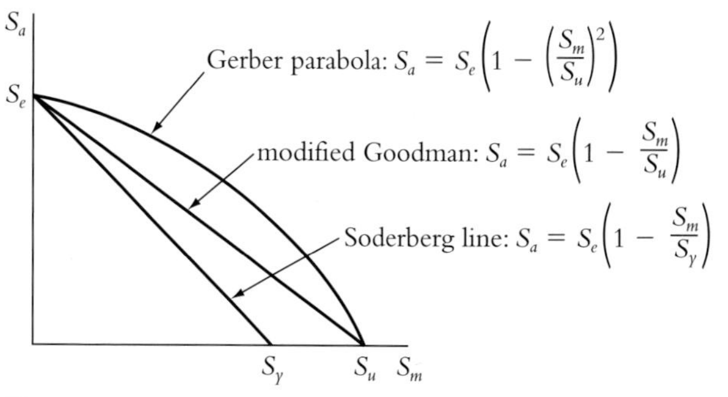

Figure 3.131 The effect of mean stress on fatigue life

Effect of stress concentrations

Ever since the first occurrences of fatigue failure, it has been recognized that such failures are most commonly associated with notch-type features in components. It is impossible to avoid notches in engineering structures, although the effects of such notches can be reduced through appropriate design. The stress concentration associated with notch-type features leads typically to local plastic strain, which eventually leads to fatigue cracking. Consequently, the estimation of stress concentration factors associated with various types of notches and geometrical discontinuities has received a lot of attention. This is typically expressed in terms of an elastic stress concentration factor (SCF), K_t, which is simply the relationship between the maximum local stress and an appropriate nominal stress:

$$K_t = \frac{\sigma^{el}_{\max}}{\sigma_{\text{nom}}}$$

It was once thought that the fatigue strength of a notched component could be predicted as the strength of a smooth component divided by the SCF. However, this is not the case. The reduction is, in fact, often less than K_t and is defined by the **fatigue notch factor,** K_f, which is defined as the ratio of the smooth fatigue strength to the notched fatigue strength:

$$K_f = \frac{S_{a,\text{smooth}}}{S_{a,\text{notch}}}$$

However, this fatigue notch factor is also found to vary with both alternating and mean stress level and thus with number of cycles to failure. Figure 3.132 shows the effect of a notch, with an SCF of 3.4, on the fatigue behaviour of a wrought aluminium alloy, where the smooth lines are for the smooth specimen and the dotted lines are for the notched specimen. The table shows how the fatigue notch factor changes with mean stress level and fatigue life. Clearly, the fatigue notch factor increases from 3.2 to 5.7 from 10^4 cycles to 10^7 cycles at 172 MPa mean stress, but remains unchanged between these lives at 2.3 for zero mean stress.

Mean stress	10^4 cycles	10^7 cycles
0 MPa	51/22 = 2.3	22/9 = 2.3
172 MPa	42/13 = 3.2	17/3 = 5.7

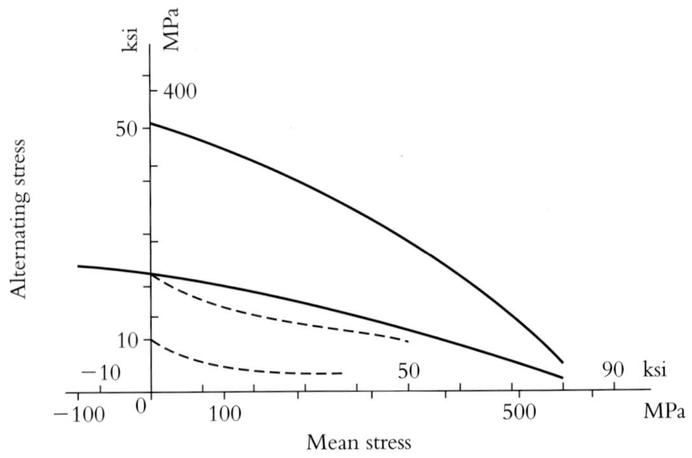

Figure 3.132 Constant life diagrams for a wrought aluminium alloy for both smooth and notched specimens

S–N design procedure for fatigue

Constant life diagrams plotted as S_a versus S_m, also called Goodman diagrams, can be used in design to give safe estimates of fatigue life and loads.

$$f = \frac{OB}{OA} = \text{factor of safety}$$

$$\text{similar triangles} \rightarrow \frac{S_e}{k_f S_u} = \frac{f S_a}{S_u - f S_m} = \frac{S_a}{\frac{S_u}{f} - S_m}$$

$$\rightarrow \frac{S_u}{f} - S_m = \frac{S_a}{\frac{S_e}{k_f S_u}}$$

$$\Rightarrow \frac{1}{f} = \frac{k_f S_a}{S_e} + \frac{S_m}{S_u}$$

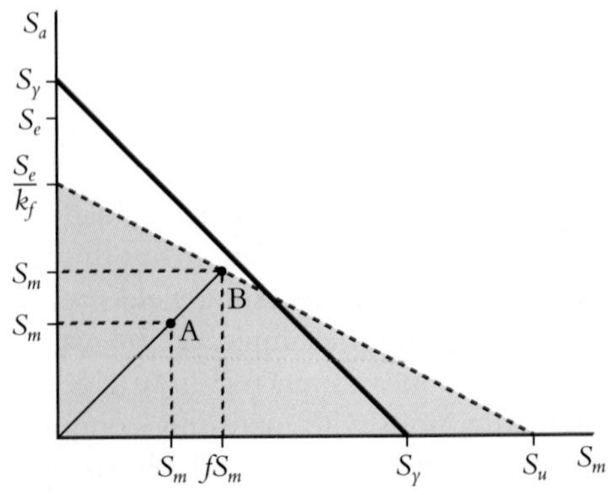

Figure 3.133

(1) The Goodman line connects the endurance limit, S_e (or long-life fatigue strength), to the UTS, S_u.

(2) The fatigue strength for zero mean stress is reduced by the fatigue notch factor, K_f. The stress concentration factor, K_t is used if K_f is not known.

(3) For static loading of a ductile component with a stress concentration, failure still occurs when the mean stress is equal to the UTS. Failure at intermediate values of mean stress is assumed to be given by the dotted line.

(4) In order to avoid yield of the whole cross section of the component, the maximum nominal stress must be less than the yield stress, S_y

$$S_m + S_a < S_y$$

This relationship gives the yield line joining S_y to S_y.

(5) The region of the diagram nearest to the origin is the 'safe' region (can also be extended to include compressive yield).

(6) A component is assessed by plotting the point corresponding to the nominal alternating stress, S_a, and the nominal mean stress, S_m, i.e. not the maximum values associated with the notch. The factor of safety is determined from the position of the point relative to the safe/fail boundary, i.e. factor of safety $F = OB/OA$.

From similar triangles:

$$\frac{S_a}{\left(\frac{S_u}{F} - S_m\right)} = \frac{S_e}{k_f S_u}$$

$$\frac{1}{F} = \frac{S_a k_f}{S_e} + \frac{S_m}{S_u}$$

A procedure similar to that described above for long life can also be used to design for a specified number of cycles. In this case the endurance limit and the fatigue notch factor are replaced by the fatigue strength and the fatigue notch factor for the specified number of cycles.

Learning summary

By the end of this section you should have learnt:

- ✓ the various stages leading to fatigue failure;
- ✓ the basis of the total life and of the damage tolerant approaches to estimating the number of cycles to failure;
- ✓ how to include the effects of mean and alternating stress on cycles to failure using the Gerber, modified Goodman and Soderberg methods;
- ✓ how to include the effect of a stress concentration on fatigue life;
- ✓ the S–N design procedure for fatigue life.

3.12 Fracture mechanics

Linear elastic fracture mechanics (LEFM)

Consider the stress concentration factor for an elliptical hole in a large, linear-elastic plate subjected to a remote, uniaxial stress.

It can be shown that the stress concentration factor is as follows:

$$K_t = \frac{\hat{\sigma}}{\sigma} = 1 + \frac{2a}{b}$$

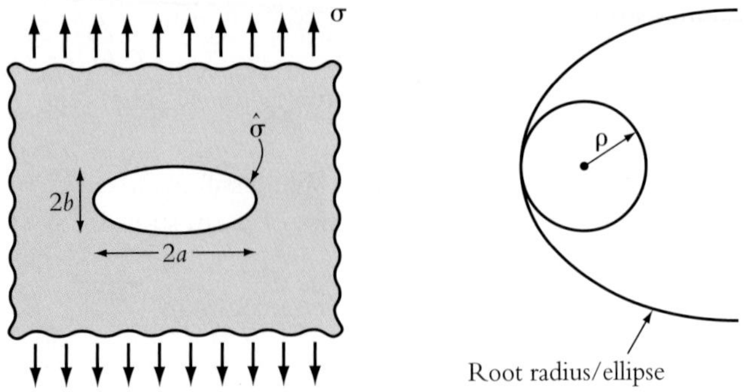

Figure 3.134 Elliptical hole in an infinite plate

Thus, as $b \to 0$, the elliptical hole degenerates to a crack, and $\frac{a}{b} \to \infty$, so that the notch stress also goes to infinity (i.e. becomes singular), $\frac{\hat{\sigma}}{\sigma} = \infty$, provided that the material behaviour remains linear elastic.

The root radius for an ellipse is given by

$$\rho = \frac{b^2}{a} \qquad \therefore \quad b = \sqrt{a\rho}$$

so that

$$\frac{\hat{\sigma}}{\sigma} = 1 + 2\sqrt{\frac{a}{\rho}}$$

and again, as the notch tip radius goes to zero, i.e. $\rho \to 0$, the notch tip stress again goes to infinity:

$$\frac{\hat{\sigma}}{\sigma} \to 2\sqrt{\frac{a}{\rho}} \to \infty$$

The singular (infinite) state of stress at a crack tip is one of the fundamental and most important aspects of fracture mechanics.

Basis of the energy approach to fracture mechanics

Griffith (1921) studied the brittle fracture of glass and adopted an energy approach to solving the problem. He reasoned that unstable crack propagation occurs only if an increment of crack growth, da, results in more strain energy being released than is absorbed by the creation of the new crack surfaces. This can be re-expressed as the change in strain energy dU, due to crack extension, being greater than the energy absorbed by the creation of the new crack surfaces. Thus, if we designate the surface energy per unit area of the crack γ_S, then the surface energy associated with a crack of length $2a$ in a body of thickness B (as shown in the Figure 3.135) is given by:

$$W_S = 4aB\gamma_S$$

Detailed stress analysis of an elliptical hole in an infinite elastic plate has established that the strain energy in such a body is

$$W_p = -\frac{\pi a^2 \sigma^2 B}{E'}$$

where σ is the remote stress (away from the hole) and where, for plane strain and plane stress, respectively,

$$E' = \frac{E}{1 - \nu^2}$$

and

$$E' = E$$

The total system energy is thus

$$W = -\frac{\pi a^2 \sigma^2 B}{E'} + 4aB\gamma_S$$

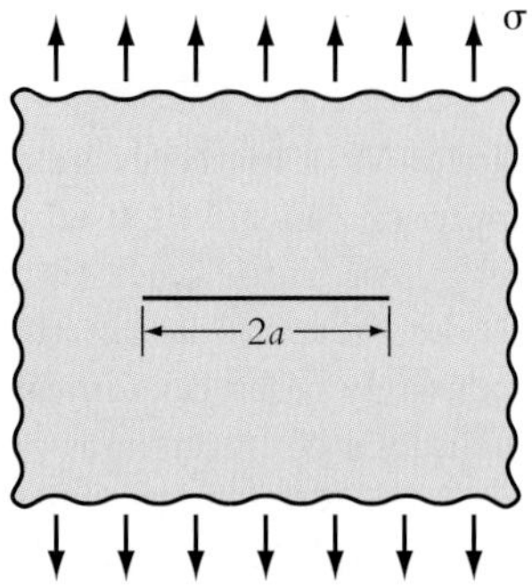

Figure 3.135 **Crack in an infinite plate**

According to Griffith, the critical condition for the onset of crack growth is

$$\frac{dU}{dA} = -\frac{\pi a \sigma^2}{E'} + 2\gamma_s = 0$$

Therefore:

$$\frac{\pi a \sigma^2}{E'} = 2\gamma_s$$

where $A = 2aB$ is the crack area and dA denotes an incremental increase in crack area. The total surface area of the two crack surfaces is $2A$. This relationship is conventionally re-expressed as

$$G = G_c$$

where G is called the strain energy release rate, the crack tip driving force or the crack extension force. G_c is a material property, which is known as the critical strain energy release rate, the toughness or the critical crack extension force. A high value of G_c means that it is difficult to cause unstable crack growth in the material whereas a low value means that it is easy to make a crack grow unstably. Thus, copper, for example, has a value of $G_c \approx 10^6\ \mathrm{J/m^2}$, whereas glass has a value of $G_c \approx 10\ \mathrm{J/m^2}$. The following relationships for plane stress and plane strain, respectively, follow from the above:

$$G = \frac{\pi a \sigma^2}{E} \text{ (plane stress)}$$

$$G = \frac{(1 - \nu^2)}{E} \pi a \sigma^2 \text{ (plane strain)}$$

Note that plane stress and plane strain are two contrasting two-dimensional assumptions that permit simplification of three-dimensional problems to two-dimensional ones. Plane stress corresponds physically to thin plate type situations while plane strain corresponds to thick plate type situations. Plane strain testing of fracture leads to lower values of G_c, so that the material property value of G_c for design purposes is taken as the plane strain value and is designated as G_{Ic}.

The critical stress that causes a crack to propagate in an unstable fashion, giving fracture, is governed by the following relationships

$$\sigma\sqrt{\pi a} = \sqrt{EG_c} \text{ (plane stress)}$$

$$\sigma\sqrt{\pi a} = \sqrt{\frac{EG_c}{1 - \nu^2}} \text{ (plane strain)}$$

Since the terms on the right-hand side of these equations are material constants, and since the term on the left-hand side is so common, it is usually abbreviated to the symbol, K, which is referred to as the **stress intensity factor**, and the equations can be re-expressed as:

$$K = K_c$$

where K_c is called the **critical stress intensity factor** or the **fracture toughness**. Thus:

$$K_c = \sqrt{EG_c}$$

In summary, $K = \sigma\sqrt{\pi a}$ is called the **stress intensity factor**

K_c is called the **fracture toughness or critical stress intensity factor**
G_c is called the **toughness** or the **critical strain energy release rate**.

Note: Most materials are not linear elastic up to failure. However, the energy approach can still be used if the plastic strain is restricted to a region very close to the crack tip; this is referred to as **small-scale yielding**. Under these conditions, the energy release rate can still be reasonably accurately based on a linear elastic analysis. Also, G_c, or G_{Ic}, now includes a component associated with plastic deformation of the crack tip as well as the creation of the surfaces. So far, we have only considered the so-called mode I loading case. There are actually three different loading modes considered in fracture mechanics.

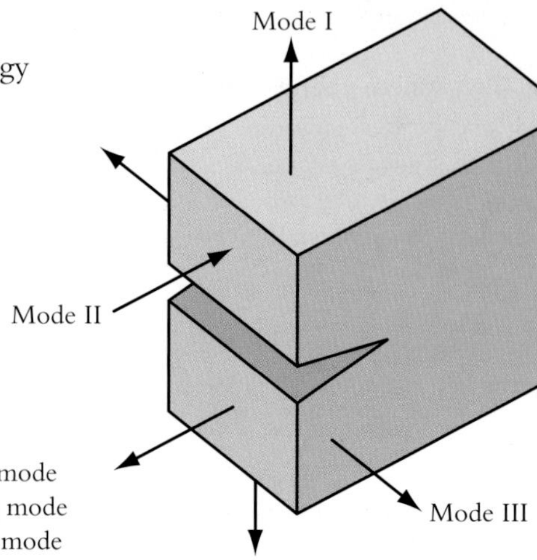

Figure 3.136 **Crack tip loading modes**

In general, the energy release rate under mixed-mode loading is given by

$$G_{\text{total}} = G_{\text{I}} + G_{\text{II}} + G_{\text{III}}$$

Elastic crack tip stress fields

Westergaard (1939) established the following equations for the elastic stress field in the vicinity of a crack tip:

$$\sigma_x = \frac{K_{\text{I}}}{\sqrt{2\pi r}}\cos\left(\frac{\theta}{2}\right)\left[1 - \sin\left(\frac{\theta}{2}\right)\sin\left(\frac{3\theta}{2}\right)\right] + \text{non-singular terms}$$

$$\sigma_y = \frac{K_{\text{I}}}{\sqrt{2\pi r}}\cos\left(\frac{\theta}{2}\right)\left[1 + \sin\left(\frac{\theta}{2}\right)\sin\left(\frac{3\theta}{2}\right)\right] + \text{non-singular terms}$$

$$\tau_{xy} = \frac{K_{\text{I}}}{\sqrt{2\pi r}}\sin\left(\frac{\theta}{2}\right)\cos\left(\frac{\theta}{2}\right)\cos\left(\frac{3\theta}{2}\right) + \text{non-singular terms}$$

$$\tau_{zx} = \tau_{zy} = 0; \sigma_z = 0 \text{ (plane stress)}$$

$$\sigma_z = \nu(\sigma_x + \sigma_y) \text{ (plane strain)}$$

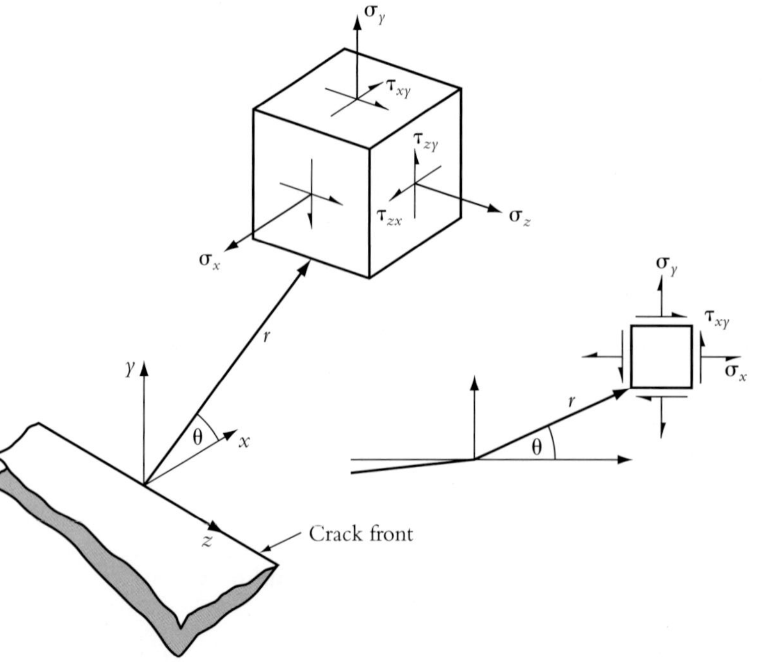

Figure 3.137 **Crack tip stress fields**

K_I is the mode-I stress-intensity factor (units $N/m^{3/2}$) which defines the magnitude of the elastic stress field in the vicinity of the crack tip. Similar expressions exist, in terms of K_{II} and K_{III}, for the modes II and III loading situations. For mixed-mode loading, the stress fields can be added together directly. It can be seen that K_I, K_{II} and K_{III} characterize the entire stress field (and hence the strain field) in the vicinity of the crack tip. It therefore seems reasonable to assume that, for mode I loading, for example, failure will occur when K_I reaches a critical value K_c (K_{Ic} under plane strain conditions).

The energy approach and the stress intensity approach are equivalent. Generally, for plane stress

$$G_{total} = G_I + G_{II} + G_{III} = \frac{1}{E}(K_I^2 + K_{II}^2 + K_{III}^2)$$

and for plane strain $\frac{(1 - \nu^2)}{E}$ is replaced by $\frac{1}{E}$.

Generally, for geometries with finite boundaries, the following expression is employed for stress intensity factor

$$K_I = Y\sigma\sqrt{\pi a}$$

and similarly for K_{II} and K_{III}, where Y is a function of the crack and component dimensions.

Material	$\sigma_y (MN/m^2)$	$K_{Ic} (MN/m^{3/2})$
mild steel	220	140–200
pressure vessel steel (HY130)	1700	170
aluminium alloys	100–600	45–23
cast iron	200–1000	20–6

Table 3.3 Typical fracture toughness values

Solutions for Y can be found in the literature for a wide range of geometries and loadings, e.g.

- H. Tada, P. Paris and G. Irwin, 1973, *The stress analysis of cracks handbook*, Hellertown, Pennsylvania: DEL Research Corporation.
- G.P. Rooke and D.J. Cartwright, 1975, *Compendium of stress intensity factors*, HMSO.
- Y Murakani (Editor), 1987, *Stress-intensity factors handbook*, Oxford: Pergamon Press, (2 volumes).

The effects of finite boundaries on expressions for stress intensity factors can be seen from the typical expressions shown below.

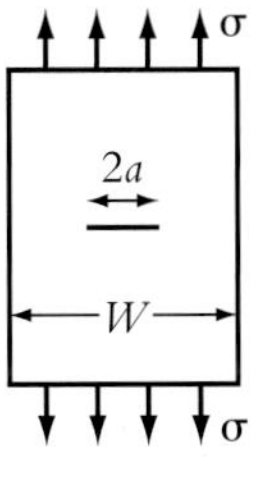

$$K_I = \sigma\sqrt{\pi a}\left(\sec\frac{\pi a}{W}\right)^{1/2}$$

$$K_{II} = \tau\sqrt{\pi a}\left(\text{small}\ \frac{a}{W}\right)$$

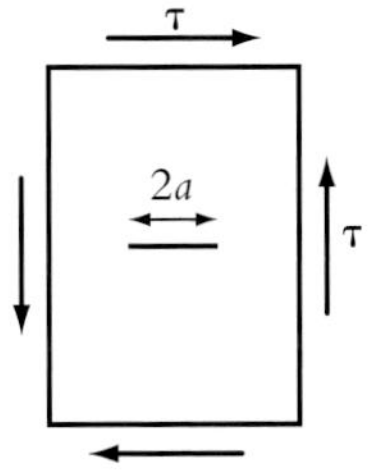

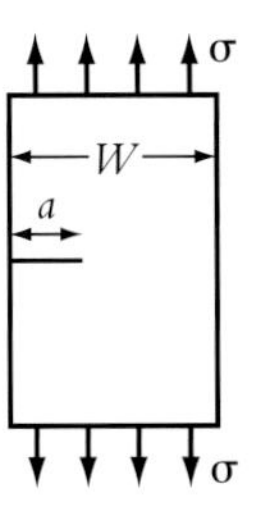

$$K_I = 1.12\sigma\sqrt{\pi a}\left(\text{small}\ \frac{a}{W}\right)$$

or $K_I = Y\sigma\sqrt{a}$

with $$Y = 1.99 - 0.41\frac{a}{W} + 18.7\left(\frac{a}{W}\right)^2 - 38.48\left(\frac{a}{W}\right)^3 + 53.85\left(\frac{a}{W}\right)^4$$

$(1.99 = 1.12\sqrt{\pi})$

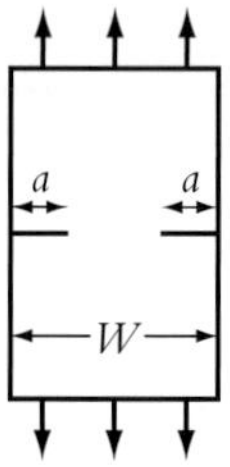

$$K_{\mathrm{I}} = 1.12\sigma\sqrt{\pi a}\left(\text{small}\,\frac{a}{W}\right)$$

or $K_{\mathrm{I}} = Y\sigma\sqrt{a}$

with $Y = 1.99 - 0.76\dfrac{a}{W} - 8.48\left(\dfrac{a}{W}\right)^2 + 27.36\left(\dfrac{a}{W}\right)^3$

$(1.99 = 1.12\sqrt{\pi})$

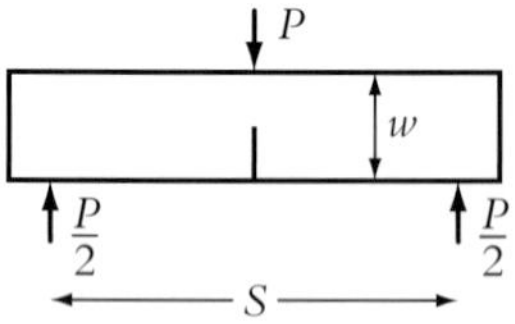

Thickness B

$$K_{\mathrm{I}} = \frac{PS}{BW^{3/2}}\left[2.9\left(\frac{a}{W}\right)^{1/2} - 4.6\left(\frac{a}{W}\right)^{3/2} + 21.8\left(\frac{a}{W}\right)^{5/2} - 37.6\left(\frac{a}{W}\right)^{7/2} + 38.7\left(\frac{a}{W}\right)^{9/2}\right]$$

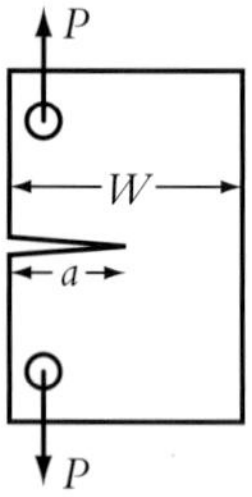

Thickness B

$$K_{\mathrm{I}} = \frac{P}{BW^{1/2}}$$

$$\left[29.6\left(\frac{a}{W}\right)^{1/2} - 185.5\left(\frac{a}{W}\right)^{3/2} + 655.7\left(\frac{a}{W}\right)^{5/2} - 1017\left(\frac{a}{W}\right)^{7/2} + 63.9\left(\frac{a}{W}\right)^{9/2}\right]$$

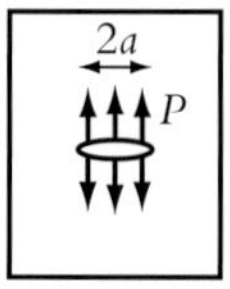

p per unit thickness

$$K_{\mathrm{I}} = p\sqrt{\pi a}$$

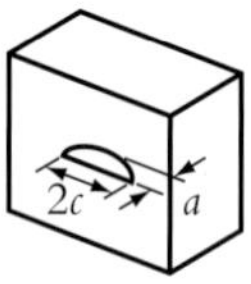

$$K_{\mathrm{Imax}} = 1.12\frac{\sigma}{\Phi}\sqrt{\pi a}$$

$$K_{\mathrm{Imin}} = 1.12\frac{\sigma}{\Phi}\sqrt{\pi a^2/c}$$

$$\Phi = \int_0^{\pi/2}\left[1 - \frac{c^2 - a^2}{c^2}\sin^2\varphi\right]\mathrm{d}\varphi$$

$$\Phi \approx \frac{3\pi}{8} + \frac{\pi}{8}\frac{a^2}{c^2}$$

Worked example

A large high carbon steel plate with a thumbnail crack for which

$$\mathbf{K_{max} = 1.2\sigma\sqrt{\pi a}}$$

has a fracture toughness of 72 MN/m$^{3/2}$ and σ_y = 1450 MN/m^2.

If $\sigma = \frac{2}{3}\sigma_y$, determine the critical initial crack size, assuming linear elastic material.

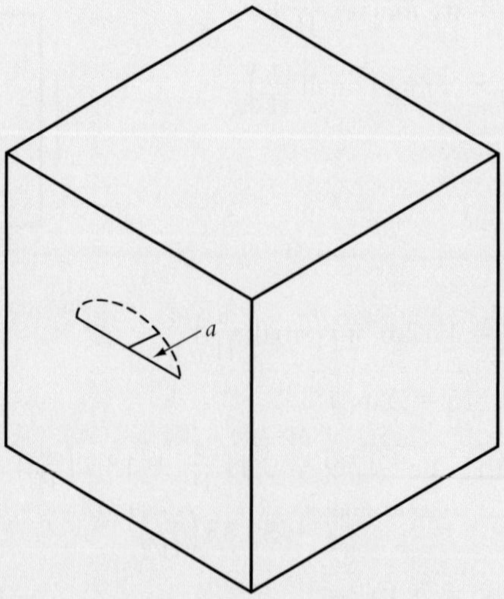

Figure 3.138

At fracture, with

$$\sigma = \frac{2}{3}\sigma_y = \frac{2x1450}{3}\frac{MN}{m^2}$$

and

$$K_{IC} = 72\frac{MN}{m^{\frac{3}{2}}}$$

Then,

$$72\frac{MN}{m^{\frac{3}{2}}} = 1.2x\left(\frac{(2x1450)}{3}\right)\sqrt{\pi a_{crit}}$$

Therefore,

$$a_{crit} = \frac{1}{\pi}\left(\frac{72x3}{1.2x2x1450}\right)^2 \text{m}$$

$$= 1.226 \times 10^{-3}\,\text{m}$$

$$= 1.226\,\text{mm}$$

If the material was mild steel, with

$$\sigma_y = 210\frac{MN}{\text{m}^2} \text{ and } K_{IC} = 200\frac{MN}{\text{m}^{\frac{3}{2}}}$$

Then

$$a_{crit} = 0.451\,\text{m} = 451\,\text{mm}$$

i.e. it is much more likely to be detected during inspection!

Fatigue crack growth

It has been shown by Paris et al. (1961) that, for a wide range of conditions, there is a log–log relationship between crack growth rate and the stress intensity factor range during cyclic loading of cracked components. This has become the basis of the damage-tolerant approach to fatigue life estimation and is widely used both in industry and in research. Essentially, it means that crack growth can be modelled and estimated based on knowledge of crack and component geometry, loading conditions and using experimentally measured crack growth data to furnish material constants. This section describes the basics of this approach.

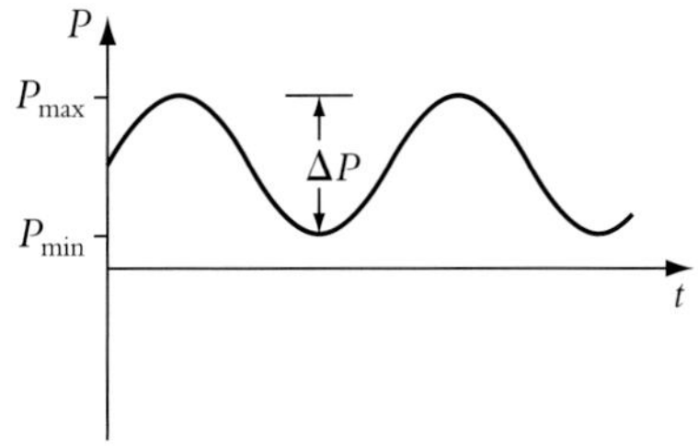

Figure 3.139 Variation of P (load) with *t* (time)

Considering a load cycle as shown above that gives rise to a load range

$$\Delta P = P_{max} - P_{min} \tag{3.90}$$

acting on a cracked body. The load range and crack geometry give rise to a cyclic variation in stress intensity factor, which is given by

$$\Delta K = K_{max} - K_{min} \tag{3.91}$$

Even though the stress intensity factor may be less than the critical stress intensity factor for unstable crack growth, stable crack growth may occur if the stress intensity factor range,

ΔK, is greater than an empirically determined material property called the **threshold stress intensity factor range**, designated ΔK_{th}. In addition, Paris showed that the subsequent crack growth can be represented by an empirical relationship as follows:

$$\frac{da}{dN} = C(\Delta K)^m \tag{3.92}$$

where C and m are empirically determined material constants. This relationship is known as the Paris equation. Fatigue crack growth data is often plotted as the logarithm of crack growth per load cycle, da/dN, and the logarithm of stress intensity factor range. There are three stages. Below ΔK_{th}, no observable crack growth occurs; region II shows an essentially linear relationship between $\log(da/dN)$ and $\log(\Delta K)$, where m is the slope of the curve and C is the vertical axis intercept; in region III, rapid crack growth occurs and little life is involved. Region III is primarily controlled by K_c or K_{Ic}.

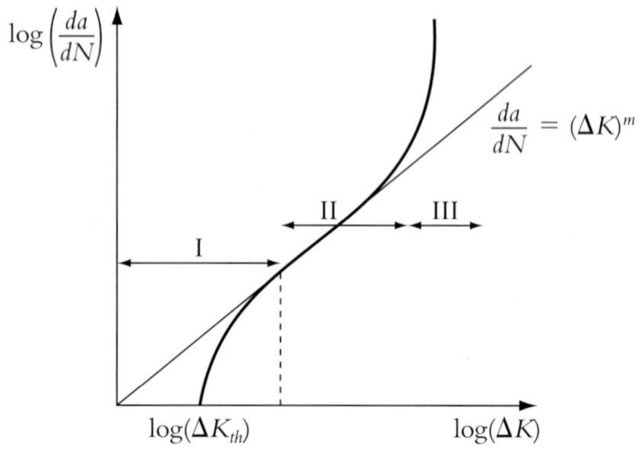

Figure 3.140 Typical (schematic) variation of log (da/dN) with log (ΔK)

The linear regime (region II) is the region which engineering components that fail by fatigue propagation occupy for most of their life. Knowing the stress intensity factor expression for a given component and loading, the fatigue crack growth life of the component can be obtained by integrating the Paris equation between the limits of initial crack size and final crack size.
For most materials, the constant C is found to be dependent on R, where R is a measure of the mean stress defined as

$$R = \frac{K_{min}}{K_{max}}$$

as shown in Figure 3.141.

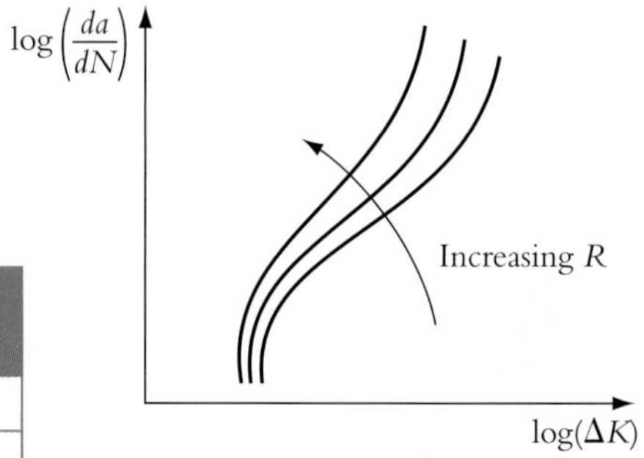

Figure 3.141 Effect of R on fatigue crack growth

Material	ΔK_{th} ($MN/m^{3/2}$)	m	ΔK ($MN/m^{3/2}$) for $da/dN = 10^{-6}$ mm/cycle
mild steel	4–7	3.3	6.2
316 stainless steel	4–6	3.1	6.3
aluminium	1–2	2.9	2.9
copper	1–3	3.9	4.3
brass	2–4	4.0	4.3–66.3
nickel	4–8	4.0	8.8

Table 3.4 Typical values of Δk_{th}, m and Δk (From Pook (1978), pp 114–35.)

The ΔK_{th} and ΔK (for $da/dN = 10^{-6}$ mm/cycle) values depend on the R-value.

Learning summary

By the end of this section you should have learnt:

- ✓ the meaning of linear elastic fracture mechanics (LEFM);
- ✓ what the three crack tip loading modes are;
- ✓ the energy and stress intensity factor (Westergaard crack tip stress field) approaches to LEFM;
- ✓ the meaning of small-scale yielding and fracture toughness;
- ✓ the Paris equation for fatigue crack growth and the effects of the mean and alternating components of the stress intensity factor.

3.13 Thermal stresses

Introduction

Changes of temperature in a body cause expansion/contraction. This phenomenon is quantified by the coefficient of thermal expansion α. In any direction, the change of length due to a temperature change, T, is $l\alpha T$ where l is the original length. Therefore, direct thermal strain is given by:

$$\varepsilon_{\text{thermal}} = \frac{l\alpha T}{l} = \alpha T$$

For isotropic materials, α is the same for all directions and there are no thermal shear strains. Uniform termperature change throughout an unrestrained body produces uniform strain but no stress, i.e. free expansion/contraction in all directions.

Generalized Hooke's law

$$\varepsilon_x = \frac{1}{E}(\sigma_x - \nu(\sigma_y + \sigma_z)) + \alpha T$$

$$\varepsilon_y = \frac{1}{E}(\sigma_y - \nu(\sigma_x + \sigma_z)) + \alpha T$$

$$\varepsilon_z = \frac{1}{E}(\sigma_z - \nu(\sigma_x + \sigma_y)) + \alpha T$$

$$\gamma_{xy} = \tau_{xy}/G$$

$$\gamma_{yz} = \tau_{yz}/G$$

$$\gamma_{zx} = \tau_{zx}/G$$

where T is the temperature at a point relative to some datum.

By introducing these thermal strains into the generalized Hooke's law we can obtain solutions to thermal stress problems, which are often very important in, for example, power and chemical plant, aeroengines and internal combustion engines (e.g. pistons and cylinder walls).

Case 1: An initially straight uniform beam

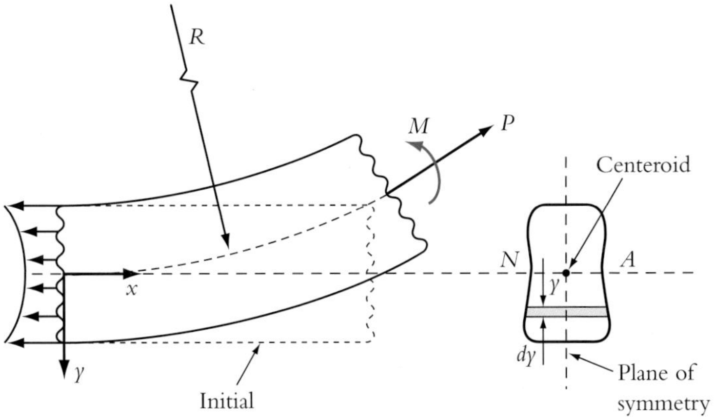

Figure 3.142

Determine deformations and stresses (small deformations)
The temperature is (assumed) purely a function of y, i.e. $T = T(y)$.

The coefficient of thermal expansion $= \alpha$. Axial force P, and pure bending, about the z-z axis, M, are also applied.

Assuming σ_z, σ_y, τ_{xz} and $\tau_{yz} = 0$ because the cross-sectional dimensions are small compared with the length.

Also

$\tau_{xy} = 0$, because M does not vary with x, $\quad \therefore \quad S = \dfrac{dM}{dx} = 0$

Compatibility

Remote from the ends, strain varies linearly with y,

i.e. $$\varepsilon_x = \bar{\varepsilon} + \frac{y}{R}$$

Where $\bar{\varepsilon}$ is the mean strain (at $y = 0$) and R is the radius of curvature.

Stress–strain

$$\varepsilon_x = \frac{\sigma_x}{E} + \alpha T\left\{= \bar{\varepsilon} + \frac{y}{R}\right\}$$

$\therefore$ $$\sigma_x = E \times \left\{\bar{\varepsilon} + \frac{y}{R} - \alpha T\right\} \quad \textbf{(3.93)}$$

Axial equilibrium

$$P = \int_A \sigma_x dA = E\int_A \left\{\bar{\varepsilon} + \frac{y}{R} - \alpha T\right\} dA = E\bar{\varepsilon}A + \frac{E}{R}\int_A y dA - E\alpha\int_A T dA$$

$$\int_A y dA = 0$$

because y is measured from an axis passing through the centroid.

$\therefore$ $$P = E\bar{\varepsilon}A - E\alpha\int_A T dA \quad \textbf{(3.94)}$$

Moment equilibrium

$$M = \int_A y\sigma_x dA = E\int_A \left\{\bar{\varepsilon} + \frac{y}{R} - \alpha T\right\} y dA = E\bar{\varepsilon}\int_A y dA + \frac{E}{R}\int_A y^2 dA - E\alpha\int_A Ty dA$$

Now

$$\int_A y^2 dA = I$$

(second moment of area, by definition)

$\therefore$ $$M = \frac{EI}{R} - E\alpha\int_A Ty dA \quad \textbf{(3.95)}$$

Worked example

A rectangular beam, width b and depth d has a temperature variation given by:

$$T = T_o\left\{1 - \frac{4y^2}{d^2}\right\}$$

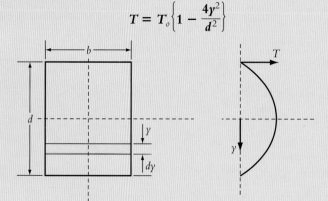

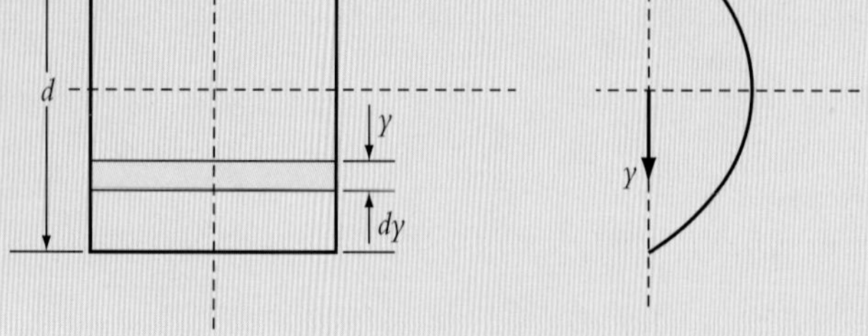

Figure 3.143

There is no restraint or applied loading (i.e. $P = M = 0$). Obtain the stress distribution.

Axial force equilibrium (equation (3.94))

$$P = E\bar{\varepsilon}A - E\alpha\int_A T\,dA$$

$$0 = E\bar{\varepsilon}\cdot bd - E\alpha\int_{\frac{-d}{2}}^{d/2} T_o\left\{1 - \frac{4y^2}{d^2}\right\}(bdy)$$

$$\therefore \qquad \bar{\varepsilon} = \frac{\alpha}{d}T_o\int_{\frac{-d}{2}}^{d/2}\left\{1 - \frac{4y^2}{d^2}\right\}dy$$

$$\text{i.e.} \qquad \bar{\varepsilon} = \frac{\alpha}{d}T_o\left[y - \frac{4y^3}{3d^2}\right]_{-d/2}^{d/2}$$

$$\therefore \qquad \bar{\varepsilon} = \frac{2}{3}\alpha\cdot T_o$$

With $M = O$ we can obtain $1/R$ from the moment equilibrium (equation (3.95)) but from symmetry we can see that $(1/R) = O$.

Using equation (3.93)

$$\sigma_x = E\left\{\bar{\varepsilon} + \frac{y}{R} - \alpha T\right\}$$

$$= E\left(\frac{2}{3}\alpha T_o + O - \alpha T_o\left(1 - \frac{4y^2}{d^2}\right)\right)$$

$$= E\alpha T_o\left(\frac{4y^2}{d^2} - \frac{1}{3}\right)$$

At $y = O$, $\qquad \sigma_x = \dfrac{-E\alpha T_o}{3}$

At $= \dfrac{\pm d}{2}$, $\qquad \sigma_x = E\alpha T_o\left(1 - \dfrac{1}{3}\right) = \dfrac{2E\alpha T_o}{3}$

$\sigma_x = 0$ when $\dfrac{4y^2}{d^2} = \dfrac{1}{3}$ i.e. at $y = \pm 0.287d$

This is the stress distribution away from the ends. At the ends, $\sigma_x = 0$ and there is a gradual transition between the two.

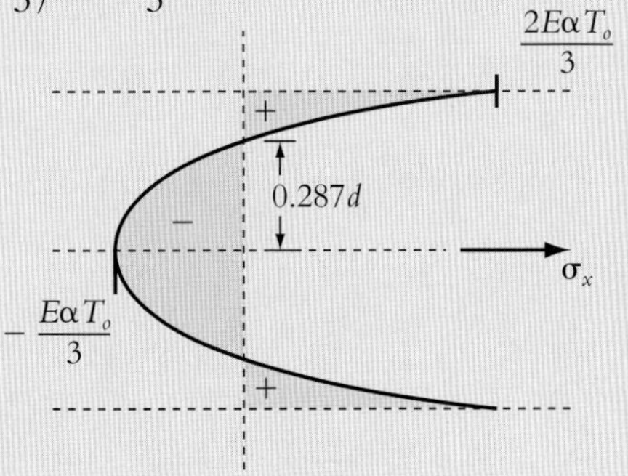

Figure 3.144

Worked example

A rectangular beam (again $b \times d$), but with:

$$T = T_o \times \frac{2y}{d}$$

It is constrained so that $\bar{\varepsilon} = 0$ and $1/R = 0$.

Determine the stresses and restraints.

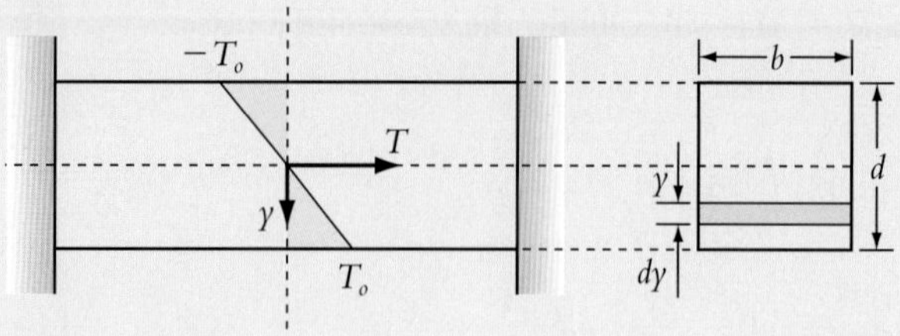

Figure 3.145

Axial force equilibrium (equation (3.94))

$$P = E\bar{\varepsilon}A - E\alpha\int_A T\mathrm{d}A$$

$$\int_A T\mathrm{d}A = \frac{2T_o b}{d}\int_{-d/2}^{d/2} y\,\mathrm{d}y = \frac{2T_o b}{d}\left[\frac{y^2}{2}\right]_{-d/2}^{d/2} = 0$$

Also,

$$\bar{\varepsilon} = 0 \qquad \therefore \quad P = 0$$

Moment equilibrium (equation (3.95))

$$M = \frac{EI}{R} - E\alpha\int_A Ty\mathrm{d}A$$

$$\int_A Ty\mathrm{d}A = \frac{2T_o b}{d}\int_{-d/2}^{d/2} y^2\mathrm{d}A = \frac{2T_o b}{d}\left[\frac{y^3}{3}\right]_{-d/2}^{d/2}$$

$$= \frac{2T_o b}{d}\left[\left(\frac{d^3}{24}\right) - \left(\frac{-d^3}{24}\right)\right] = \frac{T_o b d^2}{6}$$

Also,

$$1/R = 0, \quad \therefore\ M = \frac{-E\alpha b d^2 T_o}{6}$$

Using equation (3.93) (with $\bar{\varepsilon} = (1/R = 0)$)

$$\sigma_x = \left(\bar{\varepsilon} + \frac{y}{R} - \alpha T\right) = -E\alpha \cdot T$$

$\therefore$

$$\sigma_x = -E\alpha T_o \frac{2y}{d}$$

i.e.

$$\sigma_x = \frac{-2E\alpha T_o}{d} y$$

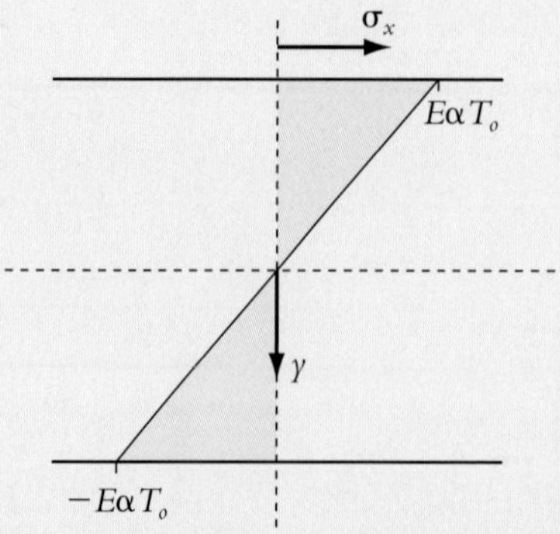

Figure 3.146

Case 2: A thin disc of uniform thickness

The equilibrium and compatibility equations are as for rotating discs, with $\omega = 0$, i.e.
Equilibrium (radial)

$$\sigma_r + r\frac{d\sigma_r}{dr} - \sigma_\theta = 0 \tag{3.96}$$

Compatibility

$$\varepsilon_\theta = \frac{u}{r} \qquad \text{(2a)} \tag{3.97}$$

$$\varepsilon_r = \frac{du}{dr} \qquad \text{(2b)} \tag{3.98}$$

i.e.

$$\varepsilon_r = \frac{d}{dr}(r\varepsilon_\theta) = \varepsilon_\theta + r\frac{d\varepsilon_\theta}{dr} \tag{3.99}$$

However, the stress–strain equations now contain thermal terms.

i.e.

$$\varepsilon_\theta = \frac{1}{E}(\sigma_\theta - \nu\sigma_r) + \alpha T \tag{3.100}$$

$$\varepsilon_r = \frac{1}{E}(\sigma_r - \nu\sigma_\theta) + \alpha T \tag{3.101}$$

Note: Because the disc is thin, $\sigma_z = 0$.
Substituting (3.100) and (3.101) into (3.99) gives

$$\left(\frac{1}{E}(\sigma_r - \nu\sigma_\theta) + \alpha T\right) = \left(\frac{1}{E}(\sigma_\theta - \nu\sigma_r) + \alpha T\right) + \frac{r}{E}\left(\frac{d\sigma_\theta}{dr} - \nu\frac{d\sigma_r}{dr}\right) + r\alpha\frac{dT}{dr}$$

$\therefore$

$$(1+\nu)(\sigma_\theta - \sigma_r) + r\frac{d\sigma_\theta}{dr} - \nu r\frac{d\sigma_r}{dr} + E\alpha r\frac{dT}{dr} = 0$$

Substitute

$$\sigma_\theta - \sigma_r = r\frac{d\sigma_r}{dr} \quad \text{from (3.96)}$$

$\therefore$

$$(1+\nu)r\frac{d\sigma_r}{dr} + r\frac{d\sigma_\theta}{dr} - \nu r\frac{d\sigma_r}{dr} + E\alpha \cdot r\frac{dT}{dr} = 0$$

i.e.

$$r\frac{d}{dr}(\sigma_\theta + \sigma_r) = -E\alpha \cdot r\frac{dT}{dr} = 0$$

Also from (3.96):

$$\sigma_\theta = \sigma_r + r\frac{d\sigma_r}{dr}$$

$\therefore$

$$r\frac{d}{dr}\left(2\sigma_r + r\frac{d\sigma_r}{dr}\right) = -E\alpha r\frac{dT}{dr}$$

$$r\left(2\frac{d\sigma_r}{dr} + \frac{rd^2\sigma_r}{dr^2} + \frac{d\sigma_r}{dr}\right) = -E\alpha r\frac{dT}{dr}$$

i.e.

$$r^2\frac{d^2\sigma_r}{dr^2} + 3r\frac{d\sigma_r}{dr} = -E\alpha r\frac{dT}{dr}$$

$\therefore$

$$\sigma_r = A - \frac{B}{r^2} + \text{P.I.}$$

From (3.96):

$$\sigma_\theta = \sigma r + r\frac{d\sigma_r}{dr}$$

Worked example

A steel disc with 300 mm outside diameter, 40 mm bore diameter and 5 mm thickness is subjected to a temperature distribution of the following form:

$$T = \left(\frac{200r^2}{150^2}\right)^{o} C$$

where r is the radius in mm. Find the hoop stress at the bore.

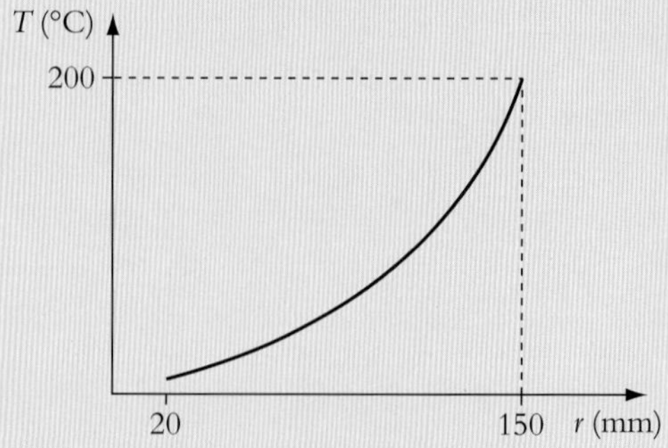

Figure 3.147

The governing differential equation is:

$$r^2\frac{d^2\sigma_r}{dr^2} + 3r\frac{d\sigma_r}{dr} = -E\alpha r\frac{dT}{dr} = -E\alpha r\frac{400r}{150^2} = 400E\alpha\frac{r^2}{150^2}$$

$$\text{P.I.} = Cr^2 \text{ (say)}$$

$\therefore$
$$r^2\cdot 2C + 3r\cdot 2Cr = -400E\alpha\frac{r^2}{150^2}$$

i.e.
$$C = \frac{-50E\alpha}{150^2}$$

$\therefore$
$$\sigma_r = A - \frac{B}{r^2} - 50E\alpha\frac{r^2}{150^2}$$

Boundary conditions:

At $r = 20\text{ mm}, \sigma_r = 0$

$\therefore$
$$0 = A - \frac{B}{20^2} - 50E\alpha\frac{20^2}{150^2} \quad \text{(a)} \qquad \textbf{(3.102)}$$

At $r = 150\text{ mm}, \sigma_r = 0$

$$0 = A - \frac{B}{150^2} - 50E\alpha\frac{150^2}{150^2} \quad \text{(b)} \qquad \textbf{(3.103)}$$

Subtracting (3.103) from (3.102) gives:

$$0 = B\left(\frac{1}{22\,500} - \frac{1}{400}\right) + 50E\alpha\left(1 - \frac{400}{150^2}\right)$$

i.e.
$$B = 20\,000E\alpha$$

Substituting B in (3.103) gives:

$$A = \frac{20\,000E\alpha}{22\,500} + 50E\alpha = 50.88E\alpha$$

So
$$\sigma_r + r\frac{d\sigma_r}{dr} = A - \frac{B}{r^2} - 50E\alpha\frac{r^2}{150^2} + r\left(\frac{2B}{r^3} - \frac{100E\alpha r}{150^2}\right)$$

i.e.
$$\sigma_\theta = A + \frac{B}{r^2} - 150E\alpha\frac{r^2}{150^2} = \left(50.88 + \frac{20\,000}{r^2} - \frac{r^2}{150}\right)E\alpha$$

At $r = 20$:

$$\sigma_\theta = \left(50.88 + \frac{20\,000}{400} - \frac{400}{150}\right)E\alpha = 98.22E\alpha$$

i.e. $$\sigma_\theta = 223.9\,\text{N/mm}^2 \text{ (at } r = 20\,\text{mm)}$$

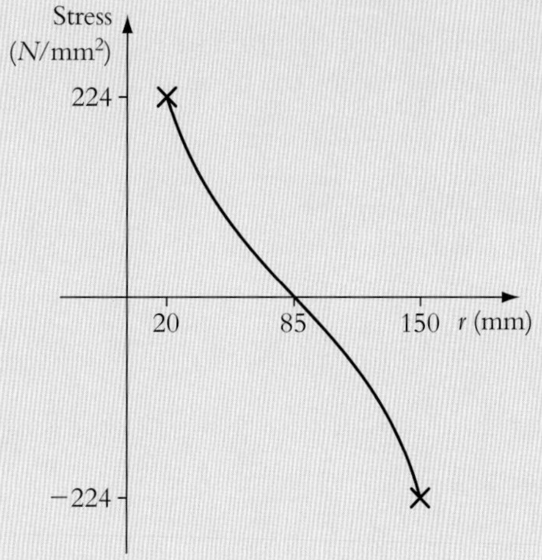

Figure 3.148

Case 3: Thin cylinders

Thin cylinders are in common use in power and chemical plant, e.g. pipes, pressure vessels, etc. Often temperature variations are approximately linear through the thickness. Considering positions remote from ends, flanges, etc.

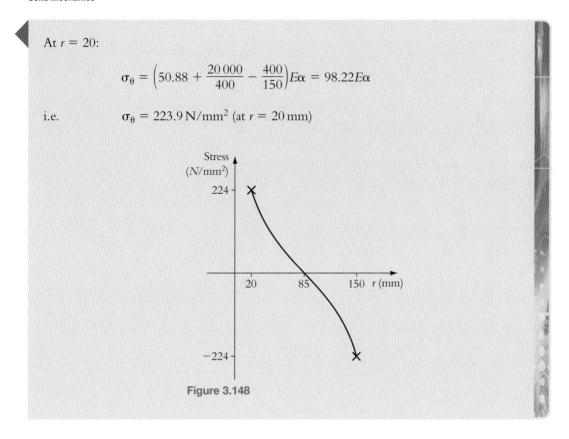

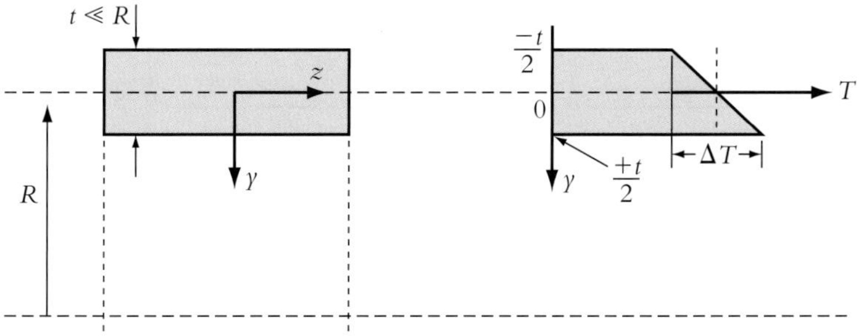

Figure 3.149

It is convenient to consider the effect of the uniform temperature change and the temperature gradient separately. If the cylinder is not restrained, then the uniform temperature change causes overall dimensional changes, but no stress. The stresses due to axial restraint are easily calculated.

For the temperature gradient we have:

$$T = \Delta T\frac{y}{t}$$

where ΔT is the temperature difference across the wall.

For a thin cylinder:

$$\sigma_r \cong 0$$

∴

$$\varepsilon_{\theta} = \frac{1}{E}(\sigma_\theta - \nu\sigma_z) + \alpha T$$

and

$$\varepsilon_z = \frac{1}{E}(\sigma_z - \nu\sigma_\theta) + \alpha T$$

Remote from the ends of the cylinder sections remain plane and circular. Therefore, from compatibility considerations (with zero mean temperature change), the hoop and axial strains must both be zero. Therefore:

$$0 = \frac{1}{E}(\sigma_{\theta} - \nu\sigma_z) + \alpha(\Delta T)\frac{y}{t}$$

and

$$0 = \frac{1}{E}(\sigma_z - \nu\sigma_\theta) + \alpha(\Delta T)\frac{y}{t}$$

Solving, gives

$$\sigma_\theta = \sigma_z = \frac{E\alpha(\Delta T)}{(1-\nu)}\left(\frac{y}{t}\right)$$

Learning summary

By the end of this section you should be able to:

- ✓ understand the cause of thermal strains and how 'thermal stresses' are caused by thermal strains;
- ✓ include thermal strains in the generalized Hooke's Law equations;
- ✓ include the temperature distribution within a solid component (e.g. a beam, a disc or a tube) in the solution procedure for the stress distribution;
- ✓ understand that stress/strain equations include thermal strain terms but the equilibrium and compatibility equations are the same whether the component is subjected to thermal loading or not.

References

Boresi, Schmidt and Sidebottom, 1993, *Advanced Mechanics of Materials*, 5th Edn, Wiley & Sons.

Griffith, A.A., 1921, 'The phenomena of rupture and flow in solids', *Philosophical Transactions of the Royal Society of London* vol. 221.

Paris, P.C., Gomez, M.P. and Anderson, W.E., 1961, 'A rational analytic theory of fatigue life', *The Trend in Engineering*, vol. 13.

Pook, L.P., 1975, 'Analysis and application of fatigue crack growth data', *J Strain Analysis*, vol. 10, no. 4, 242–64.

Westergaard, H.M., 1939, 'Bearing Pressures and Crack', *Journal of Applied Mechanics*, vol. 5, no. 2.

Unit 4

Electromechanical drive systems

UNIT OVERVIEW

- Introduction
- Characteristics of loads
- Linear and rotary inertia
- Geared systems
- Tangentially driven loads
- Steady-state characteristics of loads
- Modifying steady-state characteristics of a load using a transmission
- Sources of mechanical power and their characteristics
- Direct current motors and their characteristics
- Rectified supplies for dc motors
- Inverter-fed induction motors and their characteristics
- Other sources of power: pneumatics and hydraulics
- Steady-state operating points and matching of loads to power sources

4.1 Introduction

In many engineering situations it is necessary to design a system that involves a source of mechanical power to drive (directly or indirectly) a mechanical load of some kind, with one or more rotating shafts often being used within the means of transmitting the power from source to load. Such systems can range from the very small (for instance, focusing mechanisms within digital cameras) to the extremely large (e.g. driving the propellers of a transatlantic liner). In all cases, it is essential to understand the characteristics of the load we are trying to drive, the characteristics of the source of mechanical power that we are using to drive it, and the need for (and function of) any transmission mechanism needed to match the characteristics of the power source and load. This unit will therefore examine in depth the steady-state and transient characteristics of loads and of typical sources of mechanical power used to drive them, including electric motors and other forms of motors and engines. It is also useful to be able to alter the characteristics of the power source in order to provide control over the way that the overall system behaves, so the techniques by which electric motors can be controlled will be explored in some depth. The manner in which the power source and load interact is examined, with particular reference to how apparently incompatible characteristics can be matched, in order to achieve stable operation in the desired manner. Although this unit draws together topics from the disciplines of dynamics, drive systems, electrical engineering, fluid power and mechanical design, various texts are available that follow this kind of approach in greater depth, for example the text by Fraser and Milne (1994).

4.2 Characteristics of loads

As hinted in the introduction, in order to arrive at an optimal decision on how to drive a machine, it is necessary to have a full understanding of how that machine behaves as a mechanical load. Specifically, the following questions need to be addressed:

- How hard is it to get the machine moving from rest? (friction: 'stiction' and Coulomb friction)
- How hard is it to speed it up? (inertia)
- How much harder does it get to drive as we make it go faster? (torque–speed characteristics including viscous friction and windage)
- How hard is it to make it stop when we want to? (inertia again, with some friction considerations)
- How well do the characteristics of the machine match with those of the device we are using to drive it, and how can we overcome any mismatch?

Some of these issues can be illustrated with reference to issues familiar to automobile drivers:

- A car filled to its maximum design capacity with heavy people is much less responsive (for example, when pulling away from traffic lights) than when carrying only the driver. **The inertia of the load has increased and is no longer very well matched to the power source.**
- A car driven at a steady 50 mph uses considerably less petrol to travel a given distance than one driven at the legal motorway limit of 70 mph. **The steady-state torque required from the car engine is determined (in a large part) by wind drag which is roughly proportional to the square of the speed.** (The drag force falls to something like $(50/70)^2 = 0.51$ of its original value, though there are various other losses that show less of a reduction.)
- It is possible to tow quite a heavy trailer with a car as long as the driver allows for the poorer acceleration. However, such a trailer must include its own braking system (various mechanisms exist) because it would otherwise be difficult to stop the car/trailer combination. **The inertia of the system is no longer well matched to the braking system (nor, for that matter to the power source).**
- When a steep hill is reached, the driver must change down a gear (and probably settle for driving at a lower speed if currently travelling fast). **The engine power is limited, and in top gear the engine is badly matched to the increased load, so to obtain the required torque it is necessary to gear the engine down to obtain a high torque at low speed.**

4.3 Linear and rotary inertia

The concept of inertia is based upon Newton's second law:

$$F = ma \tag{4.1}$$

where F is the force required to cause a mass m to undergo an acceleration of a. In this context, mass is a measure of how difficult it is to accelerate something – in other words, mass is a measure of **linear inertia**.

In engineering, we are often concerned with rotational movements, and so we need to introduce the concept of **rotational inertia**, in other words how difficult it is to cause rotational acceleration in a body that is capable of rotation about an axis. This is called the **moment of inertia**, a concept which was covered in Section 6.7 of Part 1 and is used within Unit 6 of the present volume in the context of torsional vibration. It is briefly revised here for continuity.

Moment of inertia

To avoid confusion with electric current I, the symbol J will be used here for moment of inertia of a body about its axis of rotation. Moment of inertia is analogous to mass in the rotational version of Newton's second law:

$$\text{Torque} = \text{moment of inertia} \times \text{angular acceleration}$$

i.e.

$$L = J\alpha \tag{4.2}$$

For a body B rotating about a given axis OZ (Figure 4.1), J_{OZ} is calculated as follows:

$$J_{OZ} = \int_B r_z^2 dm \tag{4.3}$$

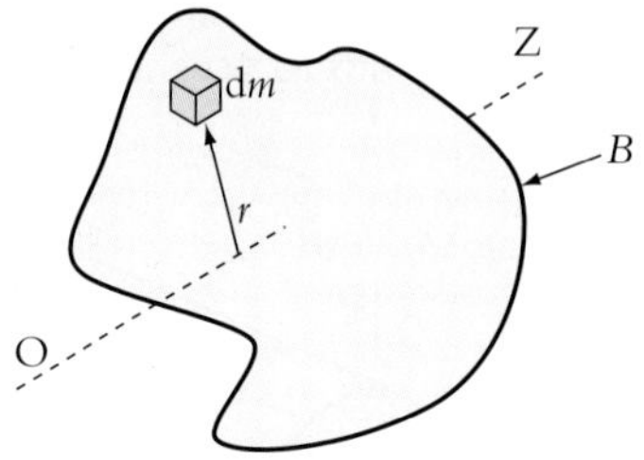

Figure 4.1 **A body with axis of rotation OZ**

By way of a specific example, a **solid cylinder** or a disc of mass m and radius r (Figure 4.2(a)) has a moment of inertia

$$J = \tfrac{1}{2}mr^2 \tag{4.4}$$

about its axis of symmetry.

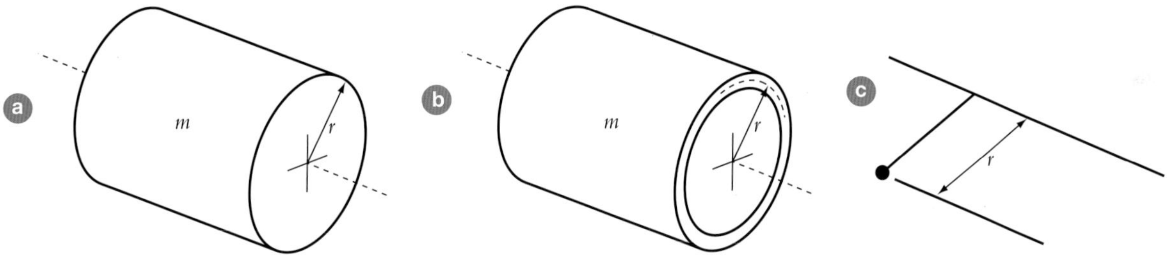

Figure 4.2 **(a) Solid cylinder, (b) thin-walled tube, (c) concentrated mass at radius *r* from axis of rotation**

By contrast, a **thin-walled tube** or a thin circular ring of mass m and radius r (Figure 4.2(b)) has a moment of inertia

$$J = mr^2 \tag{4.5}$$

about its axis of symmetry.

Finally, a **concentrated mass** m being swung at radius r about a given axis (Figure 4.2(c)) has a moment of inertia

$$J = mr^2 \tag{4.6}$$

about that axis. An example of such a system would be a golf club with a heavy head and a light handle, pivoting around the golfer's wrist.

Inertias on the same axis are simply added together; objects with portions removed are treated by subtracting the inertias of the 'missing' portions from the inertia of the overall object. For example, the moment of inertia of a hollow cylinder of outer radius r_o and inner radius r_i about its axis (Figure 4.3) is found by subtracting the moment of inertia of the 'missing' inner cylinder (of mass m_i) from the moment of inertia of the solid cylinder (of mass m_o):

$$\begin{aligned} J_{hollow} &= J_{solid} - J_{missing} \\ &= \tfrac{1}{2}m_o r_o^2 - \tfrac{1}{2}m_i r_i^2 \\ &= \tfrac{1}{2}(\pi r_o^2 \rho l)r_o^2 - \tfrac{1}{2}(\pi r_i^2 \rho l)r_i^2 \\ &= \frac{\pi}{2}\rho l(r_o^4 - r_i^4) \end{aligned} \tag{4.7}$$

which can be expressed as:

$$
\begin{aligned}
J_{\text{hollow}} &= \frac{\pi}{2}\rho l(r_o^{\,4} - r_i^{\,4}) \\
&= \frac{\pi}{2}\rho l(r_o^{\,2} - r_i^{\,2})(r_o^{\,2} + r_i^{\,2}) \\
&= \tfrac{1}{2}m_{\text{hollow}}(r_o^{\,2} + r_i^{\,2}) \qquad (4.8)
\end{aligned}
$$

Superficially this appears to give a different answer from the formula relating to thin-walled cylinders (Figure 4.2(b)), but for such a cylinder r_o and r_i tend towards being equal as the wall thickness tends to zero, and J_{hollow} approaches the familiar value of mr^2.

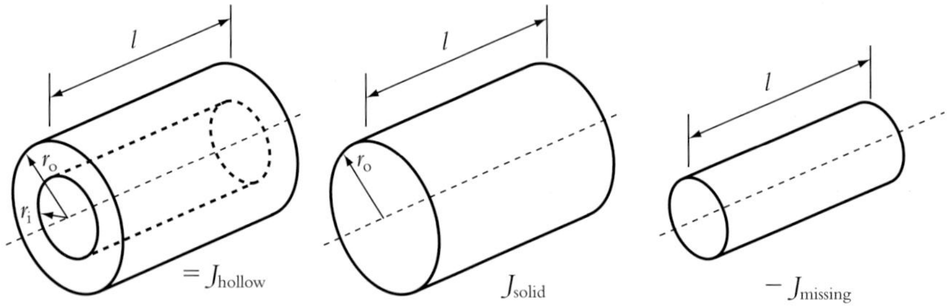

Figure 4.3: **Combining inertias**

Learning summary

By the end of this section you should have learnt:

- ✓ the similarities and differences between linear and rotational inertias and how they are analysed;
- ✓ the concept of moment of inertia;
- ✓ to calculate moment of inertia for simple components made up of cylinders and tubes.

4.4 Geared systems

In order to extend the above concepts to geared systems, the concept of a **gear ratio** must also be defined (Figure 4.4):

$$\text{Gear ratio } n = \frac{N_2}{N_1} = \frac{\omega_1}{\omega_2} = \frac{\alpha_1}{\alpha_2} = \frac{L_2}{L_1} \qquad (4.9)$$

where N_1 and N_2 are the number of teeth on the gears mounted on the input and output shafts respectively, ω_1 is the angular velocity (in rad/s), α_1 is the angular acceleration (in rad/s^2) and L_1 is the torque (in Nm) for the input shaft, with corresponding terms ω_2 etc. referring to the output shaft. Note that equation (4.9) is also valid if the angular velocities are expressed consistently in other units e.g. rev/min or rev/s; this validity does *not* extend to equations that involve calculations of actual values of power. A similar argument applies to accelerations: calculations relating inertias to accelerations must be performed using rad/s^2 as the unit of angular acceleration because equation (4.2) is not valid if units such as rev/s^2 are used.

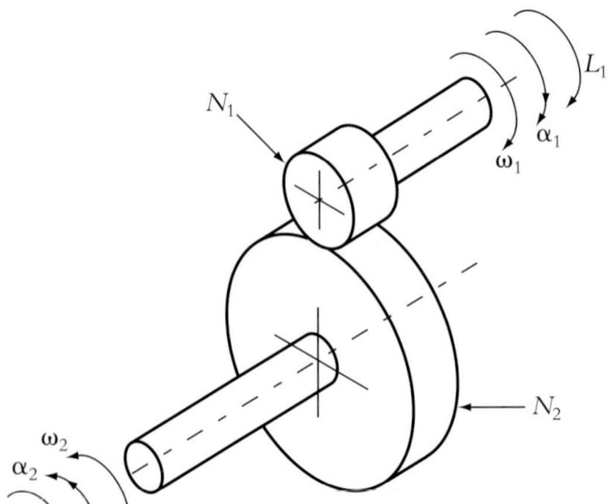

Figure 4.4 **A pair of gears showing torques, speeds etc**

The angular accelerations experienced by the different shafts are related in just the same way as are the speeds (angular velocities). The last term in equation (4.9) comes from the law of conservation of energy for a perfectly efficient geared system:

$$\text{power out} = \text{power in}$$

i.e.

$$L_2\omega_2 = L_1\omega_1 \tag{4.10}$$

so that input torque L_1 can be calculated from L_2

$$L_1 = \frac{L_2\omega_2}{\omega_1} = \frac{L_2}{n} \tag{4.11}$$

The same theory relates to other toothed drive systems, specifically sprocket/chain systems (such as those used in bicycles, motorcycles and some industrial situations) and toothed pulley/belt systems, used for the camshaft drive in cars and within numerous motion control and positioning systems in industry and within more light-duty systems such as traversing the heads of inkjet printers and moving the scanning sensor on photocopiers. All of these may be regarded as synchronous drives in that the drive ratios are exactly determined by the numbers of teeth on the driving and driven components. One difference between gears and other drive systems such as chain drives and toothed belt drives is that gears reverse the direction of drive while belts and chains normally do not.

Worked example

A machine is driven at 35 rad/s with a torque of 30 Nm. The drive is provided via a transmission with a 4:1 reduction ratio drive. What is the combination of torque and speed required at the input to the transmission system? Assume that the transmission is perfectly efficient.

For speed:

$$n = \frac{\omega_1}{\omega_2}$$

$$\text{where } n = 4, \omega_2 = 35 \text{ rad/s}$$

$\therefore$

$$\omega_1 = 4\omega_2 = 4 \times 35 = 140 \text{ rad/s}$$

For torque:

$$n = 4 = \frac{L_1}{L_2}$$

where

$$L_2 = 30\,\text{Nm}$$

$\therefore$

$$L_1 = \frac{L_2}{4} = \frac{30}{4} = 7.5\,\text{Nm}$$

Broadly similar analysis can be performed for friction-based drive systems, though these are not synchronous because the components no longer possess teeth and some degree of slippage can take place between driving and driven components. The nominal drive ratio is now determined by the ratio of the effective radii (or diameters) of the components (Figure 4.5):

$$\text{Drive ratio } n = \frac{r_2}{r_1} = \frac{d_2}{d_1} = \frac{\omega_1}{\omega_2} = \frac{\alpha_1}{\alpha_2} \tag{4.12}$$

The most commonly encountered friction-based transmissions are vee belt/pulley systems (used extremely widely, an everyday example being the drive from car engine to alternator), flat belt systems (less widely used today; examples include belt-drive turntables used for playing vinyl LP records), and systems based upon friction rollers

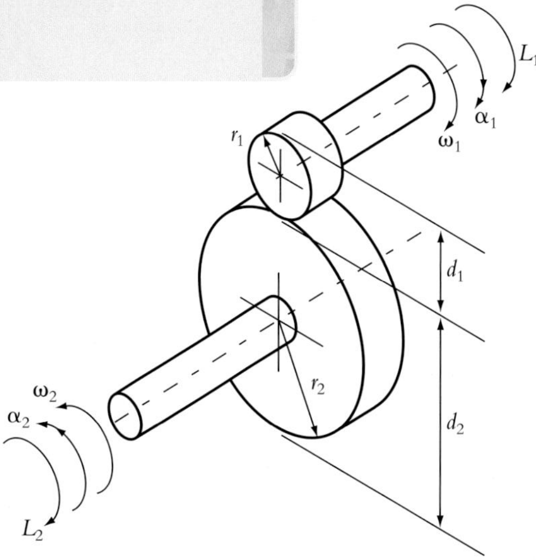

Figure 4.5 **A pair of wheels coupled by friction**

(widely used at one time in audio tape and cassette recorders, and still used within fairground and amusement rides, for example for 'gearing' rotating carriages to the rotation of the frame which carries them, or to provide the impetus to swinging 'boats' from a motorised wheel/ tyre running on the bottom of the boat). In the cases of vee belts, flat belts etc., care must be taken to understand the concept of the effective radius or pitch circle radius, since this may be significantly different from any of the measurable radii of the pulley (Figure 4.6).

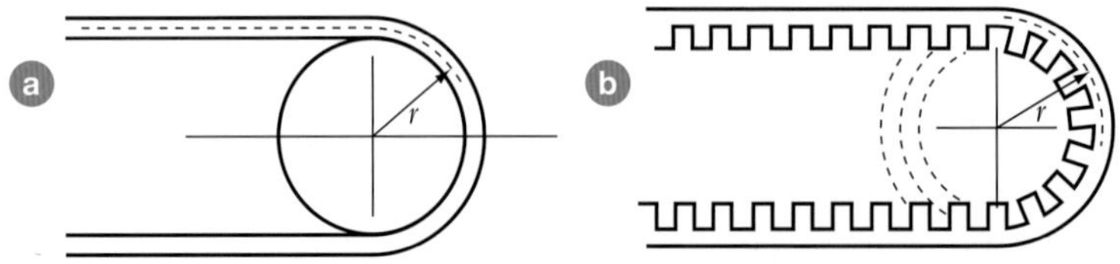

Figure 4.6 **(a) Flat belt and (b) toothed belt showing effective radius**

Worked example

A centrifugal blower needs to be driven at 3000 rev/min from a motor that runs at 1460 rev/min. If the effective diameter of the pulley on the blower is 100 mm, what should the effective diameter of the pulley on the motor be?

Drive ratio $n = \dfrac{d_2}{d_1} = \dfrac{\omega_1}{\omega_2}$

where $d_2 = 100\,\text{mm},\ \omega_1 = 1460\ \text{rev/min},\ \omega_2 = 3000\ \text{rev/min}$

$\therefore$ $d_1 = \dfrac{\omega_2}{\omega_1} d_2 = \dfrac{3000}{1460} \times 100 = 205.5\,\text{mm}$

An interesting variant of the friction-based drive is the variable-ratio drive. Various mechanisms are available, including systems based on vee pulleys of variable width (and hence of variable effective radius), and tilting spheres contacting conical rollers (Figure 4.7) so that the spheres behave as rollers with varying effective radii. Hydraulic variable-ratio drives also exist – see Section 4.12.

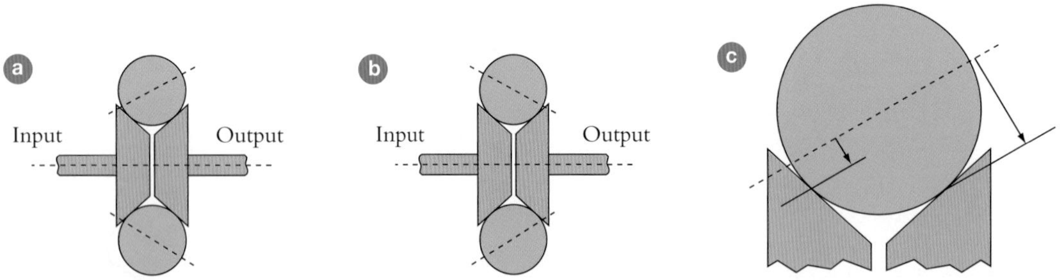

Figure 4.7 **Sphere-based variable speed drive (a) set to increase speed; (b) set to decrease speed; (c) effective radii of contact of spheres against the two cones**

Compound drive trains

In many engineering situations it is not practical to obtain the required drive ratio via a single stage of transmission. In this case, multiple stages are used, with the overall drive ratio being the product of the drive ratios of each stage in the transmission system.

For example, the drive ratio of the gear train shown in Figure 4.8 is:

$$\text{Drive ratio } n = \frac{N_2}{N_1} \times \frac{N_4}{N_3} \times \frac{N_6}{N_5} = \frac{\omega_1}{\omega_6} = \frac{\alpha_1}{\alpha_6} \qquad (4.13)$$

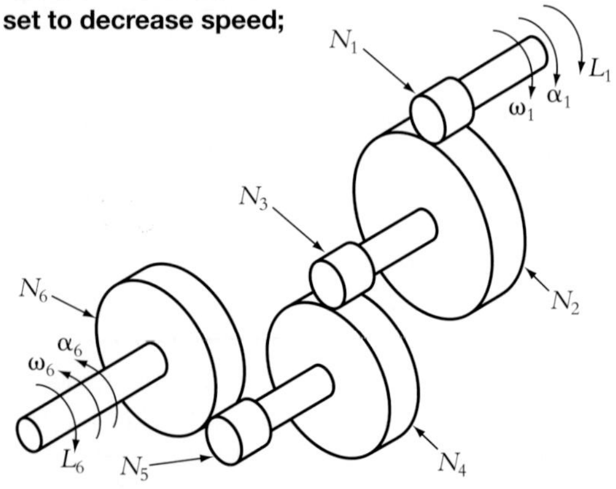

Figure 4.8 **A compound drive train**

Efficiency of geared systems

While real gears, toothed belts, chains etc. scale kinematic quantities (rotational angle, angular velocity, angular acceleration) according to the drive ratio, frictional forces and torques result in inefficiencies, so the ratios of torques do not in fact precisely follow the relationships given in equations (4.9) and (4.10). These equations may therefore be rewritten to take account of the efficiency η of the drive train:

$$\text{power out} = \text{efficiency} \times \text{power in}$$

i.e.

$$L_2\omega_2 = \eta L_1\omega_1 \quad \textbf{(4.14)}$$

so that input torque L_1 can be calculated from output torque L_2:

$$L_1 = \frac{L_2\omega_2}{\eta\omega_1} = \frac{L_2N_1}{\eta N_2} = \frac{L_2}{\eta n} \quad \textbf{(4.15)}$$

Efficiency is always less than unity, so input power is always greater than output power, with the difference being lost as heat. Typical efficiencies for geared systems may range from values close to unity (e.g. 0.95) for well-designed single-stage gear systems, down to much lower values (less than 0.5) for very high-ratio systems such as worm gear drives, which give a very compact reduction drive at the expense of involving significant sliding action and frictional losses. It must be emphasized however that kinematic quantities are unaffected by inefficiencies, so the following relationship still applies:

$$\text{Gear ratio } n = \frac{N_2}{N_1} = \frac{\omega_1}{\omega_2} = \frac{\alpha_1}{\alpha_2} \quad \textbf{(4.16)}$$

In a compound gear system, the overall efficiency of the gear train is the product of the efficiencies of each stage.

Worked example

A vehicle is driven from an electric motor via the compound gear system shown in Figure 4.9, where $N_1 = 20$, $N_2 = 100$, $N_3 = 25$, $N_4 = 75$. The efficiency of the first stage is 94 per cent and of the second stage is 96 per cent. If the final drive requires a torque of 100 Nm at 200 rev/min, what torque and speed are required from the motor?

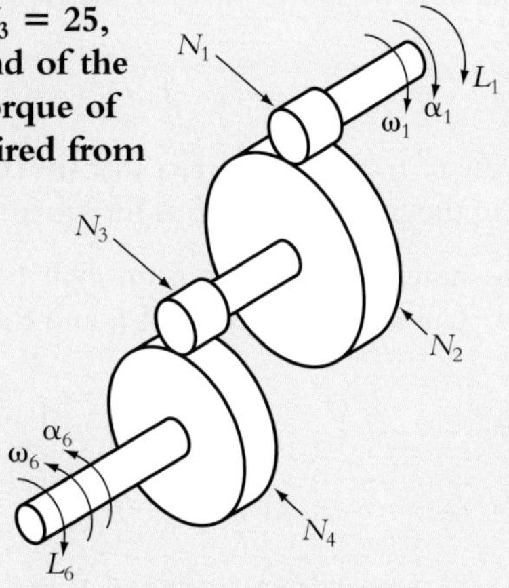

Figure 4.9 Compound gear system

Adapting equation (4.13) to the present situation:

$$\text{Drive ratio } n = \frac{N_2}{N_1} \times \frac{N_4}{N_3} = \frac{\omega_1}{\omega_4}$$

where $\omega_4 = 200$ rev/min

$$\therefore \quad \omega_1 = \frac{N_2}{N_1} \times \frac{N_4}{N_3} \times \omega_4 = \frac{100}{20} \times \frac{75}{25} \times 200$$

$$= 3000 \text{ rev/min}$$

Overall efficiency is the product of efficiency of the two stages so:

$$\eta = \eta_1 \times \eta_2 = 0.94 \times 0.96 = 0.902 \text{ or } 90.2\%$$

Adapting equation (4.15) to the present situation gives:

$$L_1 = \frac{L_4}{\eta n}$$

where $L_4 = 100$ Nm

and

$$n = \frac{N_2}{N_1} \times \frac{N_4}{N_3} = \frac{100}{20} \times \frac{75}{25} = 15$$

so that

$$L_1 = \frac{100}{0.902 \times 15} = 7.39 \text{ Nm}$$

Inertia of a geared system

Consider two shafts 1 and 2 (Figure 4.10) geared together with a perfectly efficient transmission of gear ratio n, with shaft 2 having mounted on it a moment of inertia J_2. Shaft 1 is driven by a motor, which causes the shafts to accelerate with angular accelerations α_1 and α_2 respectively.

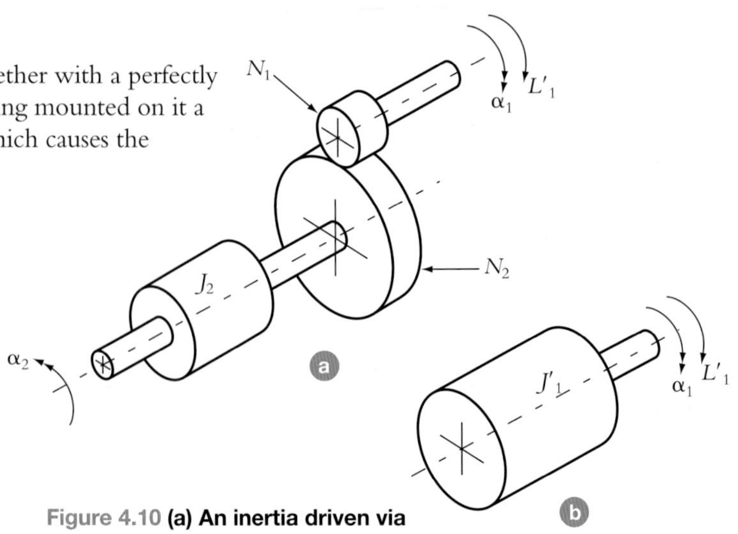

Figure 4.10 (a) An inertia driven via gears and (b) its inertia referred to the input shaft

The motor is assumed to provide a torque L_1 to shaft 1. This torque is scaled up by the gear ratio n and transmitted to shaft 2 causing inertia J_2 to accelerate with angular acceleration α_2. Because the shafts are geared together, the motor observes shaft 1 as accelerating with angular acceleration α_1. Overall, therefore, the motor is providing a torque L_1 which is causing an angular acceleration α_1 to the machine being driven. As far as the motor is concerned, this is simply an inertia load J_1' defined as:

$$J'_1 = \frac{L_1}{\alpha_1} \tag{4.17}$$

where (in general) the notation J' will be used for the referred or apparent inertia typically involving an indirectly driven inertia load. In the present case the load actually consists of inertia J_2 driven via the gear train of ratio n, experiencing a torque $L_1 n$. This causes J_2 to experience an angular acceleration of $\alpha_2 = (L_1 n)/J_2$ which is observed at shaft 1 as $\alpha_1 = \alpha_2 n = (L_1 n^2)/J_2$, It can therefore be shown that the inertia J_1' experienced by the motor is given by:

$$J'_1 = \frac{J_2}{n^2} \tag{4.18}$$

Of particular note is the n^2 term, and the fact that **inertias are scaled by the square of the gear ratio**, rather than the gear ratio itself, as for kinematic quantities as in equation (4.9).

By extension, a geared system with inertia J_1 on shaft 1 geared to shaft 2 carrying another inertia J_2 (Figure 4.11) will involve the sum of J_1 and the referred inertia due to J_2, in other words:

$$J'_1 = J_1 + \frac{J_2}{n^2} \tag{4.19}$$

is equivalent to:

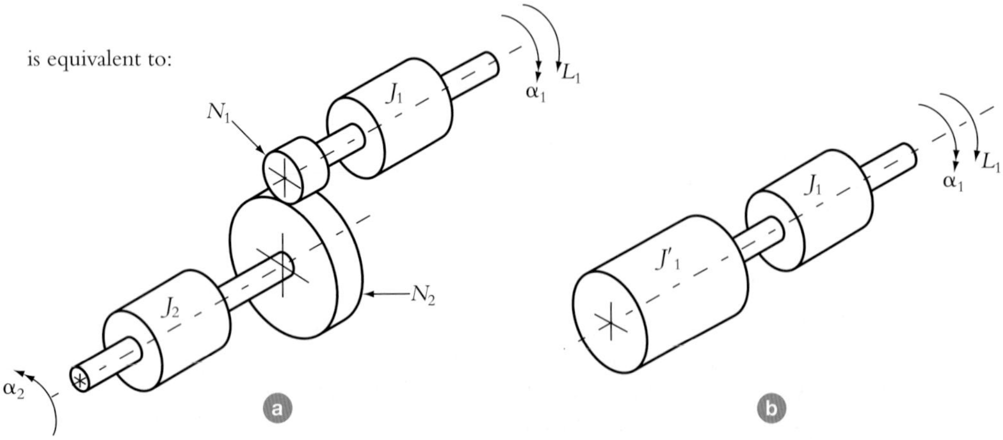

Figure 4.11 (a) A more complex system involving geared inertia; (b) summation of inertias referred to input shaft

Worked example

A machine (Figure 4.12) consists of two shafts A and B, each fitted with a flywheel of inertia $J_A = J_B = 1$ kg.m^2. Shaft B is driven from shaft A by a gear train of 3:1 reduction ratio. To a blindfolded observer, what does the system feel like if a handle is attached to shaft A (i.e. what is the inertia of the system referred to shaft A)?

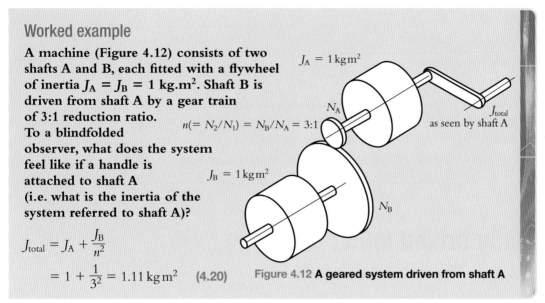

$$J_{total} = J_A + \frac{J_B}{n^2}$$

$$= 1 + \frac{1}{3^2} = 1.11\,\text{kg}\,\text{m}^2 \quad (4.20)$$

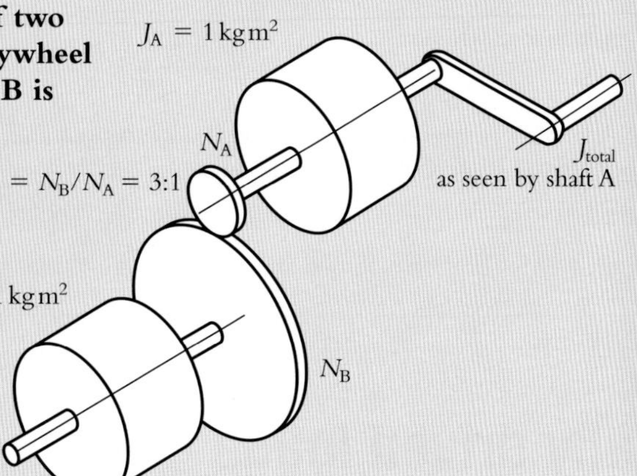

Figure 4.12 **A geared system driven from shaft A**

Worked example

The handle is instead attached to shaft B (Figure 4.13). What does the system feel like now (i.e. what is the inertia of the system referred to shaft B)?

Note that since the first shaft is now B and the second is A, the gear ratio $n = N_2/N_1$ is now 1:3 which is used in the calculation as 1/3:

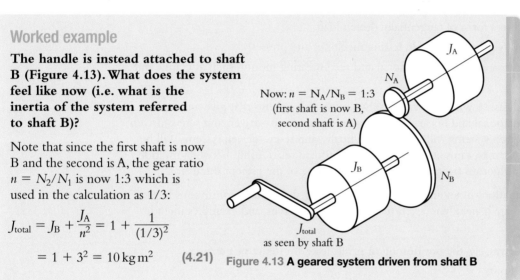

$$J_{total} = J_B + \frac{J_A}{n^2} = 1 + \frac{1}{(1/3)^2}$$

$$= 1 + 3^2 = 10\,\text{kg}\,\text{m}^2 \quad (4.21)$$

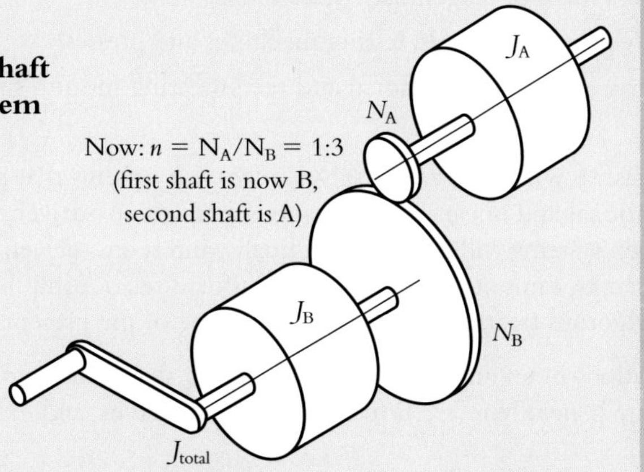

Figure 4.13 **A geared system driven from shaft B**

All the above concepts are applicable to pulley/belt and sprocket/chain drivetrains.

Influence of efficiency on the apparent inertia of geared systems

It has already been shown that practical geared systems (and similar systems) are not perfectly efficient, and this has important implications for considering the practical effects of inertia in such systems. For an imperfectly efficient version of the example shown earlier, torque L_1 results in inertia J_2 experiencing a reduced torque $\eta L_1 n$ and hence undergoing a smaller value of angular acceleration $(\eta L_1 n)/J_2$. Thus the ratio of torque to angular acceleration on shaft 1 can be shown to be:

$$\frac{L_1}{\alpha_1} = \frac{L_1 J_2}{\eta L_1 n^2} = \frac{J_2}{\eta n^2} = \frac{J_1'}{\eta} \quad (4.22)$$

Note that the true value of referred inertia J_1' is not affected by the inefficiency term – one cannot, for instance, store more rotational energy in a geared flywheel by connecting it via an inefficient gear train! However, it is useful to realize that a heavy system being accelerated via an inefficient transmission system will *appear* to have a higher value of inertia than if the transmission system were efficient. Conversely, it will appear to have a lower value of inertia when it is decelerating.

Learning summary

At the end of this section you should have learnt:

- ✓ the concept of a gear ratio;
- ✓ to relate angular velocities, angular accelerations and torques between input and output shafts of a geared system (or other system involving belts, friction drives etc.);
- ✓ the concept of 'referred inertia';
- ✓ the effect of inefficiency on the transmission of torque and on the apparent value of referred inertia.

4.5 Tangentially driven loads

There are numerous situations where rotational motion is converted to linear motion or vice versa, such as:

- tangentially driven systems involving belts, chains and ropes
- vehicles, which are also a form of tangentially driven load
- screw-driven systems (e.g. machine tools, testing machines and presses)
- mechanisms for converting between rotational and reciprocating motion such as cranks and cams.

Within the present analysis we will concern ourselves only with systems that give a constant relationship between rotational and linear motion, namely tangentially driven and screw-driven systems. Tangentially driven systems will be covered initially, and screw-driven systems will be also be covered briefly, Cranks, cams etc. involve a non-uniform relationship between rotation and displacement, and a rigorous treatment is beyond the scope of the present discussion.

There are numerous situations in engineering where a rotating shaft causes continuous linear motion. Such situations are generally referred to as **tangential drives**, and examples include:

- conveyer belts
- toothed belt linear drives, e.g. the positioning of print heads on inkjet printers
- cranes, winches and hoists
- vehicles.

Indeed, any belt or chain drive involves the tangentially driven mass of the belt or chain itself.

Tangential drives via belts, chains etc.

Consider the simplest tangential load situation where a mass m is driven by a light pulley of effective radius r via a light belt or cord (Figure 4.14). The inertia J' of this system referred to the axis of the pulley is:

$$J' = mr^2 \qquad \textbf{(4.23)}$$

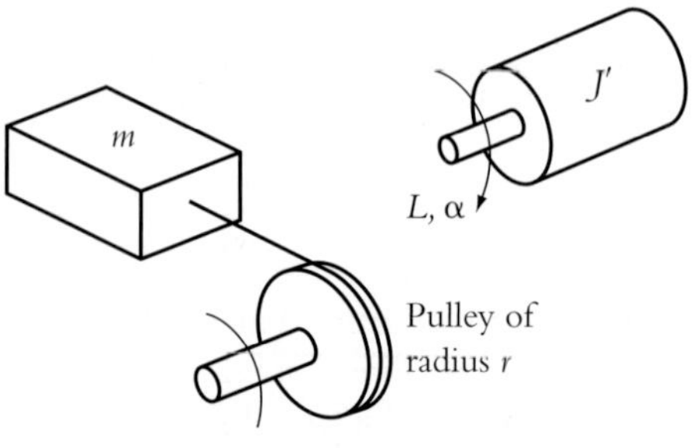

Figure 4.14 Tangentially driven mass

Practical tangential drive situations involve pulleys or sprockets, together with belts, conveyors or chains, of non-negligible mass, and each of these contributes to the total inertia. For instance, consider a conveyor belt of mass m carrying a load of mass M, mounted on a pair of rollers, one of which is connected to the drive input shaft and has moment of inertia J_1 and radius r_1, with the second being an idler of inertia J_2 and radius r_2 (Figure 4.15).

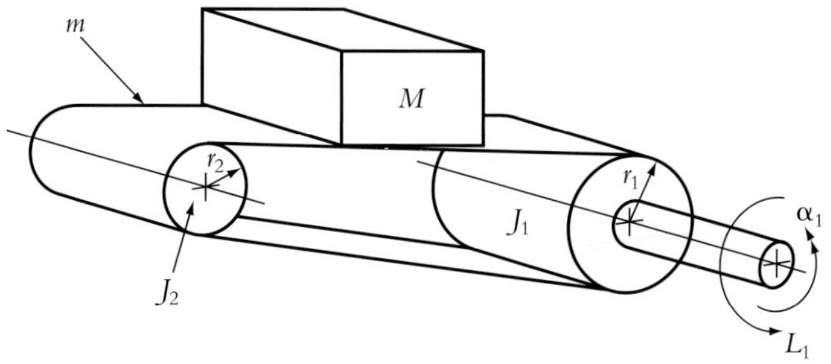

Figure 4.15 **Conveyor belt system**

Each of these contributes to the inertia J' of the system referred to the input shaft:

- the roller connected to the input shaft obviously contributes a moment of inertia J_1;
- the load of mass M is driven tangentially by the first roller of radius r_1 so contributes Mr^2;
- the conveyor belt of mass m is also driven tangentially by the first roller so contributes mr^2;
- the idler roller is driven (via the conveyor belt) from the first roller with a pulley ratio of $r_2:r_1$ so its inertia referred to the drive input shaft is $J_2(r_1/r_2)^2$.

Therefore, the total moment of inertia referred to the input drive shaft is:

$$J' = J_1 + (M + m)r^2 + J_2(r_1/r_2)^2 \qquad (4.24)$$

Worked example

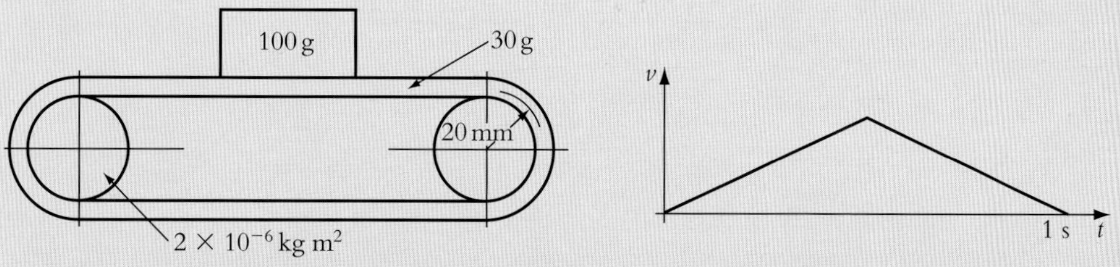

Figure 4.16 **(a) Tangentially driven probe assembly showing belt and pulleys; (b) uniform acceleration/deceleration profile**

A probe assembly of mass 100 g on an item of laboratory equipment is to move 0.2 m in 1 second. It is driven by a toothed belt that weighs 30 g and has an effective radius of 20 mm where it passes over the pulleys (Figure 4.16(a)). A uniform acceleration/deceleration (triangular) velocity profile is to be used (Figure 4.16(b)). Each pulley has a moment of inertia of 2×10^{-6} kg.m^2

(a) Calculate the inertia of the system referred to the axis of the drive pulley.

(b) Calculate the maximum acceleration of the probe and hence the maximum angular acceleration of the pulley.

(c) Calculate the torque required to cause the desired acceleration.

(a) $J' = 2 \times J_{\text{pulley}} + (M + m)r^2$

$= 2 \times 2 \times 10^{-6} + (100 \times 10^{-3} + 30 \times 10^{-3}) \times 0.02^2 = 5.6 \times 10^{-5}\,\text{kg m}^2$

(b) The probe covers 0.2 m in 1 s divided equally between acceleration and deceleration, therefore acceleration phase covers 0.1 m in 0.5 s from rest.

Applying $s = ut + \frac{1}{2}at^2$ gives

$$a = \frac{2s}{t^2} = \frac{2 \times 0.1}{0.5^2} = 0.8\,\mathrm{m\,s}$$

Angular acceleration is therefore given by

$$\alpha = \frac{a}{r} = \frac{0.8}{0.02} = 40\,\mathrm{rad\,s^{-2}}$$

(c) Torque is given by

$$L = J'\alpha = 5.6 \times 10^{-5} \times 40 = 0.002\,24\,\mathrm{Nm}$$

Vehicles

A very important type of tangentially driven load is a vehicle running on a road (or rail or other surface), and it can be useful to consider the effective inertia of a vehicle referred to (for example) the vehicle's axle, or the engine's output shaft. There are two contributions to the inertia J' referred to the axle of a vehicle (Figure 4.17), ignoring inefficiencies in the wheel/road contact, such as hysteresis in a rubber tyre:

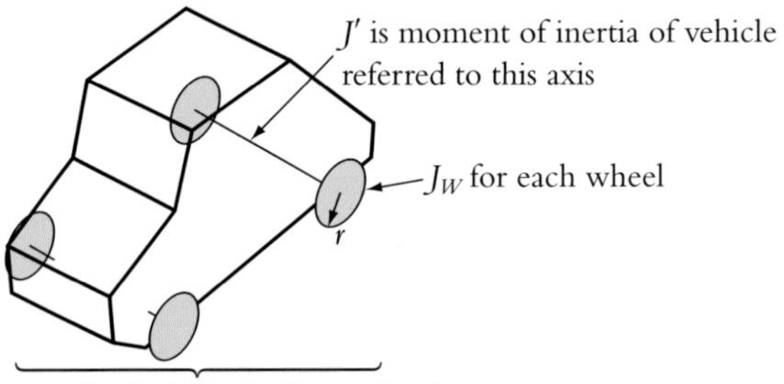

Figure 4.17 **Simplified diagram of a vehicle showing contributions to inertia referred to axle**

- mr^2 where m is the overall mass of the vehicle (**including the wheels**) and r is the effective radius of the wheels where they contact the road
- NJ_W where N is the number of wheels and J_W is the inertia of each wheel (whether the wheels are directly mounted on the axle or not – coupling between the wheels is provided via their contact with the road or rail)

This gives

$$J' = mr^2 + NJ_W \tag{4.25}$$

In practice, the contribution due to the wheels is likely to be small compared with that of the overall mass of the vehicle.

Screw drives

These provide an alternative to tangential drives where only a small movement is desired for each revolution of the input shaft. They are widely used in machine tools for accurate positioning of slideways (where they are known as leadscrews), and within the aerospace industry for actuation of landing gear.

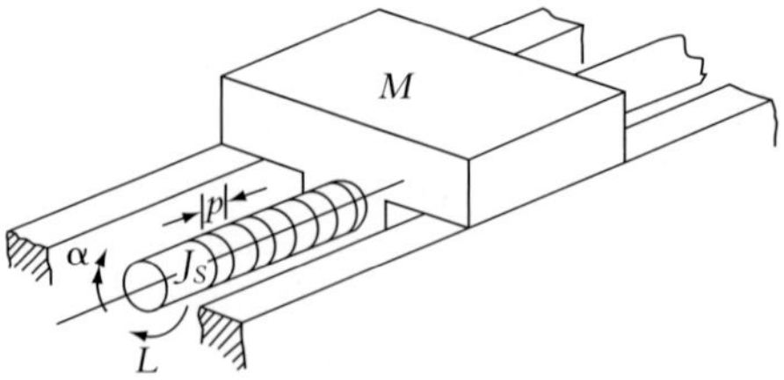

Figure 4.18 **System driven by a leadscrew**

The apparent moment of inertia of a screw-driven system (Figure 4.18), referred to the leadscrew axis, depends on the leadscrew pitch p, the mass M being driven and the efficiency η of the leadscrew as well as the moment of inertia J_s of the screw itself. The torque for a given angular acceleration α of the screw is therefore:

$$L = \left[J_s + \frac{Mp^2}{4\pi^2\eta}\right]\alpha \tag{4.26}$$

Unless the load is very heavy or the pitch is large, the biggest contribution to the total moment of inertia is often from the inertia J_s of the screw itself.

Learning summary

By the end of this section you should have learnt:

✓ to calculate the referred inertia of a tangentially driven system

✓ the inertia behaviour of screw-driven systems.

4.6 Steady-state characteristics of loads

So far, the analysis has only considered the effects of inertia. With this theory alone, it could be assumed that, once a machine has been accelerated to its running speed, it will carry on running for ever with no further input of energy. Clearly this is not true, and it is now necessary to consider the various contributions to frictional losses that tend to oppose movement when the machine is running.

Frictional effects and losses

Three types of frictional losses are typically encountered within machines:

Coulomb friction

This is based on Coulomb's well-known law for contacting bodies (Figure 4.19) that are able to move linearly with respect to each other, stating that the limiting frictional force F for two bodies experiencing contact force N is given by

$$F = \mu N \tag{4.27}$$

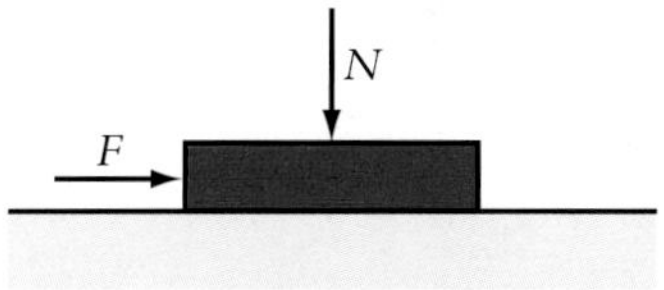

Figure 4.19 **Block in contact with a surface**

In fact, two distinct regimes can be considered:

- static friction ('stiction'), the force required to get two bodies moving with respect to each other; in effect this is the force required to break any intermolecular or interatomic bonds that have built up between the bodies over time;
- sliding friction or dynamic friction, a frictional force between two objects that are already sliding over each other; this remains (in theory) constant irrespective of the relative speed of the two objects.

In practice, the coefficients of static friction, μ_S, and dynamic friction, μ_D, tend to be noticeably different (it takes a slightly smaller force to keep two components moving with respect to each other, compared to the force required to get them to move). Typical values might be $\mu_S = 1.5\ \mu_D$. However, as stated above, μ_D may be assumed to be independent of relative speed, so the force-speed relationship may be represented graphically as shown in Figure 4.20:

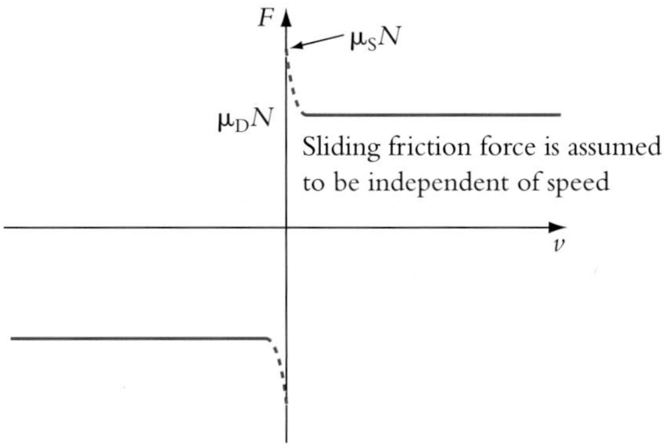

Figure 4.20 **'Stiction' and dynamic friction for linear movement**

Mathematically, this can be represented for the static situation as:

$$-F_S \leqslant F \leqslant F_S \text{ for } v = 0 \tag{4.28}$$

where

$$F_S = \mu_S N \tag{4.29}$$

and for the dynamic situation:

$$F = F_D \quad \text{for} \quad v > 0 \tag{4.30}$$

and

$$F = (-F_D) \quad \text{for} \quad v < 0 \tag{4.31}$$

where

$$F_D = \mu_D N \tag{4.32}$$

While components undergoing linear sliding certainly are encountered in engineering (for example machine tool saddles moving with respect to the stationary bed, pistons sliding in cylinders), the above concepts are equally applicable to rotational movement, and the characteristics shown earlier may be represented graphically in Figure 4.21 and mathematically as:

$$-L_S \leqslant L \leqslant L_S \quad \text{for} \quad \omega = 0 \tag{4.33}$$

$$L = L_D \quad \text{for} \quad \omega > 0 \tag{4.34}$$

and

$$L = (-L_D) \quad \text{for} \quad \omega < 0 \tag{4.35}$$

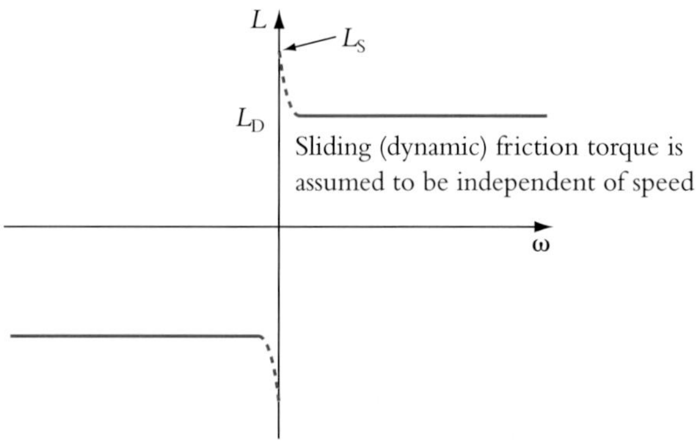

Figure 4.21 'Stiction' and dynamic friction for rotational movement

Viscous friction

Viscous friction is caused by the shearing of a Newtonian fluid, e.g. a lubricating oil that separates the two bodies that are 'sliding'. Newton's model of viscous friction states that the shear stress on a fluid is proportional to the shear strain rate of the fluid. For a linear bearing properly lubricated with a Newtonian fluid, this implies that the viscous friction force is proportional to the relative velocity of the two components. For linear motion of two plates of area A separated by distance d the force F required to maintain a relative velocity v (sometimes known as the 'drag force') is:

$$F = \frac{Av\mu}{d} \tag{4.36}$$

where here μ is used to represent the viscosity (strictly the dynamic viscosity) of the fluid.

This gives a simple linear relationship between force and velocity which can represented as:

$$F = a_1 v \tag{4.37}$$

This can be shown graphically (Figure 4.22) as:

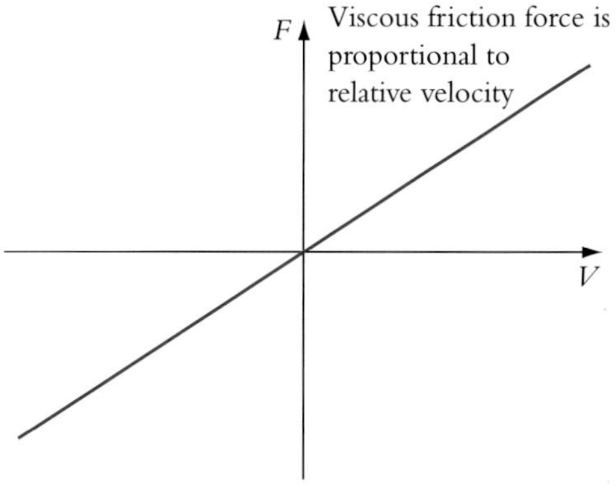

Figure 4.22 Viscous friction for linear movement

Again, it is useful to represent this in rotational form (Figure 4.23):

$$L \propto \omega \tag{4.38}$$

or

$$L = b_1 \omega \tag{4.39}$$

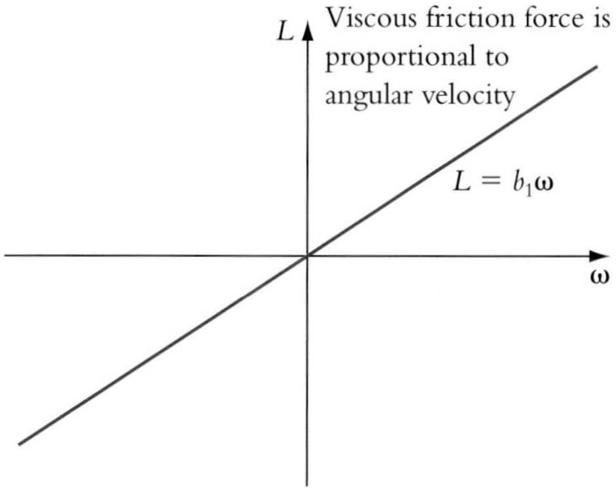

Figure 4.23 Viscous friction for rotational movement

Windage

Windage is a force caused by aerodynamic effects such as turbulence, e.g. drag on a car, windage drag on a motor due to its cooling fan and to the 'churning up' of air in its casing. The magnitude of windage force is proportional to the **square** of velocity, noting that the direction of the force opposes the direction of movement.

$$|F| \propto v^2 \tag{4.40}$$

This is represented graphically in Figure 4.24.

For a body of frontal area A moving through fluid of density ρ with velocity v the windage force it experiences is:

$$|F| = \tfrac{1}{2}\rho v^2 A C_d \tag{4.41}$$

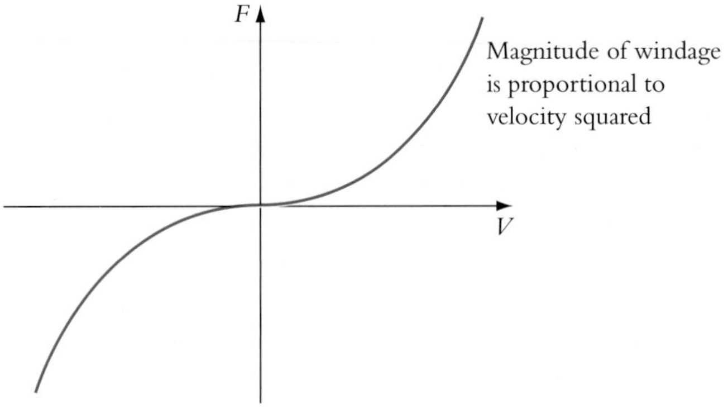

Figure 4.24 **Windage for linear movement**

where C_d is an approximately constant value called the **coefficient of drag**, which depends upon the shape of the body, and is in the order of 1 or slightly larger for 'bluff' bodies (such as cubes, plates etc.) going down to much lower values for aerodynamically designed, streamlined shapes such as aircraft. Such coefficients are widely tabulated in the fluid mechanics literature and are presented in Unit 1 of the present book. In simplified form this kind of relationship can be represented as

$$|F| = a_2 v^2 \quad (4.42)$$

In a similar manner for rotation, torque is proportional to the square of angular velocity:

$$|L| \propto \omega^2 \quad (4.43)$$

or

$$|L| = b_2 \omega^2 \quad (4.44)$$

This can be represented graphically in Figure 4.25.

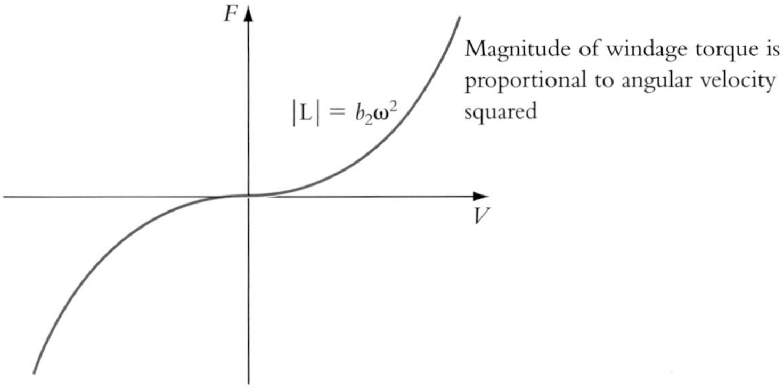

Figure 4.25 **Windage for rotational movement**

Summary of frictional effects

For linear motion, assuming motion in a positive direction:

- at rest/threshold of movement: $F = F_S$ = static friction force
- at velocity v: $F = F_D + a_1 v + a_2 v^2$

i.e. total frictional force = Coulomb friction force + viscous friction force + windage.

For rotation in a positive direction:

- at rest/threshold of movement: $L = L_S$ = static friction torque
- at angular velocity ω: $L = L_D + b_1\omega + b_2\omega^2$

i.e. total frictional torque = Coulomb friction torque + viscous friction torque + windage.

Note that for a tangentially-driven system the force-speed relationship can be transformed to a rotational characteristic by substituting $T = Fr$ and $v = \omega r$, giving (for example):

$L = F_D r + a_1 r^2 \omega + a_2 r^3 \omega^2$

Characteristics of fans, blowers, pumps etc.

Not surprisingly, windage-type effects are the predominant contribution to the torque–speed characteristics of rotodynamic fans and blowers. Broadly similar characteristics are also encountered for similar devices such as centrifugal pumps, such as those encountered in many chemical plants. It should be noted that, in approximate terms, the pressure that can be supplied by rotodynamic pumps, blowers, fans etc. is also proportional to the square of the speed. However, an important issue is that if the torque required to drive such a device is proportional to the square of the speed, then the power (which is given by torque × angular velocity) will increase with the cube of the speed. This has very important implications for energy efficiency, in that considerable savings can be made by reducing the speed of such devices to the minimum which will deliver the desired performance, rather than throttling their flow rates or allowing them to supply higher pressures and flow rates than are required.

Note that the issues surrounding the torque–speed characteristics of positive-displacement pumps, compressors etc. are rather different. The torque required to drive them is dominated by the pressure or head against which they are pumping: if this is constant (for example, if supplying a header tank), the torque required (when averaged over a cycle) will not change greatly with speed; if the pressure changes with flow rate, then the torque will also change.

Other contributions to steady-state running characteristics

While it is clearly impossible to list all possible contributions to running characteristics, some common examples include gravitational load, for example due to items being carried upwards on a sloping conveyor, winched upward using a hoist, or where a vehicle is climbing a slope. Such a contribution to load torque will be independent of speed, and can be found using simple engineering mechanics (refer to Unit 6 of *An Introduction to Mechanical Engineering: Part 1*).

Example: Torque on a high-speed rotating antenna

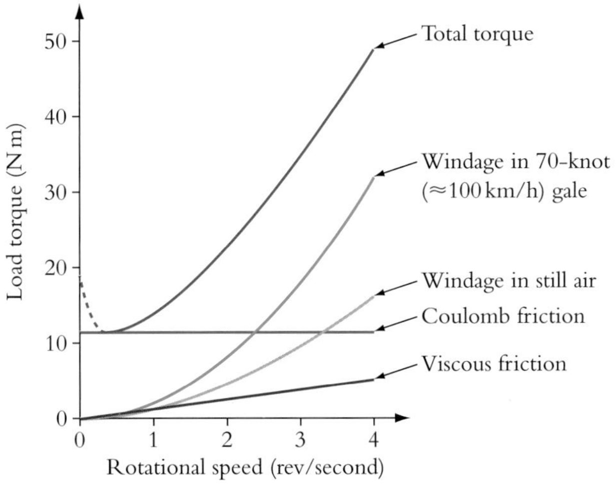

Figure 4.26 Contributions to torque–speed characteristics of a high-speed rotating antenna (Adapted from Blackburn et al., 1960)

Figure 4.26 illustrates the characteristics of a system that exhibits all of the above contributions. It is interesting to note that the windage element varies with the ambient wind speed.

Some positive aspects of friction – it wastes energy, but is it always a bad thing?

Coulomb friction, viscous friction and windage all involve conversion of mechanical power into heat. Within mechanical power transmission situations this is normally undesirable as it involves wasted energy, yet friction is usefully harnessed as an essential feature of several engineering devices:

- Brakes, which apply a frictional torque to oppose motion, are widely used to convert unwanted kinetic energy (for instance, in a vehicle) or potential energy (in the load of a crane or elevator) into heat, which clearly needs to be dissipated.
- Clutches are used to provide a means of temporarily matching differing speeds while transmitting a given value of torque. This is achieved by frictional coupling of the input shaft to the output shaft via two contacting plates. The function of a clutch will be examined further in Section 4.13.
- Worm gears (Figure 4.27) are widely used to provide a very large speed reduction ratio (and torque increase) between a rapidly rotating input shaft and a much more slowly rotating output shaft. This is achieved in much less space than would be possible via conventional (spur or helical) gears, which would require the use of compound gearing to achieve the required gear ratio. Unlike conventional gears, which involve primarily rolling friction between gear teeth, worm gears involve a primarily sliding action as the screw-like worm rotates against the flanks of the teeth of the worm wheel, driving it around. This friction results in the undesirable feature of very low efficiency (values of 30 per cent are typical for worm gear assemblies). However, friction also accounts for an important feature of most worm gear systems which contrasts with conventional gear systems: provided the helix angle of the worm is sufficiently shallow (typically corresponding to the use of as single-start or possibly two-start worm) they can normally only transmit drive in one direction through the power train, with the drive going only from the input shaft to the output shaft. In other words they are self-locking if an attempt is made to provide the drive from the output shaft to the input shaft. This makes them a particularly useful feature of devices such as hoists, cranes, torsional testing machines etc. For example, if the drive to a worm-driven hoist fails, the load will not descend as the worm gear assembly will lock up. Similarly, a worm-driven torsional testing machine will retain the torsional loading on the specimen if no input torque is applied. In both of these cases it is said that the drive system cannot be 'over-hauled'. There are exceptions to this: in some cases a multistart worm is used to provide a compact speed increase gear system. These are relatively rare in engineering – perhaps the classic example of such a system was within the speed-limiting governor used in mechanical telephone dials up to the 1970s, where a rapidly rotating centrifugal brake was operated from the slowly turning dial via a reverse-driven worm gear mechanism.

Figure 4.27 Worm mechanism

Learning summary

By the end of this section you should have learnt:

- ✓ the different contributions to steady-state running characteristics of loads;
- ✓ to express these different contributions to the load characteristics in the form of a mathematical expression, both for linear and rotational motion;
- ✓ that friction can have a beneficial role in some situations as well as having a detrimental effect on efficiency in other situations.

4.7 Modifying steady-state characteristics of a load using a transmission

In order to match the torque–speed characteristics of a load to the characteristics of the mechanical power source that will drive it, it is often necessary to drive the load via a transmission of some kind, which may take the form of gears, belts, chain drive etc. as described in Section 4.4. It is therefore very useful to find the characteristics of the load referred to the input shaft of the transmission. This is straightforwardly achieved by substituting the relations for torque and speed given in equation (4.9) (or, if inefficiencies are taken into account, the torque relations given in equation (4.15)) into the equation giving the load's torque–speed relationship. For example, consider a load with the torque–speed (or, strictly, torque–angular velocity) characteristic as follows:

$$L = L_D + b_1\omega + b_2\omega^2 \qquad (4.45)$$

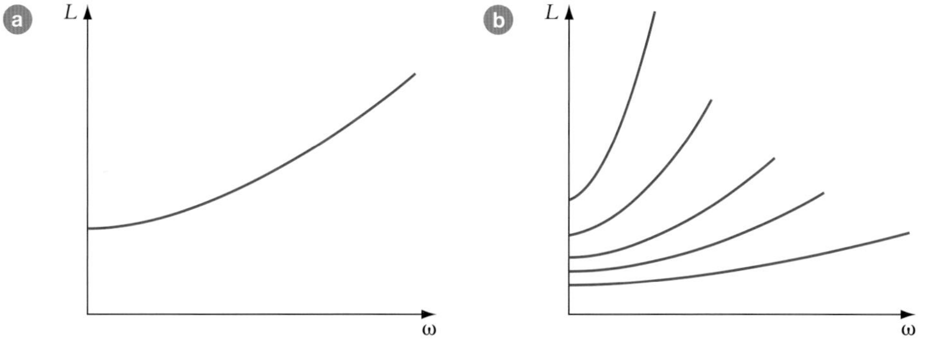

Figure 4.28 **(a) Typical torque–speed curve for a directly driven machine involving dynamic Coulomb friction, viscous friction and windage (b) family of torque-speed curves**

Such a characteristic is shown diagrammatically in Figure 4.28(a). The load is driven via a transmission with a ratio of n:1 and an efficiency of η, and it is desired to find the characteristic of the system expressed as the relationship between torque L' and speed ω' at the input shaft of the transmission. Making use of equation (4.9), ω is expressed in terms of ω':

$$\omega = \frac{\omega'}{n} \qquad (4.46)$$

Similarly, making use of equation (4.15), L' is expressed in terms of L as:

$$L' = \frac{L}{n\eta} \qquad (4.47)$$

Inserting these into equation (4.45) gives:

$$L' = \frac{1}{n\eta}(L_D + b_1\omega + b_2\omega^2)$$

$$= \frac{1}{n\eta}\left(L_D + b_1\frac{\omega'}{n} + b_2\frac{(\omega')^2}{n^2}\right)$$

$$= \frac{L_D}{n\eta} + b_1\frac{\omega'}{n^2\eta} + b_2\frac{(\omega')^2}{n^3\eta} \qquad (4.48)$$

It is instructive to observe that a family of torque–speed curves (Figure 4.28(b)) can be constructed as the ratio n is varied: a low value of n will cause the load to require high values of torque to be provided at a relatively low range of speeds, while a large value of n will require lower values of torque to be provided, but with the drive being at higher speeds.

It is crucial to note that changing the transmission ratio does *not* alter the power requirements of the load – it merely allows the characteristics of the load to be moved to provide a better match to those of the mechanical power source. The issue of matching of load and power source will be explored further in Section 4.13.

Learning summary

By the end of this section you should have learnt:

✓ to refer the torque–speed characteristics of a load to the input shaft of a transmission system in order to obtain the characteristics observed by the mechanical power source driving it;

✓ that a transmission will affect the combination of torque and speed required to drive a load but will not help to overcome a shortfall in the power available for providing the drive.

4.8 Sources of mechanical power and their characteristics

In some cases, a machine will be driven manually via a handle, pedals etc., but in most cases some other source of mechanical power will be needed. There are various categories of such systems, and it can be useful to distinguish them. Exact terminology can vary (see Chambers (2007) *Science and Technology Dictionary* for one set of formal definitions) but as a guideline:

1. The term 'motor' is usually used to describe a machine that causes motion or generates mechanical power, often drawing its energy from some other easily managed form (electrical, hydraulic, pneumatic, possibly chemical). Within the present unit, we will consider a variety of electric, pneumatic and hydraulic motors. However, the term 'motor' historically also covered internal combustion engines, leading to the formal term 'motor vehicle' and related terms such as 'motorist' etc.
2. Prime movers are devices that convert a natural source of energy into mechanical power; these can include engines which make use of burning fuel, but could also include wind turbines, water turbines etc.
3. The term 'engine' tends to be reserved for a machine that creates mechanical power from heat energy; examples are internal combustion engines, jet engines (a form of gas turbine), steam engines etc. Within the present discussion we will pay some attention to the characteristics of internal combustion engines, though reference will be made to the very different characteristics of other forms of engines.

In order to be able to design a drive system, it is important to understand how the torque available from a source of mechanical power varies with speed. With this information it is then possible to analyse the way in which the mechanical power source interacts with the load it is driving.

Characteristics of the internal combustion engine

Although there is no immediate derivation of its torque–speed characteristics, the internal combustion engine is one of the most important prime movers. Its torque–speed characteristics vary depending upon the position of the throttle (on a petrol engine) or the setting of the fuel injection pumps (diesel engine). It is usual to give only the maximum torque–speed characteristics (for a petrol engine this is known as the wide open throttle (WOT) condition); it is not usual to give the torque–speed characteristics for different settings of the throttle etc., but for illustrative purposes, possible curves of this type are included in Figure 4.29. (Note that it is also usual to present internal combustion engine characteristics such as those in Figure 4.29 in terms of a quantity known as 'brake mean effective pressure', a quantity related to torque and to the engine capacity. For simplicity, however, graphs of torque and speed are plotted directly here.)

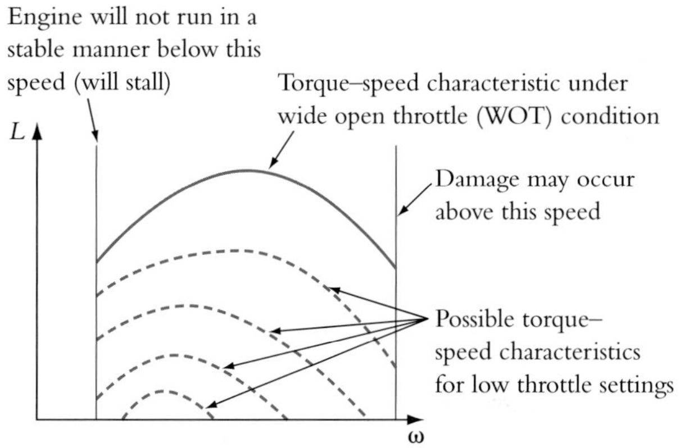

Figure 4.29 **Hypothetical set of torque–speed characteristics for an internal combustion engine at different throttle settings**

A particular feature of internal combustion engines is that they cannot run in a sustained manner at low speeds: instead they *stall*, a phenomenon very familiar to learner drivers! Rather than represent the infinite set of characteristics, it is usual to represent instead the maximum possible torque (under wide open throttle conditions etc.) for each speed, and to fill the area under the curve with a 'map' showing the *specific fuel consumption* (SFC) of the engine for each combination of torque and speed (Figure 4.30 a and b). This gives the amount of fuel consumed by the engine per unit of mechanical energy produced (usually expressed in g/kWh) It may be noted that there is an optimal combination of torque and speed (close to the maximum torque value, but well below the maximum safe running speed) at which the engine runs at maximum efficiency, consuming the least amount of fuel for a given output.

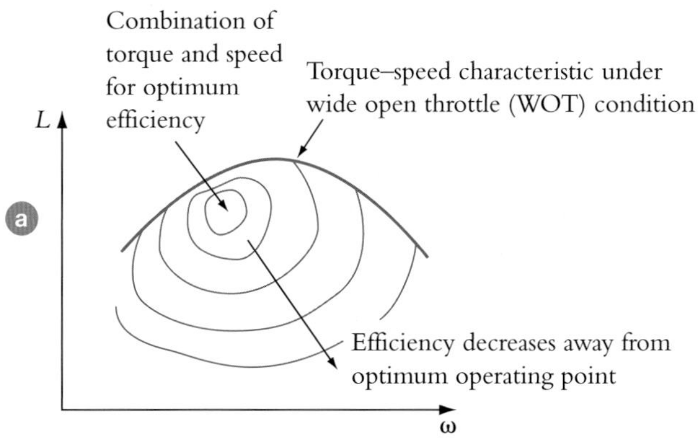

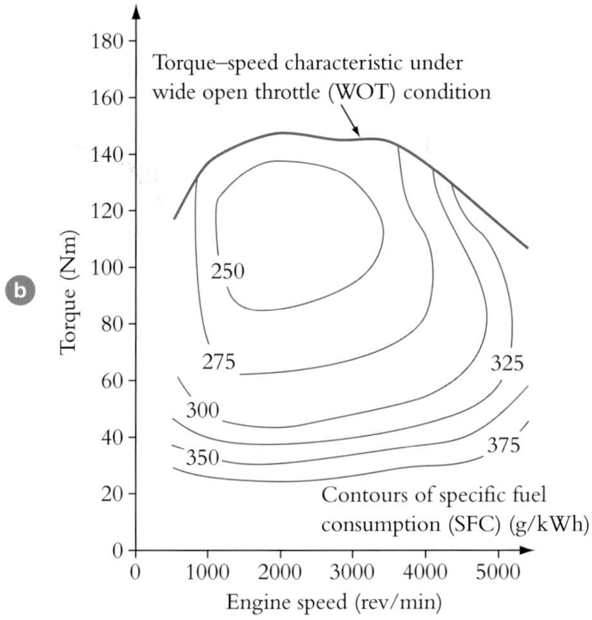

Figure 4.30 **(a) Features of a torque-speed-SFC map for an internal combustion engine; (b) torque–speed–SFC map for a 1.9-litre internal combustion engine (adapted from Shayler et al., 1999)**

The great advantage of the internal combustion engine is its ready availability and independence from other sources of power; its most obvious applications are in motor vehicles, ships and railway locomotives. In addition, diesel engines in particular are often used for powering generators for medium-sized communities and for standby pumping or power generation applications. However, rather than using a prime mover such as an engine, it is often far more appropriate to use some kind of motor which is supplied with power in the form of electricity or pressurized fluid and delivers mechanical power locally in a controllable way.

Learning summary

By the end of this section you should have learnt:

✓ the different types of mechanical power sources used within drive systems;

✓ the meaning of a torque–speed–SFC diagram for an internal combustion engine, and in particular understand the implications of its main features.

4.9 Direct current motors and their characteristics

Electric motors provide the drive for many mechanical systems, and generally provide the most convenient source of power. Each type of motor has a different torque–speed characteristic with some (e.g. dc shunt or permanent-magnet motor, or ac induction motor) providing good running characteristics, and others (e.g. dc series motors) providing much better starting characteristics. For each application the engineer must choose a machine that will provide the desired torque over the full speed range of the mechanical system. This section examines direct current (dc) motors, which are divided into five categories: shunt, series, compound, separately excited and permanent magnet motors. Shunt motors run at relatively constant speed and are ideal for use with conveyors, pumps, compressors and rolling mills. Series motors are used for traction, lifts and other applications where a large starting torque is required, and can also be used with ac supplies (when they are known as 'universal' motors). Compound motors combine the characteristics of shunt and series motors, while separately excited motors provide accurate speed control. Permanent magnet motors have similar characteristics to shunt wound motors but make use of permanent magnets to provide the stationary magnetic field, rather than generating the field electromagnetically.

In order to understand how dc motors work, it is necessary to have an understanding of electromagnetism and to be familiar with the basic equations associated with such systems. These prerequisites are covered in Part 1, Unit 5 and the reader is encouraged to refer to that material.

Because this section (along with Sections 4.10 and 4.11) overlaps the disciplines of engineering mechanics and electrical systems, there is a conflict of notation in that mechanical engineers use the symbol L to represent torque, while electrical engineers use L to represent inductance. In order to avoid confusion within the present section and Sections 4.10 and 4.11, the word 'torque' will generally be used explicitly in preference to the use of a symbol, though the symbol T will occasionally be used for torque within a formula. Also, the symbol n will be used to represent speed in rev/min.

Construction of dc motors

The basic dc motor has two parts; a stationary field and a rotating armature. The field generated either by electromagnets constructed of steel laminations, with salient poles, around which coils are wound or by permanent magnets, which replace the poles and associated coils. Figure 4.31(a) shows a four-pole motor with two N-poles and two S-poles (indicating north-seeking and south-seeking poles respectively). Direct currents flow in the coils creating a static magnetic field which links with the armature winding. The armature has circular steel laminations fixed to a central steel shaft. Slots are cut in the outer circumference as shown in Figure 4.31(b). Also fixed to the shaft, adjacent to the steel core is a commutator constructed of copper segments that are separated by thin sheets of insulation. Coils are laid in the slots in the steel laminations and their ends brazed to individual commutator segments. Carbon brushes connect the commutator to an external electric supply. Figure 4.31(c) shows a partially completed motor.

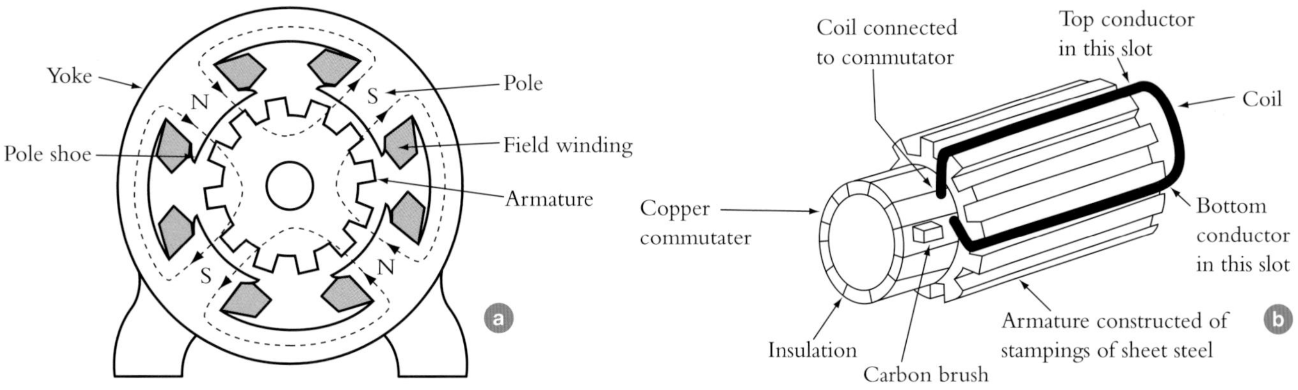

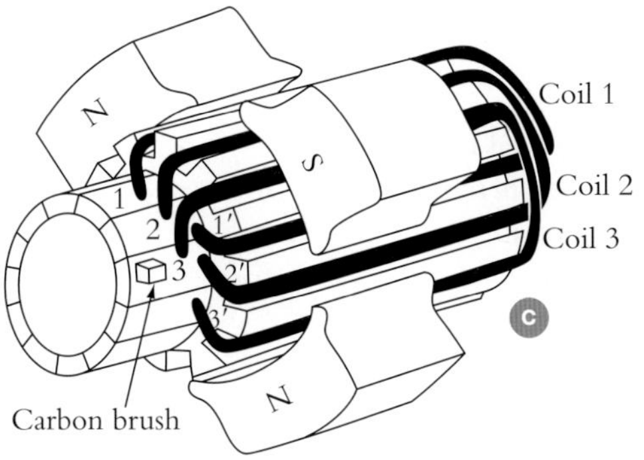

Figure 4.31 **Construction of a dc motor: (a) cross-section through a four-pole motor; (b) simplified diagram of a single armature winding; (c) partially completed motor showing relationship between armature and stator poles**

Operation of dc motors

The simplest dc motor has two poles, an armature with two conductors, 1 and 2, forming a single coil and a commutator with two segments, as shown in Figure 4.32.

An external direct voltage source supplies a current, I_f to the field winding. This establishes a magnetic flux, ϕ between the two poles. The flux passes through both air and steel. Its magnitude is a function of the field current, I_f:

$$\phi = \text{Fn}(I_f) \qquad \textbf{(4.49)}$$

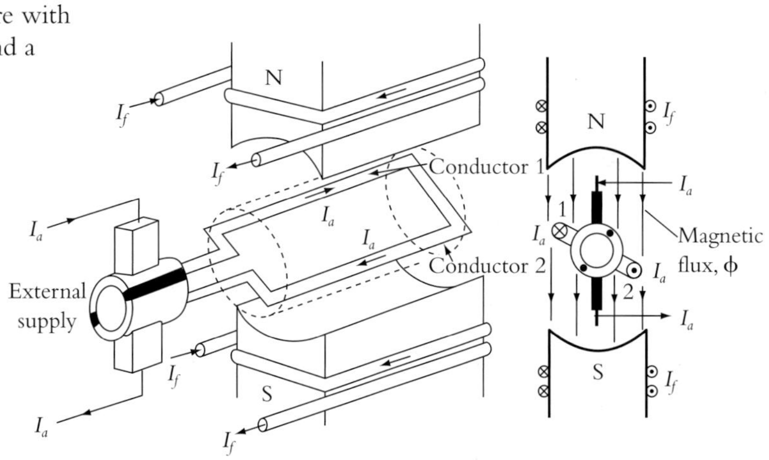

Figure 4.32 **Simplest form of dc motor showing armature and stator windings**

Because the B–H (flux density vs magnetic field strength) characteristic for steel is non-linear the relationship between ϕ and I_f is also non-linear. Another direct voltage supply feeds current, I_a to the armature winding via the carbon brushes. The armature current produces subsidiary magnetic fields that circulate around the armature conductors. Figure 4.33(a) shows a cross section of the conductors encircled by the subsidiary fields and the main field passing from the N-pole to S-pole. The subsidiary fields interact with the main field to produce the resultant field shown in Figure 4.33(b). This produces forces on the conductors, which create a torque on the armature making it rotate.

By supplying current to the armature via carbon brushes and commutator the currents flowing through the individual armature conductors reverse every time the coil is perpendicular to the main field, maintaining the torque in the same direction. Figure 4.34(a) to (e) show the change in current direction as the armature rotates. In Figure 4.34(a) and (b) the current in conductor

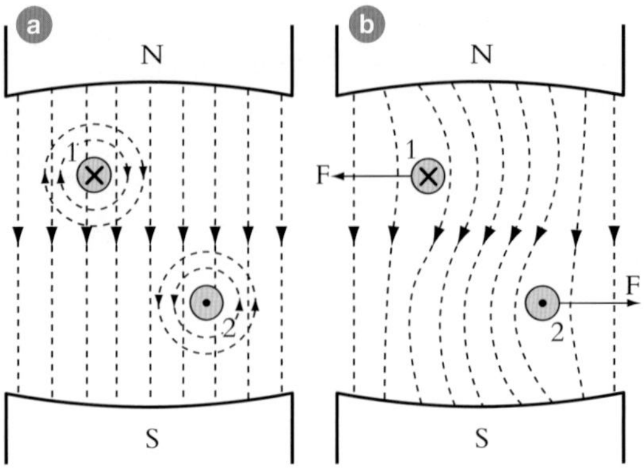

Figure 4.33 **Generation of forces on a pair of conductors carrying current into the page (⊗) and out of the page (⊙)**

1 flows 'into the page' and the current in conductor 2 flows 'out of the page' making the coil rotate anticlockwise. When the coil passes the horizontal (Figure 4.34(c)) the currents in the two conductors are reversed by the switching action of the commutator to maintain the torque and hence the rotation in the same direction.

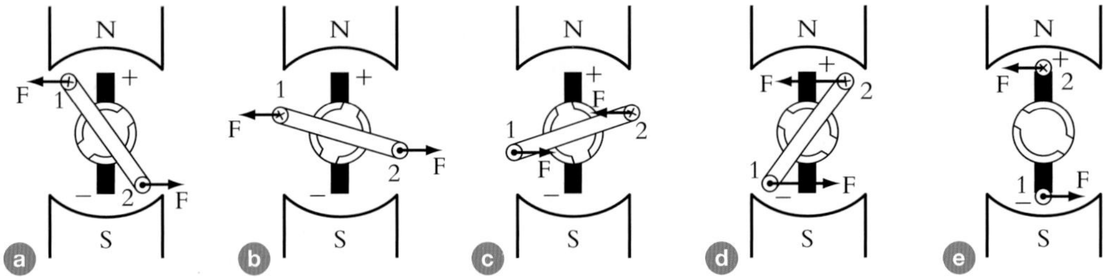

Figure 4.34 **Forces on armature windings at different armature orientations**

Emf induced in the armature winding

As the armature coil rotates, an emf is induced in it which tends to oppose the externally supplied armature current, I_a. The external supply voltage must overcome this emf if the machine is to motor and deliver mechanical power through the shaft.

Figure 4.35 shows the variation in flux linkage between the main field and the coil as the latter rotates. In Figure 4.35(a), when the coil is perpendicular to the field, there is maximum flux linkage, ψ_m. With only one coil this equals the flux between the poles, ϕ.

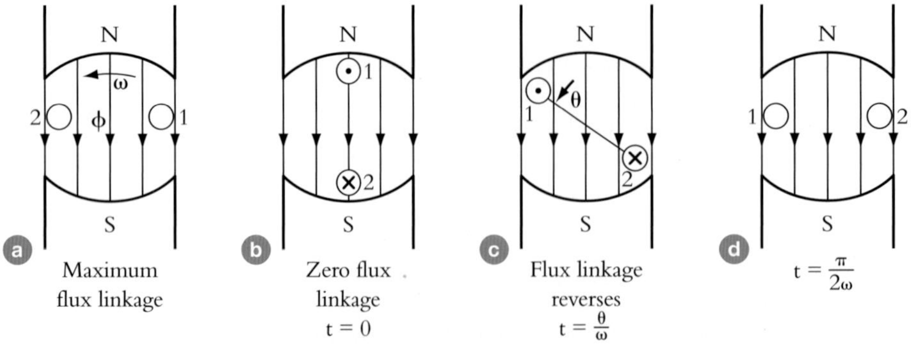

Figure 4.35 **Variation in flux linkage between the main field and armature coil**

When the coil is parallel to the field (Figure 4.35(b)) the flux linkage is zero. As the coil continues to rotate, the direction of flux linkage is reversed. In Figure 4.35(c) the coil is at an angle θ to the field. At this point

$$\text{flux linkage, } \psi = \psi_m \sin\theta$$

Taking the time when the coil is parallel to the field (Figure 4.35(b)) to be zero and assuming that the coil rotates at ω rad/s, the time, t when the coil is at angle θ:

$$t = \theta/\omega$$

and the flux linkages, ψ at that time are:

$$\psi = \psi_m \sin\omega t$$

The emf induced in the coil is, according to Faraday's law (see Part 1, Unit 5), equal to the rate of change of flux linkages.

$$e = \frac{d\phi}{dt}$$

so, the emf induced in the coil at time, $t = \theta/\omega$ is:

$$e = \psi_m\,\omega\cos\omega t$$

The coil continues to rotate to $(\pi/2)$, Figure 4.35(d), arriving at time, $t = \pi/(2\omega)$.

If ε = average emf induced in the coil as it rotates from parallel to the field (Figure 4.35(b)) to perpendicular to the field (Figure 4.35(d)) then:

$$\varepsilon = \frac{1}{\left[\frac{\pi}{2\omega}\right]}\int_0^{\pi/2\pi} \psi_m\,\omega\cos\omega t\,dt$$

$$\varepsilon = \frac{2\omega}{\pi}\psi_m\left[\sin\omega t\right]_0^{\pi/2\omega}$$

$$\varepsilon = \frac{2\omega}{\pi}\psi_m$$

$$\text{but}\quad \psi_m = \phi$$

$$\therefore\quad \varepsilon = \frac{2\omega}{\pi}\phi$$

Many machines have several pairs of poles. With two N- and two S-poles the motor shown in Figure 4.31 has two pairs of poles. For a given speed of armature rotation, additional poles reduce the time the coil links with flux from a particular pole. This affects the induced emf. For a machine with p pairs of poles the coil only links with the magnetic flux from each pole for $\left(\frac{\pi}{p\omega}\right)$ seconds. If the flux per pole is ϕ then the average emf induced in the coil, ε is given by:

$$\varepsilon = \frac{2\omega p\phi}{\pi}\ \text{[V]}$$

In practice an armature winding will have many coils connected in series and the total emf induced in the armature winding is given by:

total emf induced in armature winding	=	average emf induced in one coil	×	number of coils in series
E	=	$\frac{2\omega p\phi}{\pi}$	×	A_s

Both the number of poles ($2p$) and the number of armature coils in series (A_s) are constant for a particular machine, so it is convenient to combine these two parameters into a single design constant, k, where

$$k = \frac{2pA_s}{\pi}$$

and express the total emf induced in the armature winding, E as:

$$E = k\phi\omega \text{ [V]} \quad (4.50)$$

Torque

The electrical power delivered to the armature is converted to mechanical power, creating a torque, which makes the armature rotate.

Assuming that there are no losses in the system:

electrical power delivered to armature = mechanical power at the shaft

$$P_a = P_m \quad (4.51)$$

electrical power = armature emf × armature current

$$P_a = E \times I_a \quad (4.52)$$

mechanical power = torque × angular velocity

$$P_m = T \times \omega \quad (4.53)$$

Combining equations (4.51) and (4.53)

$$T = \frac{P_a}{\omega} \quad (4.54)$$

Torque = Electrical power delivered to the armature/Angular velocity

Substituting (4.50) and (4.52) into (4.54):

$$\text{Torque} = k\phi\omega\frac{I_a}{\omega}$$

giving:

$$\text{Torque} = k\phi I_a \text{ [Nm]} \quad (4.55)$$

In practice there will be some losses in the system due to friction at the bearings, windage and iron losses in the steel core. These reduce the mechanical torque produced at the shaft. However, as these losses are usually small, they will be ignored for the rest of this section, and it will be assumed that the torque calculated using equation (4.55) equals the mechanical torque delivered at the shaft.

Armature equivalent circuit

To analyse the operation of dc motors it is convenient to represent the armature winding by the equivalent circuit shown in Figure 4.36. E is the emf induced in the armature and R_a is the armature winding resistance, which is usually quite small. The armature terminal voltage is given by:

$$V_a = E + I_aR_a \quad (4.56)$$

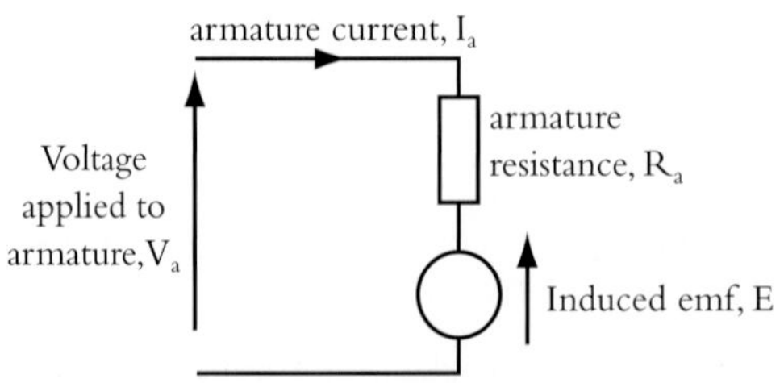

Figure 4.36 Equivalent circuit for the armature of a dc motor

Types of dc motor

There are five types of dc motor:

(1) shunt motor
(2) series motor (also used with ac supplies as 'universal motor')
(3) compound motor
(4) separately excited motor
(5) permanent magnet motor.

These are shown in Figure 4.37. The shunt motor (a) has the field and armature windings connected in parallel with the supply voltage. In the series motor (b) the two windings are connected in series. Compound motors, which have two field windings, are divided into two subcategories: the long-shunt motor (c) and the short-shunt motor (d). In the long-shunt motor the parallel field winding, R_{fp} is in parallel with the series field winding, R_{fs} and armature in series, while in the short-shunt motor, R_{fs} is in series with the parallel combination of R_{fp} and armature. The separately excited motor (e) has independent voltage supplies to the field and armature windings. The permanent magnet motor uses permanent magnets rather than field windings to provide the stationary magnetic field.

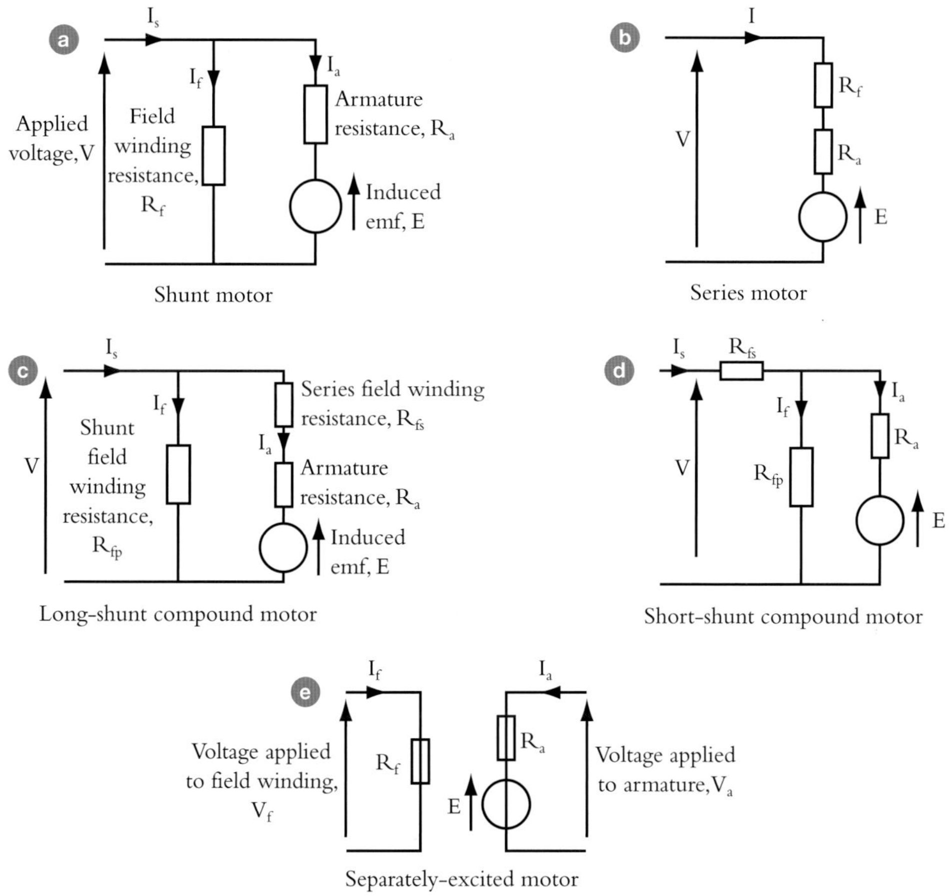

Figure 4.37 Different configurations of dc motor: (a) shunt motor; (b) series motor; (c) long-shunt compound motor; (d) short-shunt compound motor; (e) separately excited motor

The direction of rotation of any dc motor is reversed by swapping the connections to either the armature or field windings, but not both. In the case of a permanent magnet motor, only the armature connections exist; swapping the connections to this will reverse the motor.

DC shunt motor

As stated above, the dc shunt motor (Figure 4.37(a)) has the armature and field windings connected in parallel with the supply voltage. To withstand the applied voltage, the field winding is constructed of a large number of turns of thin wire with a resistance of hundreds of ohms. The armature winding has a large cross sectional area and a small resistance, usually less than 1Ω.

From equation (4.50) the emf induced in the armature winding, E is given by:

$$E = k\phi\omega$$

but from equation (4.49) ϕ is a function of the field current, I_f.

Hence

$$\frac{E}{\omega} = k\phi = \text{Fn}(I_f)$$

$k\phi$ versus I_f is normally presented as a graph, called the motor field characteristic. This graph is used to find the motor torque versus speed characteristic. The worked example demonstrates how.

Worked example

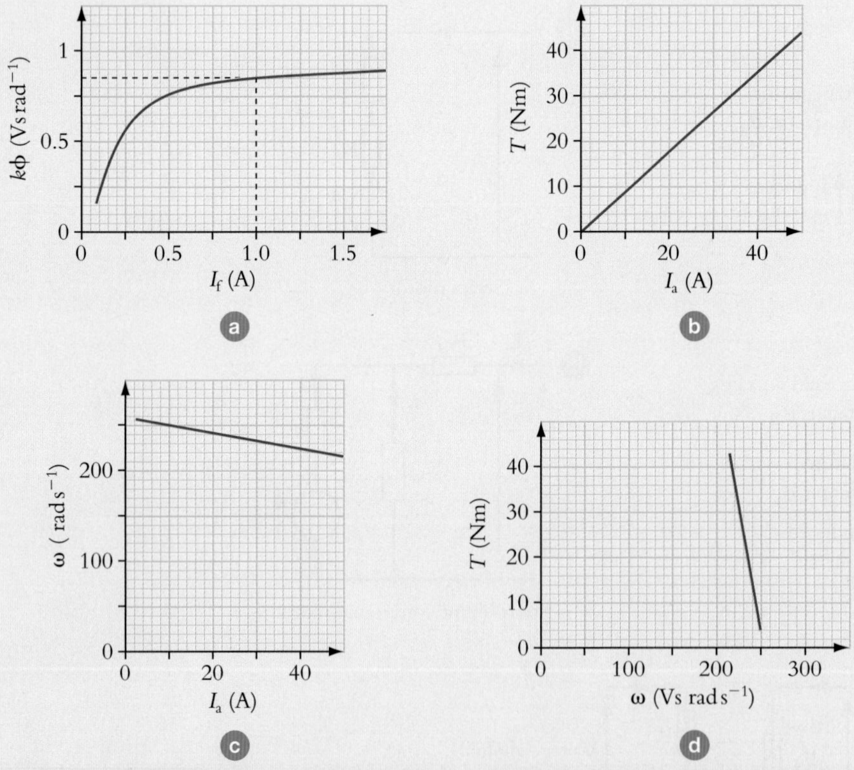

Figure 4.38 Characteristics of a dc motor: (a) motor field characteristic; (b) torque versus armature current; (c) angular velocity versus armature current; (d) torque versus angular velocity

A 220 V dc shunt motor has an armature resistance of 0.8 Ω and field winding resistance of 220 Ω. The motor field characteristic is shown in Figure 4.38(a). Calculate the field current, I_f and the corresponding value for $k\phi$.

The motor drives a load torque of 10 Nm. Calculate the armature current, I_a, and speed, n.

Field current, $I_f = \frac{V}{R_f} = \frac{220}{220} = 1\,\text{A}$

The motor field characteristic is used to find the value of $k\phi$.
When $I_f = 1\,\text{A}$, $k\phi = 0.84\,\text{Vs/rad}$.

With the field winding connected directly across the supply voltage, the field current, and hence the value of $k\phi$, is constant for all loads.

Using equation (4.55)

$$\text{Torque} = k\phi I_a$$

$$I_a = \frac{T}{k\phi}$$

when $T = 10\,\text{Nm}$,

$$I_a = \frac{10}{0.84} = 11.8\,\text{A}$$

Using equation (4.50)

$$E = k\phi\omega$$

and substituting into equation (4.56)

$$V = k\phi\omega + I_a R_a \qquad \textbf{(4.57)}$$

$$220 = 0.84\omega + 11.8 \times 0.8$$

$$220 = 0.84\omega + 9.4$$

$$\omega = \frac{220 - 9.4}{0.84} = \frac{210.6}{0.84}$$

$$\omega = 250.7\,\text{rad/s}$$

Although, we do the calculations in radians per second it is usual to express motor speed in revolutions per minute, *n*, where:

$$n = \frac{\omega}{2\pi} \times 60 = \frac{60\omega}{2\pi}\,\text{rev/min}$$

In this case, $n = 2394\,\text{rev/min}$.

By repeating these calculations using other values of torque, the following characteristics can be drawn:

- torque versus armature current – Figure 4.38(b)
- angular velocity versus armature current – Figure 4.38(c)
- torque versus angular velocity – Figure 4.38(d).

Figure 4.38(d) shows that, as the torque is increased, there is only a small drop in motor speed, making the dc shunt motor ideal for driving machine tools, pumps and compressors.

Torque–speed characteristics of dc shunt motors

From the viewpoint of the design of an electromechanical system, it is useful to express the torque–speed relationship explicitly:

$$V = k\phi\omega + I_a R_a$$

$$= k\phi\omega + \frac{R_a \times \text{torque}}{k\phi}$$

This can be rearranged as follows:

$$\frac{k\phi V}{R_a} = \frac{k^2\phi^2\omega}{R_a} + \text{torque}$$

and hence the torque can be expressed explicitly in terms of speed:

$$\text{Torque} = \frac{k\phi V}{R_a} - \frac{k^2\phi^2\omega}{R_a} \qquad (4.58)$$

Alternatively, speed may be expressed as a function of torque:

$$\omega = \frac{V}{k\phi} - \frac{R_a \times \text{torque}}{k^2\phi^2} \qquad (4.59)$$

The constant $k\phi$ (for a given value of field flux ϕ) is sometimes clustered together as the motor constant K_m giving the emf E induced in the armature, or back-emf, in terms of the angular velocity ω:

$$E = K_m\omega$$

and the torque in terms of the current I_a:

$$\text{Torque} = K_m I_a$$

This notation will be used within the unit on control theory.

Speed control of dc shunt motors

To find a way to control the speed of a dc shunt motor we should examine equation (4.57), which can be rearranged to give:

$$\omega = \frac{V - I_a R_a}{k\phi}$$

The voltage drop across the armature winding ($I_a R_a$) is usually small compared to the supply voltage, V. But, if the latter is fixed then the motor speed is dependent on the flux. However, the flux is dependent on the field current. So, with the supply voltage fixed, an increase in field circuit resistance reduces the field current, thereby reducing the flux and increasing the speed.

This is demonstrated in the following examples. The same motor is used to drive the same load torque. In the first example, the motor is operated without additional resistance. In the

Worked example

A 240 V dc shunt motor has a field winding resistance, R_f of 160 Ω, armature winding resistance, R_a of 0.2 Ω and a motor field characteristic like that shown in Figure 4.39. The load torque is 50 Nm.

Calculate the field current, armature current, supply current and motor speed.

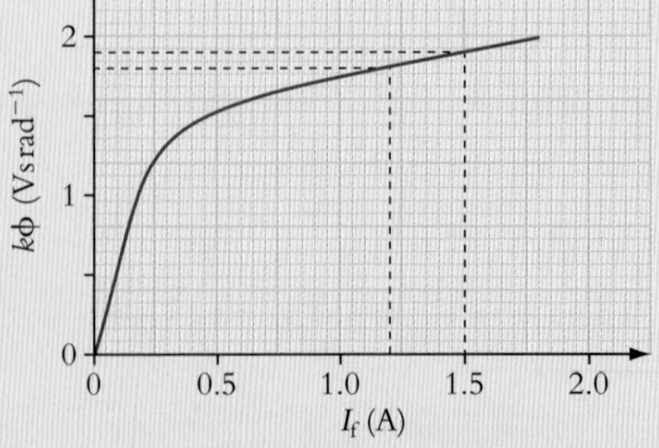

Figure 4.39 **Motor field characteristic**

Field current, $I_f = \frac{V}{R_f} = \frac{240}{60} = 1.5\,\text{A}$

From motor field characteristic, Figure 4.39, when $I_f = 1.5\,\text{A}$, $k\phi = 1.91\,\text{Vs/rad}$

Using equation (4.55)

$$\text{Torque, } T = k\phi\, I_a$$

$$\text{Armature current, } I_a = \frac{T}{k\phi}$$

$$I_a = \frac{50}{1.91}$$

$$= 26.2\,\text{A}$$

Supply current,

$$I_s = I_f + I_a$$

$$= 1.5 + 26.2 = 27.7\,\text{A}$$

From equation (4.56)

$$E = V - I_a R_a$$

$$= 240 - 26.2 \times 0.2 = 234.8\,\text{V}$$

From equation (4.50)

$$E = k\phi\omega$$

$$\omega = \frac{E}{k\phi} = \frac{234.8}{1.91} = 122.9\ \text{rad/s}$$

$$n = \frac{30\omega}{\pi} = 1174\ \text{rev/min}$$

Worked example

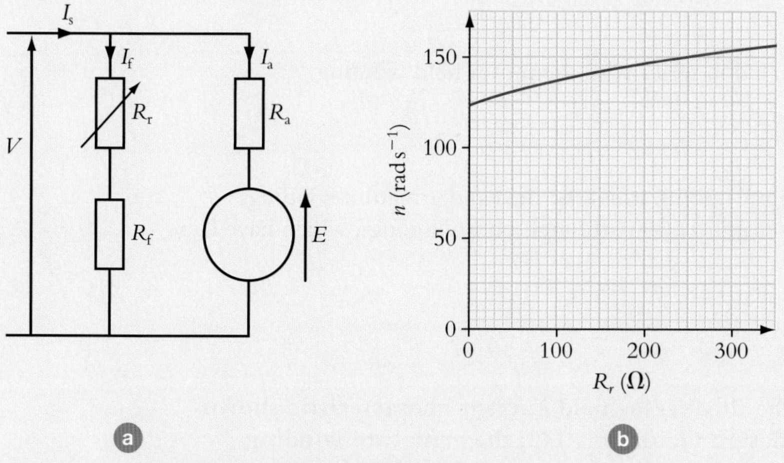

Figure 4.40 **(a) dc motor with shunt field winding regulator; (b) graph of angular velocity versus regulator resistance**

The dc shunt motor in the previous example has a shunt field winding regulator connected in series with the field winding, as shown in Figure 4.40(a). The regulator resistance, R_r is set at 40 Ω. The load torque is maintained at 50 Nm. Calculate the new field current, armature current, supply current and speed.

$$\text{Field current, } I_f = \frac{V}{R_f + R_r} = \frac{240}{160 + 40} = 1.2\text{ A}$$

From motor field characteristic, Figure 4.39, when $I_f = 1.2\text{A}$, $k\phi = 1.83\text{ Vs/rad}$

$$\text{Armature current, } I_a = \frac{\text{torque}}{k\phi}$$

$$I_a = \frac{50}{1.83} = 27.3\text{A}$$

$$\text{Supply current, } I_s = I_f + I_a$$

$$= 1.2 + 27.3 = 28.5\text{A}$$

From equation (4.56)

$$E = V - I_a R_a$$

$$= 240 - 27.3 \times 0.2 = 234.5\text{V}$$

From equation (4.50)

$$E = k\phi\omega$$

$$\omega = \frac{E}{k\phi} = \frac{234.5}{1.83} = 128.1\text{ rad/s}$$

$$n = \frac{30\omega}{\pi} = 1223\text{ rev/min}$$

The connection of a shunt field regulator in series with the field winding has increased the resistance in the field winding circuit and with it the motor speed.

By repeating these calculations using various values for the regulator resistance, the graph of angular velocity, ω, versus regulator resistance, R, can be plotted as shown in Figure 4.40(b).

The graph shows that the motor speed rises as the regulator resistance is increased.

second, additional resistance is connected in series with the motor field winding.

DC series motor

A dc series motor is illustrated in Figure 4.37(b). It has the field and armature windings connected in series, so that the same current, I flows through both windings, which have large cross sectional areas and low resistance.

Worked example

A 230 V dc series motor has the flux versus field current characteristic shown in Figure 4.41(a). The motor design factor, k is 142, the armature winding resistance, R_a is 0.12 Ω and the field winding resistance, R_f is 0.03 Ω.

Calculate the speed and torque when the current is 25 A.

From the equivalent circuit, Figure 4.37(b)

$$E = V - I(R_a + R_f)$$

$$E = 230 - I \times (0.12 + 0.03) = 230 - 0.15\, I$$

When $I = 25\,\text{A}$

$$E = 230 - 25 \times 0.15 = 230 - 3.75 = 226.25\,\text{V}$$

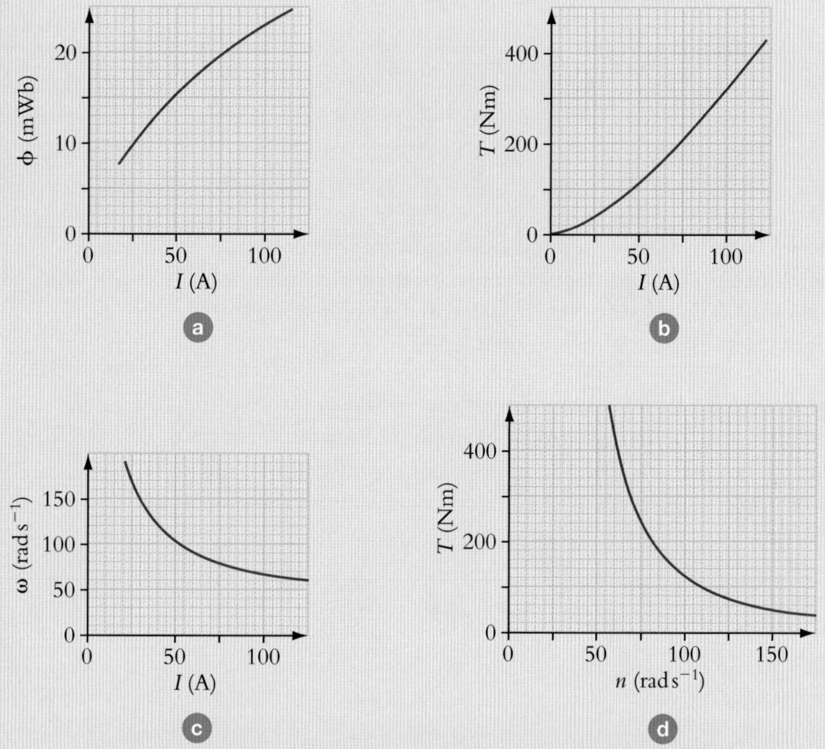

Figure 4.41 (a) Flux versus field current characteristic; (b) torque versus current; (c) angular velocity versus current; (d) torque versus angular velocity

From the motor field characteristic, Figure 4.41(a) when $I = 25\,\text{A}$, $\phi = 10\,\text{mWb}$.

Using equation (4.50)

$$\omega = \frac{E}{k\phi}$$

$$\omega = \frac{226.25}{142 \times 10 \times 10^{-3}} = 159.3\,\text{rad/s}$$

From equation (4.55)

$$\text{Torque} = k\phi I$$

When $I = 25\,\text{A}$

$$\text{Torque} = \frac{142 \times 10 \times 10^{-3} \times 25}{2\pi} = 35.4\,\text{Nm}$$

By repeating the calculations with different values of current the following graphs can be plotted:

- torque versus current – Figure 4.41(b)
- angular velocity versus current – Figure 4.41(c)
- torque versus angular velocity – Figure 4.41(d).

The following example shows how we can find the torque versus speed graph for this type of motor.

If it is assumed that ϕ is proportional to I, and if the effects of resistance are neglected, then to a first approximation:, it can be shown that

$$\text{Torque} \propto \frac{1}{n^2}$$

or:

$$\text{Torque} \propto \frac{1}{\omega^2}$$

At low speeds dc series motors create a large torque, which reduces as the speed increases, making this type of motor the ideal choice for lift and traction applications. Series motors will run at very high speeds when no load is coupled to the shaft. To prevent damaging the machines they must always be operated with permanently coupled loads.

The speed of a series motor may be controlled by a variable resistor connected in series with the field winding. This is demonstrated in the following worked examples. The same motor is used to drive a constant load torque. In the first example, the motor is operated without any additional resistance. In the second example, a resistance is connected in series with the motor windings.

Worked example

The 240 V dc series motor shown in Figure 4.42(a) has an armature resistance of 0.1 Ω and a field winding resistance of 0.05 Ω. The motor field characteristic is shown in Figure 4.42(b). The motor drives a load torque of 250 Nm. Calculate the speed.

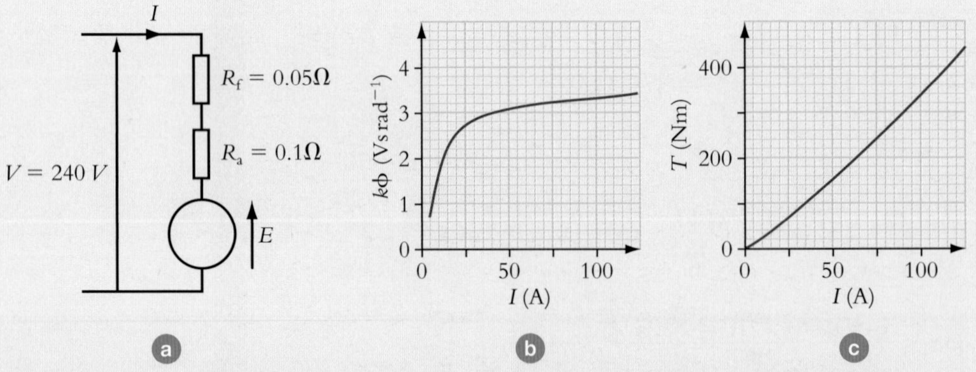

Figure 4.42 (a) dc series motor; (b) motor field characteristic; (c) graph of torque versus current

The first step to solving this problem is to draw the graph of torque versus current,. This requires selecting a number of currents on the horizontal axis of Figure 4.42(b) and for each point reading off the corresponding value of $k\phi$ and calculating torque using equation (4.55):

Torque $= k\phi\, I$

For example when $I = 50\,A$

$$\phi = 3.12\,Vs/rad$$

and

$$\text{Torque} = 3.12 \times 50 = 156\ \text{Nm}$$

These calculations are repeated for several currents and the graph of torque, T versus current, I (Figure 4.42(c)) is plotted.

From this graph:
when torque = 250 Nm

$$I = 78\,A$$

$$E = V - I\,(R_a + R_f)$$

$$= 240 - 78 \times (0.1 + 0.05) = 228.3\,V$$

From Figure 4.42(b) when $I = 78\,A$

$$k\phi = 3.25\,Vs/rad$$

Using equation (4.50)

$$\omega = \frac{E}{k\phi} = \frac{228.3}{20.4} = 70.3\ \text{rad/s}$$

$$n = \frac{60\omega}{2\pi} = 671\ \text{rev/min}$$

Worked example

The dc series motor in the previous example has a resistance, R_s of 0.05 Ω connected in series with the armature and field windings, as shown in Figure 4.43(a). The torque remains at 250 Nm. Calculate the new speed.

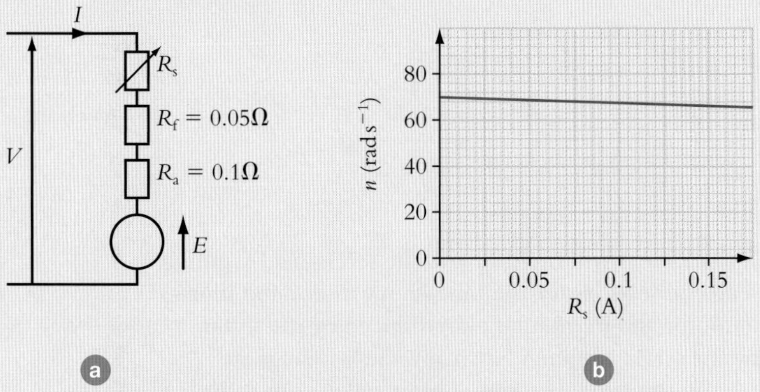

Figure 4.43 **(a) dc series motor with additional series resistance; (b) graph of motor speed vs. resistance for load of 250 Nm**

From the torque, T, versus armature current, I, characteristic for the motor, Figure 4.42(c) when Torque = 250 Nm, I = 78 A

$$E = V - I(R_a + R_f + R_s)$$
$$= 240 - 78 \times (0.1 + 0.05 + 0.05)$$
$$= 224.4\text{V}$$

From Figure 4.42(b), when $I = 78$ A

$$k\phi = 3.25\,\text{Vs/rad}$$

$$\omega = \frac{E}{k\phi} = \frac{224.4}{3.25} = 69\,\text{rad/s}$$

$$n = \frac{60\omega}{2\pi} = 659\,\text{rev/min}$$

Connecting a resistance in series with the field and armature windings has reduced the motor speed. With the same load torque further increases in the series resistance will make the motor run even more slowly, as shown in Figure 4.43(b).

Series motors can be connected to ac supplies because both the armature and field windings have relatively low inductance. They can be used for moderate- or relatively low-power high-speed ac applications such as:

- high-speed blowers for small furnaces
- vacuum cleaners
- high-speed woodworking tools (routers).

These motors can also be used for moderate-power applications requiring good power-to-weight ratios and high stall torques such as:

- power tools, e.g. hand-held power drills and garden tools
- electric mixers and food processors.

Since these motors typically run at very high speeds (e.g. 5000–10 000 rev/min) it is usual to use a large reduction drive ratio to enable them to be used to drive items such as a drill bit or mixing agitator. A cooling fan is usually incorporated; as well as providing a means of cooling the motor, the windage it creates acts as a mechanical load which limits the no-load speed of the system to a safe value.

Compound motors

There are two forms of compound motor:

- long-shunt motor
- short-shunt motor.

Both motors have two field windings, the shunt-field winding, R_{fp} and the series-field winding, R_{fs}. Figure 4.37(c) and (d) show the winding connections for the long-shunt motor and short-shunt motor respectively. In both forms of motor, the magnetic field produced by the shunt field winding is larger than that formed by the series winding. Normally, the windings are connected so that the two fields assist, i.e. act in the same direction. This is called cumulative compound. The current through the shunt winding, and hence the magnetomotive force (mmf) produced by it, are practically independent of the load, whereas, the mmf produced by the series field winding depends on the armature current and the mechanical load. When the motor is on no load, the mmf produced by the series winding is negligible and the motor performs like a shunt motor. Figure 4.44 shows a typical set of characteristics for a cumulative compound motor. As the load current rises the mmf produced by the series field increases, and the speed which is inversely proportional to the total mmf, falls. At full load torque the speed is between 0.7 and 0.9 of the no-load value. Cumulative compound motors are used where heavy loads may be suddenly applied, for example in presses, plunger pumps and hoists.

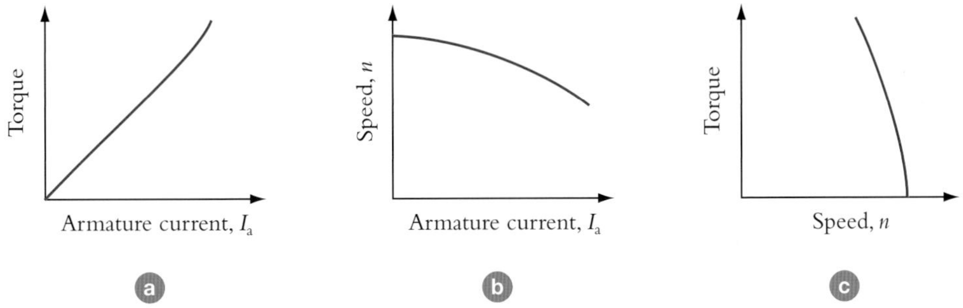

Figure 4.44 Typical characteristics for a cumulative compound motor: (a) torque vs. armature current; (b) speed vs. armature current; (c) torque vs. speed

The differential compound motor has the series winding connected to produce a field that opposes that formed by the shunt winding. This type of motor has few applications and as a consequence is rare.

Separately excited motor

The separately excited motor has the field winding and armature connected to independent voltage supplies, as shown in Figure 4.37(e), from where it can be deduced that:

$$V_f = I_f R_f \quad (4.60)$$

and:

$$V_a = E + I_a R_a \quad (4.61)$$

The induced emf, E is generally far larger than the voltage drop across the armature winding, $I_a R_a$. As a result we may ignore the latter and assume that

$$E = V_a$$

But from equation (4.50) for all dc motors

$$E = k\phi\omega$$

Hence in this case

$$V_a = k\phi\omega$$

$$\omega = \frac{V_a}{k\phi}$$

where k is the design constant and ϕ is the flux, which is dependent on the field current, I_f.

If, I_f and hence ϕ is kept constant then

$$\omega \propto V_a$$

so

$$n \propto V_a$$

That is, the motor speed is directly proportional to the armature voltage.

Worked example

A separately excited dc motor has the motor field characteristic shown in Figure 4.39. The field winding resistance is 200 Ω. The armature winding resistance is negligible. Calculate the motor speed when:

(a) the field voltage and the armature voltage both equal 220 V

(b) the field voltage is 220 V and the armature voltage is 150 V.

Field current, $I_f = \frac{220}{200} = 1.1\,\text{A}$

From motor field characteristic $k\phi = 1.8\,\text{Vs/rad}$

(a) when armature voltage, $V_a = 220\,\text{V}$

$$\omega = \frac{V_a}{k\phi}$$

$$= 122.3\,\text{rad/s}$$

$$n = \frac{60\omega}{2\pi} = 1168\,\text{rev/min}$$

(b) when armature voltage, $V_a = 150\,\text{V}$

$$\omega = \frac{150}{1.8} = 83.3\,\text{rad/s}$$

$$n = \frac{60\omega}{2\pi} = 795\,\text{rev/min}$$

Permanent magnet motors

These motors use permanent field magnets instead of laminated poles with field windings to provide the stationary magnetic field, ϕ, which is now sensibly constant in value and cannot be altered electrically. The analysis and theory of permanent magnet motors is identical to those of shunt-wound or separately excited motors with a fixed value of field excitation, so that $k\phi$ is constant. Their characteristics may be summarized as:

- no-load speed is proportional to supply voltage
- torque is proportional to current
- speed falls from its no-load value with increasing torque.

Permanent magnet dc motors are not normally used for large power applications but are very widely used within domestic and automotive applications such as:

- window winders, windscreen wipers, fans and blowers in cars
- battery-operated appliances such as electric shavers and cooling fans
- battery-operated and track-powered toys.

They are also widely used in various forms as dc servomotors within control applications. Conventional dc servomotors are simply high-quality permanent-magnet dc motors with wound armatures and commutators, but there are also variants such as:

- ironless motors (in which the armature windings form a rigid lightweight structure of their own rather than being wound onto iron laminations)

- pancake and 'printed armature' motors in which the ironless armatures are flat in shape
- brushless dc motors in which the armature windings are stationary, the field magnets form a rotor, and the switching action of the brushes and commutator is carried out by solid-state switches.

All of these behave electrically in the same manner as the conventional dc permanent magnet motor, even though their construction is totally different. The subject of dc permanent magnet motors will be revisited within Unit 5 on control.

Learning summary

By the end of this section you have learnt:

✓ the operation of dc motors;

✓ to derive the general torque equation;

✓ the different forms of dc motor;

✓ why the shunt motor produces virtually constant speed;

✓ how to vary the speed of a shunt motor;

✓ the characteristics of a series motor;

✓ how to control the speed of a series motor;

✓ why the speed of a separately excited motor is proportional to the armature voltage.

4.10 Rectified supplies for dc motors

The vast majority of power supplies around the world are alternating voltage. So to provide the power to drive a dc motor it is often necessary to convert the alternating supply to a direct voltage. To do this a rectifier is used.

The principles of rectification were examined in Part 1, Unit 5.

When an alternating (ac) voltage, Figure 4.45(a), is applied to a diode rectifier with a resistive load, like that in Figure 4.45(b), the diode only conducts when it is forward biased (i.e. the potential of the anode is positive with respect to the cathode).

Ohm's law states that:

$$v = iR$$

so while the diode is conducting, the load current has a similar waveform to that of the supply voltage but when the diode is reversed biased it cannot conduct and the current remains at zero, producing a current waveform that comprises a series of positive half cycles, as shown in Figure 4.45(c).

The voltage across the load resistor is called the output voltage. It has a waveform that has a similar shape to that of the load current. Assuming that the forward volt drop across the diode, approximately 0.7 V, is small compared to the supply voltage and may be ignored, then the average dc output voltage is given by:

$$\text{Average dc output voltage, } V_o = \frac{V_m}{\pi}$$

Where, V_m is the supply voltage maximum $= \sqrt{2} \times$ rms supply voltage.

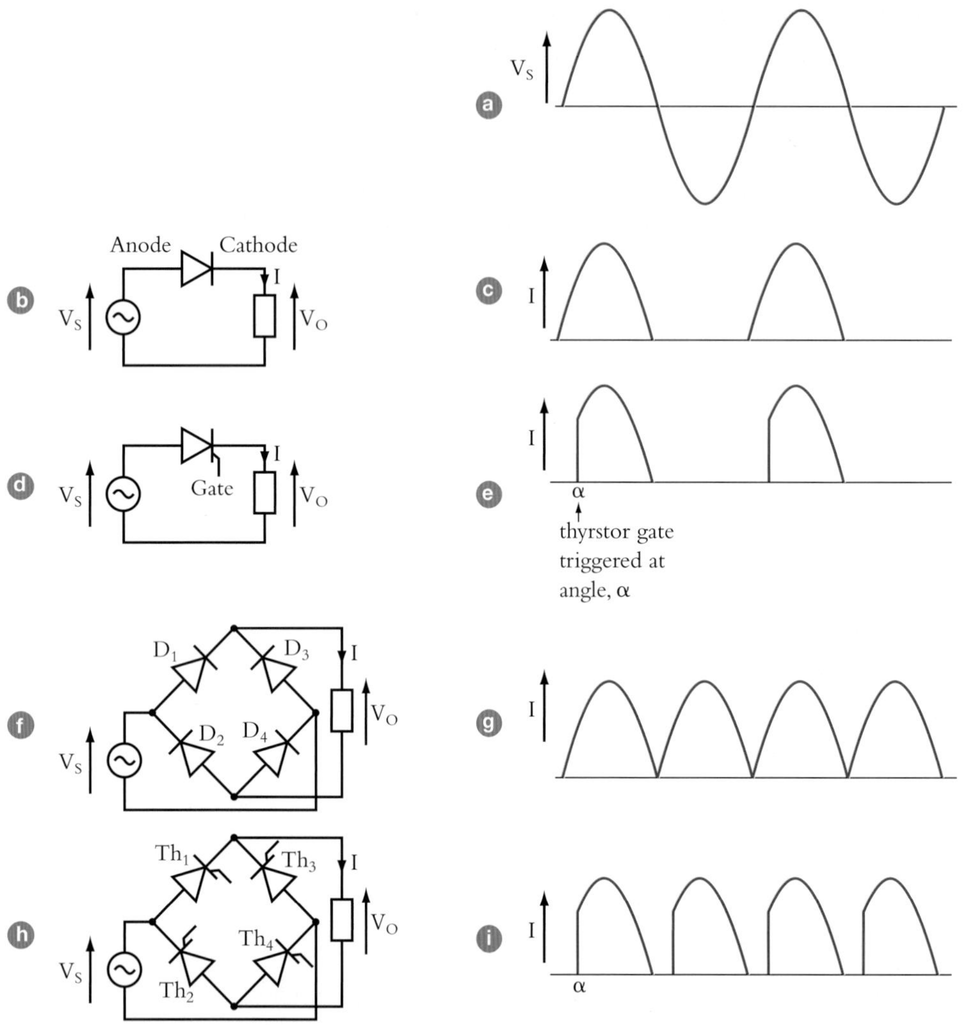

Figure 4.45 **(a) ac supply voltage waveform; (b) rectification using a single diode (half wave rectifier) and (c) resulting current waveform; (d) controlled rectification using a single thyristor and (e) resulting current waveform; (f) diode bridge rectifier (full wave rectifier) and (g) resulting current waveform; (h) thyristor bridge rectifier and (i) resulting current waveform**

Replacing the diode by a thyristor, as in Figure 4.45(d), enables the user to control the output voltage. The thyristor current is blocked in the forward direction until the thyristor gate is triggered by a small pulse from a second, independent voltage source, as shown in Figure 4.45(e), thereby reducing the mean load current and hence output voltage to:

$$\text{Average dc output voltage, } V_o = \frac{V_m}{2\pi}[1 + \cos\alpha]$$

A disadvantage of both these rectifiers is that current can only flow through the load during the positive half cycles of the voltage supply. For load current to flow during both the positive and negative voltage half cycles a bridge rectifier must be used. Figure 4.45(f) shows a diode bridge rectifier. When the supply voltage is positive, the current flows via diode D_1 to the load and back through D_4. During the negative half cycles D_2 and D_3 conduct, and D_1 and D_4 are cut off. The load current waveform and hence output voltage is shown in Figure 4.45(g). The mean voltage is given by:

$$\text{Average dc output voltage} = \frac{2V_m}{\pi}$$

This is twice as large as the output voltage for a single diode rectifier.

The corresponding output voltage waveform for a thyristor bridge rectifier (Figure 4.45(h)) is shown in Figure 4.45(i), and the mean output voltage is given by:

$$\text{Average dc output voltage} = \frac{V_m}{\pi}[1 + \cos\alpha]$$

In order to compare the diode and thyristor bridge rectifiers (Figure 4.45(f) and (h)) with other rectifier circuits described later in this unit it is useful to redraw the two circuit diagrams as in Figure 4.46(a) and (b) respectively.

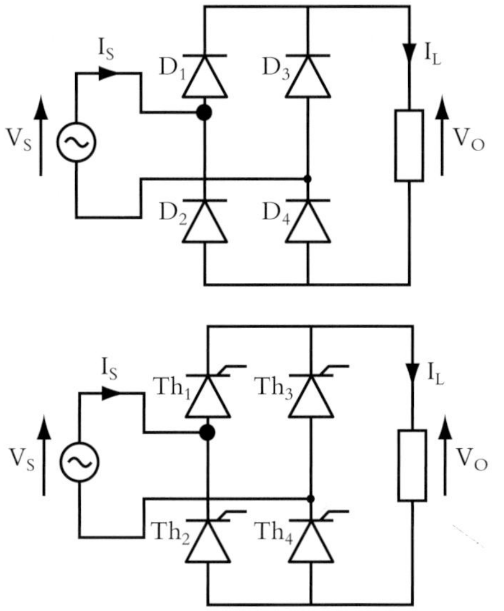

Figure 4.46 (a) Diode bridge rectifier; (b) thyristor bridge rectifier, both with resistive loads

Replacing the resistive load by a dc motor affects the load current waveform because the motor has inductance. This delays current changes and inhibits the current from following directly changes in the applied voltage.

Consider the diode bridge rectifier shown in Figure 4.46(a) but now with an inductive load.

The supply voltage and load voltage waveforms (Figure 4.47(a) and (b) respectively) are similar to those with a resistive load but the load current fluctuates, never falling to zero as illustrated in Figure 4.47(c). The peaks in the current lag the voltage maxima, their magnitude dependent on the ratio of load inductance, *L,* to resistance, *R*. The larger this ratio the smaller is the current ripple, so that with a very high ratio of *L*:*R* the load current is virtually constant.

As with a resistive load, diodes D_1 and D_4 conduct when the supply voltage is positive and D_2 and D_3 conduct when it is negative, but the shape of the diode current waveforms are different, see Figure 4.47(d) and (e). Figure 4.47(f) shows the corresponding supply current waveform.

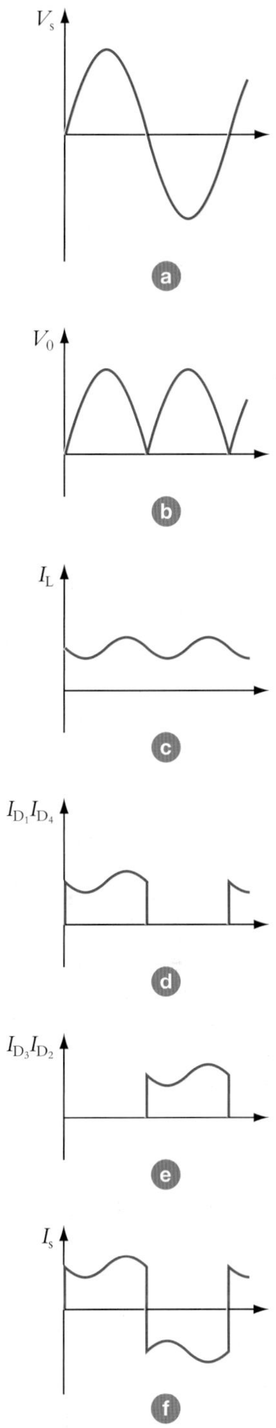

Figure 4.47 (a) ac supply voltage waveform; (b) voltage waveform after full wave rectification; (c) resulting current in dc motor; (d) current in diodes D1 and D4; (e) current in diodes D3 and D2; (f) supply current

Figure 4.48(a) to (f) show the voltage and current waveforms for a thyristor bridge rectifier, Figure 4.46(b), but now with an inductive load. Initially, thyristors Th_2 and Th_3 are conducting.

Because of the high ratio of load inductance to resistance, the load current does not fall to zero when the supply voltage reaches zero. Current flow is maintained through the two thyristors until the other thyristors are triggered at the angle, α. The output voltage will go negative during this period.

The mean output voltage can be found thus:

Average dc output voltage,

$$V_o = \frac{1}{\pi}\left[\int_{\alpha}^{\pi+\alpha} V_m \sin(\omega t)\, d(\omega t)\right]$$

$$= \frac{V_m}{\pi}[-\cos \omega t]_{\alpha}^{\pi+\alpha}$$

$$= \frac{V_m}{\pi}[-\cos(\pi + \alpha) + \cos(\alpha)]$$

$$V_o = \frac{2V_m}{\pi}\cos\alpha \qquad (4.62)$$

The output voltage becomes negative if the delay angle, α, is greater than 90°.

Three-phase bridge rectifiers have six devices and produce a smoother dc output voltage than single-phase rectifiers. Figure 4.49(a) shows a three-phase diode rectifier with a highly inductive load. The phase voltage and diode current waveforms are shown respectively in Figure 4.49(b) and (c). The diodes conduct in pairs, one in the top arm of the bridge with one in the bottom arm of another phase; D_5 with D_4, D_1 with D_4, D_1 with D_6 and so on. The dc output voltage waveform is shown in Figure 4.49(d). The average dc output voltage for a three-phase diode bridge rectifier with a highly inductive load can be found thus:

$$V_o = \frac{1}{\frac{\pi}{3}}\int_{30°}^{90°} V_{an}\sin(\omega t)\,d(\omega t) - V_{bn}\sin(\omega t - 120°)\,d(\omega t)$$

Let

$$|V_{m.ph}| = |V_{an}| = |V_{bn}| = |V_{cn}|$$

$$= \frac{3V_{m\cdot ph}}{\pi}[-\cos(\omega t) + \cos(\omega t - 120°)]_{30°}^{90°}$$

$$= \frac{3V_{m\cdot ph}\sqrt{3}}{\pi}$$

But $\quad V_{line} = \sqrt{3}\, V_{ph}$

$$V_o = \frac{3V_{m.line}}{\pi} \qquad (4.63)$$

where $V_{m\,line}$ = maximum line voltage

$= \sqrt{2} \times$ rms line voltage, V_{line}

$$V_{m\,line} = \sqrt{2}\, V_{line}$$

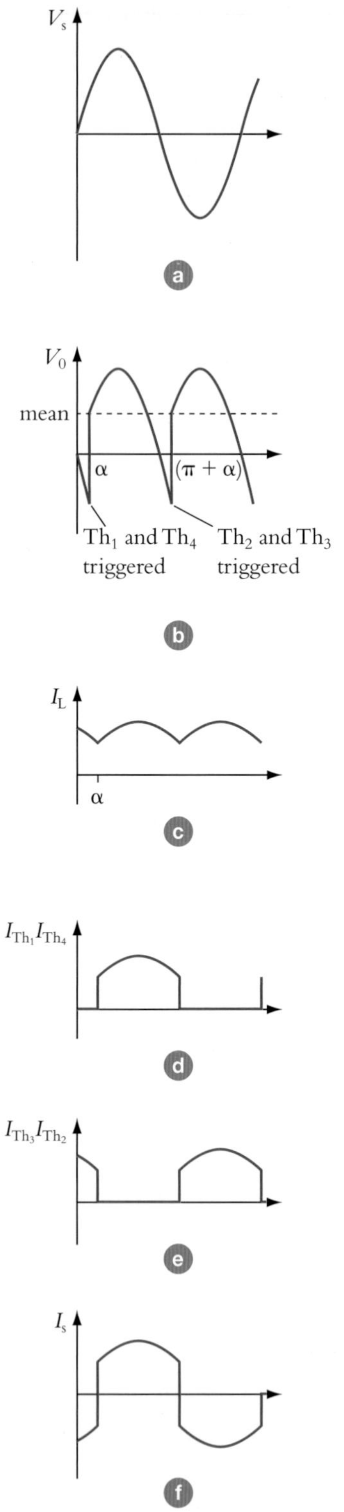

Figure 4.48 (a) ac supply voltage waveform; (b) voltage waveform from thyristor bridge rectifier; (c) resulting current in dc motor; (d) current in thyristors Th_1 and Th_4; (e) current in thyristors Th_3 and Th_2; (f) supply current

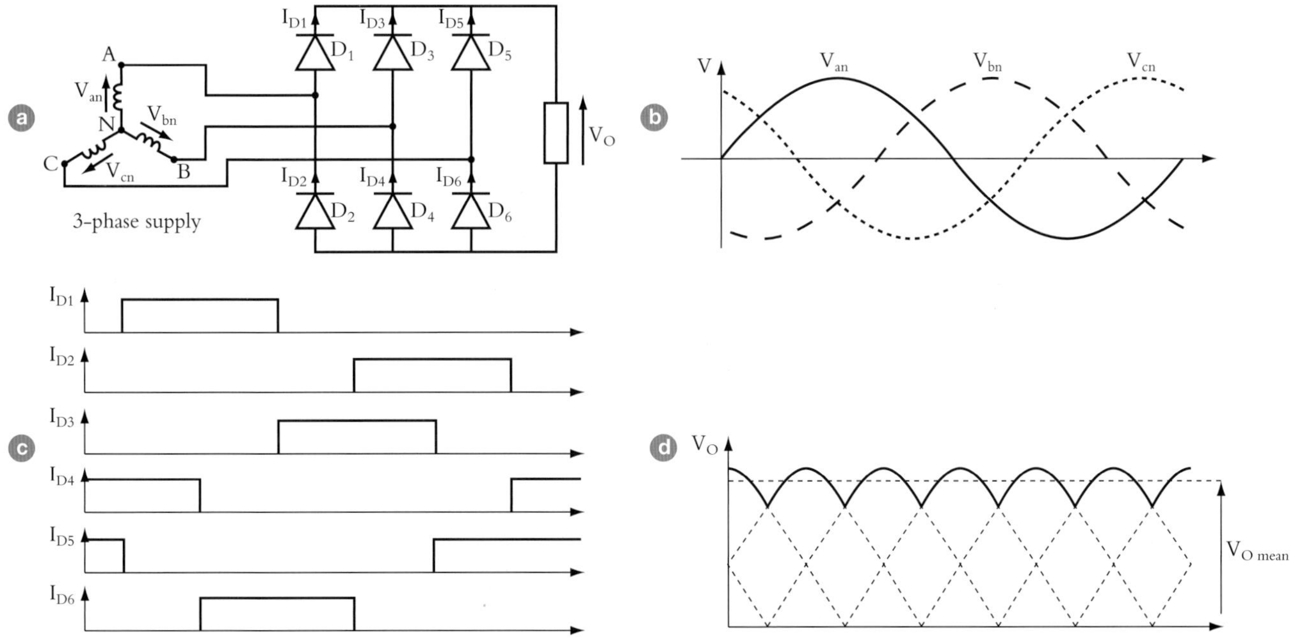

Figure 4.49 **(a) Three-phase diode rectifier with a highly inductive load; (b) phase voltage waveforms and (c) diode current waveforms; (d) dc output voltage waveform**

Replacing the diodes by thyristors (Figure 4.50(a)) allows the angle at which the devices are triggered to be delayed, thereby enabling the output voltage to be controlled. The mean dc output voltage, V_{dc} is given by:

$$V_{dc} = \frac{3V_{m.\text{line}}}{\pi} \cos\alpha \qquad (4.64)$$

where α = delay angle.

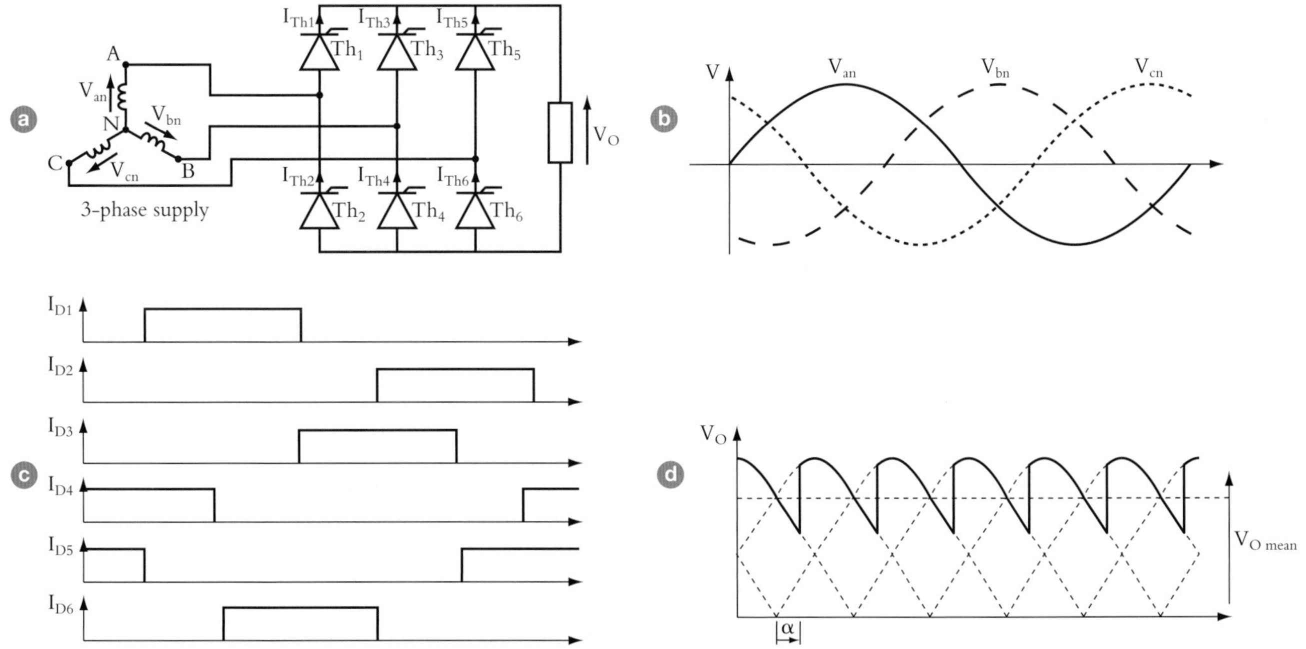

Figure 4.50 **(a) Three-phase thyristor rectifier with a highly inductive load; (b) phase voltage waveforms; and (c) thyristor current waveforms; (d) dc output voltage waveform for delay angle α**

The mean output voltage is zero when $\alpha = 90°$, and negative when the angle is greater than 90°. As the delay angle is increased, the ripple on the output voltage becomes more pronounced. Even so, the output voltage waveform is smoother than for the equivalent single-phase rectifier and should be used in preference. A dc motor connected to this type of rectifier will perform almost as well as if it were connected to a pure dc supply. Hughes (2008) has identified two reasons for this. Firstly, the motor armature inductance smoothes the armature current, and since the torque is directly proportional to this current, ripples in the output torque are limited. Secondly, the armature inertia helps to inhibit variations in motor speed caused by torque ripples.

Connecting the dc motor as a separately excited motor with the armature fed by a thyristor bridge rectifier and the field supplied from an independent diode bridge rectifier, as shown in Figure 4.51, enables the motor speed to be controlled by varying the thyristor triggering angle, α.

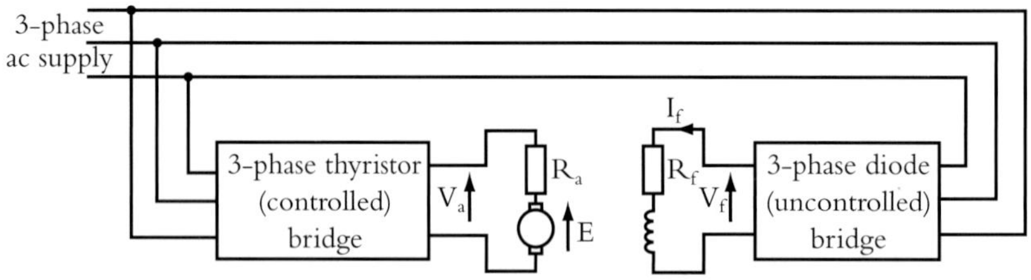

Figure 4.51 Separately excited dc motor with the armature fed by a thyristor bridge rectifier and the field supplied from an independent diode bridge rectifier

It was shown earlier, see equations (4.60)and (4.61), that for a separately excited motor:

$$\text{Field voltage, } V_f = I_f R_f$$

$$\text{Armature voltage , } V_a = E + I_a R_a$$

By ignoring the small armature resistance, R_a

$$E = V_a$$

From equation (4.50) for any dc motor:

$$\text{Induced emf, E} = k\phi\omega$$

$$\text{therefore } \omega = \frac{V_a}{k\phi}$$

The following example demonstrates that controlling the variation in the trigger angle, alters the armature voltage and hence the motor speed.

Worked example

A separately excited dc motor has the motor field characteristic shown in Figure 4.39. The field winding resistance is 500 Ω. The armature winding resistance is negligible. A three-phase 415 V (rms) ac supply is connected to the motor field winding via a diode bridge rectifier. The same ac supply feeds the motor armature via a thyristor bridge rectifier. Calculate the motor speed when the thyristor delay angle is:

(a) 0°

(b) 45°

From equation (4.63)

Field voltage = average output voltage from diode rectifier

$$V_f = \frac{3V_{\mathrm{m.line}}}{\pi}$$

where $V_{\mathrm{m\,line}}$ = maximum alternating voltage = $\sqrt{2}$ × rms line voltage.

In this case

$$V_{\mathrm{m\,line}} = \sqrt{2} \times 415 = 586.9\mathrm{V}$$

hence

$$V_{\mathrm{f}} = 560.5\mathrm{V}$$

Field current,

$$I_{\mathrm{f}} = \frac{V_{\mathrm{f}}}{R_{\mathrm{f}}} = \frac{560.5}{500} = 1.12\mathrm{A}$$

From motor field characteristic, Figure 4.39, when $I_{\mathrm{f}} = 1.12\mathrm{A}$

$$k\phi = 1.8\,\mathrm{Vs/rad}$$

Using equation (4.64):

Armature voltage = average output voltage from thyristor bridge

$$V_{\mathrm{a}} = \frac{3V_{\mathrm{m.line}}}{\pi}\cos\alpha$$

where α = delay angle in triggering the thyristors

(a) With the delay angle set at 0°

$$V_{\mathrm{a}} = \frac{3V_{\mathrm{m.line}}}{\pi}\cos\alpha$$

$$V_{\mathrm{a}} = 560.5\mathrm{V}$$

From equation (4.50)

$$\omega = \frac{V_{\mathrm{a}}}{k\phi}$$

$$\omega = \frac{560.5}{1.8} = 311.4\,\mathrm{rad/s}$$

$$n = \frac{30\omega}{\pi} = 2974\,\mathrm{rev/min}$$

(b) with the delay angle set at 45°

$$V_{\mathrm{a}} = \frac{3V_{\mathrm{m.line}}}{\pi}\cos\alpha$$

$$V_{\mathrm{a}} = 396.3\mathrm{V}$$

As the average output voltage from the diode rectifier, V_{f} is constant, I_{f} and $k\phi$ remain constant at 1.12A and 1.8Vs/rad respectively, the speed falls to:

$$\omega = \frac{V_{\mathrm{a}}}{k\phi} = \frac{396.3}{1.8} = 220.2\,\mathrm{rad/s}$$

$$n = 2102\,\mathrm{rev/min}$$

The drop in speed is directly proportional to fall in armature voltage.

Learning summary

By the end of this section you should have learnt:

- ✓ to draw the current and voltage waveforms for both diode and thyristor bridge rectifiers with an inductive load;
- ✓ to derive the dc output voltage for diode and thyristor bridge rectifiers with an inductive load;
- ✓ to calculate the speed of a separately excited dc motor supplied via bridge rectifiers.

4.11 Inverter-fed induction motors and their characteristics

The construction and operation of a three-phase ac induction motor connected to a fixed voltage, fixed frequency supply was explained in detail in Part 1, Unit 5. It will suffice to repeat the key points here.

Induction motors are the most common motors encountered in industry and are the most rugged type of motor. They have two parts: an outer stationary frame called the stator and an inner rotating part called the rotor. These are shown in Figure 4.52.

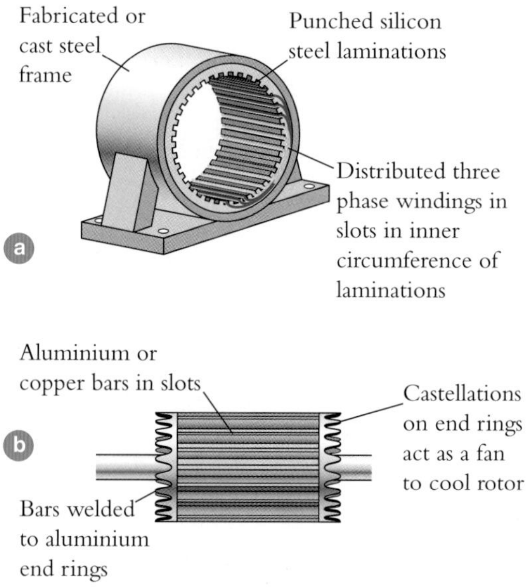

Figure 4.52 (a) Stator and (b) rotor of a simple (squirrel-cage) induction motor

The simplest form of induction motor, the two-pole motor, has three separate coils wound in the slots on the inner circumference of the stator. The coils are mutually displaced by 120°. The rotor has silicon steel laminations keyed to a central shaft. Slots are cut in the laminations into which aluminium or copper conductors are fitted. The ends of the conductors are welded to aluminium end rings. There are no electrical connections to the rotor.

When a balanced three-phase fixed frequency supply is connected to the stator windings, currents flow in the coils which produce a uniform magnetic field that rotates at constant speed in the air gap between the stator and the rotor. In the time it takes the currents to complete one cycle, the magnetic field makes one revolution of the air gap. This is illustrated in the sequence of diagrams, Figure 4.53.

The speed at which the field rotates is called the synchronous speed, n_s. Measured in revolutions per minute, it is related to the frequency of the stator currents, f, in hertz (Hz), as follows.

$$n_s = 60f$$

The rotating magnetic field interacts with the rotor, inducing emfs in the rotor conductors. The latter are short-circuited by the end rings so currents flow in the conductors. These currents create subsidiary magnetic fields around the conductors. The subsidiary fields interact with the rotating stator field to produce forces on the rotor conductors, to create a torque on the rotor and make it turn. This torque is called the electrical or driving torque and its magnitude is proportional to the magnitude of the rotating magnetic field.

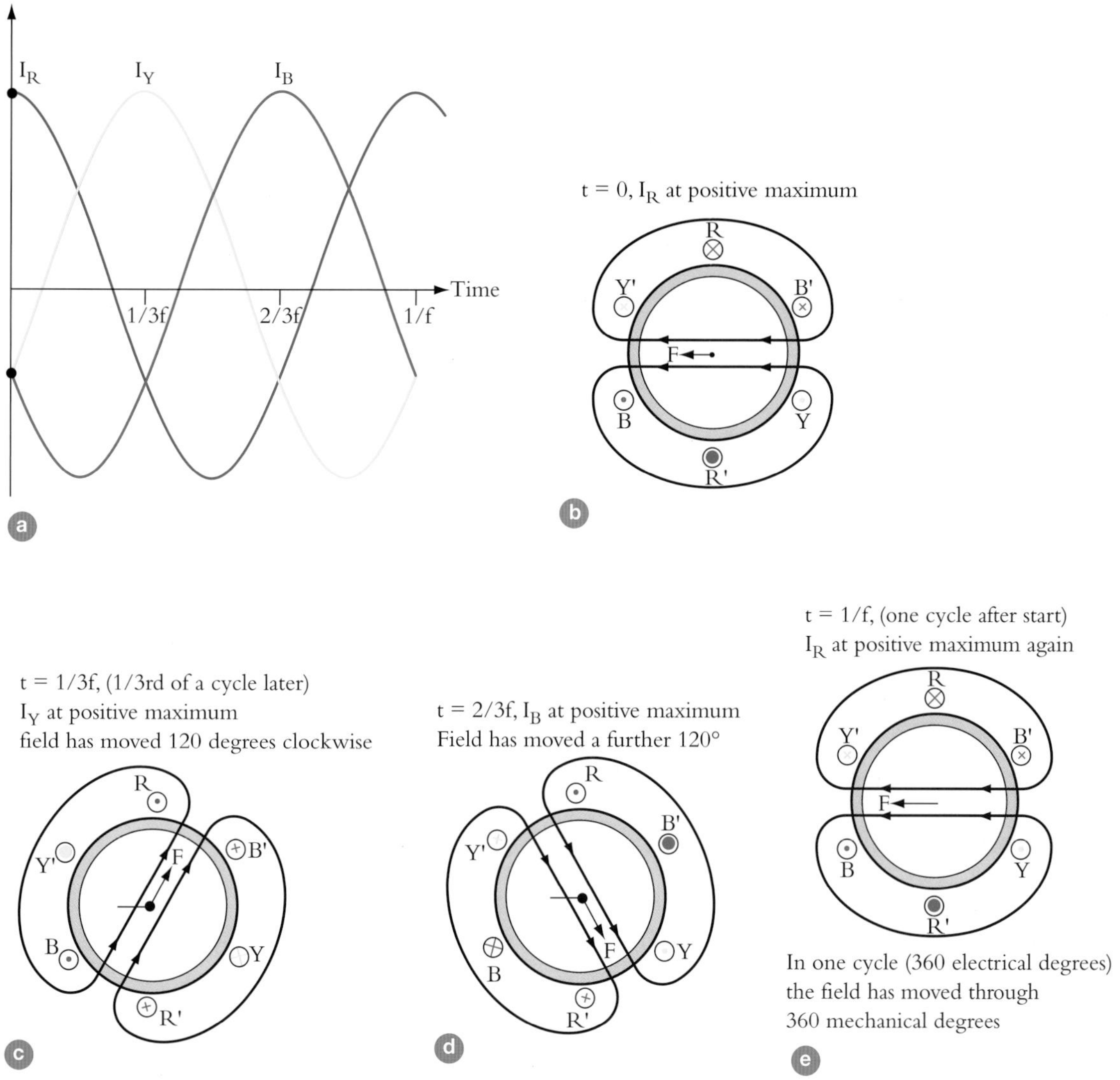

Figure 4.53 (a) Three-phase ac current waveforms; (b)–(e) sequence of energising the windings of a two-pole induction motor

The rotor accelerates until the electrical torque produced by the currents exactly equals the mechanical load torque on the shaft. At this point the rotor is running at a speed, *n*, slightly slower than the synchronous speed. This difference in speed is expressed as a ratio and is known as the 'slip' or 'per unit slip', *s*.

$$s = \frac{n_s - n}{n_s} = 1 - \frac{n}{n_s}$$

The small difference between the speed of the rotating field and that of the rotor is fundamental to the operation of the induction motor.

Figure 4.54 shows a typical torque vs. slip characteristic for a two-pole motor.

When it is operating in steady-state conditions, an increase in the load torque causes the rotor to slow down and the slip to increase. This raises the emfs induced in the rotor conductors and the rotor currents, and creates more field distortion and more driving torque. The slip increases until the driving torque equals the load torque, and steady conditions are re-established. For most motors, the steady state slip varies between around 0.01 on no load and 0.10 when the motor is driving full load. This means that a two-pole motor connected to a 50 Hz supply will run at a steady state speed somewhere between 45 rev/s (2700 rev/min) when $s = 0.1$ and 49.5 rev/s (2970 rev/min) when $s = 0.01$.

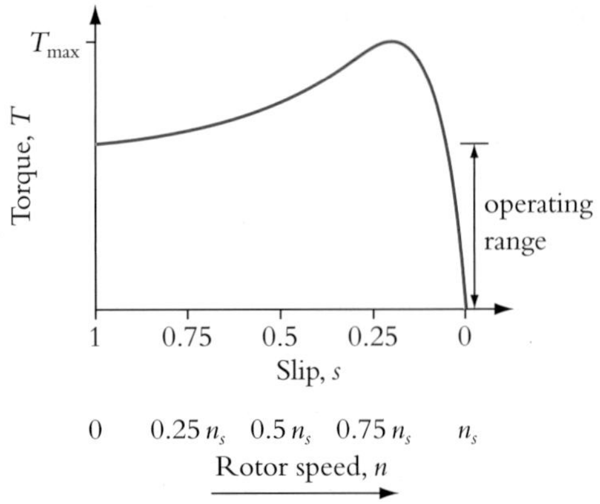

Figure 4.54 Typical torque vs. slip characteristic for a two-pole motor

For many applications these speeds are too high. Manufacturers can reduce the synchronous speed, and hence the rotor speed, by winding the stator coils in a way that effectively increases the number of poles in the rotating magnetic field. This was demonstrated pictorially in Part 1. For example, a four-pole motor, with two N-poles and two S-poles, has a synchronous speed equal to half the supply frequency.

$$n_s = 60f/2 = 30f$$

Connected to a 50 Hz supply the synchronous speed is 1500 rev/min, and the rotor speed varies between around 1350 rev/min when $s = 0.1$ and 1485 rev/min when $s = 0.01$.

Further increases in the number of poles reduce even more the synchronous speed and with it the rotor speed, as demonstrated in Table 4.1. Industrial induction motors are typically rated to run at a slip value in the order of 0.04 (4 per cent).

No. of poles	Pairs of poles	Synchronous speed, n_s as a fraction of frequency, f	Synchronous speed in rev/s when $f = 50$ Hz	Synchronous speed in rev/min when $f = 50$ Hz	Rotor speed in rev/min when $s = 0.1$	Rotor speed in rev/min when $s = 0.04$	Rotor speed in rev/min when $s = 0.01$
2	1	f	50	3000	2700	2880	2970
4	2	$f/2$	25	1500	1350	1440	1485
6	3	$f/3$	16.67	1000	900	960	990
8	4	$f/4$	12.5	750	675	720	742.5
$2p$	p	f/p	$50/p$	$3000/p$	$2700/p$	$2880/p$	$2970/p$

Table 4.1

With a fixed number of poles and fixed supply frequency the variation in rotor speed is very limited. To operate an induction motor over a wide speed range it must be connected to a variable frequency supply.

Inverters produce variable frequency supplies. They convert direct current to alternating current of a chosen frequency. To understand their operation you should consider the simple single-phase bridge inverter shown in Figure 4.55. Diagonally opposite pairs of transistors conduct in turn. When transistors Tr_1 and Tr_4 are switched on and Tr_2 and Tr_3 are off, the current flows from the positive dc rail through Tr_1, the load and Tr_4 to the negative rail, travelling through the load from left to right. When Tr_2 and Tr_3 are turned on and Tr_1 and Tr_4 are switched off, the current flows through the load in the reverse direction as it travels from the positive dc rail to the negative. Diodes are connected in parallel with the transistors to short-circuit the large back emfs generated when the magnetic field reverses, that would otherwise destroy the transistors as they are turned off.

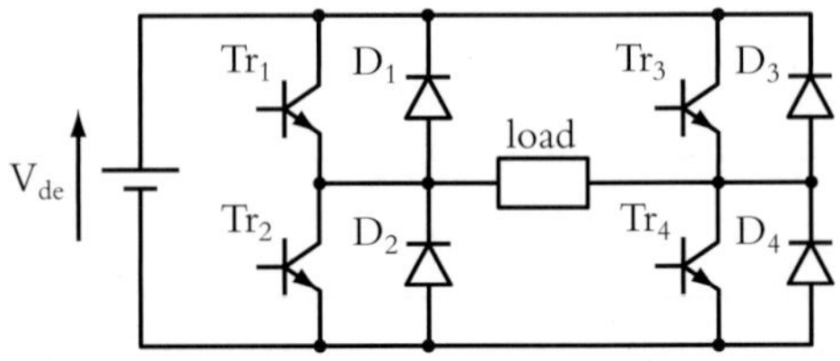

Figure 4.55 Single-phase bridge inverter circuit

By turning each diagonal pair of transistors on for slightly less than half a cycle, the load voltage waveform shown in Figure 4.56(a) is created. Having a short period with no current at the beginning and end of each half cycle reduces the risk of two transistors in the same arm conducting simultaneously and short circuiting the dc supply. The output frequency is varied by altering the conduction times of the transistors, as illustrated in Figure 4.56(b). A more comprehensive description of inverter operation can be found in Hughes (2006).

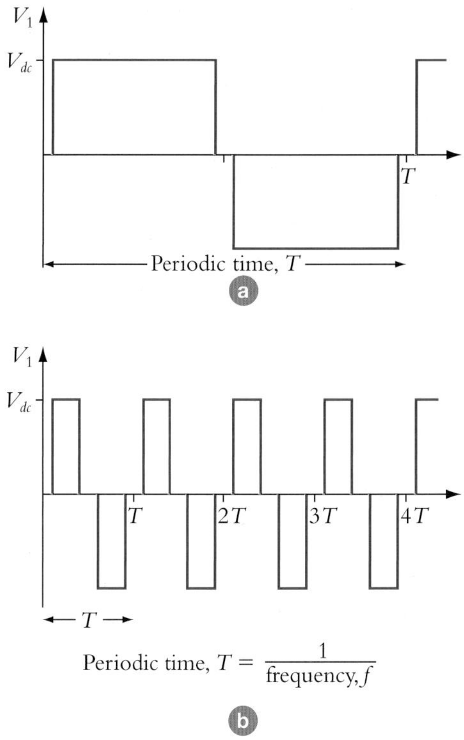

Figure 4.56 Output waveforms for simple bridge inverter circuit (a) giving a low frequency and (b) giving a higher frequency output

To control the magnitude of the load current, pulse width modulation (PWM) is often used. This involves the removal of a series of notches in each half cycle to produce a waveform like that in Figure 4.57(a). One way this is achieved is to switch Tr_4 on and off when Tr_1 is conducting and to turn Tr_2 on and off when Tr_3 is on, taking precautions to avoid two transistors on the same arm being switched on simultaneously. The mean output voltage is then equal to:

$$\text{mean output voltage} = V_1 = V_{dc} \times \frac{t_{on}}{(t_{on} + t_{off})}$$

where t_{on} = duration of the pulses, and t_{off} = time between pulses

To alter the mean output voltage the times at which Tr_2 and Tr_4 are switched on and off are adjusted to change the ratio

$$\frac{t_{on}}{(t_{on} + t_{off})}$$

In more sophisticated systems, microprocessors are used to vary the duration of the pulses and the spaces between pulses. This is shown in Figure 4.57(b). In this case, the average output voltage is given by:

$$\text{Mean output voltage} = V_{dc} \times \frac{\text{total duration of pulses in a half cycle}}{\text{duration of half cycle}}$$

$$V_{dc} \times \frac{\Sigma t_{on}}{\Sigma(t_{on} + t_{off})}$$

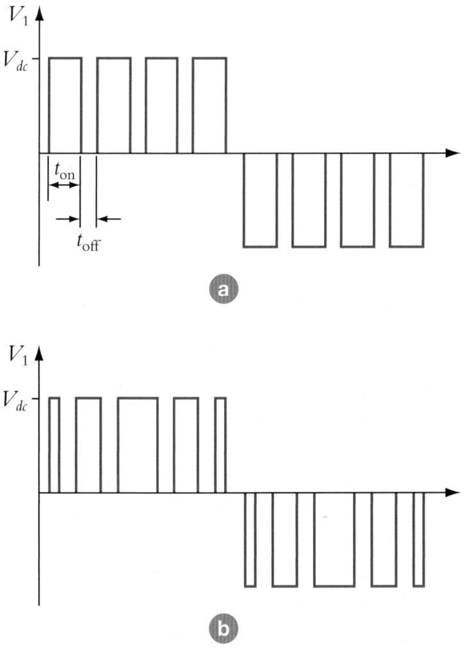

Figure 4.57 PWM output waveforms with (a) constant pulse cycle over each half-wave and (b) variable pulse cycle

For an induction motor supplied by an inverter to operate properly the number, width and spacing of the pulses must be chosen with extreme care and controlled by a microprocessor to make the load current waveform resemble as closely a possible a sine wave. Looking at the voltage waveforms in Figure 4.57, this might seem impossible. But, all repetitive waveforms with a fixed periodic time comprise an infinite series of sine and cosine waves of different magnitudes and frequencies, known as a Fourier series. Most textbooks on electric circuit theory explain how to determine the Fourier series for any waveform but we will limit our discussion to the example of the square wave illustrated in Figure 4.58a, for which the Fourier series begins:

$$V(t) = \frac{4V}{\pi}\left(\sin \omega t + \tfrac{1}{3}\sin 3\omega t + \tfrac{1}{5}\sin 5\omega t + \ldots\right) \text{ where } \omega = 2\pi f$$

The first term in the bracket is a sine wave of the same frequency as the square wave. This is called the fundamental. The other sine waves are called harmonics. They have smaller magnitudes, and their frequencies are multiples of the fundamental. Figure 4.58(b) shows the result of adding the first three terms in the series. As more terms are added, the waveform becomes closer to a square wave.

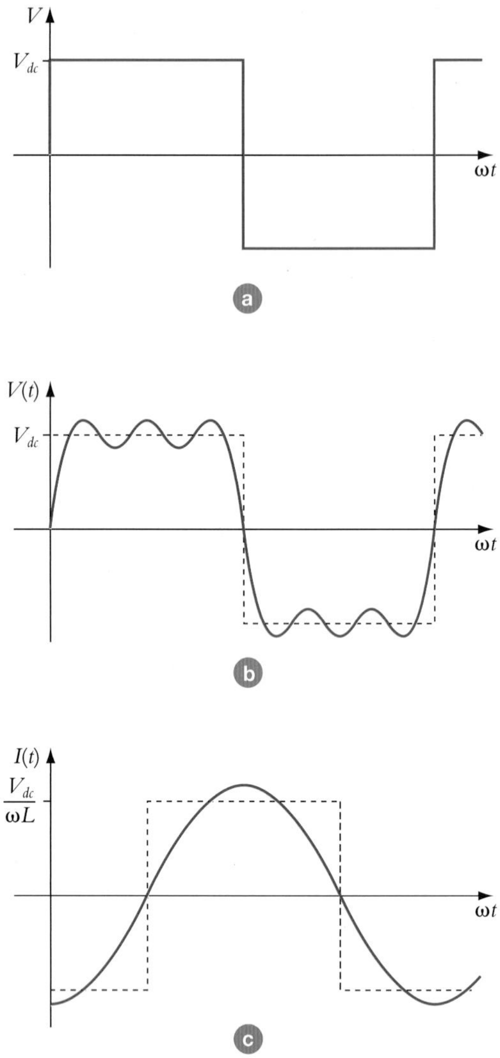

Original square wave

$$V(t) = \frac{4V_{dc}}{\pi}(\sin\omega t + \tfrac{1}{3}\sin 3\,\omega t + \tfrac{1}{5}\sin 5\,\omega t)$$

$$I(t) = \frac{4V_{dc}}{\pi\omega L}(\cos\omega t + \tfrac{1}{3}(\tfrac{1}{3}\cos 3\,\omega t) + \tfrac{1}{5}(\tfrac{1}{5}\cos 5\,\omega t)$$

Figure 4.58 (a) Voltage waveform in the form of a square wave; (b) square wave voltage waveform approximated by summation of fundamental and first two harmonics; (c) resulting current waveform through an inductive load

The impedance of a motor is a combination of resistance, R and inductive reactance, X_L, where

$$X_L = 2\pi fL$$

which is directly proportional to the frequency. As a result, the reactance and hence the impedance of the motor are both larger at the harmonic frequencies than at the fundamental frequency. So, when a non-sinusoidal voltage, like that shown in Figure 4.58(b) is applied to a motor the harmonic components of the current waveform are attenuated more than the fundamental, producing a current waveform that more closely resembles a sine wave, as illustrated in Figure 4.58(c).

In drawing these curves it was assumed that the resistance was negligible.

Even when a pulse-modulated voltage waveform, like that in Figure 4.59(a), is applied, the current waveform is approximately sinusoidal, see Figure 4.59(b). The reader needing more details is again directed to Hughes (2006).

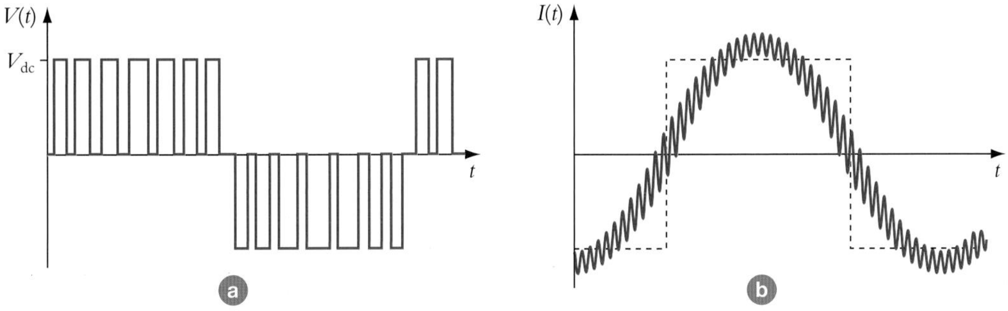

Figure 4.59 **(a) Voltage waveform synthesised via pulse width modulation, and (b) resulting current waveform through an inductive load**

Three-phase variable frequency inverters are used to supply induction motors, as shown in Figure 4.60. Three-phase inverters are similar to single-phase inverters. The main difference is that the former has three arms to the latter's two. The inverter supplies the currents to the motor stator windings to produce a rotating magnetic field in the air gap between the stator and rotor. Adjusting the frequency of the currents controls the rotational speed of the field (i.e. synchronous speed), and hence the rotor speed.

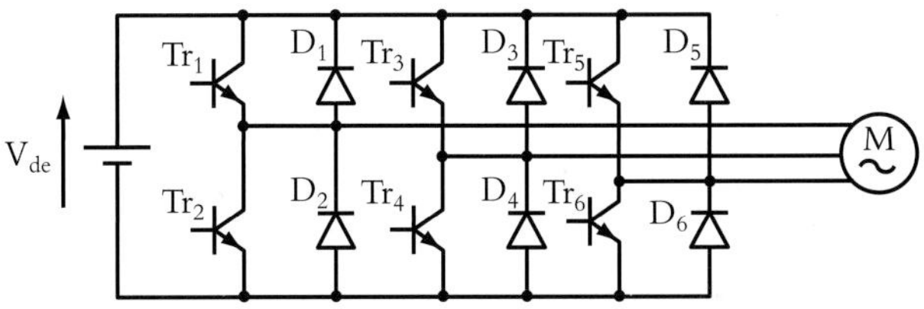

Figure 4.60 **Three-phase inverter circuit**

Figure 4.53 showed the stator currents and magnetic field distribution in a two-pole induction motor at four points in a cycle. The currents were balanced, and the field had constant magnitude and rotated at constant speed. Analysing the current distribution at each instant in the cycle (which is beyond the scope of this book) it can be shown that the currents combine to produce a magnetomotive force (mmf) of constant magnitude equal to

$$\text{mmf} = 1.5\, IN$$

where I = stator phase current, N = turns in each phase of the stator winding.

The magnitude of the mmf is independent of frequency and equals the combined mmf drops in the air gap between the stator and rotor, $H_{air}l_{air}$ and in the steel cores, $H_{steel}l_{steel}$, where H_{air} = magnetic field strength in the air gap, l_{air} = total distance flux travels through air, which is the length of the air gap between the stator and rotor multiplied by two, as the magnetic field has to cross the air gap twice, H_{steel} = magnetic field strength in the steel cores, and l_{steel} = total length of the magnetic path in the steel cores.

The reluctance of the air gaps is greater than that of the steel cores, so it will be assumed the latter can be ignored, and the mmf drop in the air gaps ($H_{air}l_{air}$) is directly proportional to the mmf produced by the stator currents.

$$H_{air}l_{air} \propto IN$$

The magnetic flux density, B in air is given by:

$$B = \mu_0 H_{air}$$

where μ_0 = constant, $4\pi \times 10^{-7}$ H/m

So the flux density is proportional to the mmf produced by the stator currents, then

$$B \propto IN$$

As the air gap between the stator and rotor is uniform it may be concluded that the air gap flux, ϕ, is also proportional to the mmf produced by the sinusoidal stator currents and will have constant magnitude.

When the magnetic field rotates it links with the stator windings. Figure 4.53(b) shows the situation in a two-pole motor at time, $t = 0$ when maximum flux links with the red phase coil, R-R′.

The field is rotating at the synchronous speed, n_s rev/min, which related to the frequency, f, by:

$$n_s = 60f$$

so the angular velocity of the field, ω equals:

$$\omega = 2\pi f$$

The flux linkage with a coil, ψ is defined as:

$$\text{flux linkage} = \text{flux linking a coil} \times \text{number of turns in the coil}$$

$$\psi = \phi N \tag{4.65}$$

Taking the maximum flux linkage with the red phase coil R-R′ at time, $t = 0$ to be ψ_m the general expression for the flux linkage with the red phase is given by:

$$\psi = \psi_m \cos \omega t$$

According to Faraday's law, the emf induced in a coil is:

$$e = N\frac{d\phi}{dt} \tag{4.66}$$

so the emf induced in the red coil will equal:

$$e_R = -\omega\psi_m \sin \omega t \tag{4.67}$$

The emfs induced in the yellow and blue coils can be determined similarly. Figure 4.53(c) shows that maximum flux linkage with the yellow phase occurs one third of a cycle (or $2\pi/3$ radians) later than the red phase, so:

$$\text{flux linkages with the yellow phase coil} = \psi_m \cos(\omega t - 2\pi/3)$$

and the induced voltage in yellow coil,

$$e_Y = -\omega\psi_m \sin(\omega t - 2\pi/3) \tag{4.68}$$

Maximum flux linkage occurs with the blue phase, two-thirds of a cycle (or $4\pi/3$ radians) later than with the red coil (Figure 4.53(d)), so:

$$\text{Flux linkages with the blue phase coil} = \psi_m \cos(\omega t - 4\pi/3)$$

and the induced voltage in blue coil,

$$e_B = -\omega\psi_m \sin(\omega t - 4\pi/3) \tag{4.69}$$

Thus from equations (4.67), (4.68) and (4.69) the rms value for induced emf in each phase, E may be found

$$E = \frac{\omega\psi_m}{\sqrt{2}} \tag{4.70}$$

But $\omega = 2\pi f$, and from equation (4.65) $\psi_m = N\phi_m$

so

$$E = \frac{2\pi f N\phi_m}{\sqrt{2}} \tag{4.71}$$

The induced emf in each phase of the stator in an ac induction motor opposes the applied phase voltage, V. If the stator winding impedance is considered to be negligible, there will be no volt drop in the stator winding and

$$|E| = |V_p|.$$

Rearranging equation (4.71) and substituting for E

$$\phi_m = \frac{\sqrt{2}\,V_p}{2\pi f N} \tag{4.72}$$

Let $k = \text{constant} = \dfrac{1}{\sqrt{2}\pi N}$

$$\phi_m = k\frac{V_p}{f} \tag{4.73}$$

The magnitude of the rotating magnetic flux is directly proportional to the applied voltage, V_p and inversely proportional to the frequency, f. To keep the magnitude of the flux constant the frequency and the voltage must be adjusted simultaneously (Hughes, 2006).

As explained earlier, the motor torque is proportional to the magnitude of the rotating magnetic flux. So with $\phi_m = k\dfrac{V_p}{f}$ (equation (4.73)) we can also say that the torque is also proportional to the applied voltage divided by the frequency (V/f).

When a variable frequency inverter is used to drive an induction motor, the output frequency of the inverter is set to produce the required speed of flux rotation in the motor, i.e. synchronous speed, and the inverter output voltage adjusted to produce full load torque.

To alter the motor speed and maintain full load torque the inverter frequency and output voltage must be adjusted simultaneously, to keep the V/f ratio, and hence the magnetic flux, constant (Hughes, 2006).

Figure 4.61(a) shows the effect on the motor torque versus speed characteristics of keeping (V/f) constant at different frequencies.

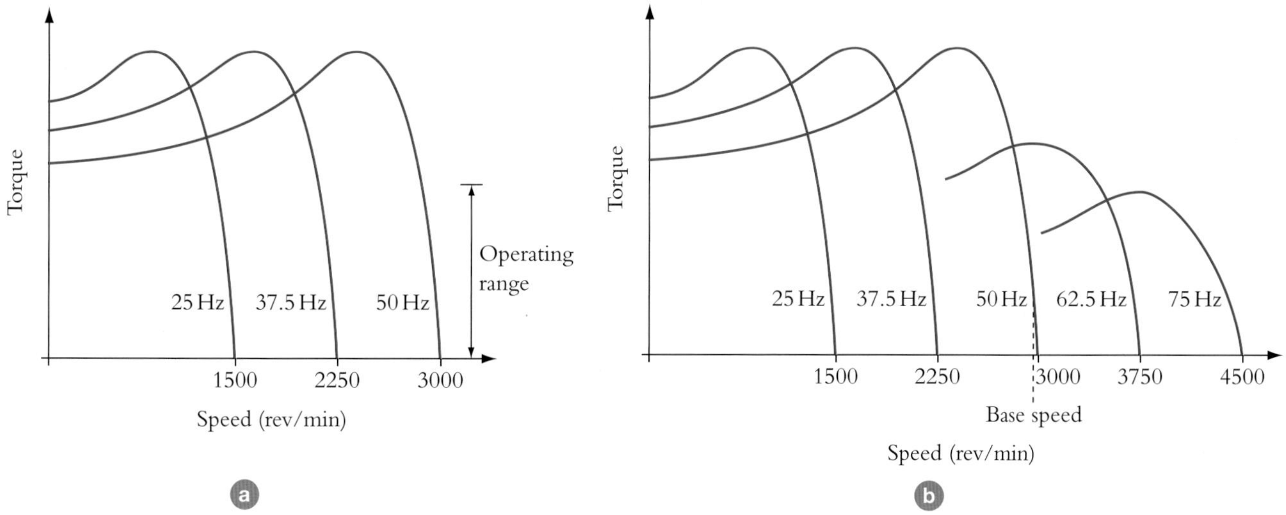

Figure 4.61(a) Torque–speed characteristics of induction motor operated at varying frequencies with constant V/f ratio; (b) reduced torque above 'base speed' where voltage is held constant as frequency increases

The simultaneous increase in inverter frequency and voltage is only possible until the voltage reaches the rated value for the inverter and motor, as specified by the manufacturers. The frequency at this point is called the 'base speed' and is usually set by manufacturers at 50 Hz or 60 Hz.

It is unsafe to operate electrical equipment above its rated voltage. If the inverter frequency is increased above the base speed, the voltage must be held constant at its rated value and the magnitude of the flux allowed to fall. The drop in flux reduces the emfs induced in the rotor conductors, the currents flowing through the conductors, the forces on the conductors and hence the torque on the rotor. As a result there is a fall in torque at frequencies above the base speed, as shown in Figure 4.61(b). The reader needing more details is again directed to Hughes (2006).

In the constant voltage region (i.e. when running at frequencies above the rated frequency the torque–speed curve is less steep. Above base speed, the maximum allowable power of the motor is constant and the allowable torque reduces inversely with the speed. Hughes (2006) presents an interesting discussion on the limits to the torque–speed envelope for induction motors. He suggests that the constant power region extends to around twice base speed (limited by the peak torque available at a given frequency) and within this region the motor may be operated at higher slips than are permitted below base speed.

All the curves in Figure 4.61(a) and (b) show a small drop in motor speed as the load torque is increased in the normal operating range. In situations where this fall in speed is unacceptable, a closed-loop speed control system, like that shown in Figure 4.62(a) must be used.

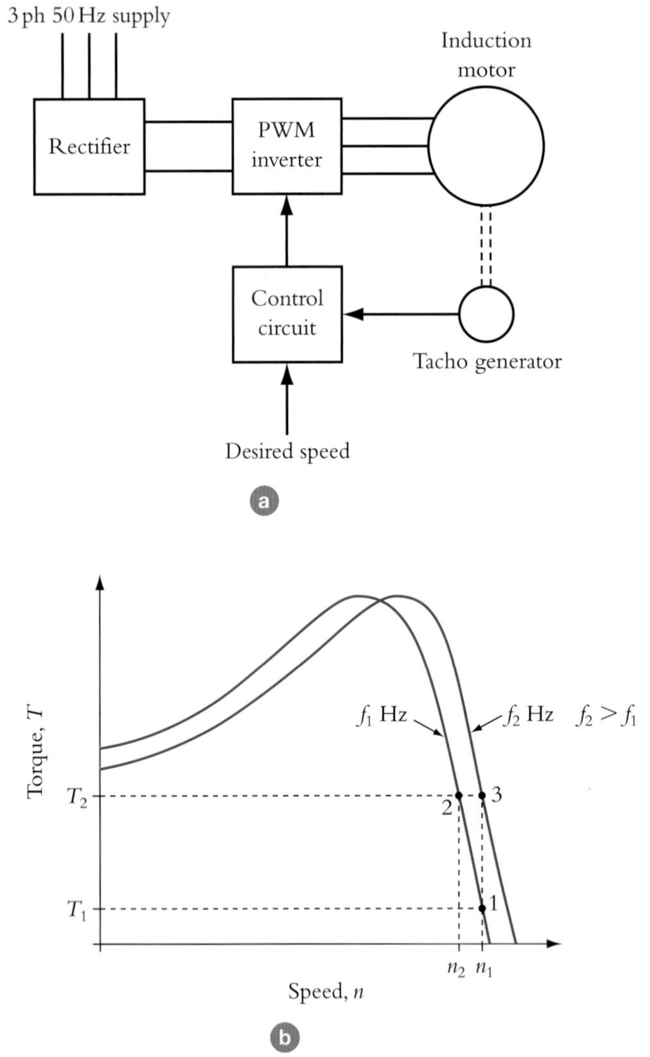

Figure 4.62 (a) Closed-loop control of inverter drive; (b) frequency changed by control loop to restore motor running speed to desired value n_1 despite torque increase from T_1 to T_2

Such systems compensate automatically for any change in motor speed by raising the inverter output frequency as the load torque is increased. Figure 4.62(b) illustrates this. Assume that the motor is running on no load, the torque is T_1 and the speed is n_1 (at operating point 1). Without closed-loop speed control an increase in load torque to T_2 will cause the speed to fall to n_2, (operating point 2) but with closed-loop control, the frequency will be increased incrementally as the load torque rises to T_2 to maintain the speed constant at n_1 (operating point 3).

Approximate characteristics of induction motors fed from inverters

The analysis of an induction motor was covered in Part 1 and the information can be used to draw a torque–speed curve such as that shown in Figure 4.54. The non-linear nature of the torque–speed equation underlying this curve makes it difficult to use this equation within drive system calculations, so it can be convenient to derive approximate torque–speed relationships relating to the operating region of the motor as shown in Figure 4.61. Within this range the torque–speed characteristic of the motor is approximately linear. It was shown in equation (5.147) in Part 1 that the torque–slip characteristics of a simple three-phase AC induction motor are given by the following equation:

$$\text{Torque} = \frac{3p}{\omega}\,\frac{E_2{}^2 as}{X_2(a^2 + s^2)} \tag{4.74}$$

where:

p = number of pairs of poles on stator

$\omega = 2\pi f$ = circular frequency of mains supply (rad/s)

E_2 is the emf induced in each phase of the rotor winding when the rotor is stationary

X_2 is the rotor reactance per phase when the rotor is stationary

$a = R_2/X_2$ = ratio of resistance to reactance of rotor winding

$s = \text{per-unit slip} = \dfrac{n_s - n}{n_s}$

where n_s and n are the synchronous speed and the running speed of the motor respectively.

Using equations (5.111), (5.112) and (5.116) from Part 1, the stator phase voltage V_1 can be expressed as:

$$V_1 = E_1 = \frac{N_1}{N_2}E_2 \tag{4.75}$$

and the rotor standstill reactance per phase, referred to the stator, is:

$$X_1 = \frac{N_1{}^2}{N_2{}^2}X_2 \tag{4.76}$$

so that equation (4.74) becomes:

$$\text{Torque} = \frac{3p}{\omega}\,\frac{V_1{}^2 as}{X_1(a^2 + s^2)} \tag{4.77}$$

Note also that rotor resistance per phase, referred to the stator, is:

$$R_1 = \frac{N_1{}^2}{N_2{}^2}R_2 \tag{4.78}$$

so that a can be re-expressed as

$$a = \frac{R_1}{X_1} \tag{4.79}$$

As all quantities are henceforward only referred to the stator, the subscripts for V_1, X_1 and R_1 can be dropped from equations (4.77) to (4.79).

The operating region of the motor typically involves slip values in the range 1–10 per cent, with the most common value being around 4 per cent. This is around a fifth of the typical value of a (which is likely to be around 0.2 or 20 per cent), so that $(a^2 + s^2)$ can be reasonably approximated as a^2. Noting the definitions of a, ω and s, and that

$$n_s = \frac{60f}{p} \text{ (in rev/min)}$$

Equation (4.77) can be approximated as:

$$\text{Torque} \approx \frac{3p}{\omega}\frac{V^2 s}{R} \tag{4.80}$$

and hence as:

$$\text{Torque} \approx \frac{1}{60}\frac{3p^2}{2\pi R}\left(\frac{V}{f}\right)^2\left(\frac{60f}{p} - n\right) \tag{4.81}$$

or, relating the torque instead to the angular velocity of the rotor, ω_r, measured in rad/s:

$$\text{Torque} \approx \frac{3p^2}{4\pi^2 R}\left(\frac{V}{f}\right)^2\left(\frac{2\pi f}{p} - \omega_r\right) \tag{4.82}$$

In practice, nearly all induction motors encountered today are squirrel-cage motors so R will be constant but unknown. By adapting an approach described (for the constant-frequency situation) by Wildi (1991), it is often easier to infer the approximate torque–speed characteristics from the design data ('nameplate data') for the motor, involving its rated power, P_{rated}, rated speed, n_{rated}, and at its rated frequency, f_{rated}, and voltage, V_{rated}, and to write an expression roughly equivalent to equation (4.81) as:

$$\text{Torque} \approx K\left(\frac{f_{\text{rated}}}{V_{\text{rated}}}\right)^2\left(\frac{V}{f}\right)^2\left(\frac{60f}{p} - n\right) \tag{4.83}$$

or as:

$$\text{Torque} \approx \frac{60K}{2\pi}\left(\frac{f_{\text{rated}}}{V_{\text{rated}}}\right)^2\left(\frac{V}{f}\right)^2\left(\frac{2\pi f}{p} - \omega_r\right) \tag{4.84}$$

where

$$K = \frac{60P_{\text{rated}}}{2\pi n_{\text{rated}}\left(\frac{60f_{\text{rated}}}{p} - n_{\text{rated}}\right)}$$

(Strictly speaking, equations (4.81) and (4.82) relate to drawing a straight-line tangent (at $s = 0$) to the torque–speed curve, while the approach of Wildi (1991) corresponds to drawing a straight line or chord between the points on the curve corresponding to zero slip and rated slip. In practice, the difference between these lines is small.)

Using these equations it is now possible to generate the approximately linear regions of the families of torque–speed curves given in Figure 4.54 and Figure 4.61:

- For the single curve shown in Figure 4.54 relating to a motor driven directly from its rated power supply $f = f_{\text{rated}}$, the approximate torque–speed characteristic is simply given by inserting the rated frequency and voltage into equation (4.83), which reduces to:

$$\text{Torque} = K\left(\frac{60f_{\text{rated}}}{p} - n\right) \tag{4.85}$$

- For the curves relating to $f < f_{\text{rated}}$ as shown in Figure 4.61(a) and the left-hand part of Figure 4.61(b), f and V are varied in proportion, as described earlier within the present section, so that their ratio remains constant, and equation (4.83) now reduces to:

$$\text{Torque} = K\left(\frac{60f}{p} - n\right) \tag{4.86}$$

- For the curves relating to $f > f_{\text{rated}}$, as shown in the right-hand part of Figure 4.61(b), the voltage remains at the rated value V_{rated} while the frequency increases, so equation (4.83) now reduces to:

$$\text{Torque} = K\left(\frac{f_{\text{rated}}}{f}\right)^2\left(\frac{60f}{p} - n\right) \tag{4.87}$$

- For completeness, it can be stated that for the situation where a motor is driven at its rated frequency at reduced voltage (for example, via a variable transformer or a so-called 'soft starter') equation (4.83) reduces to:

$$\text{Torque} = K\left(\frac{V}{V_{\text{rated}}}\right)^2\left(\frac{60f}{p} - n\right) \tag{4.88}$$

All of these approximations are valid within the approximately linear region of the motor's torque–speed curve, which applies when the actual value of torque lies within the following limit (extended from Wildi (1991) to consider varying frequency):

$$\text{Torque} \leqslant (\text{Rated torque}) \times \left(\frac{f_{\text{rated}}}{V_{\text{rated}}}\right)^2\left(\frac{V}{f}\right)^2 \tag{4.89}$$

where:

$$\text{Rated torque} = \frac{60P_{\text{rated}}}{2\pi n_{\text{rated}}} \tag{4.90}$$

It should be emphasized that this limitation on the applicability of the linearly approximated torque–speed characteristic does not necessarily correspond to the limitation on allowable torque for the motor. Once again, the reader is referred to Hughes (2006) for a discussion of the allowable operating envelope for an inverter-fed motor.

Worked example

A 25 kW, 415V (line-to-line voltage), 50 Hz, 1440 rev/min squirrel-cage induction motor is fed from a variable frequency inverter. The voltage and frequency are varied in proportion, up to the rated frequency; the voltage is held constant above this frequency. This arrangement is used to drive a blower that has a torque–speed characteristic given by $L = 5.556 \times 10^{-5} n^2$ where torque L is in Nm and n is in rev/min. Determine the inverter frequency and line-to-line voltage at blower speeds of 720 and 1560 rev/min.

For this motor the rated power is 25 000W and the rated speed is $n_{\text{rated}} = 1440$ rev/min.

It can be seen from Table 4.1 that for a rated speed of 1440 rev/min the synchronous speed is almost certainly 1500 rev/min and hence the motor is a four-pole motor with $p = 2$. Therefore:

$$K = \frac{60P_{\text{rated}}}{2\pi n_{\text{rated}}\left(\dfrac{60f_{\text{rated}}}{p} - n_{\text{rated}}\right)}$$

$$= \frac{60 \times 25000}{2\pi \times 1440 \times \left(\dfrac{60 \times 50}{2} - 1440\right)} = 2.763\ \text{Nm min/rev}$$

For the blower at $n = 720$ rev/min:

$$\text{Torque} = 5.556 \times 10^{-5} \times 720^2 = 28.8\ \text{Nm}$$

720 rev/min is below the rated speed, so $V/f = V_{\text{rated}}/f_{\text{rated}}$, and equation (4.86) applies. This can be rearranged to give:

$$f = \frac{p}{60}\left(\frac{\text{Torque}}{K} + n\right)$$

$$= \frac{2}{60}\left(\frac{28.8}{2.763} + 720\right) = 24.35\ \text{Hz}$$

The rated line voltage is 415 V so the rated phase voltage is 415/√3 = 239.6 V

The voltage is obtained by reducing the rated voltage in proportion to the frequency:

$$V = V_{\text{rated}} \times \frac{f}{f_{\text{rated}}} = 239.6 \times \frac{24.35}{50} = 116.7\text{V}$$

so the line-to-line voltage is 116.7 × √3 = 202.1 V.

It is important to check that the approximately linear operating region, and hence the applicability of the approximate equation, have not been exceeded. Noting that the rated phase voltage is 239.6 V and the actual phase voltage is 116.7 V, then using equation (4.89):

$$\text{Maximum permissible torque} = (\text{Rated torque}) \times \left(\frac{f_{\text{rated}}}{V_{\text{rated}}}\right)^2 \left(\frac{V}{f}\right)^2$$

$$= \frac{60 \times 25\,000}{2\pi \times 1440} \times \left(\frac{50}{239.6}\right)^2 \left(\frac{116.7}{24.24}\right)^2 = 165.8\,\text{Nm}$$

which is very far from being exceeded by the actual torque of 28.8 Nm.

For the blower at n = 1560 rev/min:

$$\text{Torque} = 5.556 \times 10^{-5} \times 1560^2 = 135.2\,\text{Nm}$$

1560 rev/min is above the rated speed, so $V = V_{\text{rated}}$, and equation (4.87) applies. This can be rearranged to give:

$$\text{Torque} = T = 60K\frac{f_{\text{rated}}^2}{pf} - K\left(\frac{f_{\text{rated}}}{f}\right)^2 n$$

Multiplying by f^2 and rearranging gives:

$$Tf^2 - 60K\frac{f_{\text{rated}}^2}{p}f + Kf_{\text{rated}}^2 n = 0$$

Inserting the values for the present problem gives:

$$135.2f^2 - 60 \times 2.763 \times \frac{50^2}{2}f + 2.763 \times 50^2 \times 1560 = 135.2f^2 - 207\,250f + 10\,775\,700 = 0$$

This is a quadratic in f which has the solutions:

$$f = \frac{207\,250 \pm \sqrt{207\,250^2 - 4 \times 135.2 \times 10\,775\,700}}{2 \times 135.2}$$

$$= 53.9\,\text{Hz or } 1479\,\text{Hz}$$

As the new running speed is only slightly above the rated speed of 1440 rev/min, we would expect the frequency to be slightly above the rated frequency of 50 Hz, so the first of the alternative solutions is clearly the correct one. Recall also that $V = V_{\text{rated}}$ = 239.6V.

Again it is important to check using equation (4.89) that applicability of the approximate equation has not been exceeded:

$$\text{Maximum permissible torque} = (\text{Rated torque}) \times \left(\frac{f_{\text{rated}}}{V_{\text{rated}}}\right)^2 \left(\frac{V}{f}\right)^2$$

$$= \frac{25\,000}{2\pi \times 24} \times \left(\frac{50}{239.6}\right)^2 \left(\frac{239.6}{53.9}\right)^2 = 142.7\,\text{Nm}$$

which is slightly greater than the actual torque of 135.2 Nm so the motor is (by a small margin) within its linear region and the calculations are valid.

Learning summary

By the end of this section you should have learnt:

- ✓ the principles of operation of an induction motor;
- ✓ the operation of a simple inverter;
- ✓ the principles of pulse width modulation;
- ✓ induction motor torque is proportional to the applied voltage divided by frequency (V_p/f);
- ✓ at frequencies above 'base speed', the torque falls as the frequency increases;
- ✓ approximately linear torque–speed characteristics of induction motors can be obtained for relatively low values of slip.

4.12 Other sources of power: pneumatics and hydraulics

Electric motors (particularly dc permanent magnet motors for light duties and ac induction motors for heavier duties) are by far the most common means of providing drive to mechanical systems. However, in certain circumstances it may be appropriate to use other means of providing drive, notably the use of pneumatic and hydraulic drives. Electric motors are cheap, efficient, reliable and controllable, but they can be heavy in relation to the power needed, they can overheat or burn out if stalled, and they are not ideal for providing either very large forces or long linear movements. Pneumatic and hydraulic actuators and transmission systems can in some situations overcome these problems.

Pneumatics

Pneumatic actuators can be used in preference to electrical actuators for the following reasons:

- better power-to-weight ratio (i.e. less weight for a given power output)
- better power-to-space ratio (i.e. smaller for a given power output)
- high torque of pneumatic motor at stall, controlled via pressure and with no damage to the motor being caused by stall
- long stroke of pneumatic cylinder.

A widely used form of pneumatic motor, used to provide continuous drive, is the vane motor, which operates as shown in Figure 4.63.

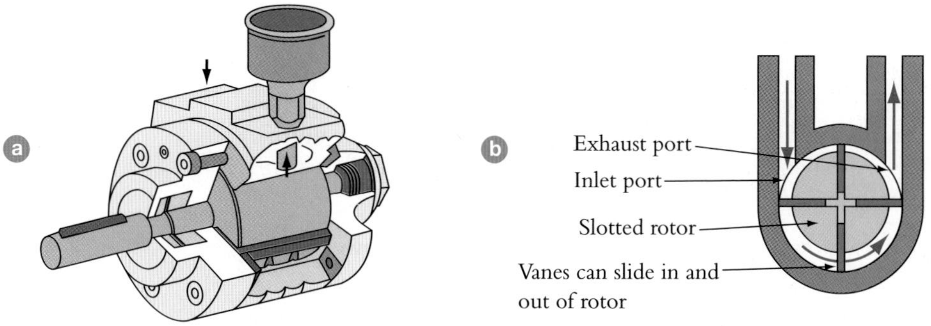

Figure 4.63 **Pneumatic vane motor: (a) cutaway view and (b) simplified diagram**

Other types of pneumatic motor are based on pistons.

The characteristics of pneumatic vane motors can be approximated as a straight-line graph joining the high stall torque to the high no-load speed (Figure 4.64), typically with a small elbow representing reduced torque on start-up.

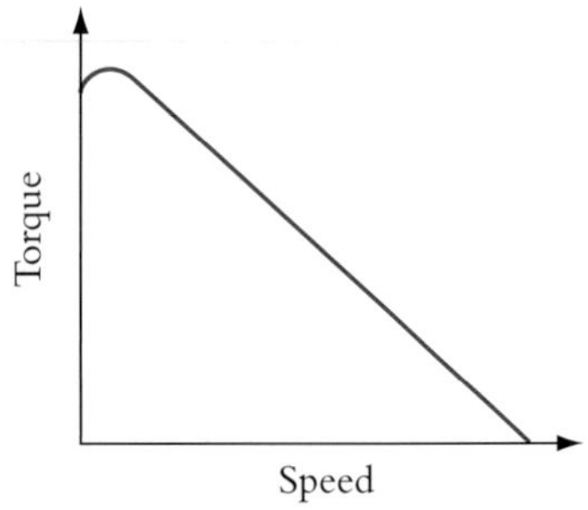

Figure 4.64 Characteristic of a pneumatic vane motor

A particular advantage of pneumatic motors is that their characteristics can easily be altered by:

- varying the pressure (which changes both the stall torque and the no-load speed) as shown in Figure 4.65(a);
- throttling the inlet or the outlet (which primarily changes the no-load speed) as shown in Figure 4.65(b).

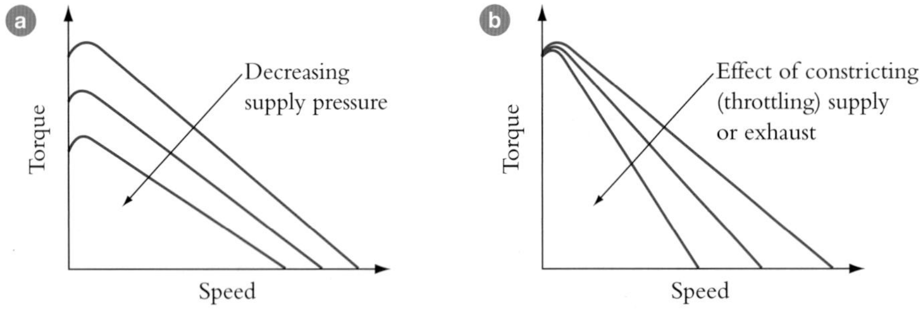

Figure 4.65 Effect of (a) varying the supply pressure and (b) throttling a pneumatic vane motor

Typical applications of rotary pneumatic motors include:

- pneumatic wrenches and similar devices for tightening bolts in assembly lines, tyre service stations etc. – these exploit both the good power-to-weight ratio and the stall-tolerant characteristics of pneumatic motors;
- bottling plant, where bottle tops need to be tightened, again exploiting the ability of pneumatic motors to provide a pressure-limited value of stall torque without damage.

Pneumatics are also particularly useful for providing linear motion, primarily via pneumatic cylinders. These provide moderately large forces over a stroke or displacement ranging from a few millimetres up to the order of a metre. When used in conjunction with solenoid-operated valves, they provide a means of obtaining rapid linear motion under electrical or computer control, and are widely used in automated production environments. The force from a pneumatic cylinder is given by:

$$F = pA \qquad \textbf{(4.91)}$$

where F is the force obtained, p is the pressure of the compressed air supply and A is the area of the piston. In practice, the available force may be a little lower than this, due to friction between the piston seals and the cylinder.

Pneumatic systems

In order to be able to use a pneumatic motor or actuator, it is of course necessary to have a source of compressed air. There is no pneumatic equivalent of the electricity supply grid, although most factories, engineering laboratories etc. are often equipped with a compressed air supply, which may be supplied centrally within the factory etc. or via local compressor units. While details of different systems vary depending upon size, cost etc., a compressed air system typically includes the following components (Figure 4.66):

- electric motor (or other mechanical power source) to provide the mechanical power to the system;
- several filters, including an inlet filter, to remove airborne particles, which may cause damage to the compressor or actuators;

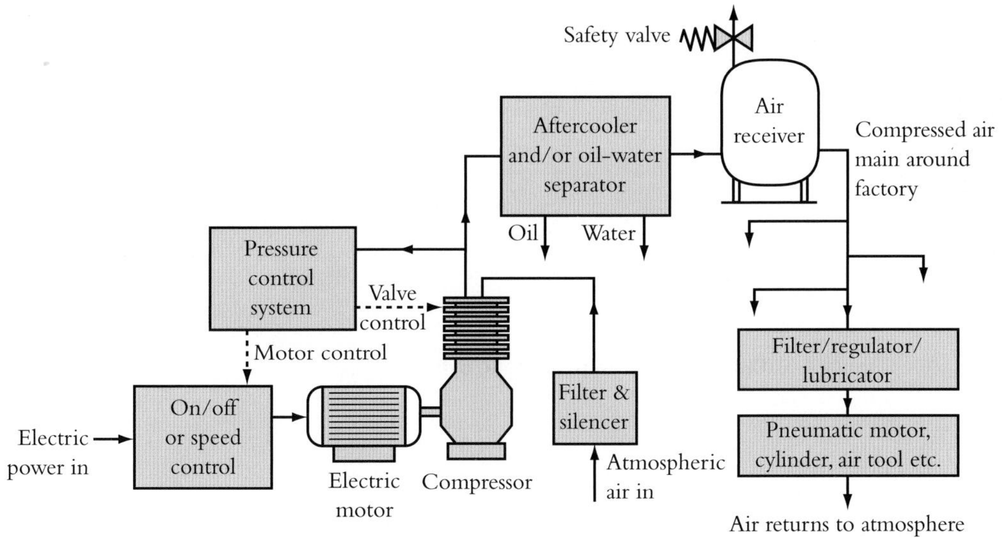

Figure 4.66 **Typical pneumatic system for a factory or similar installation**

- compressor – typically a one- or two-stage reciprocating compressor, although other types (e.g. screw compressors) may be used;
- a control system to ensure that the correct pressure is maintained without excessive wastage of energy – this may be a simple pressure switch for switching the motor on and off as required, or may involve, for instance, the lifting of the compressor's inlet valves so that it idles without compressing air, or may involve varying the speed of the motor;
- 'receiver' or pressure vessel for storing the compressed air;
- pressure relief valve, or safety valve, to prevent excessive pressure building up in the event of the failure of the control system;
- network of pipes to carry the compressed air around the factory;
- oil-mist lubricators to add lubricant to the air for use within motors.

Larger systems will also include a cooler and/or condenser to allow the dissipation of heat generated by the adiabatic compression of air, and to allow the condensation of moisture from the air as it cools after compression. An oil/water separator may also be included to avoid contaminating the drains with lubricating oil from the compressor.

Efficiency and energy utilisation issues of pneumatic systems

Pneumatic systems and actuators give a more intense, robust and lightweight source of power than is achievable with electricity alone, and provide an excellent route to high-speed, low-cost automation. However, poor design and working practices can easily lead to greater inefficiencies than those inherent in purely electrical systems. The Carbon Trust state that 'Of the total energy supplied to a compressor, as little as 8–10 per cent may be converted into useful energy that can do work at the point of use' (Carbon Trust, 2009). Reasons include the significant proportion of the energy required for compressing air which is lost as heat, and leakage from the air system. It has been stated that '35–50 per cent leakage is not uncommon' (Carbon Trust, 2008); other issues include friction in the compressor and air motor, pressure drops in filters etc. Careful searches for leaks, elimination of wasteful applications of compressed air, and the use of filters with low pressure drops, can all be used to reduce the wastage of compressed air and unnecessary use of energy. Although compressed air itself does not cause pollution, it is important for engineers to be aware of the scope for inefficiencies, and to be aware that the energy needs and associated cost and environmental issues involved in air compression can be significant.

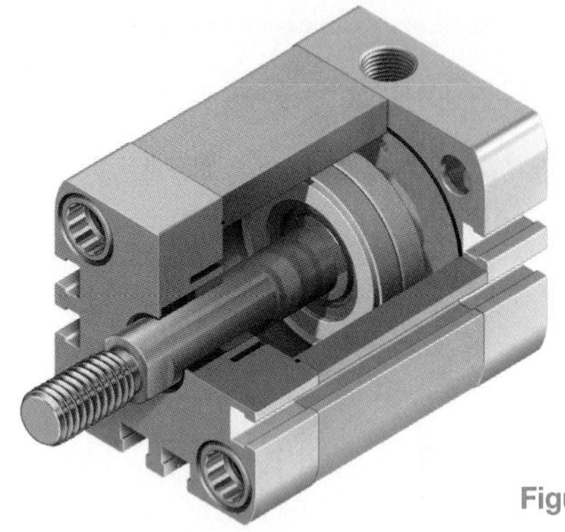

Figure 4.67 **Cutaway view of pneumatic cylinder**

Hydraulic systems

Hydraulic systems involve the use of liquids under very high pressure, and achieve even greater power-to-weight and power-to-space ratios than are achievable via pneumatic systems. There are obvious similarities with pneumatics (for example, both involve fluids under pressure, and both can be used with cylinder actuators and rotational motors) but there are major differences which can be summarized in Table 4.2:

Characteristic	Pneumatics	Hydraulics
Working fluid	air	liquid (usually oil or ester)
Pressure	6–8 bar typical	50–200 bar typical, can go up to 250–700 bar or higher
Construction of actuators	lightweight, typically from aluminium extrusions and diecastings	much more substantial, typically ferrous castings and steel tubing
Forces and torques available	medium (tens of N up to around 10 kN from linear actuators)	large (e.g. up to 10000 kN from linear actuators)
Precision	low – not usually used for precise positioning (though some pneumatic control systems provide precise control of flow etc. via the use of feedback) – more usually used for simple two-position operation	used for precise positioning e.g. within machine tools, and under control of servo valves
Stiffness	not stiff because air is compressible	very stiff as hydraulic fluid is virtually incompressible – actuator becomes rigid if inlets/outlets are blocked off, as in the case of cranes, jacks etc.
Cost	relatively low-cost	relatively high-cost due to robust construction
Power-to-weight and power-to-space ratios	very good – makes pneumatics suitable for power-intensive handheld tools, for example wrenches and grinders.	excellent – makes hydraulics suitable for providing power locally within aerospace applications, for example for operating landing gear and control surfaces in aircraft

Table 4.2

One of the major features of hydraulic systems is the ability to obtain a very large mechanical advantage (Figure 4.68). The classic example is a hydraulic jack, where a reciprocating pump with a small bore requires a relatively small force (but many operating strokes) to provide the fluid to operate a cylinder of large bore which is used to raise a heavy load by a small distance.

In the context of mechanical drive systems, a pump with a small displacement (which pumps a small amount of fluid per revolution) can be used to generate a very large force or torque when connected to a linear or rotational actuator with a large total piston area.

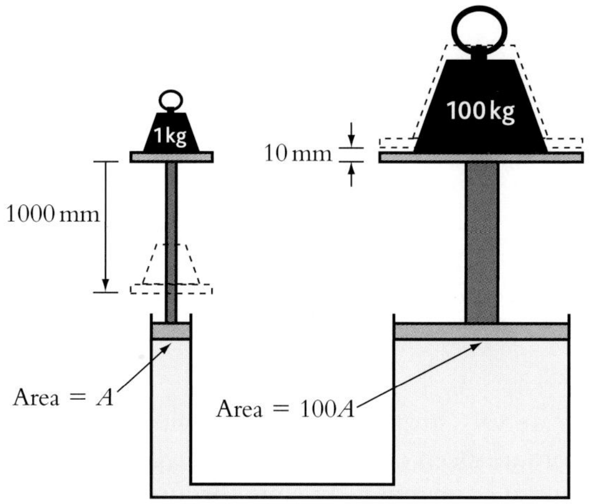

Figure 4.68 **Mechanical advantage produced by simple hydraulic system**

Hydraulic cylinders

As indicated above, these are similar in concept to pneumatic cylinder actuators but operate at much higher pressures and are consequently much more solidly built (Figure 4.69). They may be single-acting or double-acting, and provide a large force proportional to the pressure supplied to them. They are widely used where a large force is required, for example in raising and lowering the buckets of excavators and the jibs of cranes.

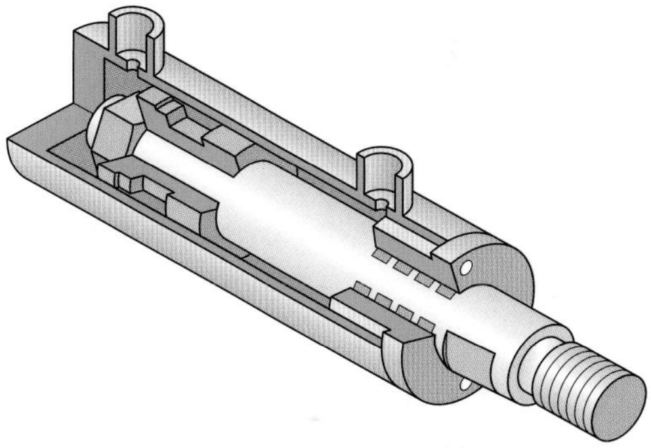

Figure 4.69 **Cutaway view of a typical hydraulic cylinder actuator (Design & Draughting Solutions Ltd – www.dds-ltd.co.uk)**

Hydraulic motors

Hydraulic motors are used in a variety of situations where a very large torque is required in a given space. Examples include providing drive to the tracks of track-laying excavators, driving rollers and other large-torque applications in industry, and within variable-speed hydraulic drives. Hydraulic motors are typically based on the use of pistons, and are generally radial (e.g. Figure 4.70, with the pistons typically acting on some form of radial cam) or axial (e.g. Figure 4.71, with the pistons acting on some form of inclined plate or swashplate).

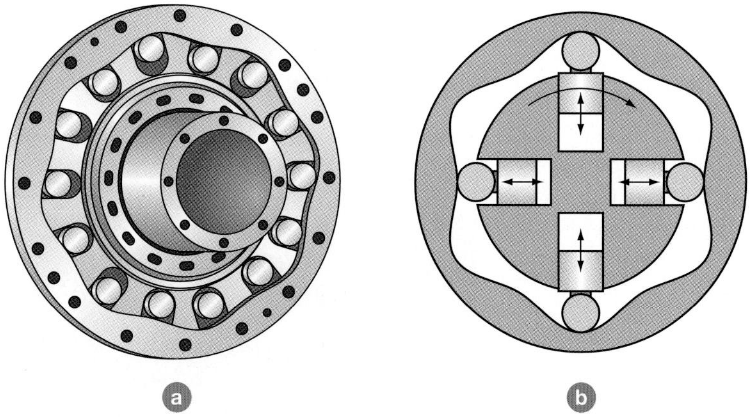

Figure 4.70 **(a) Construction and (b) operation of a radial piston hydraulic motor (Hagglunds Drives)**

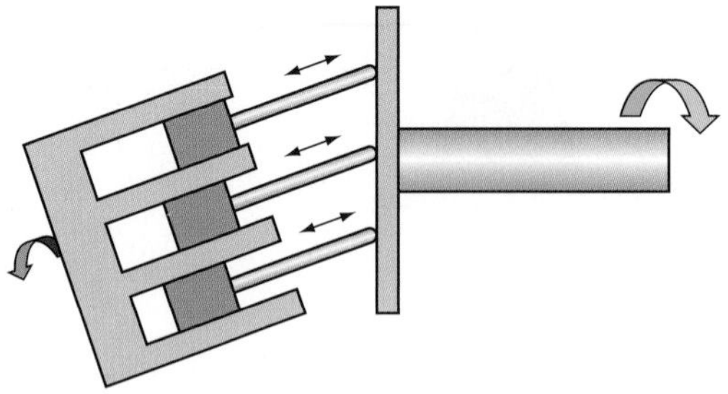

Figure 4.71 **Operation of an axial piston hydraulic motor (Hagglunds Drives)**

Because pressure drops due to dynamic head losses are very small in comparison with the pressures at which hydraulic systems operate, the torque–speed characteristic of a hydraulic motor is very nearly a horizontal line corresponding to a constant torque proportional to pressure (Figure 4.72).

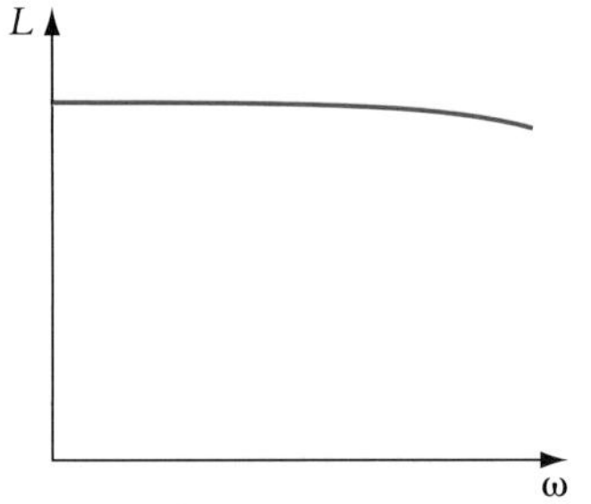

Figure 4.72 **Torque–speed characteristic of hydraulic motor**

Hydraulic power supplies

In order to power hydraulic actuators\, a pump must be used to provide a source of hydraulic fluid at high pressure. This clearly requires a mechanical power source, which is typically an electric motor (normally an induction motor) but can be a diesel engine. In a factory or laboratory environment a 'power pack' is often used, a self-contained unit consisting of the following components (Figure 4.73):

- electric induction motor
- fixed-displacement pump (e.g. vane pump)
- oil reservoir
- pressure regulator (often a relief valve)
- sometimes a cooler (either a water-cooled heat exchanger, or an air-cooled device similar to a car radiator) to dissipate the heat generated in the pump and regulator
- filters to remove particles.

In other applications it can be convenient to use variable-displacement pumps which allow the flow rate from the pump to be adjusted directly, for example to allow a piece of machinery (such as the wheels or tracks of a piece of construction machinery) to be driven with variable speed and direction.

Figure 4.73 **Hydraulic power pack showing 75 kW motor and pump, and below them the oil storage tank**

Hydraulic variable-ratio drives

One convenient application of hydraulics is within variable-ratio drives, which behave as a 'gearbox' with a ratio which is infinitely variable between given positive and negative values. The mechanism (Figure 4.74) consists simply of a variable-displacement pump driving a fixed-displacement motor. With the swashplate on the pump in its neutral position, no fluid is pumped, and the motor and output shaft do not rotate. However, if the swashplate is moved from neutral towards its positive or negative limit, the motor and output shaft will rotate at speeds up to the maximum rate in either direction.

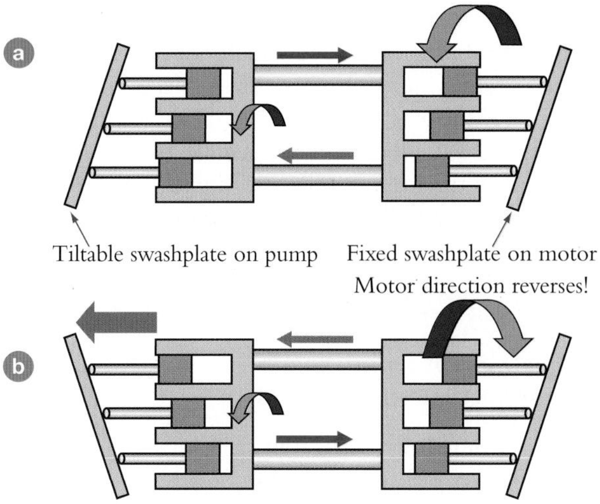

Figure 4.74 **Diagram of variable-ratio drive providing an output rotation: (a) in the forward direction; (b) in the reverse direction**

Learning summary

By the end of this section you should have learnt:

✓ the similarities and differences between pneumatic and hydraulic systems;
✓ why pneumatic or hydraulic actuators might be used in preference to electric motors;
✓ the ancillary equipment needed to power pneumatic and hydraulic systems;
✓ how hydraulics can be used to provide a variable-ratio drive;
✓ the cost and energy-efficiency issues associated with pneumatic power and compressed air.

4.13 Steady-state operating points and matching of loads to power sources

Consideration of torque–speed characteristics of loads and engines etc. provides an opportunity to introduce the concept of the steady-state operating point at which the motor or engine will drive the load without acceleration or deceleration. This concept occurs widely in all branches of engineering (mechanical, electrical, pneumatic etc.) and is closely related to the concept of 'impedance matching'. In practice, it is often necessary to use some kind of transmission to match the characteristics of the power source to the load, so that the power source and load interact in an optimal manner.

Steady-state operating points

The concept of an operating point is straightforward: if the torque–speed characteristics of the load and of the motor or engine are both known, then the combination of torque and speed at which the motor will drive the load corresponds to the point at which the characteristics cross. This is straightforward (and obvious) to determine graphically if the characteristics are plotted; it can also be found if the equations of the characteristics are known, simply by expressing torque as a function of speed for both characteristics, then equating the two expressions. Solution of the resulting equation may or may not be possible analytically, so an iterative numerical solution may be required.

An issue related to the concept of operating points concerns the scenario if, at any instant, the system is running at a speed which does *not* correspond to the operating point – for example if the system is running more slowly than the steady-state running speed. Assuming (for the purpose of the argument) that the characteristics are valid for transient (non-steady state) conditions, it will be observed that the torque available from the engine or motor exceeds that required by the load. In that case, the surplus torque is available for accelerating the load, which will speed up until the steady-state running speed is approached asymptotically.

Worked example

Example of an internal combustion engine problem

(a) A naturally aspirated diesel engine with the characteristic given in Figure 4.75 is driving a pump at 2000 rev/min; the pump consumes a power of 30 kW at this speed. How many litres of fuel are required per hour? Assume the specific gravity of the fuel (effectively, the density in kg/litre) is 0.84.

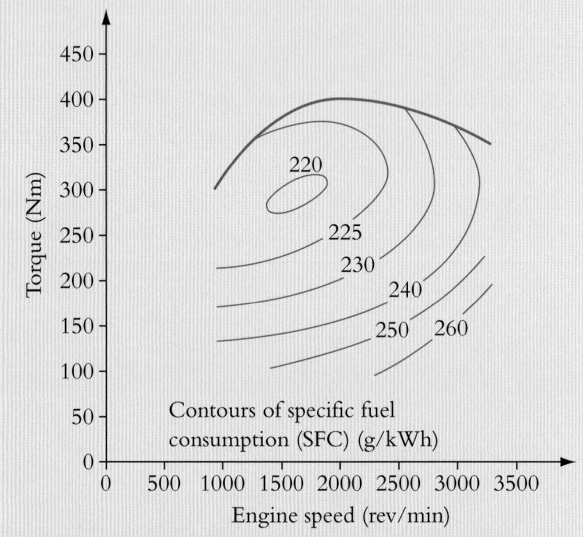

Figure 4.75 **Representative torque–speed–SFC map for a diesel engine (adapted from Heywood, *Internal Combustion Engine Fundamentals*, McGraw Hill, NY, 1988)**

(b) A sudden flood occurs, and the engine is set to maximum injection by the control system. If the combined inertia of the pump and engine is 4 kg.m^2, at what rate (rev/sec^2) does the system accelerate at the instant the throttle is opened fully? You may assume that the engine and pump characteristics both remain valid for this sudden change.

(c) If the pump has a square-law torque–speed characteristic (i.e. torque proportional to speed squared) , what is the maximum speed at which the engine can drive it?

(a) Torque = 30 000/(2π × 2000/60) = 143.2 Nm
Specific fuel consumption = 245 g/kWh
Mass of fuel consumed per hour = 30 × 245 = 7350 g = 7.35 kg
Volume of fuel per hour 7.35/0.84 = 8.75 litres

(b) Maximum engine torque at 2000 rpm = 400 Nm
Load torque = 143.2 Nm
Excess torque for acceleration = 400 − 143.2 = 256.8 Nm
Acceleration = net torque/moment of inertia = 256.8/4 = 64.2 rad/s^2 = 10.2 rev/s^2

(c) Assume equation of form $L = c\,N^2$ where L is torque in Nm, c is a constant and N is the speed in rev/min
Constant c = initial torque/(initial speed)^2 = 143.3/2000^2 = 0.000 035 8
Characteristic of pump: $L = 0.000\,035\,8\,N^2$ (Nm)
N = 3150 rpm where this characteristic crosses the maximum torque line

Matching of loads to power sources

It very unusual for the characteristics of a power source to match closely the requirements of a load without the need for any kind of transmission. One example of where a direct drive is possible is the steam railway locomotives (Figure 4.76(a)), where the pistons drive a crankshaft that forms one of the main axles carrying the wheels, often with the cranks being integral with the wheels. Another is a simple piston-engined aircraft (Figure 4.76(b)), where the propeller is mounted directly on the end of the engine's crankshaft.

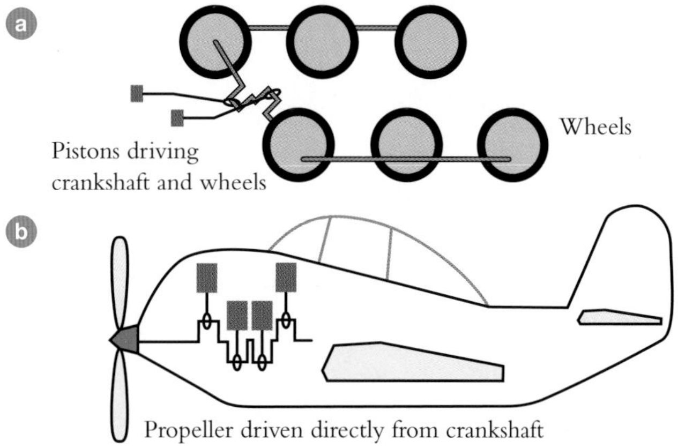

Figure 4.76 **Two examples of direct drive from engine to load (a) chassis of steam locomotive; (b) piston-engined aircraft**

Contrast these two examples, respectively, with a modern diesel locomotive, where a complex arrangement of generator and motor is used to link engine to wheels, and a 'turboprop' aircraft such as a military transport plane, where the high-speed gas turbine engine is linked to the propeller via a reduction gear.

Matching of steady-state characteristics

Probably the most common challenge in matching load to power source is where the steady-state running characteristics of the load do not cross the characteristics of the power source in a useful manner, for instance where the power source runs too fast and does not provide enough torque (see Figure 4.77).

In such a case it is straightforward to shift the torque–speed characteristic via the use of a transmission ratio, as described in Section 4.7, so that a useful operating point can be chosen.

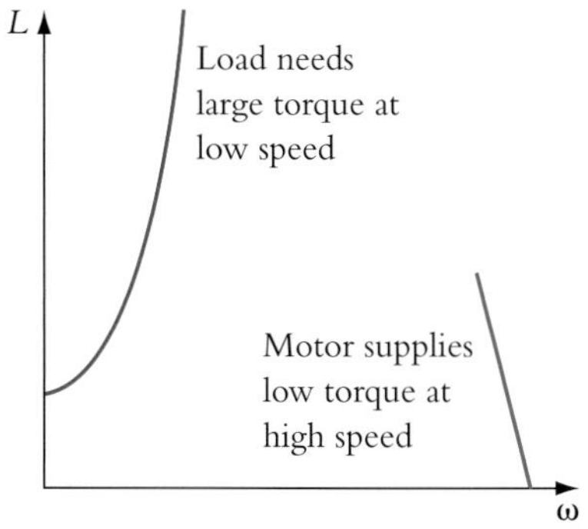

Figure 4.77 **Mismatched torque–speed characteristics**

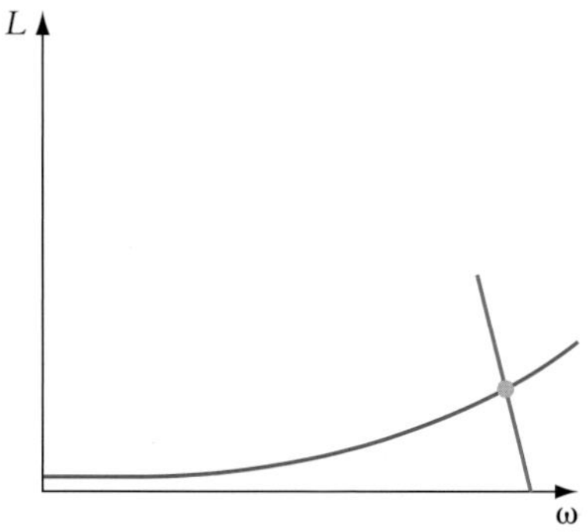

Figure 4.78 **Torque–speed characteristics which cross at a useful operating point**

Worked example

An example of such a problem might be to consider the matching of a load such as a blower to a motor (see Figure 4.79).

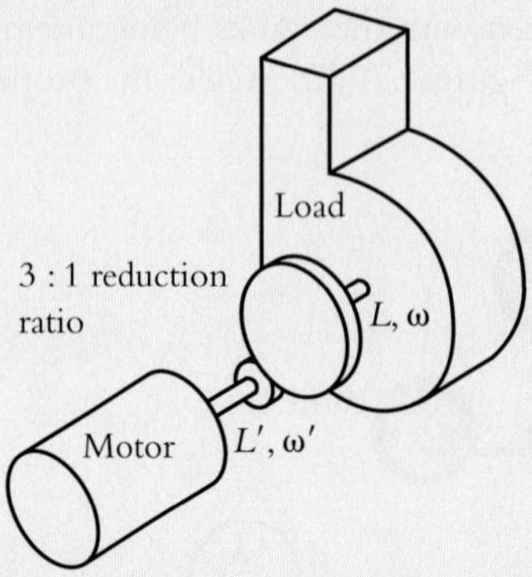

Figure 4.79 **Blower driven from motor via reduction gear**

Assume that the characteristics of a load are:

$$L = 20 + 0.2\omega + 0.025\omega^2$$

where L is the torque in Nm and ω is its angular velocity in rad/s (both measured at the shaft directly driving the load). The load is to be driven via a 3:1 reduction ratio from a 5 kW induction motor whose characteristics are approximated as:

$$L' = 5(50\pi - \omega')$$

The raw characteristics of the motor and load do not cross in a useful manner (Figure 4.80)

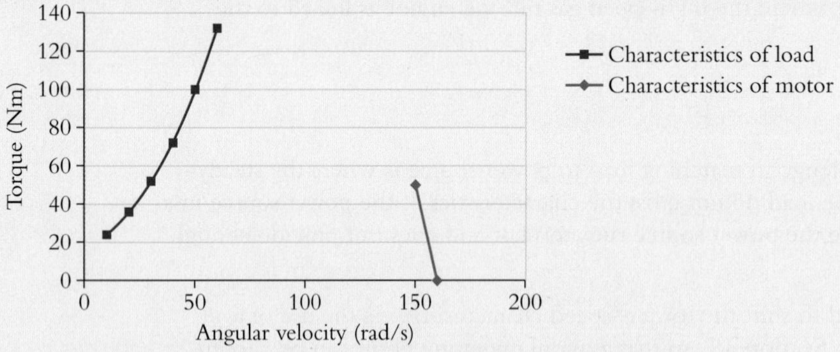

Figure 4.80 **Mismatched characteristics of motor and blower load**

This problem is, of course, tackled by connecting the motor and the load via the transmission system illustrated in Figure 4.79. The torque–speed characteristics of the load can now referred to the axis of the motor:

$$L' = L/3$$

$$\omega' = 3\omega \Rightarrow \omega = \omega'/3$$

$$L = 20 + 0.2\omega + 0.025\omega^2$$

$$L' = \tfrac{1}{3}(20 + 0.2\omega + 0.025\omega^2)$$

$$= \frac{1}{3}\left(20 + 0.2\frac{\omega'}{3} + 0.025\left(\frac{\omega'}{3}\right)^2\right)$$

$$= 6.667 + 0.022\,22\omega' + 0.000\,9259\omega'^2 \qquad \textbf{(4.92)}$$

This referred characteristic now crosses the characteristic of the motor to give a stable operating point as shown in Figure 4.81

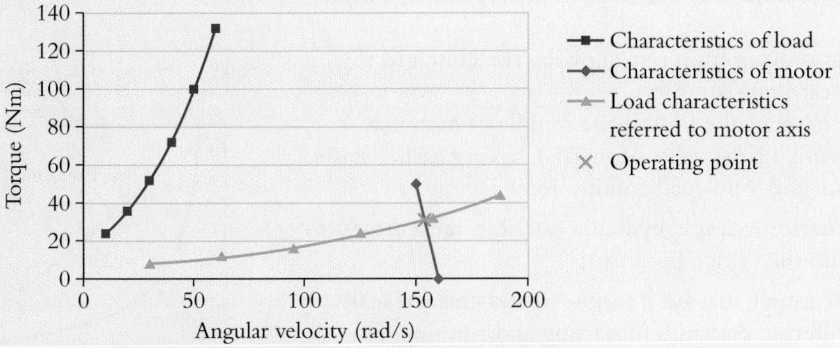

Figure 4.81 **Matching of load and motor characteristics**

This operating point can be found graphically as the point where the characteristics of power source and load (referred, for instance, to the power source axis) cross. However, if (as in the present case) the torque–speed characteristic can be expressed in the form of an equation, the operating point can be found by equating the torques for motor and load and solving to find the angular velocity. In the present case this solution involves constructing and solving a quadratic equation using the standard formula:

$$L'_{\text{load}} = L'_{\text{motor}}$$

$$6.667 + 0.022\,22\omega' + 0.000\,9259\omega'^2 = 5(50\pi - \omega')$$

$$(6.667 - 250\pi) + (0.022\,22 + 5)\omega' + 0.000\,9259\omega'^2 = 0$$

$$0.000\,9259\omega'^2 + 5.022\,22\omega' - 778.7 = 0$$

$$\omega' = \frac{-5.022\,22 \pm \sqrt{5.022\,22^2 + 4 \times 0.000\,925\,9 \times 778.7}}{2 \times 0.000\,9259}$$

$$= 150.86 \text{ rad/s (or } -5575 \text{ rad/s, which is clearly incorrect)}$$

$$L = 5(50\pi - 150.86) = 31.1 \text{ Nm} \qquad (4.93)$$

Matching of starting characteristics

Another common situation is where it is desired to start a load from rest. If the load does not have a large inertia, or if the mechanical power source has a large torque at zero speed (and is not damaged by stalling or running at low speed), this may not be a problem. However, in many practical situations, starting can be a problem. For example, internal combustion engines will not run in a stable manner at low speed (they stall easily, as any learner driver knows!), while large electric motors will burn out if they are required to provide a large torque at low speed for any length of time. Several strategies can be adopted:

- alteration of the characteristics of the mechanical power source, for example, by driving an induction motor from an inverter so that it gives a good torque as the frequency is raised from rest (see, for example, Figure 4.82). (Note: this diagram assumes that the frequency is increased slowly so that the steady-state operating points are reached at each value of speed. For rapid ramping of frequency, the torque required to cause acceleration will, of course, be significant and the locus of combinations of torque and speed will not coincide with the steady-state torque–speed curve of the load.);

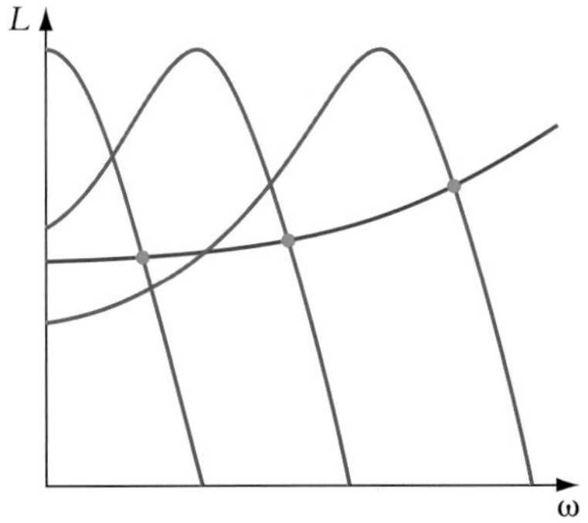

Figure 4.82 **Locus of operating points as a combined friction and windage load is accelerated slowly from rest using an inverter-driven induction motor**

- a clutch used to allow a mismatch in speed between mechanical power source and load, whilst allowing a given value of torque (equal to the slippage torque of the clutch) to be transmitted. The power source can now run at a stable speed while the load accelerates from rest under the influence of that torque. Examples include:
  - the process of getting a car moving from rest, allowing the clutch to slip gently as the speed of the car picks up
  - the engagement of a clutch on a lathe to start the spindle (which may carry a heavy workpiece with a large value of inertia), having earlier started the lathe's induction motor under no-load conditions;
- the use of variable-ratio drives (for example hydraulic variable-ratio drives) to provide a means of running up the device from rest;
- choosing a mechanical power source that has a torque–speed characteristic which is well suited to the differing demands of starting and running. This involves a high torque at low speed (without damage to the power source or any ancillary equipment, e.g. power supply, occurring due to running under stall conditions) combined with lower torques at high speeds. Good examples of motors with these characteristics (see Figure 4.83) include:
  - pneumatic motors
  - hydraulic motors
  - series-wound electric motors.

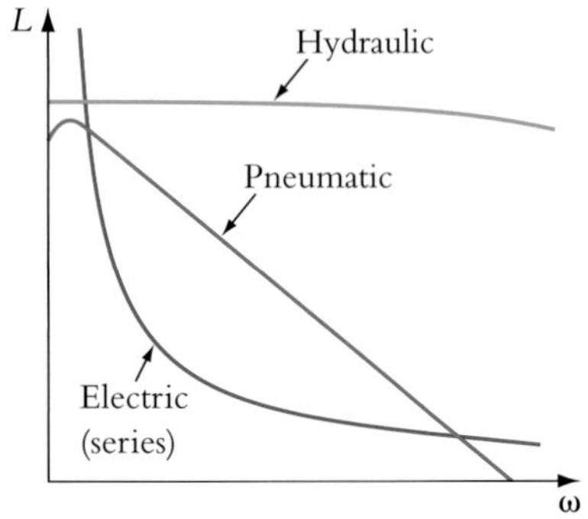

Figure 4.83 **Torque–speed characteristics of motors which are well suited to starting**

Learning summary

By the end of this section you should have learnt:

✓ how mechanical power sources and loads interact;

✓ how a transmission system may be used to match the power source to the load;

✓ to calculate the combination of torque and speed at which a load and power source will operate under steady-state conditions;

✓ the function of a clutch with particular reference to starting a mechanical load from rest;

✓ why some sources of mechanical power are better suited to starting of loads than others.

References

Blackburn, J.F., Reethof, G. and Shearer, J.L. (eds), 1960, *Fluid Power Control*, Cambridge, M.A: MIT Press.

Carbon Trust 2009, http://www.carbontrust.co.uk/energy/startsaving/tech_compressed_air.htm, accessed 23 February 2010.

Carbon Trust, 2008, 'Better business guide to energy saving – Introducing measures to help organisations save carbon'. *Carbon Trust management guide CTV034.*

Chambers dictionary of science and technology, 2007, (general editor, John Lackie), Edinburgh: Chambers.

Fraser, C. and Milne, J., 1994, *Integrated electrical and electronic engineering for mechanical engineers*, New York: McGraw-Hill.

Hughes, A., 2006, *Electric motors and drives*, 3rd edn, Oxford: Newnes, 287–9.

Shayler, P. J., Chick, J., Darnton, N. J. and Eade, D., 1999, 'Generic functions for fuel consumption and engine-out emissions of HC, CO and NOx of spark-ignition engines'. *Proc. Instn. Mech. Engrs* Vol 213 Part D, 365–378.

Wildi, T., 1991, *Electrical machines, drives and power systems*, 2nd edn., New Jersey: Prentice-Hall, 292–3.

Unit 5

Feedback and control theory

UNIT OVERVIEW

- Introduction
- Feedback and the concept of control engineering
- Illustrations of modelling and block diagram concepts
- The *s* domain: a notation borrowed from mathematics
- Block diagrams and the *s* notation: the heater controller and tensioning system
- Working with transfer functions and the *s* domain
- Building a block diagram: part 1
- Building a block diagram: part 2
- Conversion of the block diagram to the transfer function of the system
- Handling block diagrams with overlapping control loops
- The control algorithm and proportional-integral-derivative (PID) control
- Response and stability of control systems
- A framework for mapping the response of control systems: the root locus method

5.1 Introduction

There are numerous situations in engineering where it is desirable to provide power to a machine in order to make it run at a chosen (and variable) speed, flow rate, pressure etc. In many of these cases, there is a need to maintain the speed or other attribute of a system at a fixed value despite changes to the environment in which it operates. Examples including maintaining the speed of a car at a fixed value regardless of road gradient and wind, maintaining the course of an aircraft regardless of the motion of the air in which the aircraft flies, and adjusting the rotational speed of a DVD to maintain the required data rate regardless of the changing radius at which the laser beam reads the disc.

This unit initially concentrates on the concept of feedback and control, and on the mathematical concepts used in modelling a control system. The basic concepts are outlined with a minimum of mathematical overhead, then the mathematical tools required to analyse realistic situations (such as those involving dynamics) are introduced one by one until systems can be modelled, analysed and characterized in terms of their performance and stability.

The concept of automatic control is exemplified by systems such as the cruise control on a vehicle, which maintains a set speed regardless of gradient, wind speed etc., and the autopilot on an aircraft, which maintains course, altitude and speed without human intervention.

However, although control systems are now present in numerous everyday situations, they are frequently so invisible or transparent in their operation that the user may be totally unaware of them. For instance, modern car engines continuously take data from sensors to adjust the engine to optimum performance regardless of ambient temperature, televisions maintain stable pictures despite fluctuations in supply voltage and frequency, and cameras automatically compensate for varying levels of illumination. By contrast, 30 to 40 years ago, cars were equipped with a manually operated choke to compensate for a low engine temperature, televisions frequently needed adjustment by the user to prevent the picture from breaking up, and cameras needed manual setting of focus, shutter and aperture. The present unit aims to introduce some of the concepts involved in automatic control, beginning with a simplified approach involving a minimum of mathematics in order to examine:

- feedback
- block diagrams
- error
- the control algorithm.

The unit then progresses to examining some of the mathematical tools of classical control theory that are required to analyse and understand real control systems. These tools include:

- the concept of Laplace transforms to allow analysis of systems within '*s* space'
- the concept of the transfer function
- analysis of initial and final behaviour and frequency response
- techniques for determining stability of a control system
- a technique for visualizing the behaviour of a system.

This unit can only give an overview of some of the simpler topics in control engineering, and the approach taken here is to present a variety of worked examples rather than going into excessive depth on the theory or techniques. Numerous texts are available that cover control engineering in more detail and from a more rigorous mathematical perspective including the current edition of the classic text by Dorf and Bishop (2008). A slightly earlier but excellent text is that by Van de Vegte (1993).

5.2 Feedback and the concept of control engineering

Concept, history and simple examples

Feedback may broadly be defined as **the comparison of a measured physical variable with the desired value of that variable with the aim of taking action to minimize the difference between the measured and desired values**.

The original application of feedback in control of engineering plant appears to have been the centrifugal governor originally applied to steam power by James Watt in 1788. As the governor rotates and the weights fly outwards, a steam valve is closed so that engine speed is maintained at an approximately constant value. In fact, Watt's development of the steam engine governor was based closely on the use of a similar device invented a year earlier by the millwright Thomas Mead for the control of the settings (but not necessarily the speed) of grinding wheels in windmills (Bennett, 1979). Figure 5.1(a) shows one of the governors on what is thought to be the last set of engines built by James Watt and Co. in 1882–4, still in working order in the preserved Papplewick Pumping Station near Nottingham. At the time of writing, there is a similar engine (also with a centrifugal governor) in working order in Wollaton Park immediately to the north of the University of Nottingham.

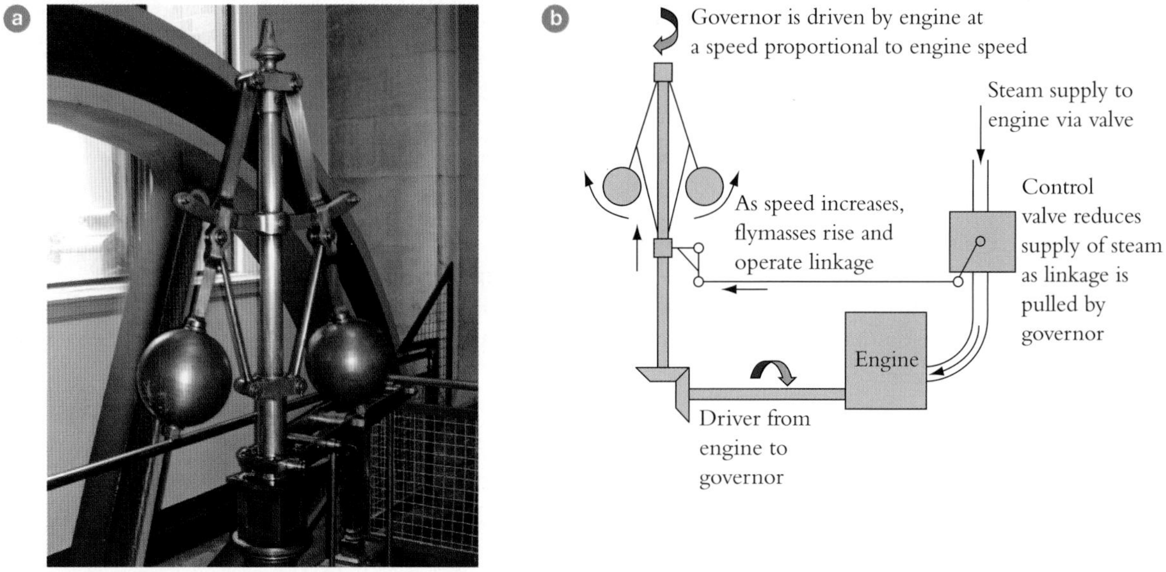
a
b
Governor is driven by engine at
a speed proportional to engine speed
Steam supply to
engine via valve
Control
valve reduces
supply of steam
as linkage is
pulled by
governor
As speed increases,
flymasses rise and
operate linkage
Engine
Driver from
engine to
governor

Figure 5.1 (a) Governor at Papplewick pumping station; (b) schematic operation of typical centrifugal governor (Papplewick Pumping Station Trust)

The process by which the governor operates (Figure 5.1(b)) is:

- if the engine speeds up beyond its desired speed, the masses on the governor move outwards under centrifugal action;
- the movement of the masses is transferred, via a collar, to a control rod that slightly closes a control valve supplying steam to the engine;
- this causes the engine to slow down;
- by means of this process, known as negative feedback (for example, where an increase in speed results in an action which causes the speed to be reduced) the engine's speed is automatically maintained at a value close to the desired value.

Conversely, if the engine speed falls unexpectedly, the governor will cause the control valve to open slightly above its current setting, causing the engine to speed up.

Since the development of the centrifugal governor, the understanding of control systems has developed to incorporate a variety of concepts and theories. These include the application to control of Laplace transforms, proportional-integral-derivative (PID) control, transfer functions, stability, root locus plots etc. These developed into the discipline we now know as classical control theory in the years during and after World War II. The history of the discipline is covered in depth in two books by Bennett (1979 and 1993); these topics form the basis of the control-related material in the present unit.

Everyday example of closed-loop feedback control: driving a vehicle at constant speed

The task of maintaining the constant speed of a vehicle provides an excellent example of closed-loop control. In the case of a car without cruise control (or with the cruise control not engaged) one aspect of the driving process is:

- a speed is chosen (for example, the driver might choose to drive as fast as possible to reach an appointment in time but without breaking the law) – this is taken to be the 'demand' for the system;
- the current speed is measured via the speedometer, which is observed by the driver;
- the driver compares the two and assesses the difference (known technically as the **error**, which is the desired value minus the measured value);

- equipped with the skill of how to drive a car, the driver takes a decision on how to correct the speed based on the value of error, and also based on an understanding of how well the speed is converging to the desired value – this decision-making process is very important even though it is probably taken for granted within a human control situation;
- the driver attempts to correct the speed by raising or lowering his foot on the accelerator (gas) pedal;
- depending on issues such as the mass of the car, the power of the engine etc., the car will respond by gradually increasing or decreasing its speed; if the driver is experienced, the car's speed will rapidly converge on the desired value, but if the driver is a learner, the speed is very likely to oscillate about the desired value until the driver hits upon the correct position of the accelerator.

This process is illustrated via a simple block diagram in Figure 5.2.

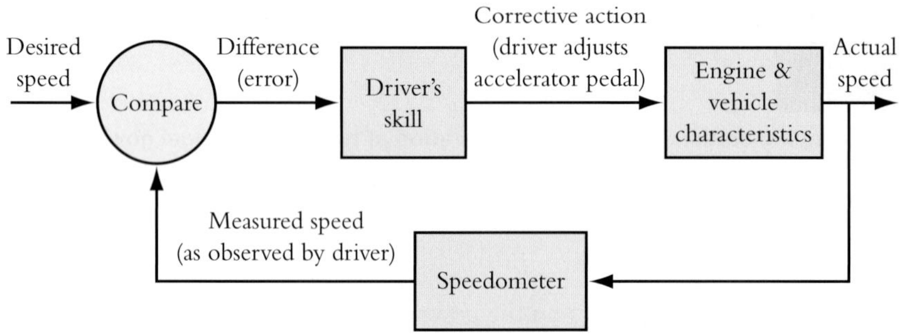

Figure 5.2 Block diagram of the process of driving a car

An obvious but essential point to note is that feedback is **negative**: for example, if the car is travelling too fast, the driver partially releases the accelerator to allow it to slow down. (Positive feedback is explored later in this section but would not be helpful in this situation – it would involve the driver pressing the accelerator ever harder as the car's speed was seen to increase!)

This process can of course be automated, and a system for doing so (known as the cruise control) is available in some modern cars. The control loop (Figure 5.3) remains essentially the same as before, except that the decision-making process requiring the skill of the driver is now replaced with the automated decision-making process (known formally as the **control algorithm**) forming the heart of the cruise control system, and the actuation of the engine's throttle takes place under the command of the cruise control rather than via the driver's operation of the pedal.

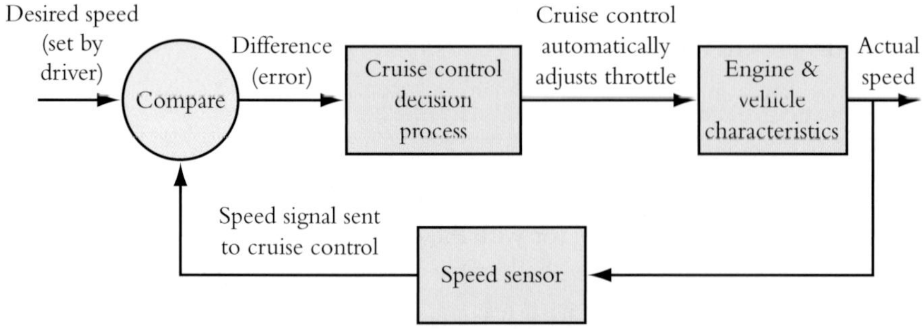

Figure 5.3 Block diagram of a cruise control system (greatly simplified)

The control algorithm is carefully designed to take account of the way in which the car's speed is changing, in order to avoid over- or under-shooting the desired speed, or causing the speed to fluctuate about its desired value.

Less constructive examples of feedback

The fact that control systems generally employ negative feedback has already been noted – for instance, if a car is going too fast, the corrective action involves sending a signal to slow the car down. Although **positive feedback** is occasionally used to beneficial effect, it is generally an undesirable phenomenon, as any discrepancy between desired and actual values of system behaviour results in a signal that makes that discrepancy greater, giving a form of **instability**. A powerful example of the destructive nature of positive feedback is the situation (a 'bank run') which occurs when a bank starts to encounter 'liquidity problems', i.e. when it starts to run out of money:

- customers begin to hear that the bank is short of cash and become worried about the security of their savings;
- they therefore withdraw their savings from the bank so that they can keep them safely elsewhere;
- this means that the bank has less cash, and more customers get worried about its problems;
- even more customers withdraw their savings ...

... and so on until either the bank fails completely (as happened numerous times in the 1930s) or the process is halted by state intervention (as in the case of Northern Rock in the UK in 2007). Other examples include the phenomenon in electronics known as thermal runaway, which relates to a device that turns electrical power into waste heat as a by-product of its normal operation, and tends to do this more vigorously as its temperature rises. Unless precautions are taken, the amount of power it converts to heat will keep increasing as its temperature rises, in turn causing it to get hotter still, until the component burns out. An even more catastrophic example was the Chernobyl nuclear disaster, which can be regarded as a complex form of thermal runaway. The more the cooling water in the nuclear reactor boiled to form steam, the more heat the nuclear reaction generated. This created yet more steam, eventually causing a disastrous explosion.

5.3 Illustrations of modelling and block diagram concepts

Simple example of feedback and control: modelling of a heated system

Real control systems usually include some form of time-dependent behaviour, but it is useful to illustrate the modelling process via a very simple control system in which time-varying effects are neglected. Consider a heater unit for a set of railway equipment, designed to maintain the equipment at an elevated temperature to avoid it icing up. The railway equipment has an exposed surface area of $A = 0.5\,\text{m}^2$ and a convective heat transfer coefficient of $h = 5\,\text{W/m}^2.°\text{C}$ (more correctly represented as $h = 5\,\text{W/m}^2.\text{K}$). A heater provides a power p of heat input into the railway equipment. It will be assumed for simplicity that the temperature of the system's surroundings is 0°C. It is easy to show that if (for example) $p = 10\,\text{W}$ and the system is allowed to settle down to a steady temperature τ, the temperature achieved (which we may regard as the output of the system) is obtained by equating the power input p to the rate of heat loss:

$$p(t) = Ah\tau(t) \tag{5.1}$$

This is solved to obtain

$$\tau(t) = \frac{p(t)}{Ah} \tag{5.2}$$

which, for the above values, gives a temperature of:

$$\tau(t) = \frac{10}{0.5 \times 5} = 4°\text{C} \qquad \textbf{(5.3)}$$

Note that the values that are properties of the system (A and h) are constant, i.e. invariant with time, while all variable quantities p and τ are expressed in the form of functions of time t, though in this example we will assume that such quantities have had time to settle down to their steady-state values, or that that the thermal capacity of the system is negligibly small. (The importance of this apparently obvious statement will become clear later as a slightly different convention is normally followed in control engineering: variable quantities will be expressed as a function of a complex variable s, rather than time; see section 5.4.) This very simple relationship between power input and temperature may be (trivially) represented using a block diagram as shown in Figure 5.4.

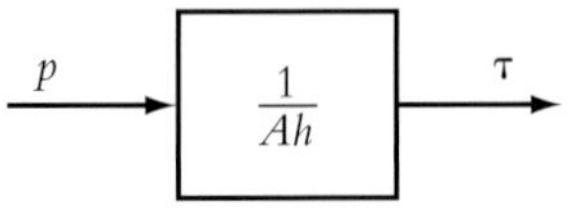

Figure 5.4

In the block diagram, the mathematical relationship between an input (in this case, power p) and an output (actual temperature τ) is shown as a 'block' which acts upon 'flows' of values or data, specifically multiplying the input by the value of the constant contained in the block in order to obtain the output.

Note: In Figure 5.4 the fact that the variables p and τ are functions of time is deliberately not shown. In fact we would never normally draw the block diagram directly with variables as a function of time; they would instead be expressed as a function of the complex variable s which will be introduced in Section 5.4.

Suppose that we now wish to maintain the temperature of the equipment at a given value $\tau_D(t)$. In order to automate this process, a very simple control system is introduced which drives the heater with a power that is proportional to the difference between the desired temperature and the actual temperature, so that for each degree Celsius of temperature difference, an extra power of 10 W will be supplied where this constant of proportionally will be termed the *gain* K of the controller. As will be explained later, this is a very simple form of the control algorithm (decision-making process) and is known as proportional control. It may be represented mathematically as:

$$p(t) = K(\tau_D(t) - \tau(t)) \qquad \textbf{(5.4)}$$

and within a block diagram as shown in Figure 5.5.

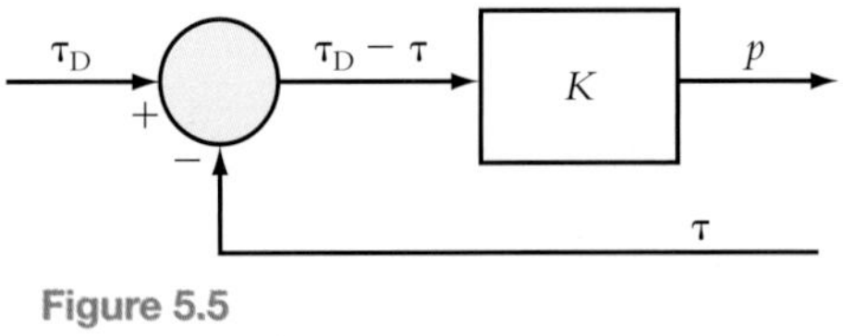

Figure 5.5

The block diagram in Figure 5.5 now involves not only the input setting τ_D (demand temperature), the output τ (actual temperature) but now also an intermediate quantity $\tau_D - \tau$, known as the 'error'. Also visible is a 'differencing junction' where τ is subtracted from τ_D. The block diagram for the complete control system can now be assembled by 'joining the loose ends' in the block diagrams for the different aspects of the system as shown in Figure 5.6.

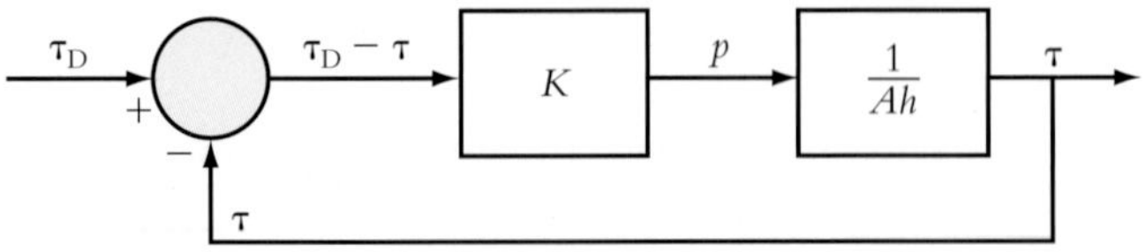

Figure 5.6

Note that, because we wish to treat τ as an output from the system as well as feeding it back, a branch is introduced into the flow; no additional symbols are involved in this. The block diagram can be simplified slightly by combining successive blocks as shown in Figure 5.7.

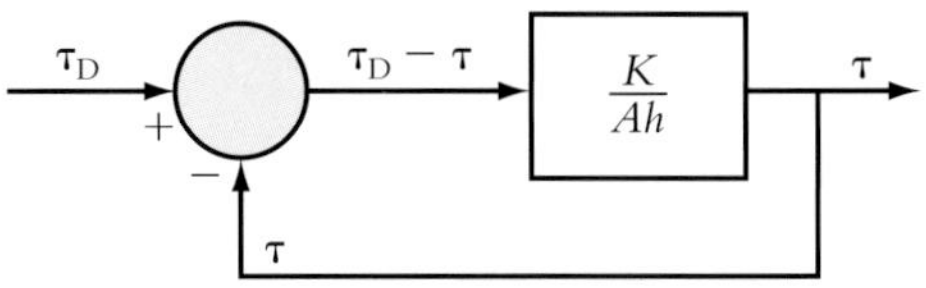

Figure 5.7

Assembling the block diagram corresponds to combining equations (5.2) and (5.4) to eliminate $p(t)$:

$$\tau(t) = \frac{1}{Ah} K(\tau_D(t) - \tau(t)) = \frac{K(\tau_D(t) - \tau(t))}{Ah} = \frac{K\tau_D(t)}{Ah} - \frac{K\tau(t)}{Ah} \tag{5.5}$$

It would appear that we have a circular relationship in which $\tau(t)$ appears twice. However, equation (5.5) can be rearranged to group the $\tau(t)$ terms together:

$$\tau(t) + \frac{K\tau(t)}{Ah} = \tau(t)\left(1 + \frac{K}{Ah}\right) = \frac{K\tau_D(t)}{Ah} \tag{5.6}$$

It is now straightforward to rearrange this to give the output $\tau(t)$ in terms of the input (demand) $\tau_D(t)$:

$$\tau(t) + \frac{K\tau_D(t)}{Ah\left(1 + \frac{K}{Ah}\right)} = \frac{K}{(Ah + K)}\tau_D(t) \tag{5.7}$$

so that the error is:

$$\tau_D(t) - \tau(t) = \tau_D(t) - \frac{K}{Ah + K}\tau_D(t) = \frac{Ah}{Ah + K}\tau_D(t) \tag{5.8}$$

The block diagram for the whole system can now be expressed in simplified form in Figure 5.8, where the contents of the 'box' are the ratio of the output to the input. This concept will be extended within the more complex analysis given in Section 5.4 and 5.5 and will be referred to as the *transfer function*.

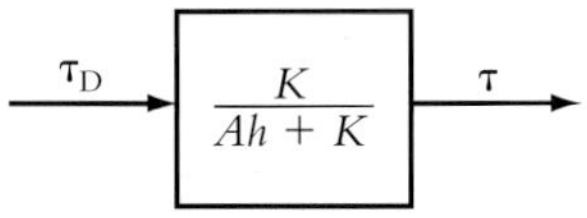

Figure 5.8

The issues of manipulating and simplifying block diagrams will be explored in detail in Section 5.9. It is important to note that, with this simple (proportional) control algorithm, the desired temperature is never actually reached. For example, with the above choice of gain $K = 10\,\text{W}/°\text{C}$, the temperature $\tau(t)$ settles down to the following value:

$$\tau(t) = \frac{10}{(0.5 \times 5 + 10)}\tau_D(t) = 0.8\tau_D(t) \tag{5.9}$$

so that a demand temperature of 5°C will result in an actual temperature of 4°C. Increasing the gain will result in the temperature being more closely approached; for example a gain of $K = 100\text{W}/°\text{C}$ will result in a temperature of

$$\tau(t) = \frac{100}{(0.5 \times 5 + 100)}\tau_D(t) = 0.976\tau_D(t) \tag{5.10}$$

being achieved, for example 4.88°C for a demand of 5°C. It will be seen later, however, that there can be a penalty (in terms of system stability) associated with making the gain of a system larger than necessary.

Example of feedback and control: modelling of a fibre tensioning system

Another introductory example involves a slightly more complex control system, again with time-related terms such as inertia being neglected, and involves only mechanical components. Consider a purely mechanical tension control system for a filament winding machine (used for producing fibre-reinforced composite containers) in which the dry fibres are held on a

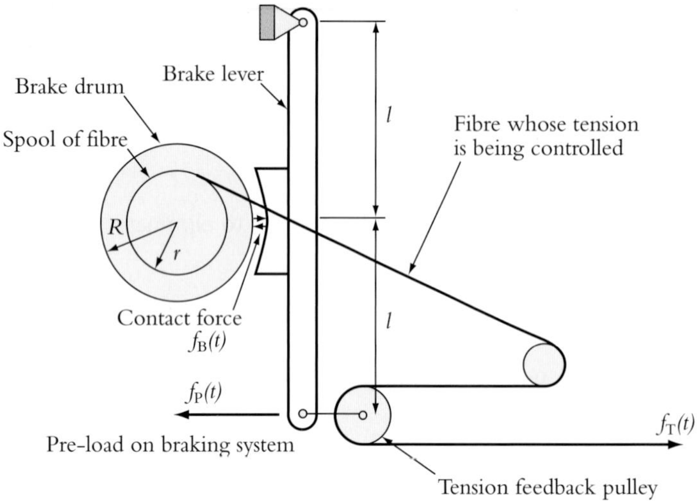

Figure 5.9: Diagram of mechanical fibre tensioning system (frictional forces omitted for clarity)

cardboard spool of radius r, braked via a drum of radius R and coefficient of friction μ with a force related to the fibre tension f_T and to a pre-load setting f_P applied to the braking system (Figure 5.9). The concept is that the brake provides a frictional force on the drum in order to tension the fibre, the pre-load contributes to the frictional force (and can be adjusted to vary the tension), and the tension feedback pulley provides a mechanism for reducing the braking force if the fibre tension becomes too large. In this system, the pre-load setting f_P is the input to the system, and the fibre tension f_T is the output.

The block diagram, and hence the relationship (transfer function) between input f_P and output f_T can be derived by considering each part of the system in turn.

By taking moments about the fulcrum O in a free-body diagram of the brake lever and pulley (Figure 5.10), the force f_B on the brake shoe is found from:

$$2l(2f_T(t) - f_P(t)) + lf_B(t) = 0$$

$$\therefore \quad f_B(t) = 2f_P(t) - 4f_T(t) \qquad \textbf{(5.11)}$$

$$\text{or} \quad f_B(t) = 2(f_P(t) - 2f_T(t))$$

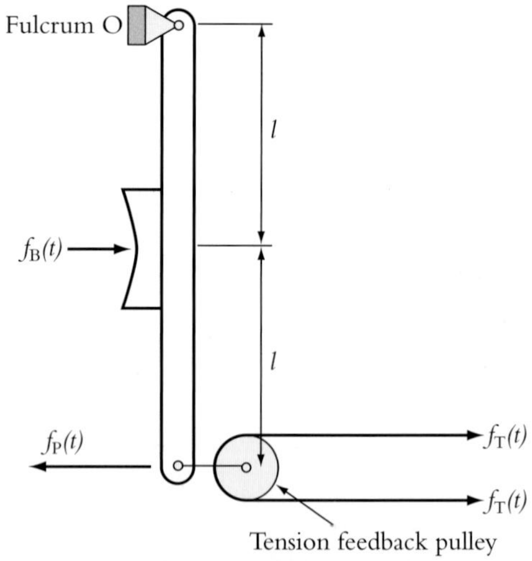

Figure 5.10 Lever and brake assembly (braking friction is assumed to have negligible moment and is omitted for clarity; reaction forces at O also omitted)

Again we note that the length l, and for that matter quantities such as R, r and μ, are all constant, i.e. invariant with time, while all variable quantities f_P, f_T and f_B are expressed in the form of functions of time t. In the present version of the example we will assume that such quantities only change very slowly, and/or that the inertia of the rotating components is negligibly small. The behaviour of the brake lever and pulley is represented as a block diagram in Figure 5.11.

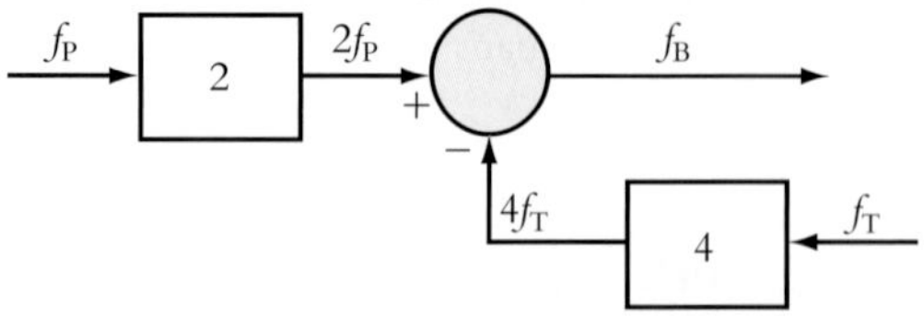

Figure 5.11 Block diagram for lever and pulley

In the block diagram, the mathematical relationships between quantities such as the input setting f_P, the output f_T (fibre tension) and intermediate quantities such as brake force f_B are shown as 'blocks' that act on 'flows' of values or data. In this particular block diagram, blocks involving multiplication by constant factors (gains) of value 2 and 4 are evident. Also visible is a 'differencing junction' where $4f_T$ is subtracted from $2f_P$ to give f_B. Again, the fact that f_P, f_T and f_B are functions of time is deliberately not shown; the correct representation of time-dependent quantities within block diagrams via the complex variable s will be introduced in Section 5.4.

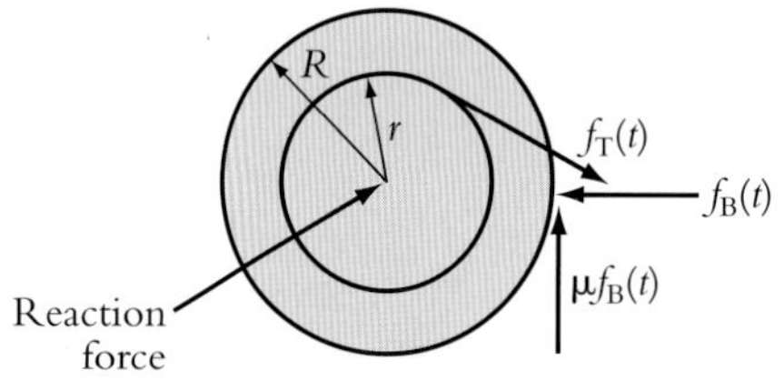

Figure 5.12 Brake drum

Provided that inertia effects are ignored, or that no acceleration or deceleration is occurring, the brake drum (Figure 5.12) is similarly analysed as follows:

$$f_T(t)r - \mu f_B(t)R = 0$$

$$\therefore \quad f_T(t) = \frac{\mu R}{r} f_B(t) \tag{5.12}$$

This analysis can be represented via a block diagram (Figure 5.13).

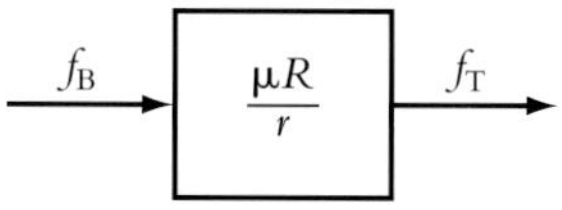

Figure 5.13 Block diagram for brake drum

By virtue of the feedback inherent in the system, there is a circular relationship involving f_B and f_T since the application of the braking force f_B results in an increase in the fibre tension f_T, but the lever system results in f_T acting to reduce the braking force f_B. This simple system could be solved algebraically without further ado, but it is instructive first to draw the complete block diagram for the system, in which the feeding-back of f_T for eventual subtraction from f_P is now clearly visible in Figure 5.14.

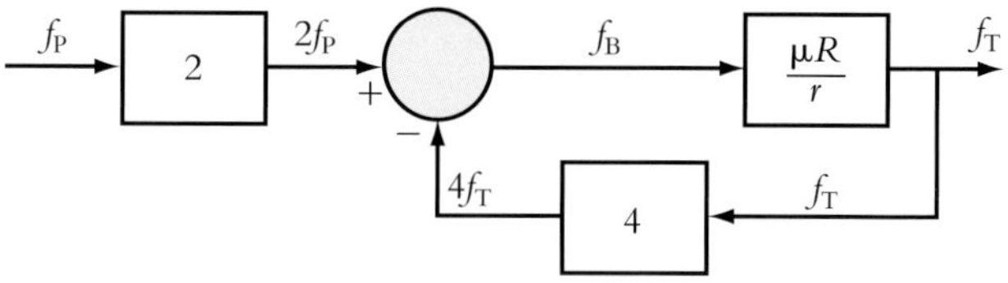

Figure 5.14 Block diagram for fibre tensioning system

Figure 5.14 may be rearranged to bring the leftmost block inside the loop as shown in Figure 5.15, and may be further simplified into the form shown in Figure 5.16 by combining the transfer functions of successive blocks by multiplication.

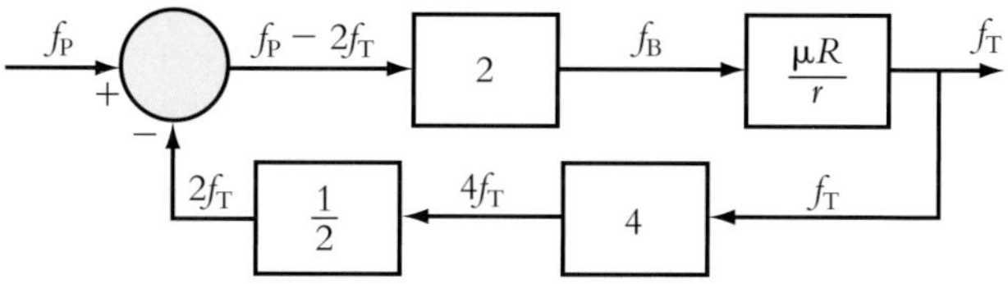

Figure 5.15 Block diagram of tensioning system after rearrangement

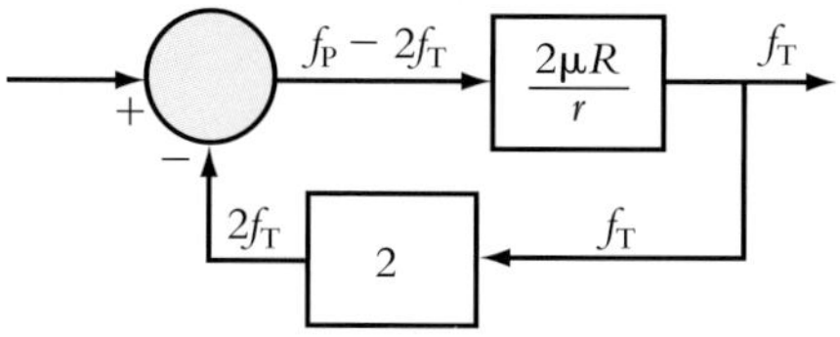

Figure 5.16 Block diagram of tensioning system after multiplying successive blocks

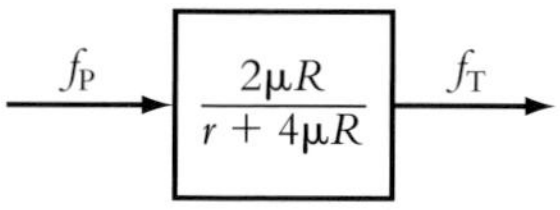

Figure 5.17 Block diagram for tensioning system represented using a single transfer function

The system may now be solved as follows:

$$f_T(t) = 2\frac{\mu R}{r}(f_P(t) - 2f_T(t)) = 2\frac{\mu R}{r} f_P(t) - 4\frac{\mu R}{r} f_T(t)$$

$$\therefore \quad f_T(t)\left(1 + 4\frac{\mu R}{r}\right) = 2\frac{\mu R}{r} f_P(t) \tag{5.13}$$

$$\therefore \quad f_T(t) = \frac{2\frac{\mu R}{r}}{\left(1 + 4\frac{\mu R}{r}\right)} f_P(t) = \frac{2\mu R}{(r + 4\mu R)} f_P(t)$$

In this case the ratio $f_T(t)/f_P(t)$ of fibre tension to pre-load force (effectively the ratio of output force to input force) is simply $2\mu R/(r + 4\mu R)$, and a block diagram which represents this is shown in Figure 5.17.

The effectiveness of the control provided by this system can now be observed for the following parameters: $r = 50$ mm, $R = 100$ mm, $\mu = 0.4$, $F_P = 2.5$ N. A 20% reduction in r (to 40 mm) as the spool of fibre unwinds results only in a 5% increase in tension from 0.95 N to 1 N – without the feedback system, the change in tension would be around 20%.

If we wish to consider the inertia of the spool and brake assembly (which would be significant, and would need to be included in the analysis in order to understand the tension in the fibre) then the simple (time domain) mathematics described above is insufficient to express the situation, and an additional tool is required to take account of rates of change within a system. This will be described in the next section.

Learning summary

By the end of this section you should have learnt:

✓ the concepts of closed-loop control and feedback, and their typical applications;

✓ the difference between positive and negative feedback;

✓ the concept of a block diagram representing the characteristics of a control system and the manner in which feedback is provided.

5.4 The *s* domain: a notation borrowed from mathematics

It is very difficult to analyse anything except the most elementary systems (such as the ones described above) in the time domain. This is because practical systems involve features such as inertia and damping, which require derivatives such as acceleration and velocity to be incorporated into the model. In such a case, the block diagram effectively becomes a graphical representation of a differential equation rather than a simple algebraic one. In order to tackle such systems, analysis within classical control engineering is generally undertaken using an adaptation of the methods of solving differential equations, using a notation based on one of two complex variables:

- the variable known as *s*, involving analysis in the *s* domain, used for modelling continuous control situations such as the ones covered in this chapter. This notation is closely related to the mathematical approach involving **Laplace transforms**, which are just a useful way of solving differential equations by converting them to polynomials in *s*;
- a different complex variable, known as *z*, involving analysis in the *z* domain, used for modelling discrete and digital control situations. This will not be considered further in the present chapter, but will be of interest in the case of more advanced control situations.

Within the present chapter, the behaviour of control systems and their components, including modes of behaviour that involve derivatives and integrals with respect to time, will henceforth be expressed in terms of the variable *s*. The concept of the *s* domain, the meaning of *s* as a variable, and the relationship to Laplace transforms, can all be discouraging or confusing to engineers whose main concern is to solve practical control problems. It is therefore convenient here to view the variable *s* and the *s* domain as a language or notation for processes involving integration and differentiation, rather than becoming too concerned about either the physical meaning of *s* or its use as a tool for solving differential equations; there are numerous introductions to the technique in the literature, such as Chapter 6 of the engineering mathematics text of Kreyszig (2006). For completeness, however, it can be stated that the Laplace transform $F(s) = \mathscr{L}\{f(t)\}$ of a time-domain function $f(t)$ is defined as:

$$F(s) = \mathscr{L}\{f(t)\} = \int_0^\infty e^{-st} f(t)\mathrm{d}t \tag{5.14}$$

where s may be interpreted as complex angular frequency, or radians per unit time. By convention, as shown in equation (5.14), variable quantities expressed in the time domain (i.e. as a function of time) are denoted by lowercase letters, e.g. $f(t)$, while the quantities expressed in the s domain are denoted by upper-case letters, e.g. $F(s)$.

Laplace transforms of a large variety of functions are tabulated in the mathematical literature (Kreyszig, 2006, pp 264–7; Healey, 1967). For the purpose of the present chapter, Table 5.1 may be considered to be the most useful:

	Function $f(t)$ (valid only for $t \geqslant 0$; assumed to by zero for $t < 0$)	Graph	Laplace transform $F(s)$
1	Unit step function (Heaviside function) $H(t)$: $f(t) = 0$ for $t < 0$ $f(t) = 1$ for $t > 0$ (Note: definitions vary for the value of $f(t)$ for $t = 0$)	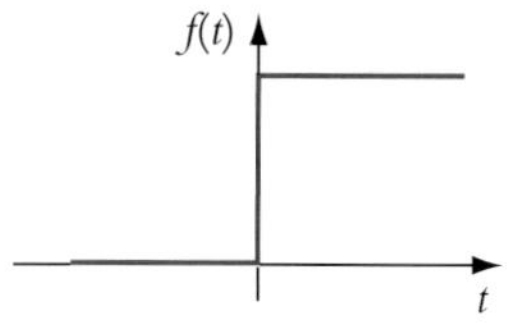	$\frac{1}{s}$
2	Unit ramp function $tH(t)$: $f(t) = 0$ for $t \leqslant 0$ $f(t) = t$ for $t > 0$	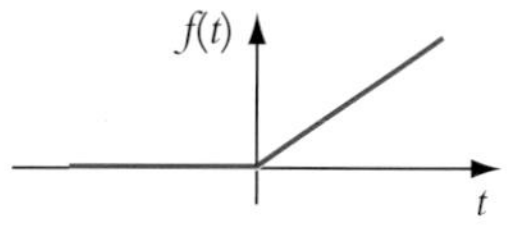	$\frac{1}{s^2}$
3	Unit impulse function (Dirac delta function) $\delta(t)$: $f(t) = 0$ for $t < 0$ $f(t) = \infty$ for $t = 0$ $f(t) = 0$ for $t > 0$ (area enclosed is unity)	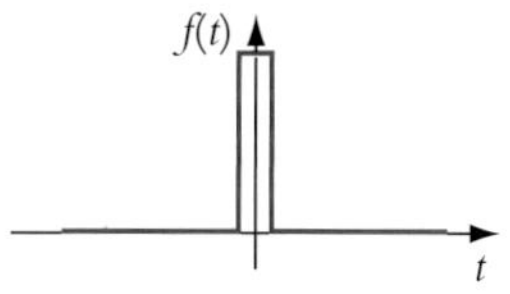	1
4	e^{-at}	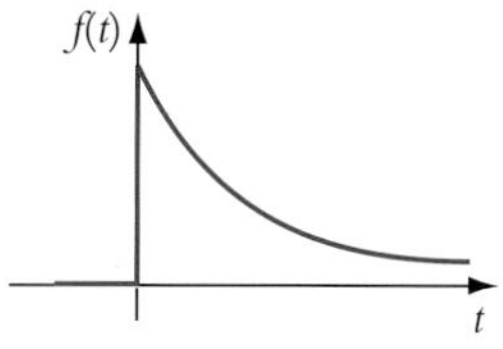	$\frac{1}{s+a}$
5	$1 - e^{-at}$	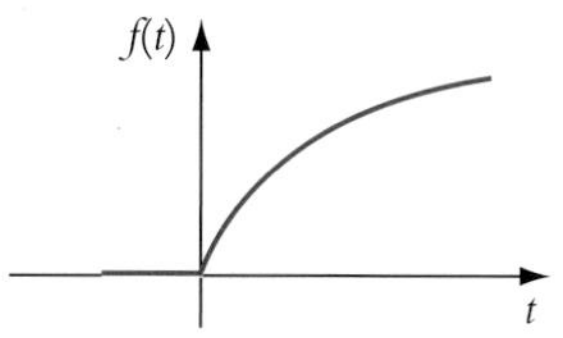	$\frac{a}{s(s+a)}$
6	$t - \frac{1}{a}(1 - e^{-at})$	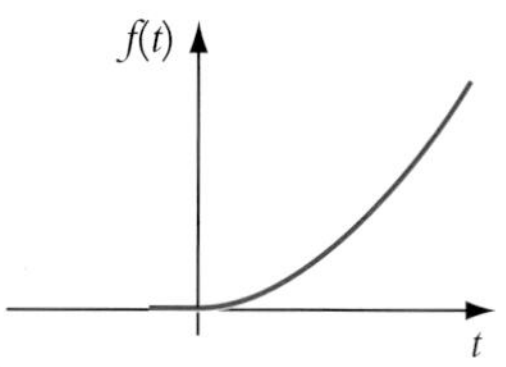	$\frac{a}{s^2(s+a)}$

Continued overleaf

7	$\sin(\omega t)$	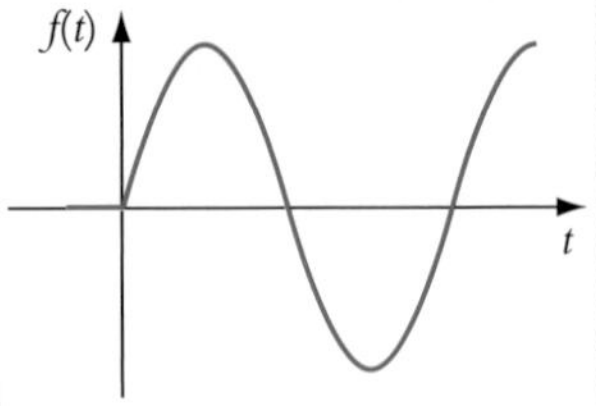	$\dfrac{\omega}{s^2 + \omega^2}$
8	$\cos(\omega t)$	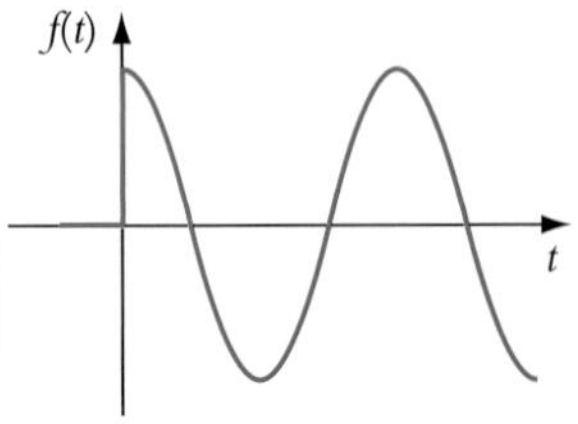	$\dfrac{s}{s^2 + \omega^2}$
9	$\dfrac{\omega}{\sqrt{1-\zeta^2}} e^{-\zeta\omega t} \sin\left(\omega t\sqrt{1-\zeta^2}\right)$	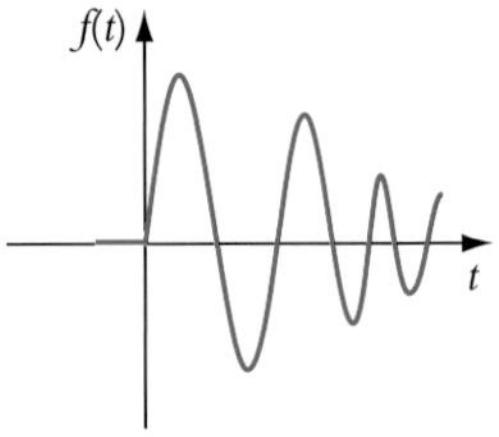	$\dfrac{\omega^2}{s^2 + 2\zeta\omega s + \omega^2}$
10	$1 - \dfrac{e^{-\zeta\omega t}}{\sqrt{1-\zeta^2}} \sin\left(\omega t\sqrt{1-\zeta^2} + \phi\right)$ where $\cos\phi = \zeta$	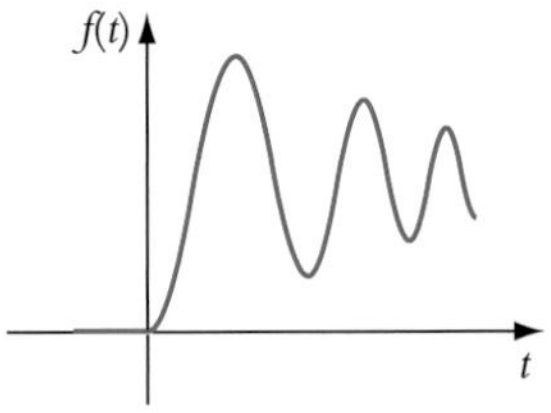	$\dfrac{\omega^2}{s(s^2 + 2\zeta\omega s + \omega^2)}$

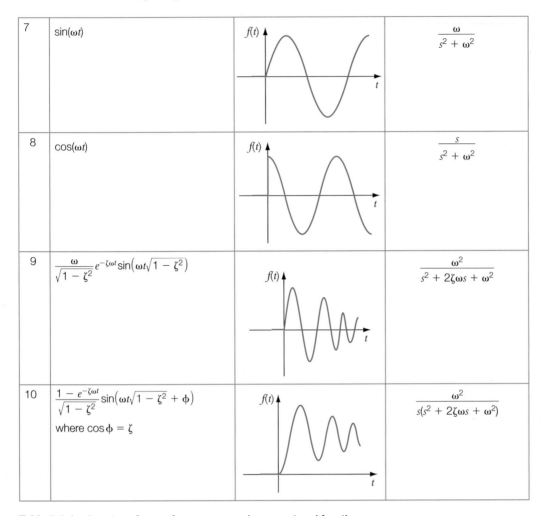

Table 5.1: Laplace transforms of some commonly encountered functions

Laplace transforms also obey various rules which include the following:

(1) Laplace transforms are **linear**, that is they obey the principle of superposition. For example:

$$\begin{aligned}\mathcal{L}\{af(t) + bg(t)\} &= \mathcal{L}\{af(t)\} + \mathcal{L}\{bg(t)\} \\ &= a\mathcal{L}\{f(t)\} + b\mathcal{L}\{g(t)\} = aF(s) + bG(s)\end{aligned} \tag{5.15}$$

In particular it can be seen from this example that:

- the original functions $f(t)$ and $g(t)$ are multiplied by the constants a and b respectively, and so are their transforms $F(s)$ and $G(s)$;
- the transform of the sum of $af(t)$ and $bg(t)$ is the sum of their transforms.

Similarly, the transform of a difference, e.g. $af(t) - bg(t)$ would be $aF(s) - bG(s)$. However, the unwary student should note that this concept does not extend to the situation where functions of t are multiplied together: the transform of $f(t)g(t)$ is *not* the product of their transforms, as this does not involve superposition or linear combination.

(2) The Laplace transform of a derivative is

$$\mathcal{L}\left\{\frac{\mathrm{d}f(t)}{\mathrm{d}t}\right\} = s\mathcal{L}\{f(t)\} - f(0) \tag{5.16}$$

so that, in turn, the transform of a second derivative is

$$\mathcal{L}\left\{\frac{\mathrm{d}^2 f(t)}{\mathrm{d}t^2}\right\} = s\mathcal{L}\left\{\frac{\mathrm{d}f(t)}{\mathrm{d}t^2}\right\} - f'(0) = s^2\mathcal{L}\{f(t)\} - f'(0) - sf(0) \qquad \textbf{(5.17)}$$

where $f'(0)$ is the value of $\dfrac{\mathrm{d}f(t)}{\mathrm{d}t}$ at $t = 0$.

(3) The Laplace transform of an integral is

$$\mathcal{L}\left\{\int_0^t f(\tau)\mathrm{d}\tau\right\} = \frac{1}{s}\mathcal{L}\{f(t)\} \qquad \textbf{(5.18)}$$

(4) The Laplace transform of a time-shifted function, where the time axis has been changed by a constant value τ, is as follows:

if $\quad \mathcal{L}\{f(t)\} = F(s) \quad$ then $\quad \mathcal{L}\{f(t-\tau)\} = e^{-s\tau}F(s) \qquad$ **(5.19)**

Of particular interest here is the observation that multiplication by s (and subtraction of the initial condition, which in a control system is typically zero) represents differentiation, while division by s represents integration. This 'shorthand' for these calculus operations is encountered often in dealing with control systems. For readers familiar with the 'operator D' method of solving differential equations (Kreyszig, 2006, pp. 59–61; Spencer *et al*, 1977, pp. 17–18 and 41), there are significant similarities between the operator D (where multiplication by D represents differentiation and division by D represents integration) and the way in which differentiation and integration in the time domain are represented respectively as multiplication and division by s.

The procedure for using Laplace transforms to solve ordinary differential equations is:

(1) with knowledge of the initial conditions, the differential equation is transformed from the time domain to the s domain;

(2) by use of partial fractions or otherwise, the transformed equation is simplified to give the solution in the s domain;

(3) the simplified equation is transformed back into the time domain to give the solution (in control engineering, this step is usually not necessary and the behaviour of the system is generally left as an expression in s).

A straightforward example illustrates the power of Laplace transforms in solving differential equations. A system (Figure 5.18) consisting of a spring of stiffness k in parallel with a viscous damper of constant c is subjected to a constant force $p(t) = q$ beginning at time $t = 0$. For traceability throughout the following derivation, the forcing term is highlighted in green, the spring term in red and the damping term in blue, up to the point where the equations are rearranged and the individual terms are no longer identifiable.

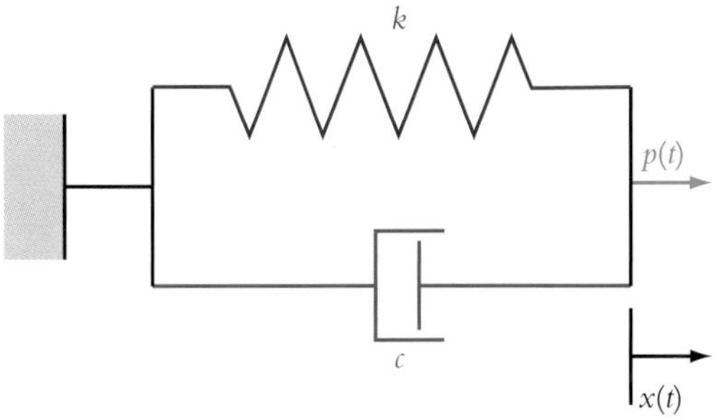

Figure 5.18 Spring–damper system

It will be assumed that the link joining the spring to the damper has no significant mass. Applying equilibrium to the link:

$$p(t) = kx(t) + c\frac{\mathrm{d}x(t)}{\mathrm{d}t}$$

At $\quad t = 0, x = 0$

$$p(t) = 0 \text{ for } t < 0 \qquad \textbf{(5.20)}$$

$$p(t) = q \text{ for } t > 0$$

Mathematically this load can be expressed in terms of the Heaviside step function $h(t)$ (see item 1 in Table 5.1):

$$p(t) = qH(t) \qquad \textbf{(5.21)}$$

so that the equation of motion now becomes

$$qH(t) = kx(t) + c\frac{dx(t)}{dt} \tag{5.22}$$

Taking Laplace transforms (see equation (5.14)) gives

$$\begin{aligned}\frac{q}{s} &= kX(s) + csX(s) - cx(0)\\ &= (k + cs)X(s) - cx(0)\end{aligned} \tag{5.23}$$

Rearrangement gives

$$X(s) = \frac{q}{s(k + cs)} + \frac{cx(0)}{k + cs} \tag{5.24}$$

Using partial fractions this can be expressed as

$$X(s) = \frac{q}{sk} - \frac{qc}{k(k + cs)} + \frac{cx(0)}{k + cs}$$

Inserting the initial condition that $x(0) = 0$, and rearranging, gives

$$X(s) = \frac{q}{k}\frac{1}{s} - \frac{q}{k}\frac{1}{(s + k/c)} \tag{5.25}$$

From items 1 in Table 5.1, it is seen that the inverse transform of $1/s$ is 1 for $t > 0$, and similarly the inverse transform of $1/(s + a)$ is e^{-at} for $t > 0$; these items are of the same form as the terms in equation (5.25). Assuming that $a = k/c$ and inserting these inverse transforms into equation (5.25) gives the well-known solution to the problem of a spring-damper system under a step load:

$$x(t) = \frac{q}{k}.1 - \frac{q}{k}e^{-\frac{kt}{c}} = \frac{q}{k}\left(1 - e^{-\frac{kt}{c}}\right) \tag{5.26}$$

Instead of using partial fractions, it is often possible to make use of more complicated inverse transforms to get straight to the answer. For example, equation (5.25) can be expressed as:

$$X(s) = \frac{q}{k}\left(\frac{k/c}{s(s + k/c)}\right) \tag{5.27}$$

From item 5 in Table 5.1, it is seen that the inverse transform of

$$\frac{a}{s(s + a)}$$

is

$$1 - e^{-at}$$

Assuming as before that $a = k/c$ and substituting this inverse transform into equation (5.27) once again recovers the solution equation (5.26), this time without any further working being required.

Learning summary

By the end of this section you should have learnt:

- ✓ the concept of the Laplace transform and its role in the solution of differential equations;
- ✓ the basic rules relating to application of Laplace transforms to problems involving integration, differentiation and time-shifting;
- ✓ how to transform a range of expressions from the time domain into the s domain.

5.5 Block diagrams and the *s* notation: the heater controller and tensioning system examples revisited

Heater controller

The real strength of the *s* notation, when applied to control engineering, comes when it is used in conjunction with block diagrams to model systems whose behaviour involves differentiation or integration with respect to time. Revisiting the simple example of the heated system modelled earlier, consider the heated equipment to have a mass of $m = 50\,\text{kg}$ and a specific heat capacity of $C_p = 500\,\text{J/kg K}$. Now, it is assumed that the heating power supplied and the rate of heat loss are no longer equal (so the system is no longer in thermal equilibrium), and the net rate of heat gain causes the temperature to increase:

$$p(t) - Ah\tau(t) = mC_p\frac{d\tau(t)}{dt} \tag{5.28}$$

Assuming initial conditions (i.e. assuming that $\tau(t)$ is initially zero), and making use of equation (5.16), the Laplace transform of this expression is

$$P(s) - AhT(s) = mC_psT(s) \tag{5.29}$$

This can be rearranged as

$$T(s) = \frac{P(s) - AhT(s)}{mC_ps} \tag{5.30}$$

The block diagram (Figure 5.19) for this aspect of the system is now slightly more complex than before and shows the circular relationship inherent in equation (5.29), in which $T(s)$ is defined via a function involving $T(s)$ itself.

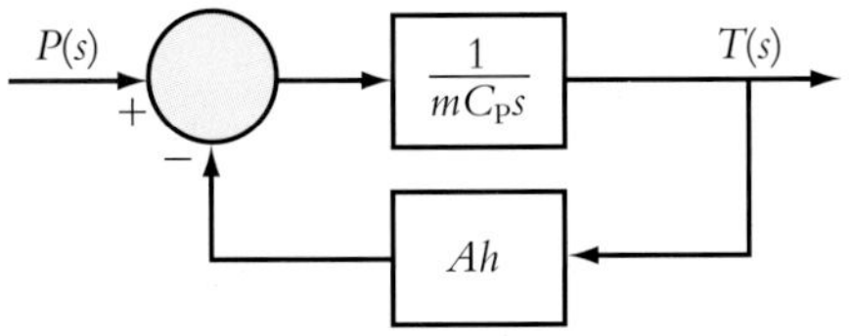

Figure 5.19 Block diagram relating to heat balance of heating system

This circular relationship can be eliminated from the mathematics by straightforward manipulation of equation (5.30):

$$\begin{aligned} T(s) + \frac{AhT(s)}{mC_ps} &= \frac{mC_psT(s) + AhT(s)}{mC_ps} \\ &= \frac{mC_ps + Ah}{mC_ps}T(s) = \frac{P(s)}{mC_ps} \\ \Rightarrow T(s) &= \frac{1}{mC_ps + Ah}P(s) \end{aligned} \tag{5.31}$$

We could use this to simplify the block diagram relating to this aspect of the problem, but for illustrative purposes the loop in the block diagram will be left in place and will instead be tackled within Section 5.9.

The controller still follows the same approach as in the simpler version of the problem, i.e. the power supplied to heat the equipment is proportional to the error, which is the difference between desired and achieved temperatures:

$$p(t) = K(\tau_D(t) - \tau(t)) \quad (5.32)$$

This expression is very straightforward to transform to the s domain:

$$P(s) = K(T_D(s) - T(s)) \quad (5.33)$$

and is represented by substantially the same block diagram as before (Figure 5.20). Compare this with Figure 5.5; it is now (correctly) expressed in terms of the complex variable s.

Assembly of the complete block diagram now results in a diagram which involves two nested loops (Figure 5.21).

This block diagram will be simplified to a straightforward transfer function (ratio of output to input in the s domain) in Section 5.9.

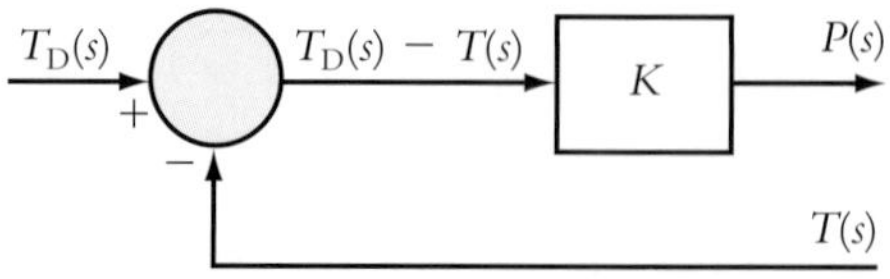

Figure 5.20 Block diagram relating to controller

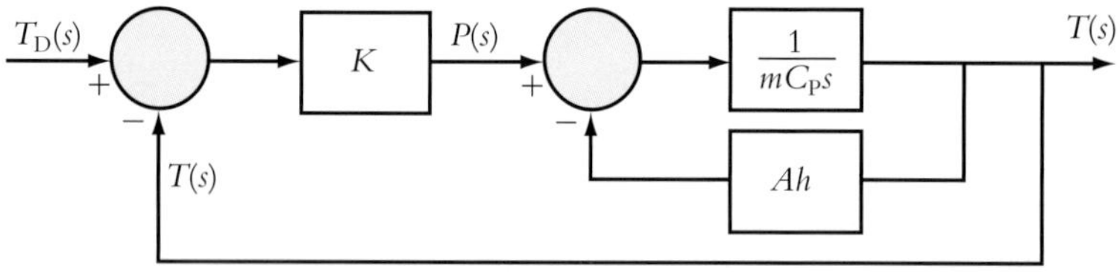

Figure 5.21 Complete block diagram of heater controller system

Fibre tensioning system

Revisiting also the fibre tensioning system example, now consider the spool/brake drum assembly to have inertia J. Also, consider that the fibre is being drawn through the system at a velocity $u(t)$ (which is equal to $r\dot{\theta}(t)$) and acceleration $du(t)/dt$, so that the angular acceleration $\ddot{\theta}(t)$ of the drum is

$$\ddot{\theta}(t) = \frac{1}{r}\frac{du(t)}{dt} \quad (5.34)$$

The analysis starts in almost the same manner as the static analysis of the system undertaken earlier. Equation (5.11) is still applicable:

$$f_B(t) = 2(f_P(t) - 2f_T(t)) \quad (5.35)$$

Equation (5.35) straightforwardly transforms to the s domain to give

$$F_B(s) = 2(F_P(s) - 2F_T(s))$$

or:

$$F_B(s) = 2F_P(s) - 4F_T(s) \quad (5.36)$$

The block diagram for the lever is now unchanged from the version in Figure 5.11 other than being expressed in s space (Figure 5.22).

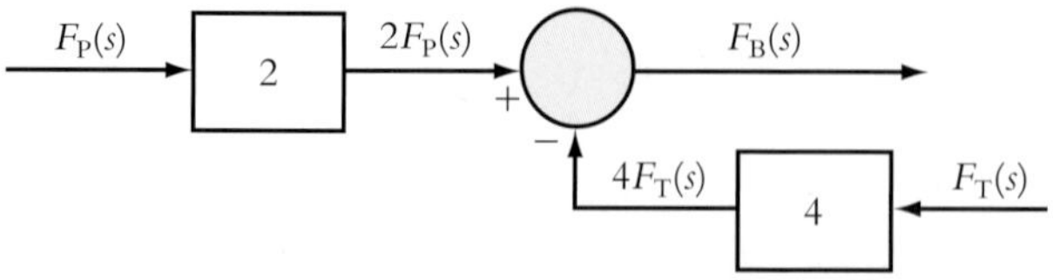

Figure 5.22 Block diagram for lever system

However, introducing inertia into equation (5.12) results in an expression of the form $L = J\ddot{\theta}$:

$$f_T(t)r - \mu f_B(t)R = J\ddot{\theta} = \frac{J}{r}\frac{du(t)}{dt} \quad (5.37)$$

Rearranging gives

$$\therefore \qquad f_T(t) = \frac{\mu R}{r} f_B(t) + \frac{J}{r}\frac{du(t)}{dt} \tag{5.38}$$

Making use of equation (5.16), the Laplace transform of this expression (assuming that $u(t)$ and $du(t)/dt$ are both zero initially, i.e. at $t = 0$) is now

$$\therefore \qquad F_T(s) = \frac{\mu R}{r} F_B(s) + s\frac{J}{r} U(s) \tag{5.39}$$

The block diagram (Figure 5.23) for the brake/drum assembly is therefore somewhat more complex than before and now involves a summing junction.

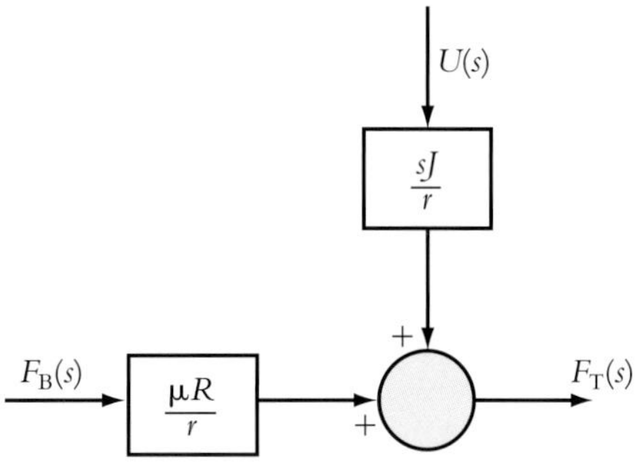

Figure 5.23 Block diagram of brake/drum assembly including inertia

By combining the block diagrams for the lever and for the brake, and joining the arrows relating to $F_B(t)$ and $F_T(t)$, the overall block diagram for the system can now be assembled in Figure 5.24.

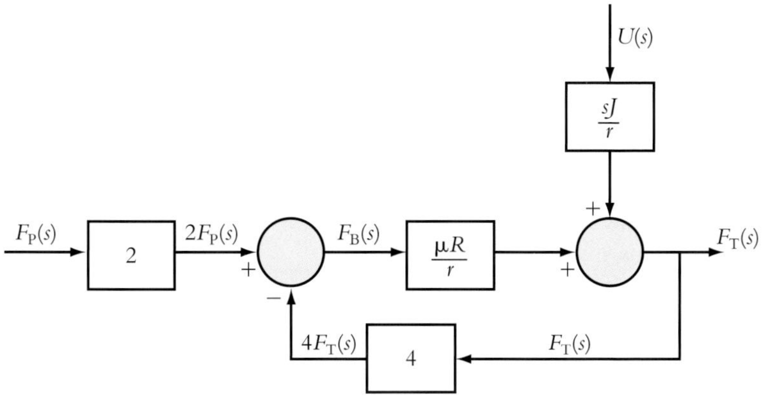

Figure 5.24 Block diagram of tensioning system

It will no longer be possible to express the transfer function purely as a ratio of output to input, as effectively there are now two inputs to the system (pre-load $f_P(t)$ and filament speed $u(t)$).

By way of introduction to the manipulation and simplification of block diagrams, note that proceeding along the main flow of the block diagram leads to the following (circular) expression, involving $F_T(s)$ on both sides of the equation:

$$F_T(s) = (2F_P(s) - 4F_T(s))\frac{\mu R}{r} + U(s)\frac{sJ}{r}$$

$$= 2F_P(s)\frac{\mu R}{r} - 4F_T(s)\frac{\mu R}{r} + U(s)\frac{sJ}{r} \tag{5.40}$$

Rearranging to take $F_T(s)$ to the left-hand side of the equation gives

$$F_T(s) + 4F_T(s)\frac{\mu R}{r} = 2F_P(s)\frac{\mu R}{r} + U(s)\frac{sJ}{r} \quad \textbf{(5.41)}$$

Multiplying throughout by r gives

$$F_T(s)(r + 4\mu R) = 2F_P(s)\mu R + U(s)sJ \quad \textbf{(5.42)}$$

Finally, dividing throughout by $(r + 4\mu R)$ gives the transfer function:

$$F_T(s) = \frac{2\mu R}{(r + 4\mu R)}F_P(s) + \frac{Js}{(r + 4\mu R)}U(s) \quad \textbf{(5.43)}$$

Learning summary

By the end of this section you should have learnt:

✓ the use of block diagrams involving the s domain;

✓ the construction of the block diagram of a system involving dynamic behaviour such as mechanical or thermal inertia.

5.6 Working with transfer functions and the *s* domain

Open- and closed-loop transfer functions

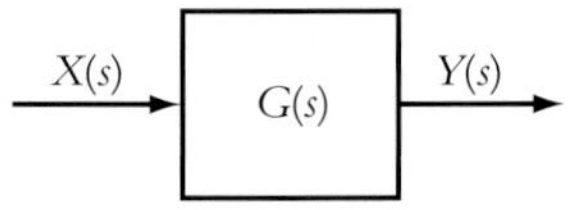

Figure 5.25 A block diagram representing a transfer function G(s)

It has already been mentioned in passing that a transfer function gives the relationship between the output and input. In general, the transfer function of a simple system (e.g. that given in Figure 5.25) is expressed as a function of s relating the Laplace transform $Y(s)$ of the output $y(t)$ of a system divided by the Laplace transform $X(s)$ of the input $x(t)$, for example:

$$G(s) = \frac{Y(s)}{X(s)} \quad \textbf{(5.44)}$$

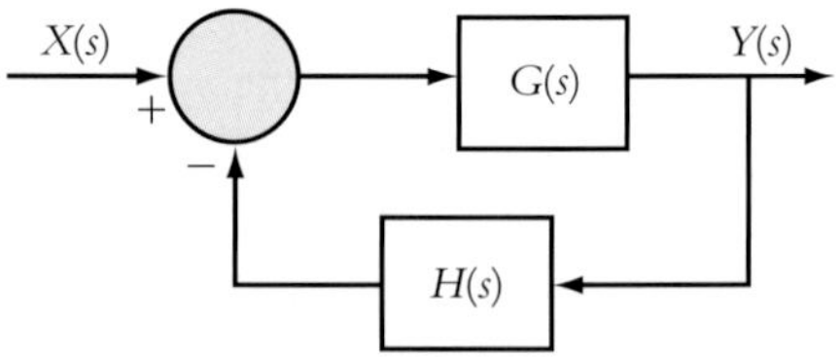

Figure 5.26 A typical feedback loop

While this is straightforward for a simple system having a block diagram consisting of a single box, practical control systems are more complex and, like the example of the tensioning system, nearly always involve some form of (usually negative) feedback. Such systems typically have a block diagram of the form shown in Figure 5.26.

In such a system:

- $G(s)$ typically represents the transfer function of the item of equipment being controlled, along with the transfer function of the controller;
- $H(s)$ typically represents the transfer function of the sensor or transducer being used to measure the output of the system and compare it with the input or demand signal.

Two versions of the transfer function for this system can be defined:

- Open-loop transfer function. There are two conventions for this, which can be confusing for the unwary student. For the typical system shown in Figure 5.26, the open-loop transfer function is sometimes defined simply as $G(s)$, the relationship between output $Y(s)$ and input $X(s)$ *for the condition that the feedback does not actually take place* (Figure 5.27). However, the open-loop transfer function is also sometimes defined as the relationship between the input and the value which would be subtracted from the input, again for the condition that no feedback actually takes place. In other words, it is sometimes defined as $G(s)H(s)$, the product of the transfer functions of the two boxes through which the signal passes on its way back to the differencing junction:

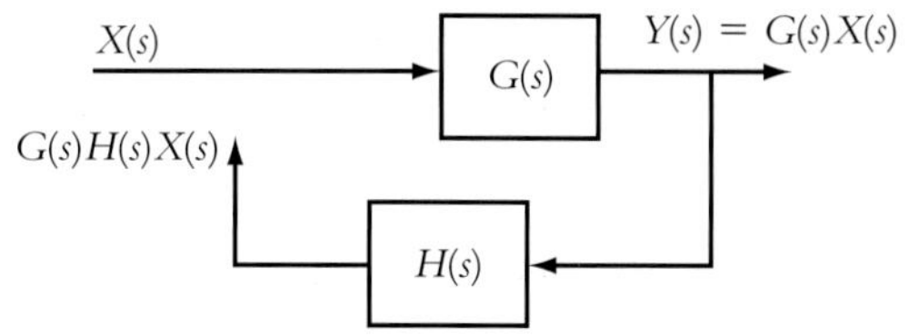

Figure 5.27 A control system operating in open loop mode

The latter version of the open-loop transfer function is of particular interest in predicting the stability of a control system from experimental measurements without actually causing the control system to go unstable, and is the one which will be used later in this chapter.

- Closed-loop transfer function. This gives the relationship between the output of the closed-loop control system (Figure 5.26) and its input. The closed-loop transfer function can be derived by noting that:

$$Y(s) = G(s)[X(s) - H(s)Y(s)] \tag{5.45}$$

Rearranging gives

$$Y(s) + G(s)H(s)Y(s) = Y(s)[1 + G(s)H(s)] = G(s)X(s) \tag{5.46}$$

and hence

$$\frac{Y(s)}{X(s)} = \frac{G(s)}{1 + G(s)H(s)} \tag{5.47}$$

which is the closed-loop transfer function of the system, the ratio of output $Y(s)$ to input $X(s)$ in the s domain. The closed-loop transfer function is of particular interest in terms of understanding the error of the system (how the actual output from the system compares with the desired output).

In order to determine the behaviour of the output of the system, two further mathematical tools are needed: the **initial-value theorem** and the **final-value theorem**.

Initial-value theorem

It is convenient to assume within classical control modelling that the initial state of a given control system is known. In other words, the starting point for future behaviour is unambiguously defined. It can usually be assumed that everything is initially at rest and in its 'neutral' position (i.e. all values defining the state are zero). It can be useful to check that the transfer function, in conjunction with the inputs to the system, correctly calculates this state. This theorem may be stated as:

$$\lim_{t \to 0^+} f(t) = \lim_{s \to \infty} sF(s) \tag{5.48}$$

Note that this finds the value of $f(t)$ *immediately after* time $t = 0$ – in other words, an infinitesimal amount of time is allowed for the analysis to begin and for inputs to be 'switched on' (via the unit step and unit impulse functions given in items 1 and 3 of Table 5.1, for example).

As an example, consider the spring–damper system described earlier, consisting of a spring of stiffness k and a damper of constant c in parallel, subjected to a load which is zero for $t < 0$ and of value q for $t > 0$. Mathematically this load can be expressed in terms of the Heaviside step function $H(t)$:

$$p(t) = qH(t) \tag{5.49}$$

The Laplace transform of the Heaviside step function is given in row 1 of Table 5.1, so that equation 5.49 transforms to:

$$P(s) = \frac{q}{s} \tag{5.50}$$

From earlier work it was shown (equation (5.25)) that the displacement of the spring–damper system under this load of q applied at $t = 0$ can be expressed as

$$X(s) = \frac{q}{sk} - \frac{q}{k\left(\frac{k}{c} + s\right)} \tag{5.51}$$

Inserting this expression into the initial-value theorem (equation (5.48)) gives

$$\begin{aligned} \lim_{t \to 0^+} x(t) = \lim_{s \to \infty} sX(s) &= \lim_{s \to \infty} s\left(\frac{q}{sk} - \frac{q}{k\left(\frac{k}{c} + s\right)}\right) = \lim_{s \to \infty}\left(\frac{sq}{sk} - \frac{sq}{k\left(\frac{k}{c} + s\right)}\right) \\ &= \lim_{s \to \infty}\left(\frac{q}{k} - \frac{q}{\frac{k^2}{sc} + k}\right) \\ &= \frac{q}{k} - \frac{q}{\frac{k^2}{\infty \times c} + k} \\ &= \frac{q}{k} - \frac{q}{k} = 0 \end{aligned} \tag{5.52}$$

Note that the very large terms involving s swamp the smaller terms such as k/c, so that the latter may be neglected; the zero initial displacement of the system is correctly recovered.

Now consider a simplified version of this system, which omits the damper. In this case, the equation describing the spring is very simple:

$$kx(t) = p(t);\ x(0) = 0 \tag{5.53}$$

Dividing by k, taking Laplace transforms and noting the zero initial conditions gives

$$X(s) = \frac{1}{k}P(s) \tag{5.54}$$

Once again the load is represented as a Heaviside step function (item 1 in Table 5.1):

$$p(t) = qH(t) \tag{5.55}$$

so that

$$P(s) = \frac{q}{s} \tag{5.56}$$

Combining these expressions gives the expression for displacement under the applied load:

$$X(s) = \frac{1}{k}\frac{q}{s} = \frac{q}{sk} \tag{5.57}$$

Inserting this into the initial-value theorem correctly recovers the displacement of the spring immediately after application of the load:

$$\begin{aligned} \lim_{t \to 0^+} x(t) = \lim_{s \to \infty} sX(s) &= \lim_{s \to \infty} s\left(\frac{q}{sk}\right) \\ &= \frac{q}{k} \end{aligned} \tag{5.58}$$

It could usefully be argued that this example is unrealistic as the displacement has occurred instantaneously on application of the load; in practice, of course, there would be some inertia, but that is not modelled here.

Final-value theorem

Although the initial-value theorem provides a useful check that initial conditions are satisfied, a potentially more useful tool is the **final-value theorem**, which is a method of calculating the long-term value of the response of a system. The theorem may be stated as

$$\lim_{t \to \infty} f(t) = \lim_{s \to 0} sF(s) \tag{5.59}$$

Note that this is only applicable where the final value is constant and finite.

A straightforward example is the calculation of the displacement of the spring–damper system after it has had a chance to settle down. Recall that

$$X(s) = \frac{q}{sk} - \frac{q}{k\left(\frac{k}{c} + s\right)} \tag{5.60}$$

Inserting this expression into the final-value theorem correctly recovers the steady-state displacement of the system:

$$\begin{aligned} \lim_{t \to \infty} x(t) &= \lim_{s \to 0} sX(s) = \lim_{s \to 0} s\left(\frac{q}{sk} - \frac{q}{k\left(\frac{k}{c} + s\right)}\right) \\ &= \frac{q}{k} - 0 \times \frac{q}{k\left(\frac{k}{c} + 0\right)} = \frac{q}{k} \end{aligned} \tag{5.61}$$

A more sophisticated example will be given in Section 5.12.

Learning summary

By the end of this section you should have learnt:

✓ to simplify a block diagram containing a feedback loop to give a simplified block diagram involving a single transfer function;

✓ to find the values of an expression in the s domain at times $t = 0$ and $t = \infty$.

5.7 Building a block diagram: part 1

The strength of the techniques of control modelling lies in the fact that models of complete systems can be assembled from the building blocks consisting of the models of individual components. For each component, it is necessary to understand how the output and input are related (in the time domain). This relationship, often in the form of a differential equation, is then transformed into the s domain, and is then rearranged to give the transfer function of that component. Some typical examples of components follow.

Mass

The transfer function giving the displacement of a mass in terms of a force applied to it is obtained by transforming Newton's second law:

$$f(t) = ma(t) = m\frac{d^2x(t)}{dt^2} \tag{5.62}$$

Using equation (5.17) and taking Laplace transforms gives

$$F(s) = ms^2X(s) - mx'(0) - msx(0) \tag{5.63}$$

where $x'(0)$ is the value of dx/dt at $t = 0$. If zero initial conditions are assumed (i.e. displacement and velocity are zero up to the instant when $t = 0$), this simplifies to

$$F(s) = ms^2X(s) \tag{5.64}$$

This may be represented in block diagram form as shown in Figure 5.28.

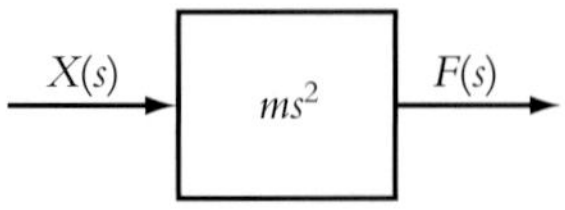

Figure 5.28 Block diagram for a mass

In this context (i.e. for this form of the block diagram) the transfer function of the mass is therefore the force divided by the displacement (where both are expressed as functions of s):

$$\frac{F(s)}{X(s)} = ms^2 \tag{5.65}$$

Sometimes, it is more convenient to consider the conversion to acceleration and the effect of the mass as separate 'boxes' in the block diagram as shown in Figure 5.29.

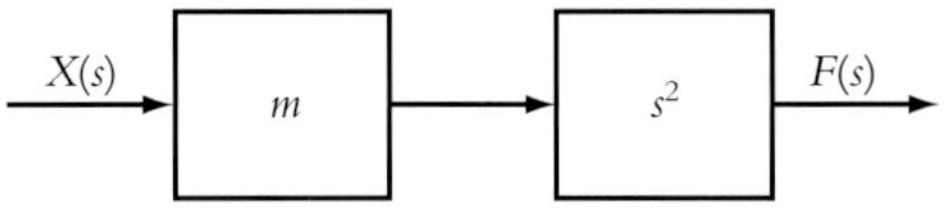

Figure 5.29 Alternative block diagram for a mass

Alternatively, equation (5.65) may be rearranged to give the displacement caused by a particular force:

$$X(s) = \frac{1}{ms^2}F(s) \tag{5.66}$$

The resulting block diagram is shown in Figure 5.30.

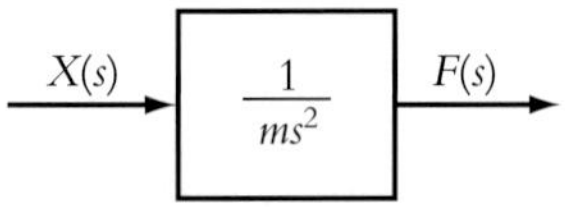

Figure 5.30 Another block diagram for a mass

In this context, the transfer function is now:

$$\frac{X(s)}{F(s)} = \frac{1}{ms^2} \tag{5.67}$$

Spring

The mathematical model of a spring is a very straightforward statement of proportionality (Hooke's law):

$$f(t) = kx(t) \tag{5.68}$$

Taking Laplace transforms gives an equally simple result:

$$F(s) = kX(s) \tag{5.69}$$

so that the block diagram of a spring giving the displacement in terms of the force can be drawn (Figure 5.31).

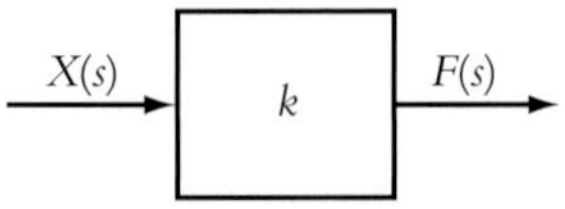

Figure 5.31 Block diagram for a spring

The transfer function is

$$\frac{F(s)}{X(s)} = k \tag{5.70}$$

Conversely, the displacement may be expressed in terms of the force by simple rearrangement:

$$X(s) = \frac{1}{k}F(s) \tag{5.71}$$

so that the transfer function is now

$$\frac{X(s)}{F(s)} = \frac{1}{k} \tag{5.72}$$

The resulting block diagram is shown in Figure 5.32.

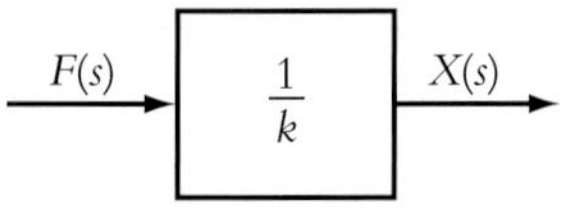

Figure 5.32 Alternative block diagram for a spring

Dashpot or damper

A dashpot or damper is a component that gives a reaction force proportional to the velocity $v(t)$ of one end relative to the other. It is described mathematically as follows:

$$f(t) = cv(t) = c\frac{dx(t)}{dt} \tag{5.73}$$

Taking Laplace transforms:

$$F(s) = csX(s) - cx(0) \tag{5.74}$$

Assuming zero initial conditions, and rearranging, gives the transfer function

$$\frac{F(s)}{X(s)} = cs \tag{5.75}$$

or alternatively

$$\frac{X(s)}{F(s)} = \frac{1}{cs} \tag{5.76}$$

The block diagrams for these two alternative representations are shown in Figure 5.33(a) and (b):

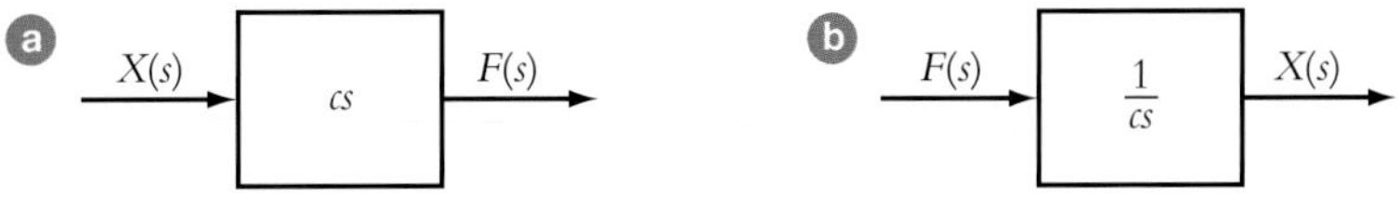

Figure 5.33 Alternative block diagrams for a damper

Flywheel or rotational inertia

Just as Newton's law for linear motion relates force to acceleration via the equation $F = ma$, the torque l applied to a flywheel is related to that flywheel's angular acceleration α $(= \ddot{\theta} = \dot{\omega})$ by the following equation, which also involves the moment of inertia J, the rotational equivalent of mass:

$$l(t) = J\alpha(t) = J\ddot{\theta} = J\frac{d^2\theta(t)}{dt^2} \tag{5.77}$$

(Note that $l(t)$ is used for torque rather that $L(t)$ to be consistent with our notation that lowercase italic letters represent variables in the time domain, while uppercase italic letters represent variables transformed to the s domain.)

Taking Laplace transforms by applying equation (5.16) to equation (5.77) gives the following equation in the s domain:

$$L(s) = Js^2\Theta(s) - J\theta'(0) - Js\theta(0) \tag{5.78}$$

where $\Theta(s) = \mathcal{L}\{\theta(t)\}$, i.e. the Laplace transform of the angular position $\theta(t)$. Setting initial conditions (angular position $\theta(0)$ and angular velocity $\theta'(0)$) to zero, and dividing by $\Theta(s)$ to rearrange the equation, gives the transfer function:

$$\frac{L(s)}{\Theta(s)} = Js^2 \tag{5.79}$$

This can be represented via the block diagram shown in Figure 5.34.

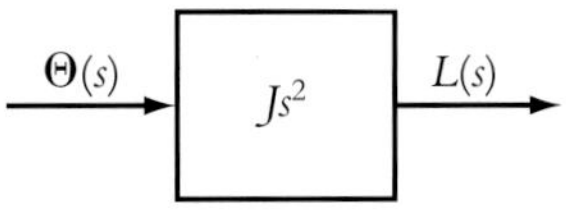

Figure 5.34 Block diagram for an inertia load

Alternatively, the same concept may be expressed in terms of angular velocity $\omega(t)$:

$$l(t) = J\alpha(t) = J\dot{\omega} = J\frac{d\omega(t)}{dt} \quad (5.80)$$

Applying equation (5.16) gives:

$$L(s) = Js\Omega(s) - J\omega(0) \quad (5.81)$$

Again assuming initial conditions, this gives:

$$\frac{L(s)}{\Omega(s)} = Js \quad (5.82)$$

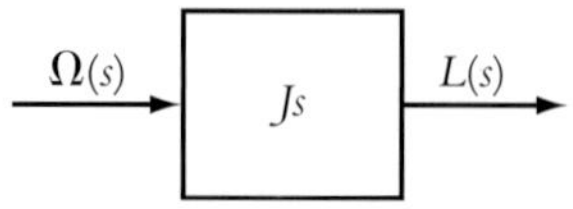

Figure 5.35 Alternative block diagram for a rotational inertia load

where $\Omega(s)$ is the Laplace transform of angular velocity $\omega(t)$. The block diagram is shown in Figure 5.35.

Rotational load with viscous characteristics

This is the rotational equivalent of a dashpot, and again may be represented in terms of the angular position or angular velocity:

$$l(t) = b\omega(t) = b\frac{d\theta(t)}{dt} \quad (5.83)$$

This gives the transfer function

$$\frac{L(s)}{\Omega(s)} = b \quad (5.84)$$

or

$$\frac{L(s)}{\Theta(s)} = bs \quad (5.85)$$

DC motor

For steady-state purposes, the characteristics of a motor are very simple: torque $l(t)$ is proportional to current $i(t)$ supplied:

$$l(t) = K_m i(t) \quad (5.86)$$

giving the transfer function

$$\frac{L(s)}{I(s)} = K_m \quad (5.87)$$

It should be noted that a constant magnetic field is assumed within the motor (for instance, assuming the motor is of the permanent magnet type), and hence it is possible to use the single constant K_m which is equivalent to the term $k\phi$ used within Unit 4. As a first approximation (and especially under no-load conditions), the angular velocity $\omega(t)$ is proportional to the motor supply voltage $v(t)$:

$$\omega(t) = \frac{v(t)}{K_m} \quad (5.88)$$

Note that, for an ideal motor, the same constant K_m appears in both equations (5.86) and (5.88). Equation (5.88) leads to the transfer function

$$\frac{\Omega(s)}{V(s)} = \frac{1}{K_m} \quad (5.89)$$

A more sophisticated model of the motor takes account of the fact that the speed under load conditions depends on both the supply voltage $v(t)$ and the torque $l(t)$ which the motor is providing:

$$\omega(t) = \frac{v(t)}{K_m} - \frac{l(t)R}{K_m^2} \quad (5.90)$$

where R is the armature resistance of the motor. Note that, strictly, the torque in this expression includes any component of torque used for acceleration of the motor armature as well as any provided as the useful mechanical output from the motor.

It is usual to plot torque vs. speed rather than vice versa, so Equation (5.90) may be represented as a set of torque vs. speed characteristics for a range of supply voltages (Figure 5.36).

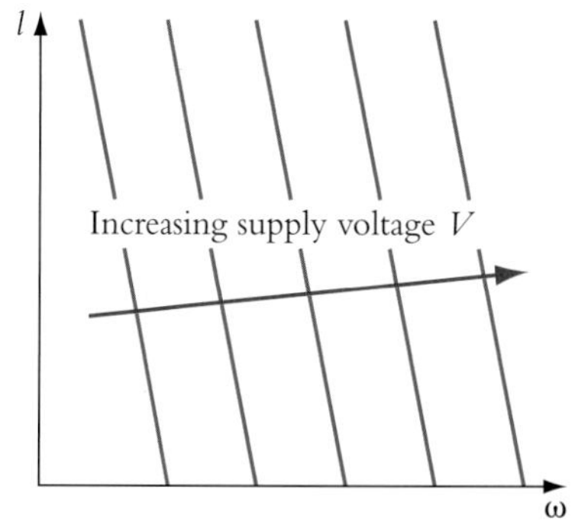

Figure 5.36 Typical set of torque-speed characteristics for a dc motor

The Laplace transform of equation (5.90) (for zero initial conditions) is simply

$$\Omega(s) = \frac{V(s)}{K_{\mathrm{m}}} - \frac{L(s)R}{K_{\mathrm{m}}^{2}} \tag{5.91}$$

This situation can be represented via a block diagram (Figure 5.37).

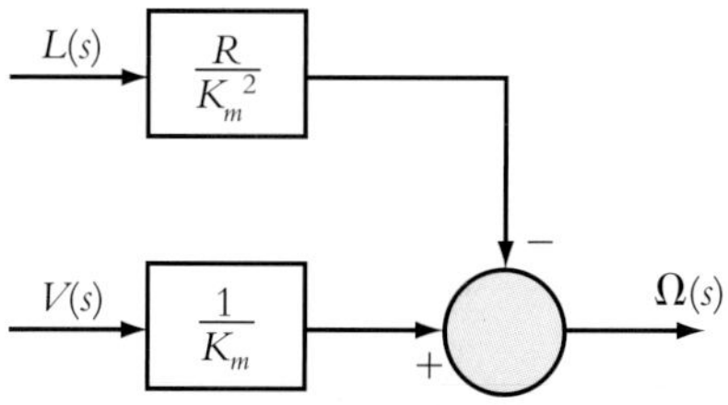

Figure 5.37 Block diagram for a dc motor

Alternatively the torque available for driving and/or accelerating the load (and the armature) can be represented in terms of the speed:

$$L(s) = V(s)\frac{K_{\mathrm{m}}}{R} - \Omega(s)\frac{K_{\mathrm{m}}^{2}}{R} \tag{5.92}$$

so that the block diagram can now be drawn as in Figure 5.38.

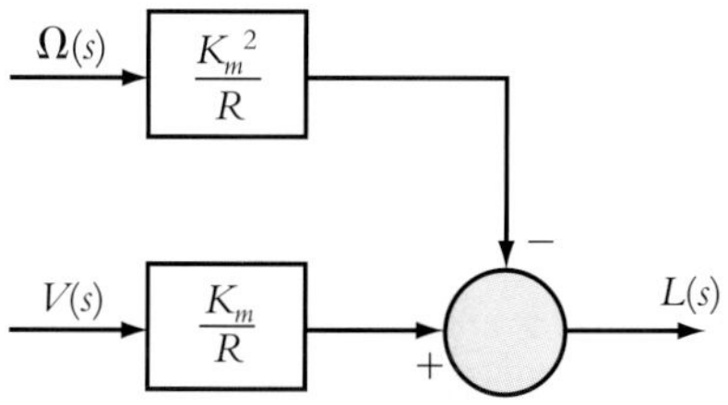

Figure 5.38 Alternative block diagram for a dc motor

Amplifiers and other components with gain

The situation is frequently encountered in control engineering where a signal is amplified (made larger) using some form of amplifier. An everyday example is the amplifier used in a home entertainment system, which takes a small signal (from a DVD player, MP3 player etc.) and makes it much larger in order to drive the loudspeakers. In simple terms, an amplifier multiplies a signal (e.g. a voltage) by a given constant K and therefore involves a transfer function which can be expressed as

$$\frac{V_{\mathrm{out}}(s)}{V_{\mathrm{in}}(s)} = K \tag{5.93}$$

The concept of a gain or constant of proportionality can be extended to numerous other situations, including sensors with calibration constants. A good example is a tachogenerator, which is a speed sensing device that produces a dc voltage output signal proportional to angular velocity. Such a device will have a certain gain in volts per unit of angular velocity. Other examples are simple components such as resistors (which allow a current to flow in proportion to voltage). A simple proportional gain (where the error signal is multiplied by a constant to give the corrective action) will form one possible component of the control algorithm to be examined in Section 5.11.

Pumps and tanks

These components are widely encountered within courses on control theory as they relate to quantities (flow rate, liquid level) that are easily visible. The behaviour of a positive-displacement pump is very straightforward as a first approximation: the volume flow rate of fluid $q(t)$ delivered is proportional to the angular velocity $\omega(t)$ with a constant of proportionality K_{p}, giving the transfer function

$$\frac{Q(s)}{\Omega(s)} = K_{\mathrm{p}} \tag{5.94}$$

The behaviour of a tank requires an understanding of its differential equation:

$$q_{\mathrm{in}}(t) - q_{\mathrm{out}}(t) = A\frac{\mathrm{d}h(t)}{\mathrm{d}t} \tag{5.95}$$

where $h(t)$ is the liquid level, A is the cross-sectional area of the tank, and $q_{in}(t)$ and $q_{out}(t)$ are respectively the rates of inflow to, and outflow from, the tank (Figure 5.39).

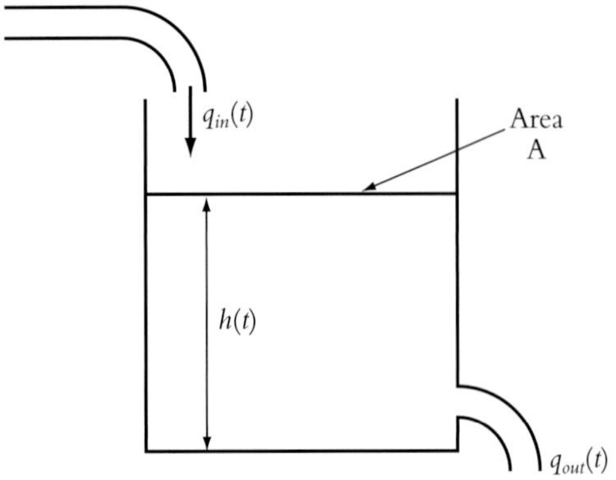

Figure 5.39 Water tank showing inflow $Q_{in}(t)$ and outflow $Q_{out}(t)$

This results in the following expression in the s domain:

$$Q_{in}(s) - Q_{out}(s) = AsH(s) - Ah(0) \qquad \textbf{(5.96)}$$

which, for zero initial conditions, can be represented as

$$H(s) = \frac{1}{As}(Q_{in}(s) - Q_{out}(s)) \qquad \textbf{(5.97)}$$

This can be represented in block diagram form (Figure 5.40).

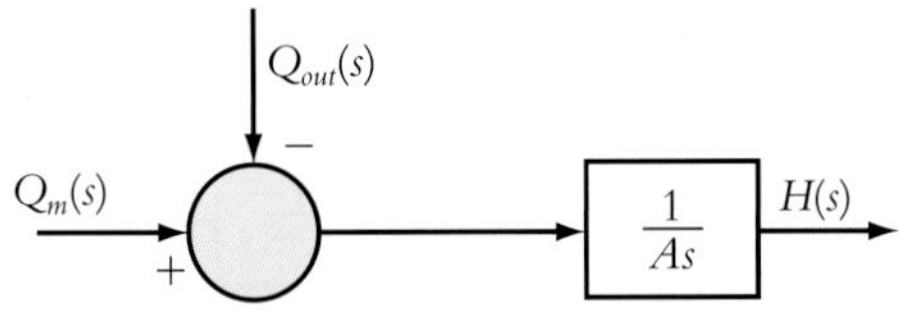

Figure 5.40 Block diagram of tank

Flow restrictions and linearization

Components such as pipe bends, orifices, valves etc. are often encountered in control systems that involve fluid flow. It would be tempting to treat such flow restrictions as being analogous to resistors, until it is realized that, strictly speaking, they are not linear devices. For example, an orifice plate or valve has a pressure drop $\Delta p(t)$ across it that is approximately proportional to the *square* of the volume flow rate $q(t)$:

$$\Delta p(t) = C(q(t))^2 \qquad \textbf{(5.98)}$$

where C is a constant depending upon the nature of the flow restriction and the density of the fluid. It would appear superficially that it is impossible to represent such a component within the present (linear) framework. However, in practice, many control systems operate over a fairly narrow range of speeds, depths, flow rates etc., and it is possible to simplify the system's behaviour by considering *small perturbations* about a given operating point. This results in behaviour that can be assumed to be linear within a small range around the operating point. For the present (simplified) purposes, it is sufficient to assume that the flow rate $q(t)$ from a tank via a flow restriction is proportional to the height ('head') $h(t)$ of liquid above the restriction:

$$q(t) = \frac{h(t)}{R} \qquad \textbf{(5.99)}$$

where R is the linearized flow resistance of the restriction referred to the head of liquid. Taking Laplace transforms results in the following expression:

$$Q(s) = \frac{H(s)}{R} \qquad \textbf{(5.100)}$$

where $H(s) = \mathcal{L}\{h(t)\}$

In general, linearization is used to cope with all manner of situations where the true behaviour of a device is non-linear, but where small perturbations around an operating point are to be considered. Examples might include

- the relationship between liquid depth h (head) and flow rate q, described above (Figure 5.41(a));
- the relationship between torque and speed of a centrifugal blower (Figure 5.41(b)).

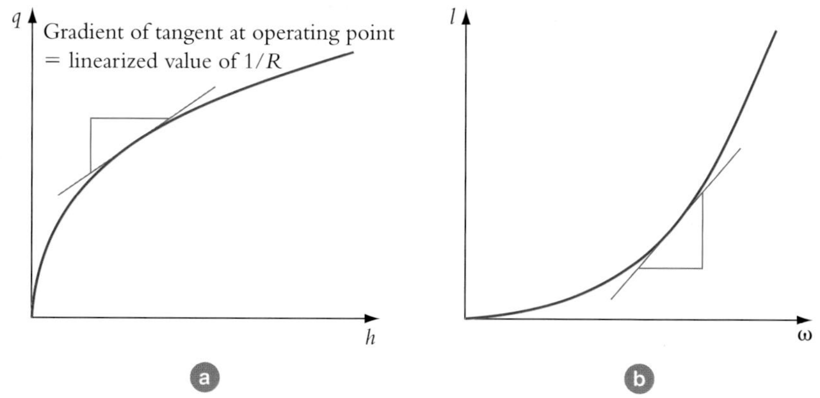

Figure 5.41 **Linearization of (a) flowrate–head characteristics for a tank; (b) torque–speed characteristics for a blower**

Summing and differencing junctions

Within control systems, the effects of inputs and other influences are often combined via summing and differencing junctions; such junctions have already been used freely within the earlier examples. As another example, two flows may sum to form a larger flow (Figure 5.42(a)). The summing of the flows is represented via summing junction in a block diagram (Figure 5.42(b)).

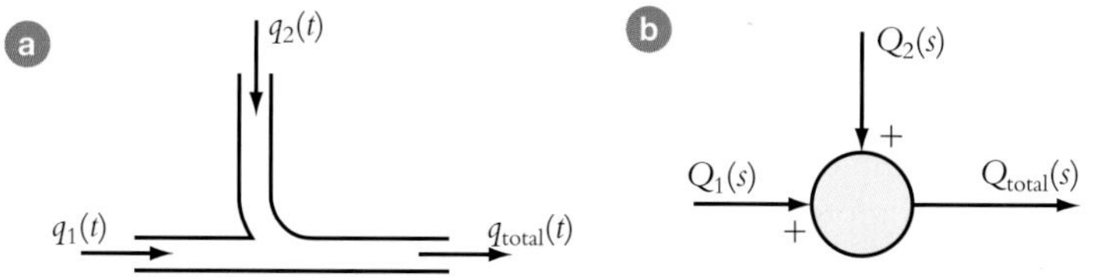

Figure 5.42 **(a) Summation of two flows at a pipe junction; (b) block diagram of pipe junction represented as a summing junction**

A very frequently encountered situation in control loops is the concept of negative feedback where the measured value of a signal is compared with the desired value, and the difference between them is known as the error. This has already been encountered in the examples given above. In the introductory example of the cruise control, the actual speed $v(t)$ may be compared with the desired or demand speed $v_D(t)$, so that, in the s domain, the feedback quantity $V(s)$ is subtracted from the demand $V_D(s)$ to obtain the error signal $E(s)$ (Figure 5.43).

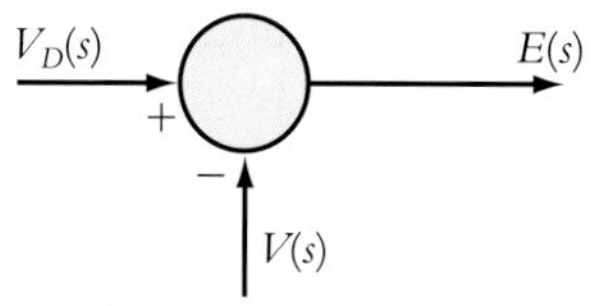

Figure 5.43 **Typical differencing junction**

Transfer function of a controller

In order to correct the error consisting of the difference between desired and actual values of the system response, some form of controller is used. The detailed behaviour (and transfer function) of typical systems will be explored in Section 5.11, but for the present time it will be represented simply by the symbol $G_c(s)$ (Figure 5.44).

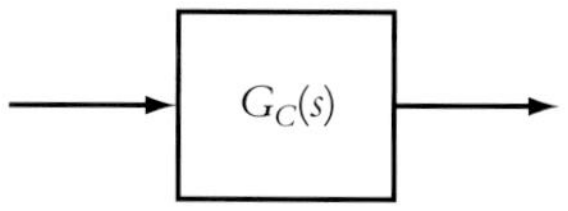

Figure 5.44 **Block diagram of a controller**

Learning summary

By the end of this section you should have learnt:

- ✓ the concept of a transfer function;
- ✓ the transfer functions of a range of simple components;
- ✓ the concept of linearization.

5.8 Building a block diagram: part 2

By this stage, the reader should have sufficient information to be able to construct the transfer functions (and hence draw the block diagrams) for each of the components. The remaining challenge lies in correctly assembling the diagram for the whole system from its components. In constructing block diagrams, it is conventional to assume that all variable quantities or signals (such as displacement, speed etc.) are functions of s rather than time. Although each component may itself be trivial to model, the construction of the block diagram of a system may require careful thought, and it is not always obvious which way around a particular relationship (such as the relationship between torque and speed of a motor) should be applied.

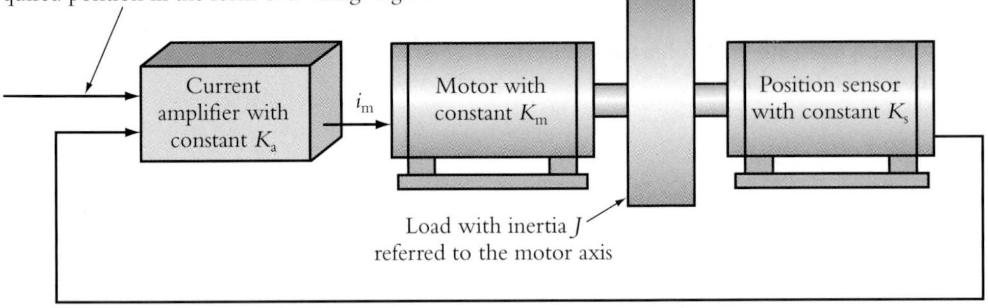

Figure 5.45 Over-simplistic positioning system

Two examples of block diagram construction have already been presented within Sections 5.3 and 5.5, but it is instructive to explore the assembly of the block diagram components from Section 5.7 via two further examples of varying complexity. A simple (and flawed) example is a positioning system (Figure 5.45) consisting of a motor with constant K_m driving a load of inertia J, with the motor being driven via current amplifier that provides the motor with a current of K_a amperes per volt of input. Position is assumed to be sensed via a device that gives an output of K_s volts per radian of rotation of the motor. The input to the system is a demand voltage $v_D(t)$ which is proportional to the desired position. The voltage, v_P, corresponding to the actual position is subtracted from v_D to give an error voltage, which is amplified by the servo-amplifier to give the input current i_m to the motor which will create a torque to rotate the load with the aim of correcting the positioning error:

$$i_m(t) = K_a(v_D(t) - v_P(t))$$

Taking Laplace transforms and assuming that all voltages etc. are initially zero:

$$I_m(s) = K_a(V_D(s) - V_P(s))$$

This may be represented within a block diagram (Figure 5.46).

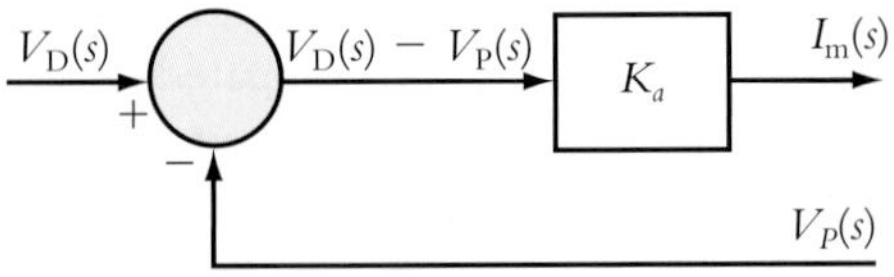

Figure 5.46 Block diagram of differencing junction and amplifier for positioning system

The motor will be assumed to be acting as a source of torque $l(t)$ dependent upon the current $i(t)$ by adapting equations (5.86) and (5.87) to the present situation and rearranging so that

$$L(s) = K_m I_m(s) \qquad \textbf{(5.101)}$$

The block diagram corresponding to this is straightforward and is shown in Figure 5.47

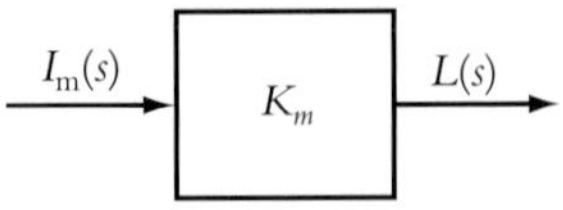

Figure 5.47 Block diagram of motor

The relationship between torque $l(t)$ and angular position $\theta(t)$ (and hence their transforms $L(s)$ and $\Theta(s)$) illustrates one of the pitfalls of constructing a block diagram from its components. Although the block diagram for an inertia load J is straightforward and was given earlier as Figure 5.34, it is necessary here to reverse the flow and invert the transfer function in order to obtain $\Theta(s)$ in terms of $L(s)$ (Figure 5.48).

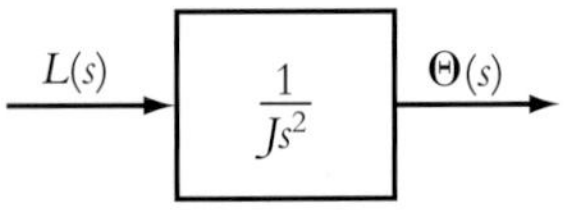

Figure 5.48 Rearranged block diagram of inertia

The relationship between angular position $\theta(t)$ and the position-related voltage $v_P(t)$ is a straightforward linear one involving the sensor constant K_s and, for brevity, we will proceed straight to the block diagram (Figure 5.49).

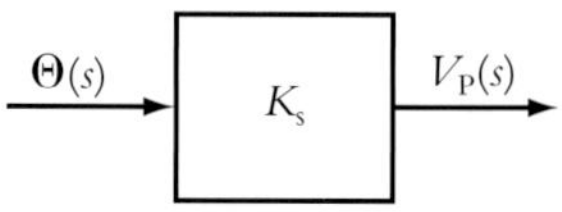

Figure 5.49 Block diagram of sensor

The complete block diagram for this system is shown in Figure 5.50.

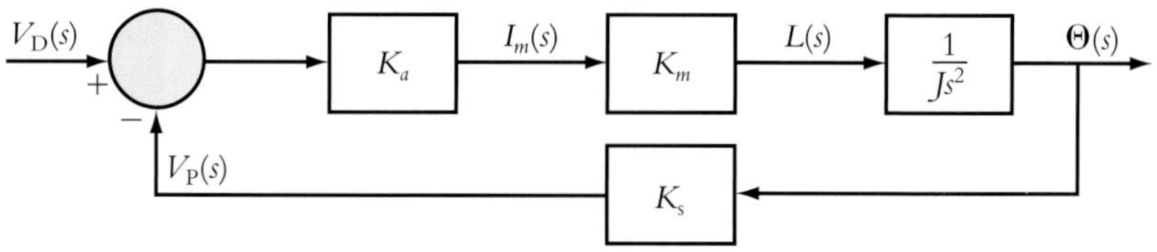

Figure 5.50 Block diagram for (over-simplistic) position control system

Interestingly, this would not form a very useful control system as its response would be both undamped and oscillatory (the control system is merely equivalent to an undamped rotational spring acting upon the inertia load). This further illustrates the need for an effective control algorithm, and will be explored further in Sections 5.11 and 5.13.

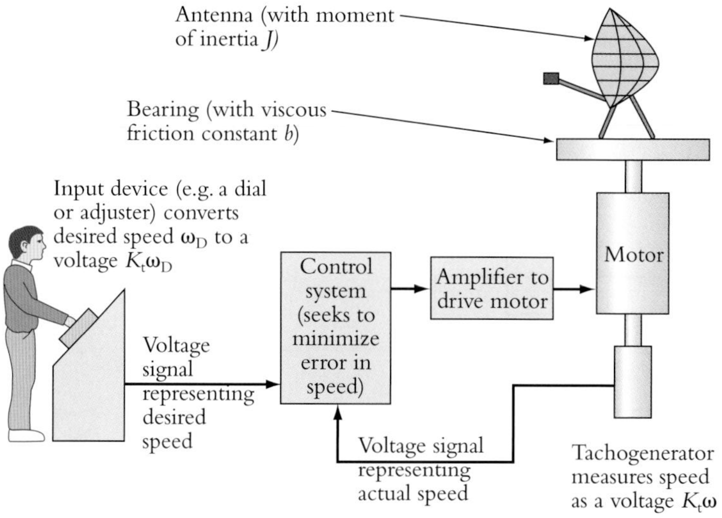

Figure 5.51 Radar antenna on viscous bearing, with tachometer control

A more complex example is as follows. A radar antenna (or aerial) may be represented as a viscous load with torque–speed constant b, i.e. in order to drive it at constant angular velocity ω, a constant torque l is required:

$$l = b\omega \tag{5.102}$$

It also has a moment of inertia J. It is driven by an electric motor with motor constant K_m and armature resistance R, at a speed that must be closely controlled. The desired angular velocity is ω_D (which is shown for illustrative purposes in Figure 5.51 as being manually entered via an adjuster which creates a variable voltage proportional to ω_D), and the actual angular velocity ω is measured via a tachogenerator with a constant of K_t Vs/rad. The system is controlled via negative feedback with a controller with transfer function $G_c(s)$ and via a voltage amplifier with gain K_a.

Consider firstly the antenna itself. In order to drive it at a given value of angular velocity ω, a constant torque l is required, as expressed in equation (5.102), so it is straightforward to treat the relationship between antenna torque and antenna speed in the time domain as

$$l(t) = b\omega(t) \tag{5.103}$$

After taking Laplace transforms this results in

$$L(s) = b\Omega(s) \tag{5.104}$$

which leads to the block diagram shown in Figure 5.52.

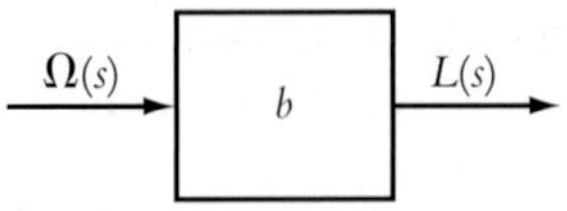

Figure 5.52 Block diagram of pure viscous load

However, if the torque l_m available from the motor for driving it exceeds the torque l required to overcome viscous friction and maintain a steady speed, it will accelerate at the rate $(l_m - l)/J$, so that the equation of motion of the antenna is

$$J\ddot{\theta}(t) = J\frac{\mathrm{d}\omega(t)}{\mathrm{d}t} = l_m(t) - l(t) \tag{5.105}$$

This can be represented in the s domain as

$$Js\Omega(s) = L_m(s) - L(s) \tag{5.106}$$

which can be rearranged to give

$$\Omega(s) = \frac{1}{Js}(L_m(s) - L(s)) \tag{5.107}$$

The corresponding block diagram is therefore as shown in Figure 5.53.

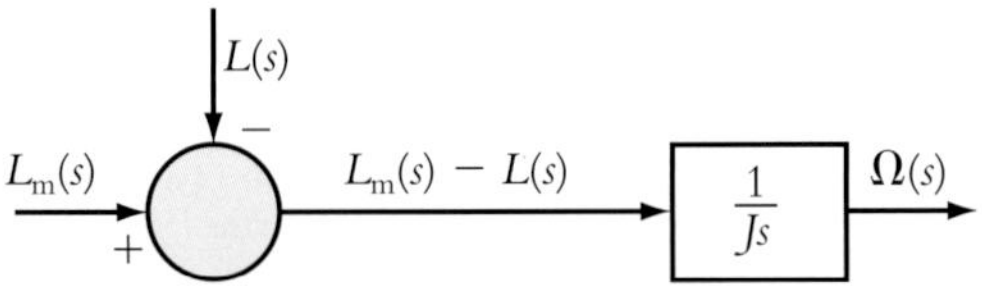

Figure 5.53 Block diagram of inertia load with external drag *L*(*s*)

Inserting equation (5.104) into equation (5.107) gives

$$\Omega(s) = \frac{1}{Js}(L_m(s) - b\Omega(s)) \tag{5.108}$$

The corresponding block diagram (Figure 5.54) for the antenna, treated as a combined frictional and inertia load, can be created either directly from equation (5.108) or by joining together the relevant flows from Figure 5.52 and Figure 5.53.

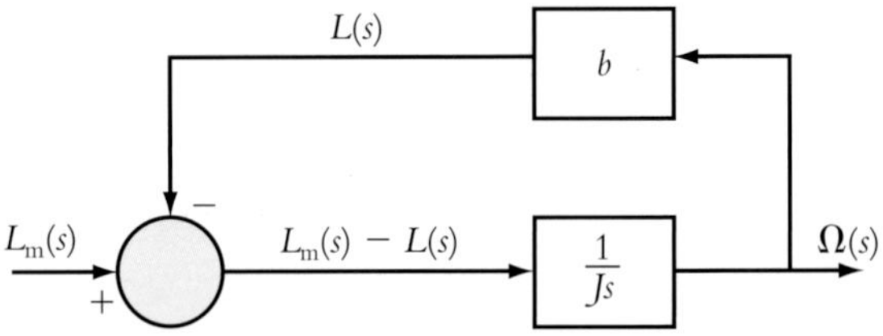

Figure 5.54 Block diagram of antenna treated as combined frictional and inertia load

However, the antenna is driven from a motor that will provide a torque l_m that depends on the speed ω at which it is rotating:

$$l_m(t) = v(t)\frac{K_m}{R} - \omega(t)\frac{K_m^2}{R} \tag{5.109}$$

or in the s domain:

$$L_m(s) = V(s)\frac{K_m}{R} - \Omega(s)\frac{K_m^2}{R} \tag{5.110}$$

This is shown in block diagram form in Figure 5.55.

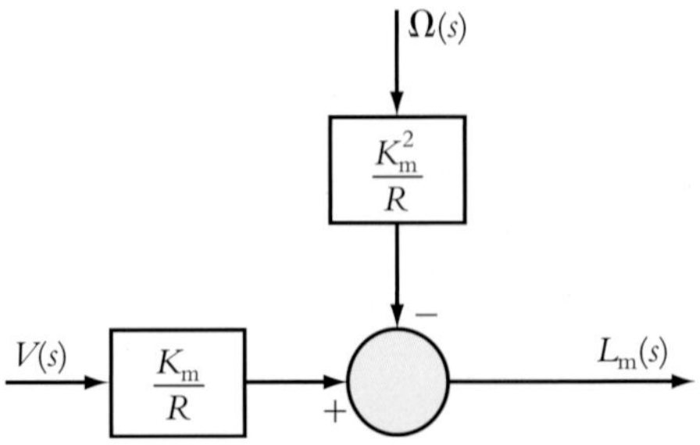

Figure 5.55 Block diagram for motor

The block diagram for the antenna–motor assembly is therefore as shown in Figure 5.56.

But recall that this system is driven via a voltage amplifier from the output of a controller, which acts upon the error signal. This error signal is in turn the difference between the required speed $\Omega_D(s)$ (converted into a voltage) and the voltage signal $K_t\Omega(s)$ from the tachogenerator. This control system driving the antenna and its motor may be represented as shown in Figure 5.57.

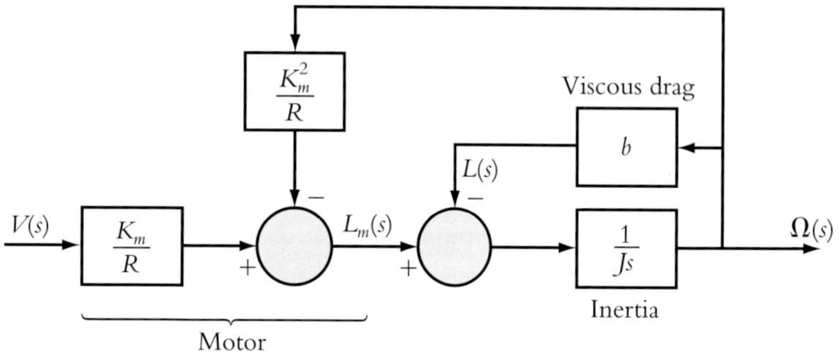

Figure 5.56 Block diagram for antenna–motor assembly

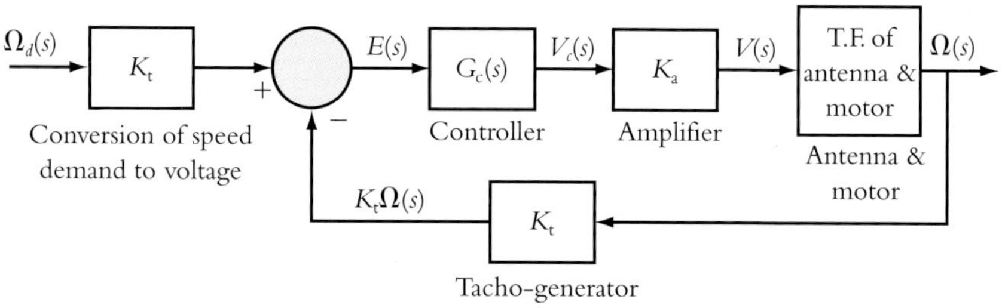

Figure 5.57 Block diagram for control system

Inserting the block diagram for the antenna and motor results in the block diagram for the whole system (Figure 5.58). The behaviour of this system will be explored in Sections 5.12 and 5.13, though it is first necessary to convert the rather complex block diagram to a single transfer function.

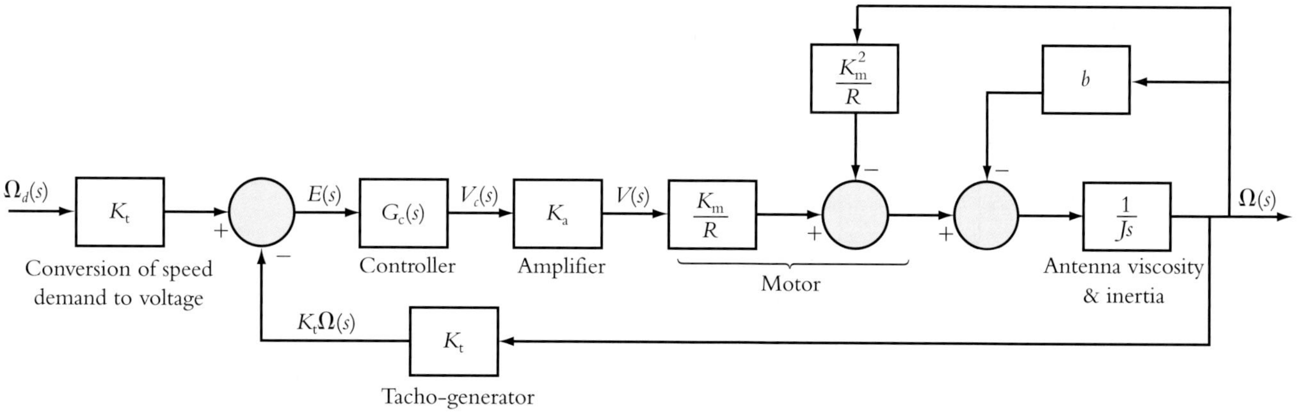

Figure 5.58 Block diagram for complete antenna system

Learning summary

By the end of this section you should have learnt:

- ✓ to construct a block diagram for systems or subsystems based on the transfer functions of each of its components or features;
- ✓ to assemble the block diagram for a complex system from the diagrams for its various subsystems;
- ✓ to appreciate that distinguishing the input and output of particular processes is not always straightforward.

5.9 Conversion of the block diagram to the transfer function of the system

This process is broadly based upon two rules:

(1) the multiplication of the transfer functions of successive blocks in a sequence;

(2) the simplification of the block diagram shown in Figure 5.59(a) into a single transfer function shown in Figure 5.59(b), as demonstrated in equations (5.45) to (5.47).

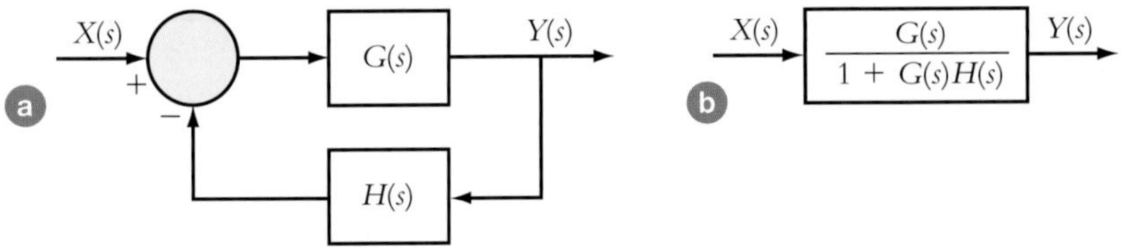

Figure 5.59 Block diagram of a typical control loop: (a) before and (b) after reduction to single transfer function

Essentially, this pair of actions is applied repeatedly to the feedback loops within a complex control system, starting with the innermost loop and working outwards.

Taking first the example of the block diagram for the controlled heat transfer system (Figure 5.21), the innermost loop can be tackled by replacing $G(s)$ with the transfer function $1/mC_ps$ of the main flow, and $H(s)$ with the transfer function Ah of the feedback flow. Substituting into equation (5.47) gives:

$$\frac{G(s)}{1+G(s)H(s)} = \frac{\dfrac{1}{mC_ps}}{1+Ah\times\dfrac{1}{mC_ps}} = \frac{1}{mC_ps+Ah} \qquad \textbf{(5.111)}$$

giving the same result as the more lengthy manipulation within equation (5.31).

The overall block diagram then simplifies to Figure 5.60(a) and hence to Figure 5.60(b).

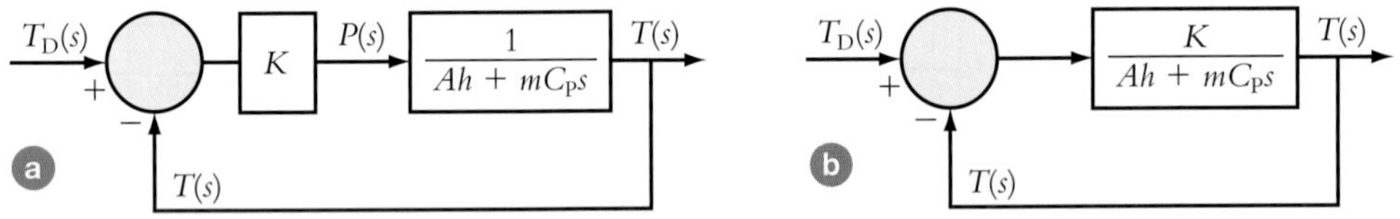

Figure 5.60 Simplified versions of block diagram for heating system

Now, the process of simplifying the structure of the block diagram is repeated. $G(s)$ is replaced with $K/(Ah+mC_ps)$, and $H(s)$ with 1, since there is no transfer function associated with the feedback flow, which simply feeds the output directly back to the differencing junction. This manipulation results in the transfer function for the system:

$$\frac{T(s)}{T_D(s)} = \frac{G(s)}{1+G(s)H(s)} = \frac{K/(mC_ps+Ah)}{1+K/(mC_ps+Ah)} = \frac{K}{mC_ps+Ah+K} \qquad \textbf{(5.112)}$$

The error (i.e. the difference between desired and actual values of output) is:

$$T_D(s) - T(s) = T_D(s) - T_D(s)\frac{K}{mC_ps+Ah+K} = T_D(s)\frac{mC_ps+Ah}{mC_ps+Ah+K} \qquad \textbf{(5.113)}$$

This expression will be used further in examining steady-state error in section 5.12.

Dealing now with the radar antenna example, the innermost loop in Figure 5.58 can be considered by replacing $G(s)$ with the transfer function of the main flow $1/(Js)$ and $H(s)$ with the transfer function of the feedback flow *b*. The identity given in equation (5.47) is then used to enable the innermost loop to be replaced with a single block containing the transfer function:

$$\frac{1/(Js)}{1 + b/(Js)} = \frac{1}{Js + b} \qquad (5.114)$$

The resulting block diagram for the antenna and motor is now as shown in Figure 5.61.

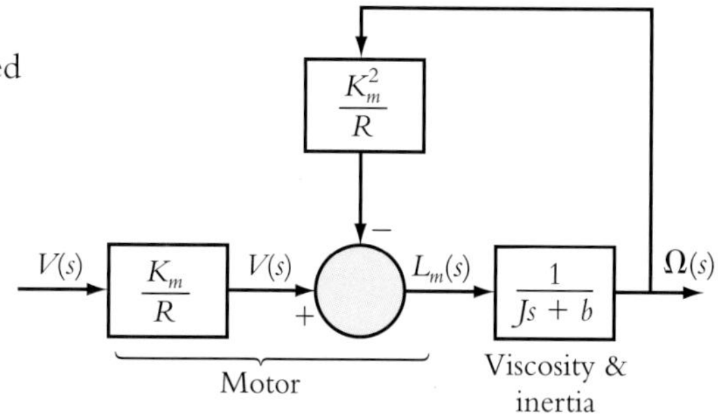

Figure 5.61 Block diagram for antenna and motor after simplification of inner loop

Applying the same principle to the next loop involves replacing *G*(s) with $1/(Js+b)$ and *H*(*s*) with K_m^2/R, again giving a single block with the transfer function

$$\frac{1/(Js + b)}{1 + K_m^2/[R(Js + b)]} = \frac{1}{Js + b + K_m^2/R} = \frac{R}{RJs + Rb + K_m^2} \qquad (5.115)$$

giving a further simplified block diagram for the combination of antenna and motor (Figure 5.62).

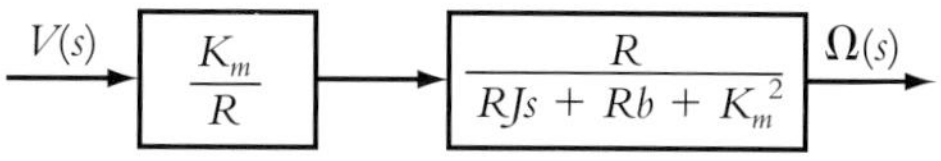

Figure 5.62 Block diagram for antenna and motor after simplification of outer loop

The resulting block diagram for the whole system is now as shown in Figure 5.63.

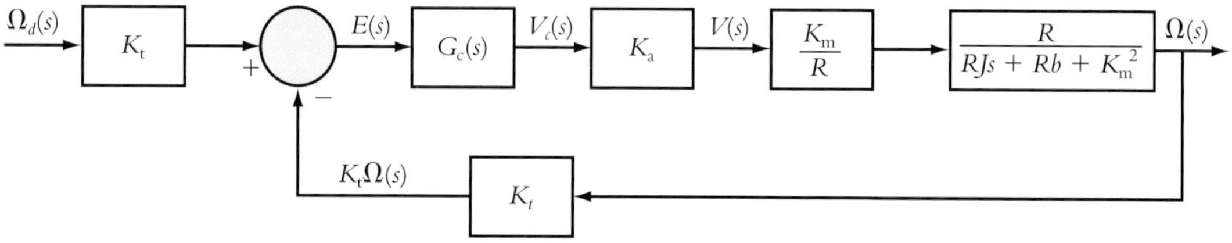

Figure 5.63 Block diagram for complete antenna system before combining blocks

The transfer functions of the four innermost blocks may be multiplied out to give the following much simpler block diagram for the system (Figure 5.64).

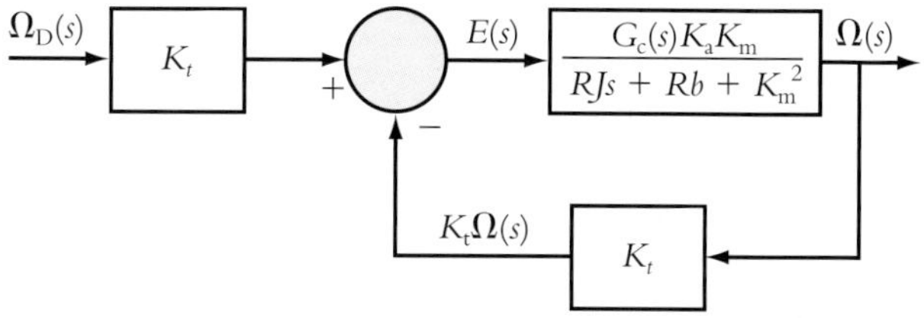

Figure 5.64 Simplified block diagram for complete antenna system

Substituting yet again for *G*(*s*) and *H*(*s*) within equation (5.47) for the remaining loop allows us to replace that loop with the following transfer function (Figure 5.65):

$$\frac{G_c(s)K_aK_m}{RJs + Rb\ 1\ K_m^2} \Big/ 1 + \frac{K_tG_c(s)K_aK_m}{RJs + Rb + K_m^2} = \frac{G_c(s)K_aK_m}{RJs + Rb + K_m^2 + K_tG_c(s)K_aK_m} \qquad (5.116)$$

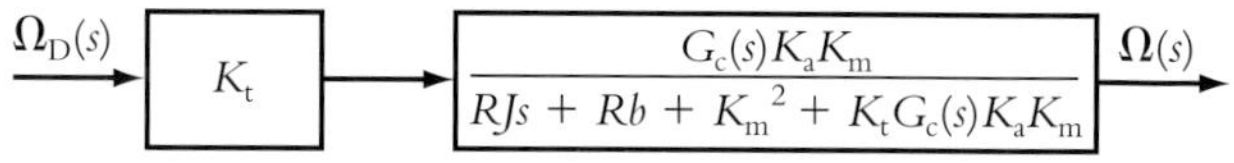

Figure 5.65 Elimination of loop for complete antenna system

Finally, multiplying by the transfer function of the leftmost block results in the transfer function for the whole system:

$$\frac{\Omega(s)}{\Omega_D(s)} = \frac{K_t G_c(s) K_a K_m}{RJs + Rb + K_m^2 + K_t G_c(s) K_a K_m} \tag{5.117}$$

enabling the block diagram to be reduced to a single box (Figure 5.66)

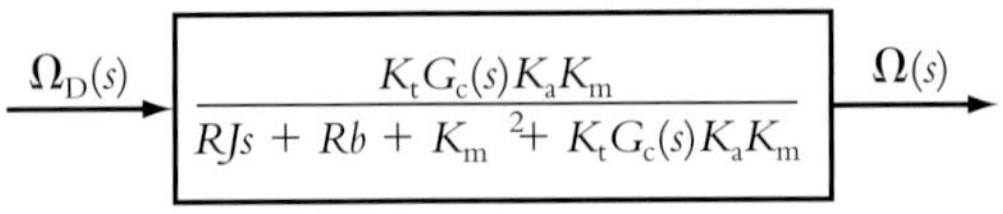

Figure 5.66 Final block diagram for complete antenna system

An expression can therefore be written for the error, which is the difference between the desired value of angular velocity and the achieved value:

$$\Omega_D(s) - \Omega(s) = \left(\frac{1 - K_t G_c(s) K_a K_m}{RJs + Rb + K_m^2 + K_t G_c(s) K_a K_m}\right)\Omega_D(s)$$

$$= \frac{RJs + Rb + K_m^2 + K_t G_c(s) K_a K_m - K_t G_c(s) K_a K_m}{RJs + Rb + K_m^2 + K_t G_c(s) K_a K_m}\Omega_D(s) \tag{5.118}$$

$$= \frac{RJs + Rb + K_m^2}{RJs + Rb + K_m^2 + K_t G_c(s) K_a K_m}\Omega_D(s)$$

This expression will be used in Section 5.12 when examining the steady-state response of the system.

Learning summary

By the end of this section you should have learnt:

- ✓ the relationship between the transfer function of a closed-loop system and the transfer functions of its main and feedback paths;
- ✓ the procedure for simplifying a block diagram with one or more loops in order to reduce it to a block diagram involving a single transfer function.

5.10 Handling block diagrams with overlapping control loops

A problem appears to arise when dealing with a block diagram including loops that are not simply nested one inside the other, but overlap in the manner shown in Figure 5.67.

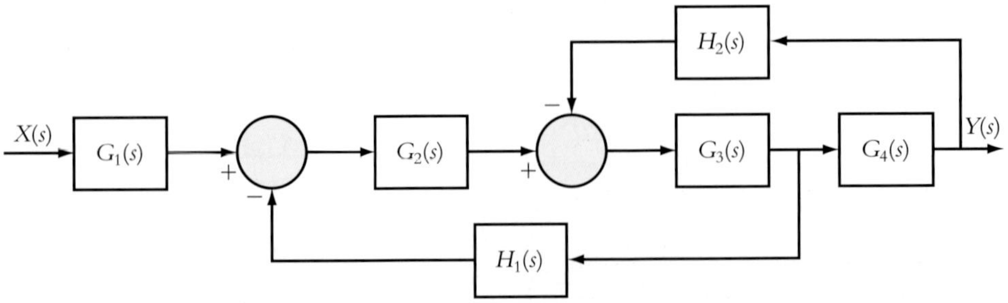

Figure 5.67 Block diagram with overlapping control loops

This is handled via a simple trick. The point at which one of the loops branches away from the main flow of the diagram is moved, and appropriate blocks are introduced into the feedback loop to compensate for the change in the topology of the diagram. For example, if a branch leading to a feedback loop is moved 'upstream' from the original branching point, the blocks that now lie outside the loop are now replicated within the loop itself so that the open-loop transfer function for that loop stays the same. In the example given above, the point at which the upper loop leaves the main flow is moved to coincide with the point at which the lower loop leaves. However, this would involve omitting the block containing $G_4(s)$ from that control path. That block is therefore reintroduced into the loop so that the overall transfer function for that loop stays the same (Figure 5.68).

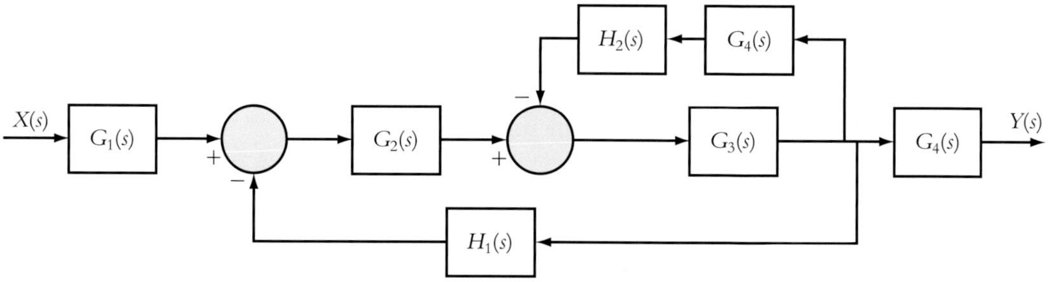

Figure 5.68 Block diagram rearranged to avoid overlapping of loops

It is now straightforward to reduce this block diagram to a single transfer function in the usual way. Replacing $G(s)$ in equation (5.47) with $G_3(s)$ and $H(s)$ with $H_2(s)H_4(s)$, the inner loop involving G_3, H_2 and G_4 can be expressed as a single transfer function, so that the block diagram can be drawn as shown in Figure 5.69.

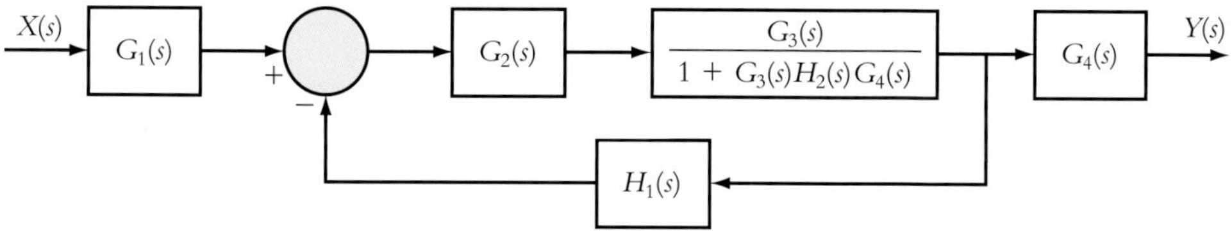

Figure 5.69 Block diagram redrawn to eliminate inner loop

By a similar process the remaining loop is eliminated (Figure 5.70).

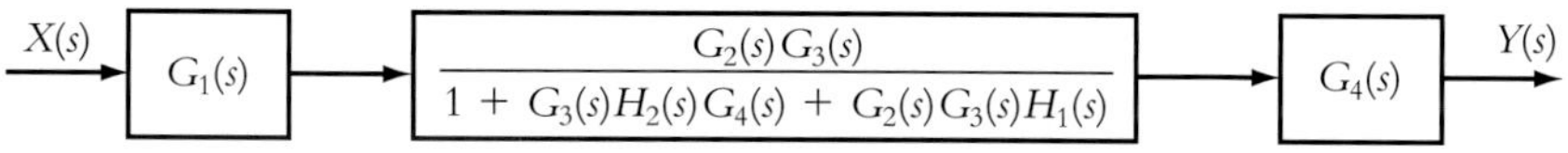

Figure 5.70 Block diagram redrawn to eliminate outer loop

The block diagram is now in the form of a straightforward sequence of blocks whose transfer functions can be multiplied together to give the transfer function of the overall system (Figure 5.71).

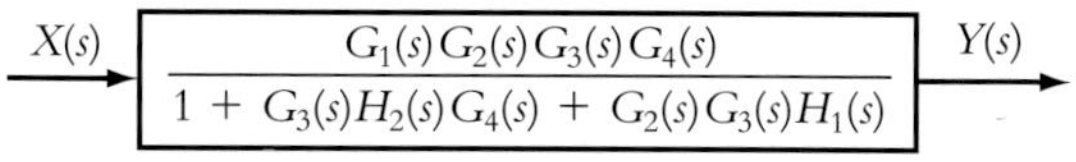

Figure 5.71 Block diagram redrawn in the form of a single transfer function

5.11 The control algorithm and proportional-integral-derivative (PID) control

So far, only the simplest assumptions have been made regarding how the *error* (which may be colloquially expressed as 'the difference between what you want and what you get') is processed. The rule by which this error is processed in order to obtain the required corrective action is known as the *control algorithm* and is central to the behaviour of feedback and control systems. Ideally, the control system should:

- take more corrective action if the error is larger;
- take more corrective action if the error has persisted for a long time;
- take more corrective action if the error is getting worse (and less if it is getting better).

A very widely used control algorithm is the **three-term** or **PID** (proportional-integral-derivative) controller, which formalizes the three kinds of decisions defined verbally above. Specifically, it sets the corrective action to a value that depends on the sum of

- a multiple of the current value of the error;
- a multiple of the integral of the error, i.e. the value of the error integrated or accumulated over the time since the beginning of the process;
- a multiple of the derivative of the error i.e. its rate of change with time.

Mathematically the control algorithm is defined as:

$$G_c(s) = K_c\left(1 + T_D s + \frac{1}{T_I s}\right)$$
$$= K_c + K_c T_D s + \frac{K_c}{T_I s} \quad \textbf{(5.119)}$$

where:

K_c is the gain of the controller
T_D is the derivative time constant (a larger time constant leads to more derivative action)
T_I is the integral time constant (a larger time constant leads to *less* integral action).

This is sometimes written alternatively as

$$G_c(s) = K_c + K_D s + \frac{K_I}{s} \quad \textbf{(5.120)}$$

In general, the terms within PID control have the following effects:

- proportional control alone tends to result in **steady-state error**, which means that the system output never reaches the desired value, typically tending asymptotically to it;
- integral action in conjunction with proportional control tends to eliminate steady-state error, at the expense of tending to make the system's response oscillatory or even unstable;
- derivative action can in some circumstances damp down the tendency towards oscillatory behaviour without reintroducing steady-state error. However, there can be pitfalls in differentiating real signals as this procss can tend to accentuate noise. It can be preferable to measure a derivative directly, for example as an angular velocity rather than obtaining it by numerical differentiation.

As an example, consider the positioning system described earlier. It is straightforward to show, from Figure 5.50, that the open-loop transfer of the position controller system is

$$\frac{V_P(s)}{V_D(s)} = \frac{K_a K_m K_s}{Js^2} \quad \textbf{(5.121)}$$

For illustrative purposes, the values $K_a = 10$, $K_m = 1\,\text{A/V}$ and $K_s = 1\,\text{Vs/rad}$ will be assumed. It can be demonstrated, either analytically by reversing the application of Laplace transforms to find the time domain solution, or by numerical simulation, for instance, using the Simulink simulation package (www.mathworks.com/products/simulink), that the response of this system to a demand of 10V (representing a request to move by 10 radians) results in an oscillatory response which continues forever at constant amplitude (Figure 5.72), which is clearly not a useful response. And we have not yet considered what happens when there is some external torque trying to move the load away from its desired position.

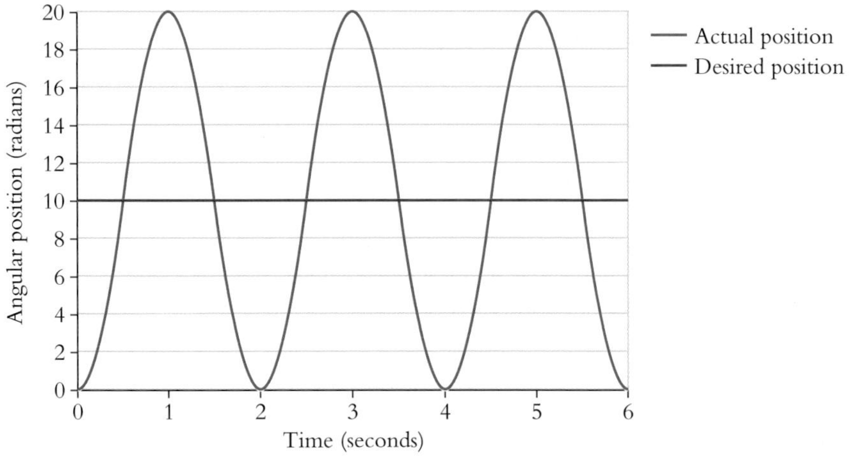

Figure 5.72 Response of positioning system under proportional control only

Now let us replace the simple servo-amplifier driving the motor with a controller/amplifier of gain K_c and integral and derivative time constants T_I and T_D, and additionally introduce a torque load $L_L(s)$ which is applied externally to the shaft carrying the flywheel. The resulting block diagram is shown in Figure 5.73.

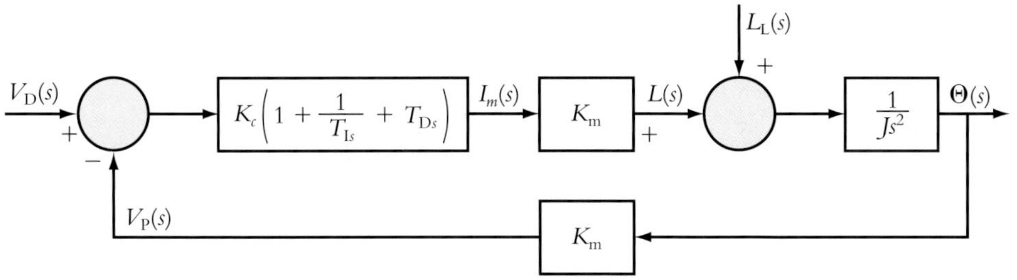

Figure 5.73 Block diagram of positioning system with PID control and an external load

The open loop transfer function can no longer be expressed as a simple ration of output to input, but for completeness the following expression can be derived:

$$V_P(s) = V_D(s)\frac{K_cK_mK_sT_D\left(s^2 + \frac{1}{T_D}s + \frac{1}{T_IT_D}\right) + sL_L(s)}{Js^3} \qquad (5.122)$$

Along with the values of system constants assumed earlier, the values $T_D = 0.5$ second and $T_I = 1$ second will be assumed for illustrative purposes, and a demand of 10V and an external load of 20 Nm will be imposed at $t = 0$. Proportional control only (omitting both integral and derivative term) again results in a sinusoidal response of constant amplitude similar to that shown in Figure 5.72, this time oscillating about a mean value of 12 radians. Proportional plus integral control only (i.e. omitting the derivative term) results in an exponentially growing sinusoidal response (Figure 5.74). Proportional plus derivative control gives a well-damped response rapidly settling to a steady value of 12 radians, indicating a steady-state error of 2 radians. None of these responses is really satisfactory; however, using all three terms (proportional, integral and derivative control) results in a response that overshoots a little but rapidly settles to the desired value of 10 radians without any steady-state error.

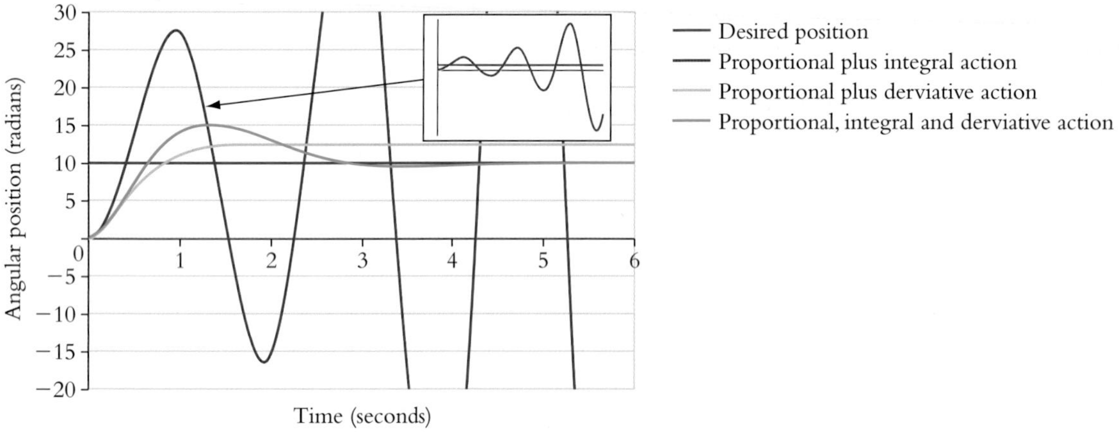

Figure 5.74 Response of positioning system with proportional plus integral control alone (blue line and inset), proportional plus derivative control (cyan line) and PID control (green line)

The application of different control algorithms to the radar antenna problem is considered in a broadly similar manner in the next section with reference to steady-state error and dynamic response, and the response of both of these systems is revisited yet again in Section 5.13 on the root locus method.

Learning summary

By the end of this section you should have learnt:

- ✓ to assemble a collection of transfer function boxes relating to separate aspects of a control system to form a complete block diagram for that system;
- ✓ to rearrange and simplify that system to find the transfer function of the system;
- ✓ the importance of the control algorithm;
- ✓ the purpose of the three terms in a PID controller;
- ✓ typical effects of the three contributions to PID control.

5.12 Response and stability of control systems

In order to understand the actual behaviour of a given control system, the chosen control algorithm must be substituted for $G_c(s)$. Once this has been performed, a number of issues can be examined:

- Is the system stable?
- Is its response oscillatory, i.e. is it overdamped or underdamped?
- Is there a steady-state error, i.e. does the actual output from the system converge to the desired outcome, or is there a difference that persists indefinitely?

These issues will be examined by continuing to examine the heater and radar antenna problems.

Steady-state error

Steady-state error is the difference between demand and achieved output which persists as time becomes infinite, and is typically evaluated for one of two situations:

- a step input at $t = 0$, corresponding to a demand issued at $t = 0$ for the system to attain a steady value;

- a ramp input beginning at $t = 0$, corresponding to a demand for the output to increase linearly from zero with time, i.e. for a constant rate of motion or rate of change of output.

As a link with earlier work it will be recalled that the error for the heated system under closed-loop control was given in equation (5.113). If a step input is assumed, consisting of a demand for a temperature τ_1 imposed at $t = 0$, the demand can be expressed in the s domain (by using line 1 of Table 5.1).

$$T_D(s) = \tau_1 \mathcal{L}\{H(t)\} = \frac{\tau_1}{s} \tag{5.123}$$

Inserting this into equation (5.113) and applying the final value theorem (equation (5.59)) gives the error as time tends to infinity:

$$\text{Steady-state error} = \lim_{s \to 0} s \frac{\tau_1}{s} \frac{mC_p s + Ah}{mC_p s + Ah + K} = \frac{Ah}{Ah + K} \tau_1 \tag{5.124}$$

which recovers an expression for error equivalent to equation (5.8) which was derived via the simple, time-independent analysis, confirming that this simple proportional control algorithm does not achieve a degree of control which eliminates steady-state error.

If, by contrast, a linearly increasing demand of τ_2°C/s is imposed, beginning at $t = 0$, the demand is modelled in the s space by making use of line 2 of Table 5.1:

$$T_D(s) = \tau_2 \mathcal{L}\{tH(t)\} = \frac{\tau_2}{s^2} \tag{5.125}$$

Attempting to calculate the steady-state error gives

$$\text{Steady-state error} = \lim_{s \to 0} s \frac{\tau_2}{s^2} \frac{mC_p s + Ah}{mC_p s + Ah + K} = \infty$$

from which we can conclude that this simple control system will not only fail to achieve the desired (constantly increasing) temperature, but indeed will fail even to provide a steady offset from the desired temperature and the error will become infinitely large with time.

Considering now the antenna problem, if proportional action K_c alone is assumed within equation (5.119) so that $G_c(s) = K_c$, $T_D = 0$ and $T_I = \infty$, and that expression for $G_c(s)$ is further substituted into equation (5.118) which gives the error in the antenna angular velocity, the following expression is obtained for the error as a function of s:

$$\Omega_D(s) - \Omega(s) = \Omega_D(s) \left(\frac{RJs + Rb + K_m{}^2}{RJs + Rb + K_m{}^2 + K_t K_c K_a K_m} \right) \tag{5.126}$$

Applying the final-value theorem enables the steady-state error to be calculated for the situation where the demand $\Omega_D(s)$ takes the form of a Heaviside step function at $t = 0$ and of magnitude Ω_1:

$$\lim_{s \to \infty} (\Omega_D(s) - \Omega(s)) = \lim_{s \to 0} s \frac{\Omega_1}{s} \left(\frac{RJs + Rb + K_m{}^2}{RJs + Rb + K_m{}^2 + K_t K_c K_a K_m} \right) \tag{5.127}$$

$$= \left(\frac{Rb + K^2{}_m}{Rb + K^2{}_m + K_t K_c K_a K_m} \right) \Omega_1$$

which is a non-zero value.

The effect of proportional action alone is illustrated in Figure 5.75. For illustrative purposes, the value of R will be taken as 1 Ω, J as 1 kgm², b as 1 Ns/rad, K_m and K_t both as 1 Vs/rad, K_a as 1 and K_c as 10. It is seen that the response to a demand beginning at $t = 0$ for a speed of $\Omega_1 = 10$ rad/s is an actual response tending asymptotically to 8.333 rad/s – a steady-state error of 1.667 rad/s correctly predicted by equation (5.127).

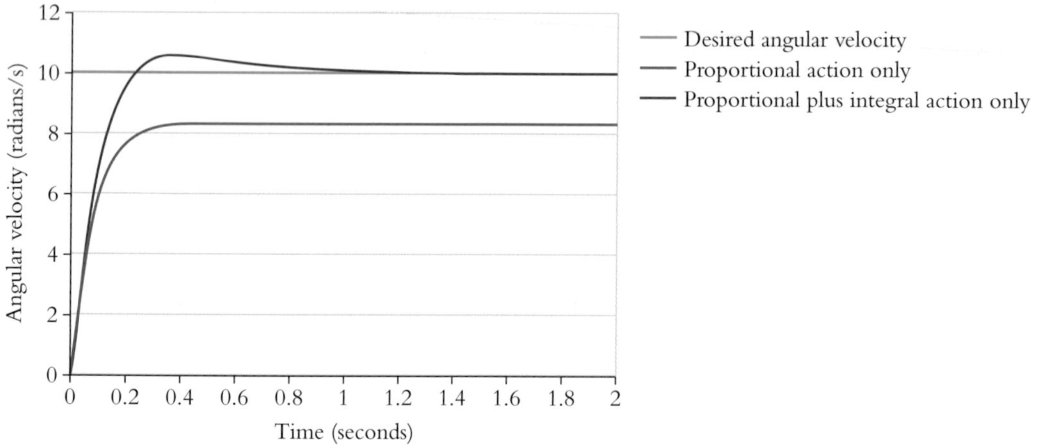

Figure 5.75 Response of radar antenna under proportional control alone and under proportional plus integral action

By contrast, substituting the following expression (corresponding to proportional plus integral action, but no derivative action) into equation (5.118):

$$G_c(s) = K_c\left(1 + \frac{1}{T_I s}\right) \quad \textbf{(5.128)}$$

results in the following expression for error:

$$\Omega_D(s) - \Omega(s) = \Omega_D(s)\frac{RJs + Rb + K_m{}^2}{RJs + Rb + K_m{}^2 + K_t K_c[1 + 1/(T_I s)]K_a K_m}$$

$$= \Omega_D(s)\left(\frac{T_I s(RJs + Rb + K_m{}^2)}{T_I s(RJs + Rb + K_m{}^2) + K_t K_c(1 + T_I s)]K_a K_m}\right) \quad \textbf{(5.129)}$$

$$= \Omega_D(s)\left(\frac{T_I RJs^2 + T_I(Rb + K_m{}^2)s}{T_I RJs^2 + T_I(K_t K_c K_a K_m + Rb + K_m{}^2)s + K_t K_c K_a K_m}\right)$$

Therefore, substituting for $\Omega_d(s)$ as before and applying the final-value theorem gives the following expression for steady-state error:

$$\lim_{s \to \infty}(\Omega_D(s) - \Omega(s))$$

$$= \lim_{s \to \infty} s\frac{\Omega_1}{s}\left(\frac{T_I RJs^2 + T_I(Rb + K_m{}^2)s}{T_I RJs^2 + T_I(K_t K_c K_a K_m + Rb + K_m{}^2)s + K_t K_c K_a K_m}\right) \quad \textbf{(5.130)}$$

$$= 0$$

In other words, inclusion of the integral term has eliminated steady-state error. This is also shown in Figure 5.75 for an integral time constant T_I of 2.8 seconds, chosen to achieve a rapid settling time without oscillation (corresponding to critical damping) in response to a step input. A more detailed consideration of the form of the response (oscillatory or otherwise), and the effect of including derivative action, are included in the discussion on dynamic response of second-order systems contained in the next subsection.

Dynamic response

So far, we have concentrated on what happens immediately after our system experiences its input and in the (infinitely) long term. No attempt has yet been made to quantify how the control system behaves in the short and medium term in response to an input. We are likely to be particularly interested in the following questions (refer to Figure 5.76 for typical responses):

- Does the system respond monotonically to an input or does its value overshoot and oscillate – in other words, is its response overdamped, underdamped or critically damped?
- If the response is underdamped, at what frequency does the output oscillate?
- Does the system settle down to a steady value or does the output grow increasing large (possibly in an oscillatory manner) with time? In other words, is its response stable?

These issues are covered in significant depth in Unit 6, and an understanding of these types of damping will be assumed here.

In practice, it is found that many systems typically exhibit behaviour analogous to a second-order (mass–spring–damper) system, and it is instructive to answer the above questions by relating the system's response to that of a second-order system.

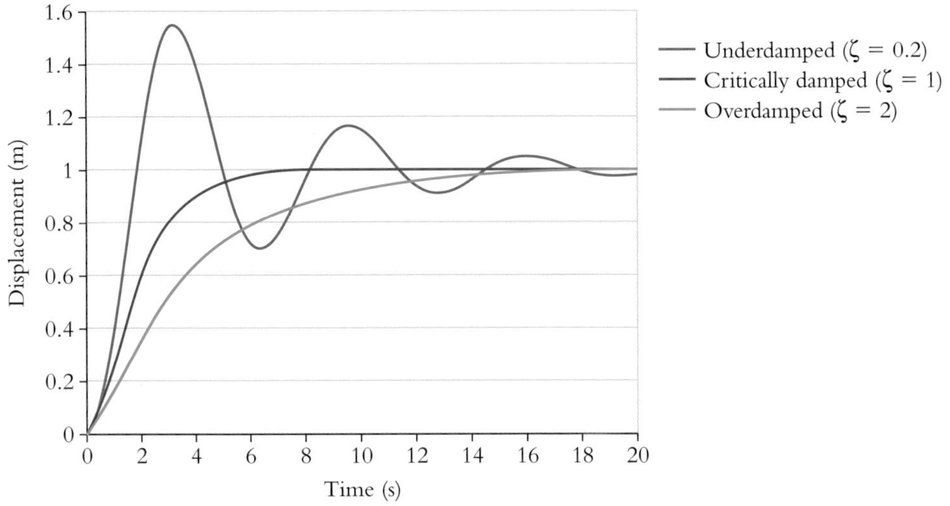

Figure 5.76 Graphs of underdamped, critically damped and overdamped responses of a second-order system

The key to answering the above questions lies in the closed-loop transfer function, and in particular the **characteristic equation** which is obtained by setting the denominator of the transfer function (the **characteristic polynomial**) to zero. However, before examining a true control system it is instructive to examine the simple mass–spring–damper system (Figure 5.77(a)) against which we will compare a control system's behaviour.

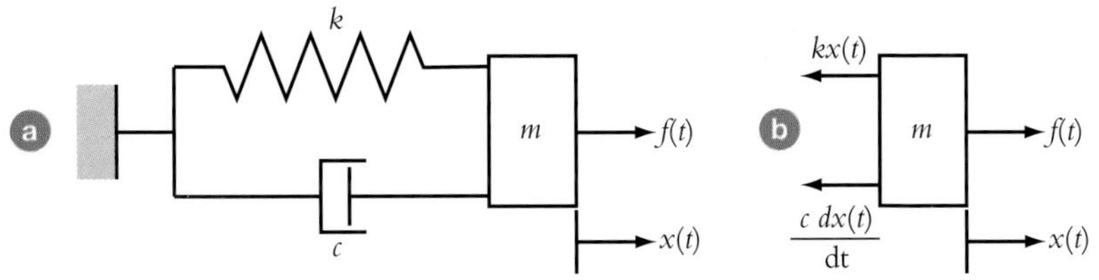

Figure 5.77 (a) Mass–spring–damper system; (b) free-body diagram of mass

Consider the mass to be subjected to an externally applied force $f(t)$. From the free-body diagram of the mass under displacement $x(t)$ (Figure 5.77(b)) and noting that Newton's second law states that net force is equal to mass × acceleration, the equation of motion of the mass can be written down as

$$f(t) - kx(t) - c\frac{dx(t)}{dt} = m\frac{d^2x(t)}{dt^2} \quad (5.131)$$

This can be rearranged as

$$f(t) = m\frac{d^2x(t)}{dt^2} + c\frac{dx(t)}{dt} + kx(t) \tag{5.132}$$

Taking Laplace transforms, and assuming zero initial conditions, gives

$$F(s) = s^2mX(s) + scX(s) + kX(s) \tag{5.133}$$

This can be rearranged to give the transfer function of the system:

$$\frac{X(s)}{F(s)} = \frac{1}{ms^2 + cs + k} \tag{5.134}$$

However, it is instructive to rearrange it again into as

$$\frac{X(s)}{F(s)} = \frac{1/m}{s^2 + \frac{c}{m}s + \frac{k}{m}} \tag{5.135}$$

The characteristic equation can now be obtained by setting the denominator equal to zero:

$$s^2 + \frac{c}{m} + \frac{k}{m} = 0 \tag{5.136}$$

However, it is well known that, for a simple mass–spring damper system, the following quantities can be defined:

- the natural frequency ω_n where

$$\omega_n = \sqrt{\frac{k}{m}} \tag{5.137}$$

- the damping ratio ζ (pronounced 'zeta') where

$$\zeta = \frac{c}{\sqrt{4mk}} \tag{5.138}$$

The value of ζ gives an indication of the response of the system, giving a more precise way to characterize the behaviours shown in Figure 5.76:

- If $\zeta = 0$, the system is undamped and will continue to oscillate indefinitely if it is perturbed.
- If $0 < \zeta < 1$, the system is said to be **underdamped** and will exhibit overshoot and a decaying oscillatory response when perturbed.
- If $\zeta = 1$, the system is said to be **critically damped**, and will exhibit a response that tends towards its steady-state response rapidly but without oscillation (or with only a small amount of overshoot).
- If $\zeta > 1$, the system is said to be **overdamped** and will tend asymptotically to its steady-state value, and may take a long time to reach a value that can be regarded as close enough to the value desired.

An everyday example of a second-order system that could fall into any of these categories would be a door that can open in either direction (such as a door between sections of a factory, or between a restaurant and a kitchen) and spring-loaded to return it to its closed position while allowing people or vehicles to push it open in either direction without operating a latch (Figure 5.78).

If the sprung door is underdamped, it will swing backwards and forwards after someone has passed through it; if it is overdamped, it will take a long time to close, and if it is critically damped it will close rapidly without swinging backwards and forwards. Common sense indicates that in this case, as in many practical situations, a critically damped response gives the best compromise between obtaining a quick response and avoiding prolonged oscillations, and this reasoning is generally applicable also to control systems.

Figure 5.78 An everyday example of a second-order system

Note that the natural frequency is *not* necessarily quite the same as the frequency at which oscillations will occur if the system is perturbed and allowed to undergo free vibration – it is only equal to the free oscillation frequency for the undamped case ($\zeta = 0$). If the system is underdamped, the actual frequency will be $\omega_n\sqrt{(1 - \zeta^2)}$ while if it is critically damped or overdamped, no oscillation will occur at all.

Returning to the analysis of the second-order (mass–spring–damper) system, the characteristic equation can now be expressed in terms of the natural frequency and damping ratio as

$$s^2 + 2\zeta\omega_n s + \omega_n^2 = 0 \tag{5.139}$$

Second-order characteristic equations are frequently encountered in control systems, and it is often useful to determine by inspection of the characteristic equation what its response will be.

As an example, consider again the motor-driven, viscous-damped antenna described earlier. If proportional plus integral action is used, the closed-loop transfer function becomes

$$\frac{\Omega(s)}{\Omega_D(s)} = \frac{K_t K_c(1 + 1/T_I s))K_a K_m}{RJs + Rb + K_m^2 + K_t K_c(1 + 1/(T_I s))K_a K_m}$$

$$= \frac{K_t K_c(T_I s + 1)K_a K_m}{T_I RJs^2 + RbT_I s + K_m^2 T_I s + K_t K_c(T_I s + 1)K_a K_m} \tag{5.140}$$

$$= \frac{K_t K_c K_a K_m(T_I s + 1)}{T_I RJs^2 + (Rb + K_m^2 + K_t K_c K_a K_m)T_I s + K_t K_c K_a K_m}$$

The characteristic function (the denominator) is clearly that of a second-order system. The transfer function can then be rearranged to get the characteristic function (the denominator) into the form in which it appears in equation (5.135):

$$\frac{\Omega(s)}{\Omega_D(s)} = \frac{K_t K_c K_a K_m(T_I s + 1)/(T_I RJ)}{s^2 + [(Rb + K_m^2 + K_t K_c K_a K_m)/(RJ)]s + K_t K_c K_a K_m/(T_I RJ)} \tag{5.141}$$

The characteristic equation can therefore be written down by setting the denominator to zero:

$$s^2 + [(Rb + K_m^2 + K_t K_c K_a K_m)/(RJ)]s + K_t K_c K_a K_m/(T_I RJ) = 0 \tag{5.142}$$

From inspection of this characteristic equation, and by comparison with equation (5.139), it is seen that

$$\omega_n = \sqrt{\frac{K_t K_c K_a K_m}{T_I RJ}} \tag{5.143}$$

and

$$\zeta = \frac{Rb + K_m^2 + K_t K_c K_a K_m}{2RJ\omega_n} = \frac{Rb + K_m^2 + K_t K_c K_a K_m}{2}\sqrt{\frac{T_I}{RJK_t K_c K_a K_m}} \tag{5.144}$$

It can be seen therefore that introducing integral action has the effect of turning the system from a first-order to a second-order one. Also, increasing the amount of integral action (specifically, reducing T_I) has the effect of reducing the damping ratio and making the system more oscillatory. For example, with the parameters used earlier but with $T_I = 1$, the system becomes overdamped ($\zeta = 1.9$) and approaches the demand speed slowly (Figure 5.79), whereas with $T_I = 0.01$, the system is noticeably underdamped ($\zeta = 0.19$) and the speed will fluctuate before settling down to its desired value.

If proportional-integral-derivative (PID) action is used, the corresponding analysis becomes rather messy but eventually results in the following expressions for natural frequency and damping ratio:

$$\omega_n = \sqrt{\frac{K_t K_c K_a K_m}{T_I RJ + T_I T_D K_t K_c K_a K_m}} \tag{5.145}$$

and

$$\zeta = \frac{Rb + K_{\mathrm{m}}^2 + K_t K_c K_a K_{\mathrm{m}}}{2(RJ + T_D\, K_t K_c K_a K_{\mathrm{m}})\omega_n} = \frac{Rb + K_{\mathrm{m}}^2 + K_t K_c K_a K_{\mathrm{m}}}{2}\sqrt{\frac{T_I}{(RJ + T_D K_t K_c K_a K_{\mathrm{m}})K_t K_c K_a K_{\mathrm{m}}}} \tag{5.146}$$

In fact, as this is a first-order system in open loop, it can be seen that derivative action actually has no benefit in terms of increasing damping and can actually make stability worse by reducing the value of ζ; this is not the case for systems with a second-order open-loop transfer function, such as the positioning system analyzed earlier.

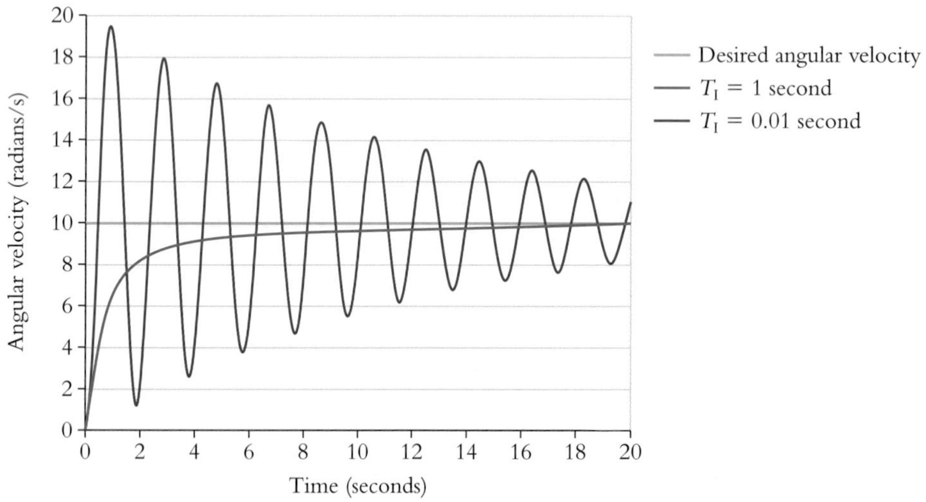

Figure 5.79 Response of radar antenna system for two values of integral time constant

Frequency response for harmonic excitation

This will only be covered briefly, though it forms a more significant part of texts on control theory. The response of a control system to a sinusoidally varying (harmonic) input can be determined by inserting the complex quantity $j\omega$ into the system's transfer function $G(s)$ where j is the symbol used by engineers for the imaginary unit (the square root of −1, often termed i by mathematicians). The result is a complex quantity whose real and imaginary parts $\mathrm{Re}(G(j\omega))$ and $\mathrm{Im}(G(j\omega))$ will generally vary with the value of circular frequency ω. The result can either be left directly in complex form, or can be converted into magnitude and phase (or modulus and argument) as follows:

$$|G(j\omega)| = \sqrt{(\mathrm{Re}(G(j\omega)))^2 + (\mathrm{Im}(G(j\omega)))^2} \tag{5.147}$$

and

$$\arg(G(j\omega)) = \tan^{-1}\left(\frac{\mathrm{Im}(G(j\omega))}{\mathrm{Re}(G(j\omega))}\right) \tag{5.148}$$

Note that the argument must be placed in the correct quadrant.

This calculation process may be performed for both the open- and closed-loop situations. The results may be plotted in two different forms:

Bode plot

This is actually a pair of graphs, presenting modulus and argument of $G(j\omega)$ plotted against $j\omega$. Logarithmic axes are used for both variables of the modulus plot and for the frequency axis of the argument plot. This is useful in order to identify resonances. Figure 5.80 gives a Bode plot for a second-order system, showing its single resonant frequency.

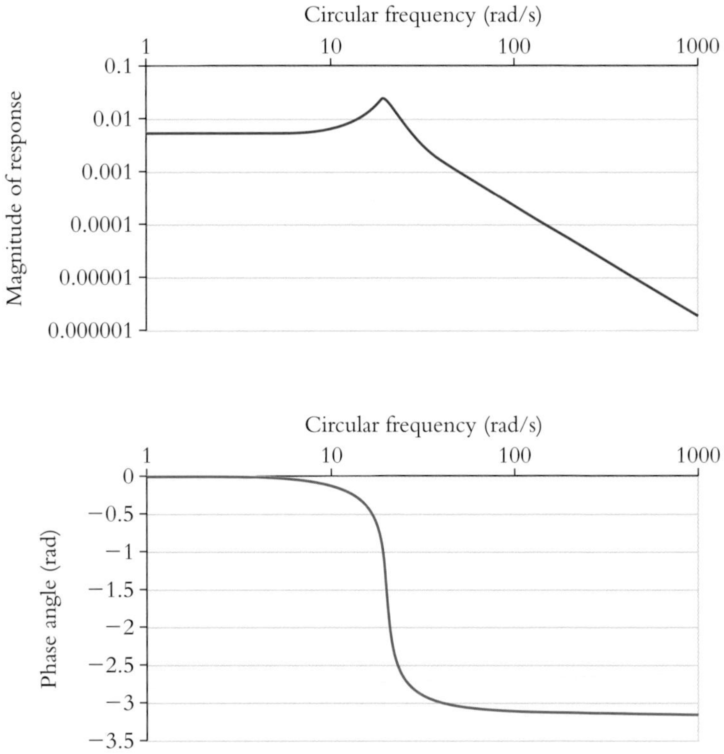

Figure 5.80 Example of Bode plot for a second-order system

Nyquist plot

This is a single graph plotted on an Argand diagram (i.e. a representation of the complex plane) that presents the locus of the points $(\mathrm{Re}(G(j\omega)), \mathrm{Im}(G(j\omega)))$ as ω is varied from zero to infinity. The result is typically a spiral-like plot that approaches the origin as ω tends to infinity. Although it contains less data than the Bode plot (the dependency upon frequency is lost, unless specific points are marked with values of ω), it is particularly helpful in understanding stability issues. Figure 5.81 shows plots for a typical second-order system and a higher-order system.

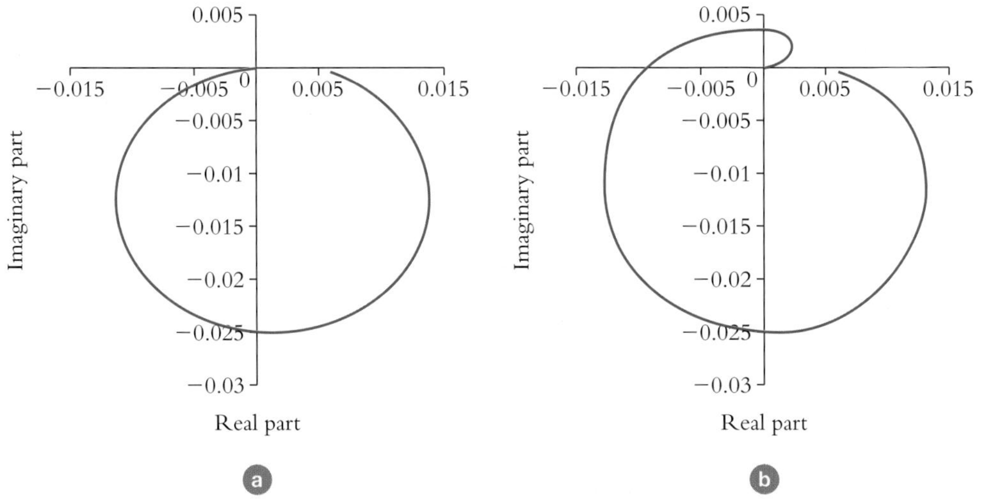

Figure 5.81 Nyquist plot for (a) typical second-order system; (b) typical higher-order system

Stability

Another very important issue is the stability of a control system. It has already been shown that a system may respond in an underdamped, critically damped or overdamped manner. However, it is also very important to determine whether or not a system exhibits **stability**, i.e. whether, for a bounded input (such as a unit impulse or step function), the output is also bounded, rather than (for example) undergoing oscillations which grow in amplitude with time rather than decaying. Such behaviour can occur owing to the use of excessive gain in a controller, or to time delays or phase lags in the controller or elsewhere in the loop.

A good example of unstable behaviour of a feedback system is the unpleasant 'howling' that occurs with a public address (PA) system at a rock concert or in a lecture theatre, when the microphone is placed too close to the loudspeaker, or if the volume control (gain) on the amplifier is turned up too far. The result is an oscillatory response that builds up until the maximum amplitude of the amplifier is reached.

There are three approaches to determining the stability of a control system, two of which will be considered here (the root locus method is dealt with separately in section 5.13).

Routh–Hurwitz criterion

This is perhaps the easier criterion to apply theoretically, though its physical meaning is not obvious. The closed-loop transfer function is determined as shown in Section 5.9, and the characteristic polynomial is found. It is then necessary to discover whether or not the characteristic function (obtained by equating the characteristic polynomial to zero) has any roots with positive real parts, which would result in a response that would become larger (rather than smaller) with time, and the Routh–Hurwitz criterion is a rapid way to find this information without going to the trouble of actually finding the roots. The coefficients of the characteristic polynomial are used as the basis of figures that are inserted into a table known as the Routh array. Let the characteristic function take the form:

$$a_n s^n + a_{n-1} s^{n-1} + \ldots + a_2 s^2 + a_1 s + a_0 \quad \textbf{(5.149)}$$

The following table is completed:

s^n	a_n	a_{n-2}	a_{n-4}	a_{n-6}	...
s^{n-1}	a_{n-1}	a_{n-3}	a_{n-5}	a_{n-7}	...
s^{n-2}	b_1	b_2	b_3	...	
s^{n-3}	c_1	c_2	c_3	...	
s^{n-4}	d_1	d_2	...		
...	...	...	...		

where

$$b_1 = \frac{a_{n-1}a_{n-2} - a_n a_{n-3}}{a_{n-1}}, \quad b_2 = \frac{a_{n-1}a_{n-4} - a_n a_{n-5}}{a_{n-1}} \text{ etc,}$$

and

$$c_1 = \frac{b_1 a_{n-3} - a_{n-1} b_2}{b_1}, \quad c_2 = \frac{b_1 a_{n-5} - a_{n-1} b_3}{b_1} \text{ etc,}$$

or more generally:

$$b_i = \frac{a_{n-1}a_{n-2i} - a_n a_{n-2i-1}}{a_{n-1}}$$

$$c_i = \frac{b_1 a_{n-2i-1} - a_{n-1} b_{i+1}}{b_1}$$

Diagrammatically, the terms are obtained as shown in Figure 5.82.

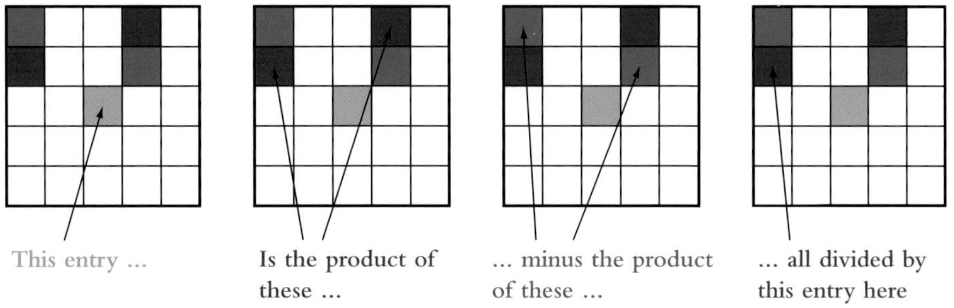

Figure 5.82 Sequence of calculating a new term in the Routh array

The number of changes of sign in the left-hand column is equal to the number of roots of the characteristic polynomial having positive real parts. In the context of control system stability, if *any* of the roots have positive real parts, the system will be unstable. We can therefore conclude that if there are *any* changes in sign of the left-hand column of the Routh array, the system will be unstable.

This can be illustrated via the following example.

Worked example

Find how many positive roots the following characteristic equation has:

$$s^6 - 4s^5 - 7s^4 + 44s^3 - 214s^2 + 140s + 400 = 0 \qquad (5.150)$$

The Routh array is as follows:

s^6	1	-7	-214	400
s^5	-4	44	140	0
s^4	$\frac{(-7)\times(-4) - 1\times 44}{(-4)} = 4$	$\frac{(-214)\times(-4) - 1\times 140}{(-4)} = -179$	$\frac{400\times(-4) - 1\times 0}{(-4)} = 400$	0
s^3	$\frac{44\times 4 - (-4)\times(-179)}{4} = -135$	$\frac{140\times 4 - (-4)\times 400}{4} = 540$	0	0
s^2	$\frac{(-179)\times(-135) - 4\times 540}{(-135)} = -163$	$\frac{400\times(-135) - 4\times 0}{(-135)} = 400$	0	0
s	$\frac{540\times 163 - (-135)\times 400}{(-163)} = 208.7$	0	0	0

There are four changes of sign in the left-hand column indicating four roots with positive real parts. In fact, this example of the characteristic function was generated by expanding the left-hand side of the following equation:

$$(s + 1)(s - 2)(s + 4)(s - 5)(s - 1 + 3j)(s - 1 - 3j) = 0 \qquad (5.151)$$

so the roots are $-1, 2, -4, 5, 1 - 3j$ and $1 + 3j$, with the four roots $2, 5, 1 - 3j$ and $1 + 3j$ all having positive real parts. A system with this characteristic equation would clearly be extremely unstable!

Nyquist stability criterion

This approach is more intuitive though it involves more numerical calculation. Moreover it can be applied experimentally without actually causing the control system to go unstable. The Nyquist plot for the **open-loop** (not closed-loop) transfer function (including the contribution from the feedback loop, i.e. $G(s)H(s)$ within Figure 5.27), is plotted as shown in Figure 5.83. If the curve encloses the point in the complex plane corresponding to the real number -1 (i.e. the point one unit to the left of the origin, on the negative real axis) as shown in Figure 5.83(b), the system will be unstable. The closer the curve comes to enclosing this point, the less stable the system will be (meaning in practice that disturbances will die away more slowly).

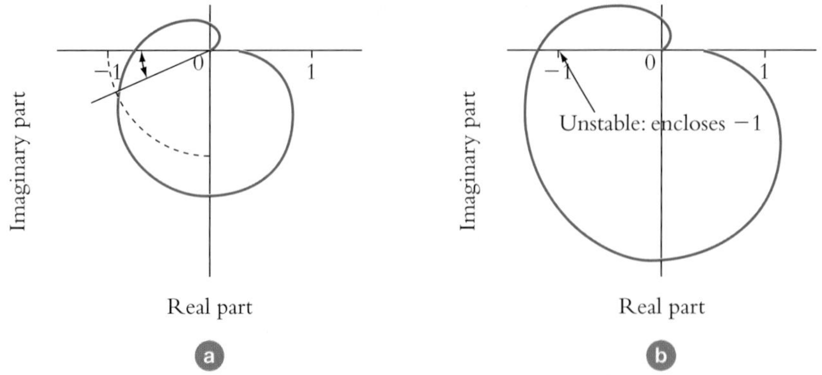

Figure 5.83 Nyquist plots of (a) the open-loop transfer function of a stable high-order system, showing the phase margin; (b) the open-loop transfer function of an unstable system showing how the plot encloses the point −1

The physical interpretation to this is as follows. If the feedback signal to the differencing junction is of opposite sign to the demand signal (or, more accurately, contains a component which is of opposite sign), the result from the differencing junction is actually an amplification of the input signal, and the net effect is equivalent to positive rather than negative feedback. If the amplitude of this 'positive feedback' signal is less than the input, the effect is not catastrophic as the feedback signal is still attenuated with respect to the input and each pass through the system attenuates it further. However, if the opposite-sign (180° out-of-phase) component of the feedback signal equals or exceeds the amplitude of the input signal, the feedback signal will be further amplified by the system, and a larger signal will be fed back. This implies that the signal will continue to grow and that instability will result.

The practical applicability of the Nyquist stability criterion is that the open-loop Nyquist plot can be constructed for a real (but theoretically uncharacterized) piece of control equipment, based on experimental response (magnitude and phase) to a sinusoidal input. The Nyquist stability criterion provides a very useful tool for predicting the closed-loop behaviour of a piece of apparatus without taking any risk of allowing it to become unstable. The amount by which the feedback signal would have to be scaled to enclose −1 on the plot gives a measure of the amount by which the gain of the controller could be increased without instability occurring. This is known as the *gain margin* and is usually expressed in dB (decibels, defined as 20 $\log_{10}$ of the factor by which the gain must be increased to cause instability) rather than as the factor itself.

Another useful quantity is the *phase margin*. This is the angle between the negative real axis of the Nyquist plot and the point at which the curve intersects the unit circle (Figure 5.83(a)). It represents the amount of additional phase lag needed in the control system to cause instability.

Learning summary

By the end of this section you should have learnt:

- ✓ the issues of response, damping and stability;
- ✓ the meaning of the terms 'characteristic polynomial' and 'characteristic equation';
- ✓ to determine the natural frequency and damping of a second-order system from the characteristic equation;
- ✓ the meaning of a Bode plot and a Nyquist plot;
- ✓ the meanings of gain margin and phase margin;
- ✓ the principle of the Nyquist stability criterion;
- ✓ to apply the Routh–Hurwitz stability criterion to a given characteristic polynomial.

5.13 A framework for mapping the response of control systems: the root locus method

The response of second-order systems is straightforward to analyse by drawing on the analogy with a mass-spring-damper system and inferring the natural frequency and damping ratio, but for third-order (and higher-order) systems, a more sophisticated method of conceptualizing the response is required. A useful tool is the **root locus plot** which is used for understanding the different aspects of a control system's behaviour, as represented by the various roots of the characteristic equation (which was introduced earlier in the present section as being the equation obtained by setting the denominator of the transfer function to zero).

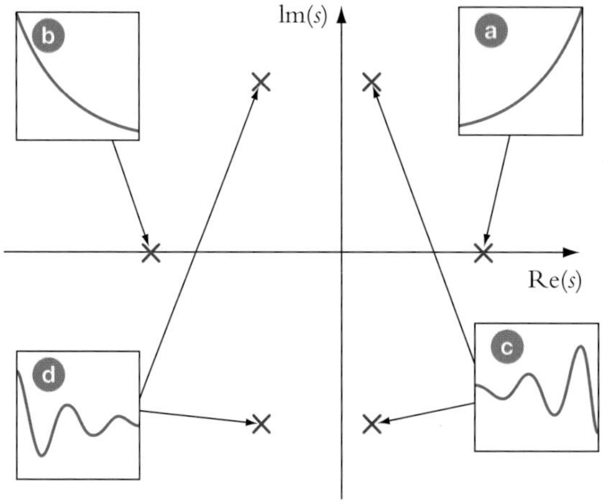

Figure 5.84 **Typical responses corresponding to positions of roots in the complex plane**

The root locus method involves plotting the locations of the roots in the complex plane (i.e. plotting their imaginary parts versus their real parts) as the gain of the control system is varied, and inferring the stability and nature of the response of the system from the position of the roots on the resulting diagram. Before looking in detail at the generation (and interpretation) of root locus plots, it is instructive to explore the physical meaning of different types of root, which can be represented in the complex plane, shown as (a) to (d) in Figure 5.84:

(a) positive real roots correspond to an exponentially growing (but non-oscillatory) transient response to a disturbance (which is clearly an inherently unstable response since it does not converge to a steady value);

(b) negative real roots correspond to an exponentially decaying, non-oscillatory response that dies away to zero with time and can form part of a stable response;

(c) complex roots, which always occur in 'conjugate pairs' with equal positive and negative imaginary parts, may have positive real parts and be of the form $s = \alpha \pm i\beta$ giving an oscillatory response of exponentially growing amplitude – clearly another example of an unstable response;

(d) complex roots may have negative real parts and be of the form $s = -\alpha \pm i\beta$ giving an oscillatory response of exponentially decaying amplitude – this dies away with time, and can again contribute to a stable response.

A simple example of a root locus plot, relating to a non-control example of a second-order system, is the familiar example of a mass–spring–damper system discussed earlier in the present section. It can be seen from the denominator in equation (5.134) that this system has a characteristic function $ms^2 + cs + k$, which leads to the characteristic equation (5.136) given again here for convenience:

$$s^2 + \frac{c}{m}s + \frac{k}{m} = 0 \qquad \textbf{(5.152)}$$

For illustrative purposes let us choose mass $m = 1$ kg, damping constant $c = 10$ Ns/m and allow spring stiffness k to vary from zero to infinity so that the characteristic equation becomes

$$s^2 + 10s + k = 0 \qquad \textbf{(5.153)}$$

With very small k, the roots are real (for $k = 0$, the roots are $s = -10$ and $s = 0$), and the response to an initial input (such as a step or impulse) is heavily damped, decaying slowly but without oscillation. As the spring constant is made larger (in this case 25 N/m) the system becomes critically damped, and the roots become equal but still real with the value, −5. A slightly larger value of the spring constant causes the roots to break away from the real axis to form a conjugate pair of complex roots, and the response becomes oscillatory. Further increases to the spring constant make the imaginary parts of the roots larger and the oscillations are faster but they do not die away any more slowly with time. The locus of the roots as the value of k varies is shown in Figure 5.85.

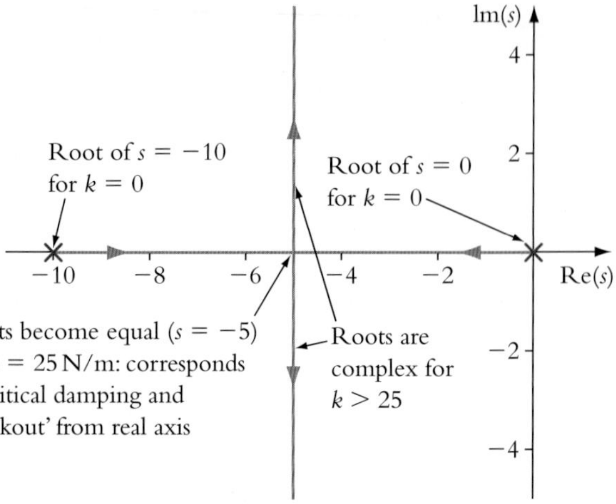

Figure 5.85 Root locus plot for a mass-spring-damper system with varying spring constant *k*

The same concepts can be applied to control systems, whose behaviour may or may not be oscillatory. In practice, it is found that the response of a system can vary between any or all of the kinds of behaviour shown in Figure 5.84 as the gain of the controller in a closed-loop control system is varied.

In order to understand the relevance of the root locus method to control systems, recall that a typical control system has an open-loop transfer function $G(s)H(s)$ as shown earlier in Figure 5.27. Furthermore, the effect on the transfer function of closing the loop has also been demonstrated (Figure 5.26 and equations (5.45) to (5.47)). Now let us assume that the transfer function of the main (forward) path involves a variable gain term K, arising for example from varying the gain of a controller lying within that forward path, and two polynomials $Z_G(s)$ and $P_G(s)$ so that

$$G(s) = K\frac{Z_G(s)}{P_G(s)} \qquad (5.154)$$

Similarly the transfer function of the feedback path can be expressed in terms of two more polynomials $Z_H(s)$ and $P_H(s)$:

$$H(s) = K\frac{Z_H(s)}{P_H(s)} \qquad (5.155)$$

so that the open-loop transfer function is

$$G(s)H(s) = K\frac{Z_G(s)Z_H(s)}{P_G(s)P_H(s)} = \frac{Z(s)}{P(s)} = K\frac{(s-z_1)(s-z_2)\ldots(s-z_{nz})}{(s-p_1)(s-p_2)\ldots(s-p_{np})} \qquad (5.156)$$

where $P(s) = P_G(s)P_H(s)$, $Z(s) = Z_G(s)Z_H(s)$. Also, $s = z_1, s = z_2 \ldots s = z_{nz}$ are the roots of the equation $Z(s) = 0$ and are known as the **zeros** of the open-loop transfer function, while $s = p_1, s = p_2, \ldots s = p_{np}$ are the roots of the equation $Z(s) = 0$ and are known as the **poles** of the open-loop transfer function.

By inserting equations (5.154) and (5.155) into equation (5.47) giving the closed-loop transfer function in terms of $G(s)$ and $H(s)$, the following is obtained:

$$\text{Closed loop transfer function} = \frac{G(s)}{1 + G(s)H(s)} = \frac{K\frac{Z_G(s)}{P_G(s)}}{1 + K\frac{Z_G(s)Z_H(s)}{P_G(s)P_H(s)}} \qquad (5.157)$$

$$= \frac{KZ_G(s)P_H(s)}{P_G(s)P_H(s) + KZ_G(s)Z_H(s)} = \frac{KZ_G(s)P_H(s)}{P(s) + KZ(s)}$$

The closed-loop response of the system is governed by the roots of the characteristic equation, which once again is obtained by setting the denominator of the closed-loop transfer function to zero:

$$P(s) + KZ(s) = (s-p_1)(s-p_2)\ldots(s-p_{np}) + K(s-z_1)(s-z_2)\ldots(s-z_{nz}) = 0 \qquad (5.158)$$

Now consider three possible situations:

(1) When K is very small (i.e. K tends to zero), $P(s)$ is much larger than $KZ(s)$ so the characteristic equation can be approximated as

$$(s - p_1)(s - p_2)\ldots(s - p_{np}) = 0 \tag{5.159}$$

whose roots are simply the open-loop poles of the system.

(2) When K is very large (i.e. K tends to infinity) and s is not large, $P(s)$ is much smaller than $KZ(s)$ so the characteristic equation can be approximated as

$$(s - z_1)(s - z_2)\ldots(s - z_{nZ}) = 0 \tag{5.160}$$

whose roots are simply the open-loop zeros of the system. However, there will also be some very large complex roots to the characteristic equation which (for very large complex s) can also be approximated by expanding it out and truncating the expansions of the two sets of brackets after the terms involving $s^{n_P - 1}$ and $s^{n_Z - 1}$ respectively:

$$s^{n_P} - s^{n_P - 1}p_1 - s^{n_P - 1}p_2\ldots - s^{n_P - 1}p_{np} + K(s^{n_Z} - s^{n_Z - 1}z_1 - s^{n_Z - 1}z_2\ldots - s^{n_Z - 1}z_{nZ}) = 0 \tag{5.161}$$

or

$$s^{n_P - 1}(s - p_1 - p_2\ldots - p_{np}) + Ks^{n_Z - 1}(s - z_1 - z_2\ldots - z_{nZ}) = 0 \tag{5.162}$$

Noting that s is assumed to be large, equation (5.156) can be further approximated (via a binomial expansion of the form $(1 - a)^{-1} = 1 + a + a^2\ldots \approx 1 + a$ for small a) and rearranged to give

$$s^{n_P - n_Z}\left(1 - \frac{p_1}{s} - \frac{p_2}{s}\ldots - \frac{p_{nP}}{s} + \frac{z_1}{s} - \frac{z_2}{s}\ldots - \frac{z_{nZ}}{s}\right) = -K \tag{5.163}$$

Taking the $(n_P - n_Z)$th root:

$$s\left(1 - \frac{p_1}{s} - \frac{p_2}{s}\ldots - \frac{p_{nP}}{s} + \frac{z_1}{s} - \frac{z_2}{s}\ldots - \frac{z_{nZ}}{s}\right)^{\frac{1}{n_P - n_Z}} \approx (-K)^{\frac{1}{n_P - n_Z}} \tag{5.164}$$

Noting that $(1 + a)^{1/n} = 1 + a/n + a^2/n^2 + \ldots \approx 1 + a/n$ for small a, and re-expressing the $(n_P - n_Z)$th roots of $(-K)$ in terms of the $(n_P - n_Z)$th roots of (-1):

$$s\left(1 - \frac{1}{n_P - n_Z}\left(\frac{p_1}{s} + \frac{p_2}{s}\ldots + \frac{p_{nP}}{s} + \frac{z_1}{s} - \frac{z_2}{s}\ldots - \frac{z_{nZ}}{s}\right)\right) = K^{\frac{1}{n_P - n_Z}}(-1)^{\frac{1}{n_P - n_Z}} \tag{5.165}$$

There are of course $(n_P - n_Z)$ possible values for the $(n_P - n_Z)$th root of (-1). Rearranging equation (5.165) and expressing the various $(n_P - n_Z)$th roots of (-1) in polar form:

$$s = \frac{p_1 + p_2\ldots p_{nP} - z_1 + z_2\ldots z_{nZ}}{n_P - n_Z} + K^{\frac{1}{n_P - n_Z}}\left(\cos\left(\frac{(2k + 1)\pi}{n_P - n_Z}\right) + j\sin\left(\frac{(2k + 1)\pi}{n_P - n_Z}\right)\right) \tag{5.166}$$

where $k = 0 \ldots (n_P - n_Z - 1)$. In other words, for large K, the roots of the characteristic equation tend asymptotically to large multiples of the $(n_P - n_Z)$th root of -1, shifted by a term sometimes referred to as the centroid of the poles and zeros.

(3) When K is neither very small nor very large, the roots of the characteristic equation will take some intermediate values which may be real or complex. The purpose of the root locus method is to show graphically how the roots move between the open-loop poles and the open-loop zeros or the large complex values (asymptotes).

The rules for plotting the root locus of an arbitrary closed-loop transfer function were originally proposed in a paper by Evans (1948). They are given in detail, though in slightly different forms and orders, in any comprehensive text on classical control theory (e.g. Dorf and Bishop (2008), who use a slightly different convention for poles and zeros, or Van de Vegte (1993)) and the reader is referred to such a text. A summary of the rules, with minor differences from the approach presented here, is given on the RoyMech website, along with a number of examples. The rules are summarized here (without further proof) to act as a starting point, noting that this version of the rules does not cover all contingencies.

(1) The root locus plot is symmetric about the real axis, because complex roots occur in conjugate pairs.

(2) The number of paths on the root locus is equal to the number of open-loop poles n_P.

(3) The paths start (for $K = 0$) at the open-loop poles and are normally marked on the complex plane with X symbols.

(4) n_Z of these terminate at the open-loop zeros, which are marked on the complex plane with O symbols.

(5) The remaining $n_P - n_Z$ paths do not terminate, but tend towards asymptotes oriented at angles θ to the real axis where

$$\theta = \frac{(2k+1)\pi}{n_P - n_Z} \quad \text{where } k = 0, 1 \ldots (n_P - n_Z - 1) \tag{5.167}$$

so that (for example) if there are five poles and two zeros there will be three asymptotes oriented at $\pi/3$, π and $5\pi/3$ radians (60°, 180° and 300°) to the real axis (Figure 5.86a). Similarly, if there are five poles and one zero there will be four asymptotes oriented at $\pi/4$, $3\pi/4$, $5\pi/4$ and $7\pi/4$ radians (45°, 135°, 225° and 315°) to the real axis (Figure 5.86b).

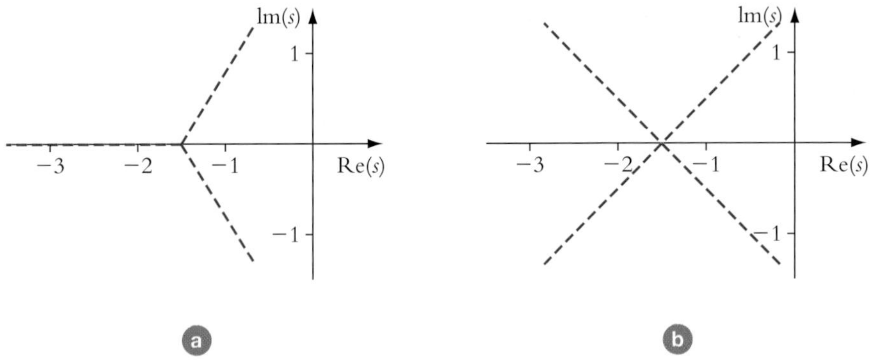

Figure 5.86 Typical asymptotes for (a) $n_P = 5$, $n_Z = 2$ and (b) $n_P = 5$, $n_Z = 1$

(6) The asymptotes radiate outwards from the point on the real axis sometimes known as the centroid of the poles and zeros:

$$s = \frac{p_1 + p_2 \ldots + p_{n_P} - z_1 + z_2 \ldots - z_{n_Z}}{n_P - n_Z} \tag{5.168}$$

(7) Root loci can be drawn on the real axis in the regions to the left of an odd number of poles and zeros. If there are an even number of poles and/or zeros on the real axis to the right of a given point, no root locus is present at that point.

(8) Where paths come together along the real axis (or converge back onto the real axis), the values of s at the 'breakout' and 'breakin' points (collectively known as 'singular' points, where there are multiple roots at a single point) are given by rearranging equation (5.158) to give K in terms of s then solving the equation:

$$\frac{dK}{ds} = 0 \tag{5.169}$$

Note that not all solutions to equation (5.169) will necessarily correspond to breakaway points.

(9) Where a path starts at a complex pole, the angle at which it leaves the pole, plus the sum of each of the angles to the real axis (horizontal) direction of the present pole from each of the other poles, add up to an angle $(2k + 1)\pi$ radians where k is a real integer (in other words, add up to an odd multiple of π). This is best illustrated via the example shown in Figure 5.87.

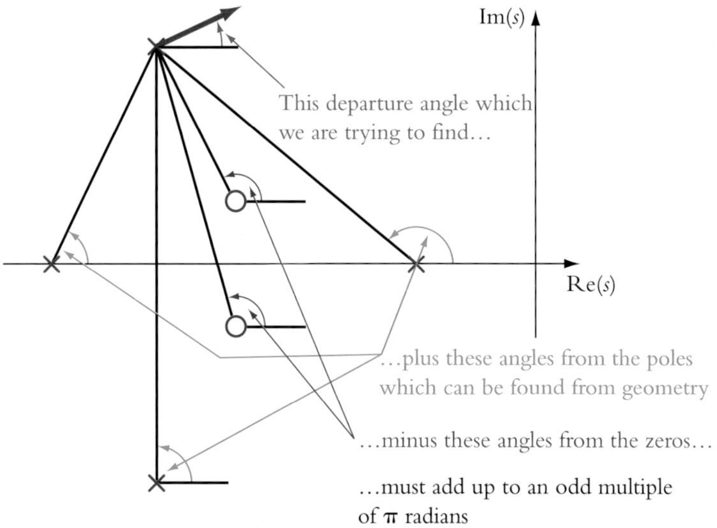
Im(s)
This departure angle which we are trying to find...
Re(s)
...plus these angles from the poles which can be found from geometry
...minus these angles from the zeros...
...must add up to an odd multiple of π radians

Figure 5.87 Finding the angle of departure from a complex pole

The angles of arrival of the paths at complex zeros are found by a process that is effectively the opposite of the above (the angle of arrival at a zero, plus the angles from the other zeros, minus the angles from the poles adds up to an odd multiple of π).

(10) The loci cross the imaginary axes at points that can be determined by replacing s by $j\omega$ and solving. It may or may not first be necessary to find the limiting value of K for stability (for example, by applying the Routh–Hurwitz stability criterion) before solving to find the imaginary roots. This can be a little tedious so will not be explored in detail here, though the worked example includes a simple illustration.

(11) Once we have plotted the root locus, the value of K can be chosen to ensure that the roots stay well away from the right-hand part of the complex plane. If a given point on the root locus is chosen, the corresponding value of K can be found by the product of distances of that point from the poles divided by the product of distances of that point from the zeros:

$$K = \frac{|s - p_1| \times |s - p_2| \times \ldots \times |s - p_{np}|}{|s - z_1| \times |s - z_2| \times \ldots \times |s - z_{nz}|} \qquad \textbf{(5.170)}$$

If there are no zeros, the denominator is unity.

With the availability of computing packages such as MATLAB, root locus plots can be drawn automatically and accurately without the need to apply the rules given above. An online root locus plotting Java applet is available (see References). It is nonetheless useful to be able to obtain a graphical visualization of a system's behaviour without the need for computational effort.

The power of the root locus method lies in the fact that an experienced control system designer can choose a suitable point on the root locus plot that will give a good compromise between stability and responsiveness; the corresponding value of K can then be calculated from rule 11 above.

Worked example

Plot the root locus of a closed-loop system having the following open-loop transfer function:

$$G(s)H(s) = \frac{K}{(s^2 + 2s + 2)(s + 2)} \quad (5.171)$$

This can be expressed as

$$G(s)H(s) = \frac{K}{(s + 1 + j)(s + 1 - j)(s + 2)} \quad (5.172)$$

which has three poles at $s = (-1 + j)$, $s = (-1 - j)$ and $s = -2$, and no zeros. Following the rules above:

(1) Symmetry about the real axis will be assumed and is satisfied by the open-loop poles.

(2) There are three open-loop poles so there are three paths on the root locus plot.

(3) The paths start at the open-loop poles $s = (-1 + j)$, $s = (-1 - j)$ and $s = -2$, which are marked on the diagram with X symbols (Figure 5.88(a))

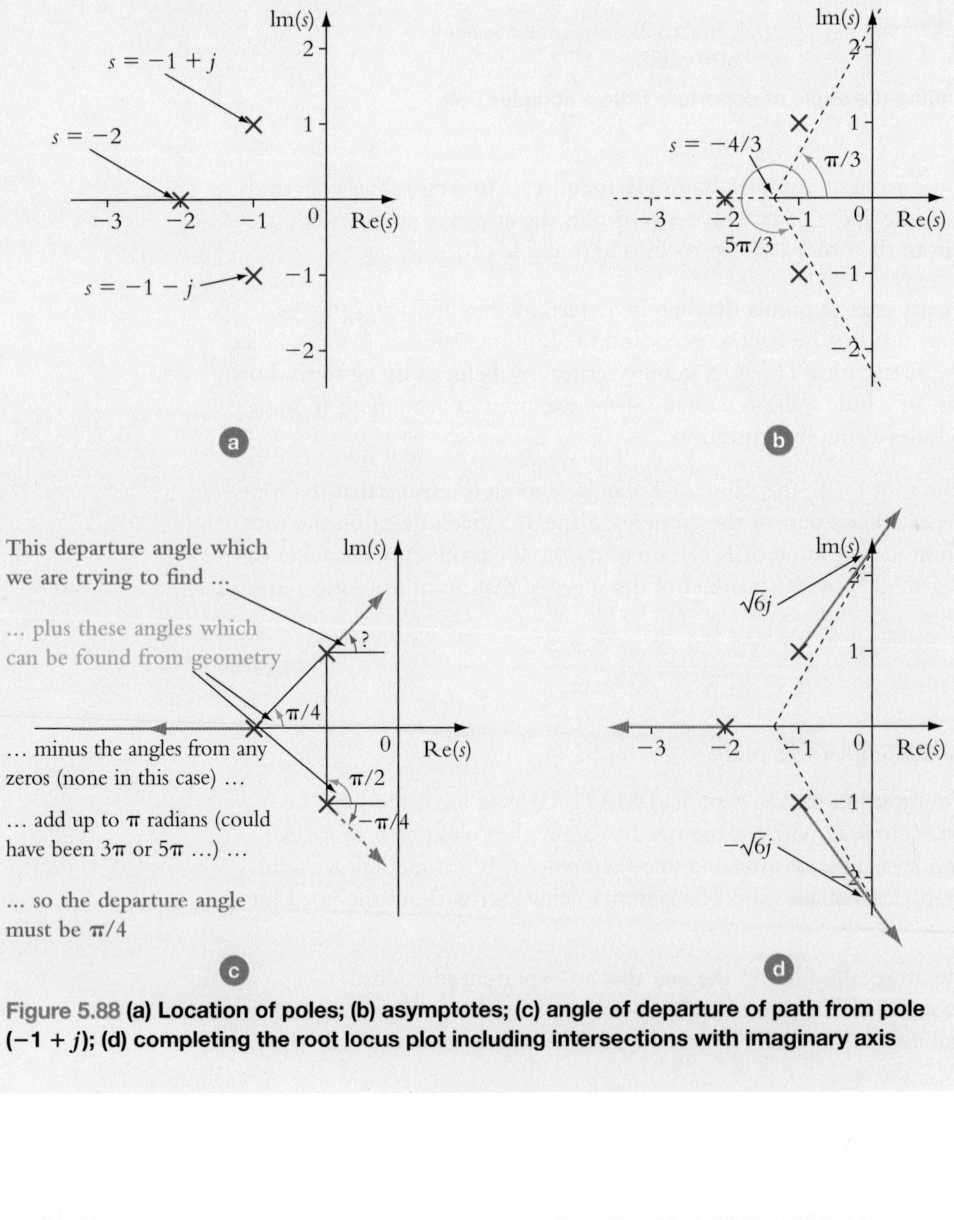

Figure 5.88 (a) Location of poles; (b) asymptotes; (c) angle of departure of path from pole (−1 + j); (d) completing the root locus plot including intersections with imaginary axis

(4) There are no zeros to be marked on the plot, so all of the paths end in asymptotes rather than at defined positions.

(5) $n_P = 3, n_Z = 0$, so there are three asymptotes at angles $\pi/3$, π and $5\pi/3$ radians (or $60°, 180°$ and $300°$) to the real axis (Figure 5.88(b))

(6) The asymptotes radiate outwards from the point:

$$s = \frac{(-1 + j) + (-1 - j) + (-2)}{3} = -\frac{4}{3} \tag{5.173}$$

as shown in Figure 5.88(b).

(7) The only point at which a path can be present on the real axis is to the left of the root $s = -2$, because only in this region is there an odd number of poles and zeros (in this case, only one pole and no zeros) to its right on the real axis. By enforcing this rule it is seen that this particular path starts at the pole $s = -2$ and extends infinitely along the negative real axis (Figure 5.88(c)).

(8) There is no breakout or breakin as the paths in this case either lie on the real axis or away from it.

(9) The angle of departure of from the pole $s = (-1 + j)$ (measured relative to the real axis), plus the total of the angles of that pole from the other poles, minus the total of the angles to that pole from the zeros, must add up to an odd multiple of π. It can be seen from Figure 5.88(c) how this is applied in the present case, with the total of the angles simply being π itself. The other angles are (in this example) simple to find geometrically, so the angle of departure is found as

$$\pi - \frac{\pi}{4} - \frac{\pi}{2} = \frac{\pi}{4}$$

By symmetry about the real axis, the angle of departure from the pole $s = (-1 - j)$ must be $-\pi/4$.

(10) Two of the paths cross the imaginary axis after leaving the complex poles. At this stage it is noted that (by inserting the open-loop transfer function into (5.47), expanding the bracketed terms and rearranging) the characteristic equation of the closed-loop system can be expressed as

$$s^3 + 4s^2 + 6s + 4 + K = 0 \tag{5.174}$$

Making the substitution $s = j\omega$ gives

$$-j\omega^3 - 4\omega^2 + 6j\omega + 4 + K = 0 \tag{5.175}$$

Taking the real part of equation (5.175) and rearranging gives

$$6\omega = \omega^3 \tag{5.176}$$

ω is therefore either zero (not a valid solution) or $\pm\sqrt{6}$ rad/s, so the loci intersect the imaginary axis at the points $s = \pm\sqrt{6}j = \pm 2.44j$. Now taking the imaginary part of equation (5.175), rearranging and inserting this value of ω gives the value of gain for limiting stability:

$$K = 4\omega^2 - 4 = 4 \times 6 - 4 = 20 \tag{5.177}$$

At this stage, the root locus plot can be completed as shown in Figure 5.88(d), with the two complex paths leaving the two complex poles at $\pm\pi/4$, crossing the imaginary axis at $\pm 2.44j$, and converging on the asymptotes at $\pi/3$ and $5\pi/3$. A control engineer may then choose (by eye and experience) a point on the plot for which the gain can then be calculated from the distances of that point to the poles and zeros. For example, if point (–0.5 + 1.65*j*) on the locus is chosen by eye (Figure 5.89), *K* is calculated as

$$K = \frac{\text{product of all distances from chosen } s \text{ to poles}}{\text{product of all distances from chosen } s \text{ to zeros}}$$

$$= \frac{|s + 1 - j| \times |s + 1 + j| \times |s + 2|}{1} = \frac{|0.5 + 0.65j| \times |0.5 + 2.65j| \times |1.5 + 1.65j|}{1}$$

$$= 4.93 \qquad \textbf{(5.178)}$$

noting that (in this case) the denominator is simply 1 as there are no zeros.

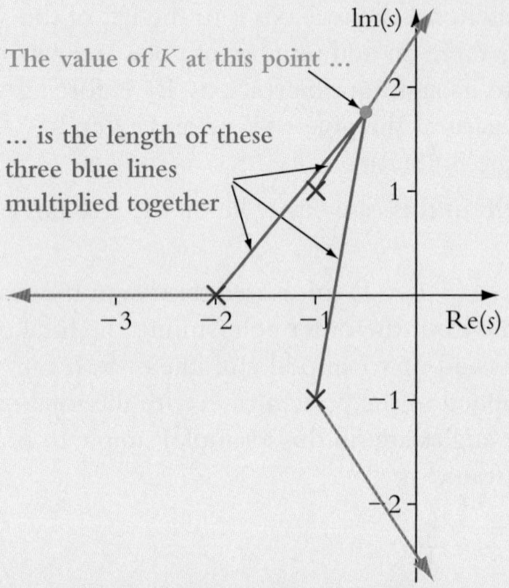

Figure 5.89 Finding the gain of a system corresponding to a point on the root locus plot

Applications of the root locus method

The root locus method can be used to explore further two of the examples from earlier in the present unit, namely the position controller and the antenna system, with particular reference to the stability and oscillatory response of the systems.

It has already been shown from Figure 5.50 that the open-loop transfer of the position controller system with a simple proportional amplifier is

$$\frac{V_P(s)}{V_D(s)} = \frac{K_a K_m K_s}{Js^2} \qquad \textbf{(5.179)}$$

The root locus is very easy to plot (for varying amplifier gain) as we can immediately see:

- there are two poles, both of value $s = 0$, so there are two paths;
- as the poles are equal, they also form the breakout point;
- there are no zeros, so there are $2 - 0 = 2$ asymptotes;
- the centroid of the poles is $s = 0$ so the asymptotes radiate from the origin at an angle $\pm\pi/2$ to the real axis.

Without further ado, the root locus plot can be drawn as shown in Figure 5.90, showing the poles of the closed-loop transfer function as the amplifier gain is increased.

This illustrates that the root locus lies in the imaginary axis – in other words, the response is always oscillatory and is on the margin of stability, with any input resulting in a position that oscillates at a frequency dependent upon the amplifier gain, with the oscillations never dying away. However, if as before the simple amplifier is replaced with a PID controller/amplifier with gain K_c and proportional and integral time constants T_I and T_D respectively (and this time ignoring any additional external load), the open-loop transfer function now becomes

$$\frac{V_P(s)}{V_D(s)} = \frac{K_cK_mK_s\left(1 + \frac{1}{T_Is} + T_Ds\right)}{Js^2} = \frac{K_cK_mK_sK_D\left(s^2 + \frac{1}{T_D}s + \frac{1}{T_IT_D}\right)}{Js^3}$$

Once again, assuming, that $T_D = 0.5$ second and $T_I = 1$ second, and assuming also the system parameters chosen in Section 5.11, the open-loop transfer function becomes:

$$\frac{V_P(s)}{V_D(s)} = \frac{K_cK_mK_sT_D(s^2 + 2s + 2)}{Js^3} = \frac{K_cK_mK_sT_D(s + 1 + j)(s + 1 - j)}{Js^3}$$

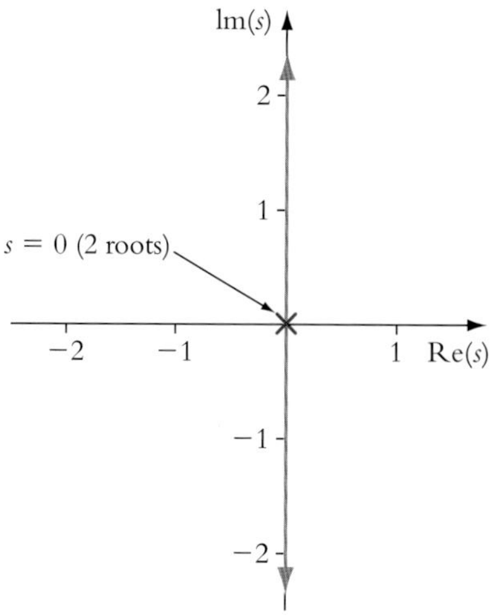

Figure 5.90 Root locus plot for positioning system with proportional control

This system now has:

- three poles, all at s = 0
- two zeros, at $s = (-1 + j)$ and $s = (-1 - j)$
- two paths that arrive at the zeros at angles that can be found in a similar way to that for angle of departure from poles: sum of angles from zeros − sum of angles from poles comes to $(2k + 1)\pi$ for integer k, i.e. an odd multiples of π. In the case of the root at $s = (-1 - j)$, try $k = -1$ so angle $= -\pi - (\pi/2 - 3 \times 3\pi/4) = 3\pi/4$, and by symmetry the angle of arrival at the other complex zero is $-3\pi/4$. It can also be shown that the angles of departure from the three coincident poles at the origin are π and $\pm\pi/3$, though this is not obvious from the rules already described
- one asymptote at angle π radians, radiating from the point $s = 2$.

The root locus plot can now be drawn as shown in Figure 5.91.

It is seen that, as the gain is increased, the roots move into the right-hand half of the plane and the system becomes unstable; however, for larger values of K the system becomes stable again as two of the closed-loop poles tend to the open-loop zeros in the left-hand half of the complex plane (indicating decaying oscillatory responses) and the third closed-loop pole tends to $-\infty$, indicating a rapidly decaying monotonic response. This example shows the usefulness of being able to visualize what is going on within a system that may initially seem to exhibit puzzling or counterintuitive behaviour, in this case showing unstable behaviour at small gains and becoming stable for larger gains. The practical implication of this is shown in Figure 5.92, which shows the response for gains of 1 and 10 respectively.

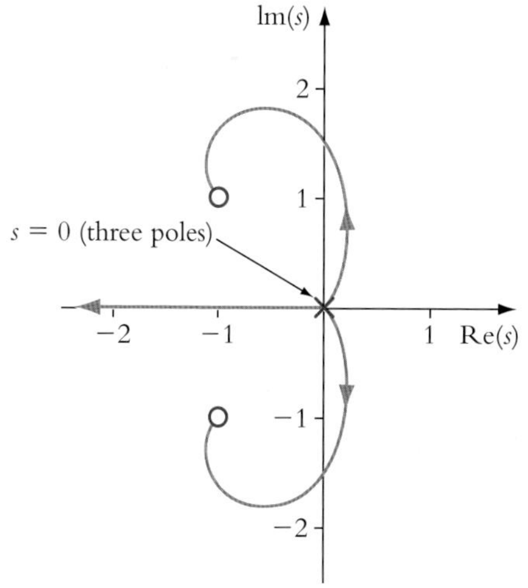

Figure 5.91 Root locus plot for positioning system with PID control

The root locus plot can also be drawn for the radar antenna problem. The block diagram in Figure 5.64 may be rearranged slightly by placing the common factor of K_t inside the feedback loop as shown in Figure 5.93.

After multiplying out the terms on the forward path, and substituting for proportional plus integral action, the following open-loop transfer function is obtained:

$$\text{Open loop transfer function} = \frac{G_c(s)K_tK_aK_m}{RJs + Rb + K_m^2}$$

$$= \frac{K_c(1 + 1/(T_Is))K_tK_aK_m}{RJs + Rb + K_m^2} = \frac{K_cK_tK_aK_m(s + 1/T_I)}{RJs(s + (Rb + K_m^2)/RJ)}$$

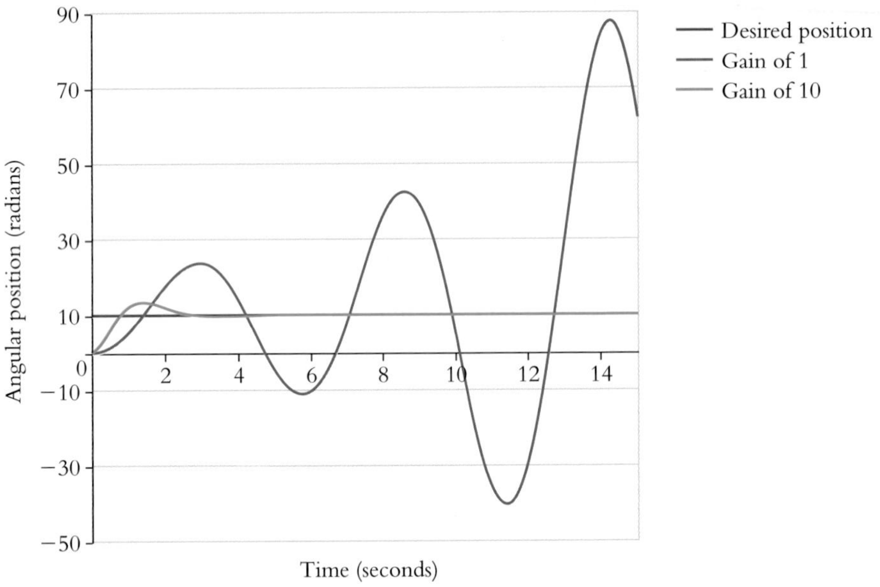

Figure 5.92 Response of positioning system with PID control for two values of gain

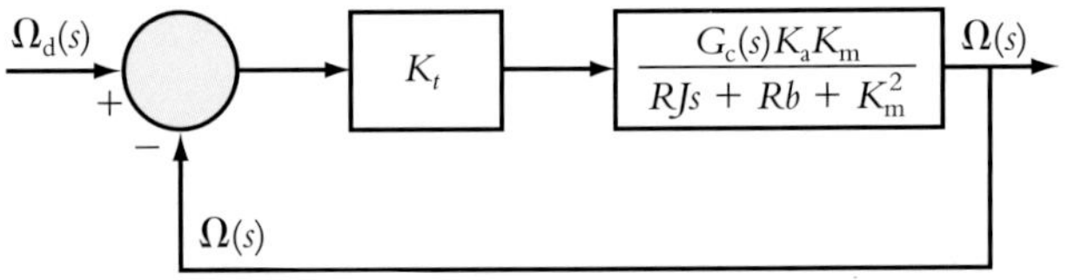

Figure 5.93 Rearranged block diagram for complete antenna system

This is now a second-order system with poles at $s = 0$ and $s = -(Rb + K_m{}^2)/RJ$, and a zero at $s = -1/T_I$. Using the same system constants as before, but with $T_I = 0.2$ seconds (as this involves more interesting behaviour than other values which are arguably more suitable), the poles are now at $s = 0$ and $s = -2$, and the zero is at $s = -5$. The root locus plot shown in Figure 5.94 can be drawn.

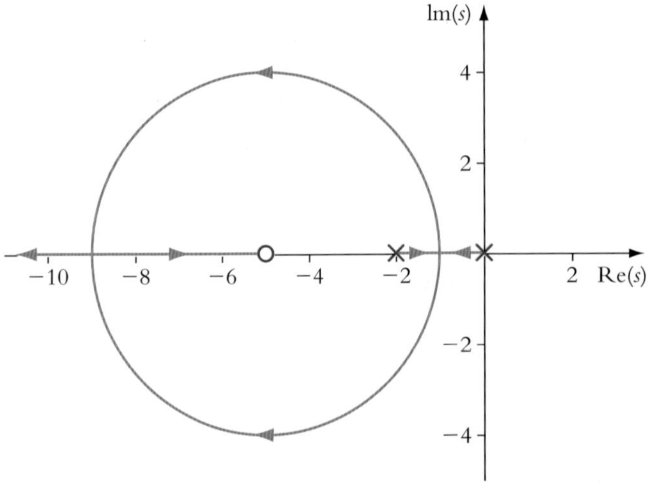

Figure 5.94 Root locus plot for antenna system

This shows that, as the gain is increased from zero, the behaviour is initially overdamped, then becomes (slightly) underdamped, then becomes overdamped again. Such behaviour would not be obvious from examining the open-loop transfer function! The practical implication of this is shown via plots for $K_a = 0.1$ ($\zeta = 1.48$), 4 ($\zeta = 0.67$) and 20 ($\zeta = 1.1$) as shown in Figure 5.95.

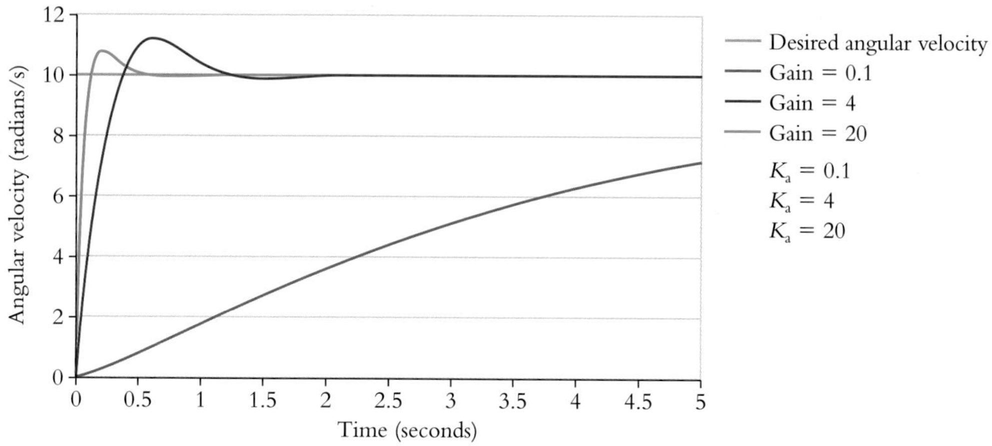

Figure 5.95 Responses of antenna system for various values of K_a

Learning summary

By the end of this section you should have learnt:

✓ to draw root locus plots and interpret them in terms of system stability and response.

References

Bennett, S., 1979, *A History of Control Engineering, 1800–1930*. Peter Peregrinus/IEE.

Bennett, S., 1993, *A History of Control Engineering, 1930–1955*, 1st edn, Peter Peregrinus.

Dorf, R.C. and Bishop, R.H., 2008, *Modern Control Systems*, 11th edn, London: Pearson.

Evans, W. R., 1948, 'Graphical analysis of control systems', *Trans. AIEE*, 67, 547-51.

Healey, M., 1967, *Tables of Laplace, Heaviside, Fourier and Z transforms,* Edinburgh: Chambers.

http://www.roymech.co.uk/Related/Control/root_locus.html, accessed 9 February 2010.

http://users.ece.gatech.edu/bonnie/book/OnlineDemos/InteractiveRootLocus/applet.html, accessed 9 February 2010

Kreyszig, E. (2006), *Advanced engineering mathematics*, 9th int. edn, Hoboken, NJ: Wiley.

Simulink™ Details can be found at: http://www.mathworks.com/products/simulink/ (accessed 27 February 2010)

Spencer, A. J. M., Parker, D. F., Berry, D. S., England, A. E. H., 1977, *Engineering mathematics*, Vol. 1., Wokingham: Van Nostrand Reinhold.

Van de Vegte, J., 1993 *Feedback Control Systems*, 3rd edn, London: Pearson.

Unit 6

Structural vibration

UNIT OVERVIEW

- Introduction
- Natural frequencies and mode shapes
- Response of damped single-degree-of-freedom systems
- Response of damped multi-degree-of-freedom systems
- Experimental modal analysis
- Approximate methods
- Vibration control techniques

6.1 Introduction

Many people encounter vibration for the first time through the music produced by stringed and percussive instruments, or through the speakers of their radios or the earphones of their iPods. All of these produce their sounds by vibration. While these provide pleasure to the user, vibration in most engineering structures is something to avoid.

This unit will describe some basic causes and effects of vibration and show you how to model the vibration of structures mathematically.

A recurring theme is **resonance**. A key observation is that if the frequency of an alternating force coincides with a resonant frequency, large amplitude vibrations will result.

Large alternating displacements often produce large alternating stresses. It is no surprise, therefore, that **resonance-induced fatigue** is a major cause of in-service failures – normally when the effects of vibration have not been considered adequately at the design stage.

In design, it is common to calculate the stresses due to static loads and to use a so-called 'factor of safety' to make allowance for effects that have been excluded from the analysis. An alternating force that induces resonance can easily produce stresses that are 100 times those that a static force of the same magnitude would give. Typical factors of 'safety' range from 1.5 to 4!

Figure 6.1 **Wind-induced torsional vibration of the Tacoma Narrows bridge shortly before it collapsed**

The most dramatic failure caused by resonant vibration was probably that of the Tacoma Narrows Bridge in New York State. It collapsed in truly spectacular fashion in 1940, after only four months' use. From its opening, bending waves were set up if the wind speed rose and people would drive across just for the fun of the experience (Disney World didn't exist in 1940!). However, on 7th November a wind speed of 42 mph produced a severe torsional vibration. At its most extreme, the sides of the bridge were moving up and down by over 8 m. Take a look at the video at www.tinyurl.com/2ca5mh. Seeing is believing.

Designers now understand about the vertical bending and torsional vibration that can be caused by wind or by loads moving across bridges. Or at least they thought they did, until the opening of the Millennium Bridge in London showed that there was still more to learn. It was built for pedestrians and cyclists, and the designers had taken account of the vertical forces from their feet as they walked across. A big crowd had gathered for the opening ceremony and when they started to walk across, the bridge began to sway from side-to-side. No structure is completely rigid and the crowd found that the bridge swayed slightly from side to side. To help steady themselves, they moved their feet apart as they walked and in doing so exerted a small alternating horizontal force onto the bridge with each step. The problem was that everyone responded in the same way at the same time and the combined effect of the forces from all the feet was sufficient to excite a horizontal bending mode of the bridge. Take a look at the video at www.tinyurl.com/33xeyle. The problem was cured by a combination of viscous dampers and tuned-mass absorbers.

Section 6.2 explains how to calculate the resonant frequencies of structures and see how structures deform when they vibrate; their so-called 'mode shapes'. Structures only vibrate if they are subjected to some form of excitation and Section 6.3 considers the response of structures that exhibit only one resonant frequency. Section 6.4 extends this to structures with many resonant frequencies. Even the most sophisticated computer analysis has to make assumptions and simplifications that may not reflect the actual properties of the real structure. Section 6.5 describes how a structure's resonant frequencies and mode shapes can be measured experimentally. At the preliminary stage of a design, approximate values can help determine whether vibration is likely to be an issue, and Section 6.6 introduces some useful techniques for obtaining these. Finally, Section 6.7 introduces two common techniques for controlling the potentially damaging effects of vibration.

Figure 6.2 **The crowd of pedestrians who had gathered for the opening of the Millennium Bridge in London caused a pronounced side-to-side vibration when they began to walk across**

6.2 Natural frequencies and mode shapes

Introduction

Anyone who has tapped a bottle or cup will know that it 'rings' at a characteristic frequency. We hear a sound because the tapping has excited one or more natural frequencies. As noted in Section 6.1, resonance-induced fatigue is a major cause of premature structural failure, and this section discusses several ways of modelling structures so that their natural frequencies can be calculated.

Before doing so, however, we introduce the classification of structure in terms of **degrees of freedom**. These can be thought of as 'possible motions'. The number of degrees of freedom for a structure:

- is related to the number of resonant frequencies it has;
- affects the number of motion coordinates needed to define its position.

A rigid body can have up to six degrees of freedom: three translations and three rotations. In addition to the six rigid body freedoms, a flexible body can have other possible motions, such as bending and twisting, or flexibility between components.

A key step is to decide how many degrees of freedom are appropriate. With a few exceptions (such as a spacecraft), structures are constrained so that their freedom to move is limited. Even so, this can still leave perhaps thousands of possible ways in which a structure might vibrate (its so-called **modes of vibration**). In most cases, we are only interested in a few of these (possibly in only one mode) and we will look for a mathematical model that concentrates on these and omits other possible motions. If, for example, we wanted to study the way an aircraft's wing vibrates when it hits an air pocket, we would ignore the fact that this also caused a light fitting in one of the toilets to vibrate! On the other hand, if we were designing the light fitting, we would want a mathematical model that described how it responded to motion of the panel that it was attached to. In this case, we would probably just model the fitting and the panel, but wouldn't include every other part of the aircraft.

In summary, we can say that we aim to model the particular motions we are interested in. We will start by looking at structures that can be characterized by just one type of motion. We will then move on to those that can exhibit two or more types of motion.

The mathematical descriptions we will develop relate the motion of the structure to the forces that cause it. Central to this process is **Newton's second law of motion**. This links **cause** and **effect**. It applies only to rigid bodies and allows us to relate the applied forces (and/or moments) to the motion that they produce.

There are two forms: one for translation, one for rotation. These were covered in Unit 6 in Part 1 and you may want to review that unit before proceeding.

For linear translational motion, Newton's second law can be written as:

$$\overrightarrow{\begin{array}{c}\text{Resultant force in the direction of}\\ \text{the acceleration}\end{array}} = \text{Mass} \times \overrightarrow{\begin{array}{c}\text{Absolute acceleration of}\\ \text{the centre of mass}\end{array}}$$

It's best to think of it in words, not as '$F = ma$'. In problems, you should write it down as it's said, noting always that it's a vector equation. Put the forces (resolved into the vector direction that you have chosen for acceleration) on the left of the equals sign and the mass and acceleration terms on the right. Don't mix the two together. 'Mass × acceleration' is *not* itself a force. It describes the effect that the actual forces have on the mass in question.

Newton's second law for rotational motion can be expressed as:

$$\overrightarrow{\begin{array}{c}\text{Resultant torque (moment)}\\ \text{about the axis in the}\\ \text{direction of the acceleration}\end{array}} = \begin{array}{c}\text{Moment of}\\ \text{inertia about}\\ \text{the axis}\end{array} \times \overrightarrow{\begin{array}{c}\text{Absolute}\\ \text{angular}\\ \text{acceleration}\end{array}}$$

It is important to note that 'the axis' about which the torque is taken *must* be one of the following.

- a fixed axis
- an 'instantaneous centre of rotation' (this is a point on a body that is stationary for an instant, such as the point of contact of a tyre rolling along the road)
- the centre of mass.

No other axis is allowed.

As with the linear equation, think of the words and write it down as it's said.

Structures with one dominant mode of vibration

Initially, we will look at structures that can be modelled with just one degree of freedom. This is the simplest model for a vibrating structure. The model normally takes the form of a rigid body restrained by one or more massless springs. It works well for structures with one resonance or where one resonance dominates the vibration behaviour. It gives good insight into vibration behaviour and is often used as a first approximation for more complicated structures.

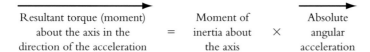

We will look at a number of different examples where single-degree-of-freedom dynamic models are appropriate, but the general approach to the analysis will be the same in all cases and will involve the following steps:

(1) Convert the physical structure into a dynamic mass–spring model.

(2) Draw a free-body diagram. Displacing the body from its equilibrium position will create forces in the restraining springs that will try to return the body to equilibrium. The free-body diagram can be thought of as a 'snapshot' of the state of the system when it has moved away from equilibrium by a chosen amount. Drawing the free-body diagram is the key step in the analysis and should be tackled systematically. There are three stages:

   (i) Start with the system in equilibrium and draw it as a free body. To create the 'free' body, draw it without any of the restraining springs. The forces exerted by the springs will be added in stage (iii).

   (ii) Select a motion coordinate to describe how the system will deflect from its equilibrium position and mark it on the diagram.

   (iii) Apply a positive deflection in the chosen motion coordinate, identify the forces (and/or moments) that result and draw them on the diagram. It is critical that the positive directions of the forces due to a positive deflection are shown correctly.

(3) Apply the appropriate form of Newton's second law of motion to give the equation of motion for the system.

The following pages show how steps (1), (2) and (3) can be applied to a number of examples. You should adopt the same systematic approach using the three steps when setting up solutions.

At this stage, we will concentrate on finding the **natural frequency** of the systems. This is the frequency at which a system will vibrate when displaced from equilibrium and then released. Later, we will consider the effects of external excitation and damping, but these are omitted for the moment.

When we look at the effects of excitation, we will find that, for most engineering structures, there is a maximum response if the excitation frequency coincides with the natural frequency (which is why we need to know what its value is). This effect is called **resonance** and the term **resonant frequency** is often used instead of natural frequency (although, strictly speaking, the two are different as we shall see later).

Example 1: Simple mass–spring system

If a mass m [kg] is suspended from a spring of stiffness k [N/m], it will move down under the effect of gravity and stretch the spring by a distance x_{eq} before reaching its static equilibrium position (Figure 6.3).

Once it is in equilibrium, the force in the spring (given by the stiffness multiplied by the change of length) exactly balances the gravitational force on the mass and hence:

$$mg = kx_{eq}$$

What then happens if the mass is given a *further* downward displacement, x, away from equilibrium and then released (Figure 6.4)?

The displacement x produces an additional force in the spring of kx, as shown in Figure 6.4(b). Since we know from the static equilibrium case that $mg = kx_{eq}$, the resultant force on the mass is kx, as shown in Figure 6.4(c).

Remembering that the displacement and the force are both vectors, we see that a positive downward displacement from equilibrium produces a positive upward resultant force on the mass. This acts to

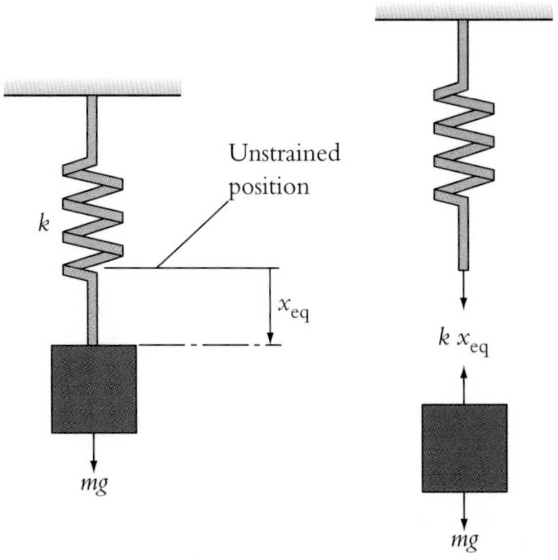

Figure 6.3

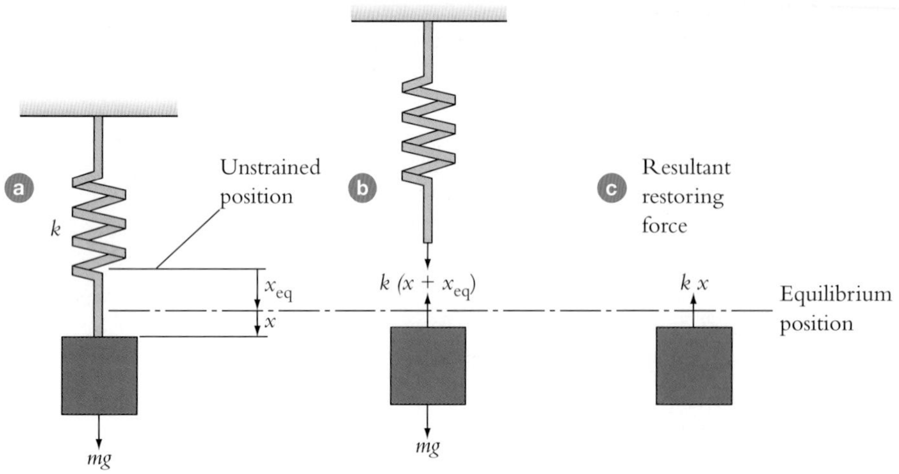

Figure 6.4

return it to its equilibrium position. At any instant, when x was negative (meaning that the mass was *above* its equilibrium position), the 'upward' resultant force, kx, would also be negative, telling us that the force was actually acting downwards at this instant. Since the force always acts to return the mass to equilibrium, the term **restoring force** is often used.

So, what happens when the mass is displaced downwards (x positive) away from equilibrium and then released? Initially, there will be an upward resultant force on the mass, so it will accelerate upwards. Since downward movement has been chosen as positive, upward velocity and upward acceleration are both negative. When the mass reaches its equilibrium position (point A in Figure 6.5), x is zero and the resultant force and acceleration are zero as well. At this point, the velocity has its maximum upward (i.e. negative) value. This upward velocity carries the mass past this point and it moves above the equilibrium position. So x now becomes negative and the resultant force then acts downwards and slows the mass down (acceleration is downwards). At point B, the mass reaches its maximum upward (i.e. negative) displacement and is instantaneously at rest. Since x is still negative, the resultant force continues to act downwards. The mass passes back through the equilibrium position (point C), when it has its maximum downward (i.e. positive) velocity. Then x becomes positive once more and the upward resultant force slows the mass down until it reaches its original starting position (point D).

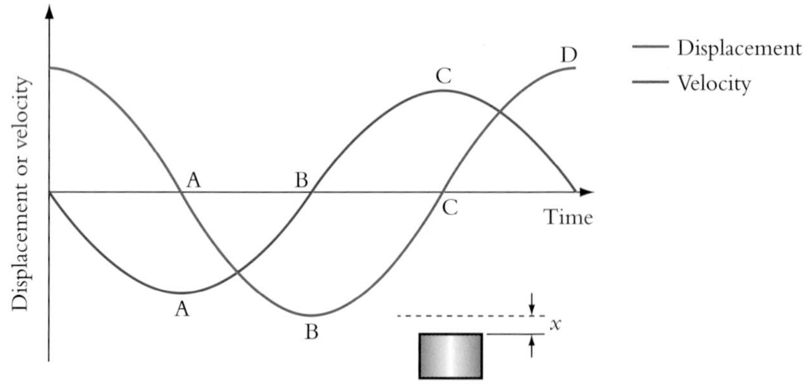

Figure 6.5

We see that the resultant force on the mass depends only on the displacement measured from the equilibrium position. Here, and in all other problems, the static forces in springs exactly balance any gravitational forces on the mass under equilibrium conditions. Because they always cancel each other, we ignore them and start at the equilibrium position and consider displacements away from that position.

Note: Throughout this unit, we will assume that displacements are small and that the spring stiffnesses are constant. This means that the force in a spring varies linearly with displacement. Such systems are called **linear systems**. If displacements are not small, the stiffness may vary with displacement, but this is not considered here.

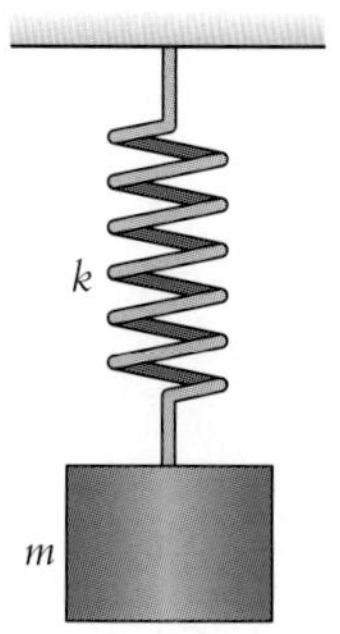

Figure 6.6 **Physical system**

To create a mathematical model of the problem, the first step is to replace the physical system (Figure 6.6) with a dynamic mass-spring model (Figure 6.7). In this simple case, the mass is represented by a square block and the coil spring by a zig-zag representation.

Step 1 Dynamic mass–spring model

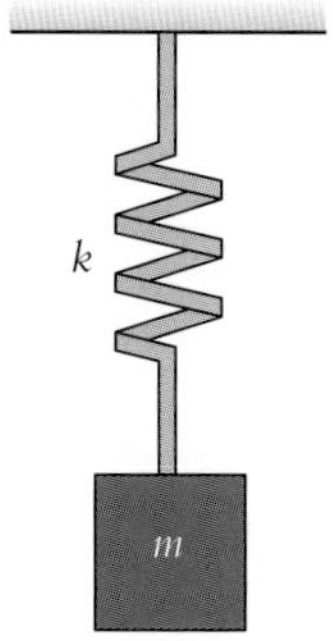

Figure 6.7

Step 2 Free-body diagram

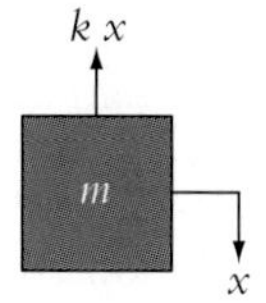

Figure 6.8

(i) Remove the spring to leave the mass by itself.

(ii) Mark the chosen positive direction for displacement.

(iii) Give the mass a positive displacement, write down the expression for the force and add it to the diagram to show its positive direction.

Step 3 Equation of motion

Applying Newton's second law, we write down the resultant force in the chosen direction of motion (downwards in this example) and equate to the mass multiplied by the acceleration. Hence,

$$-kx = m\ddot{x}$$

This can be rearranged to give an equation in the form of a second-order ordinary differential equation with constant coefficients.

$$m\ddot{x} + kx = 0 \tag{6.1}$$

Any sinusoidal function satisfies this equation, but from the earlier description of what happens to the mass when it is displaced downwards and then released, a suitable mathematical form would be $x(t) = X\cos\omega t$. This describes a sinusoidally varying displacement at frequency ω, with maximum deflections of X above and below the equilibrium position. The frequency of the vibration is called the **natural frequency** and is the characteristic quantity that we are trying to find. Sinusoidal displacement like this is often referred to as **simple harmonic motion**.

Differentiating $x(t) = X\cos\omega t$, we get $\dot{x} = -\omega X\sin\omega t$ and $\ddot{x} = -\omega^2 X\cos\omega t$.

Substitute for $x(t)$ and its derivatives into the equation of motion.

$$-m\omega^2 X\cos\omega t + kX\cos\omega t = 0$$

Cancelling $X\cos\omega t$,

$$-m\omega^2 + k = 0$$

The natural frequency for this system is therefore given by $\sqrt{\frac{k}{m}}$. The symbol, ω_n, is normally used for the natural frequency. When substituting numerical values into the equation of motion, ω must have the units of **rad/s** to make the equation consistent. However, the value would normally be quoted (in a report, for example) using the units of Hz (hertz, or cycles per second). The two are linked by the equation:

$$f_n[\text{Hz}] = \frac{\omega_n}{2\pi}\,[\text{rad/s}]$$

You will find that other systems have different equations of motion, but provided that we can assume that displacements remain small, all will have the same form, namely:

$$M\ddot{z} + Kz = 0$$

where z is the chosen motion coordinate. Following the above analysis, we find that the natural frequency is given by

$$\omega_n = \sqrt{\frac{K}{M}} \text{ [rad/s]}$$

Hence, as soon as you've derived the equation of motion and obtained the coefficients of the displacement and acceleration, you can immediately write down the expression for the natural frequency of a system.

Example 2: Vertical vibration of a block on a flexible cantilever beam

The system can be made to vibrate by lifting the block up (causing the supporting beam to bend) and then releasing it, resulting in an up and down vibration. The block will be treated as a rigid mass and the supporting beam as a massless spring. The **bending stiffness** of the beam is given by $k_B = \frac{3EI_2}{L^3}$. Note that the symbol I_2 in this formula is the second moment of area for the beam. Since displacements are assumed to be small, we will neglect the rotation of the block caused by the bending of the beam and consider only the vertical translation of the mass. The example on page 402 shows one way of including the effect of the rotation in the analysis.

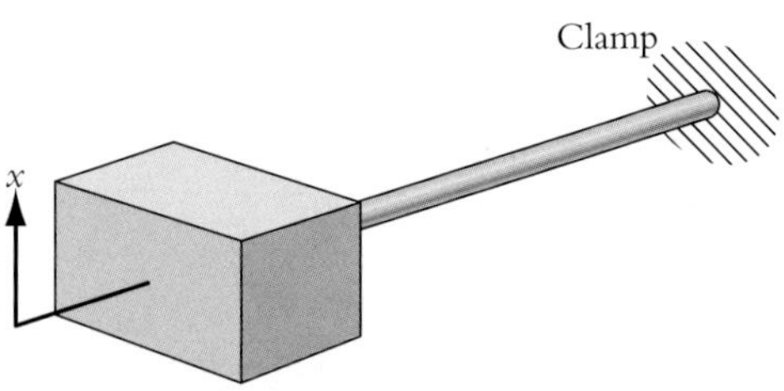

Figure 6.9 **Physical system**

Step 1 Dynamic model

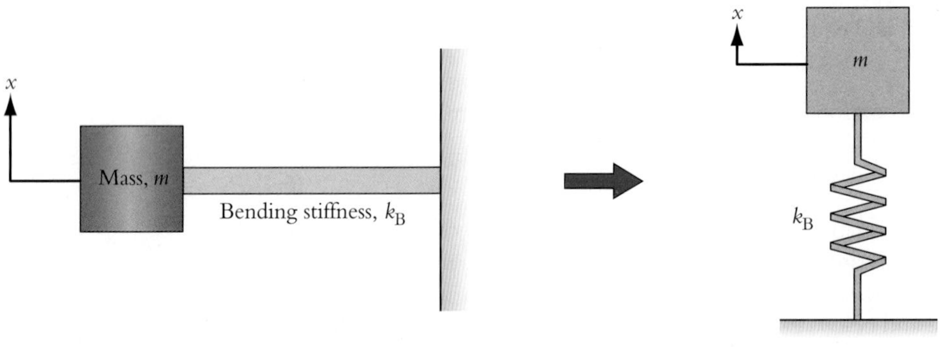

Figure 6.10

Dynamically, this case is identical to the first example and following the same procedure of free-body diagram and equation of motion, the natural frequency will be found to be

$$\omega_n = \sqrt{\frac{k_B}{m}} \tag{6.2}$$

Example 3: Torsional vibration of a block on a cantilever beam

The block and beam from the previous example can also be made to vibrate by rotating the block about the axis of the support beam and then releasing it, resulting in torsional vibration. In this case, the support beam can be modelled as a spring with **torsional stiffness** $k_T = \frac{GJ}{L}$ (with units of Nm/rad). The block can again be modelled as a rigid body, but with a moment of inertia I about the beam axis, which we will assume to be a fixed axis. Methods of calculating the moment of inertia of rigid bodies are covered in Unit 6 of Part 1.

Step 1 Dynamic model

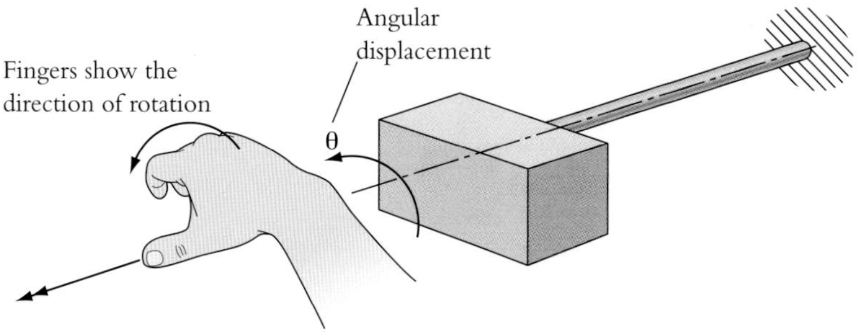
Angular
displacement
Fingers show the
direction of rotation
θ
Thumb points in
the vector direction

Figure 6.11

It is convenient to use rotational vector notation in this example. As shown in Figure 6.11, the angular displacement vector acts along the axis of rotation in a direction given by the 'right-hand rule'.

Step 2 Free-body diagram

If the block rotates by an angle θ, the induced torque in the support beam will be $k_T\theta$.

This will act in a direction to oppose the imposed rotation and to return the block to its equilibrium position.

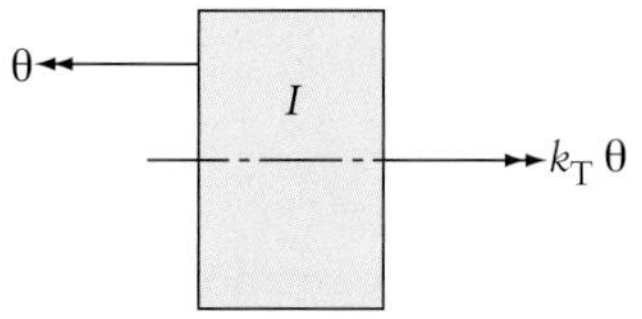
θ
I
$k_T\,\theta$

Figure 6.12

Step 3 Equation of motion

Applying Newton's second law for rotation, we write down the resultant torque in the chosen direction of motion (to the left in this example) and equate to the moment of inertia multiplied by the angular acceleration:

$$-k_T\theta = I\ddot{\theta}$$

Rearranging,

$$I\ddot{\theta} + k_T\theta = 0$$

From the coefficients in the equation of motion, the natural frequency for this system is:

$$\omega_n = \sqrt{\frac{k_T}{m}} \tag{6.3}$$

Example 4: Rocker system

This example consists of a rigid, massless bar with a fixed pivot at one end and a large mass attached at the other. The rocking motion about the pivot is restrained by two springs, one attached to the mass and the other that is connected to the bar at point B.

Note: We will use the angular displacement of the bar about the fixed pivot as the motion coordinate and assume that this displacement is small. This means that $\cos\theta \approx 1$ and $\sin\theta \approx \tan\theta \approx \theta$. Taken together with the earlier assumption that stiffness values are constant, the linear dependence of the spring forces on the motion coordinate, θ, will be maintained. As a result, the model will give an equation of motion that is a second order ordinary differential equation with constant coefficients.

Step 1 Dynamic model

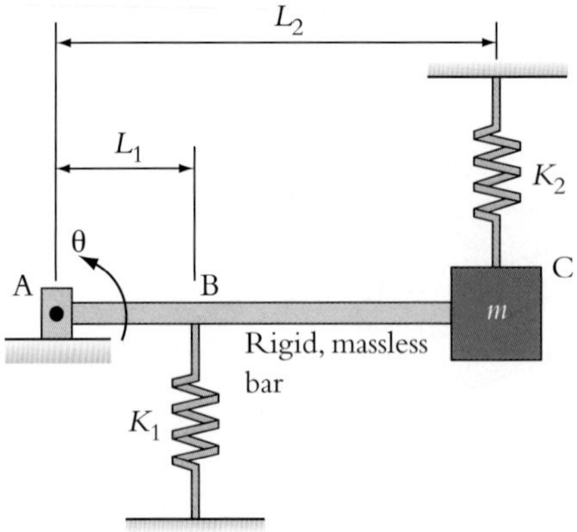

Figure 6.13

Step 2 Free-body diagram

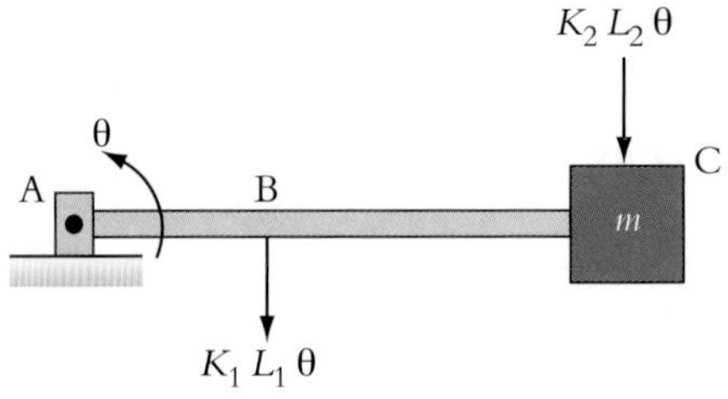

Figure 6.14

For a positive rotation of the bar, point B will move up by $L_1\theta$. This puts the spring K_1 into tension, resulting in a downward force on the bar. Point C will move up by $L_2\theta$, putting the spring K_2 into compression and resulting in a downward force on the bar.

Step 3 Equation of motion

Taking anti-clockwise moments of the forces about the fixed pivot at A, noting that the moment of inertia of the mass about the pivot is mL_2^2, we get:

$$-K_1L_1\theta \times L_1 - K_2L_2\theta \times L_2 = mL_2^2\ddot{\theta}$$

Rearranging,

$$mL_2^2\ddot{\theta} + (K_1L_1^2 + K_2L_2^2)\theta = 0$$

From the coefficients of $\ddot{\theta}$ and θ, the expression for the natural frequency is:

$$\omega_n = \sqrt{\frac{K_1L_1^2 + K_2L_2^2}{mL_2^2}} \qquad (6.4)$$

Lumped mass–spring systems

Many structures exhibit more than one mode of vibration, each with a different frequency. For example, the Tacoma Narrows bridge had become known for its bending vibration almost from the day it opened, but it was a torsional vibration excited by a particular speed of cross wind that cause its collapse.

Some structures can be approximated by several rigid bodies connected by massless springs. The analysis approach is an extension of the one used for single-degree-of-freedom systems, except that each of the rigid elements will have a separate free-body diagram. Each will also have its own equation of motion.

Example 1: Single-axle caravan

In this example, the body of the caravan is assumed to behave like a rigid mass and is connected to the axle (also assumed to be a rigid mass) by road springs in the suspension. The axle is separated from the road by flexible tyres. The mass of the tyres and the suspension springs is assumed to be negligible in comparison with the body and axle parts. A suitable dynamic model for studying vertical vibration is shown below.

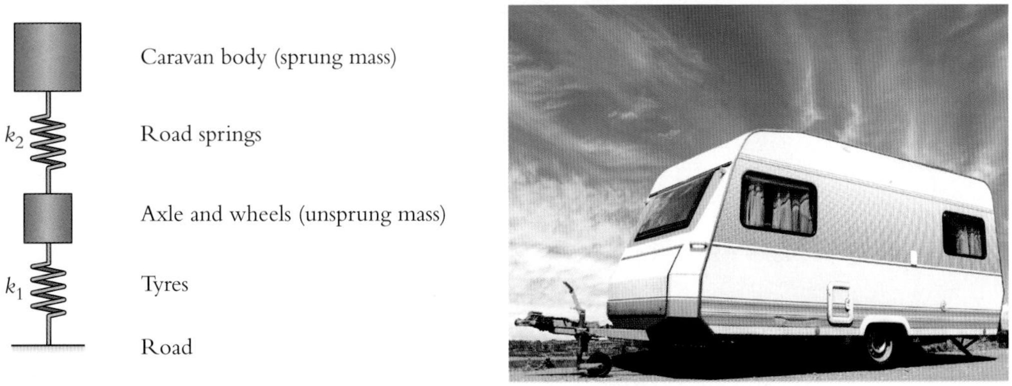

Figure 6.15

Since the body and the axle can move separately from each other, this model has two degrees of freedom (two independent possible motions). Two coordinates are needed to describe how the system moves: axle displacement and body displacement.

We will see that the model will predict two natural frequencies, each of which will have a characteristic pattern of displacement, called a **mode shape**.

Step 1 Dynamic mass–spring model

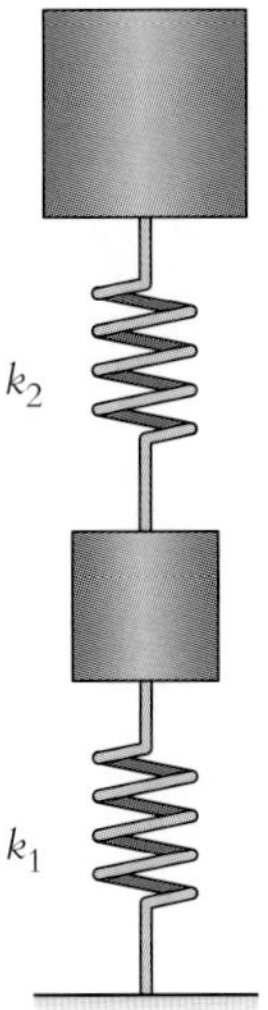

Figure 6.16

Step 2 Free-body diagrams

Upward (positive) displacement of mass m_1 will put the spring k_1 into tension, resulting in a downward force on the underside of the mass. The force in spring k_2 is given by the stiffness multiplied by the change of length due to the displacements x_1 and x_2. Since both have been chosen to be positive upwards, the net change of length is the difference between them, that is $(x_1 - x_2)$. The force in the spring is therefore $k_2(x_1 - x_2)$. This is positive whenever $x_1 > x_2$, that is, whenever the spring is in compression. The compressed spring will exert forces in the direction shown in Figure 6.17.

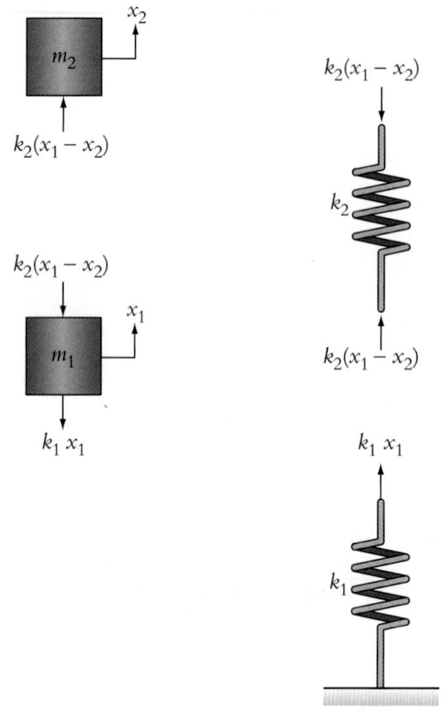

Figure 6.17

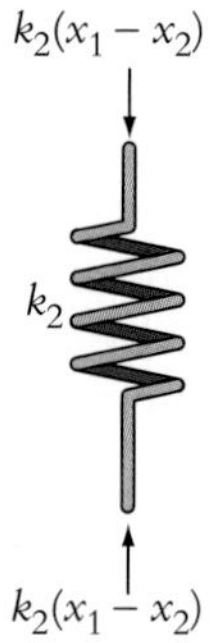

Figure 6.18 Case 1 Spring in compression $x_1 > x_2$

We could have written the net change of length of the spring as $(x_2 - x_1)$ to give the force expression as $k_2\,(x_2 - x_1)$. This is positive whenever $x_1 < x_2$, or whenever the spring is in tension. With the spring in tension, the directions of the forces are reversed.

The two cases are equivalent to each other, so both are correct.

Note: The free-body diagram shows a 'snapshot' of the system when x_1 and x_2 are both positive. The arrows on the diagram need to show the positive direction of the forces that each spring exerts on the masses. The expression for each force is given by

spring force = stiffness × change of length.

Take each spring in turn and ask yourself the following questions:

- What is the change of length of the spring?
- Is the spring in tension or compression?

For the change of length, look at the positive displacements of the two ends of the spring. If the direction of movement at both ends is the same (as it is in this example), the change of length will be the difference between the individual displacements. If, however, the chosen positive directions for the two ends of the spring are in opposite directions, the change of length will be the sum of the individual displacements. You need to be systematic when setting up the free-body diagrams.

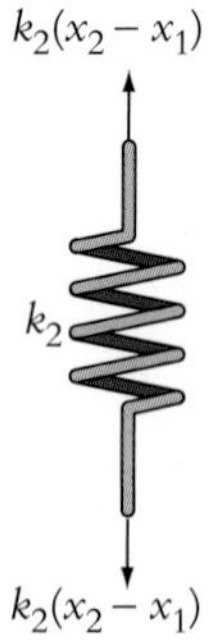

Figure 6.19 Case 2 Spring in tension $x_1 < x_2$

Step 3 Equations of motion

For mass m_1,

$$-k_1x_1 - k_2(x_1 - x_2) = m_1\ddot{x}_1$$

and for mass m_2,

$$+\,k_2(x_1 - x_2) = m_2\ddot{x}_2$$

or
$$m_1\ddot{x}_1 + (k_1 + k_2)x_1 - k_2x_2 = 0 \tag{6.5}$$

and
$$m_2\ddot{x}_2 - k_2x_1 + k_2x_2 = 0 \tag{6.6}$$

In matrix form,

$$\begin{bmatrix} m_1 & 0 \\ 0 & m_2 \end{bmatrix} \begin{Bmatrix} \ddot{x}_1 \\ \ddot{x}_2 \end{Bmatrix} + \begin{bmatrix} (k_1 + k_2) & -k_2 \\ -k_2 & k_2 \end{bmatrix} \begin{Bmatrix} x_1 \\ x_2 \end{Bmatrix} = \begin{Bmatrix} 0 \\ 0 \end{Bmatrix}$$

or

$$[\mathrm{M}]\{\ddot{x}\} + [K]\{x\} = \{0\}$$

where [M] and [K] are called the mass and stiffness matrices for the system.

As with single-degree-of-freedom systems, we can check for errors in the equations at this stage. In particular,

(1) the terms in the leading diagonals of [M] and [K] are *always positive;*

(2) the off-diagonal terms may be positive or negative;

(3) [M] and [K] are *often symmetric* about the leading diagonal.

The equations are **coupled**: each involves both x_1 and x_2. Physically, the coupling tells us that if one mass moves, the other mass will also move; push the top mass down and the lower mass moves as well. Put another way, motion of one mass cannot occur independently of the other. Mathematically, the coupling means that the equations must be solved simultaneously.

For free vibration of the system at one of its natural frequencies, the motion of each mass will be sinusoidal. Use as substitutions, $x_1(t) = X_1 \cos \omega t$ and $x_2(t) = X_2 \cos \omega t$, where X_1 and X_1 are the amplitudes of vibration of each mass.

Substituting into equations (6.5) and (6.6) and cancelling $\cos \omega t$, which is a common factor in all terms, gives

$$-m_1\omega^2X_1 + (k_1 + k_2)X_1 - k_2 = 0$$

and

$$-m_2\omega^2X_2 - k_2X_1 + k_2X_2 = 0$$

or in matrix form

$$\begin{bmatrix} (k_1 + k_2) - m_1\omega^2 & -k_2 \\ -k_2 & k_2 - m_2\omega^2 \end{bmatrix} \begin{Bmatrix} X_1 \\ X_2 \end{Bmatrix} = \begin{Bmatrix} 0 \\ 0 \end{Bmatrix} \tag{6.7, 6.8}$$

or
$$([K] - \omega^2[M])\{X\} = \{0\} \tag{6.9}$$

or just

$$[Z]\{X\} = \{0\}$$

where

$$[Z] = [K] - \omega^2[M]$$

Equation (6.9) is in the form of an eigenvalue problem; commonly presented in maths textbooks in the form:

$$([A] - \lambda[B])\{X\} = \{0\}$$

In our vibration problem, the eigenvalues give the natural frequencies and the eigenvectors give the corresponding mode shapes.

There are many methods of solving eigenvalue problems. In practice, most engineers will turn to software such as Matlab, which has the necessary functions. A Matlab script for doing this is provided on the book's website: www.hodderplus.co.uk/mechanicalengineering. For problems with only two or three degrees of freedom, however, the following procedure can be used.

For a non-trivial solution of equation (6.9),

$$\det[Z] = 0$$

Multiplying out the determinant in this case gives

$$m_1 m_2 \omega^4 - (m_1 k_2 + m_2(k_1 + k_2))\omega^2 + k_1 k_2 = 0 \qquad (6.10)$$

Equation (6.10) is called the **Frequency equation**. For this problem, it is a quadratic in ω^2 and will have two roots, ω_{n1}^2 and ω_{n2}^2, where ω_{n1} and ω_{n2} are the two natural frequencies of the system and this gives us the first part of the information we are looking for.

To find the corresponding mode shapes, we substitute each root back into equation (6.7) or (6.8) to get the relationship between X_1 and X_2. Since equations (6.7) and (6.8) form a pair of homogeneous equations, we cannot find unique solutions for X_1 and X_2 separately.

One way of resolving this is to make the amplitude of one mass equal to unity and then find the amplitude of the other mass relative to this. For example, if we choose $X_2 = 1$, equation (6.8) becomes:

$$(k_1 + k_2 - \omega^2 m_1)X_1 - k_2 \cdot 1 = 0$$

Hence,

$$\begin{Bmatrix} X_1 \\ X_2 \end{Bmatrix} = \begin{Bmatrix} \dfrac{k_2}{k_1 + k_2 - \omega^2 m_1} \\ 1.0 \end{Bmatrix} \qquad (6.11)$$

The vector $\begin{Bmatrix} X_1 \\ X_2 \end{Bmatrix}$ is the required **mode shape**, which tells us the relative amplitude and phase of the two masses.

Here is a numerical example for the caravan problem. Let $m_1 = 70\,\text{kg}$, $m_2 = 350\,\text{kg}$, $k_1 = 400\,\text{N/m}$ and $k_2 = 40\,\text{N/m}$

Natural frequencies

Substituting into equation (6.10) and solving gives $\omega_{n1}^2 = 103.7\,\text{s}^{-2}$ and $\omega_{n2}^2 = 6296.3\,\text{s}^{-2}$.

Hence, $\omega_{n1} = 10.18\,\text{rad/s} = 1.62\,\text{Hz}$ and $\omega_{n2} = 79.35\,\text{rad/s} = 12.63\,\text{Hz}$

Mode shapes

Mode 1

Put $\omega_{n1}^2 = 38.1\,\text{s}^{-2}$ into equation (6.11) to give:

$$\begin{Bmatrix} X_1 \\ X_2 \end{Bmatrix} = \begin{Bmatrix} 0.092 \\ 1.0 \end{Bmatrix}$$

In this mode, the axle and the body vibrate in phase with each other (X_1 and X_2 are both positive values) and the amplitude of the displacement of the body is about 10 times that of the axle. This is because the tyres are relatively stiff compared with the springs in the suspension. It's a low frequency vibration (1.62 Hz) and is commonly called the 'bounce' mode of the vehicle. A Matlab script is included on the book's website; that shows an animated version of the mode shapes, and Figure 6.20 is a still from the animation showing the positions of the two masses at the instant that mass m_2 is at its uppermost displaced position.

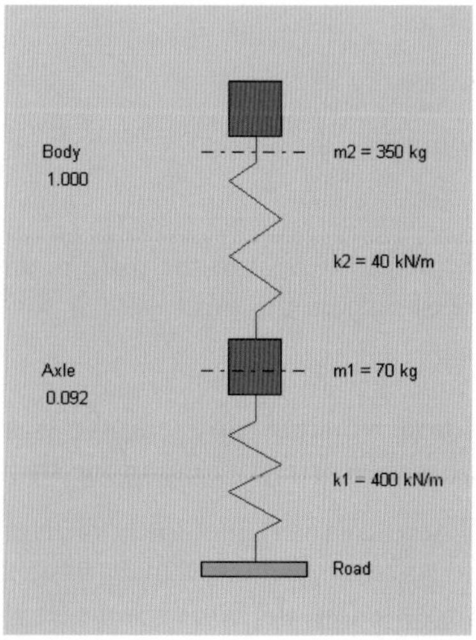

Figure 6.20

Since the axle movement is small compared with that of the axle in this mode, we could have made the assumption that the tyres were effectively rigid, implying that the axle didn't move. This would have given a single-degree-of-freedom system with the body mass supported only by the suspension springs. The natural frequency for this system is 1.70 Hz, which is similar to the figure of 1.62 Hz obtained above.

Mode 2

Put $\omega_{n2}^2 = 261.8\,\text{s}^{-2}$ into equation (6.11) to give:

$$\begin{Bmatrix} X_1 \\ X_2 \end{Bmatrix} = \begin{Bmatrix} -54.09 \\ 1.0 \end{Bmatrix}$$

In this mode, the axle and the body vibrate out of phase with each other (X_1 and X_2 have opposite signs), and the amplitude of the displacement of the axle is over 50 times that of the body. It's called a 'wheel-hop' mode and is at a much higher frequency (12.6 Hz) than the bounce mode.

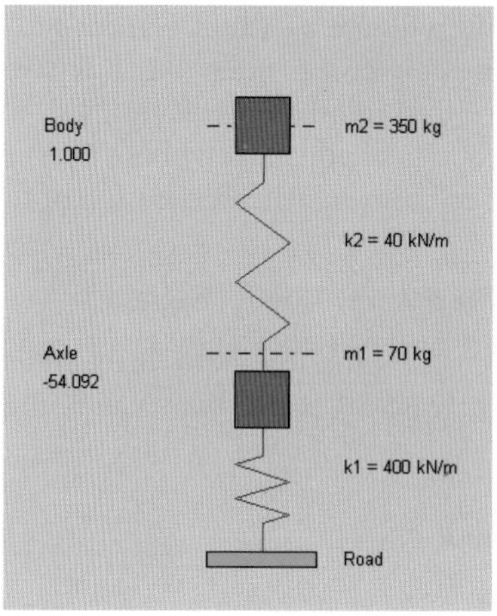

Figure 6.21

Example 2: 2D vehicle model (coupled bounce and pitch)

The previous worked example showed that the bounce mode of the caravan could be estimated by assuming that the tyres were rigid. The second example applies this observation to a four-wheeled vehicle to study the combined effects of bounce and pitch.

Step 1 The following assumptions will be made in developing the dynamic model:

(1) Roll motion is not considered – pitch and vertical translation only.

(2) The body is rigid, with mass, m, and moment of inertia, I_G about its centre of mass.

(3) The tyres are very stiff, so that the axles do not move.

(4) k_A and k_B are the combined stiffnesses for the front and rear springs respectively.

(5) The shock absorbers are ignored.

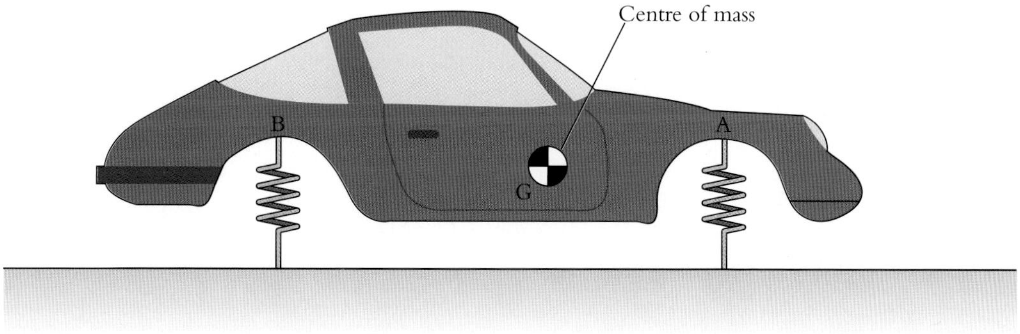

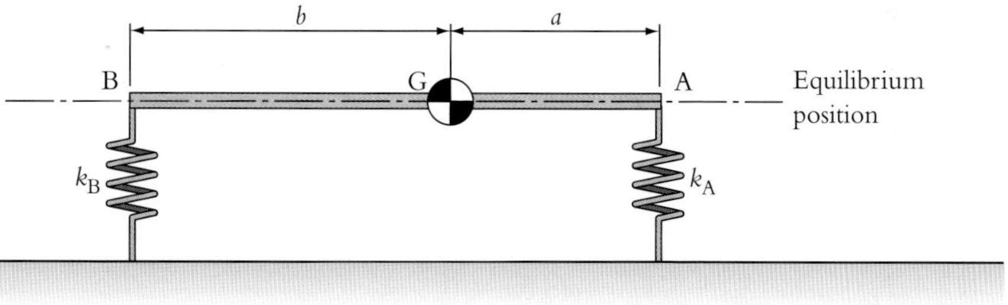

Figure 6.22

A and B (the attachment points for the suspension springs) along with the centre of mass, G, are marked on the dynamic model, which is shown in its equilibrium position.

We will use x_G (displacement of G) together with θ (pitch angle) as motion coordinates. These are convenient since they link directly with the two equations of motion for the body. The vertical translation equation will need the absolute acceleration of the centre of mass and the rotational equation will use G as the axis (there is no fixed axis on the car body) and we will need the angular acceleration of the body.

Step 2 Free-body diagram

To draw the free-body diagram, it has been assumed that x_G and θ are such that both springs increase in length. For small θ, the increases in spring lengths are:

$$x_A = x_G - a\theta \tag{6.12}$$

and

$$x_B = x_G + b\theta$$

With both springs in tension, the positive directions of the forces on the body are as shown in Figure 6.23.

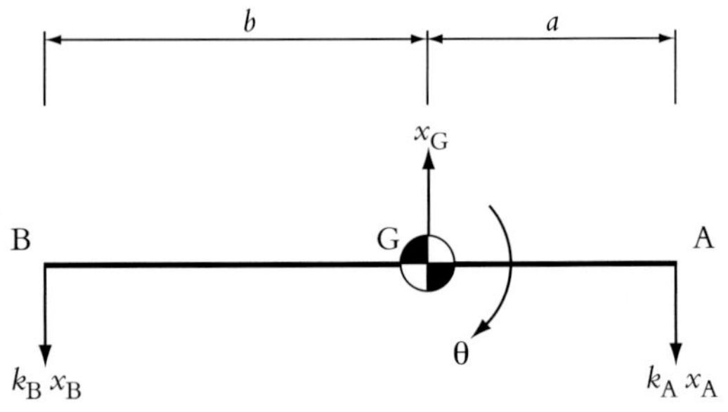

Figure 6.23

Step 3 Equations of motion

For vertical translation of the centre of mass,

$$-k_A(x_G - a\theta) - k_B(x_G + b\theta) = m\ddot{x}_G$$

For rotation about the centre of mass,

$$+k_A(x_G - a\theta)\cdot a - k_B(x_G + b\theta)\cdot b = I_G\ddot{\theta}_G$$

Rearranging,

$$m\ddot{x}_G + (k_A + k_B)x_G + (bk_B - ak_A)\theta = 0$$

and

$$I_G\ddot{\theta} + (bk_B - ak_A)x_G + (a^2k_A + a^2k_B)\theta = 0$$

In matrix form,

$$\begin{bmatrix} m & 0 \\ 0 & I_G \end{bmatrix}\begin{Bmatrix} \ddot{x}_G \\ \ddot{\theta} \end{Bmatrix} + \begin{bmatrix} k_A + k_B & bk_B - ak_A \\ bk_B - ak_A & a^2k_A + b^2k_B \end{bmatrix}\begin{Bmatrix} x_G \\ \theta \end{Bmatrix} = \begin{Bmatrix} 0 \\ 0 \end{Bmatrix}$$

For a solution, put $x_G(t) = X_G \cos \omega t$ and $\theta(t) = \Theta \cos \omega t$.

This gives,

$$\begin{bmatrix} k_A + k_B - m\omega^2 & bk_B - ak_A \\ bk_B - ak_A & a^2k_A + b^2k_B - I_G\omega^2 \end{bmatrix}\begin{Bmatrix} X_G \\ \Theta \end{Bmatrix} = \begin{Bmatrix} 0 \\ 0 \end{Bmatrix} \tag{6.13}$$

or

$$[Z]\begin{Bmatrix} X_G \\ \Theta \end{Bmatrix} = \begin{Bmatrix} 0 \\ 0 \end{Bmatrix} \tag{6.14}$$

To find the natural frequencies, solve the frequency equation given by $\det[Z] = 0$.

The mode shapes can be found by substituting the natural frequencies into either equation (6.13) or (6.14). Using equation (6.13) for example:

$$(k_A + k_B - m\omega^2)X_G + (bk_B - ak_A)\Theta = 0$$

Let $\Theta = 1$ rad to give:

$$\begin{Bmatrix} X_G \\ \Theta \end{Bmatrix} = \begin{Bmatrix} (ak_A - bk_B)/(k_A + k_B - m\omega^2) \\ 1.0 \end{Bmatrix} \tag{6.15}$$

Worked example

$$\mathbf{m = 900\,kg} \qquad \mathbf{I_G = 1000\,kg\,m^2}$$

$$\mathbf{k_A = 25\,kN/m} \qquad \mathbf{k_B = 10\,kN/m}$$

$$\mathbf{a = 1\,m} \qquad \mathbf{b = 2\,m}$$

Natural frequencies

The frequency equation becomes

$$\begin{vmatrix} 35 \times 10^3 - 900\omega^2 & -5 \times 10^3 \\ -5 \times 10^3 & 25 \times 10^3 + 40 \times 10^3 - 1000\omega^2 \end{vmatrix} = 0$$

or

$$\begin{vmatrix} 35 - 0.9\omega^2 & -5 \\ -5 & 65 - \omega^2 \end{vmatrix} = 0$$

Expanding, $$0.9\omega^4 - 93.5\omega^2 + 2250 = 0$$

Roots are $\omega_{n1}{}^2 = 37.8\,\text{s}^{-2}$ and $\omega_{n2}{}^2 = 66.0\,\text{s}^{-2}$

Hence, $\omega_{n1} = 0.98\,\text{Hz}$ and $\omega_{n2} = 1.29\,\text{Hz}$

Mode shapes

Mode 1: Put $\omega_{n1}{}^2 = 37.8\,\text{s}^{-2}$ into equation (6.15) to give $\begin{Bmatrix} x_G \\ \Theta \end{Bmatrix} = \begin{Bmatrix} 5.43 \\ 1.0 \end{Bmatrix} \begin{matrix} \text{m} \\ \text{rad} \end{matrix}$

To visualize the mode shape, $\begin{Bmatrix} X_A \\ X_B \end{Bmatrix}$ is more convenient. This ratio can be obtained by using equations (6.12). Thus,

$$\begin{Bmatrix} X_A \\ X_B \end{Bmatrix} = \begin{Bmatrix} X_G - a\Theta \\ X_G + b\Theta \end{Bmatrix} = \begin{Bmatrix} 5.43 - 1 \times 1 \\ 5.43 + 2 \times 1 \end{Bmatrix} = \begin{Bmatrix} 4.43 \\ 7.43 \end{Bmatrix}$$

Normalizing to make $X_B = 1.0$, we get:

$$\begin{Bmatrix} X_A \\ X_B \end{Bmatrix} = \begin{Bmatrix} 0.60 \\ 1.0 \end{Bmatrix}$$

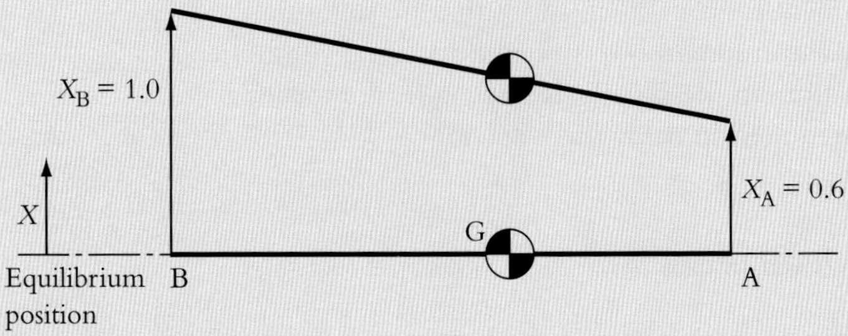

Figure 6.24

This 'snapshot' of the mode at its extreme position shows a predominant bounce motion, with a little pitching.

Mode 2: Put $\omega_{n2}^2 = 66.0\,\mathrm{s}^{-2}$ into equation (6.15) to give $\begin{Bmatrix} x_G \\ \Theta \end{Bmatrix} = \begin{Bmatrix} -0.205 \\ 1.0 \end{Bmatrix} \begin{matrix} \mathrm{m} \\ \mathrm{rad} \end{matrix}$

or from equation (6.12), $\begin{Bmatrix} X_A \\ X_B \end{Bmatrix} = \begin{Bmatrix} -0.67 \\ 1.0 \end{Bmatrix}.$

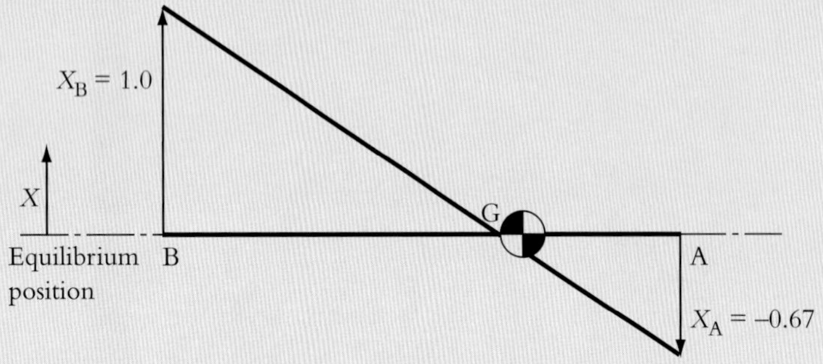

Figure 6.25

In this case, we see mainly pitching motion, with only a small displacement at the centre of mass.

If the car was driven over a road with a gradually undulating profile (such as produced by ground subsidence), the car will tend to bounce up and down with mode 1 dominating the response. Alternatively, if the front wheels hit an obstruction in the road (for example, the speed bumps used in traffic calming schemes in urban areas), it's likely that a pitching response dominated by mode 2 would result. We will see in Section 6.4 that the response of a structure can be written as a combination of its modes of vibration, so in the general case, the response will be a mixture of modes 1 and 2.

Torsional systems

Most mechanical power is transmitted via rotating shafts. Examples include the turbine–alternator sets in power stations, the tail rotor drive in a helicopter and the propeller shaft in a ship's propulsion system. One of the problems for the vibration engineer is that power surges or sudden changes in load can induce transient torsional oscillations that are superimposed on the uniform rotation. Also, the drive torque can sometimes contain fluctuating components that can set up steady state torsional oscillations in the system. In each case, the torsional natural frequencies and modes shapes are required to study the behaviour.

Example: Main drive shaft of a gas turbine engine

There are two drive shafts in the V2500 engine, which powers the majority of Airbus 320 aircraft (and several others) currently in service. The main shaft transmits power from a five-stage turbine at the rear to the fan and the first three compressor stages. The whole assembly rotates freely, supported by three bearings. To study the torsional behaviour, the dynamic model will assume that the fan and turbine assemblies can be treated as rigid elements and that the mass of the shaft can be neglected.

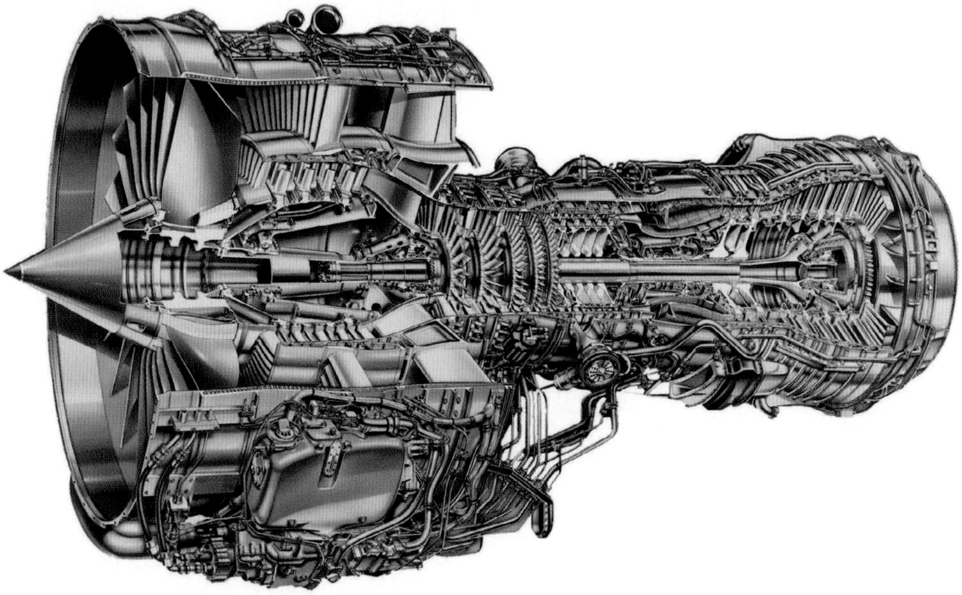

Figure 6.26 **V2500 twin-shaft aero engine**

Step 1 Dynamic mass–spring model

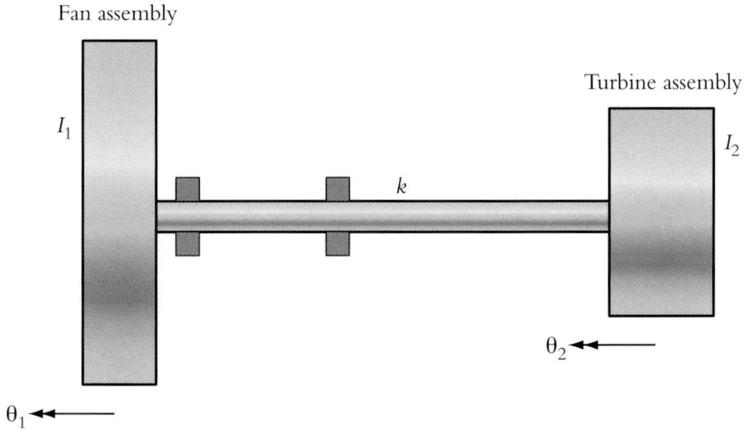

Figure 6.27

The engine rotates anti-clockwise when viewed from the front, so this has been chosen for the positive direction of rotation for the two motion coordinates, θ_1 and θ_2.

Step 2 Free-body diagrams

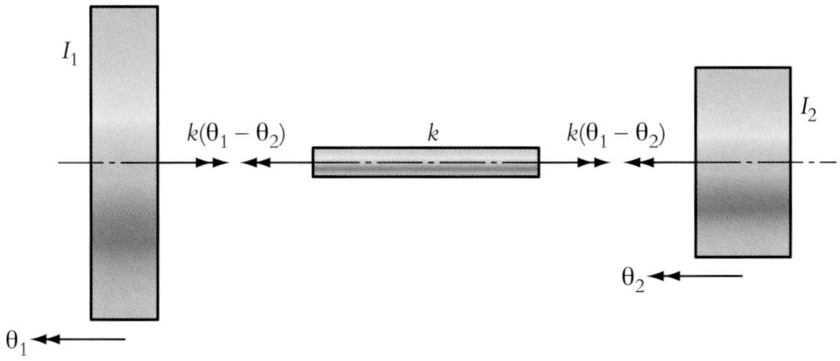

Figure 6.28

If both motion coordinates are given a positive rotation, the twist in the shaft will be the difference between the two. Hence the expression for the torque in the shaft can be written as: $k(\theta_1 - \theta_2)$. This is positive whenever $\theta_1 > \theta_2$ and results in the torque directions shown in Figure 6.28.

Step 3 Equations of motion

For the fan, $$-k(\theta_1 - \theta_2) = I_1\ddot{\theta}_1$$

For the turbine, $$+k(\theta_1 - \theta_2) = I_2\ddot{\theta}_2$$

Rearranging, $$I_1\ddot{\theta}_1 + k\theta_1 - k\theta_2 = 0$$

and $$I_2\ddot{\theta}_2 - k\theta_1 + k\theta_2 = 0$$

In matrix form,

$$\begin{bmatrix} I_1 & 0 \\ 0 & I_2 \end{bmatrix}\begin{Bmatrix} \ddot{\theta}_1 \\ \ddot{\theta}_2 \end{Bmatrix}\begin{bmatrix} k & -k \\ -k & k \end{bmatrix}\begin{Bmatrix} \theta_1 \\ \theta_2 \end{Bmatrix} = \begin{Bmatrix} 0 \\ 0 \end{Bmatrix}$$

Substitute $\theta_1(t) = \Theta_1 \cos \omega t$ and $\theta_2(t) = \Theta_2 \cos \omega t$ to give

$$\begin{bmatrix} k - I_1\omega^2 & -k \\ -k & k - I_2\omega^2 \end{bmatrix}\begin{Bmatrix} \Theta_1 \\ \Theta_2 \end{Bmatrix} = \begin{Bmatrix} 0 \\ 0 \end{Bmatrix} \qquad \begin{matrix}(6.16)\\(6.17)\end{matrix}$$

Expanding the determinant of the coefficient matrix, the frequency equation is

$$I_1 I_2 \omega^4 - k(I_1 + I_2)\omega^2 = 0$$

This has, as its roots, $\omega_{n1}{}^2 = 0$ and $\omega_{n2}{}^2 = \dfrac{k(I_1 + I_2)}{I_1 I_2}$

To find the mode shapes, substitute roots into equation (6.16) or (6.17). Using equation (6.17) and putting $\Theta_2 = 1$,

$$\begin{Bmatrix} \Theta_1 \\ \Theta_2 \end{Bmatrix} = \begin{Bmatrix} (k - I_2\omega^2)/k \\ 1.0 \end{Bmatrix} \qquad (6.18)$$

For mode 1, $\omega_{n1} = 0$ and $\begin{Bmatrix} \Theta_1 \\ \Theta_2 \end{Bmatrix} = \begin{Bmatrix} 1 \\ 1 \end{Bmatrix}$. This describes a **rigid body mode**, which in this case tells us that the rotor is able to rotate continuously in one direction, with no twisting of the shaft ($\Theta_1 = \Theta_2$).

In all cases, a rigid body mode is characterized by a natural frequency equal to zero and by a mode shape in which there is no deformation of any of the parts; in other words, the system moves as if it were a single rigid body.

Any structure that is capable of moving without deformation (this is true of any structure not connected to the ground) *will* have one (or more) rigid body modes with $\omega_n = 0$. It follows that the frequency equation will not contain a constant term. Since you can tell in advance that this should be the case, it's a useful check that the frequency equation is correct.

For mode 2, $\omega_{n2} = \sqrt{\dfrac{k(I_1 + I_2)}{I_1 I_2}}$ and $\begin{Bmatrix} \Theta_1 \\ \Theta_2 \end{Bmatrix} = \begin{Bmatrix} -I_2/I_1 \\ 1.0 \end{Bmatrix}$

This mode has a non-zero frequency and the fan and turbine vibrate 180° out of phase with each other, resulting in significant alternating twisting of the shaft.

Shaft whirl and critical speeds

Shaft whirl is a potentially destructive, self-sustaining flexural vibration observed in rotating shafts. It occurs if the rotational frequency of the shaft coincides with a resonant frequency for flexural vibration. These shaft speeds are called **critical speeds**. A given shaft will be designed to operate with some maximum speed. Ideally, if this maximum design speed is less than the lowest critical speed, whirl will not be a problem. Unfortunately, this is not always possible and it is vital to be able to calculate what the critical speeds will be. For example, the need to minimize the mass of the drive shafts in an aero engine means that they may have critical speeds within the operating range.

To examine the characteristics of shaft whirl behaviour, we will model the shaft as a 'beam' with a circular cross section in order to find its natural frequencies in flexure. The analysis below shows that shafts potentially have an infinite number of flexural natural frequencies, which means that they have an infinite number of critical speeds.

Flexural vibration of uniform beams

Apart from shafts, many structures exhibit beam-like vibrational behaviour. Examples include aircraft wings, helicopter rotor blades and suspension bridge decks (all of which can vibrate in response to aerodynamic interaction) and tall buildings, which vibrate significantly during earthquakes. While these are more complex than uniform beams, they exhibit many of the same characteristics and this section will provide some insight into this behaviour.

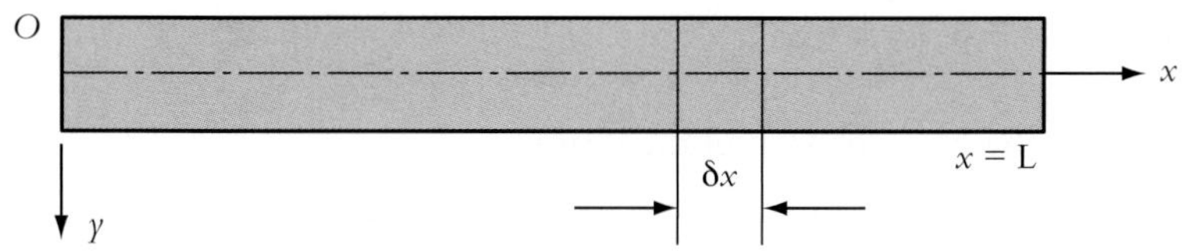

Figure 6.29

Unlike previous cases, a beam does not consist of discrete masses connected by massless springs. Both mass and stiffness are distributed along the length. A different approach is required and we start by considering the free-body diagram of an infinitesimal element of the beam of length δx.

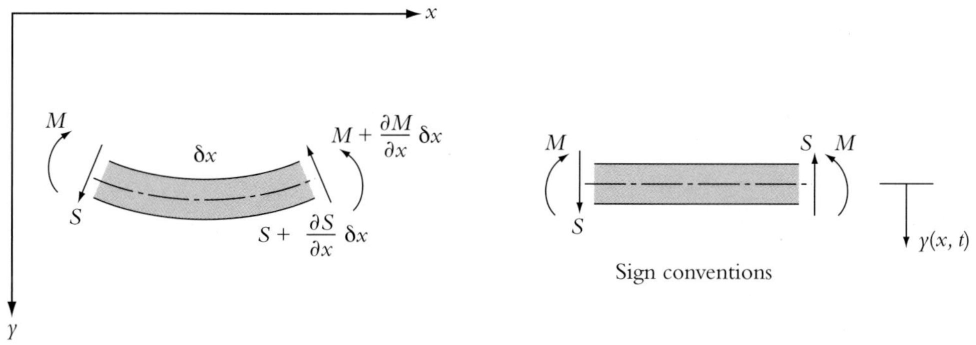

Figure 6.30

The equations derived and used in this section depend on the use of a standard sign convention (shown in Figure 6.30) that defines the positive directions of the transverse displacement, $y(x, t)$, and of the shear force and bending moment. It is the same sign convention as is used in Unit 3.

Bending moment – curvature relationship is

$$M = -EI\frac{\partial^2 y}{\partial x^2} \tag{6.19}$$

Equation for vertical motion (downward motion is positive):

$$S - \left(S + \frac{\partial S}{\partial x}\delta x\right) = (\rho A \delta x)\frac{\partial^2 y}{\partial t^2}$$

$$\therefore \qquad \frac{\partial S}{\partial x} = -\rho A \frac{\partial^2 y}{\partial x^2} \qquad \textbf{(6.20)}$$

To simplify the analysis, we will neglect the rotational inertia of the element and the shear deformation of planes of cross section. Both assumptions tend to give an over-estimate of the natural frequencies of the beam. While this error is normally small for the first few modes, it increases progressively when higher frequencies are evaluated.

The equation for rotational motion about an axis through the centre of mass of the element is:

$$S\frac{\delta x}{2} + \left(S + \frac{\partial S}{\partial x}\delta x\right)\frac{\delta x}{2} - M + \left(M + \frac{\partial M}{\partial x}\delta x\right) = 0$$

$$\therefore \qquad S = -\frac{\partial M}{\partial x} \qquad \textbf{(6.21)}$$

Substituting for M from (6.19) into equation (6.21) and then for S in equation (6.20) we get

$$EI\frac{\partial^4 y}{\partial x^4} = -\rho A \frac{\partial^2 y}{\partial t^2} \qquad \textbf{(6.22)}$$

This is the governing differential equation for the free vibration of the beam.

Equation (6.22) is a partial differential equation giving the deflection, y, which is a function of space x and time t. The objective in solving the equation will be to find the natural frequencies and the corresponding mode shapes.

For free vibration at a natural frequency, the displacement of any point on the beam in the y-direction will be sinusoidal, but the amplitude of the vibration will vary along the length. We can therefore use as a substitution,

$$y(x, t) = Y(x)\cos\omega t$$

Substituting into (6.22), we get

$$EI\frac{d^4 Y}{dx^4}\cos\omega t = \rho A \omega^2 Y(x)\cos\omega t$$

$$\therefore \qquad \frac{d^4 Y}{dx^4} = \frac{\rho A \omega^2}{EI} Y(x)$$

For a uniform cross section, A and I are constant and it is convenient to introduce the so-called **wavenumber**, λ, defined by

$$\lambda^4 = \frac{\rho A \omega^2}{EI} \qquad \textbf{(6.23)}$$

to give

$$\frac{d^4 Y}{dx^4} = \lambda^4 Y(x)$$

Take as solution:

$$Y(x) = Ae^{\alpha x}$$

Thus,

$$\alpha^4 Ae^{\alpha x} = \lambda^4 Ae^{\alpha x}$$

or

$$\alpha^4 = \lambda^4$$

so that

$$\alpha = \pm\lambda \quad \text{or} \quad \pm\mathbf{i}\lambda$$

The complete solution for $Y(x)$ is therefore

$$Y(x) = A_1 \mathrm{e}^{\lambda x} + A_2 \mathrm{e}^{-\lambda x} + A_3 \mathrm{e}^{\mathrm{i}\lambda x} + A_4 \mathrm{e}^{-\mathrm{i}\lambda x}$$

which may be rewritten to give the more convenient form,

$$Y(x) = C_1 \sin \lambda x + C_2 \cos \lambda x + C_3 \sinh \lambda x + C_4 \cosh \lambda x \quad (6.24)$$

This is a **general equation** giving the deflected shape of any beam of uniform cross section. The constants $C_1 - C_4$ need to be determined from the boundary conditions at the ends of the beam. Four basic types of support are shown in Table 6.1 and other types of boundary conditions are considered later.

Descriptive terms	Diagrammatic	Boundary conditions
Built-in clamped encastré		$y = 0 \quad \frac{\partial y}{\partial x} = 0$
Simple support hinged pinned	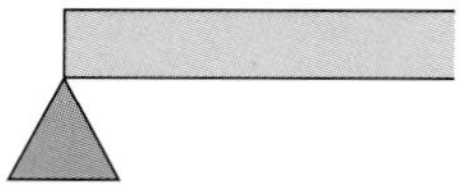	$y = 0$ $M = 0 \quad \therefore \frac{\partial^2 y}{\partial x^2} = 0$
Free		$M = 0 \quad \therefore \frac{\partial^2 y}{\partial x^2} = 0$ $S = 0 \quad \therefore \frac{\partial^3 y}{\partial x^3} = 0$
Massless slider	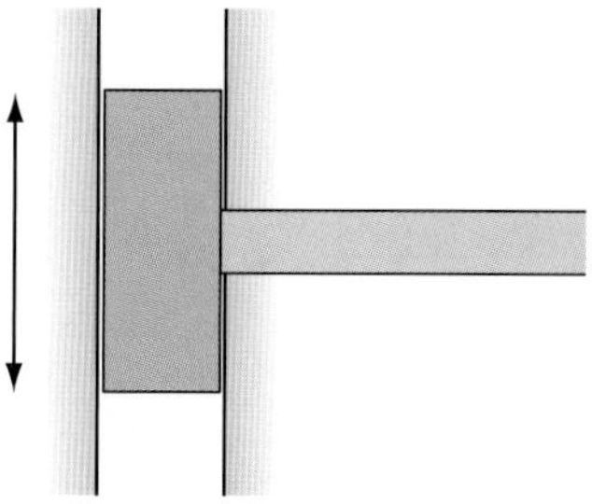	$\frac{\partial y}{\partial x} = 0$ $S = 0 \quad \therefore \frac{\partial^3 y}{\partial x^3} = 0$

Table 6.1 Boundary conditions for basic types of support

The following steps should be used to find the solutions for particular cases:

Step 1 Start by identifying the four boundary conditions. Use $y(x,t) = Y(x) \cos \omega t$, with equation (6.24) to express the boundary condition in terms of $Y(x)$ and its derivatives.

Step 2 Since each of the boundary condition equations depends on $C_1 - C_4$, they can be assembled in the form

$$[Z]\{C\} = \{0\} \quad (6.25)$$

where $\{C\}$ is a vector of the constants $C_1 - C_4$ and $[Z]$ is a coefficient matrix.

Step 3 For a valid solution, det $[Z] = 0$.

This gives the **frequency equation** and its roots will give the natural frequencies of the beam.

Step 4 When each root is substituted back into equation (6.25), the solution vector $\{C\}$ will define the corresponding **mode shape** when the values are put into equation (6.24).

Pinned–pinned	$\sin \lambda L = 0$
Clamped–clamped & free–free*	$\cos \lambda L \cosh \lambda L - 1 = 0$
Clamped–pinned & free–pinned*	$\tan \lambda L - \tanh \lambda L = 0$
Clamped–free	$\cos \lambda L \cosh \lambda L + 1 = 0$

Table 6.2 Frequency equation for particular end conditions

r	1	2	3	4	5	>5
Pinned–pinned	π	2π	3π	π	5π	$r\pi$
Clamped–clamped & free–free*	4.730	7.853	10.996	14.137	17.279	$\approx (r + 0.5)\pi$
Clamped–pinned & free–pinned*	3.927	7.096	10.210	13.351	16.493	$\approx (r + 0.25)\pi$
Clamped–free	1.875	4.694	7.855	10.996	14.137	$\approx (r - 0.5)\pi$

Table 6.3 Numerical values of roots, $\lambda_r L$, of frequency equations.

*A free-free beam will also have two rigid body modes and a free-pinned beam one rigid body mode, each corresponding to $\lambda L = 0$.

Selecting the values of $\lambda_r L$ from Table 6.3 for the beam of interest, the natural frequencies can be found from equation (6.23). That is: $\omega_r = \left(\frac{\lambda_r L}{L}\right)^2 \sqrt{\frac{EI}{\rho A}}$

Example 1: Simply supported beam

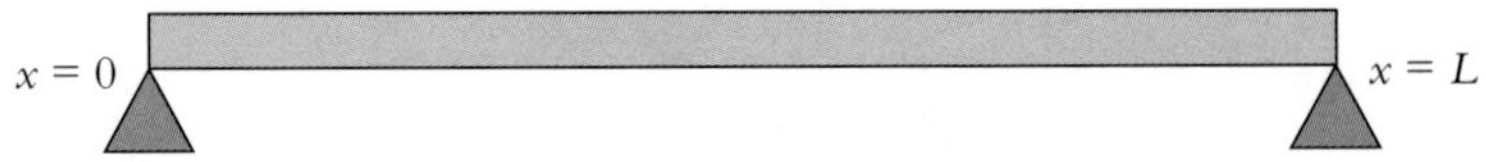

Figure 6.31

Step 1
The boundary conditions at $x = 0$ and at $x = L$ are:

$$y = 0 \text{ and } M = 0, \text{ the latter implying that } \frac{\partial^2 y}{\partial x^2} = 0.$$

Since $y(x,t) = Y(x) \cos \omega t$, the boundary conditions become

$$Y = 0 \quad \text{and} \quad \frac{d^2Y}{dx^2} = 0$$

From equation (6.24),

$$Y(x) = C_1 \sin \lambda x + C_2 \cos \lambda x + C_3 \sinh \lambda x + C_4 \cosh \lambda x$$

$$\frac{d^2Y}{dx^2} = -\lambda^2 C_1 \sin \lambda x - \lambda^2 C_2 \cos \lambda x + \lambda^2 C_3 \sinh \lambda x + \lambda^2 C_4 \cosh \lambda x$$

Note that $\frac{d}{d\theta} \sinh \theta = \cosh \theta$ and $\frac{d}{d\theta} \cosh \theta = \sinh \theta$

Hence, at $x = 0$

$$Y(0) = C_1 \times 0 + C_2 \times 1 + C_3 \times 0 + C_4 \times 1 = 0$$

$$\left(\frac{d^2Y}{dx^2}\right)_{x=0} = -\lambda^2 C_1 \times 0 - \lambda^2 C_2 \times 1 + \lambda^2 C_3 \times 0 + \lambda^2 C_4 \times 1 = 0$$

and at $x = L$

$$Y(L) = C_1 \sin\lambda L + C_2 \cos\lambda L + C_3 \sinh\lambda L + C_4 \cosh\lambda L = 0$$

$$\frac{d^2Y}{dx^2} = -\lambda^2 C_1 \sin\lambda L - \lambda^2 C_2 \cos\lambda L + \lambda^2 C_3 \sinh\lambda L + \lambda^2 C_4 \cosh\lambda L = 0$$

Step 2

Assembling the four equations in matrix form gives:

$$\begin{bmatrix} 0 & 1 & 0 & 1 \\ 0 & -\lambda^2 & 0 & \lambda^2 \\ \sin\lambda L & \cos\lambda L & \sinh\lambda L & \cosh\lambda L \\ -\lambda^2\sin\lambda L & -\lambda^2\cos\lambda L & \lambda^2\sinh\lambda L & \lambda^2\cosh\lambda L \end{bmatrix} \begin{Bmatrix} C_1 \\ C_2 \\ C_3 \\ C_4 \end{Bmatrix} = \begin{Bmatrix} 0 \\ 0 \\ 0 \\ 0 \end{Bmatrix} \quad \begin{matrix} (a) \\ (b) \\ (c) \\ (d) \end{matrix} \qquad (6.26)$$

Step 3

This is the particular form of equation (6.25) for a simply supported beam. Expanding the determinant of the coefficient matrix and equating to zero gives the **frequency equation**, so-called because the wavenumber is directly related to the frequency via equation (6.23).

$$-4\lambda^4 \sin\lambda L \sinh\lambda L = 0$$

To identify the roots of the frequency equation, we note that a simply supported beam cannot deflect without bending. In other words, it has no freedom to move as a rigid body and will not, therefore, have a rigid-body mode of vibration. From equation (6.23), it follows that $\lambda = 0$ will not be a root of the frequency equation. Not only is $\lambda \neq 0$, but $\sinh\lambda L \neq 0$.

The frequency equation therefore reduces to

$$\sin\lambda L = 0$$

which has roots $\lambda_r L = r\pi$ for $r = 1, 2, 3, \ldots$

From equation (6.23), the natural frequencies are $\omega_r = \left(\frac{r\pi}{L}\right)^2 \sqrt{\frac{EI}{\rho A}}$ for $r = 1, 2, 3, \ldots$

Step 4

To find the corresponding mode shapes, substitute the roots into equation (6.26) and solve for the constants $C_1 - C_4$.

From (6.26a), $$C_2 + C_4 = 0$$

From (6.26b), $$\lambda_r^2(-C_2 + C_4) = 0$$

Since $\lambda_r \neq 0$, it follows that $$C_2 = C_4 = 0$$

From (6.26c), $$\sin\lambda_r L \cdot C_1 + \sinh\lambda_r L \cdot C_3 = 0$$

Since $\sin\lambda_r L = 0$ (the frequency equation) and $\sinh\lambda_r L \neq 0$, $C_3 = 0$.

The only non-zero constant is C_1. Its value is arbitrary and we normally choose $C_1 = 1$. From equation (6.24), the mode shape expression becomes:

$$Y_r(x) = \sin\lambda_r x = \sin\frac{r\pi x}{L}, \text{ with } r = 1, 2, 3, \ldots$$

The first three mode shapes are as follows (there are animated versions on the book's supporting website):

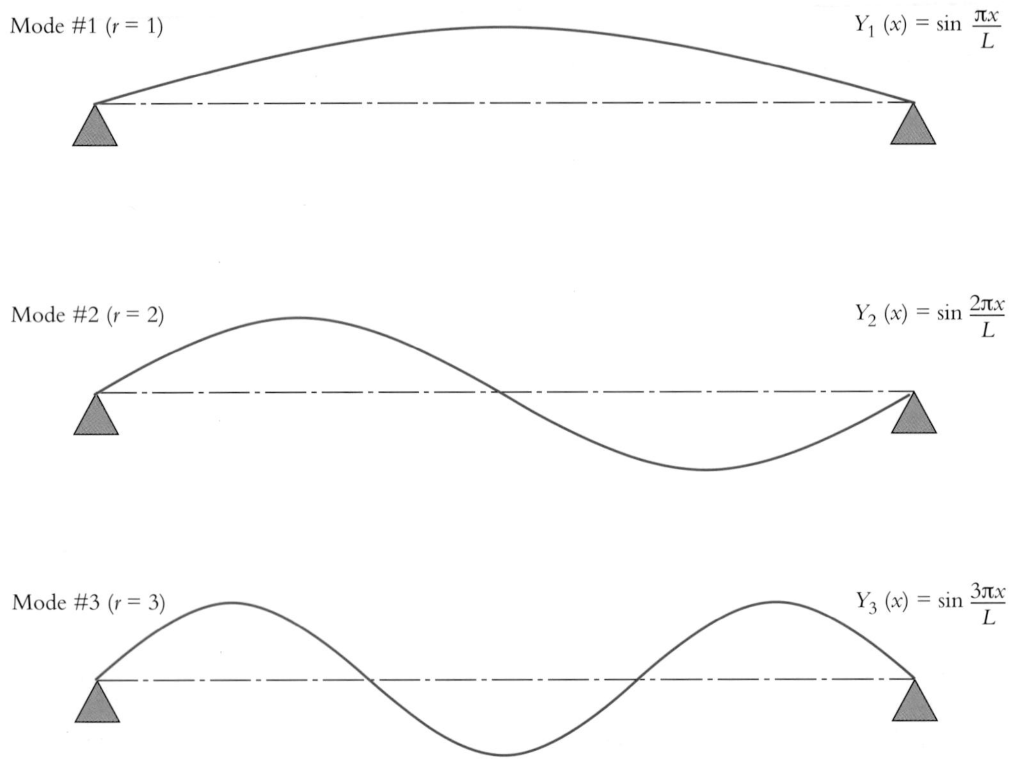

Figure 6.32

Example 2: Vibration of a cantilever (clamped–free) beam

Figure 6.33

Consider a cantilever that is clamped at $x = 0$ and free at $x = L$.

Step 1

The boundary conditions are:

At $x = 0$, $\quad y = 0$ and $\frac{\partial y}{\partial x} = 0$

At $x = L$, $\quad M = 0 \quad \therefore \quad \frac{\partial^2 y}{\partial x^2} = 0$

and $\quad S = 0 \quad \therefore \quad \frac{\partial^3 y}{\partial x^3} = 0$

Since $y(x,t) = Y(x)\cos \omega t$, the end conditions become:

At $x = 0$, $\quad Y = 0$ and $\frac{dY}{dx} = 0$

At $x = L$, $\quad \frac{d^2Y}{dx^2} = 0$ and $\frac{d^3Y}{dx^3} = 0$

Step 2

Substituting from equation (6.24) we get (in matrix form),

$$\begin{bmatrix} 0 & 1 & 0 & 1 \\ \lambda & 0 & \lambda & 0 \\ -\lambda^2 \sin\lambda L & -\lambda^2 \cos\lambda L & \lambda^2 \sinh\lambda L & \lambda^2 \cosh\lambda L \\ -\lambda^3 \sin\lambda L & -\lambda^3 \cos\lambda L & \lambda^3 \sinh\lambda L & \lambda^3 \cosh\lambda L \end{bmatrix} \begin{Bmatrix} C_1 \\ C_2 \\ C_3 \\ C_4 \end{Bmatrix} = \begin{Bmatrix} 0 \\ 0 \\ 0 \\ 0 \end{Bmatrix} \begin{matrix} (a) \\ (b) \\ (c) \\ (d) \end{matrix} \qquad (6.27)$$

This is the particular version of equation (6.25) for a cantilever beam.

Step 3

The **frequency equation** is given by setting the determinant of the coefficients of $C_1 - C_4$ to zero. After some manipulation (and noting that a cantilever has no rigid-body modes), this gives

$$1 + \cos\lambda L \cosh\lambda L = 0$$

There are no closed-form solutions to this equation, so the roots $\lambda_r L$ must be obtained numerically and are given in Table 6.3 As before, the natural frequencies can be found using equation (6.23), which is the definition of the wavenumber, λ.

Step 4

The **mode shapes** are obtained by substituting $\lambda = \lambda_r$ into equation (6.27) and solving for the constants $C_1 - C_4$.

From (6.27a) and (6.27b) $\qquad C_3 = -C_1$ and $C_4 = -C_2$

Thus from (6.27c) or (6.27d)

$$C_2 = -\frac{\sin\lambda_r L + \sinh\lambda_r L}{\cos\lambda_r L + \cosh\lambda_r L} C_1$$

$$= \sigma_r C_1$$

This gives C_2, C_3 and C_4 in terms of C_1, an arbitrary constant.

If we choose $C_1 = 1$, the mode shape becomes

$$Y_r(x) = \sin\lambda_r x - \sinh\lambda r x + \sigma_r(\cos\lambda_r x - \cosh\lambda r x)$$

When each value of λ_r is used in this equation, a different deflected shape is obtained.

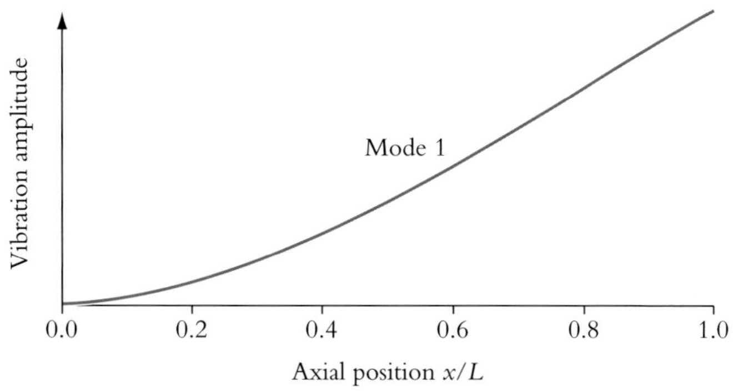

Figure 6.34

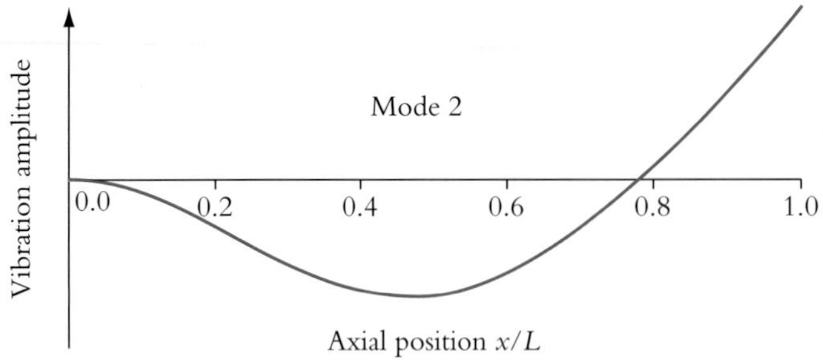

Figure 6.35

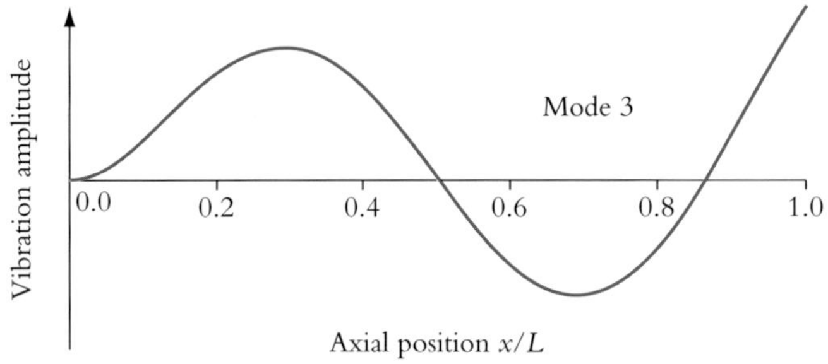

Figure 6.36

Other boundary conditions

Example: Cantilever beam with a mass at the free end

Figure 6.37

Step 1

The boundary conditions at the clamped end are identical to the previous case. So, $Y = 0$ and $\frac{\mathrm{d}Y}{\mathrm{d}x} = 0$ at $x = 0$.

However, at $x = L$, $S \neq 0$ and $M \neq 0$. To look at the effect that the mass has on the vibration of the beam, we use two of the basic principles of mechanics. These are

(1) compatibility of displacements
(2) equilibrium of forces and moments.

Consider first the shear force reaction between the beam and the mass.

The principle of **compatibility** determines that the displacement at the end of the beam is the same as the displacement of the mass.

The principle of **equilibrium of forces** determines that the shear force on the beam is equal and opposite to the force on the mass.

The free-body diagram is:

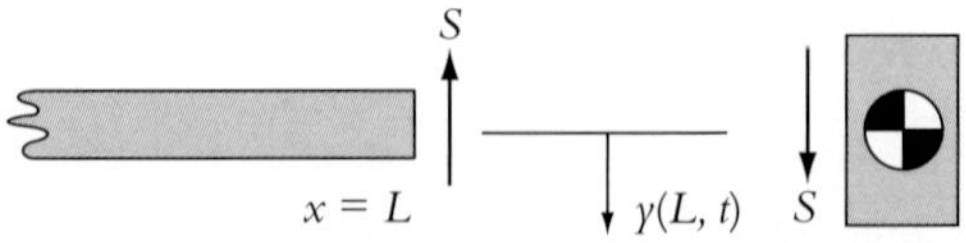

Figure 6.38

Note the positive directions for the displacement coordinate and the shear force on the beam defined by the sign convention adopted here.

We can write separate statements relating to the beam and the mass, which can then be combined using the statement of equilibrium of forces.

For the beam

$$S(t) = EI\left(\frac{\partial^3 y}{\partial x^3}\right)_{x=L}$$

But $y(x,t) = Y(x)\cos\omega t$ ⟶

$$S(t) = EI\left(\frac{d^3 Y}{dx^3}\right)_{x=L}\cos\omega t$$

For the mass

$$S(t) = m\left(\frac{\partial^2 y}{\partial x^2}\right)_{x=L}$$

$$S(t) = m \times (-\omega^2 Y(L)\cos\omega t)$$

Equating the shear force expressions and noting that $\omega^2 = \dfrac{EI\lambda^4}{\rho A}$,

$$\left(\frac{d^3 Y}{dx^3}\right)_{x=L} + \frac{m(\lambda L)^4}{\rho A L^4}Y(L) = 0 \qquad (6.28)$$

Now consider the bending moment reaction between the beam and the mass.

The principle of **compatibility** determines that the slope at the end of the beam is the same as the rotation of the mass.

The principle of **equilibrium of moments** determines that the bending moment on the beam is equal and opposite to the bending moment on the mass.

The free-body diagram is:

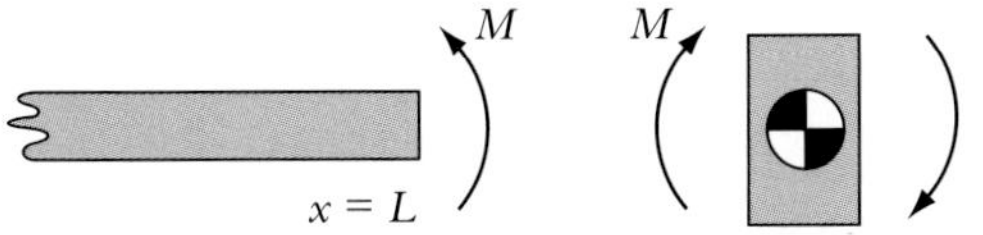

Slope $\theta = \left[\dfrac{\partial y}{\partial x}\right]_{x=L}$

Figure 6.39

For the beam

$$M(t) = -EI\left(\frac{\partial^2 y}{\partial x^2}\right)_{x=L}$$

But $y(x,t) = Y(x)\cos\omega t$

$$M(t) = -EI\left(\frac{d^2 Y}{dx^2}\right)_{x=L}\cos\omega t$$

For the mass

The rotational equation of motion for the mass is

$$M(t) = I_M\left(\frac{\partial^2\theta}{\partial t^2}\right)_{x=L}$$

$$\text{Slope, } \theta(t) = \frac{\partial y}{\partial x} = \left(\frac{dY}{dx}\right)\cos\omega t$$

$$\text{Therefore } \frac{\partial^2\theta}{\partial t^2} = -\omega^2\left(\frac{dY}{dx}\right)\cos\omega t$$

Substituting into the equation of motion,

$$M(t) = -I_M\,\omega^2\left(\frac{dY}{dx}\right)_{x=L}\cos\omega t$$

Equating the bending moment expressions,

$$\left(\frac{d^2 Y}{dx^2}\right)_{x=L} - \frac{I_M(\lambda L)^4}{\rho A L^4}\left(\frac{dY}{dx}\right)_{x=L} = 0 \qquad (6.29)$$

Collecting the four boundary condition equations together, we have

$$Y(0) = 0 \qquad \text{(a)}$$

$$\left(\frac{\mathrm{d}Y}{\mathrm{d}x}\right)_{x=0} = 0 \qquad \text{(b)}$$

$$\left(\frac{\mathrm{d}^3Y}{\mathrm{d}x^3}\right)_{x=L} + \frac{m(\lambda L)^4}{\rho A L^4} Y(K) = 0 \qquad \text{(c)}$$

$$\left(\frac{\mathrm{d}^2Y}{\mathrm{d}x^2}\right)_{x=L} - \frac{I_M(\lambda L)^4}{\rho A L^4}\left(\frac{\mathrm{d}Y}{\mathrm{d}x}\right)_{x=L} = 0 \qquad \text{(d)}$$

Step 2
To set up equation (6.25) for this system, substitute for $Y(x)$ and its derivatives from equation (6.24).

Steps 3 and 4 then follow as in the previous examples.

Other structures with distributed mass and stiffness

The models presented in the earlier parts of this section are based on a range of assumptions that may or may not be acceptable in particular cases. An aircraft wing, for example, will exhibit bending modes of vibration, so the study of uniform beams gives some insight into the behaviour that can be expected. However, a wing does not have a uniform cross section, so the analysis will not be able to predict the natural frequencies of this structure accurately.

For this and many other structures, engineers will invariably turn to the finite element method. There is an introduction to finite elements in Unit 3 related to obtaining displacements and stresses due to static loads. This involves creating and solving an equation of the form:

$$[K]\{X\} = \{P\}$$

$[K]$ is a global stiffness matrix formed by combining the individual stiffness matrices of the finite elements chosen to discretize the structure, and $\{P\}$ is a vector of applied loads. $\{X\}$ is the solution vector giving the displacements at the nodes.

As well as solving static problems, most commercial finite element codes can also solve vibration problems by calculating a mass matrix in addition to the stiffness matrix. This can be used to set up and solve an eigenvalue problem in the form:

$$([K] - \omega^2[M])\{X\} = \{0\}$$

This can be solved to give the natural frequencies and mode shapes for the structure.

Learning summary

By the end of this section you should have learnt:

- ✓ a systematic approach for setting up the equations of motion for single-degree-of-freedom and lumped mass–spring systems using the following three steps

  Step 1 convert the physical structure into a dynamic mass–spring model
  Step 2 draw free-body diagram(s)
  Step 3 apply the appropriate form(s) of Newton's second law of motion to give the equation(s) of motion for the system;
- ✓ to obtain the natural frequency of single-degree-of-freedom systems from the coefficients in the equation of motion;
- ✓ the equations of motion for lumped mass–spring systems and how to obtain the natural frequencies and the corresponding mode shapes;
- ✓ to set up the boundary condition equations for shaft/beam vibration problems and obtain the frequency equation and an expression for the mode shapes.

6.3 Response of damped single-degree-of-freedom systems

Introduction

So far, we have looked at free vibration of systems and at their natural frequencies and mode shapes. This section will analyse the response of single-degree-of-freedom systems to external excitation. This takes the form either of applied forces and/or moments or of imposed displacement on part of the system.

We will also introduce the effects of damping. **Damping** is the phenomenon that dissipates energy in a structure. If there was no damping and a structure was set vibrating and then left, the mathematics would suggest that it would carry on vibrating forever. This, of course, is impossible and the structure would stop vibrating sooner or later. This is because all real structures dissipate energy to a greater or lesser extent. There are many ways this can take place and mechanisms include hysteresis effects in the material, friction between parts, aerodynamic interaction with the surrounding fluid and noise radiated from the surfaces. Most real damping mechanisms are difficult to handle mathematically and we will consider one theoretical damping model, called **viscous damping** and will only consider discrete dampers. These can be pictured as a piston–cylinder device in which a viscous fluid is displaced from one side of the piston to the other through a constriction such as an orifice. Vehicle shock absorbers are normally based on this idea.

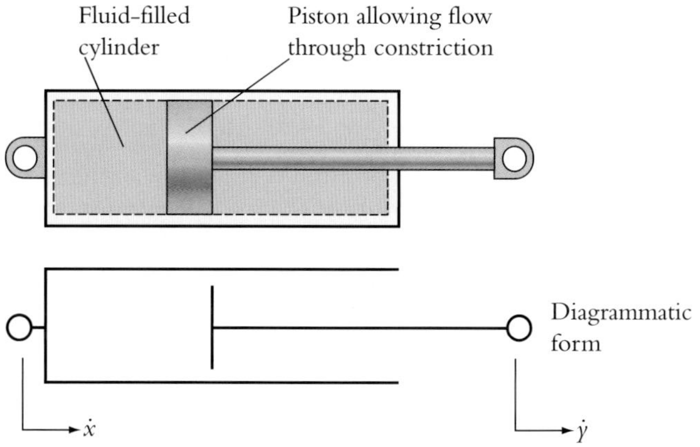

Figure 6.40

In the viscous damping model, we assume that the force in the damper is proportional to the relative velocity between the ends and acts to oppose the imposed motion. The constant of proportionality is called the damping coefficient (normally given the symbol c) and has units of N/(m/s) [or Ns/m].

Hence the force opposing the motion is $c\,(\dot{x} - \dot{y})$.

Note that dampers do not impose any stiffness on the structure; they only transmit a force if there is relative motion between the ends. If there is no motion, there is no force.

In most engineering structures, the level of damping is low. As a result, any discrepancies between the assumed viscous damping model and the actual damping mechanism are generally small, so that the error introduced by our mathematical model is also small.

Equations of motion

In the examples that follow, the steps leading to the equation of motion are the same as presented in Section 6.2 with the addition of any damping forces and any external excitation. We will then look at the solution to the equation of motion for a few types of excitation.

Example 1: Mass-spring-damper system

Step 1 Dynamic mass–spring model

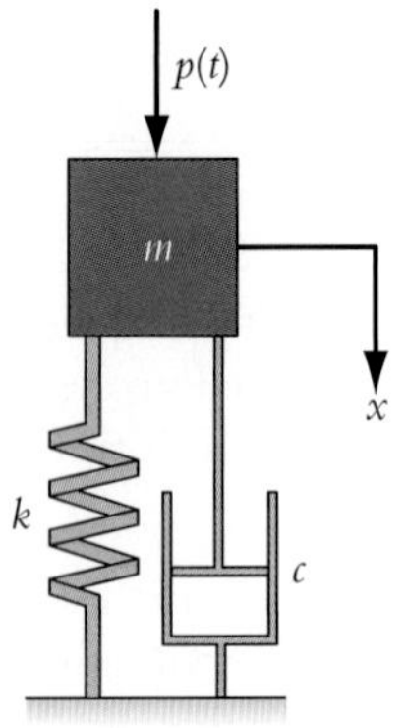

Figure 6.41

Step 2 Free-body diagram

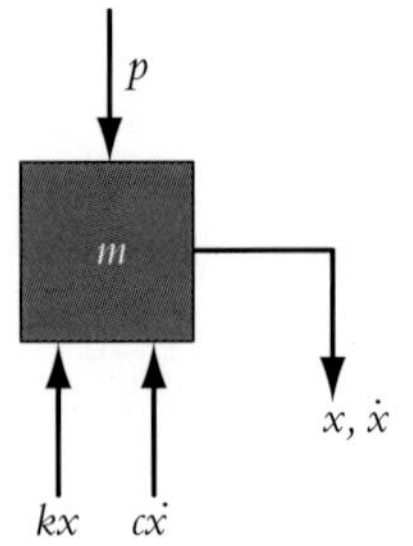

Figure 6.42

Positive (downward) displacement of the mass from its equilibrium position will put the spring into compression, resulting in an upward force acting on the underside of the mass. Positive (downward) velocity of the mass will compress the damper. The damper will oppose this motion, resulting in an upward force acting on the underside of the mass.

Step 3 Equation of motion

$$p - kx - c\dot{x} = m\ddot{x}$$

or

$$m\ddot{x} + c\dot{x} + kx = p(t) \tag{6.30}$$

Example 2: Rocker system

Step 1 Dynamic mass–spring model

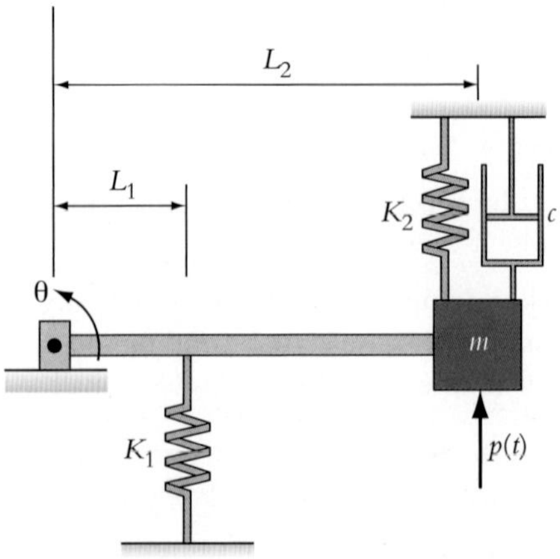

Figure 6.43

Step 2 Free-body diagram (for small displacements)

This is similar to the rocker system given as an example in Section 6.2, but with the addition of the damper and the applied force.

For the damper, a positive (anti-clockwise) angular velocity will produce upward velocity of the mass and cause the damper to compress. The damper will oppose this motion, resulting in a downward force acting on the top of the mass.

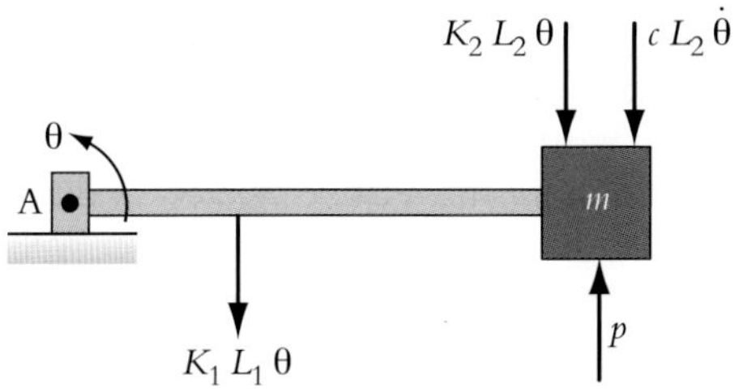

Figure 6.44

Step 3 Equation of motion
Taking moments about the fixed pivot at A, the equation of motion is:

$$-K_1L_1\theta \cdot L_1 - K_2L_2\theta \cdot L_2 - cL_2\dot{\theta} \cdot L_2 + p \cdot L_2 = I_A\ddot{\theta}$$

Again modelling the rocker as a mass on the end of a massless bar, the moment of inertia of the rocker about the pivot is $I_A = mL_2^2$.

Hence, on rearranging,

$$mL_2^2\ddot{\theta} + cL_2^2\dot{\theta} + (K_1L_1^2 + K_1L_2^2)\theta = p(t)L_2 \qquad (6.31)$$

Example 3: Single-axle caravan

We will make the following simplifying assumptions:

(1) The tyres are very stiff compared to the suspension springs (typically they are about ten times stiffer).

(2) The tyres do not lose contact with the road. Taken with (1), this means that vertical motion of the axle will follow the road profile exactly.

(3) The caravan body behaves as a rigid mass.

(4) Only vertical motion of the body is considered; pitching and rolling are ignored.

Step 1 Dynamic mass–spring model

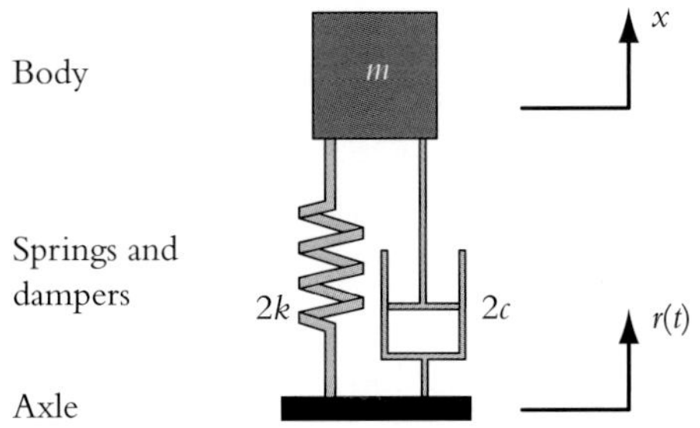

Figure 6.45

Step 2 Free-body diagram

This is an example where the excitation is in the form of a prescribed **displacement** instead of a force. With the assumption that the tyres are rigid, the road profile over which the caravan travels exactly defines the motion (displacement and velocity) of the axle. Any difference between the relative displacement or velocity of body and axle will give rise to forces in the springs and dampers that will react onto the underside of the body.

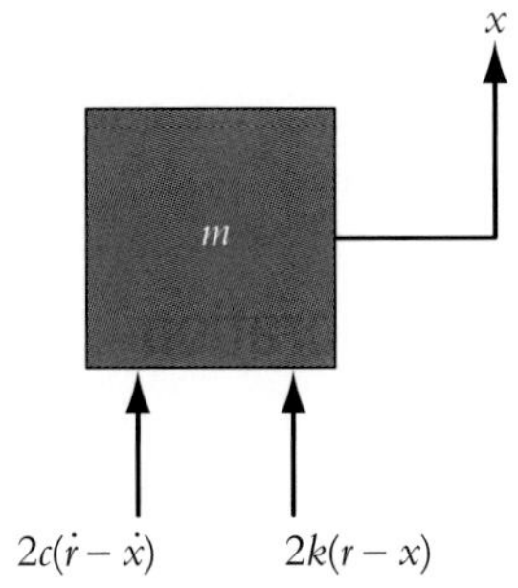

Figure 6.46

Recall that the free-body diagram is a snapshot of a system when all motions (both displacements and velocities) are positive. It shows the positive directions of the forces that the springs and dampers exert on the mass. The force in the spring is given by ***Spring force*** = ***Stiffness*** × ***Change of length***. In this case, the change of length of the spring is the difference between the body and axle displacements, which we can write as $(r - x)$. This is positive whenever the axle displacement is greater than the body displacement. In this situation, the spring will be in compression and will produce an upward force on the body.

For the damper, ***Damper force* = *Damping coefficient* × *Relative velocity***. The relative velocity can be written as $(\dot{r} - \dot{x})$. With positive (upward) velocities of both axle and body, the expression for the relative velocity will be positive if the axle velocity is greater than the body velocity. In this situation, the damper will be compressing (i.e. getting shorter). The damper will resist this with the result that there will be an upward reaction force acting on the caravan body.

Step 3 Equation of motion

$$2k(r - x) + 2c(\dot{r} - \dot{x}) = m\ddot{x}$$

or

$$m\ddot{x} + 2c\dot{x} + 2kx = 2c\dot{r}(t) + 2kr(t) \tag{6.32}$$

For any given vehicle speed, the shape of the road profile, $r(t)$, can be measured as an explicit function of time, so that we know exactly what the displacement of the axle will be. Differentiating the displacement gives the axle velocity $\dot{r}(t)$, again as an explicit function of time. As a result, the excitation terms on the right-hand side of the equation of motion are completely defined.

Collecting the equations of motion for the three examples, we have:

Mass–spring–damper system	$m\ddot{x} + c\dot{x} + kx = P(t)$
Rocker system	$mL_2^2\ddot{\theta} + cL_2^2\dot{\theta} + (K_1L_1^2 + K_1L_2^2)\theta = L_2P(t)$
Single-axle caravan	$m\ddot{x} + 2c\dot{x} + 2kx = 2c \cdot \dot{r}(t) + 2k \cdot r(t)$

While each of the equations is different in detail, you will see that they all share a common mathematical form, that of a **second-order ordinary differential equation with constant coefficients.** All linear, single-degree-of-freedom systems have this form, which can be written generically as:

$$M\ddot{z} + C\dot{z} + Kz = F(t) \tag{6.33}$$

in which

z	is the response coordinate
M	is the coefficient of $\ddot{z}$
C	is the coefficient of $\dot{z}$
K	is the coefficient of z and
$F(t)$	is the excitation function; independent of z.

The exact form of the three coefficients and of the excitation function will depend uniquely on the system being analysed.

The subsequent mathematical manipulation required to solve the equation of motion depends on the nature of the excitation function and on the amount of damping in the system. Mathematics textbooks provide comprehensive coverage of such problems, and Unit 4 presented the use of Laplace transforms for this purpose. In this unit, however, we consider three specific cases, namely 'free' vibration and harmonic (sinusoidal) and arbitrary periodic excitation.

Free vibration

If no externally applied forces or moments act on a structure, it can vibrate freely – hence the term 'free' vibration. We have previously used this situation to find the natural frequency of undamped systems. As mentioned above, the presence of damping means that any vibration will stop sooner or later, and consequently the term 'transient response' is often used in place of free vibration.

For $F(t) = 0$, we can use a solution in the form, $z(t) = Ae^{\lambda t}$

Substituting into the equation of motion gives,

$$M\lambda^2 Ae^{\lambda t} + C\lambda Ae^{\lambda t} + KAe^{\lambda t} = 0$$

For a non-trivial solution,

$$M\lambda^2 + C\lambda + K = 0$$

so that

$$\lambda_{1,2} = \frac{-C \pm \sqrt{C^2 - 4KM}}{2M} \tag{6.34}$$

The complete solution is then

$$z(t) = A_1 e^{\lambda_1 t} + A_2 e^{\lambda_2 t} \tag{6.35}$$

The integration constants, A_1 and A_2, are found from the 'initial conditions' specified in the problem. These will normally be the displacement and velocity at $t = 0$.

It can be seen from equation (6.34) that the roots $\lambda_{1,2}$ can be either real or complex, depending on the amount of damping present. While equation (6.35) gives a mathematically correct description of the response, other forms relate more closely to engineering intuition and are thus easier to interpret. Four cases with different damping levels are considered.

Zero damping

For zero damping, the system will oscillate with a sinusoidal waveform, although this is not obvious from equation (6.35).

For $C = 0$, $\qquad \lambda_{1,2} = \pm\dfrac{\sqrt{-4KM}}{2M} = \pm i\sqrt{\dfrac{K}{M}}$

The term $\sqrt{\dfrac{K}{M}}$ is called the **undamped natural frequency** and is given the symbol ω_n.

Previously, we used the term 'natural frequency' for this. As we shall see later, a damped system will have a frequency at which it will vibrate freely, so the words 'undamped' or 'damped' are used to distinguish between the two.

Returning to the general case, equation (6.35) becomes

$$z(t) = A_1 e^{i\omega_n t} + A_2 e^{-i\omega_n t}$$

This still doesn't look much like a sinusoidal waveform. However, by using the complex number identities,

$$e^{i\theta} = \cos\theta + i\sin\theta \quad \text{and} \quad e^{-i\theta} = \cos\theta - i\sin\theta$$

and the fact that A_1 and A_2 are a complex conjugate pair, it can be shown that

$$z(t) = B\cos\omega_n t + C\sin\omega_n t \tag{6.36}$$

The constants B and C are found from the initial conditions of the problem.

High damping

If the damping level is high so that $C^2 > 4KM$, the two roots, $\lambda_{1,2}$, are both **real** and **negative**. The response, as given by equation (6.35), is the sum of two decaying exponential functions. The constants A_1 and A_2 are found from the initial conditions as usual.

Critical damping

A special case for the roots of equation (6.34) occurs if $C^2 = 4KM$. This value of damping is known as 'critical damping', which is thus given by

$$C_{\text{crit}} = 2\sqrt{KM} \tag{6.37}$$

This is an important relationship and represents the boundary between two types of response. As we shall see in the next section, for subcritical damping the system will oscillate about its equilibrium position before coming to rest. For high damping, the free response decays exponentially back to its equilibrium position without oscillation.

From equation (6.34),

$$\lambda_1 = \lambda_2 = -\frac{C_{\text{crit}}}{2M} \equiv -\omega_n$$

In this situation, in order to maintain distinct parts to the solution, the response is given by

$$z(t) = A_1 e^{-\omega_n t} + A_2 t e^{-\omega_n t} \tag{6.38}$$

Note the 't' in the second term, which is needed to retain two distinct terms in the solution.

Light damping

The vast majority of engineering structures possess damping levels much less than critical. From common experience, we know that a structure left to vibrate freely will come to rest eventually. The light damping case is thus the norm. Mathematically, this corresponds to the case where $C^2 < 4KM$ and the roots of equation (6.34) are a complex conjugate pair:

$$\lambda_{1,2} = -\frac{C}{2M} \pm i\frac{\sqrt{4KM - C^2}}{2M} \tag{6.39}$$

It is convenient to introduce the **damping ratio**, $\gamma = \dfrac{C}{\text{critical damping}} = \dfrac{C}{2\sqrt{KM}}$

Using also the expression for the undamped natural frequency, ω_n, equation (6.39) becomes

$$\lambda_{1,2} = -\gamma\omega_n \pm i\omega_n\sqrt{1 - \gamma^2} \tag{6.40}$$

Substitution into equation (6.35) gives a mathematically correct (but not very convenient) solution.

$$z(t) = A_1 e^{(-\gamma\omega_n + i\omega_n\sqrt{1-\gamma^2})t} + A_2 e^{(-\gamma\omega_n - i\omega_n\sqrt{1-\gamma^2})t} \tag{6.41}$$

Again, making use of the complex exponential identities and the fact that A_1 and A_2 are a complex conjugate pair, equation (6.41) can be rewritten as

$$z(t) = e^{-\gamma\omega_n t}\left(B_1 \cos \omega_n\sqrt{1-\gamma^2}\, t + B_2 \sin \omega_n\sqrt{1-\gamma^2}\, t\right) \tag{6.42}$$

Equation (6.42) matches our intuition since it describes a sinusoidal waveform (indicated by the terms in the brackets) with an exponentially decaying term at the front that will cause the amplitude of the sinusoid to decrease. Note that the constants B_1 and B_2 are both real. As an alternative, equation (6.42) can be rewritten as

$$z(t) = C_0 e^{-\gamma\omega_n t} \cos\left(\omega_n\sqrt{1-\gamma^2}\, t - \psi\right) \tag{6.43}$$

As with the previous cases, the two constants can be found from the initial conditions.

Note that equations (6.42) and (6.43) both show that the frequency of vibration is $\omega_n\sqrt{1-\gamma^2}$. This is known as the **damped natural frequency** and is less than the undamped natural frequency, ω_n.

Worked example

When at rest in equilibrium, the mass receives an impulse J of magnitude 5 Ns applied at time, $t = 0$. Find the response for $t > 0$.

Data: $k = 500\,\text{N/m}$
$c = 20\,\text{Ns/m}$
$m = 10\,\text{kg}$

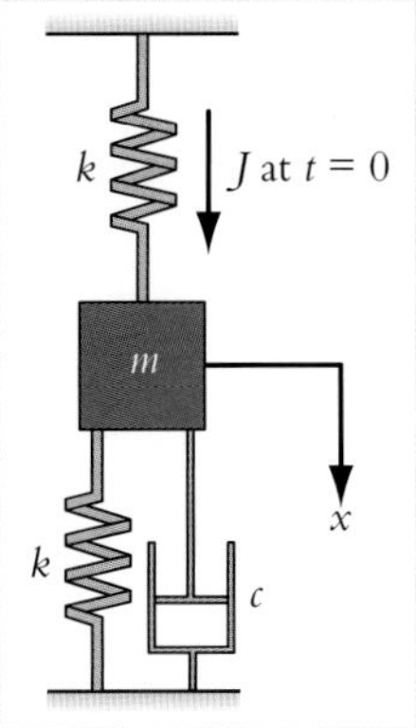

Figure 6.47

Free-body diagram

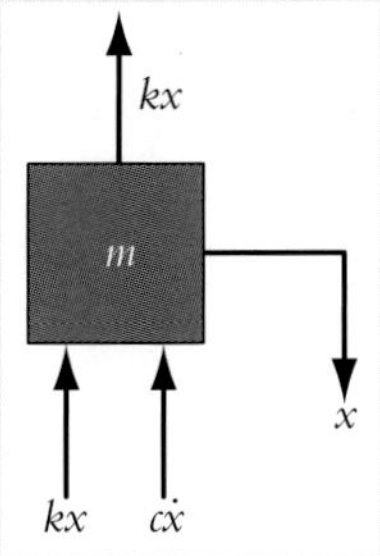

Figure 6.48

Equation of motion

$$-2kx - c\dot{x} = m\ddot{x}$$

or

$$m\ddot{x} + c\dot{x} + 2kx = 0$$

Note that the impulse J doesn't appear on the free-body diagram or in the equation of motion since it doesn't exist for $t > 0$.

From the coefficients in the equation of motion, we find that $\omega_n = \sqrt{\frac{2k}{m}} = 10\,\text{rad/s}$ and $\gamma = \frac{c}{2\sqrt{2km}} = 0.1$. Since the damping ratio is less than 1.0 in this problem, we can identify

that it is a case of light damping. From equation (6.42), we can write the expression for the response as:

$$x(t) = e^{-\gamma\omega_n t}(B_1 \cos \Omega_n t + B_2 \sin \Omega_n t)$$

where $\Omega_n = \omega_n\sqrt{1 - \gamma^2} = 9.9\,\text{rad/s}$.

First initial condition: $x = 0$ at $t = 0$ $\quad \therefore \quad B_1 = 0$

Hence, $x(t) = B_2 e^{-\gamma\omega_n t} \sin \Omega_n t$

The second initial condition is the velocity immediately after the impulse, $\dot{x}_0$. This can be found from the change in momentum caused by the impulse.

That is: $J = m(\dot{x}_0 - 0)$

Therefore, $\dot{x}_0 = \frac{J}{m}$

Differentiating the expression for displacement gives

$$\dot{x} = B_2[\Omega_n e^{-\gamma\omega_n t} \cos \Omega_n t - \gamma\omega_n e^{-\gamma\omega_n t} \sin \Omega_n t]$$

$$\dot{x}_0 = \frac{J}{m} \text{ at } t = 0, \quad \therefore \quad \frac{J}{m} = B_2[\Omega_n - 0]$$

Hence, $B_2 = \frac{J}{m\Omega_n}$ and $x(t) = \frac{J}{m\Omega_n} e^{-\gamma\omega_n t} \sin \Omega_n t$

Substituting the numerical values gives $x(t) = 0.0505\, e^{-t} \sin 9.9t$ [m]

Estimating damping

While the mass and stiffness of a structure can normally be calculated, the structural damping is very difficult to predict. However, equations (6.42) and (6.43) show that the rate of decay of the free vibration of a structure depends directly on the damping ratio, and this gives a method of measuring damping.

In the above worked example, suppose we didn't know the damping value, but had done an experiment to measure the transient displacement from the impulse. Here is the measured response waveform.

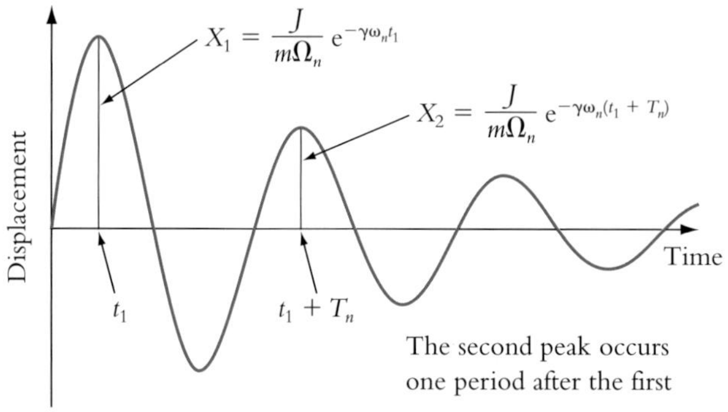

Figure 6.49

We know that the expression for the displacement is $x(t) = \frac{J}{m\Omega_n} e^{-\gamma\omega_n t} \sin \Omega_n t$

The amplitude of the first peak at $t = t_1$ is $X_1 = \frac{J}{m\Omega_n} e^{-\gamma\omega_n t_1}$

The second peak occurs one period later and has an amplitude $X_2 = \frac{J}{m\Omega_n} e^{-\gamma\omega_n (t_1 + T_n)}$

The ratio of the peaks is therefore $\frac{X_1}{X_2} = e^{\gamma\omega_n T_n}$

The period of the damped oscillation is $T_n = \frac{2\pi}{\omega_n\sqrt{1 - \gamma^2}} \approx \frac{2\pi}{\omega_n}$ for light damping.

Taking logs of both sides of the expression for the ratio of the peaks, we get:

$$\ln\left(\frac{X_1}{X_2}\right) = 2\pi\gamma$$

In this case, $X_1 = 0.0431$ m and $X_2 = 0.0229$ m, so that $\gamma = 0.101$ and $c = 20.1$ Ns/m, which compare well with the exact values of 0.1 and 20.0 respectively.

Note that the ratio of *any* two successive peaks is a constant, so there is good scope for making several estimates of the damping ratio from a measured response waveform.

Harmonic excitation

Solution of the equation of motion

$$M\ddot{z} + C\dot{z} + Kz = f(t)$$

The complete solution for $z(t)$ consists of a **particular integral** and the solution to the **complementary function**. The latter is simply the solution to the equivalent free vibration problem considered earlier in this section. For non-zero damping, all these solutions decay to zero with time and they describe the **transient response** of the structure due to the sudden start of the excitation. In practice, the transient response normally decays quickly. As an example, the graph below shows the complete solution for the response of a single-degree-of-freedom system due to the start of a cosine wave excitation. Near the start, we see the decaying transient response superimposed on the particular integral solution.

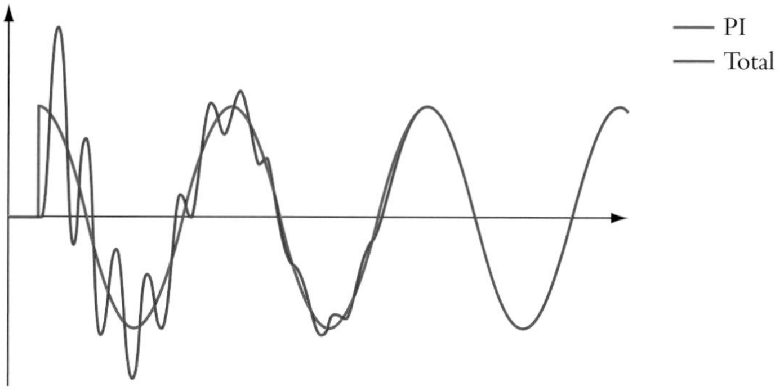

Figure 6.50

After the transient response has decayed, we are left with the particular integral, which continues for as long as the excitation remains. It's normally called the **steady-state response** and, in most cases, this is all we are interested in.

Method 1 Direct Substitution
Consider harmonic excitation of the form, $f(t) = F\cos\omega t$

For pure sinusoidal excitation, the response is also sinusoidal. The response has the same frequency as the excitation, but the two are likely to have a phase difference. The key information we want from the analysis is the **amplitude of the response** and its **phase relative to the excitation**. A suitable expression for the response is therefore

$$z(t) = Z\cos(\omega t + \alpha) \quad (6.44)$$

where Z is the amplitude of the vibration and α is the phase angle.

To find Z and α, we substitute for $z(t)$ and its derivatives in the equation of motion, expand the various trigonometric terms and equate the coefficients of $\cos\omega t$ and $\sin\omega t$. This gives

$$Z = \frac{F}{\sqrt{(K - M\omega^2)^2 + \omega^2 C^2}} \quad (6.45)$$

and $$\alpha = \tan^{-1}\left(\frac{-\omega C}{K - M\omega^2}\right) \quad (6.46)$$

Note from equation (6.46) that the phase angle α is negative, indicating that the response of a system *lags* the excitation.

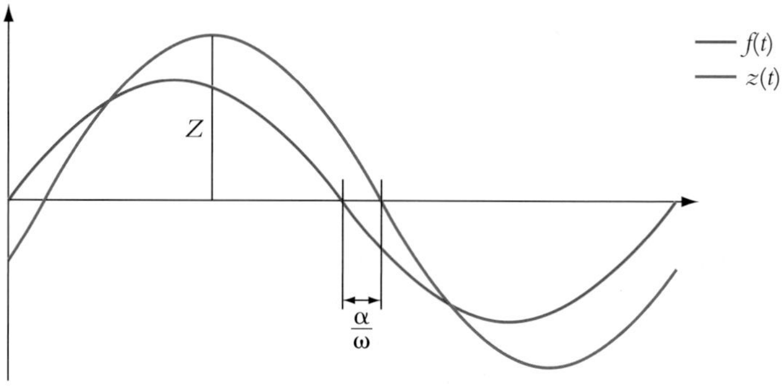

Figure 6.51

Since the steady-state response to sinusoidal excitation is also sinusoidal and with the same frequency as the excitation, the key parameters to be identified are the **amplitude** and **phase angle**. Instead of drawing conventional time waveforms, a convenient way of showing the amplitude and phase relationship between excitation and response is with a **phasor diagram** (Figure 6.52).

Method 2 Complex algebra

This provides a mathematically convenient way of finding the amplitude and phase angle of the response. It has the added advantage that the same mathematical approach can be extended to more complicated (and therefore more realistic) structures, to more general forms of excitation and to experimental testing and digital data analysis procedures.

The substitutions used are always the same; that is,

Put
$$f(t) = Fe^{i\omega t} \tag{6.47}$$

and
$$\begin{aligned} z(t) &= Ze^{i(\omega t + \alpha)} \\ &= (Ze^{i\alpha})e^{i\omega t} \\ &= Z^{\star} e^{i\omega t} \end{aligned} \tag{6.48}$$

where $Z^{\star}$ is a complex number $= Z(\cos\alpha + i\sin\alpha)$

Differentiating $z(t)$, $\dot{z} = i\omega Z^{\star} e^{i\omega t}$ and $\ddot{z} = -\omega^2 Z^{\star} e^{i\omega t}$

When these are substituted into the equation of motion, we get

$$(-M\omega^2 + i\omega C + K)Z^{\star} e^{i\omega t} = Fe^{i\omega t}$$

Hence
$$Z^{\star} = \frac{F}{(K - M\omega^2) + i\omega C}$$

or
$$Z^{\star} = \frac{F}{(K - M\omega^2)^2 + \omega^2 C^2}[(K - M\omega^2) - i\omega C]$$

or
$$Z^{\star} = a + ib$$

Hence
$$Z = \frac{F}{\sqrt{(K - M\omega^2)^2 + \omega^2 C^2}}$$

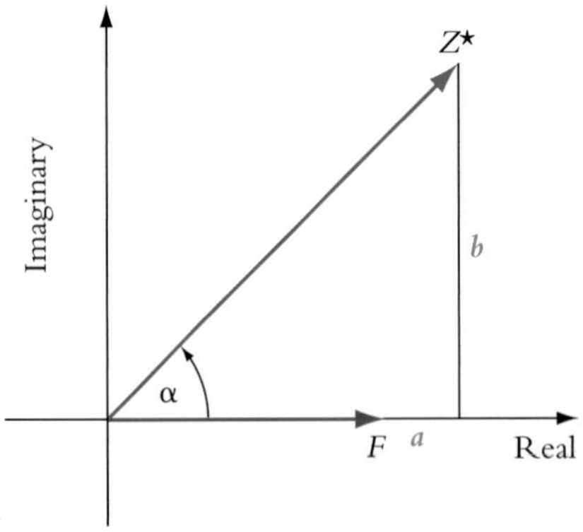

Figure 6.52

and
$$\alpha = \tan^{-1}\left(\frac{-\omega C}{K - M\omega^2}\right)$$

Plotting $Z^{\star}$ and F on the complex (Argand) plane (Figure 6.52), gives the same phasor diagram discussed before (Figure 6.53). Note that in practice, the imaginary part of $Z^{\star}$ is always negative. As a result, α is also negative, meaning that the response **lags** behind the excitation (Figure 6.54).

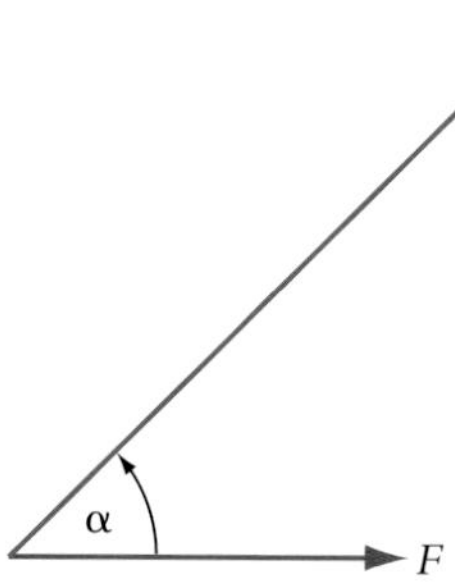

Figure 6.53

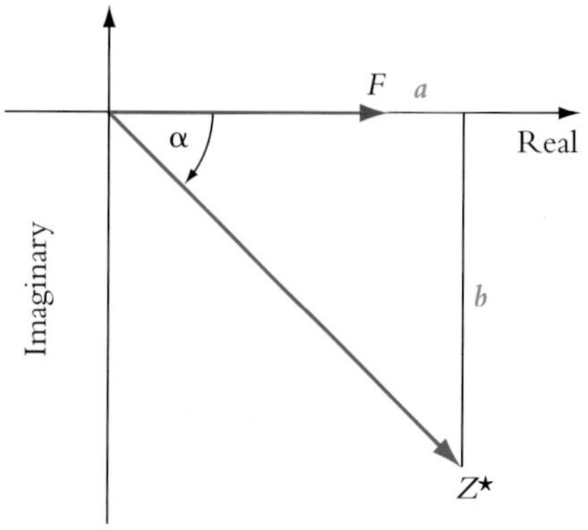

Figure 6.54

Frequency response function

To see how the excitation frequency affects the response, we will consider the **frequency response function** (FRF). By definition, this is the **response/unit applied force**. Note that it is used exclusively for **force** excitation.

Start with the general form of the equation of motion,

$$M\ddot{z} + C\dot{z} + Kz = f(t)$$

Dividing by M and noting that $\dfrac{C}{M} = 2\gamma\omega_n$, we get

$$\ddot{z} + 2\gamma\omega_n\dot{z} + \omega_n^2 z = \frac{f(t)}{M}$$

With $f(t) = Fe^{i\omega t}$ and $z(t) = Z^\star e^{i\omega t}$, $Z^\star = \dfrac{F}{M}\dfrac{1}{(\omega_n^2 - \omega^2) + i2\gamma\omega_n\omega}$

The FRF is therefore:

$$H(\omega) = \frac{Z^\star}{F} = \frac{1}{M}\frac{1}{(\omega_n^2 - \omega^2) + i2\gamma\omega_n\omega} \tag{6.49}$$

An alternative expression that emphasizes the frequency dependence is obtained by dividing top and bottom by ω_n^2 and noting that $M\omega_n^2 = K$ gives

$$H(\omega) = \frac{Z^\star}{F} = \frac{1}{K}\frac{1}{\left(1 - \dfrac{\omega^2}{\omega_n^2}\right) + i2\gamma\dfrac{\omega}{\omega_n}}$$

The FRF magnitude is

$$|H(\omega)| = \frac{1}{K}\frac{1}{\sqrt{\left(1 - \dfrac{\omega^2}{\omega_n^2}\right)^2 + 4\gamma^2\dfrac{\omega^2}{\omega_n^2}}} \tag{6.50}$$

and the phase angle is

$$\alpha = \tan^{-1}\left(\frac{-2\gamma\dfrac{\omega}{\omega_n}}{1 - \dfrac{\omega^2}{\omega_n^2}}\right) \tag{6.51}$$

It is clear from these expressions that the response of the structure (its amplitude and phase angle) depends on the ratio between the excitation frequency, ω, and the undamped natural frequency, ω_n and on the damping ratio, γ.

Figure 6.55 shows how the FRF amplitude and phase angle vary with both frequency ratio and damping ratio.

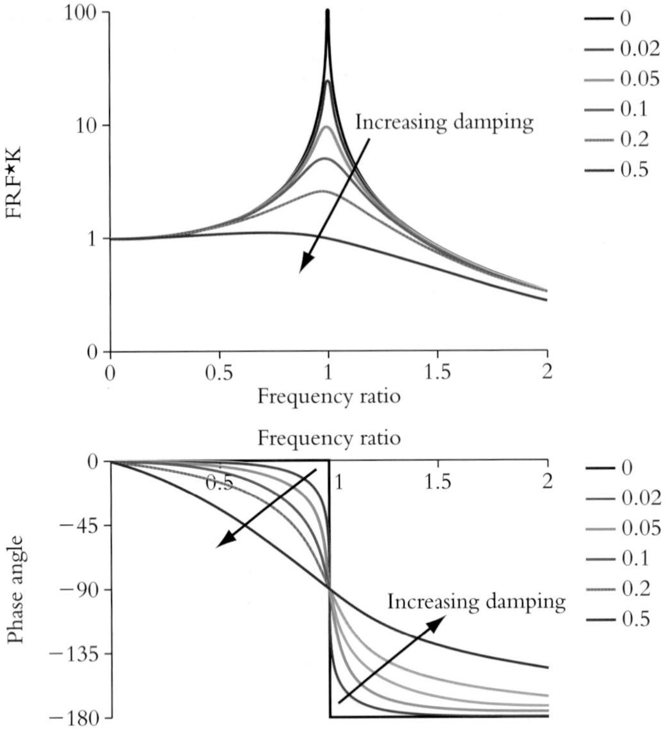

Figure 6.55 **Amplitude (top) and phase angle (bottom) of FRF plotted against frequency ratio**

The graphs have been plotted for several values of damping ratio. Most structures have damping ratios of less than 0.1 and it can be seen that the resonant peak (the maximum response near the frequency ratio of 1.0) is large. Increasing damping will reduce the height of this resonant peak. In the phase angle plot, we see that for low values of damping ratio, the phase angle is close to 0° (i.e. the response is roughly in phase with the excitation) for frequencies below the undamped natural frequency and close to −180° (roughly out-of-phase) above the undamped natural frequency.

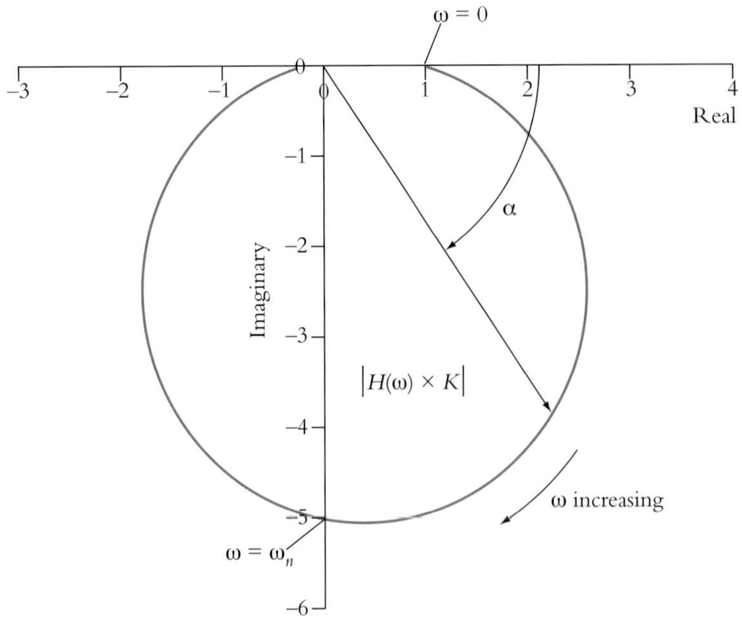

Figure 6.56 **Imaginary part of the FRF plotted against the real part (γ = 001)**

The frequency giving the maximum response is called the **resonant frequency**. With low damping, the resonant frequency, the undamped natural frequency and the damped natural frequency are all virtually identical. For higher damping (look at the curve for $\gamma = 0.2$), the resonant frequency is less than the undamped natural frequency.

Each point on the real vs. imaginary plot shown above gives the phasor diagram we met before (the arrow illustrates one particular frequency). Although this is not shown explicitly, we move around the plot as the frequency ratio varies. Comparing this with Figure 6.56, when $\omega = 0$, the amplitude is 1.0 and the phase angle is 0°. When $\omega = \omega_n$, the phase angle is −90° and the response is purely imaginary. As ω tends to infinity, the amplitude tends to zero and the phase angle tends to −180°.

Worked example – Use of the frequency response function

Find the waveform of the steady-state vertical displacement of a reciprocating air compressor (mass, $m = 4000$ kg) that operates at a crank speed, Ω, of 300 rev/min. The machine is supported on a set of four resilient mounts that have an overall vertical stiffness of 2.5 MN/m and a damping ratio of 0.04.

The effect of the masses of the reciprocating pistons is to produce a vertical force on the compressor given by

$$s(t) = 1.3\,\Omega^2 \cos \Omega t + 3.0\,\Omega^2 \cos 2\Omega t = S_1 \cos \omega_1 t + S_2 \cos \omega_2 t$$

The compressor can be modelled as a rigid mass and its mounts as a spring–damper combination as shown in Figure 6.57. Each sinusoidal term in the excitation will produce a steady-state response that is sinusoidal with the same frequency as the excitation. For example, the term $S_1 \cos \omega_1 t$ will produce a response in the form $X_1 \cos(\omega_1 t + \alpha_1)$. We can use the frequency response function to do this.

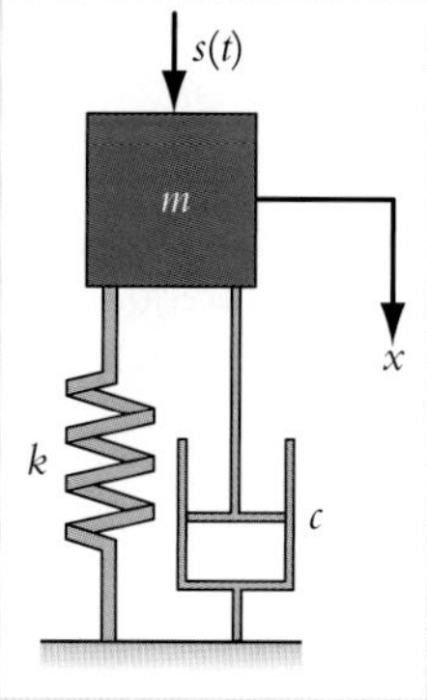

Figure 6.57

Once the response to each excitation term has been found, the total response is obtained simply by adding the two together.

For each excitation term, we have $X_j^\star = H(\omega_j) \times S_j$ so that $X_j = |X_j^\star| = |H(\omega j)| \times S_j$ where $j = 1$ or 2. We can use the equations (6.50) and (6.51) to work out the FRF magnitude and phase angle values.

Using the data given,

$$S_1 = 1283\,\text{N}, |H(\omega_1)| = 6.81 \times 10^{-4}\,\text{mm/N}, \alpha_1 = -170^\circ, X_1 = 0.87\,\text{mm}$$

$$S_2 = 2961\,\text{N}, |H(\omega_2)| = 7.52 \times 10^{-5}\,\text{mm/N}, \alpha_2 = -178^\circ, X_2 = 0.22\,\text{mm}$$

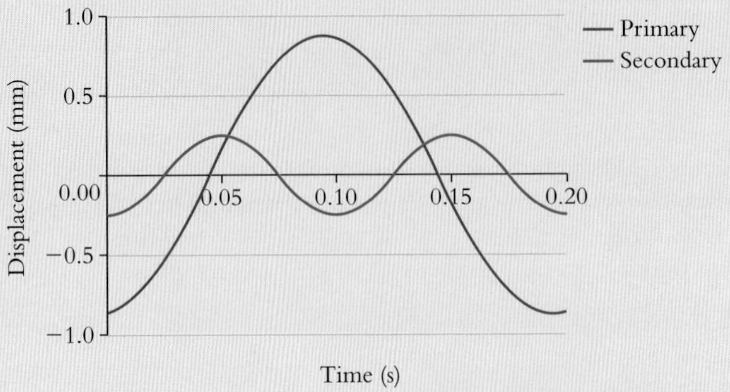

Figure 6.58 **Showing $X_1 \cos(\omega_1 t + \alpha_1)$ and $X_2 \cos(\omega_2 t + \alpha_2)$**

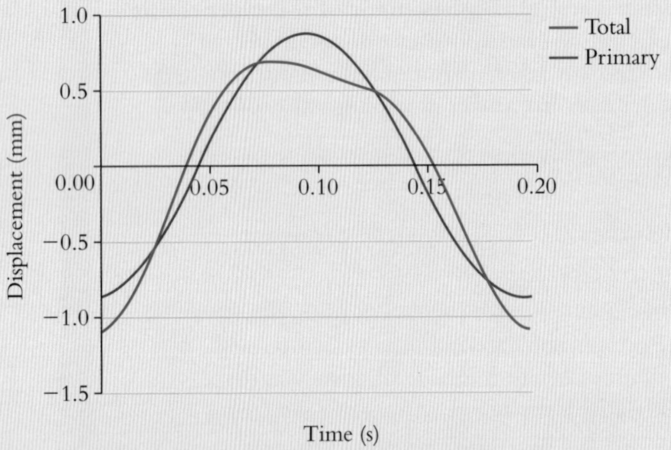

Figure 6.59 **Showing the total response by adding the two waveforms together**

The term 'primary' relates to the fact that the vibration frequency, ω_1, is equal to the rotational frequency of the crank. The 'secondary' component has a frequency that is twice the rotational frequency of the crank.

Periodic excitation

Structures are rarely subjected to pure sinusoidal excitation in their operating environment and it may therefore seem strange that so much time is devoted to it in vibration textbooks. In many situations, however, *periodic* excitation is present, a simple example being the air compressor problem considered in the previous section.

If we consider the system shown in Figure 6.60, the normal mathematical approach for solving the differential equation

$$m\ddot{x} + c\dot{x} + kx = p(t)$$

is to 'guess' an appropriate expression for the particular integral. This was successful in the case of sinusoidal excitation, but it can be difficult (or even impossible) for many other excitation functions. The method used here is an extension of the approach used for the air compressor problem.

For a periodic waveform that repeats exactly every T seconds, Fourier series analysis or its digital equivalent, the discrete Fourier transform (DFT), can split it into a series of sine and cosine components. In this case, we will get

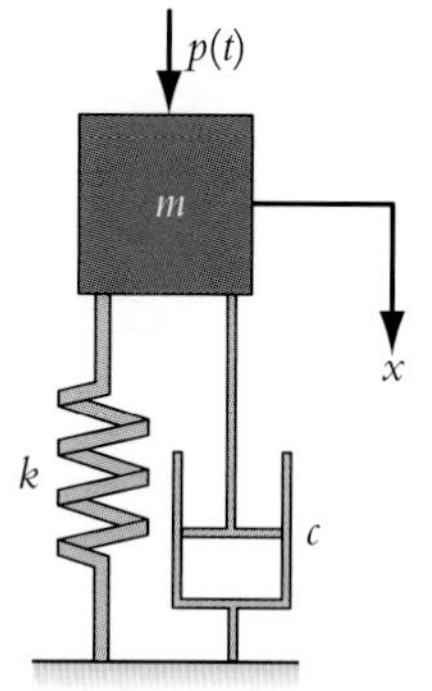

Figure 6.60

$$p(t) = A_0 + \sum_{j=1}^{\infty}[A_j \cos(j\omega_0 t) + B_j \sin(j\omega_0 t)]$$

where the repetition frequency, $\omega_0 = \frac{1}{T}[\text{Hz}] = \frac{2\pi}{T}[\text{rad/s}]$.

The term in the square brackets represents a general sinusoidal wave with frequency $j\omega_0$. The steady-state response to this general term will also be sinusoidal with the same frequency and we can calculate this using the frequency response function in much the same way that we did for the air compressor problem. If we do this for all of the terms in the series, we can then add all the individual responses together to give the overall response.

The process therefore is as follows:

Step 1 Use Fourier series analysis to decompose the excitation waveform into a set of sinusoids.

We can use the complex number representation of the waveforms to good effect here. For a typical pair of components we write

$A_j \cos(j\omega_0 t) + B_j \sin(j\omega_0 t)$ as $P_j^{\star} e^{\mathbf{i}j\omega_0 t}$ where $P_j^{\star} = A_j + \mathbf{i}B_j$

(If the DFT is used, this complex number form is calculated directly.)

Step 2 The response to this excitation can be calculated from the frequency response function, so that

$$X_j^{\star} = H(j\omega_0) \times P_j^{\star}$$

The response term, $X_j^{\star}$, will have the form $C_j + \mathrm{i}D_j$ which can be interpreted as

$$C_j \cos(j\omega_0 t) + D_j \sin(j\omega_0 t)$$

Step 3 Having done this for all frequencies in the excitation, all of the individual sinusoidal responses can be summed together.

As an example, Figure 6.61 shows a periodic excitation in the form of a series of rectangular force pulses with a repetition period of 0.25 s (i.e. a repetition frequency of 4 Hz). The response of a single-degree-of-freedom system is shown in Figure 6.62.

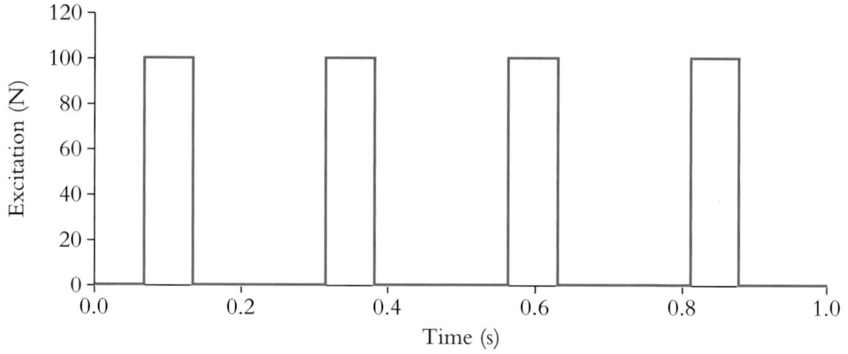

Figure 6.61 **The 4 Hz force pulse waveform**

Figure 6.63 gives the results of Step 1 and shows the A_j (top) and B_j (bottom) Fourier coefficients of the rectangular pulses plotted against a horizontal frequency axis. In addition to the A_0 at 0 Hz on the top graph, the components appear at 4 Hz, 8 Hz, 12 Hz, ... (i.e. the $j\omega_0$ terms from the Fourier series, which are at integer multiples of the repetition frequency) and some are positive, some negative. In this case, the coefficients up to $j = 12$ are shown.

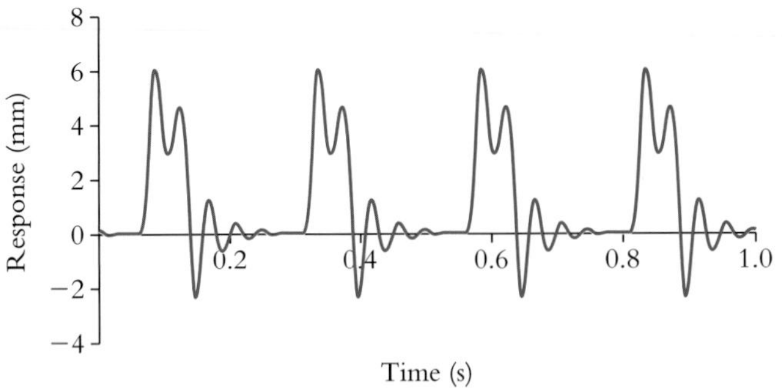

Figure 6.62 **Response of single-degree-of-freedom system**

[Note that since cos(0) = 1 and sin(0) = 0, the A_0 term can be seen as a special case of the general summation when $j = 0$.]

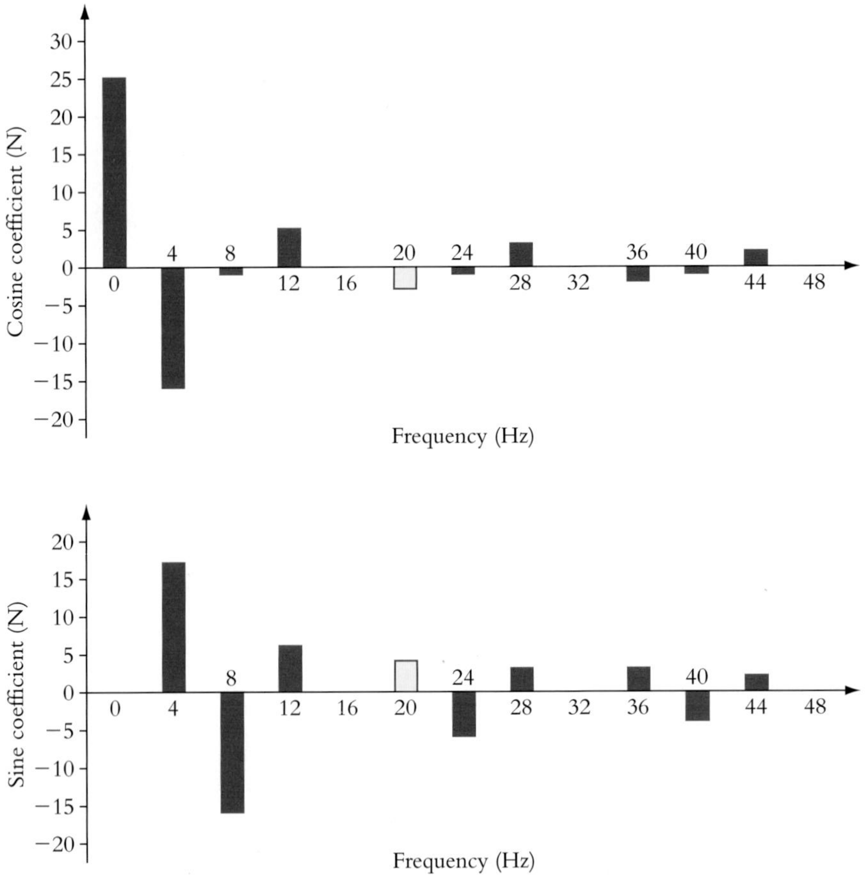

Figure 6.63 **Cosine (top) and sine (bottom) coefficients of the Fourier series expansion of the excitation waveform up to $j = 12$**

Figure 6.64 shows the frequency response function for the system in the example, with the real and imaginary parts plotted against frequency. The system has an undamped natural frequency of 25 Hz and a damping ratio of 0.2.

In Step 2, we perform the complex multiplications between the FRF and excitation components. As an example, component 5 (at 20 Hz) gives the following data.

$$P_5^\star = -2.98 + \mathbf{i}3.77\ \text{N}$$

$$H(20\ \text{Hz}) = 0.0629 - \mathbf{i}0.0559\ \text{mm/N}$$

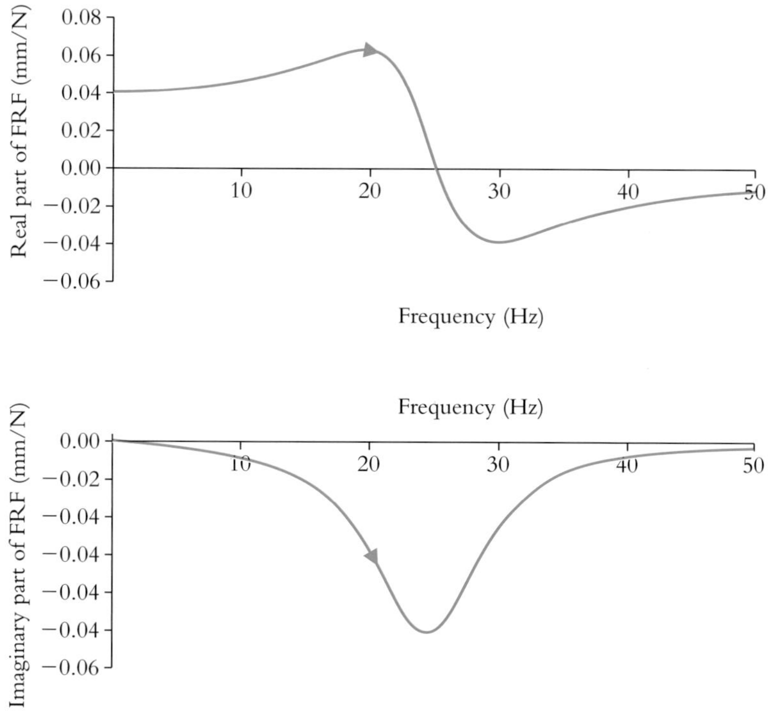

Figure 6.64 **Real and imaginary parts of the frequency response function, natural frequency = 25 Hz, damping ratio = 0.2**

These points are highlighted in Figures 6.63 and 6.64. After the appropriate complex multiplication we get

$$P_5^{\star} = -0.376 - \mathbf{i}0.045\,\text{mm}$$

Repeating this process by hand for a large number of components would be very laborious, but software (such as Matlab) that is able to do the necessary complex arithmetic using the Discrete Fourier Transform is widely available.

Component 5 is highlighted in Figure 6.65, which shows all of the C_j (top) and D_j (bottom) terms plotted against frequency. These are the cosine and sine coefficients of the response waveform, which are added together in Step 3. The resulting waveform is shown in Figure 6.62.

Figure 6.66 shows how the response waveform builds up as more terms are added. Figure 6.66(a) shows the constant term and the first three sinusoidal components (up to $j = 5$, or 20 Hz) and Figure 6.66(b) includes up to $j = 13$ (52 Hz). The waveform does not change much once the $j = 6$ term (24 Hz) has been added. This is because the excitation components are smaller above this frequency (Figure 6.63) and the FRF magnitude is also reducing (Figure 6.64).

Notes

(1) The above procedure will work for any periodic excitation.

(2) The frequency response function used here is the same expression that was derived using pure sinusoidal excitation. We are thus able to use the result from a mathematically convenient excitation case (but one which is *not* often experienced in practice) to find the response to a more complicated (and practical) excitation.

(3) Most engineering structures have more than one resonant frequency. Provided the frequency response function can be calculated (as explained in Section 6.4) or measured (as in Section 6.5), the same procedure can be used.

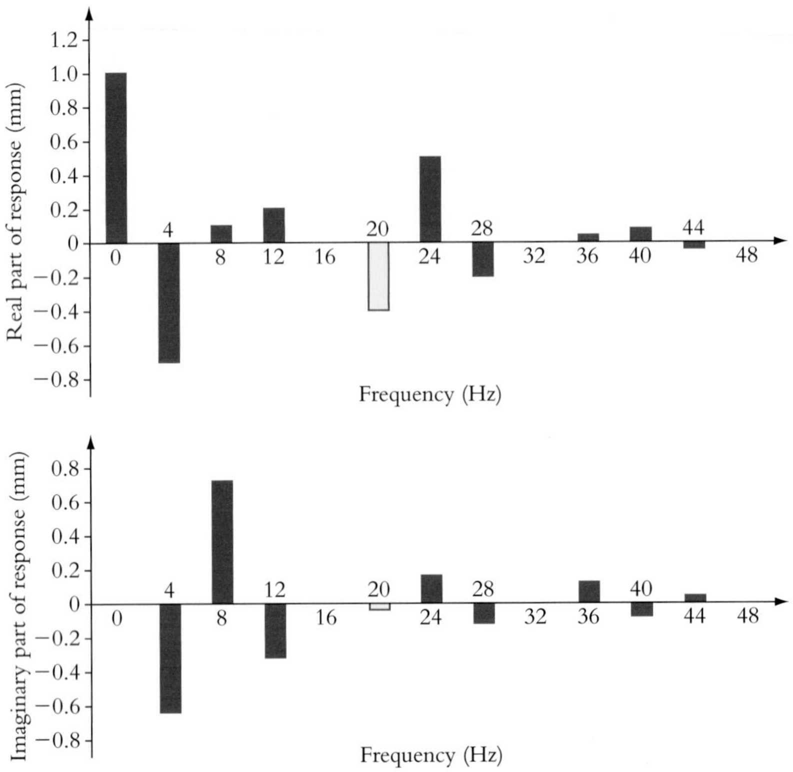

Figure 6.65 **Real (cosine) and imaginary (sine) coefficients of the response of the system**

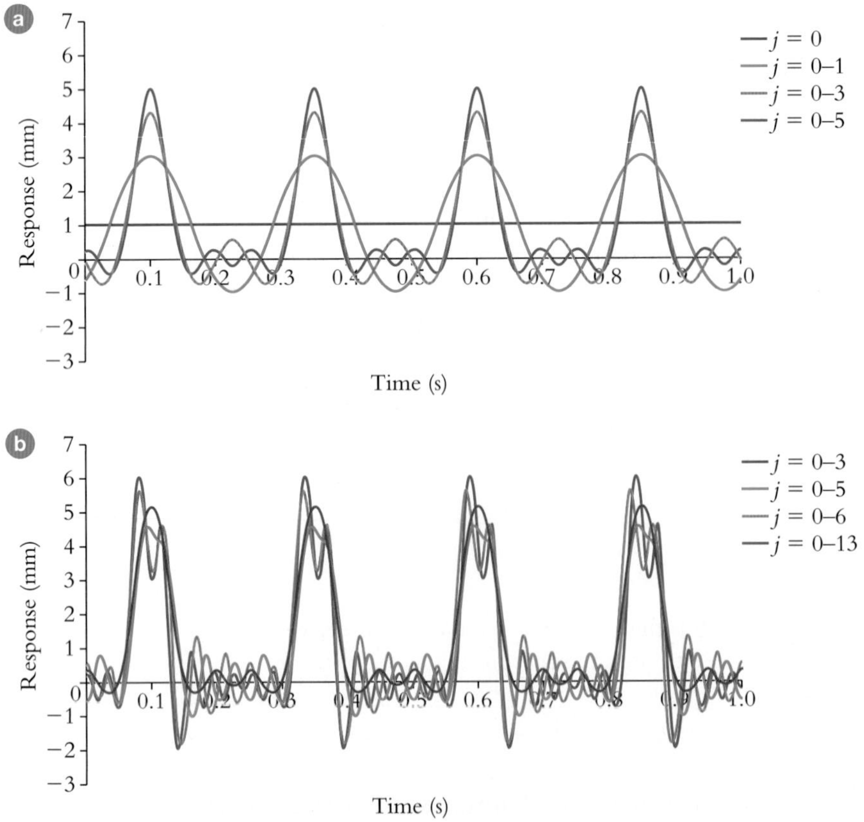

Figure 6.66 **The summation process in Step 3: (a) terms up to j = 5; (b) terms up to j = 13**

As an example, Figure 6.67(a) shows the magnitude plot of the frequency response function of a structure that has five resonant frequencies in the range from 0 to 350 Hz. To find the response of this system to the same 4 Hz pulse sequence, the 3-step procedure is repeated, but using the new frequency response function instead. The result is shown in Figure 6.67(b).

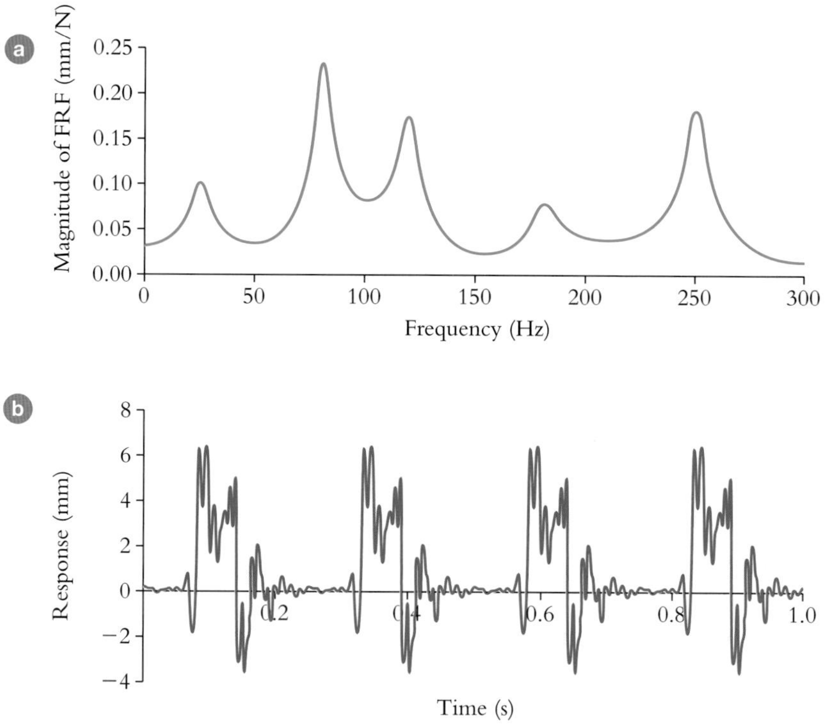

Figure 6.67 (a) Magnitude plot of FRF of a five-degrees-of-freedom system and (b) its response to the 4 Hz force pulses

Apart from some obvious differences in detail between this and the single-degree-of-freedom response waveform, the general shape and amplitude are similar for both. The main reason for this is that the original single-degree-of-freedom system is the same as the lowest natural frequency of the five-degrees-of-freedom system. The single-degree-of-freedom system is therefore a good approximation to the more complicated five-degrees-of-freedom structure. In Section 6.6 we will discuss how we can create single-degree-of-freedom approximations to more complicated structures.

Overall, the method is very powerful as it can be applied to any structure subjected to any periodic excitation.

Learning summary

By the end of this section you should have learnt:

✓ to obtain the response of single-degree-of-freedom mass–spring-damper systems for the cases of 'free' vibration and for harmonic (sinusoidal) and arbitrary periodic excitation.

6.4 Response of damped multi-degree-of-freedom systems

Introduction

Section 6.2 covered free, undamped vibration problems in the form,

$$[M]\{\ddot{x}\} + [K]\{x\} = \{0\}$$

These were solved with the substitution, $\{x(t)\} = \{X\}\cos\omega t$, and led to the eigenvalue problem,

$$([K] - \omega^2[M])\{X\} = \{0\} \tag{6.52}$$

Solving this gave the natural frequency, ω_r, and modal vectors, $\{X\}$, for each mode.

Section 6.3 discussed the forced response of damped single-degree-of-freedom systems. This section extends this to damped systems with many degrees of freedom. The inclusion of damping and excitation adds extra forces to the free-body diagrams and produces a set of equations in the form,

$$[M]\{\ddot{x}\} + [C]\{\dot{x}\} + [K]\{x\} = \{p(t)\} \tag{6.53}$$

These equations are coupled, implying that they need to be solved simultaneously. It is, however, possible to uncouple the equations to give a set of individual second-order ordinary differential equations, each with only one response variable. These look just like single-degree-of-freedom equations and can be solved using the methods described in Section 6.3.

The steps involved are as follows:

Step 1 Uncouple the equations of motion to give a set of individual second-order ordinary differential equations

Step 2 Solve each equation

Step 3 Combine the individual solutions to give the actual response of the system.

Orthogonality of modes

A property of the modal vectors of the undamped system holds the key to uncoupling equation (6.53).

For a typical mode, r, we can rearrange equation (6.52) to give

$$[K]\{X\}_r = \omega_r^2[M]\{X\}_r$$

Premultiplying by the transpose of a different mode shape vector, $\{X\}_s$, we get

$$\{X\}_s^T[K]\{X\}_r = \omega_r^2\{X\}_s^T[M]\{X\}_r \tag{6.54}$$

Similarly, by starting with mode s, we can get

$$\{X\}_r^T[K]\{X\}_s = \omega_s^2\{X\}_r^T[M]\{X\}_s \tag{6.55}$$

Since $[K]$ and $[M]$ are symmetric, it follows that

$$\{X\}_r^T[K]\{X\}_s = \{X\}_s^T[K]\{X\}_r \quad \text{and} \quad \{X\}_r^T[M]\{X\}_s = \{X\}_s^T[M]\{X\}_r$$

Subtracting (6.55) from (6.54) gives

$$0 = (\omega_r^2 - \omega_s^2)\{X\}_r^T[M]\{X\}_s$$

Thus, provided that $\omega_r \neq \omega_s$, $\quad \{X\}_r^T[M]\{X\}_s = 0 \quad$ (6.56)

from which it follows that $\quad \{X\}_r^T[K]\{X\}_s = 0 \quad$ (6.57)

For $r = s$, $$\{X\}_r^T[M]\{X\}_r = M_r \tag{6.58}$$

which is called the **modal mass** (a scalar) for mode r

and $$\{X\}_r^T[K]\{X\}_r = K_r \tag{6.59}$$

which is the **modal stiffness** for mode r. Together, these equations describe the property of **orthogonality** for different modes of the system and they show that each individual mode shape is unique and independent of the others.

Modal matrix and modal scaling

The so-called **modal matrix**, $[\Phi]$, is obtained by assembling all of the modal vectors, column by column. That is,

$$[\Phi] = [\{X\}_1 \quad \{X\}_2 \quad \ldots \quad \{X\}_r \quad \ldots \quad]$$

Thus, column r of the modal matrix is the modal vector for mode r, with each element giving the amplitude of the displacement in each coordinate for that mode.

$$\{X\}_r = \begin{Bmatrix} X_{1r} \\ X_{2r} \\ X_{3r} \\ \vdots \end{Bmatrix}$$

Written out long-hand, the modal matrix is:

$$[\Phi] = \begin{bmatrix} X_{11} & X_{12} & X_{13} & \cdots \\ X_{21} & X_{22} & X_{23} & \cdots \\ X_{31} & X_{32} & X_{33} & \cdots \\ \vdots & \vdots & \vdots & \ddots \end{bmatrix}$$

It is important to get a clear understanding of the notation being used here since, for example, X_{13} is *not* the same as X_{31}. In particular, note that

the first suffix (i.e. the row number) = response coordinate number
and the second suffix (i.e. the column number) = mode number.

Thus, X_{13} is the response in coordinate 1 for the third mode of vibration and X_{31} is the response in coordinate 3 for the first mode of vibration.

Using equations (6.56) to (6.59), it is easy to show that

$$[\Phi]^T[M][\Phi] = \begin{bmatrix} \ddots & & \\ & M_r & \\ & & \ddots \end{bmatrix}$$

and

$$[\Phi]^T[K][\Phi] = \begin{bmatrix} \ddots & & \\ & K_r & \\ & & \ddots \end{bmatrix}$$

These are the **modal mass** and **modal stiffness matrices** and it will be seen that they are both **diagonal matrices**. It is these results that enable the equations of motion to be uncoupled.

Modal vectors have an arbitrary scaling. In equation (6.11), for example, the mode shape vectors were found by setting the amplitude in one of the coordinates to 1.0. The particular scaling adopted also affects the numerical values of the modal masses and stiffnesses. As a result, these contain an arbitrary factor too.

It is common to choose to scale the vectors so that all of the modal masses are equal to unity. The benefit of this will become apparent later. Starting with an arbitrarily scaled vector $\{X\}_r$ and the resulting modal mass M_r, the modal vector can be scaled (or **normalized**) to **unit modal mass** using the expression,

$$\{U\}_r = \frac{\{X\}_r}{\sqrt{M_r}} \tag{6.60}$$

There is an example of this later.

One advantage of using this method of scaling is that, in addition to having all of the modal masses equal to unity, the modal stiffness values are numerically equal to the square of the relevant natural frequency, that is, $M_r = 1.0$ and $K_r = \omega_r{}^2$ for all modes.

Forced response of an undamped multi-degree-of-freedom structure

Before considering a damped multi-degree-of-freedom structure, we will consider the equivalent undamped problem. The coupled equations of motion are

$$[M]\{\ddot{x}\} + [K]\{x\} = \{p(t)\} \tag{6.61}$$

For step 1 of the solution process, we uncouple the equations by introducing the substitution

$$\{x\} = [\Phi]\{q\} \tag{6.62}$$

where $\{q\}$ is a vector of **modal coordinates**.

Substituting into equation (6.61) and premultiplying by $[\Phi]^T$ gives

$$[\Phi]^T[M][\Phi]\{\ddot{q}\} + [\Phi]^T[K][\Phi]\{q\} = [\Phi]^T\{p(t)\}$$

Using the orthogonality properties for modal vectors and assuming that they have been scaled to unit modal mass, we get

$$[I]\{\ddot{q}\} + \begin{bmatrix} \ddots & & \\ & \omega_r^2 & \\ & & \ddots \end{bmatrix} \{q\} = [\Phi]^T\{p(t)\} = \{f(t)\} \text{ say.} \tag{6.63}$$

Just as equation (6.61) describes the equations of motion in terms of the physical coordinates, equation (6.63) describes the equations of motion in terms of the modal coordinates. Equation (6.62) can be seen mathematically as a vector space transformation, taking the equations from **physical space** (described by the physical displacement coordinates in $\{x\}$) to **modal space** (described by the modal coordinates $\{q\}$).

Because the two matrices on the left-hand side of equation (6.63) are both diagonal matrices, the individual equations are *not* coupled and each involves only one mode. The equation for a typical mode r has the form

$$\ddot{q}_r + \omega_r{}^2 q_r = f_r(t) \tag{6.64}$$

Note that if the vectors have *not* been scaled to unit modal mass, the substitution of equation (6.62) into (6.61) still results in uncoupled equations, but each then has the form

$$M_r\ddot{q}_r + K_r q_r = f_r(t)$$

In either case, there will be as many individual equations as there are modes of the structure, with each one involving only one of the modal coordinates.

Each equation describes a single mode of vibration and will tell us how much that mode contributes to the overall response. This has the natural frequency ω_r and is subjected to excitation defined by $f_r(t)$. Solutions for free vibration, harmonic and periodic excitation were described in Section 6.3, and solutions using Laplace transforms are presented in Unit 4. The fact that these are modal rather than physical space equations does not affect the mathematics.

The solutions to the modal space equations are the responses of each individual mode of vibration expressed in terms of the modal coordinates, $q_r(t)$. Finding the expressions for these responses is step 2 of the solution process.

Once we have them, we can obtain the response in each of the physical coordinates by reusing equation (6.62),

$$\{x\} = [\Phi]\{q\}$$

This is step 3 of the solution process.

Written out longhand, these equations are

$$\begin{Bmatrix} x_1(t) \\ x_2(t) \\ \vdots \\ x_j(t) \\ \vdots \end{Bmatrix} = \begin{bmatrix} u_{11} & u_{12} & \cdots & u_{12} & \cdots \\ u_{21} & u_{12} & \cdots & u_{12} & \cdots \\ \vdots & \vdots & \ddots & \vdots & \\ u_{j1} & u_{j2} & & u_{j1} & \cdots \\ \vdots & \vdots & & \vdots & \end{bmatrix} \begin{Bmatrix} q_1(t) \\ q_2(t) \\ \vdots \\ q_r(t) \\ \vdots \end{Bmatrix}$$

The response in a typical coordinate, j, is therefore

$$x_j(t) = \sum_r u_{jr} q_r(t) \qquad \textbf{(6.65)}$$

It can be seen that equation (6.65) states that the response of the structure at any instant is a weighted sum of the individual modal vectors. The modal coordinates, $q_r(t)$, define how much each mode shape contributes to the structure's overall response at each instant.

Worked example

Find the steady-state response of each mass in the following two-degrees-of-freedom system to a sinusoidal force of amplitude 10 N and frequency 6 Hz applied to the 1 kg mass.

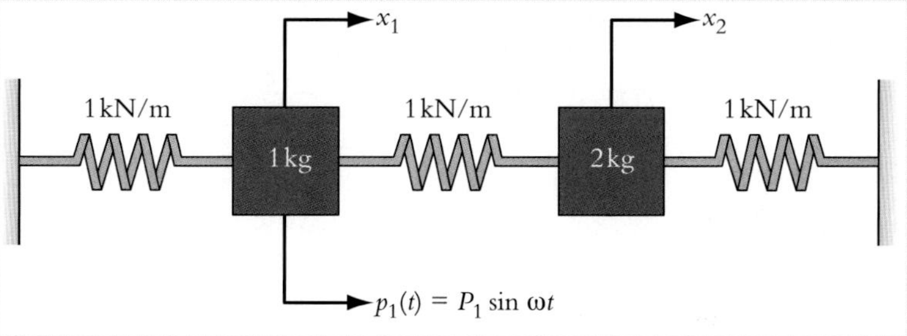

Figure 6.68

Using the method presented in Section 6.2, the equations of motion, natural frequencies and modal vectors are obtained first.

$$\begin{bmatrix} 1 & 0 \\ 0 & 2 \end{bmatrix} \begin{Bmatrix} \ddot{x}_1 \\ \ddot{x}_2 \end{Bmatrix} + \begin{bmatrix} 2000 & -1000 \\ -1000 & 2000 \end{bmatrix} \begin{Bmatrix} x_1 \\ x_2 \end{Bmatrix} = \begin{Bmatrix} p_1(t) \\ 0 \end{Bmatrix}$$

The eigenvalues are $\omega_1^2 = 634\,\mathrm{s}^{-2}$ and $\omega_2^2 = 2366\,\mathrm{s}^{-2}$ giving natural frequencies of 4.0 Hz and 7.7 Hz respectively.

The corresponding mode shape vectors are

$$\{X\}_1 = \begin{Bmatrix} X_{11} \\ X_{21} \end{Bmatrix} = \begin{Bmatrix} 0.732 \\ 1.0 \end{Bmatrix} \quad \text{and} \quad \{X\}_2 = \begin{Bmatrix} X_{12} \\ X_{22} \end{Bmatrix} = \begin{Bmatrix} -2.732 \\ 1.0 \end{Bmatrix}$$

These vectors are scaled by making $X_2 = 1.0$ for each mode. Using equations (6.58) and (6.59), the modal mass and modal stiffness values are:

$$M_1 = 2.54, M_2 = 9.46, K_1 = 1608 \text{ and } K_2 = 22\,392.$$

Rescaling to unit modal mass using equation (6.58) gives the new modal matrix:

$$[\Phi] = \begin{bmatrix} 0.460 & -0.888 \\ 0.628 & 0.325 \end{bmatrix}$$

Using the new modal matrix, the reader is left to confirm that $[\Phi]^T[M][\Phi] = \begin{bmatrix} 1 & 0 \\ 0 & 1 \end{bmatrix}$ and that $[\Phi]^T[K][\Phi] = \begin{bmatrix} 634 & 0 \\ 0 & 2366 \end{bmatrix}$. As mentioned earlier, the new modal stiffnesses of $K_1 = 634$ and $K_2 = 2366$ are numerically equal to the squares of the corresponding natural frequencies.

From $\{f(t)\} = [\Phi]^T\{p(t)\}$, the modal forces are

$$\begin{Bmatrix} f_1(t) \\ f_2(t) \end{Bmatrix} = \begin{bmatrix} 0.460 & 0.628 \\ -0.888 & 0.325 \end{bmatrix} \begin{Bmatrix} p_1(t) \\ 0 \end{Bmatrix} = \begin{Bmatrix} 0.460 \times p_1(t) \\ -0.888 \times p_1(t) \end{Bmatrix}$$

The modal equation for mode 1 is

$$\ddot{q}_1 + \omega_1^{\ 2} q_1 = f_1(t)$$

or

$$\ddot{q}_1 + 634 q_1 = 0.460 p_1(t) \tag{6.66}$$

For the steady-state response, put $p_1(t) = 10e^{i\omega t}$ and $q_1(t) = Q_1^\star e^{i\omega t}$ into (6.66) with an excitation frequency of 6 Hz to give:

$$Q_1^\star = \frac{4.60}{634 - (6 \times 2\pi)^2} = -0.005\,84$$

Similarly for mode 2,

$$Q_2^\star = \frac{-8.88}{2366 - (6 \times 2\pi)^2} = -0.009\,40$$

From equation (6.65),

$$\begin{Bmatrix} X_1^\star e^{i\omega t} \\ X_2^\star e^{i\omega t} \end{Bmatrix} = \begin{bmatrix} 0.460 & -0.888 \\ 0.628 & 0.325 \end{bmatrix} \begin{Bmatrix} Q_1^\star e^{i\omega t} \\ Q_2^\star e^{i\omega t} \end{Bmatrix}$$

Hence,

$$X_1^\star = 0.460 \times Q_1^\star - 0.888 \times Q_2^\star = 5.66 \text{ mm}$$

and

$$X_2^\star = 0.628 \times Q_1^\star + 0.325 \times Q_2^\star = -6.72 \text{ mm}$$

Notice that $X_1^\star$ is positive, showing that mass 1 moves in phase with the force applied to it, while the negative sign for $X_2^\star$ shows that mass 2 moves 180° out of phase with the applied force.

Repeating the exercise across a range of frequencies, the modal responses and the displacements of the two masses are as shown in Figures 6.69 and 6.70.

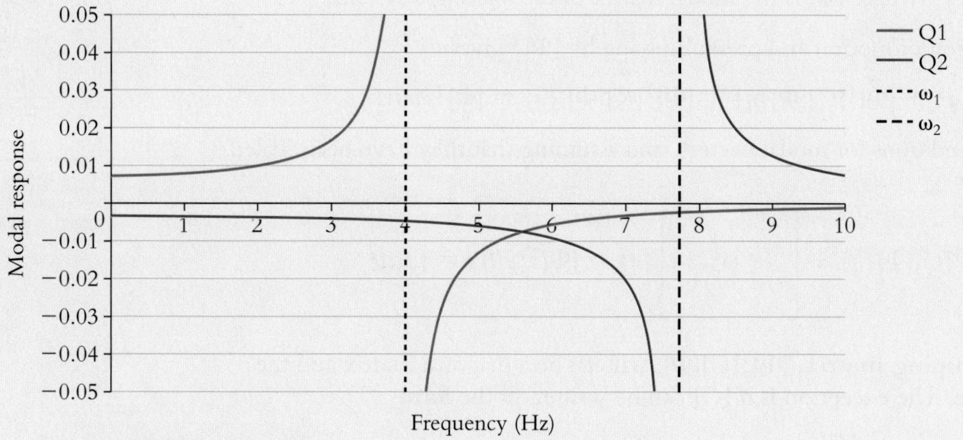

Figure 6.69 **Modal responses**

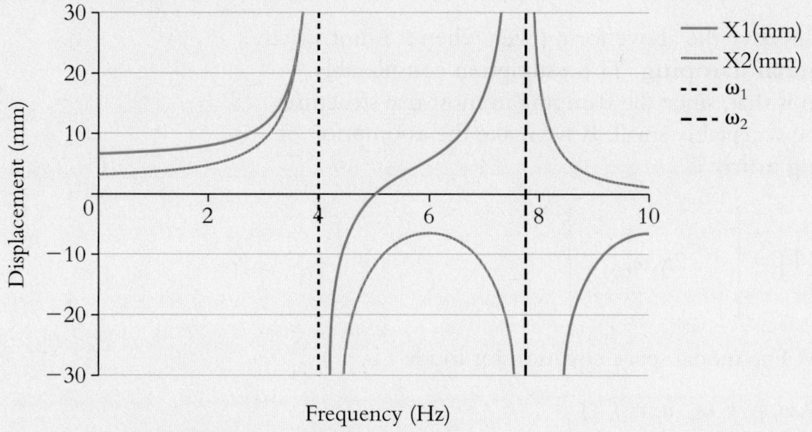

Figure 6.70 **Displacement of masses 1 and 2**

We see from the modal responses that each produces an infinite response if the excitation frequency coincides with the relevant natural frequency (4.0 Hz and 7.7 Hz, in this case). At frequencies in the vicinity of each natural frequency, the relative displacements of the two masses will closely resemble the proportions given by the relevant mode shape. In the vicinity of ω_1, for example, $|Q_1^\star| \gg |Q_2^\star|$ so the first mode will dominate the response. The chosen excitation frequency of 6 Hz is about midway between the two natural frequencies and we find that the two modes make similar contributions at this frequency.

Looking at the displacements of the masses, it's interesting to note that at about 5.5 Hz, mass 1 has a displacement amplitude of zero. This is perhaps surprising when you consider that the force is applied to this mass. The ability to design a dynamic system that has zero displacement at the point of force application is the basis for devices called **tuned vibration absorbers**. These were used, for example, to help cure the resonant vibration problem on London's Millennium Bridge.

Forced response of a damped structure

As with single-degree-of-freedom systems, the presence of damping complicates the analysis. The equations of motion are

$$[M]\{\ddot{x}\} + [C]\{\dot{x}\} + [K]\{x\} = \{p(t)\}$$

In the undamped case, the equations could be uncoupled by transforming into modal space using the modal matrix $[\Phi]$. We can try the same approach here.

For step 1, put $\{x\} = [\Phi]\{q\}$, where $[\Phi]$ is the modal matrix of the *undamped* system.

Substituting into the equation of motion and premultiplying by $[\Phi]^T$ gives

$$[\Phi]^T[M][\Phi]\{\ddot{q}\} + [\Phi]^T[C][\Phi]\{\dot{q}\} + [\Phi]^T[K][\Phi]\{q\} = [\Phi]^T\{p(t)\}$$

Using the orthogonality conditions for modal vectors, and assuming that they have been scaled to unit modal mass, we get

$$[I]\{\ddot{q}\} + [\Phi]^T[C][\Phi]\{\dot{q}\} + \begin{bmatrix} \ddots & & \\ & \omega_r^2 & \\ & & \ddots \end{bmatrix}\{q\} = [\Phi]^T\{p(t)\} = \{f(t)\}$$

In general, the **modal damping matrix**, $[\Phi]^T[C][\Phi]$, will *not* be a diagonal matrix and the equations will *not* uncouple. The exception is if $[C]$ can be written in the form

$$[C] = a_1[M] + a_2[K]$$

where a_1 and a_2 are constants. In this case, the equations *will* uncouple.

Engineers commonly assume that $[C]$ does have the above form (even when it is not strictly true), an assumption known as **proportional damping**. The assumption considerably simplifies the analysis and the justification is that, since the damping in most real structures is low, the error introduced is found to be acceptably small. If we make the assumption of proportional damping, the modal damping matrix is

$$[\Phi]^T[C][\Phi] = \begin{bmatrix} \ddots & & \\ & 2\gamma_r\omega_r & \\ & & \ddots \end{bmatrix}$$

where γ_r is the damping ratio for mode r. The modal space equation for mode r is

$$\ddot{q}_r + 2\gamma_r\omega_r\dot{q}_r + \omega_r{}^2 q_r = f_r(t)$$

Steps 2 and 3 of the solution are then the same as before. An example of this is used in the next section to find an expression for the frequency response function.

Frequency response function

The frequency response function for a multi-degree-of-freedom system can be calculated using a simple extension of the method used in Section 6.3 for a single-degree-of-freedom system. We will consider the case of sinusoidal excitation applied to coordinate k and find the response in coordinate j.

Section 6.4 showed that the equations of motion can be uncoupled to give a set of equations describing the behaviour of each mode. For mode r,

$$\ddot{q}_r + 2\gamma_r\omega_r\dot{q}_r + \omega_r{}^2 q_r = f_r(t)$$

The modal force vector, $\{f(t)\}$, is calculated from $[\Phi]^T\{p(t)\}$. In this case, only the kth coordinate has excitation; all the other elements in the vector $\{p(t)\}$ are zero.

$$\{p(t)\} = \begin{Bmatrix} 0 \\ \vdots \\ 0 \\ p_k(t) \\ 0 \\ \vdots \\ 0 \end{Bmatrix}$$

From $\{f(t)\} = [\Phi]^T\{p(t)\}$, we find that $f_r(t) = u_{kr}p_k(t)$.

For harmonic excitation, put $f_r(t) = u_{kr}P_k e^{i\omega t}$ and $q_r(t) = Q_r^\star e^{i\omega t}$ into the modal equation of motion.

$$Q_r^\star = \frac{u_{kr}P_k}{(\omega_r^2 - \omega^2) + i2\gamma_r\omega_r\omega}$$

or

$$q_r(t) = \frac{u_{kr}}{(\omega_r^2 - \omega^2) + i2\gamma_r\omega_r\omega\ P_k e^{i\omega t}}$$

Combining the individual modal responses gives the response in coordinate j to be:

$$x_j(t) = X_j^\star e^{i\omega t} = \sum_r u_{jr}q_r(t)$$

$$= \sum_r \frac{u_{jr}u_{kr}}{(\omega_r^2 - \omega^2) + i2\gamma_r\omega_r\omega\ P_k e^{i\omega t}}$$

The frequency response function giving the response in coordinate j due to a unit sinusoidal force in coordinate k is therefore:

$$H_{jk}(\omega) = \frac{X_j^\star}{P_k} = \sum_r \frac{u_{jr}u_{kr}}{(\omega_r^2 - \omega^2) + i2\gamma_r\omega_r\omega} \quad (6.67)$$

Learning summary

By the end of this section you should have learnt:

✓ to uncouple the equations of motion of a proportionally damped multi-degree-of-freedom structure by making use of the orthogonality properties of the modal matrix;

✓ to formulate response expressions, including the frequency response function.

6.5 Experimental modal analysis

Introduction

Methods for predicting the natural frequencies and mode shapes of different types of structure were discussed in Section 6.2. In each case, approximations were required to derive the model. Moreover, none of the models was able to predict the damping level in the structure.

Experimental modal analysis tests a structure as it is, without the need for modelling approximations. For each mode of vibration within a chosen frequency range, it identifies the natural frequency, damping and mode shape. These are called the modal parameters.

As an example, Figure 6.71 illustrates the first bending and first torsional mode shapes obtained from a test on a free–free beam. There are animated versions of these mode shapes on the book's supporting website. While many structures (such as helicopter rotor blades and lighting columns) have beam-like characteristics, few meet the idealized conditions incorporated into these theoretical models.

In this particular test, each mode shape has been defined in terms of the amplitude and relative phase of the vertical motion at 18 points on the beam (9 along each edge). Each point vibrates sinusoidally about the undisplaced (equilibrium) position, shown by the dotted lines in the figure. Figure 6.71 itself shows 'snapshots' of the beam displaced when it is at one extreme of this motion. As seen previously, some points on the beam (indicated by the chain-dashed lines) are seen to be stationary. These are called nodal lines and are unique to each mode shape.

A common way of performing the tests is to measure a set of frequency response functions and then find the modal parameters using a curve fitting procedure.

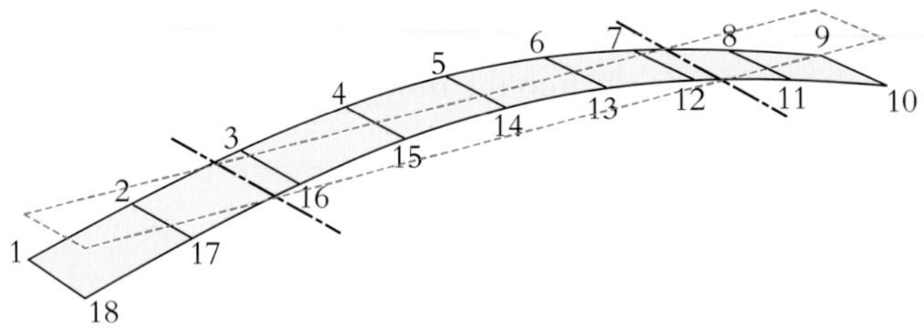

First bending mode

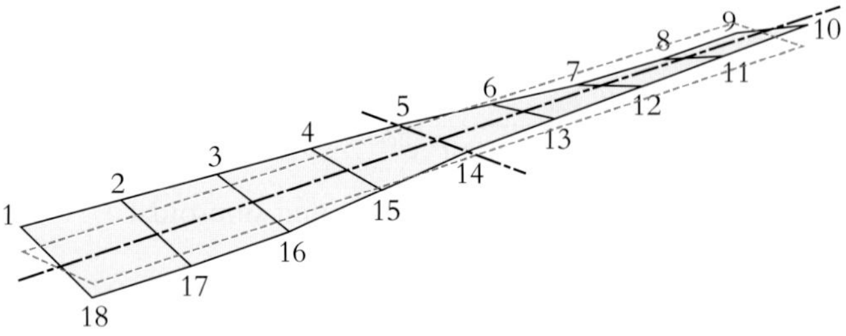

First torsional mode

Figure 6.71

Frequency response function testing

The frequency response function (FRF) defines the response per unit applied force across a range of frequencies. This was discussed in Section 6.3 where the FRF for a single-degree-of-freedom system was derived. A sinusoidal force input was used and an expression for the response displacement (expressed in terms of the steady-state amplitude and phase) was found. The FRF for an excitation frequency ω was shown (equation (6.49)) to be:

$$H(\omega) = \frac{Z^{\star}}{F} = \frac{1}{M}\frac{1}{(\omega_n^2 - \omega^2) + i2\gamma\omega_n\omega} \qquad \textbf{(6.68)}$$

where $Z^{\star}$ is a complex number containing amplitude and phase information about the response, F is the force amplitude, M is the coefficient of acceleration in the equation of motion and ω is the excitation frequency. The expression also contains two of the modal parameters; ω_n, which is the undamped natural frequency and γ, the damping ratio.

The amplitude and phase spectra are shown in Figure 6.72, plotted against the frequency ratio, ω/ω_n. These were discussed in detail in Section 6.3.

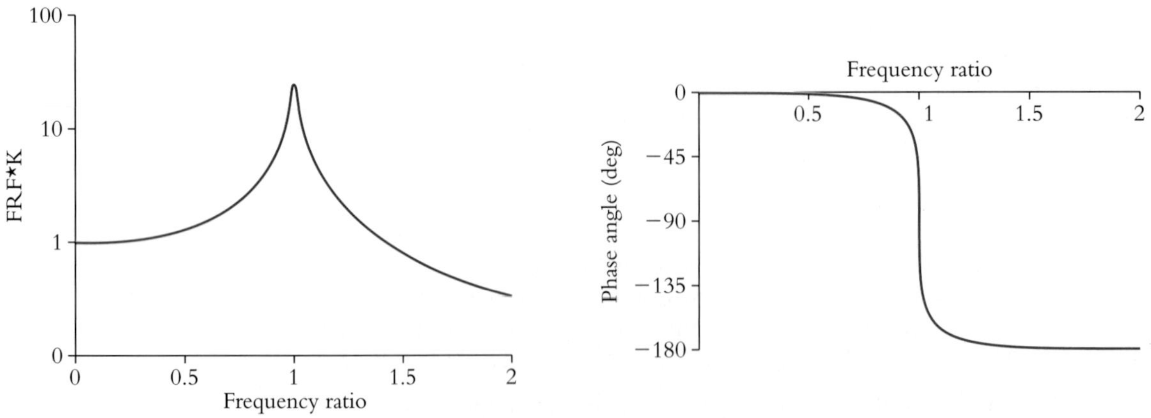

Figure 6.72

If a sinusoidal force was applied to a structure and the steady-state amplitude and phase of the response was measured, one point on each graph would be found. The complete FRF plot could be measured by repeating this test at many different excitation frequencies. This is called a swept-sine test.

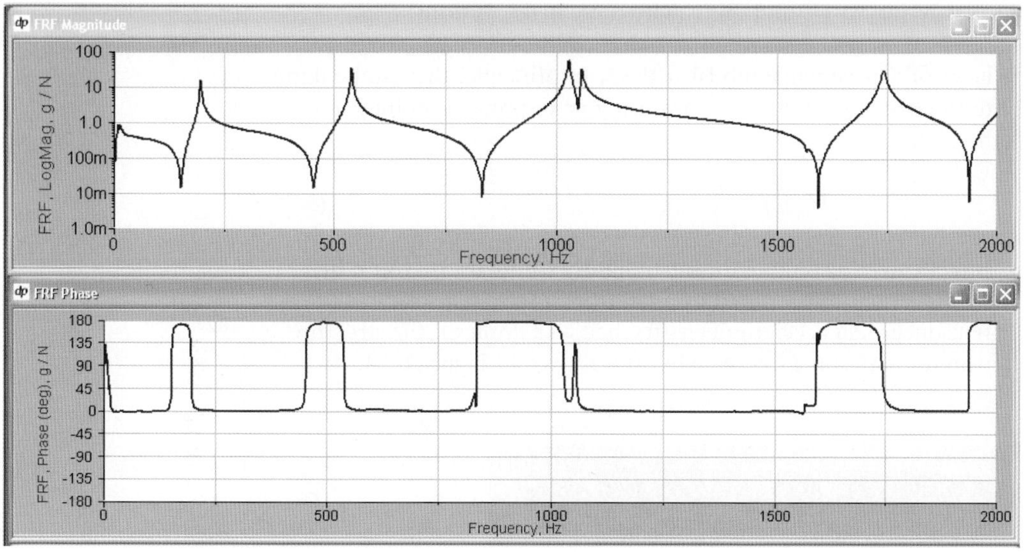

Figure 6.73

For a single-degree-of-freedom system, there is one resonant peak. For a multi-degree-of-freedom system, there would be several resonant peaks, but the swept-sine test would still work.

Figure 6.73 is an example of an FRF of a multi-degree-of-freedom system. This one has five resonant peaks, indicating five modes of vibration in the frequency range. Note that each peak (such as the one highlighted in the box) has a related phase shift of 180°. This is the same characteristic seen in the FRF for the single-degree-of-freedom system in Figure 6.72.

Since a single-degree-of-freedom system has only one coordinate to describe it, the response in that coordinate tells us all we need to know. For a multi-degree-of-freedom system, the displacement varies from point to point on the structure. At each resonance, the relative amplitude and phase of the different points is called the mode shape and is different from one resonance to the next.

Section 6.4 showed – in equation (6.67) – that the FRF for a multi-degree-of-freedom system giving the response in coordinate j due to a unit sinusoidal force in coordinate k is:

$$H_{jk}(\omega) = \frac{X_j^{\star}}{P_k} = \sum_r \frac{u_{jr}u_{kr}}{(\omega_r{}^2 - \omega^2) + \mathrm{i}2\gamma_r\omega_r\omega} \tag{6.69}$$

where the summation is taken over all of the modes, r, of the structure, and ω_r and γ_r are the undamped natural frequency and damping ratio for mode number r. u_{jr} and u_{kr} are the mode shape values in coordinates j and k for mode r.

This can be rewritten as:

$$H_{jk}(\omega) = \sum_r \frac{(A_{jk})_r}{(\omega_r^2 - \omega^2) + \mathrm{i}2\gamma_r\omega_r\omega} \tag{6.70}$$

The terms $(A_{jk})_r = u_{jr}u_{kr}$ are called 'residues' and are simply the product of the mode shape values of the structure at the response and excitation positions for each mode.

You can see that each of the five modes – i.e. each term in the summation in equation (6.70) – is described by an expression like equation (6.68). In particular, if there was only one mode, so that the summation in equation (6.70) had only one term, and the residue was equal to $\frac{1}{M}$, then it would be identical to equation (6.68). When deriving equation (6.69), the mode shapes were scaled to give modal masses equal to 1.0 and this is why there is no $\frac{1}{M}$ term.

Measuring FRFs lies at the heart of experimental modal analysis. In principle, this can be done by applying a known force to the structure at point k, measuring its response at point j and then dividing one by the other. That is:

$$H_{jk}(\omega) = \frac{X_j^*}{F_k} \qquad (6.71)$$

An 'impact test' is a simple and effective technique for small, lightly damped structures. In a typical set-up for such a test, the structure is excited by a tap from a special instrumented hammer containing a force transducer. An accelerometer attached to a point on the structure measures the response. A set-up for a test on a free–free beam is shown in Figure 6.74.

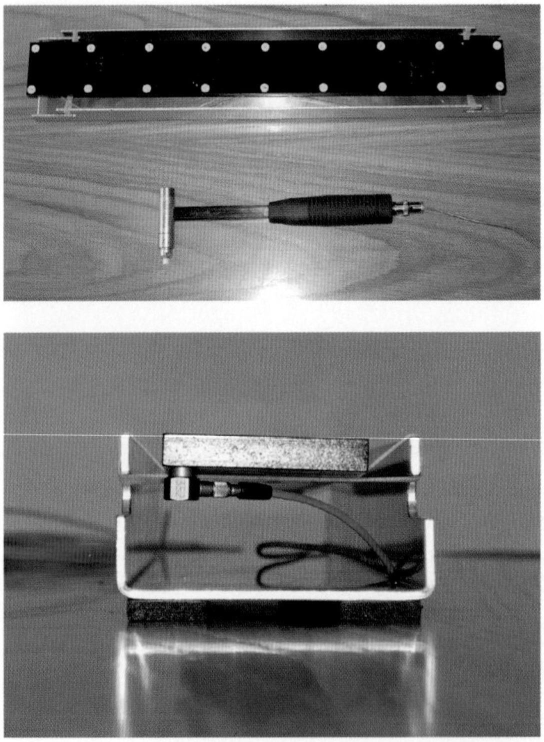

Figure 6.74 **(Top) Beam supported in a cradle using rubber bands to simulate a free–free condition An instrumented hammer is next to the beam; (Bottom) Accelerometer measuring the vertical response at one corner (Point #1)**

The impact excites a wide range of frequencies simultaneously and the resulting transient response is picked up by the accelerometer. Typical force and response signals are shown in Figure 6.75.

The tap from the hammer excites many modes of the structure simultaneously. Each mode behaves like a separate single-degree-of-freedom system and will give a transient response consisting of a decaying sinusoidal vibration at that mode's natural frequency. The overall

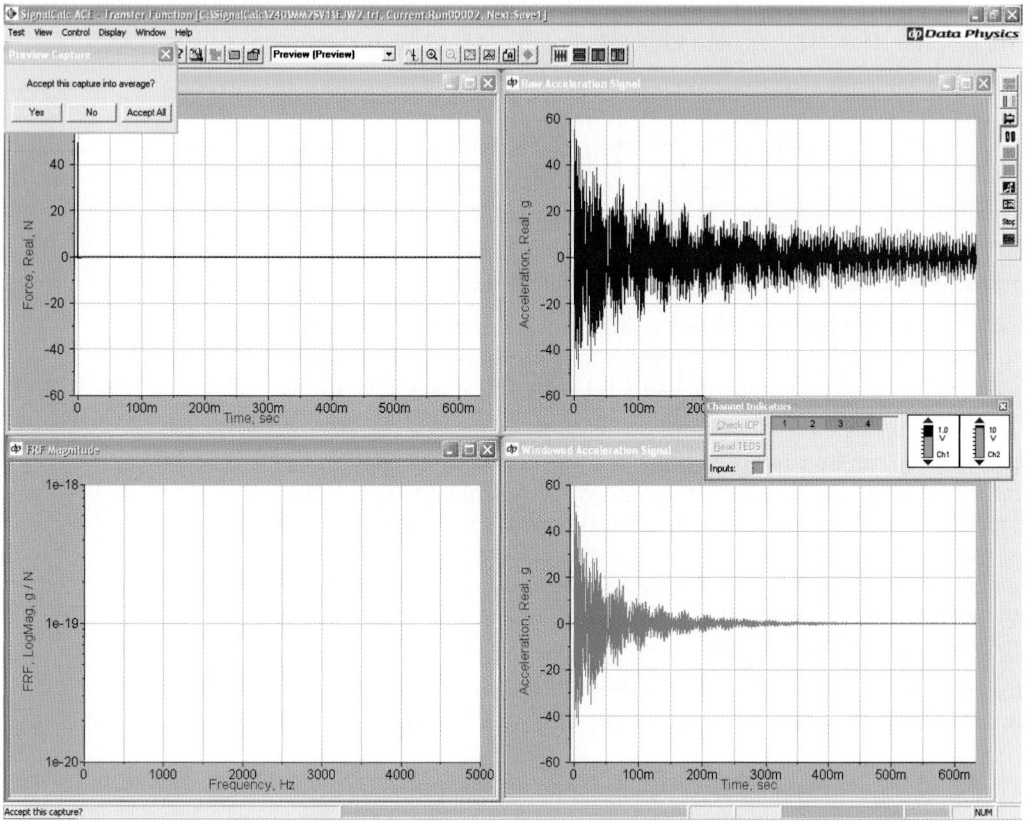

Figure 6.75 **Typical force and response signals**

response is the combination of the individual decaying oscillations from all of the modes in the range. Provided that the structure behaves as a linear system, the principle of superposition applies, and the individual modal responses will not interact with one another.

The force and response are both time domain signals, but they are clearly not sine waves as used in the theoretical derivation of the FRF. Once digitized, the force and response signals can, however, be analysed using the discrete Fourier transform to identify the individual frequency components they comprise. For each pair of frequency components, dividing response by force gives the FRF at that frequency.

$$H_{jk}(\omega) = \frac{X_j(\omega)}{F_k(\omega)} \tag{6.72}$$

Doing this for all of the frequencies gives the complete FRF. An example was shown in Figure 6.73.

Mode identification

It is apparent from equation (6.70) that the frequency response function embodies the modal properties of the system (natural frequency, damping and mode shape). It follows that if you can find values of $(A_{jk})_r$, ω_r and γ_r for each mode that make equation (6.70) match the measured FRF, you have a means of identifying the modal parameters. The matching can be done by curve fitting an analytical fit function such as equation (6.70) to measured FRFs. This is the basis for many commercial parameter identification programs.

In the earlier example of a free–free beam, the vertical coordinate at 18 measurement points were selected. If we used every combination of force and response coordinate, it would be possible to measure a total of 324 (i.e. 18 × 18) separate FRFs. These tests could be put into matrix form as:

$$\begin{Bmatrix} x_1 \\ x_2 \\ \vdots \\ x_{18} \end{Bmatrix} = \begin{bmatrix} H_{1,1} & H_{1,2} & \cdots & H_{1,18} \\ H_{2,1} & H_{2,2} & \cdots & H_{2,18} \\ \vdots & \vdots & \ddots & \vdots \\ H_{18,1} & H_{18,2} & \cdots & H_{18,18} \end{bmatrix} \begin{Bmatrix} F_1 \\ F_2 \\ \vdots \\ F_{18} \end{Bmatrix} \tag{6.73}$$

In this matrix, $H_{1,2}$ is the response in coordinate 1 due to a force applied in coordinate 2, etc.

Fortunately, it's only necessary to measure one row or one column of the matrix of FRFs. If we were using an impact test, we might place the accelerometer to measure the response at coordinate 1 due to forces applied in turn at coordinates 1 to 18. Doing this would give us the top row of the matrix.

Equation (6.70) shows that each FRF can be computed if the residue, natural frequency and damping values are known for each mode. Conversely, the right-hand side of equation (6.70) can be thought of as a parametric fitting curve in which the modal parameters must be selected to give the best fit between the computed FRF and the measured one. General software packages such as Matlab can be programmed to do the fitting and there are several dedicated commercial programs on the market too. A variety of algorithms are used to do the curve fitting, but a description of these is outside the scope of this book.

The frequency and damping of a mode are global properties; they are the same regardless of the measurement positions. However, the residue for each mode, $(A_{jk})_r$ will be different for each of the FRFs.

Once the curve fitting has been done for all of the FRFs, there will be residue values for each of the measurement coordinates and for each mode. In the beam example, if we had measured the top row of the FRF matrix, the curve fitting process would have given us the following vector of residues for each mode:

$$\{A_{1,1} \quad A_{1,2} \quad \cdots \quad A_{1,18}\}_r \tag{6.74}$$

Recalling that the residues are simply the products of mode shape values, equation (6.74) can be rewritten as:

$$\{(u_1)_r \cdot (u_1)_r \quad (u_1)_r \cdot (u_2)_r \quad \cdots \quad (u_1)_r \cdot (u_{18})_r\} \tag{6.75}$$

You can see that $(u_1)_r$ is a common factor in every term in this case. This value can be found provided that one of the measurements has the force applied at the same coordinate that the response is measured at. In the example being used, this is coordinate #1. Once you have the value of $(A_{1,1})_r$, the common factor can be found from equation (6.76).

$$(u_1)_r \cdot (u_1)_r = (A_{1,1})_r$$

or

$$(u_1)_r = \sqrt{(A_{1,1})_r} \tag{6.76}$$

Hence, the complete mode shape vector can be found from the corresponding residue vector using equation (6.75).

$$(u_k)_r = \frac{(A_{1,k})_r}{\sqrt{(A_{1,1})_r}} \tag{6.77}$$

Learning summary

By the end of this section you should have learnt:

✓ how the expression for the frequency response function for a multi-degree-of-freedom structure can be used to identify the natural frequencies, damping and mode shapes of structures from experimental tests.

6.6 Approximate methods

Introduction

Often the lowest natural frequency of a structure is the most important. For example, if the first critical speed of a shaft is above its operating speed range, whirl is avoided. It is often the case that the first mode gives the largest displacement for a given excitation.

This section presents two methods for estimating the lowest natural frequency and shows how single-degree-of-freedom approximations of more complex systems can be created.

Dunkerley's method

Stanley Dunkerley developed a method of estimating the lowest critical speed of shafts. His formula can be used to estimate the lowest natural frequency of any system.

$$\frac{1}{\omega_n^2} \approx \frac{1}{\omega_1^2} + \frac{1}{\omega_2^2} + \cdots + \frac{1}{\omega_r^2} + \cdots \quad (6.78)$$

where ω_r is the natural frequency of a subsystem that has only the rth mass present. The following examples show how the formula is used.

Example 1: Two-degree-of-freedom system

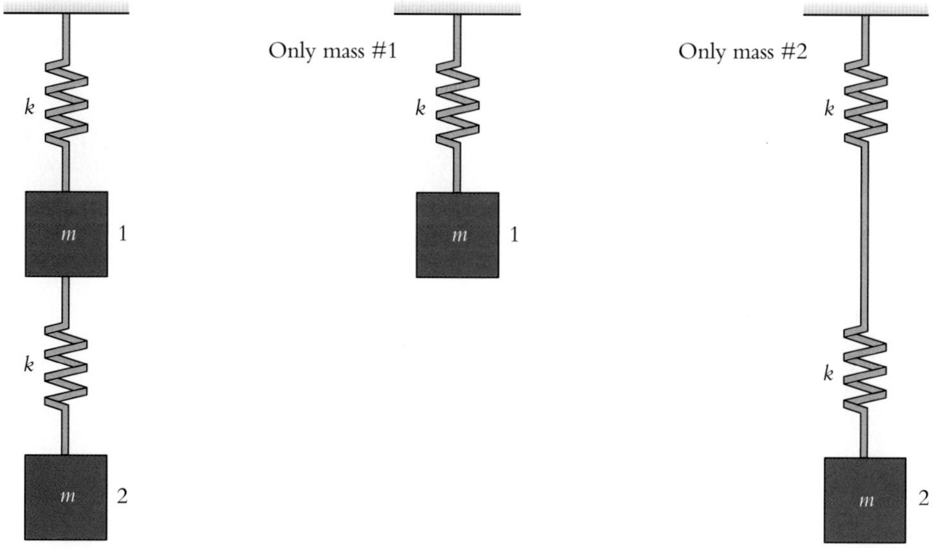

Figure 6.76 **Two-degree-of-freedom system and its two subsystems**

The system on the left of Figure 6.78 has two masses, so there will be two subsystems, one with only mass 1 and the other with only mass 2. The two subsystems are also shown in Figure 6.76.

Subsystem with only mass 1 When we remove mass 2 to create this subsystem, we also ignore the lower spring, since this is assumed to be massless. The natural frequency of the subsystem is given by:

$$\omega_1^2 = \frac{k}{m}$$

Subsystem with only mass 2 Removing mass 1 leaves mass 2 supported by two springs in series, each of stiffness k. These are equivalent to a single spring of stiffness $\frac{k}{2}$. The natural frequency of the subsystem is given by:

$$\omega_2^2 = \frac{\frac{k}{2}}{m} = \frac{k}{2m}$$

With two subsystems, Dunkerley's formula will have two terms in the series on the right-hand side. Substituting the subsystem results into the formula, we get

$$\frac{1}{\omega_n^2} \approx \frac{m}{k} + \frac{2m}{k} = \frac{3m}{k}$$

Hence,

$$\omega_n = \sqrt{\frac{k}{3m}} \tag{6.79}$$

As a numerical example, if $m = 2\,\text{kg}$ and $k = 200\,\text{N/m}$, equation (6.79) gives the lowest natural frequency to be 0.918 Hz. The exact answer is 0.984 Hz.

Example 2: Shaft with added masses

The phenomenon of shaft whirl, and a method for calculating the whirl frequencies for plain shafts with uniform cross-sections, were presented in Section 6.2. While this could be a good approximation for the drive shaft on a vehicle, for example, it would not be appropriate for the steam turbine shaft shown in Figure 6.77. This carries several large-bladed discs that add significant mass to the system and affect its natural frequencies and hence its whirl speeds. Whirl can be avoided if the rotational frequency of the shaft remains below its lowest natural frequency in flexure and Dunkerley's method is able to give an estimate of this.

Figure 6.77 **Steam turbine shaft and discs**

We will look at a simplified example consisting of a uniform shaft carrying three added masses.

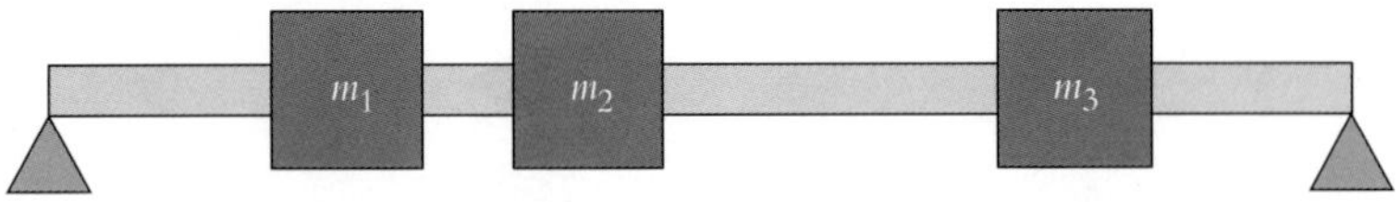

Figure 6.78

Subsystem with only mass 1 present is:

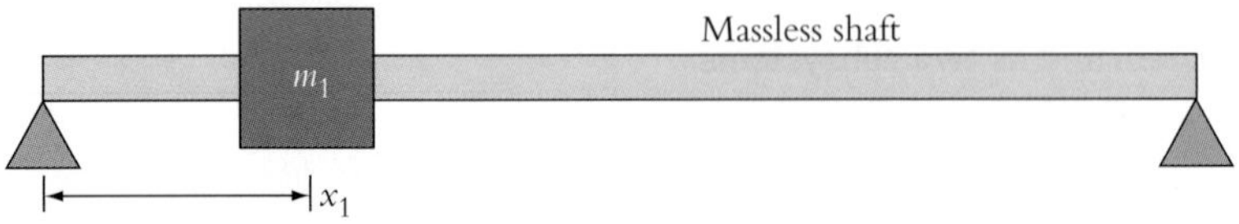

Figure 6.79

Since we are assuming that only mass 1 is present, we model the shaft as a massless spring. Hence, the natural frequency for the system can be written as:

$$\omega_1^2 = \left(\frac{\text{shaft stiffness at mass position, } x_1}{m_1}\right)$$

For a uniform, simply supported massless shaft, the stiffness at the position of mass 1 is

$$k_1 = \frac{3EIL}{x_1{}^2(L - x_1)^2}$$

Hence,

$$\omega_1{}^2 = \frac{3EIL}{m_1 x_1{}^2(L - x_1)^2}$$

This can be repeated with mass 2 only, and with mass 3 only, to give $\omega_2{}^2$ and $\omega_3{}^2$. There is also a fourth system in which all three of the added masses are removed to leave only the mass of the shaft present. This is:

Figure 6.80

For a uniform, simply supported shaft, the first natural frequency is

$$\omega_0 = \left(\frac{\pi}{L}\right)^2 \sqrt{\frac{EI}{\rho A}}$$

Combining all of the frequencies using Dunkerley's formula gives:

$$\frac{1}{\omega_n{}^2} \approx \frac{1}{\omega_1{}^2} + \frac{1}{\omega_2{}^2} + \frac{1}{\omega_3{}^2} + \frac{1}{\omega_0{}^2}$$

For all systems, Dunkerley's formula always gives an **underestimate** of the true frequency. This is very useful, since it provides a conservative estimate for whirl calculations. Specifically, to avoid the possibility of whirl, the lowest critical speed (i.e. the lowest natural frequency) of the shaft must be greater than the top speed of the shaft. Therefore, if the top speed is below the value predicted by Dunkerley's formula, it should also be below the actual critical speed.

Rayleigh's method

Lord Rayleigh observed that for an undamped system vibrating freely at one of its natural frequencies, energy is conserved so that

maximum kinetic energy = maximum strain energy

This is the basis for his method. The strain and kinetic energies can be found for any structure, provided that we know the deflected shape (i.e. the mode shape). Since we do not normally know the exact mode shape, it is necessary to make an estimate of it. This is an important step since the accuracy depends on making a good guess.

For systems with lumped mass and massless springs, the instantaneous kinetic energy for mass i is

$$(T)_{\text{mass}\, i} = \tfrac{1}{2} m_i \dot{x}_i^2$$

For sinusoidal vibration, $x_i(t) = X_i \sin \omega t$ and $\dot{x}_i(t) = \omega X_i \cos \omega t$. Hence the maximum kinetic energy for mass i is

$$(T_{\text{MAX}})_{\text{mass}\, i} = \tfrac{1}{2} m_i (\omega X_i)^2 = \tfrac{1}{2} \omega^2 m_i X_i^2 \qquad (6.80)$$

where X_i is the amplitude of vibration for mass i based on the assumed mode shape.

The maximum strain energy in the spring

$$(U_{\text{MAX}})_{\text{spring}\, j} = \tfrac{1}{2} k_j (\text{maximum change of length})^2 \qquad (6.81)$$

The totals for the complete system are given by summing the contributions of all the masses and all the springs. Equating the strain and kinetic energies, and rearranging, gives an expression for ω^2. The result can also be calculated from the mass and stiffness matrices for the system. That is

$$\omega^2 = \frac{\{\phi\}^T[K]\{\phi\}}{\{\phi\}^T[M]\{\phi\}}$$

where $\{\phi\}$ is the assumed mode shape vector.

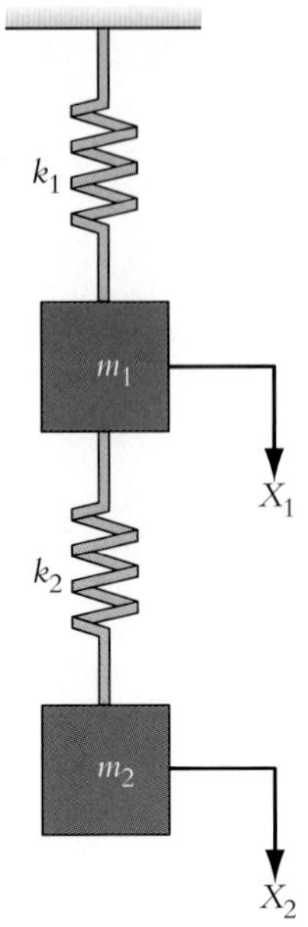

Figure 6.81

Example 1: Two-degree-of-freedom system

From equation (6.78) the combined maximum kinetic energy for both masses is

$$T_{\text{MAX}} = \tfrac{1}{2}\omega^2 m_1 X_1^{\,2} + \tfrac{1}{2}\omega^2 m_2 X_2^{\,2}$$
$$= \tfrac{1}{2}\omega^2(m_1 X_1^{\,2} + m_2 X_2^{\,2})$$

From equation (6.81) the combined maximum strain energy for the two springs is

$$U_{\text{MAX}} = \tfrac{1}{2}k_1 X_1^{\,2} + \tfrac{1}{2}k_2(X_1 - X_2)^2$$

Equating T_{MAX} and U_{MAX}, we get

$$\omega_n^{\,2} = \frac{k_1 X_1^{\,2} + k_2(X_1 - X_2)^2}{m_1 X_1^{\,2} + m_2 X_2^{\,2}} \tag{6.82}$$

If we can estimate the mode shape, we will have values of X_1 and X_2 that can be substituted into this equation.

As a numerical example, take the case of $m_1 = m_2 = 2$ kg and $k_1 = k_2 = 200$ N/m. For the first mode of vibration, we would expect the two masses to vibrate in phase with each other and for $X_2 > X_1$. Given that the springs have equal stiffness, we might guess that they both stretch by the same amount, which would suggest that $X_1 = 1$ and $X_2 = 2$ would be a reasonable estimate.

Substituting these values into equation (6.82) gives a value for ω_n of 1.007 Hz. The exact answer for this problem is 0.984 Hz, so the error is 2.3%.

The lowest mode shape often resembles the **static deflection shape**, so this is invariably a good 'guess'.

Here, noting that the bottom spring supports one mass, but the top spring carries the weight of both masses, the static extension of the top spring will be twice that of the lower spring. As a result, the static deflection shape is $X_1 = 2$, and $X_2 = 3$. This gives $\omega_n = 0.987$ Hz, which is an error of only 0.4%.

If we knew the exact mode shape, we would find that this gave the exact answer.

Because Rayleigh's method imposes a deflection shape on the system, it effectively constrains it to vibrate in a different way from the true mode shape. As a result, Rayleigh's method will always give an overestimate of the natural frequency, unless you happen to guess the exact mode shape by good fortune.

Thus,

$$\omega_{\text{Rayleigh}} \geqslant \omega_{\text{Exact}}$$

The technique to adopt is to try several possible mode shapes. The lowest of the predicted frequencies will be the most accurate.

Rayleigh's method for shafts and beams

Rayleigh's method can also be applied to shafts and beams. In this case, the expressions for the maximum kinetic and strain energies are

$$T_{\text{MAX}} = \frac{1}{2}\omega^2 \int_0^L \rho A[Y(x)]^2\,dx \tag{6.83}$$

$$U_{\text{MAX}} = \frac{1}{2}\int_0^L EI\left(\frac{d^2Y}{dx^2}\right)^2 dx \tag{6.84}$$

where $Y(x)$ is the mode shape function, which defines the amplitude of vibration of the shaft/beam along its length. Non-uniform cross sections (where A and I are functions of x) can be analysed, as can systems of interconnected beams, and systems that include discrete masses and springs. In each case, we sum the contribution of each element to the overall strain and kinetic energies.

We need to guess $Y(x)$ in order to evaluate the integrals.

Example 1: Uniform cantilever beam

The exact answer for this problem can be found by substituting the first root of the frequency equation given in Table 6.3 into equation (6.23). This gives:

$$\omega_n = \frac{3.52}{L^2}\sqrt{\frac{EI}{\rho A}}$$

The main criterion for choosing the mode shape is to ensure that it satisfies the displacement and slope conditions at the ends of the beam/shaft.

For a cantilever beam, $Y = \frac{dY}{dx} = 0$ at $x = 0$. The exact mode shape (see Section 6.2) is:

$$Y_1(x) = \sin\lambda_1 x - \sinh\lambda_1 x - \frac{\sin\lambda_1 L + \sinh\lambda_1 L}{\cos\lambda_1 L + \cosh\lambda_1 L}(\cos\lambda_1 x - \cosh\lambda_1 x)$$

It would be impossible to 'guess' this function, particularly when we don't know the natural frequency (which is needed to work out the wavenumber, λ_1).

Choice 1 Quadratic function $\quad Y(x) = Cx^2$

This expression satisfies the displacement and slope conditions at the clamped end and gives a shape that is similar to the actual mode shape.

The maximum kinetic energy is given by

$$T_{MAX} = \frac{1}{2}\omega^2\int_0^L \rho A[Cx^2]^2\,dx$$

$$= \frac{1}{2}\omega^2\rho A \times \frac{CL^5}{5}$$

$$= \omega^2\frac{\rho A C^2 L^5}{10}$$

The maximum strain energy

$$U_{MAX} = \frac{1}{2}\int_0^L EI\left(\frac{d^2Y}{dx^2}\right)^2 dx$$

$$= \frac{1}{2}\int_0^L EI(2C)^2\,dx$$

$$= 2EIC^2L$$

Equating gives $\quad \omega_n^{\,2} = 20\frac{EI}{\rho AL^4}$

Prediction is $\omega_n = \frac{4.47}{L^2}\sqrt{\frac{EI}{\rho A}}$, which is significantly (27%) higher than the exact value.

Choice 2 Static deflection shape $Y(x) = C\left[\left(\frac{x}{L}\right)^4 - 4\left(\frac{x}{L}\right)^3 + 6\left(\frac{x}{L}\right)^2\right]$

This expression gives $\omega_n = \frac{3.54}{L^2}\sqrt{\frac{EI}{\rho A}}$, an error of less than 1%.

Choice 3 A trigonometric function

Suitable functions can often be found by inspection, avoiding the need to calculate the static deflection shape. For a cantilever beam, a suitable function would be:

$$Y(x) = 1 - \cos\left(\frac{\pi x}{2L}\right)$$

Using this, the prediction is $\omega_n = \frac{3.66}{L^2}\sqrt{\frac{EI}{\rho A}}$, which is 4% high.

Figure 6.82 compares each of the three approximate mode shapes with the exact expression. The quadratic function is the least accurate shape and gives the least accurate estimate of the natural frequency. Notice that there is greater bending in the quadratic function, resulting in an overestimate of the strain energy. Also the displacement amplitudes are less, resulting in an underestimate of the kinetic energy. Since we effectively divide the strain energy by the kinetic energy (albeit without the ω^2 term), these two factors explain why the quadratic function gives such a high estimate. The static deflection shape is closest to the exact mode shape and this is why it gives the most accurate estimate of the natural frequency. The trigonometric function is a good compromise and can be found for different beams by inspection.

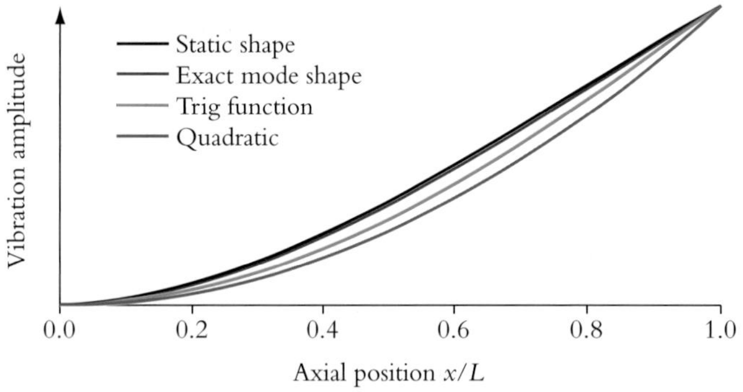

Figure 6.82 Comparison of mode shape estimate for a cantilever beam

Example 2: Beams and shafts with added masses

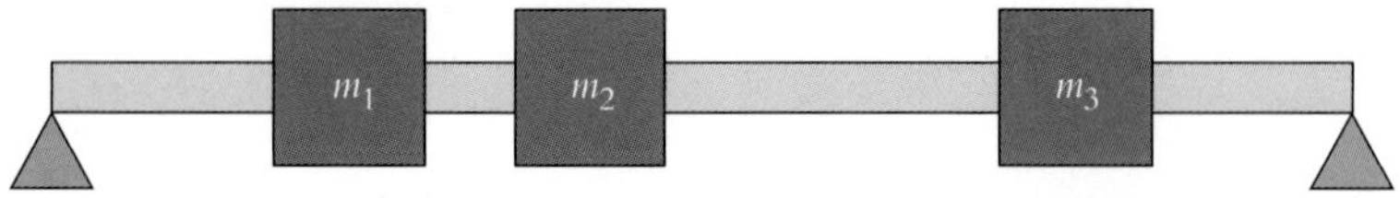

Figure 6.83

The added masses don't change the strain energy, but they add extra kinetic energy.

For mass at $x = x_p$ the maximum velocity is $\omega Y(x_p)$

Contribution of mass p to the kinetic energy is $\frac{1}{2}m_p[\omega Y(x_p)]^2$

These contributions need to be added to the kinetic energy of the shaft itself. Hence, the total kinetic energy becomes

$$T_{\text{MAX}} = \frac{1}{2}\omega^2\int_0^L \rho A[Y(x)]^2\,dx + \sum_{\text{ALL MASSES}} \frac{1}{2}m_r\omega^2[Y(x_p)]^2$$

Single-degree-of-freedom dynamic models of complex systems

The approach is based on the observation that if *the total strain and kinetic energies of two different dynamic systems are identical, an exact analogue relationship will exist between the two systems.* Such systems are called **dynamically equivalent systems**.

In situations where the real system has many degrees of freedom, but where only one mode of vibration is of interest, the concept provides a powerful method for producing an approximate single-degree-of-freedom model to describe that mode of vibration. Not only are single-degree-of-freedom models easy to analyse, but examples in this section have confirmed that they can often give good predictions of the general behaviour of more complex systems. We will consider only undamped systems here.

The approximate model consists of a simple mass–spring system, in which the displacement of the mass represents the displacement of some chosen point on the real structure.

Using the concept of dynamically equivalent systems, the mass and spring stiffness of the approximate model are chosen so that the maximum strain and kinetic energies of real and model systems are the same.

To do this, we assume that the lowest mode of vibration is dominant and therefore defines the deformation pattern in the structure. This is a good assumption if the system is vibrating sinusoidally near its lowest mode of vibration, but can also work well in other cases. An example of the latter was shown in Figure 6.67 where the response of a five-degrees-of-freedom system to a sequence of rectangular pulses was shown for comparison with that of an equivalent single-degree-of-freedom model.

As with Rayleigh's method, the accuracy of the model relies on having a good estimate of the real structure's mode shape.

Example 1: Lumped mass system

In this example, we will find an approximate single-degree-of-freedom model to analyse the motion of the top mass of the three-degree-of-freedom system shown in Figure 6.84.

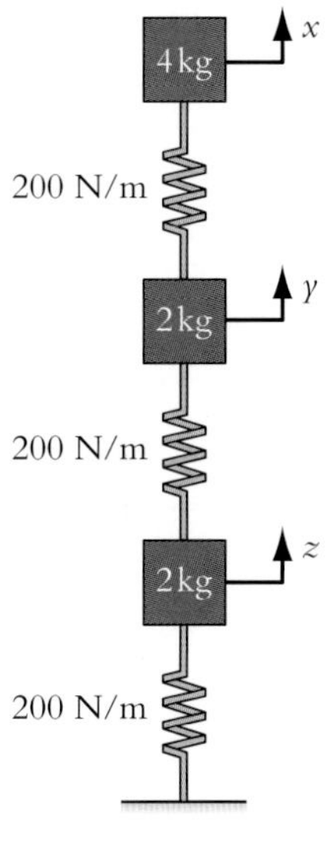

Figure 6.84

The first step is to establish a link between the displacement of the mass in the approximate single-degree-of-freedom model and some chosen point on the real system. In this case, the obvious choice is to link the displacement of the approximate model to the displacement of the top mass, x, since it's the motion of this mass that we want to predict.

However, since we are assuming that the motion of the real system is given by our chosen mode shape, not only does the approximate model predict the behaviour of the coordinate used as the link with the real system, but we can also work out the motion of the other coordinates since they are all linked by the assumed mode shape. In this example, it means that once we've found $x(t)$ we can get $y(t)$ and $z(t)$, since each will be related to $x(t)$ in proportion to the mode shape $\begin{Bmatrix} X \\ Y \\ Z \end{Bmatrix}$.

The approximate model is shown in Figure 6.85, with the motion coordinate chosen as x to match the coordinate of the top mass in the actual system.

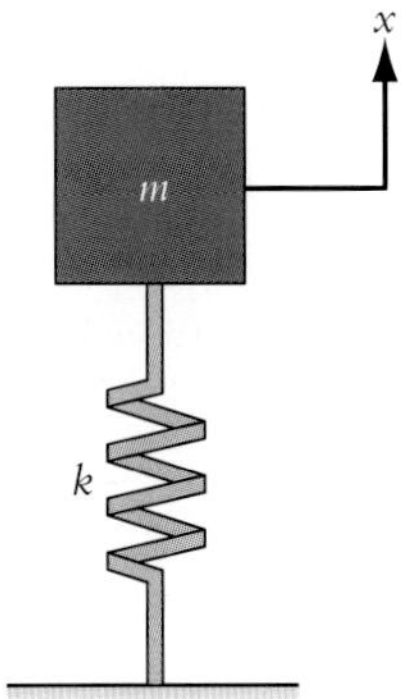

Figure 6.85

To find the values for the mass and stiffness in the model, we equate the maximum kinetic and strain energies in the real and approximate systems.

Equating maximum kinetic energies in the real (LHS of the equation) and model systems (RHS of the equation), we get:

$$\tfrac{1}{2}4(\omega X)^2 + \tfrac{1}{2}2(\omega Y)^2 + \tfrac{1}{2}2(\omega Z)^2 = \tfrac{1}{2}m(\omega X)^2$$

Hence,

$$m = \frac{4X^2 + 2Y^2 + 2Z^2}{X^2} \tag{6.85}$$

Equating maximum strain energies in the real and model systems, we get:

$$\tfrac{1}{2}200(X - Y)^2 + \tfrac{1}{2}200(Y - Z)^2 + \tfrac{1}{2}200 \cdot Z^2 = \tfrac{1}{2}kX^2$$

Hence,
$$k = \frac{200[(X - Y)^2 + (Y - Z)^2 + Z^2]}{X^2} \qquad \textbf{(6.86)}$$

To get values for m and k, we need an estimate for the mode shape. For the lowest mode shape, we expect all three masses in the real system to vibrate in phase with each other and, since the top mass is not connected to ground, that $X > Y > Z$.

Choice 1 Since all springs have the same stiffness, we might guess that they all deflect by the same amount. This implies that the mode shape would be:

$$\begin{Bmatrix} X \\ Y \\ Z \end{Bmatrix} = \begin{Bmatrix} 3 \\ 2 \\ 1 \end{Bmatrix}$$

Substituting these values into equations (6.83) and (6.84) gives, $m = 5.11$ kg and $k = 66.7$ N/m.

Choice 2 The static deflection shape for this system is $\begin{Bmatrix} X \\ Y \\ Z \end{Bmatrix} = \begin{Bmatrix} 9 \\ 7 \\ 4 \end{Bmatrix}$ and substituting these values into equations (6.83) and (6.84) gives, $m = 5.60$ kg and $k = 71.6$ N/m.

To decide which of these models is the more accurate, we can compare the natural frequencies predicted by each one.

Choice	m [kg]	k [N/m]	ω_n [Hz]
1	5.11	66.7	0.575
2	5.60	71.6	0.569

It can be seen that the second choice (the static deflection shape) is the more accurate since it gives the lower estimate of the natural frequency.

Example 2: Forced response of a cantilever beam

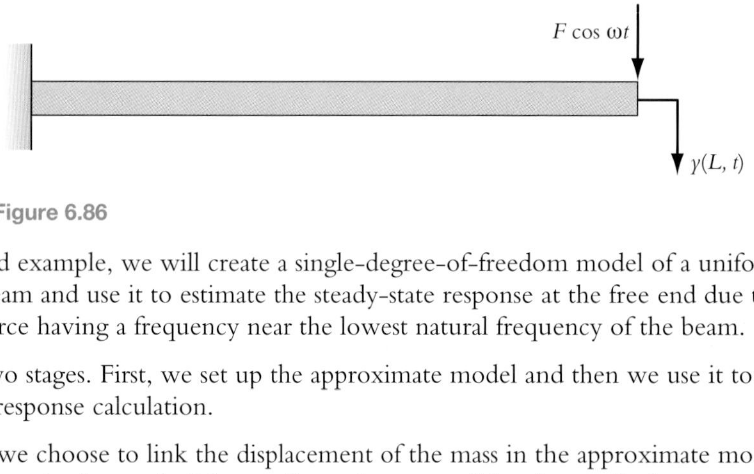

Figure 6.86

In the second example, we will create a single-degree-of-freedom model of a uniform cantilever beam and use it to estimate the steady-state response at the free end due to a sinusoidal force having a frequency near the lowest natural frequency of the beam.

There are two stages. First, we set up the approximate model and then we use it to do the steady-state response calculation.

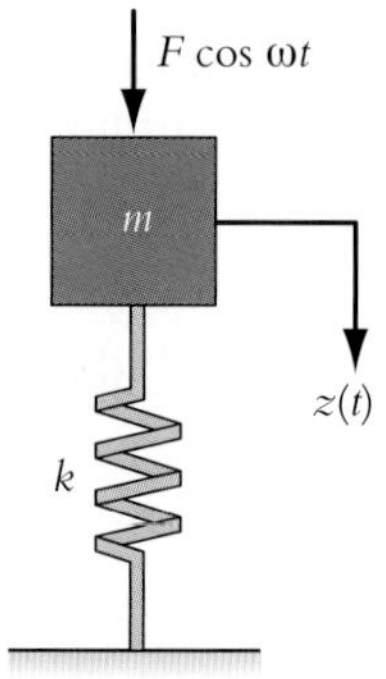

Figure 6.87

In this case, we choose to link the displacement of the mass in the approximate model to the displacement at the free end of the cantilever. In terms of the chosen displacement variables,

$$z(t) = y(L, t)$$

For steady-state, sinusoidal vibration, the link can be written as

$$Z \cos \omega t = Y(L) \cos \omega t \quad \text{or} \quad Z = Y(L)$$

To proceed, we need to choose an expression for $Y(x)$, the amplitude of vibration for the cantilever. Two possibilities are considered.

Choice 1 $$Y(x) = Cx^2$$

Linking this with the approximate model, $Z = Y(L) = CL^2$

Hence, $$C = \frac{Z}{L^2} \quad \text{so that} \quad Y(x) = \frac{Z}{L^2}x^2$$

This expression for $Y(x)$ is then used to calculate the maximum kinetic and strain energies in the beam. By equating these to the equivalent expressions for the single-degree-of-freedom model, we get the required mass and stiffness values.

For the kinetic energies, $$T_{\mathrm{MAX}} = \frac{1}{2}\omega^2 \int_0^L \rho A[Y(x)]^2\,\mathrm{d}x = \frac{1}{2}m\omega^2 Z^2$$

This gives the mass for the approximate model to be $m = 0.2\rho AL$. Note that ρAL is the mass of the beam.

For the strain energies, $$U_{\mathrm{MAX}} = \frac{1}{2}\int_0^L EI\left(\frac{\mathrm{d}^2 Y}{\mathrm{d}x^2}\right)^2 \mathrm{d}x = \frac{1}{2}kZ^2$$

This gives the spring stiffness for the model to be $k = 4\dfrac{EI}{L^3}$

Applying the natural frequency test to the approximate model, we find that $\omega_1 = \dfrac{4.47}{L^2}\sqrt{\dfrac{EI}{\rho A}}$, which is the same (poor) result obtained with Rayleigh's method using this choice for $Y(x)$.

Choice 2 Static deflection shape, $Y(x) = C\left[\left(\frac{x}{L}\right)^4 - 4\left(\frac{x}{L}\right)^3 + 6\left(\frac{x}{L}\right)^2\right]$

Linking this expression at $x = L$ with the coordinate for the approximate model, we get

$$Z = Y(L) = C[1^4 - 4 \times 1^3 + 6 \times 1^2] = 3C$$

Hence, $C = \dfrac{Z}{3}$ and $Y(x) = \dfrac{Z}{3}\left[\left(\frac{x}{L}\right)^4 - 4\left(\frac{x}{L}\right)^3 + 6\left(\frac{x}{L}\right)^2\right]$.

Using this expression to calculate the mass and stiffness values for the model gives the following.

$$m = 0.257\rho AL \quad \text{and} \quad k = 3.20\frac{EI}{L^3}$$

The natural frequency test in this case gives $\omega_1 = \dfrac{3.53}{L^2}\sqrt{\dfrac{EI}{\rho A}}$, which is lower than the result from the first choice, confirming that the static deflection shape is the better approximation.

Forced response analysis

Having established the single-degree-of-freedom model, we can now use it to estimate the steady-state response at the free end of the beam due to a sinusoidal force. The equation of motion for the approximate model is:

$$m\ddot{z} + kz = F(t)$$

Making the standard substitutions, $F(t) = F\mathrm{e}^{\mathrm{i}\omega t}$ and $z(t) = Z^\star \mathrm{e}^{\mathrm{i}\omega t}$, we get

$$Z^\star = \frac{F}{(k - m\omega^2)}$$

Note that since there is no damping in this model, the expression for $Z^\star$ is real.

This meets the objective of finding the steady-state amplitude of the deflection at the free end of the cantilever. However, since we have assumed that the deflected shape of the beam can be defined by the mode shape (that is, the function $Y(x)$), the expression for $Z^{\star}$ can also tell us the vibration amplitude at *any* point along its length. In the case of choice 2, this gives:

$$Y(x) = \frac{Z^{\star}}{3}\left[\left(\frac{x}{L}\right)^4 - 4\left(\frac{x}{L}\right)^3 + 6\left(\frac{x}{L}\right)^2\right]$$

$$= \frac{F}{3(k - m\omega^2)}\left[\left(\frac{x}{L}\right)^4 - 4\left(\frac{x}{L}\right)^3 + 6\left(\frac{x}{L}\right)^2\right]$$

Learning summary

By the end of this section you should have learnt:

✓ to use Dunkerley's method and Rayleigh's method to obtain estimates of the lowest natural frequency of structures;

✓ to create a single-degree-of-freedom approximation of a more complex structure and use it to estimate the structure's response.

6.7 Vibration control techniques

Introduction

Vibration isolators (also known as anti-vibration mounts) are used for reducing the vibration transmitted from a source. They work by introducing flexibility between a device and its support.

Case 1 In some cases, the source of vibration is within the device and *the objective is to minimize the force transmitted to the support*. Examples are a ship's engine that can transmit vibration to the deck structure or a passing train that can produce ground-borne vibration.

Case 2 In other cases, the source may be remote, but causes the support for a device to vibrate. Here, *the objective is to minimize the displacement transmitted to the device*. Examples are a satellite mounted in a launch vehicle or the need to protect sensitive laser instruments from ground-borne vibration.

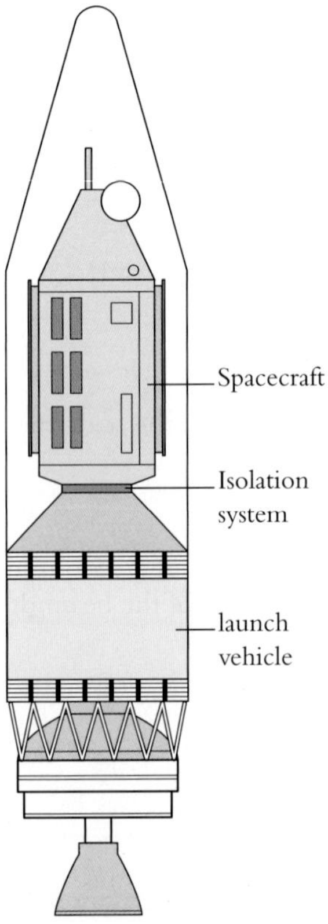

Figure 6.88 Satellite isolation system

Types of isolator

Elastomeric isolator

This is the most common type of isolator. Elastomers can be moulded from many different combinations of many different materials, including natural rubber, neoprene, butyl and silicone. A typical mount made with these materials generally employs the elastomer in shear but many utilize compressive strain also. The mounts may employ the elastomer in a manner that provides both shear and compressive loading for effective isolation performance in both the horizontal as well as the vertical direction. It is relatively easy to design various degrees of damping, shape, load–deflection characteristics and transmissibility characteristics into elastomeric isolators. The inherent damping of elastomers is useful in preventing problems at resonance that would be difficult to restrain if coil springs were used.

For isolation from shocks, elastomers offer some significant advantages because of the fact that they can generally absorb more shock energy per unit weight than other forms of isolator system.

Pneumatic vibration isolators

Pneumatic isolators are air-filled, reinforced rubber bellows with mounting plates on top and bottom. Isolators such as these can provide very low natural frequencies with small static deflections. To provide a 1 Hz natural frequency, a steel coil spring isolator would need to be about 600 mm long and capable of deflecting about 250 mm. It would thus be difficult to install and would also present some lateral stability problems. Unlike most isolators, which are passive devices, pneumatic isolators are also used with position feedback in active control systems. This is used, for example, to maintain the height of a table for mounting optical equipment that is sensitive to the slightest movement.

Figure 6.89

Coil spring isolators

Springs may be loaded in tension but it is frequently more convenient to load them in compression. Coil springs are used primarily for the isolation of low-frequency vibration. Consequently, they operate with a relatively large static deflection and lateral stability may be a problem. It can be shown that a coil spring will be stable if

$$\frac{\text{lateral stiffness}}{\text{axial stiffness}} > \frac{\text{static deflection}}{\text{working height}}$$

Coil springs possess practically no damping, and the transmissibility at resonance is extremely high. This can be overcome by the addition of friction dampers in parallel with the load-carrying spring and these types of isolators are widely used. Another method of adding damping to a spring is by the use of an air chamber with an orifice for metering the airflow. For applications where all metal isolators are desired because of temperature extremes or other environmental factors, damping can be added to a load-carrying spring by the use of metal mesh inserts.

High-frequency vibrations can be transmitted through the coils to the isolated unit. To overcome this, one or both ends of the spring can be fitted with elastomeric pads. Conventionally, the pad is attached to the bottom of the spring assembly, which has the added advantage of providing a non-slip surface that frequently eliminates the need to bolt the isolator to the floor.

Transmissibility analysis

The isolators are invariably very much more flexible than the device they support, so the first approximation is to use a single-degree-of-freedom model in which the device to be isolated is treated as a rigid body and the isolators are represented by a spring–damper combination. The steady-state response to harmonic excitation provides a way of characterizing the isolation performance at different frequencies.

Case 1 Source of vibration within a device transmitting vibration to the support

We will assume that the device generates an excitation force, of amplitude P and frequency ω and use the method from Section 6.2.

Step 1 Dynamic mass–spring model

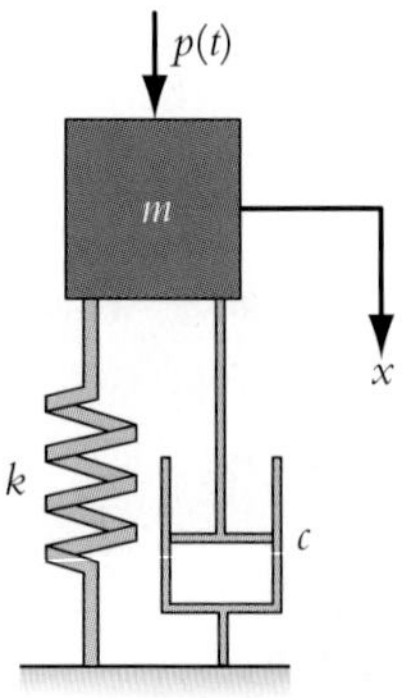

Figure 6.90

Step 2 Free-body diagram

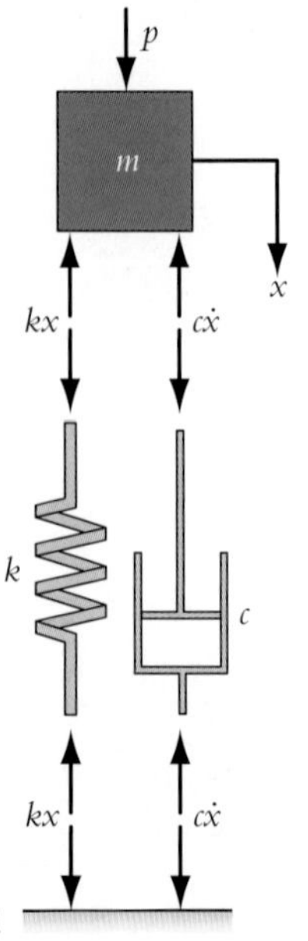

Figure 6.91

Step 3 Equations of motion

For the device: $p - kx - c\dot{x} = m\ddot{x}$

or $m\ddot{x} + c\dot{x} + kx = p$ (6.87)

The transmitted force is $q(t) = kx + c\dot{x}$ (6.88)

Substitute $p(t) = P\mathbf{e}^{\mathbf{i}\omega t}$, $q(t) = Q^{\star}\mathbf{e}^{\mathbf{i}\omega t}$, and $x(t) = X^{\star}\mathbf{e}^{\mathbf{i}\omega t}$ into equations (6.87) and (6.88) to give

$$X^{\star} = \frac{P}{(k - m\omega^2) + \mathbf{i}c\omega} \quad \text{and} \quad Q^{\star} = (k + \mathbf{i}c\omega)X^{\star}$$

Eliminating $X^{\star}$ we get

$$\frac{Q^{\star}}{P} = \frac{(k + \mathbf{i}c\omega)}{(k - m\omega^2) + \mathbf{i}c\omega}$$

Here, only the magnitude of the transmitted force is of interest and we can define **force transmissibility** as

$$T_{\mathrm{F}} = \left|\frac{Q^{\star}}{P}\right| = \sqrt{\frac{(k^2 + c^2\omega^2)}{(k - m\omega^2)^2 + c^2\omega^2}} \tag{6.89}$$

Case 2 Vibration from the support transmitted to the device

The support vibration is defined by the displacement, $y(t) = Y\cos\omega t$.

Step 1 Dynamic mass–spring model

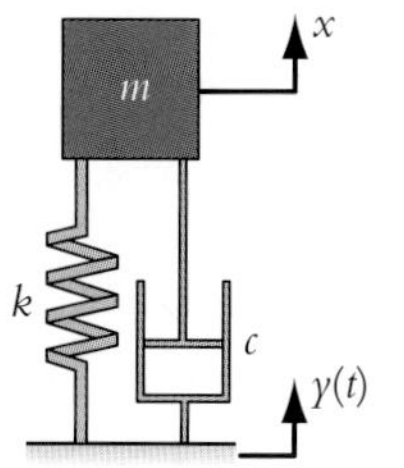

Figure 6.92

Step 2 Free-body diagram

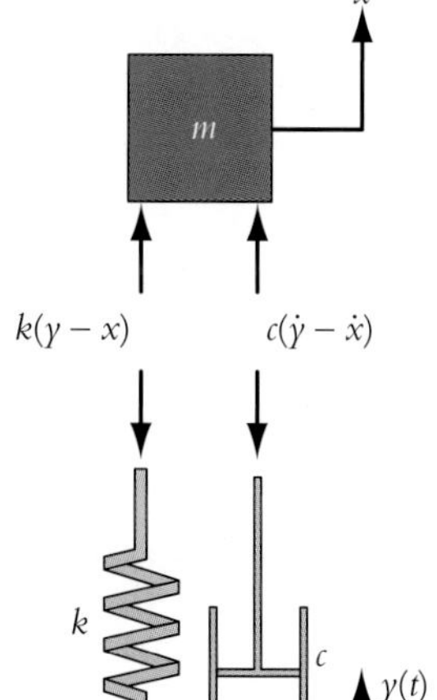

Figure 6.93

Step 3 Equations of motion

For the device: $$k(y - x) + c(\dot{y} - \dot{x}) = m\ddot{x}$$

or $$m\ddot{x} + c\dot{x} + kx = c\dot{y} + ky \qquad (6.90)$$

Substitute $y(t) = Ye^{i\omega t}$ and $x(t) = X^{\star}e^{i\omega t}$ into equation (6.90) to give

$$X^{\star} = \frac{(k + \mathbf{i}c\omega)Y}{(k - m\omega^2) + \mathbf{i}c\omega}$$

Again, only the magnitude of the transmitted displacement is of interest and we can define **displacement transmissibility** as

$$T_D = \left|\frac{X^{\star}}{Y}\right| = \sqrt{\frac{(k^2 + c^2\omega^2)}{(k - m\omega^2)^2 + c^2\omega^2}} \qquad (6.91)$$

Note that the force and displacement transmissibility expressions for this mass–spring–damper system are identical. The same is true of other physical systems when analysing the displacements or forces applied to a given pair of coordinates (the mass and the support in this case). Note also that equations (6.89) and (6.91) only apply to this mass–spring–damper system. Other physical systems will have different transmissibility expressions.

The reader is left to confirm that equations (6.89) or (6.91) can be rearranged by introducing the expressions for the natural frequency and damping ratio for the system, namely $\omega_n = \sqrt{\frac{k}{m}}$ and $\gamma = \frac{c}{2\sqrt{km}}$. The alternative expression for the transmissibility ratio is

$$T_{D,F} = \frac{\sqrt{1 + 4\gamma^2\frac{\omega^2}{\omega_n^2}}}{\sqrt{\left(1 - \frac{\omega^2}{\omega_n^2}\right)^2 + 4\gamma^2\frac{\omega^2}{\omega_n^2}}} \qquad (6.92)$$

This form of the transmissibility expression emphasizes the importance of the ratio of the excitation frequency to the natural frequency in determining the effectiveness of the isolation system in reducing vibration.

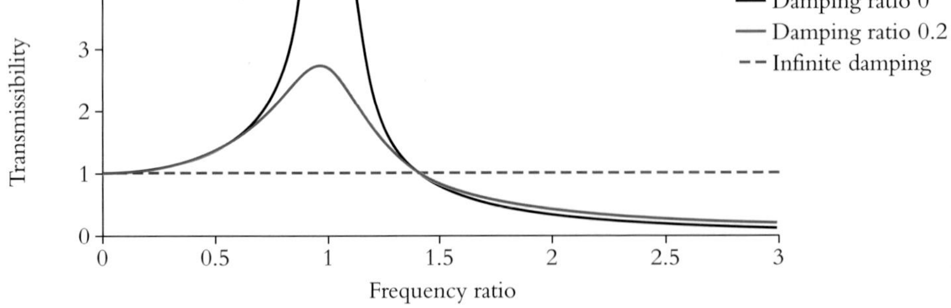

Figure 6.94 **Transmissibility curves for a simple mass–spring–damper system**

The transmissibility graphs in Figure 6.94 are plotted for three different damping ratios. With zero damping, the transmissibility is infinite if the excitation frequency coincides with the natural frequency of the system. The main effect of increasing the damping ratio ($\gamma = 0.2$ is shown in Figure 6.94) is to reduce the maximum transmissibility in the vicinity of the natural frequency. At higher frequencies, the difference between the curves is small. Theoretically infinite damping would result in no relative movement across the damper, effectively giving a rigid connection between the device and its support. This, of course, is the original situation without isolators.

It's easy to show from equation (6.92) that $T = 1$ when $\omega/\omega_n = \sqrt{2}$ and that it is independent of the value of the damping ratio. This explains why the three curves intersect at this point on the graph.

Notice that if $\omega/\omega_n < \sqrt{2}$, the transmitted displacement (or force) is higher than the input. In this case, it would be better to have no isolators and fix the device rigidly to the support! However, if $\omega/\omega_n > \sqrt{2}$, the transmissibility is less than 1.0, resulting in vibration reduction. The aim in selecting isolators is to ensure that the system operates in this **isolation region**.

It is convenient to define an isolation efficiency to describe the reduction in displacement (or force) compared with the applied displacement (or force). This is:

$$\text{Isolation efficiency} = \frac{\text{Reduction in displacement (or force)}}{\text{Original displacement (or force)}} \times 100\% = (1 - T) \times 100\%$$

Isolator selection

This section considers the task of selecting isolators to achieve a desired reduction in vibration. There are two constraints governing the selection: the lowest excitation frequency to be encountered, ω_{MIN}, and the maximum allowable transmissibility, T_{MAX}.

The combination of T_{MAX} and $\frac{\omega_{MIN}}{\omega_n}$ is marked on Figure 6.95 and represents the worst case in respect of isolation efficiency, since for any excitation frequency greater than ω_n, the transmissibility will be lower than T_{MAX}. It is apparent from Figure 6.95 that ω_n must be very much less than ω_{MIN} to ensure that T_{MAX} will be less than 1.0 for all excitation frequencies greater than or equal to ω_{MIN}.

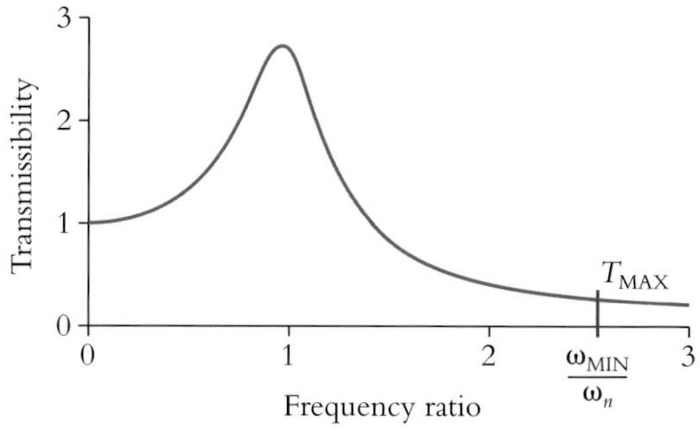

Figure 6.95 **Design constraints for isolator selection**

Three variables affect the system's dynamics; m, k and c, and m and k together determine ω_n. The stiffness, k, is given by the set of isolators selected.

At first sight, the mass of the device may not seem to be a variable for this problem. However, m can be increased by mounting the machine on an inertia base. This will reduce ω_n and increase $\frac{\omega_{MIN}}{\omega_n}$ and have the beneficial effect of reducing the transmissibility. Figure 6.96 gives an example of an inertia base installation.

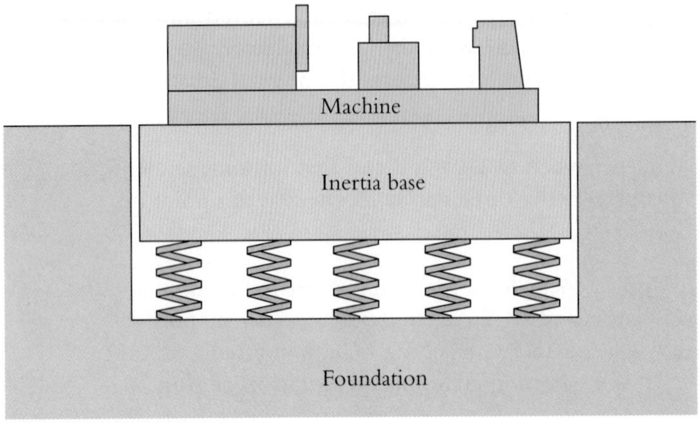

Figure 6.96 **Example of inertia base used to increase the isolated mass**

In most vibration situations, it is desirable to increase damping since this limits the amplifying effect of resonance. Here, since the system is designed to operate well above the resonant frequency, low damping is desirable since increasing γ will increase the transmissibility in the isolation region. Low damping is easy to achieve and most commercial isolators give a damping ratio that is less than 0.1.

It is normal to base the design on the assumption of zero damping and, as can be seen from the transmissibility graph in Figure 6.94, the error involved in the isolation region is small.

It is also normal to treat each isolator independently of the others. *In this case, m is the effective mass supported by the isolator in question.*

For the simple mass–spring model with zero damping,

$$T = \left| \frac{1}{1 - \dfrac{\omega^2}{\omega_n^{\,2}}} \right|$$

$$= \frac{1}{\dfrac{\omega^2}{\omega_n^{\,2}} - 1}$$

for the isolation region where $\omega > \omega_n$.

If $T = T_{\mathrm{MAX}}$ at $\omega = \omega_{\mathrm{MIN}}$,

$$\omega_n^{\,2} = \frac{T_{\mathrm{MAX}}\,\omega_{\mathrm{MIN}}^{\;2}}{1 + T_{\mathrm{MAX}}}$$

Since $\omega_n^{\,2} = \dfrac{k}{m}$, the required isolator stiffness is

$$k = m\omega_n^{\,2} = \frac{mT_{\mathrm{MAX}}\,\omega_{\mathrm{MIN}}^{\;2}}{1 + T_{\mathrm{MAX}}} \qquad (6.93)$$

If selecting isolators from a manufacturer's catalogue, it is unlikely that one with precisely this stiffness will be found. The stiffness given by equation (6.93) therefore gives the *maximum* value consistent with the design requirements.

It might appear from this that *any* isolator would be suitable provided its stiffness was less than this value. This is not the case, however, since there are also constraints imposed by static considerations. In the case of coil spring isolators, there could be installation, coil bottoming or lateral stability problems if the static deflection is too large. With elastomeric isolators, there are strength limitations under static load.

Manufacturers often express these constraints by specifying a **maximum static deflection**. Therefore, after selecting an isolator to satisfy the maximum stiffness limit, it is necessary to check that the static deflection limit is not exceeded.

The actual static deflection, X_0, is given by

$$X_0 = \frac{mg}{k_{\text{isolator}}} \tag{6.94}$$

Alternatively, combining (6.93) and (6.94) gives

$$X_0 = \frac{g}{\omega_{\text{MIN}}^2}\left(1 + \frac{1}{T_{\text{MAX}}}\right) \tag{6.95}$$

This represents the *minimum* static deflection consistent with the design requirements.

These equations can be incorporated into a procedure for selecting suitable isolators for an application. The procedure is as follows:

Step 1 Find the centre of mass of the machine.

Step 2 Select the number and position of attachment points for isolators.

Step 3 Estimate the load supported by each isolator.

Step 4 For each isolator position in turn:

(a) calculate the maximum stiffness from equation (6.93);

(b) select an isolator with a lower stiffness;

(c) check that this does not exceed any static deflection limit using equation (6.94);

(d) although this will give a satisfactory selection, it is often worth repeating (b) and (c) with other isolators having even lower stiffness; the lower the stiffness, the greater the isolation efficiency, so the limiting factor becomes the maximum allowable static deflection.

Tuned vibration absorbers

Tuned vibration absorbers were mentioned in Section 6.1 as part of the solution to the resonant vibration on the Millennium Bridge. They were also mentioned in Section 6.4 where it was observed that it was possible to design a dynamic system that has zero displacement at the point of force application. This section explains how this can be achieved.

Let's look at the case of a structure that is subjected to a sinusoidal force whose frequency is very close to a natural frequency and consider the worst-case scenario that the structure has no damping to limit the resonant amplitude. We can model the mode in question as the undamped single-degree-of-freedom system shown in Figure 6.97(a). For the original system without the tuned vibration absorber, the response amplitude is given by equation (6.45) as

$$X_0 = \left|\frac{P_1}{k - m\omega^2}\right|$$

As a numerical example, we will take $k = 1\,\text{MN/m}$, $m = 1000\,\text{kg}$ and $P_1 = 1\,\text{kN}$ to get the response spectrum shown in Figure 6.97(b). As expected, the model predicts a very large response when the excitation frequency is close to the natural frequency. The aim of adding the tuned vibration absorber is to reduce this response to zero.

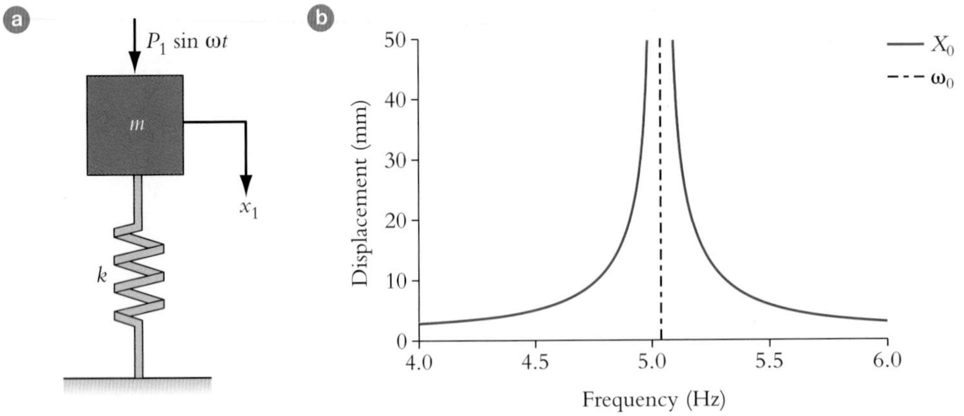

Figure 6.97

The tuned absorber itself is a secondary mass–spring system in which the ratio of stiffness to mass is identical to the original system. This is attached to the original mass as shown in Figure 6.98(a). In the numerical example shown here, the ratio of the absorber mass to the original mass, μ, has been taken to be 0.01.

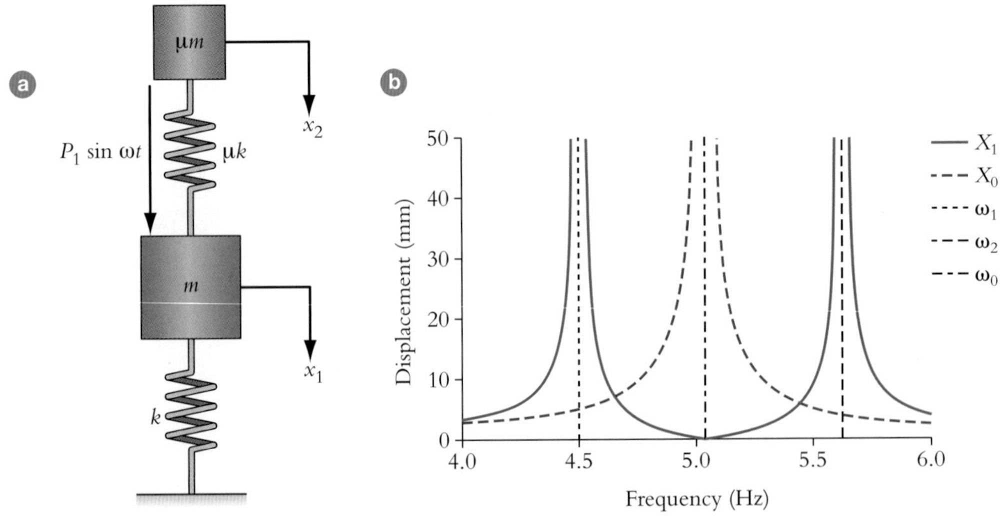

Figure 6.98

This can be analysed in exactly the same way as the worked example in Section 6.4 to give expressions for the response of each mass. It is seen from Figure 6.98(b) that the response amplitude X_1 is zero when the excitation frequency is equal to the original natural frequency. The effect of adding the tuned absorber is to create a two-degrees-of-freedom system whose natural frequencies lie either side of the natural frequency of the original system. A limitation of this absorber is that it is only effective if the excitation frequency is constant and at a frequency very close to the natural frequency of the original system. If the excitation frequency deviates from this, there is a strong chance of exciting one of the new natural frequencies, since it can be seen from Figure 6.98(b) that these are only about 0.5 Hz (10%) away from the original natural frequency. The separation of the new natural frequency can be increased by choosing a larger value for the mass ratio, μ. However, there may be limits to the amount of mass that can be added, as for example in the case of the Millennium Bridge (Section 6.1).

This problem can be overcome by using a damped vibration absorber that uses a damper in parallel with the absorber spring, as in Figure 6.99(a). Note that the original structure remains undamped. The response spectrum with the damped absorber is shown in Figure 6.99(b) using a damping coefficient of 100 Ns/m. This has greatly reduced resonant peaks and still gives a very low response at the original natural frequency.

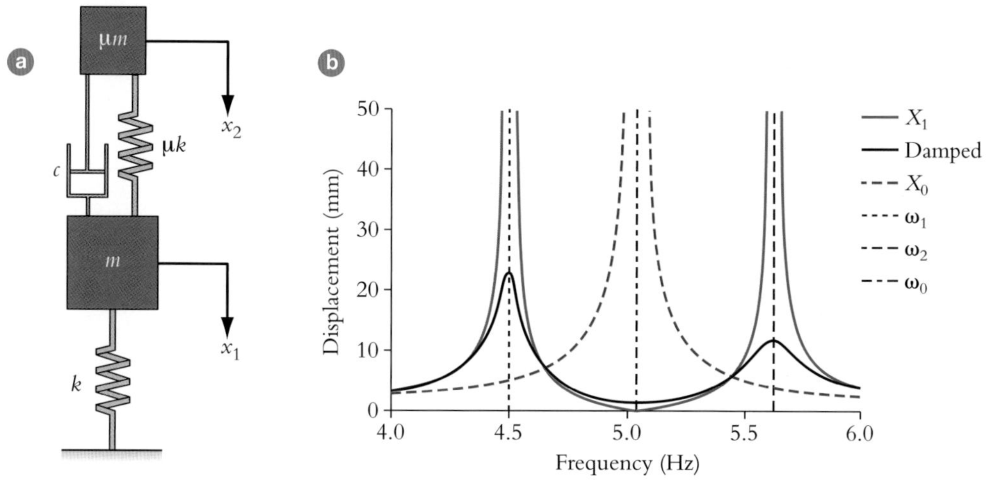

Figure 6.99

The modal approach to solving the equations of motion cannot be used for this problem since they cannot be uncoupled. The equations of motion are:

$$\begin{bmatrix} m & 0 \\ 0 & \mu m \end{bmatrix} \begin{Bmatrix} \ddot{x}_1 \\ \ddot{x}_2 \end{Bmatrix} + \begin{bmatrix} c & -c \\ -c & c \end{bmatrix} \begin{Bmatrix} \dot{x}_1 \\ \dot{x}_2 \end{Bmatrix} + \begin{bmatrix} k(1+\mu) & -\mu k \\ -\mu k & \mu k \end{bmatrix} \begin{Bmatrix} x_1 \\ x_2 \end{Bmatrix} = \begin{Bmatrix} p_1(t) \\ 0 \end{Bmatrix}$$

It will be seen that the damping matrix is not in the form $[C] = a_1[M] + a_2[K]$, meaning that we do not have proportional damping in this case. Since the equations of motion cannot be uncoupled, they must be solved as a simultaneous pair. For sinusoidal excitation, we can still make the usual substitutions, namely: $p_1(t) = P_1 e^{i\omega t}$, $x_1(t) = X_1^\star e^{i\omega t}$ and $x_2(t) = X_2^\star e^{i\omega t}$. This leads to the equations:

$$\begin{bmatrix} (k(1+\mu) - m\omega^2) + \mathbf{i}\omega c & -\mu k - \mathbf{i}\omega c \\ -\mu k - \mathbf{i}\omega c & (\mu k - \mu m\omega^2) + \mathbf{i}\omega c \end{bmatrix} \begin{Bmatrix} X_1^\star \\ X_2^\star \end{Bmatrix} = \begin{Bmatrix} P_1 \\ 0 \end{Bmatrix}$$

The responses are then given by

$$\begin{Bmatrix} X_1^\star \\ X_2^\star \end{Bmatrix} = \begin{bmatrix} (k(1+\mu) - m\omega^2) + \mathbf{i}\omega c & -\mu k - \mathbf{i}\omega c \\ -\mu k - \mathbf{i}\omega c & (\mu k - \mu m\omega^2) + \mathbf{i}\omega c \end{bmatrix}^{-1} \begin{Bmatrix} P_1 \\ 0 \end{Bmatrix}$$

In practice, this problem would be solved using software such as Matlab, which has the capability to invert complex matrices. A Matlab script based on this example is included on the book's supporting website.

The most common application of tuned vibration absorbers is for overhead power lines. When wind blows across the cables, vortex shedding can induce vertical oscillations, which can cause fatigue in the cable. This is most prevalent where the cable is supported by the transmission pylons, and vibration absorbers are attached here to suppress the oscillations.

Figure 6.100 **Tuned vibration absorber for overhead power lines (circled)**

The absorber consists of a pair of masses, each one on the end of a short cantilever made from cable. Bending of the cantilevers provides the required spring stiffness, and friction between the cable strands provides damping.

Learning summary

By the end of this section you should have learnt:

- ✓ how to derive the transmissibility expression that gives the ratio of transmitted force to applied force (or of transmitted displacement to applied displacement) as a function of excitation frequency;
- ✓ to select suitable isolators to achieve a given isolation efficiency;
- ✓ how tuned vibration absorbers can suppress resonant vibration in situations where sinusoidal excitation coincides with a natural frequency of the structure.

Questions

Fluid Dynamics

1. Show that the two-dimensional velocity field given by $u = ay$ and $v = bx$ satisfies the continuity equation. Using the Navier-Stokes equations, derive the expression for the pressure p.

2. The three-dimensional velocity profile for an incompressible flow is given by

$$u = x^3 + 2z^2 \quad \text{and} \quad w = y^3 - 2yz$$

Derive a general form of the third velocity component, v from the 3-D continuity equation:

$$\frac{\partial u}{\partial x} + \frac{\partial v}{\partial y} + \frac{\partial w}{\partial z} = 0.$$

3. By neglecting the viscous and gravity terms, derive the Bernoulli equation from the steady Navier-Stokes equations. (*Hint*: choose the x-axis along the streamline.)

4. For an axial flow inside a circular tube, the Reynolds number which causes the transition to turbulence is approximately 2300, based on the tube diameter and average flow velocity. If the tube diameter is 4 cm and the fluid is petrol at 20°C, find the volumetric flow rate which causes the transition. The density and viscosity of the petrol are $\rho = 680\ \text{kg/m}^3$ and $\mu = 2.92 \times 10^{-4}$ kg/ms, respectively.

5. Determine the shape factor of the laminar boundary layer when the velocity profile is given by:

$$\frac{u}{U_0} = \sin\left(\frac{\pi\eta}{2}\right) \quad \text{where } \eta = \frac{y}{\sigma}$$

6. A smooth flat plate 2.4 m long and 900 mm wide moves at 6 m/s through still air. Assuming that the boundary layer is laminar, calculate the boundary layer thickness at the end of the plate, the wall-shear stress halfway along, and the power required to move the plate. What power would be required if the boundary layer is 'tripped' to turbulent at the leading edge?

7. A flow straightener is formed from perpendicular flat metal strips to give 25 mm square passages, which are 150 mm long. Water of kinematic viscosity $1.21 \times 10^{-6}\ \text{m}^2/\text{s}$ approaches the straightener at 1.8 m/s. Neglecting the thickness of the metal, calculate the displacement thickness of the boundary layer and the velocity of the main stream at the outlet of the straightener. Using the Bernoulli equation, deduce the pressure drop through the straightener.

8. Atmospheric boundary layers are very thick but follow formulae very similar to those of the flat-plate boundary layer. Consider the wind blowing at 10 m/s at the height of 80 m above a smooth beach, which is considered to be the thickness of the atmospheric boundary layer. Estimate the wall-shear stress on the beach ground. What will be the wind velocity over your nose if:

   (a) you are standing up 1.7 m off the ground?

   (b) you are lying on the beach with your nose 0.17 m above the ground?

9. An aircraft travels at a cruising speed of 250 m/s at 10 km altitude. It has a smooth wing 7 m long and 55 m wide, which can be approximated by a flat plate. Estimate the power required to overcome the friction drag of this wing.

10. A man weighing 800 N jumps from an aircraft flying at 1000 m altitude. The parachute's diameter is 8 m and the time of descent is 2.9 minutes. Calculate the drag coefficient of the parachute, assuming that terminal velocity is attained within the first few seconds of the fall.

11. An electric transmission line consists of aluminium cable of density 2670 kg/m^3 with 15 mm diameter. The maximum cross-winds expected on the line are 150 km/h. Calculate the magnitude and direction of the maximum loading on the line.

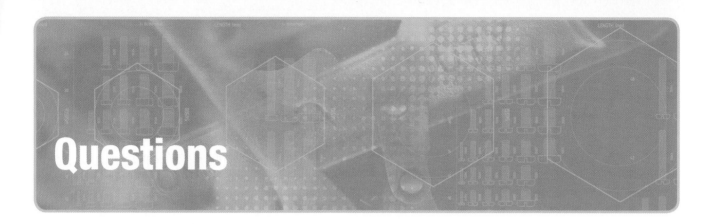

12. A four-cup wind speed indicator of 0.7 m diameter has a rusty centre bearing and the starting torque is very high. These cups are made of hemispherical shells of diameter 50 mm. Determine the starting torque of this instrument if a free stream velocity of 4 m/s is required to initiate the rotation. You can ignore the connecting rods and side winds on the cups.

13. A 60° cone with a 30 mm base diameter is mounted on a 200 mm long and 5 mm diameter rod, with its apex pointing to the free stream of 45 m/s. Estimate the total drag on this structure and the bending moment at the root of the rod.

14. Icebergs can be driven at substantial speeds by the wind. Let the iceberg be modelled by a large flat cylinder, with its diameter D much greater than the height L. Only the top 1/8th of the iceberg is exposed above the sea surface. Derive the expression for the steady iceberg speed V when it is driven by the wind of velocity U.

15. A horizontal circular duct with a 0.3 m diameter, 20 m in length draws air at atmospheric pressure into a centrifugal fan, which discharges the air into a rectangular duct 0.20 m × 0.25 m, 50 m long and to the atmospheric pressure. Show that the total pressure rise across the fan is 700 N/m^2. The flow rate of the fan is 0.5 m^3/s and the friction factor f of the duct is 0.04. You should account for the minor losses at the inlet and outlet of the duct.

16. Kerosene of relative density 0.82 and kinematic viscosity 2.3×10^{-6} m^2/s is to be pumped through 185 m of galvanised iron pipe ($\varepsilon = 0.15$ mm) at 40 litres/s into a storage tank. The pressure at the inlet end of the pipe is 370 kPa and the liquid level in the storage tank is 20 m above that of the pump. Determine the size of pipe necessary, neglecting losses other than those due to pipe friction.

17. The viscous sublayer is normally less than one per cent of the pipe diameter, therefore it is very difficult to probe it with a finite-sized instrument. To generate a thick viscous sublayer, one can use glycerine as a flow medium. Assume a smooth 305 mm diameter pipe, 12.2 m long, with the bulk velocity of 18.3 m/s and the flow temperature of 20 °C. Compute the sublayer thickness of the pipe flow and the power required if the pumping efficiency is 70 per cent.

18. Ethyl alcohol at 20°K flows at $V_b = 3$ m/s through 0.1 m diameter drawn tube. Obtain:

    (a) the head loss per 100 m of tube;

    (b) the wall-shear stress; and

    (c) the local velocity at $r = 20$ mm.

    By what percentage is the head loss increased due to the surface roughness of the tube?

19. A ship travels at a constant speed at sea surface. A propeller is required to propel this ship to maintain the steady motion, so that the thrust R from this propeller exactly matches the resistance of the ship. Using the π-theorem, derive a functional relationship for R and other relevant physical quantities.

20. Instead of the ship in Question 19, consider a submarine cruising below sea surface. How would this modify your answer? Over the submarine surface there are a hatch cover, hand rails etc. How could these be considered in a dimensional analysis?

21. Derive a relationship for the power to rotate a disk at a constant speed in a homogeneous fluid as a function of relevant physical quantities.

22. The flow through a circular pipe may be metered from the speed of rotation of a propeller, having its axis along the pipe centreline. Derive a relationship between the volumetric flow rate and the rotational speed of the propeller in terms relevant physical quantities.

23. A propeller of 75 mm diameter, installed in a 150 mm diameter pipe carrying water at 42.5 litres/s was found to rotate at 20.7 rps. If a geometrically similar propeller of 375 mm diameter rotates at 10.9 rps in air through a pipe of 750 mm diameter, estimate the volumetric flow rate.

24. A torpedo-shaped object with a 900 mm diameter is to move in air at 60 m/s. Its drag is to be estimated from tests in water using a half-scale model. Determine the necessary speed of the model and the drag of the full-scale object if the measured drag of the model is 1140 N.

25. Tests on a 371.3 mm diameter centrifugal water pump at 2134 rpm yield the data given below. What is the Best efficiency point (BEP) and the specific speed of this pump? Estimate the maximum discharge.

Q m^3/s	0	.0566	.113	.170	.227	.283
H m	104	104	104	101	91.5	67.1
bhp kW	101	120	154	191	248	248

26. A pump from the same family as in Questions 25 is built with $D = 457.2$ mm and $n = 1{,}500$ rpm. Find the discharge, water pressure rise and the brake horsepower at BEP.

27. A large centrifugal pump with a non-dimensional specific speed of 0.183 revolutions is to discharge 2 m^3/s of liquid against a total head of 15 m. The kinematic viscosity of the liquid may vary between three and six times that of water. Determine the range of speeds and test heads for a one-quarter scale-model investigation to be carried out in water.

Thermodynamics

Refrigeration

1. (a) What pressure is required in a freezer compartment evaporator coil using refrigerant R134a if the temperature is to be maintained at –20°C?

   (b) What is the enthalpy and entropy when the R134a is:

   (i) all liquid?

   (ii) all vapour?

   (iii) What does the change in entropy between the two states tell you?

   (c) Assuming a mass flow rate of R134a of 20g/s, and making use of the SFEE, what heat transfer rate occurs in the evaporator if the refrigerant leaves at saturated gas condition (i.e. all vapour, with no superheat) and enters at saturated liquid condition (i.e. all liquid and no sub-cooling)?

Air conditioning

2. An air conditioning unit draws in 0.5kg/s of atmospheric air at 30°C and specific humidity 0.012.

   (a) What is the mass flow rate of water vapour in the air?

   (b) Using the formula relating specific humidity to partial pressure of the water vapour and the atmospheric pressure, and assuming one atmosphere atmospheric pressure, what is the partial pressure of the water vapour?

   (c) Find the p_g of water vapour. What is the relative humidity of the incoming air?

   (d) Find the dew point corresponding to the pressure p_g. Sketch a T–v diagram showing the process that the water vapour in atmospheric air must go through in order to form dew. If the air is to be issued at 20°C and 50% humidity, determine the moisture removal rate, and the heat transferred in the cooler and heater sections.

Vapour power cycles

3. A turbine receives steam at 550°C and 200 bar.

   (a) Assuming an isentropic expansion through the turbine to a pressure of 40 bar, what is the final temperature?

   (b) The turbine actually expands to 40 bar and 330°C. What is the isentropic efficiency of that turbine?

4. A power station uses the Rankine cycle between a boiler with feedwater at 200 bar and a condenser which is at a pressure corresponding to the saturated water vapour at 30°C.

   (a) What are the pressure and specific enthalpy in the condenser?

   (b) What are the temperature and specific enthalpy of the saturated steam?

   (c) If the mass flow rate of the water is 100 kg/s, what is the heat input to the steam in the boiler?

   (d) What is the work done in the pump?

   (e) The Rankine cycle is modified with a superheat steam pipe circuit prior to entering the high pressure turbine. The cycle is also modified with reheat after exit from the high pressure turbine up to 550°C, before exhausting in the low pressure turbine at the pressure of the condenser. Given the isentropic efficiency of the low pressure turbine is 85%, what are:

   (i) specific enthalpy at turbine entry?

   (ii) specific enthalpy at turbine exit?

   (iii) the steam quality at exit?

   (f) Knowing the pump work, the work produced by the high pressure turbine and the low pressure turbine work, what is:

   (i) specific steam consumption?

   (ii) work ratio?

Combustion chemistry

5. (a) Write the reaction equation for the stoichiometric combustion of butane in oxygen.

   (b) If the mass of butane consumed is 20 g, how many moles of butane is this?

   (c) For this 20 g of butane, what mass of O_2 is consumed in the reaction?

   (d) How many moles of oxygen is this?

   (e) What are the masses of the gases in the products?

6. (a) Write the stoichiometric reaction equation for butane in air.

   (b) What is the air to fuel ratio by volume and by mass?

   (c) What are the proportions by mass and by volume of the product gases?

   (d) Determine how much air is required for the stoichiometric combustion of propane burning in air. Determine the AFR by mass and by volume.

   (e) The propane is burned with 30% excess air. Determine the volume and mass fractions of reactants and products.

Heat transfer

7. (a) What is the thermal resistance of a brick wall (k = 1.5W/mK), 10 cm thick, on a room wall of height 2.3 m and width 3.1 m?

   (b) Calculate the thermal resistance of the wall given that it has three layers: brick, insulation, brick. Each layer is 10 cm thick, k = 1.2W/mK for the inner brick, 0.5W/mK for the insulation and 2W/mK for the outer brick.

   (c) Given that the convective heat transfer to the wall's inner surface is 9 W/m^{2}K, and on the outer wall is 95 W/m^{2}K, find the thermal resistance for the inside and outside convective conditions, and hence for the overall wall.

   (d) What is the heat transfer if the inner air is at 18°C and the outer air is at 1°C?

   (e) Sketch the temperature profile through the wall.

   (f) The ground at night has air blowing over it at 2°C, with a heat transfer coefficient of 20W/m^{2}K. It also sees the night sky, which is at −55°C. The view factor is 1 and the emissivity of the ground is 0.9. Given the ground has a temperature of 10°C at a depth of 3m and conductivity of 2 W/m^2, calculate the surface temperature of the ground.

Combustion energy

8. Calculate the energy release by combusting 10g of butane (C_4H_{10}) given the enthalpy of formation of liquid butane at standard conditions is −147.6kJ/mol. Assume the products contain water as vapour.

9. Butane is burned by a stove with 100% excess air at 25°C to heat up a pan of water. If the combusted gases escape at 800K, how much heat is extracted from the flame per kmol of fuel?

10. What mass of butane is required to heat up 400g of water from 25°C to boiling point?

Heat exchangers

11. (a) A heat exchanger operates with an oil with c_p of 1.67 kJ/kgK and density 910 kg/m^3, and water with c_p of 4.2 kJ/kgK and density 1000 kg/m^3. The oil volume flow rate is 3158 l/h and the water flow rate is 2000 l/h. What is the capacity rate of the oil and of the water?

    (b) The water enters at 50°C and leaves at 70°C and the oil enters at 120°C and leaves at 85°C. The heat exchanger is of the shell and tube type with counter-current flow. What is the logarithmic mean temperature difference?

    (c) What is the surface area of the heat transfer surface within the heat exchanger if it has an overall heat transfer coefficient of 1100W/m^{2}K? If this is done with a tube having diameter 12 mm, what is the length of tube in the exchanger?

12. A compact heat exchanger is constructed with a cross-flow matrix arrangement with unmixed streams for cooling oil in an air stream. The oil enters at 70°C and is required to leave at 30°C. The oil density is 700kg/m^3. The flow rate of oil is 5 litres per minute and it has a specific heat capacity c = 1.7 kJ/kgK at 10°C and 2.5 kJ/kgK at 100°C.

    (a) What is a suitable average specific heat capacity of the oil over the working range?

    (b) What is the rate of heat transfer?

13. A compact heat exchanger has oil exchanging heat with water and is used to cool engine oil. The oil mass flow rate is 50 g/s and c_p = 1800 J/kgK. The water mass flow rate is 60 g/s. Given that the ingoing oil is at 130°C and the ingoing water is at 15°C:

    (a) What are c_{min} and c_{max} for this situation?

    (b) What is the maximum potential heat transfer in this situation, q_{max}?

    (c) What is the NTU if the area is 0.5 m^2 and the overall heat transfer coefficient is 800 W/m^{2}K?

    (d) What is the effectiveness, ε, and hence the actual heat transfer?

Compressors

14. A two-stage reciprocating air compressor, running at four cycles per second, delivers air at 18 bar from atmospheric air at 1.03 bar at 11°C. The dimensions of the first stage are 100 mm diameter and 100 mm

stroke, with a clearance length of 5 mm. Taking the polytropic index as 1.25 for compression and expansion processes:

(a) What is the swept volume and clearance volume of the first stage?

(b) What is the intermediate pressure for minimum work?

(c) What is the expansion of the clearance volume in the first stage?

(d) What is the volumetric efficiency and hence the mass flow rate of air?

Solid Mechanics

1. A helicopter rotor shaft, 50 mm in diameter, transmits a torque of 2.4 kNm and an upward tensile lifting force of 125 kN. Determine the maximum tensile stress, maximum compressive stress and the maximum shear stress in the shaft.

2. For the purposes of analysis, a segment of a crankshaft in a vehicle is represented as shown in Figure Q1. The load is P = 1 kN, and the dimensions are b_1 = 80 mm, b_2 = 120 mm and b_3 = 40 mm. The diameter of the shaft is d = 20 mm. Determine the maximum tensile, compressive and shear stress at point A, located on the surface of the shaft at the z axis.

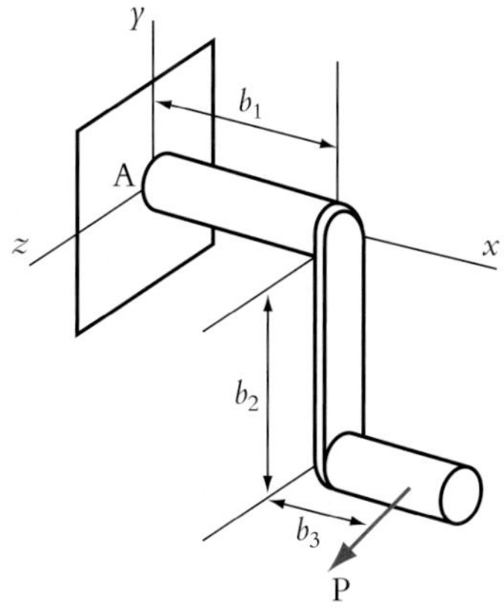

Figure Q1

3. Using the Tresca and von Mises yield criterion in turn, calculate the pressure to cause yielding in a steel cylinder, which has an 80 mm diameter and is 1 mm thick. The cylinder is closed at each end; end effects should be neglected. Assume σ_y = 250 MPa.

4. What additional torque can be applied about the axis of the cylinder in question 3 if yielding is to occur with an internal pressure of 4.0 MPa.

5. A steel beam of rectangular section 10 mm × 30 mm is subjected to pure bending in a plane parallel to the 30 mm faces. Ideal elastic–plastic behaviour may be assumed. Calculate the bending moment necessary for:

(a) the onset of yield;

(b) the complete yield through the section.
Assume σ_y = 250 MPa.

6. The web and flanges of a straight I section steel beam are 80 mm wide and 10 mm thick. The beam is loaded in pure bending in the plane of the web until the whole of each flange has yielded but the whole of the web remains elastic. Calculate the residual curvature in the unloaded beam. Assume ideal elastic–plastic behaviour. Assume σ_y = 250 MPa and E = 200 GPa.

7. Figure Q7 shows a simply supported beam carrying two concentrated loads at the positions indicated. Given that the beam has a rectangular cross-section as shown, calculate the deflection of the beam at a position 3 m from the left-hand end and at a position 5 m from the right-hand end.
(E steel = 200 GPa.)

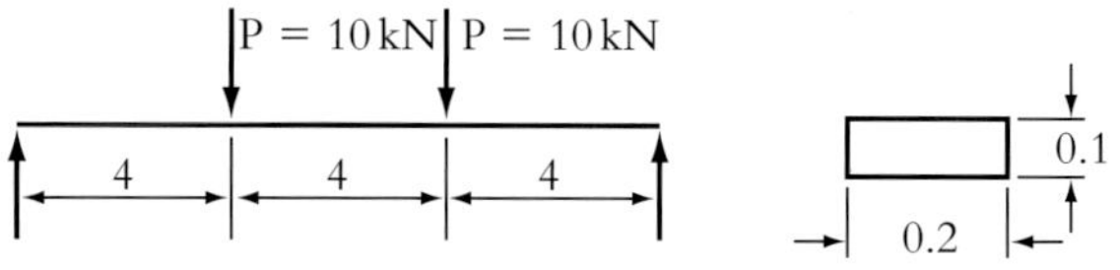

All dimensions in metres

Figure Q7

8. Find the slope at point A and the deflection at point B of the beam shown in Figure Q8.
(EI = 4 × 10^6 Nm^2.)

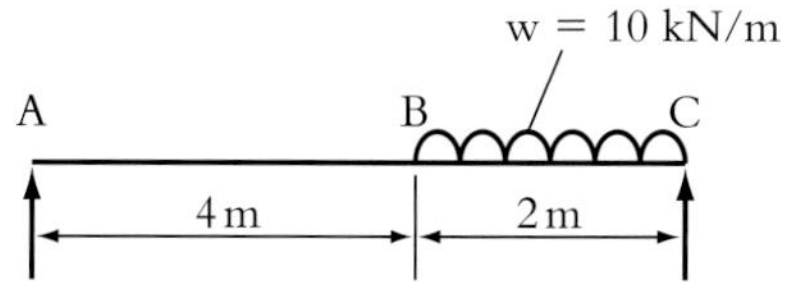

Figure Q8

9. A steel beam with both ends built in has an effective span of 7.5 m and carries a concentrated load of 60 kN at a point 4.5 m from the left-hand end. Calculate the position and magnitude of the maximum deflection due to this load, given that it occurs at a point between the left-hand end and the concentrated load. (E = 208 × 10^9 Nm^2 ; I = 85 × 10^6 mm^4.)

10. Find the critical load for a steel column of rectangular cross-section 50 × 25 mm and length 3 m for each of the following assumptions:

(a) the ends are hinged;

(b) the ends are built in;

(c) one end is hinged and restrained from moving horizontally, the other end is clamped.

11. A new tripod for a surveying instrument will consist of three equal tubular legs, hinged at the top and resting on points at the bottom. When the instrument is set up for use, the top hinge points are equally spaced on a 100 mm diameter horizontal circle, while the pointed ends are equally spaced on a 1 m diameter circle. The top hinge points are 1.3 m above the level ground, on which the pointed ends rest. The greatest expected instrument weight is 80 N. Assuming a factor of safety of 10 against elastic buckling, select the lightest safe aluminium tube (E = 70 GPa) from the following stock list.

Thickness (mm)	Outer diameters available (mm)
1	5, 10, 15, 20, 25, 30
2	10, 15, 20, 25, 30, 40
3	15, 20, 25, 30, 40, 50

12. A hollow steel column, 9 m long and of circular cross-section, 90 mm outer diameter and 75 mm inner diameter is subjected to an end thrust of 30 kN; the line of action of the thrust is parallel to the unstrained line of the strut, but does not coincide with it. Under load the maximum deviation of the strut from the straight is 80 mm. Find the eccentricity of the load and the maximum compressive and tensile stresses. The ends may be assumed to be hinged.
Assume E = 200 GPa

13. Figure Q13 shows the cross-scetion of a solid beam which carries a vertical shear force of 100 kN.

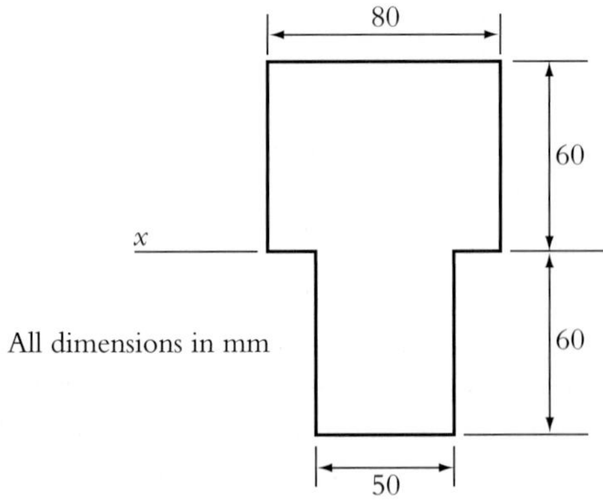

Figure Q13

Determine:

(a) the shear stress just above and just below X–X;

(b) the shear stress at the neutral axis of the section.

(c) Sketch the shear stress distribution through the section and state where the maximum shear stress occurs.

14. Show that the difference between the maximum and mean shear stress in the web of an I beam is $Sd^2/24I$, where d is the height of the web.

15. The outer dimensions of a channel girder section are 120 mm (web) × 50 mm (flanges); the web and flanges are 5 mm thick. Determine the position of the shear centre of the section.

16. Find the shear centre of the beam cross-section shown in Figure Q16 (it is much smaller than R).

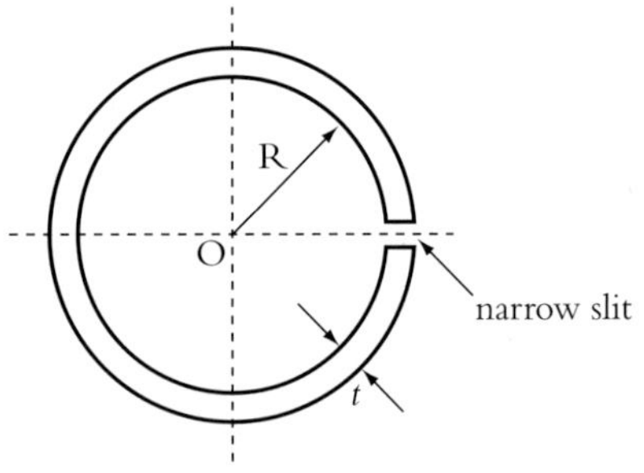

Figure Q16

17. A heat exchanger consists of a cylindrical vessel which contains 35 U-shaped tubes of 20 mm bore and 30 mm outside diameter. The ends of these tubes are welded into one of the flat end plates. The total length of the tubes within the vessel is 7 m. Calculate the hoop and axial stresses on the inside and outside of the straight parts of the tubes, remote from the bend and the ends, due to pressures of 10 bar in the vessel (i.e. outside the tubes) and 100 bar inside the tubes.

18. A compound tube is made up of an inner steel tube, 4.5 in internal diameter and 6.5 in external diameter, on which is shrunk an outer steel tube that is 9 in external diameter. If the radial pressure at the common surface is 4000 lbf/in^2, find the maximum and minimum circumferential stresses in both tubes due to shrinkage and plot the distribution of hoop stress across the wall of the compound tube.

19. A circular saw that is 5 mm thick and has a 900 mm diameter has a bore of 100 mm. The steel, of which the saw is made, has a density of 7800 kg m^{-3}, and $\eta = 0.3$. Find the maximum speed permitted if the hoop stress is restricted to 240 MPa. What is the maximum value of the radial stress?

20. A bronze gyro-wheel has a moment of inertia of 0.4 kg m^2 and can be run with safety at 3000 rpm. A larger steel wheel of similar shape is required to give an angular momentum of 40 × 10^3 kg m^2 s^{-1} at 1800 rpm. In what ratio must all the linear dimensions

be increased, and what elastic limit is required for the steel so that both wheels have the same factor of safety. (Density of bronze = 8480 kg m^{-3}; Elastic limit of bronze = 78 MPa.)

21. Calculate the principal second moments of area and the directions of the principal axes for the section shown in Figure Q21.

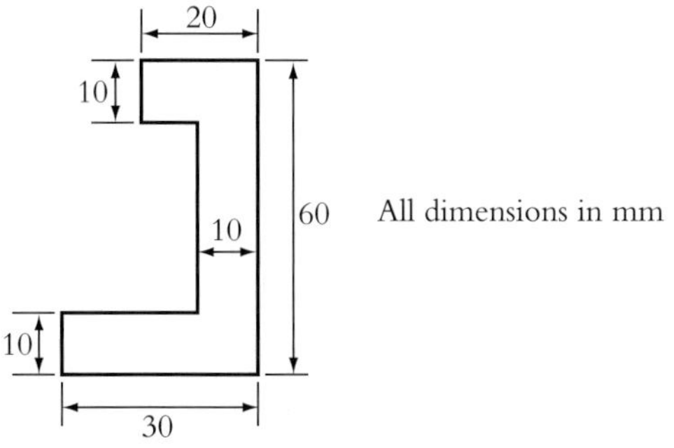

Figure Q21

22. A 50 mm × 30 mm × 5mm angle is used as a cantilever of 500 mm, with the 30 mm leg horizontal and uppermost. A vertical load of 1000 N is applied at the free end. Determine the position of the neutral axis and the maximum tensile and compressive bending stresses.

23. Calculate the maximum tensile stress and the position of the neutral axis for the section shown in Figure Q23, when a bending moment of 225 Nm is applied about the x-axis in the sense shown.

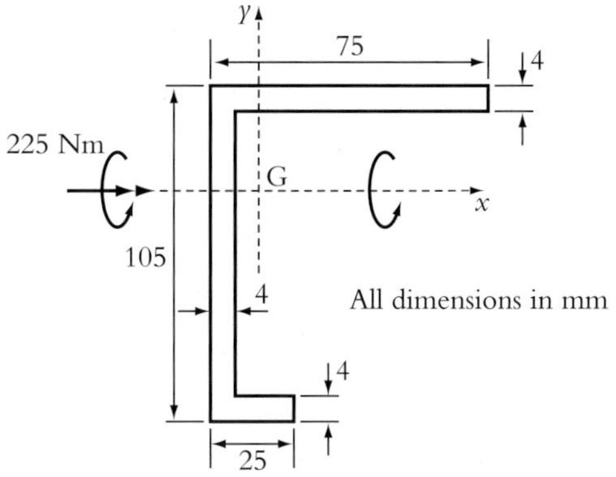

Figure Q23

24. Using strain energy, derive an expression for the deflection of the load point of the beam shown in Figure Q24.

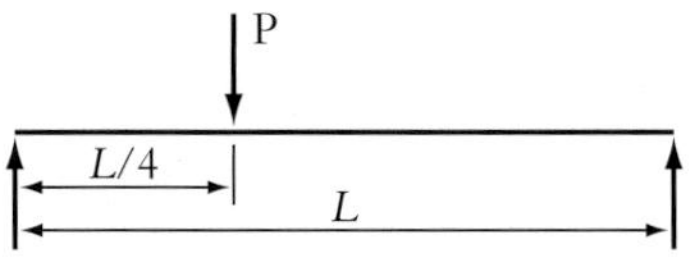

Figure Q24

25. Calculate the deflection beneath the force for the cantilevered bracket shown in Figure Q25. (E = 200 GPa; G = 80 GPa.)

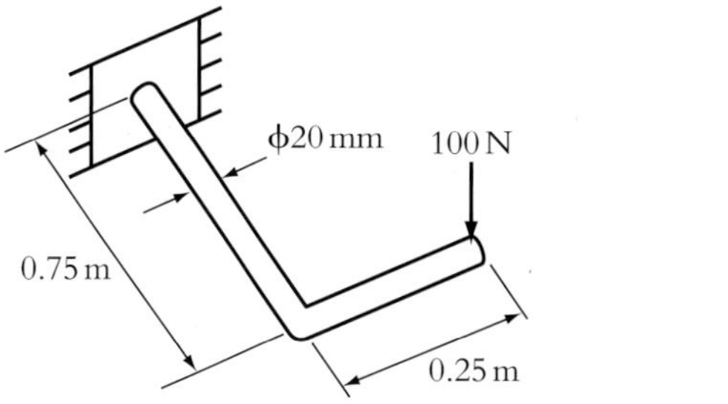

Figure Q25

26. Derive an expression for the increase in distance between the ends A and B of a thin bar of uniform cross-section consisting of of a semi-circular portion CD and two straigth portions AC and BD as shown in Figure Q26.

If the bar is of diameter 6 mm, R is 40 mm and is to have a spring stiffness, $\frac{P}{\delta}$ of 1000 kg/m, show that the necessary length for L is approximately 210 mm. The bar is made from mild steel with Young's modulus E = 210 GPa.

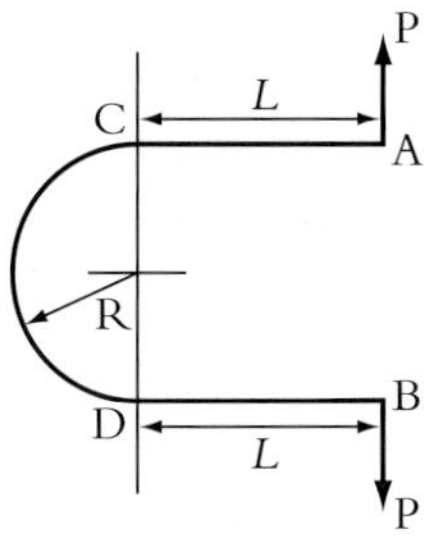

Figure Q26

27. A circular connecting rod is made from a steel having the following properties:

- UTS: 950 MN/m^2
- YP: 800 MN/m^2
- Fatigue limit: 500 MN/m^2

The rod is subjected to a fully reversed axial load of 180kN. Determine the minimum rod diameter, allowing a factory for safety of 2, if the rod end produces a fatigue strength reduction factor of 2.1, where the stress concentration factor is 2.5.

28. A mild steel cantilever beam of circular cross-section is subjected to a load at its free end which varies cyclically from P to –3P as shown in Figure Q28. Determine the maximum value of P if the fatigue strength reduction factor for the fillet is 1.85 and a safety factor of 2.0 is assumed. (Hint: use the

Goodman Diagram and apply the fatigue strength reduction factor to the stress amplitude only.)

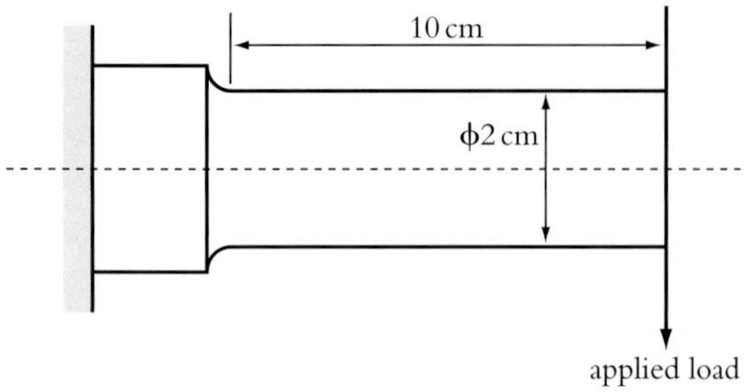

Figure Q28

29. A circular steel shaft having a transverse oil hole is subjected to a torsional load which varies from –100 Nm to 400 Nm (i.e. values in opposite senses). Determine the necessary shaft diameter, assuming that the hole causes a fatigue strength reduction factor of 1.75 and making use of a factor of safety of 1.5. Assume the following properties for the steel:

- Ultimate tensile strength: 400 NM/m^2
- Fatigue endurance limit: 260 NM/m^2

30. A rectangular section mild steel beam with the cross-sectional dimensions shown in Figure Q30, has a temperature given by:

$$T = 50\cos\left(\frac{\pi y}{40}\right)°C$$

Find the factor of safety of the beam against yield, at a section remote from the ends. Assume $\sigma_y = 250$ MPa and E = 200 GPa.

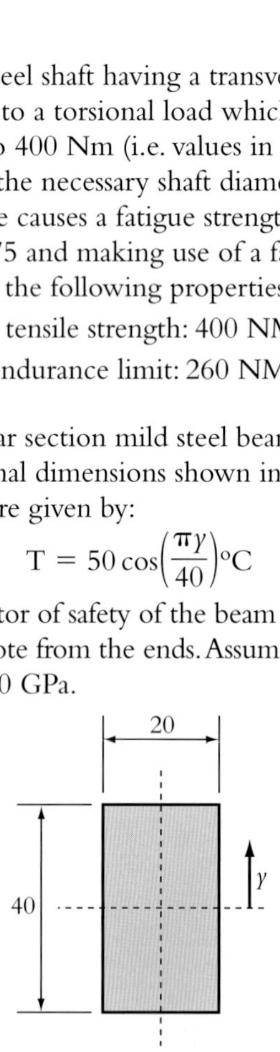

Figure Q30

31. A thin circular steel plate with a 400mm outside diameter and a 200mm insider diameter has a temperature distribution which varies approximately linearly with *r*. At the bore, T = 100°C, and at the outside T = 30°C. Find the growth of the bore and the growth of the outside diameter compared with their room temperature (20°C) values.

Electromechanical Drive Systems

1. A machine may be regarded as consisting of the system shown in Figure Q1, where $J_1 = J_2 = 0.1$ kgm^2, $N_1 = N_3 = 20$, $N_2 = N_4 = 40$. Find the moment of inertia of the system referred to the input (bottom) shaft, ignoring the inertia of the gears themselves.

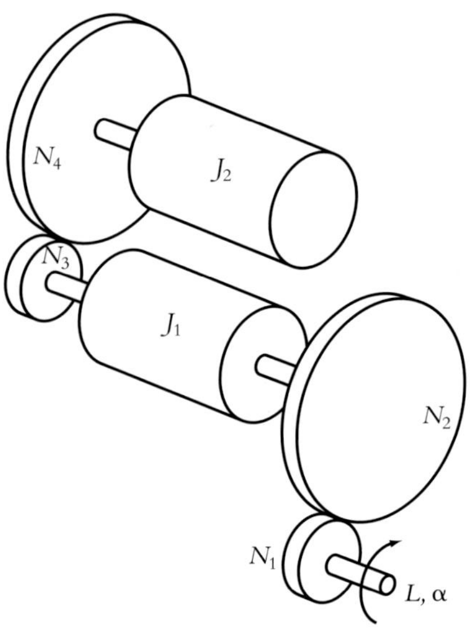

Figure Q1

(Hint: there are two ways to do this. Either work down the system, building up the contributions to the referred inertia for each axis, or treat each inertia separately and refer it directly to the input shaft. Both methods are equivalent and give identical answers.)

2. Part of an *x-y-z* positioning system consists of a carriage which has a total mass of 6 kg and is carried on a sliding ball-assisted carriage (Figure Q2). There is no viscous friction and windage is negligible but the coefficient of (Coulomb) friction for the ball-slide may be regarded as 0.05. The carriage is attached to a continuous toothed belt, which passes over two pulleys of effective (pitch) diameter 20 mm, one of which is connected directly to the motor that is used for positioning the carriage. The total mass of the belt is 50 g. Each pulley may be regarded as a solid cylinder with mass 10 g and diameter 18 mm.

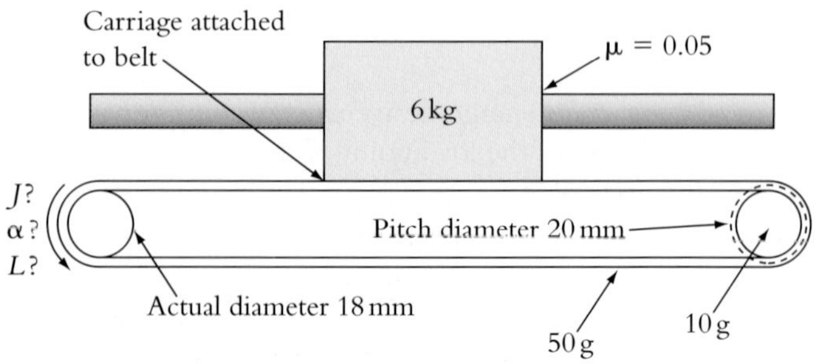

Figure Q2

(a) What is the moment of inertia of the system referred to the pulley that is connected to the motor?

(b) The carriage is required to move from one end of the 600 mm-long bed to the other in 1 second. Assuming that this is achieved by a period of constant acceleration over 300 mm, followed by constant deceleration over the remaining 300 mm, find the maximum linear acceleration of the carriage and hence find the maximum angular acceleration of the shaft.

(c) Find the frictional force (ignore the frictional effect of the weight of the belt), the frictional torque and the total torque required from the motor in order to achieve the necessary acceleration while overcoming the friction.

3. A mass of 100 kg is moved in a straight line using a lead screw of pitch 5 mm. As it is a ball-assisted high-quality leadscrew, its efficiency is high (95%), and it may be approximated as a solid cylinder of density 7800 kgm^{-3}, diameter 25 mm and length 2 m. What torque must be applied to the leadscrew to accelerate the mass from rest to 0.2 m/s over a distance of 0.5 m? Assume constant acceleration.

(Hint: work out the linear acceleration of the mass using one of the kinematic formulae, then noting that 2π radians of rotation equates to 0.005 m of movement, work out the angular acceleration needed. Work out the moment of inertia of the screw itself, then use equation 4.26 to obtain the torque.)

4. A 'penny-farthing' (Figure Q4, a primitive type of bicycle, having a very large front wheel driven directly by the pedals) is being ridden by a person of weight 80 kg. The penny-farthing has a total mass of 20 kg, of which 8 kg is concentrated in the rim of the front wheel. The front wheel has a radius of 0.65 m.

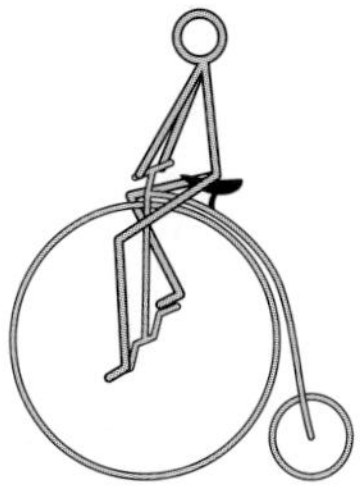

Figure Q4

(a) Calculate the inertia of the system referred to the axis of the wheel. Ignore the contribution to the rotational inertia of the front wheel spokes; also ignore the rotational inertia of the very small rear wheel.

(b) Calculate how much torque is required at the pedals to obtain an acceleration of 0.5 m/s^2:
   (i) when the penny-farthing is being ridden on a level surface;
   (ii) when it is climbing a gradient of 1 in 50 or 2% (equivalent to an angle of inclination of 0.02 radian).

5. A vehicle has mass 1600 kg. This total includes four wheels, each of mass 15 kg and diameter 0.7 m, which may be regarded for inertia purposes as solid discs.

(a) What is the moment of inertia of each wheel about its own axis?

(b) What is the moment of inertia of the vehicle referred to the axis of the wheels?

6. A mixing head for stirring a slurry has the following torque-speed characteristic when mixing is taking place:

$$L = 100 + \omega + 0.1\omega^2$$

where L is the torque in Nm and ω is the angular velocity in rad/s.
It is driven via a 50:1 worm gear of efficiency 45% from an induction motor with the following torque-speed characteristic in its operating region:

$$L' = 0.1(100\,\pi - \omega')$$

where L' and ω' are again in Nm and rad/s respectively.

(a) What is the torque-speed characteristic of the mixing head, referred to the axis of the motor i.e. expressed as L' vs ω'?

(b) At what speed will the motor run when driving the mixing head? Express your answer in rad/s and rev/min.

(c) At what speed (in rad/s and rev/min) will the mixing head itself rotate?

(d) What power does the motor produce when driving the mixing head under load?

7. A floating container is pulled along a channel of water using a chain, which is hauled using a winch of diameter 300 mm. The drag force F on the container is related to its speed by the following equation:

$$F = 500\,v^2$$

where F is in N and v is in m/s.
The winch is to be driven from a petrol engine with the following characteristics:

$$L' = 1 + 0.06\omega' - 0.001\omega'^2$$

which is valid within the range $100 < \omega < 450$ rad/s.

(a) Refer the force-speed characteristics of the container to the axis of the winch.

(b) Hence, refer the force-speed characteristics of the container to the axis of the engine for the following combinations of gear ratio and efficiency:
(i) 30:1, 65%
(ii) 40:1, 55%

(c) In both cases, find the combination of torque and angular velocity at which the engine will run when driving the winch. Hence, decide whether either of these combinations is feasible, and which will make the more effective use of the system.

8. A 220V dc series motor has the flux *vs.* current characteristic:

Flux (mWb)	16.5	22.3	24.4	25.0
Current (A)	20	30	45	60

The armature resistance is 0.1 Ω and the field resistance is 0.2 Ω.

The motor runs at a speed of 10 rev s^{-1} and draws a current of 50 A.

Calculate:

(a) emf induced in the armature;

(b) design constant, *k*.

The current is reduced to 30A.

(c) Calculate the new speed.

9. A 240V dc shunt motor has an armature resistance of 0.6 Ω and field winding resistance of 160 Ω. The motor field characteristic (ϕ versus field current) is shown in Figure Q9.

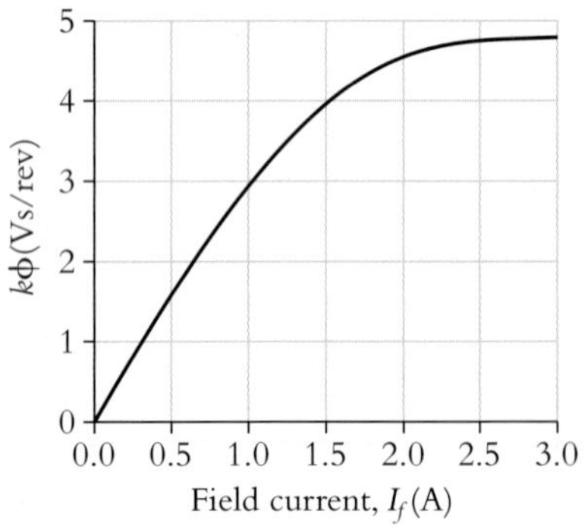

Figure Q9

Calculate:

(a) field current.

Given that the motor drives a constant load torque of 10 Nm, calculate:

(b) armature current;

(c) speed.

10. A separately excited motor is rated at 25 kW, 400V, 70 A and 1000/1500 rev/min. It is operated with the rated field voltage at speeds below 1000 rev/min and with field weakening above 1000 rev/min.

(a) Calculate the motor constant and the armature resistance R_a.

(b) Calculate the field voltage, armature voltage and armature current if the motor is running at its rated power and at its maximum speed of 1500 rev/min, taking account of armature resistance.

(c) The motor is used to drive a load with the torque-speed characteristic $T = 25 + \omega$, where the units of T are Nm and those of ω are rad/s. It is desired to drive this load at 800 rev/min. Calculate the required armature and field voltages and the armature current drawn, taking account of the effects of armature resistance.

11. A 220 Vdc series motor with an armature resistance of 0.4 Ω and a field resistance of 0.267 Ω drives a load at a speed of 520 rev min^{-1}. The supply current is 15A.

(a) Calculate the load torque.

(b) The load torque is doubled and the supply current rises to 35A. Calculate:
(i) the new speed;
(ii) the mechanical output power.

12. A 25 kW, 415 V, 50 Hz, 1440 rev/min cage induction motor is fed from a variable-frequency supply. The voltage and frequency are varied in proportion, with rated frequency corresponding to rated voltage. This arrangement is used to drive a blower which has a torque-speed characteristic given by $T = 6 \times 10^{-3}\,\omega^2$, where T is in Nm and ω is in rad/s. Determine the frequency and line-to-line voltage at a blower speed of 1300 rev/min.

13. (a) Explain the reasons why the use of a pneumatic motor may in some situations be preferable to the use of an electric motor. Illustrate your argument with proper graphs and examples where appropriate. Explain the shortcomings of pneumatic motors.

(b) Draw a diagram illustrating the various features of a compressed air system, such as may be found in a typical factory. Briefly identify the function of each component and explain how the air pressure in the system is controlled.

Feedback and Control Theory

1. (a) Figure Q1(a) shows a multi-loop system. Determine the overall transfer function of the system shown in Figure Q1(a), relating the output $Y_o(s)$ to the input $X_i(s)$

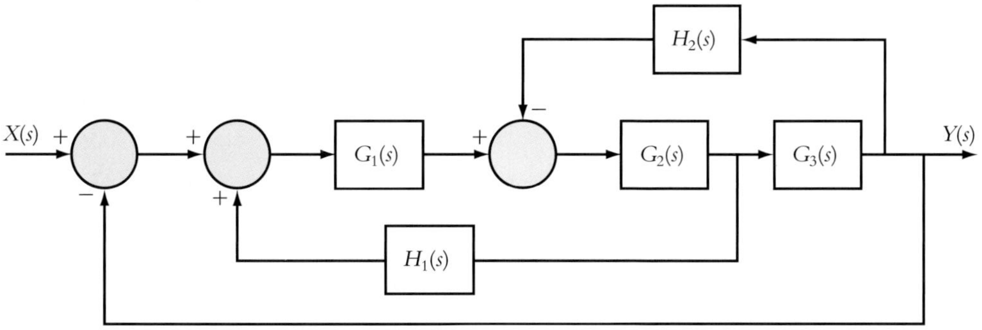

Figure Q1(a)

(b) Figure Q1(b) and Figure Q1(c) show two block diagrams for the same system. Determine the transfer functions G(s) and H(s) of the block diagram shown in Figure Q1(c), that are equivalent to those of the block diagram of Figure Q1(b).

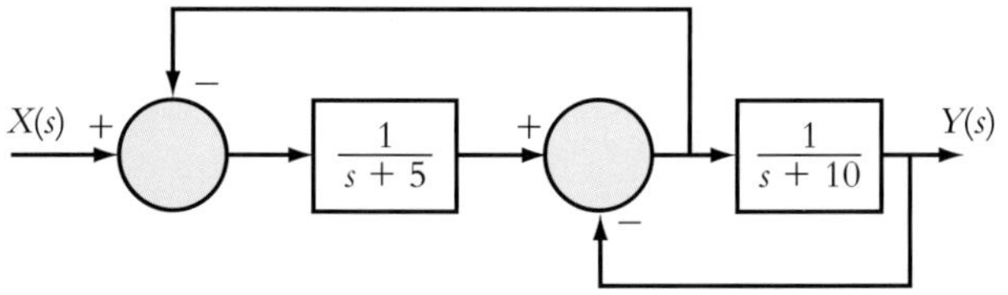

Figure Q1(b)

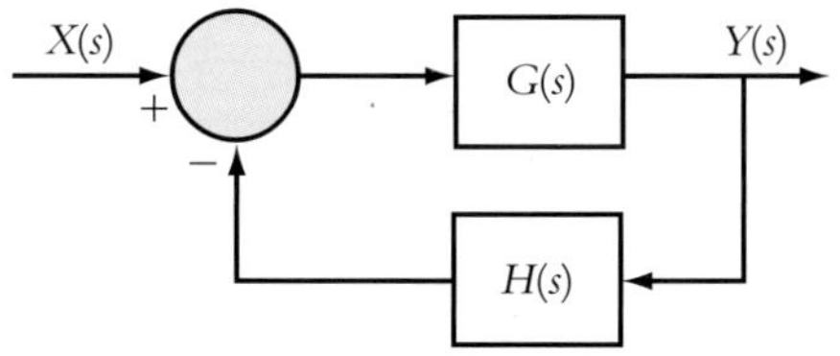

Figure Q1(c)

2. Figure Q2 shows a system for controlling the level of liquid in a tank. The tank has a fixed cross-sectional area A, fixed linearized flow resistance R for the outflow (relating the outflow to the liquid level) and variable liquid level $h(t)$. For this system, the difference between the actual level h in the tank and the desired level h_i is used to form the error signal $\varepsilon(t)$, where the actual level h is measured by a transducer. This error signal is fed to a controller that drives a variable speed pump such that the volumetric inflow rate in the tank is given by $Q_i(s) = G_c(s)\,E(s)$, where $G_c(s)$ is the transfer function of the controller and $E(s)$ is the Laplace transform of the error signal $\varepsilon(t)$. In addition, there is an uncontrolled disturbance inflow to the tank which is given by $q_d(t)$.

For the case when the controller only contains proportional action with gain K:

(a) Draw the block diagram of the system, taking h_i to be the input and h to be the output.

(b) Determine the overall transfer function relating h to h_i and q and show that the system is of first order.

(c) Show the location of the closed loop pole of the system in the s-plane.

(d) With $q_d = 0$, if $A = 2$, $R = 10$ and $K = 0.3$ and a constant demand level $h_i = \overline{h_i}$, in consistent units, find the steady-state response in terms of $\overline{h_i}$.

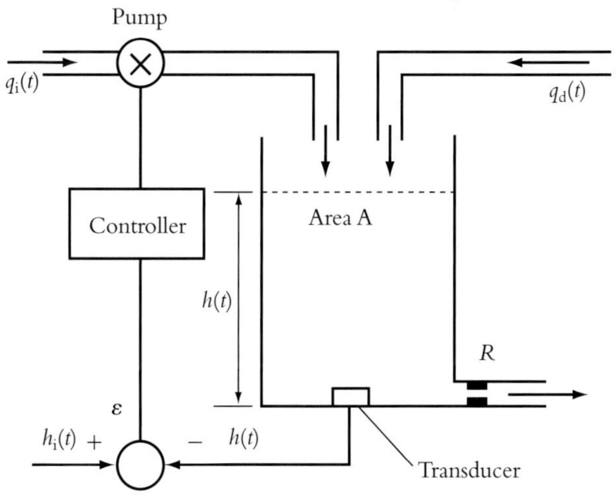

Figure Q2

3. Figure Q3 shows a schematic of a system for controlling the angular velocity Ω of a cable drum of radius R under conditions where the cable tension T is variable. The drum, which has a moment of inertia J_D is driven via n:1 speed reduction gearing by a servo-motor with rotating

parts having a moment of inertia J_M. Viscous friction of coefficient C resists the rotation of the drum. The feedback signal V_o is derived from a tacho-generator and is subtracted from the demand signal $K_G\Omega_D$ to form the error signal. The controller is a simple proportional controller with gain K and delivers current I to the servomotor that develops a drive torque L. The transfer functions for the tacho-generator and servo-motor, respectively, are as follows:

$$\frac{V_o(s)}{\Omega(s)} = \frac{K_G}{1 + T_G s}; \frac{L(s)}{I(s)} = \frac{K_M}{1 + T_M s}$$

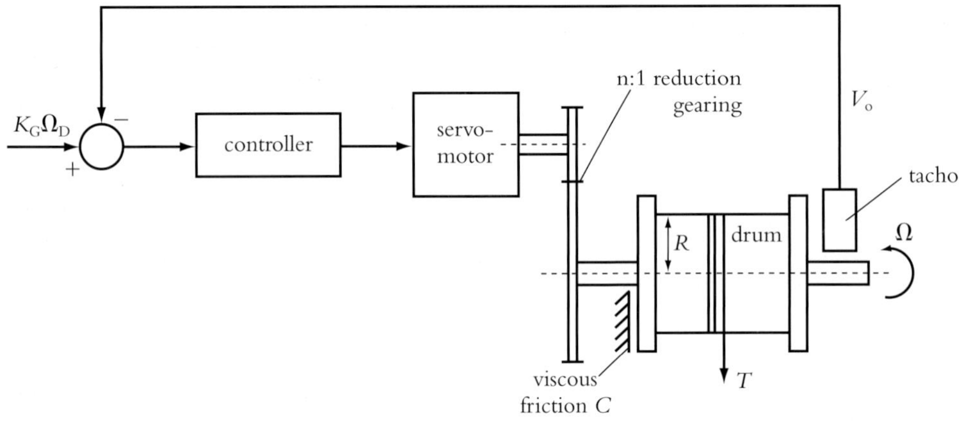

Figure Q3

(a) Draw a block diagram for the system and derive the overall transfer function relating the drum speed Ω to the demand speed Ω_D, and the cable tension T.

(b) For the case when $T_G = 0$, determine the order of the system and derive expressions for appropriate parameters which characterize completely the transient behaviour.

4. The block diagram in Figure Q4 represents a system for controlling the speed of a rotor that is driven by a field-controlled electric motor. A tacho-generator is used to provide the feedback signal. T_M, T_L, and T_T are time constants associated with the motor, rotor and tacho-generator respectively, and K is a gain associated with the motor. For a particular system, the time constants have the following numerical values: $T_M = 0.5$, $T_L = 1$ and $T_T = 0.1$.

(a) Find the minimum value of K for which the steady state error in rotor speed $(\Omega_i - \Omega_o)$ following a step change in input is 2% or less.

(b) Find the maximum value of K for which the system is stable. For this maximum value of K determine the frequency at which the system would oscillate if disturbed.

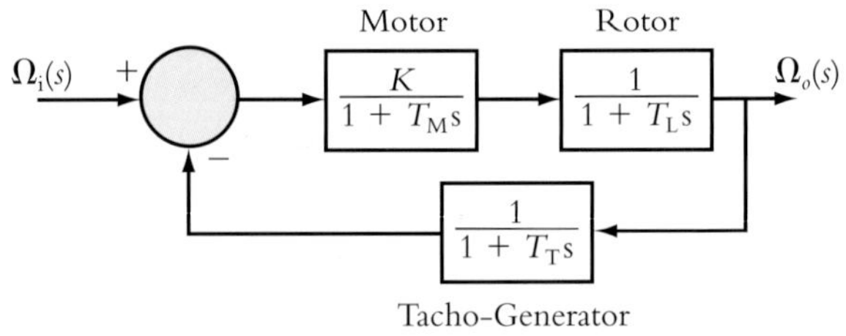

Figure Q4

5. The block diagram in Figure Q5 represents a system for controlling the rotational speed of an inertia load that is driven by a diesel engine. K and T_C are respectively the gain and time constant of the fuel injector. The demand speed is $\omega_i(t)$ and the actual load speed is $\omega_o(t)$ (these are expressed in the Laplace space as $\Omega_i(s)$ and $\Omega_o(s)$ respectively).

(a) Derive an expression for the maximum value of K for which the closed loop system is stable.

(b) Show that the steady state error in speed $\omega_i - \omega_o$ following a step change in demand is zero and derive an expression for the steady state error when a ramp input is applied.

(c) With reference to your answers to parts (a) and (b), explain whether it is better for the fuel injector time constant T_C to be shorter or longer.

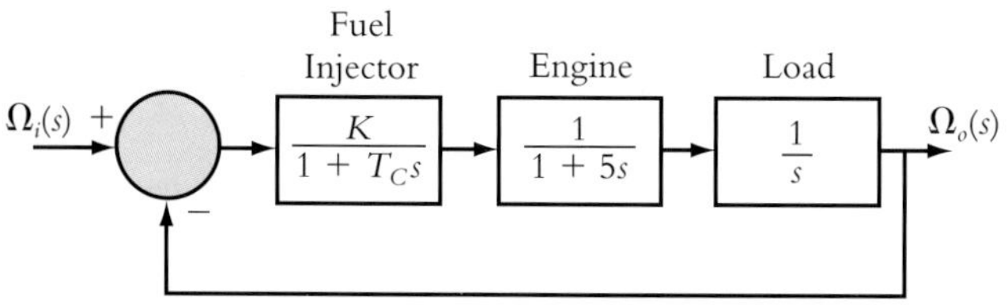

Figure Q5

6. (a) Figure Q6 shows the block diagram of a system for controlling the temperature of liquid in a tank. For the case where the controller is a proportional controller with transfer function $G(s) = K$, draw the root locus plot for the system.

(b) To reduce steady state errors, the controller is modified to incorporate integral action so that the controller transfer function is now given by:

$$G_C(s) = K\left(\frac{s + 4}{s}\right)$$

Draw the root locus plot for the modified system and comment on the practical significance of the differences between the plots for the original and modified systems.

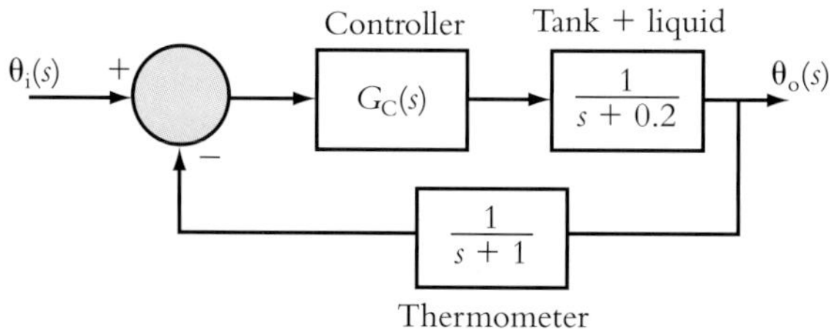

Figure Q6

7. Figure Q7 shows a schematic of a system for controlling the angular position (expressed in the Laplace domain) $\Theta_o(s)$ of a satellite dish. The dish assembly has a moment of inertia J_D and is driven via n:1 speed reduction gearing by a servo motor with rotating parts having a moment of inertia J_M. A viscous friction of coefficient c resists the motion of the dish assembly. The feedback signal $V_o(s)$ is derived from a potentiometer and is subtracted from the demand signal $K_P\Theta_i(s)$, to form the error signal. The controller is a simple proportional controller with gain K and delivers current I to a servo-motor that develops a drive torque $L_M(s)$. The transfer functions for the potentiometer and servo-motor, respectively, are as follows:

$$\frac{V_o(s)}{\Theta_o(s)} = K_P; \quad \frac{L_M(s)}{I(s)} = \frac{K_M}{1 + T_M s}$$

(a) Draw a block diagram for the system and derive the overall transfer function relating the angular position of the dish $\Theta_o(s)$ to the demand rotation $\Theta_i(s)$.

(b) For the case when $T_M = 0$, determine the order of the system and derive expressions for appropriate parameters which characterize completely the transient behaviour.

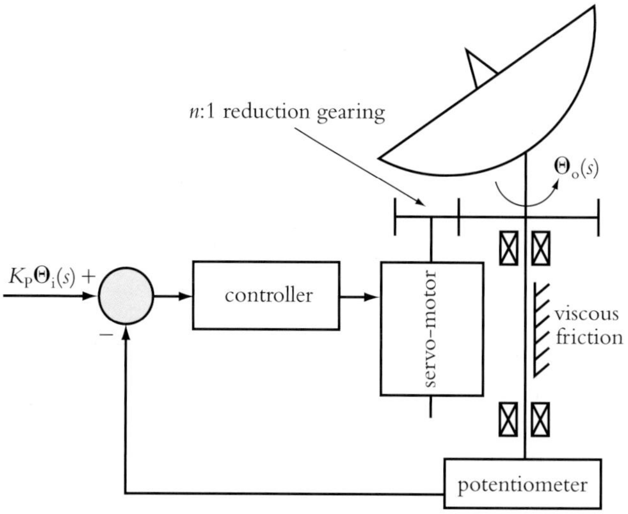

Figure Q7

8. Find the numbers of positive roots of the following equations using the Routh-Hurwitz criterion:

(a) $s^3 + 4s^2 + s - 6 = 0$

(b) $s^4 - s^3 - 2s^2 - 2s + 4 = 0$

(c) $s^5 + 4s^4 - 14s^3 - 46s^2 + 25s + 150 = 0$

9. Plot the closed loop root loci for the situations where the open loop transfer function is given by following transfer functions combined with a variable gain K:

(a) $\dfrac{s - 1}{(s^2 + 2s + 2)(s + 2)}$

(b) $\dfrac{s}{(s + 1)(s + 2)(s + 3)(s + 4)}$

Hence, for each case, determine the value of gain K for marginal stability, and (if appropriate) the natural frequency at marginal stability.

Structural Vibration

Natural frequencies and mode shapes

1. Derive the equation of motion and hence find the natural frequencies for the system shown in Figure Q1.

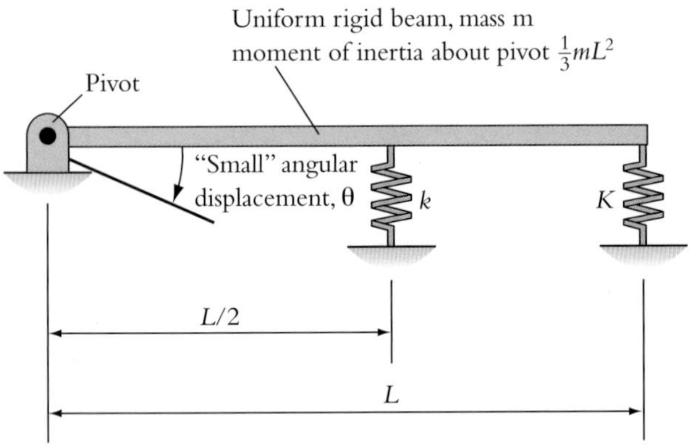

Figure Q1

2. A wheel (radius r, mass m, moment of inertia about its centre I) can roll without slipping on a horizontal plane. It is restrained by a horizontal spring (stiffness k) attached at one end to the centre of the wheel and at the other end to a rigid vertical wall, as in Figure Q2. Derive the equation of motion and hence find the natural frequency for the system.

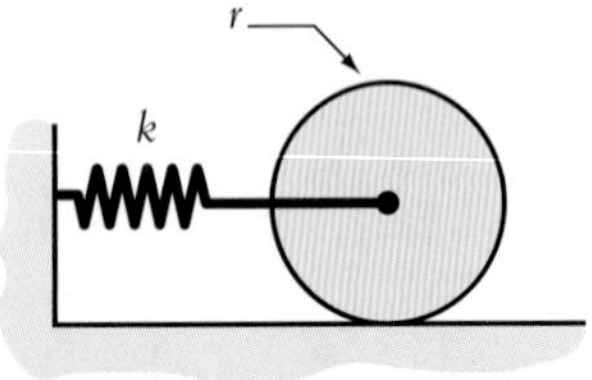

Figure Q2

What would the natural frequency be if there was no friction between the wheel and the plane?

3. Derive the equations of motion for the system in Figure Q3. Assume that all displacements are small.

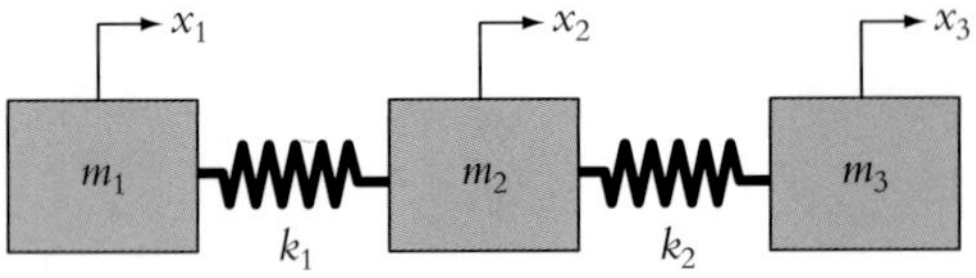

Figure Q3

4. Find the natural frequencies and mode shapes for the system in Figure Q3 when $k_1 = 10\,\text{kN/m}$, $k_2 = 30\,\text{kN/m}$, $m_1 = m_2 = 5\,\text{kg}$ and $m_3 = 10\,\text{kg}$.

5. Derive the equations of motion for the system in Figure Q5. Assume that all displacements and angles are small.

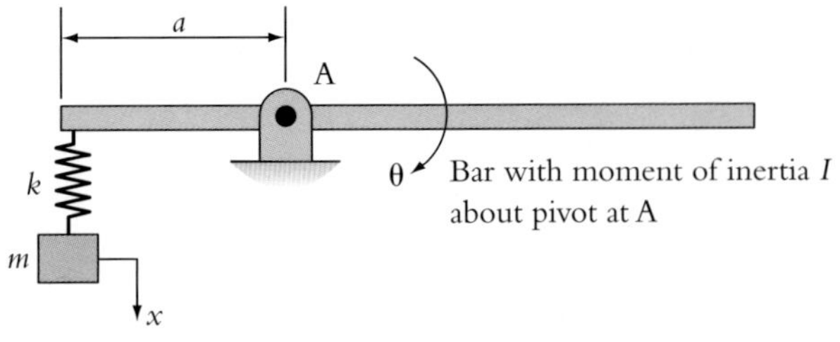

Figure Q5

6. Derive the frequency equation for flexural vibration of a uniform beam that is free at both ends and find an expression for the mode shape function.

7. A 25 mm diameter shaft, 1.5 m long, is held by two roller bearings at one end (giving a 'clamped' boundary condition) and by a self-aligning ball bearing at the other end (giving a 'pinned' boundary condition). Using the roots of the appropriate frequency equation given in Table 6.3 on page 398, find the first three critical speeds of the shaft.

Response of damped single-degree-of-freedom systems

8. If a heavily damped structure is given an initial displacement Z_O and then released from rest, find the constants of integration and sketch the graph of $z(t)$ against time.

9. A critically damped structure is subjected to an impulse such that it acquires an instantaneous initial velocity, V_O, while the displacement remains zero. Show that the subsequent displacement is given by $z(t) = V_O\, t\, e^{-\omega_n t}$.

10. The rigid beam shown in Figure Q10 has a moment of inertia of 10 kgm^2 about the pivot at A. End C is displaced downwards by 10 mm from its equilibrium position and then released from rest. Find the maximum upward displacement of C from its equilibrium position and the elapsed time at which this occurs.

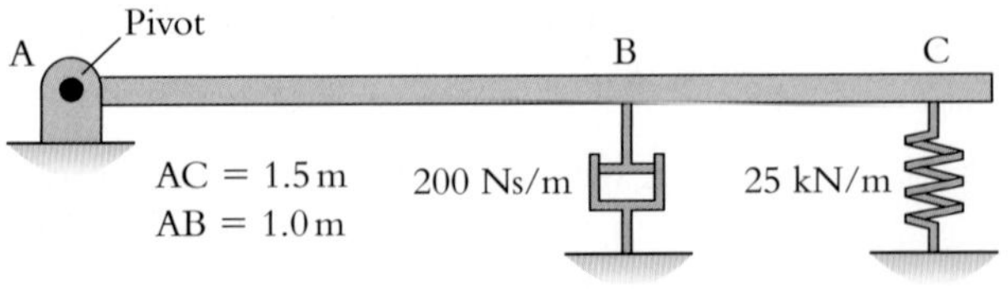

Figure Q10

11. Figure Q11 shows a rocker arm with moment of inertia I_O, about the pivot at O. Rubber blocks at ends A and B can each be modelled as a spring (stiffness, k) in parallel with a viscous damper (damping coefficient, c). The base of the block at A is attached to a rigid foundation. The base of the block at B is attached to a follower, which is driven by a cam that gives the follower a sinusoidal displacement of amplitude Y, and frequency ω. Derive an expression for the steady-state amplitude of the displacement at A, assuming that the angular displacement of the bar is small.

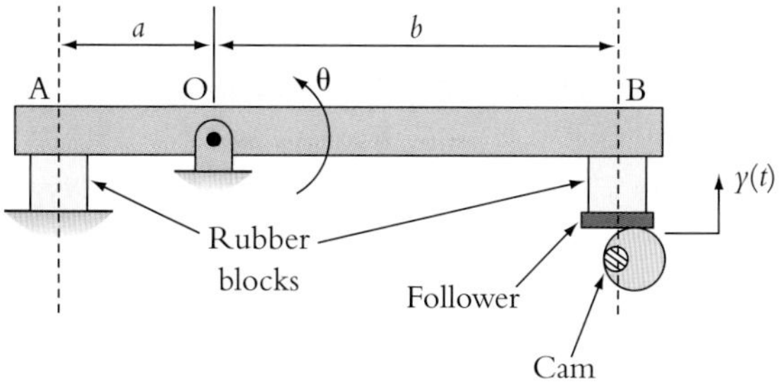

Figure Q11

12. For an undamped system with n degrees of freedom, show that the steady-state response in coordinate j due to a sinusoidal force of amplitude P and frequency ω applied in coordinate k is given by

$$x_j(t) = \left\{ \sum_{r=1}^{n} \frac{u_{jr}\, u_{kr}}{\omega_r^2 - \omega^2} \right\} P \sin \omega t$$

Approximate methods

13. Use Dunkerley's Method and Rayleigh's Method to estimate the lowest natural frequency of the torsional system shown in Figure Q13.

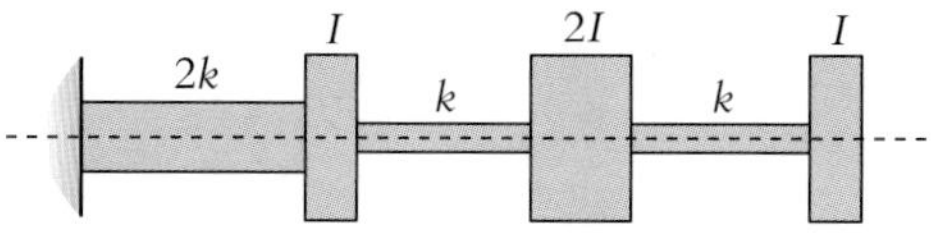

Figure Q13

14. A shaft with universal joints at each end has a length of 6 m, a second moment of area of 0.00025 m^4 and a mass/unit length of 75 kg/m. It carries three discs, which can be regarded as point masses of 100, 150, and 200 kg located 1.2, 3 and 4.8 m from the left-hand end. Estimate the lowest critical speed using Dunkerley's and Rayleigh's Methods. Take $E = 207$ GN/m^2, $\rho = 7800$ kg/m^2.

15. Find a single-degree-of-freedom approximate model to analyse the motion of the 1 kg mass in the system in Figure Q15 when it is vibrating near its lower natural frequency.

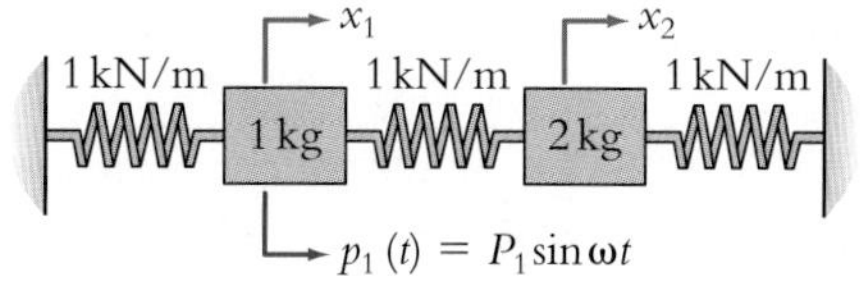

Figure Q15

16. Use the model from Question 15 to estimate the steady-state response of each mass in the two-degree-of-freedom system due to a sinusoidal force of amplitude 10 N and frequency 3.8 Hz applied to the 1 kg mass.

Vibration control techniques

17. Derive the displacement transmissibility for the system in Figure Q17.

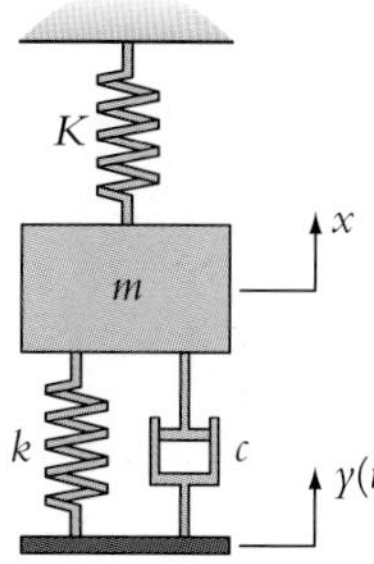

Figure Q17

18. A compressor of mass 300 kg is to be installed on four isolators, two at each end. The centre of mass is 0.4 m from end A and 0.2 m from end B. In the end elevation, the isolators are located symmetrically with respect to the centre of mass.

Isolators are available with stiffnesses of 40, 70, 110, 180 and 290 kN/m and each has a maximum allowable static deflection of 10 mm. Select suitable isolators for the installation so that the isolation efficiency is at least 70% at the normal running speed of 870 rev/min. Estimate the actual isolation efficiency for each of the isolators you select. Neglect damping.

Index

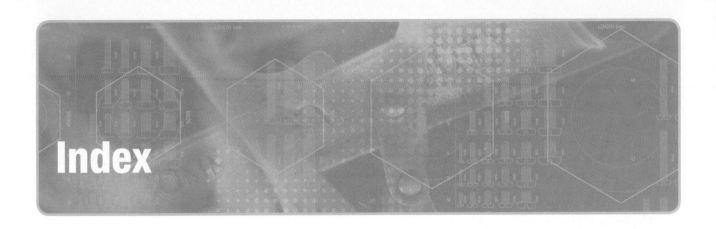